D1674804

Molecular System Bioenergetics

Edited by
Valdur Saks

1807–2007 Knowledge for Generations

Each generation has its unique needs and aspirations. When Charles Wiley first opened his small printing shop in lower Manhattan in 1807, it was a generation of boundless potential searching for an identity. And we were there, helping to define a new American literary tradition. Over half a century later, in the midst of the Second Industrial Revolution, it was a generation focused on building the future. Once again, we were there, supplying the critical scientific, technical, and engineering knowledge that helped frame the world. Throughout the 20th Century, and into the new millennium, nations began to reach out beyond their own borders and a new international community was born. Wiley was there, expanding its operations around the world to enable a global exchange of ideas, opinions, and know-how.

For 200 years, Wiley has been an integral part of each generation's journey, enabling the flow of information and understanding necessary to meet their needs and fulfill their aspirations. Today, bold new technologies are changing the way we live and learn. Wiley will be there, providing you the must-have knowledge you need to imagine new worlds, new possibilities, and new opportunities.

Generations come and go, but you can always count on Wiley to provide you the knowledge you need, when and where you need it!

William J. Pesce
President and Chief Executive Officer

Peter Booth Wiley
Chairman of the Board

Molecular System Bioenergetics

Energy for Life

Edited by
Valdur Saks

WILEY-VCH Verlag GmbH & Co. KGaA

The Editor

Prof. Valdur Saks
J. Fourier University
Laboratory of Bioenergetics
2280 rue de la Piscine
38041 Grenoble Cedex 9
France

■ All books published by Wiley-VCH are
carefully produced. Nevertheless, authors,
editors, and publisher do not warrant the
information contained in these books,
including this book, to be free of errors.
Readers are advised to keep in mind that
statements, data, illustrations, procedural
details or other items may inadvertently be
inaccurate.

Library of Congress Card No.: applied for

British Library Cataloguing-in-Publication Data
A catalogue record for this book is available
from the British Library

**Bibliographic information published by
the Deutsche Nationalbibliothek**
Die Deutsche Nationalbibliothek lists this
publication in the Deutsche National-
bibliografie; detailed bibliographic data are
available in the Internet at http://dnb.d-nb.de

© 2007 WILEY-VCH Verlag GmbH & Co.
KGaA, Weinheim

Printed in the Federal Republic of German
Printed on acid-free paper

Composition   Asco Typesetters, Hong Kong
Printing   betz-druck GmbH, Darmstadt
Bookbinding   Litges & Dopf GmbH,
Heppenheim
Wiley Bicentennial Logo   Richard J. Pacifico

ISBN 978-3-527-31787-5

Contents

8   **Integration of Adenylate Kinase and Glycolytic and Glycogenolytic**
         Circuits in Cellular Energetics   *265*
         Petras P. Dzeja, Susan Chung, and Andre Terzic

9   **Signaling by AMP-activated Protein Kinase**   *303*
         Dietbert Neumann, Theo Wallimann, Mark H. Rider,
         Malgorzata Tokarska-Schlattner, D. Grahame Hardie, and Uwe Schlattner

Preface

When Thucydides wrote his famous book *The History of the Peloponnesian War* in Athens in 460–400 BC, he supplied it with the following commentary: "My work is not a piece of writing designed to meet the taste of an immediate public, but was done to last forever." We could borrow this comment for our book, whose intention is to survey the results of more than 200 years of research in cellular bioenergetics, from the discovery that respiration is a process of biological oxidation by Lavoisier and Laplace in the 18th century, to the latest achievements of molecular system bioenergetics in explaining the mechanistic basis of its regulation. Respiration is *la raison de être* of cells, and a knowledge of the mechanisms associated with biological oxidation, free energy transduction, and regulation of cellular respiration implies understanding how cells obtain energy for life. The scope of cellular energetics, however, is much broader than studying the mechanisms involved in respiratory regulation, because all biochemical processes are inseparable from free energy transduction, thus including the whole metabolism and the description of all kinds of work performed by cells.

At present, the biological sciences are witnessing a radical change in paradigms. Reductionism – which used to be the philosophical basis of biochemistry and molecular biology, when everything from genes to proteins and organelles were studied in their isolated state – is making way for systems biology, which favors the study of integrated systems at all levels: cellular, organ, organism, and population. Reductionism was justified in the initial stages of biological research, giving a wealth of information on system components. It is timely to put them together and analyze them in interaction, to understand the principles of functioning of the whole. These paradigmatic changes also concern bioenergetics. Unraveling the mechanisms of regulation of cellular energetics only appears to be possible by using an integrated approach based on computational models, which we call molecular system bioenergetics.

Not everybody, however, was taken by surprise following these paradigmatic changes. There have been pioneers in the systemic approach to biochemistry and metabolism. The roots of systems biology can be traced back to the works of Pasteur, Claude Bernard, Meyerhof, Cori, and Krebs, among others, and its theoretical basis can be found in the work of Norbert Wiener, the founder of cybernetics. Regarding cellular energetics, an important line of research of metabolic

Molecular System Bioenergetics: Energy for Life. Edited by Valdur Saks
Copyright © 2007 WILEY-VCH Verlag GmbH & Co. KGaA, Weinheim
ISBN: 978-3-527-31787-5

networks, energy transfer, and metabolic regulation has always existed, particularly emphasizing the significance of structure–function relationships for the integrated regulation of metabolic networks.

The aim of this book is to describe the "state of the art" of these investigations by the authors who have been most actively committed to developing integrated approaches to studying energy metabolism. In our pursuit we follow the historical unfolding of these developments, emphasizing the most important achievements in the area during the last 100 years. We begin with basic information about mitochondrial structure and function and a general description of the theoretical bases of cellular metabolism and bioenergetics in open thermodynamic systems, including Schrödinger's principle of negentropy. This is followed by an analysis of mitochondrial behavior and processes in cells, taking into account macromolecular crowding and compartmentation phenomena. The most important part of the book is devoted to the description of compartmentalized energy transfer in muscle and brain cells. The experimental data gathered during the last several decades in many laboratories allow us to explain the mechanistic bases of two highly significant phenomena for physiological energetics: the Frank–Starling law of the heart, related to the regulation of respiration, and acute ischemic cardiac contractile failure. This approach explains the regulation of substrate supply in the heart as well.

The last part of the book analyzes the applied aspects of cellular bioenergetics – the problems of mitochondrial pathology, exercise physiology, cancer research, and a new important area of research, the "metabolic syndrome" – related to the medical and socioeconomic problems of modern societies. Because these diseases, which are related to cellular and whole-body derailment of basic metabolism, are reaching epidemic proportions in the "civilized" world, the regulation of sugar and fat metabolism, once considered an old-fashioned discipline of physiological biochemistry, is back in the limelight again. Amazingly, these old disciplines and fields of metabolic regulation often are rediscovered and uncovered by young scientists, making it even more important to analyze the achievements of bioenergetics in historical perspective.

This effort intends to be a useful manual and valuable reference book for a large scientific audience, namely, in biology and medicine. We hope it will also be a resourceful tool for undergraduate and doctoral students, post-doctoral students, and research workers in bioenergetics, biophysics, biochemistry, cell biology, pharmacology, physiology, pathophysiology, and medicine.

Grenoble, May 2007   *Valdur Saks*

List of Contributors

Irina Agarkova
University Hospital of Zurich
Department of Cardiovascular
Surgery Research
Rämistrasse 100
8091 Zurich
Switzerland

Mayis K. Aliev
Cardiology Research Center
Institute of Experimental
Cardiology
Laboratory of Cardiac Pathology
3rd Cherepkovskaya Street 15A
121552 Moscow
Russia

Fabrizio Andreelli
Bichat Claude Bernard Hospital
Diabetology-Endocrinology-
Nutrion Department
46 Rue Henri Huchard
75877 Paris Cedex 18
France

Robert H. Andres
University of Bern
Department of Neurosurgery
Inselspital
Freiburgerstrasse 10
3010 Bern
Switzerland

Tiia Anmann
National Institute of Chemical and
Biological Physics
Laboratory of Bioenergetics
Akadeemia tee 23
12618 Tallinn
Estonia

Miguel A. Aon
Johns Hopkins University
Institute of Molecular Cardiobiology
720 Rutland Avenue
1059 Ross Building
Baltimore, MD 21205
USA

Stephan J.L. Bakker
University of Groningen and
University Medical Center Groningen
Department of Internal Medicine
Hanzeplein 1
9700 RB Groningen
The Netherlands

Cécile Batandier
Université Joseph Fourier
Bioénergétique Fondamentale et
Appliquée
INSERM E-0221
BP 53X
38041 Grenoble Cedex
France

Molecular System Bioenergetics: Energy for Life. Edited by Valdur Saks
Copyright © 2007 WILEY-VCH Verlag GmbH & Co. KGaA, Weinheim
ISBN: 978-3-527-31787-5

Dieter Brdiczka
University of Konstanz
Faculty of Biology
Untere Bohlstrasse 45
78465 Konstanz
Germany

Susan Chung
Mayo Clinic College of Medicine
Departments of Biochemistry and
Molecular Biology
200 First Street SW
Rochester, MN 559095
USA

Jolita Ciapaite
Vytautas Magnus University
Faculty of Nature Sciences
Centre of Environmental
Research
Vileikos 8
44404 Kaunas
Lithuania
and
VU University
Faculty of Earth and Life Sciences
Department of Molecular Cell
Physiology
Institute for Molecular Cell
Biology
De Boelelaan 1085
1081 HV Amsterdam
The Netherlands

Sonia Cortassa
Johns Hopkins University
Institute of Molecular
Cardiobiology
720 Rutland Avenue
1059 Ross Building
Baltimore, MD 21205
USA

Dominique Detaille
Université Joseph Fourier
Bioénergétique Fondamentale et
Appliquée
INSERM E-0221
BP 53X
38041 Grenoble Cedex
France

Anne Devin
Université de Bordeaux 2
Institut de Biochimie et de Génétique
Cellulaires du CNRS
1 Rue Camille Saint Saëns
33077 Bordeaux Cedex
France

Petras P. Dzeja
Mayo Clinic College of Medicine
Departments of Medicine, Molecular
Pharmacology, and Experimental
Therapeutics
Division of Cardiovascular Diseases
200 First Street SW
Rochester, MN 559095
USA

Jüri Engelbrecht
Tallinn University of Technology
Institute of Cybernetics
Centre for Nonlinear Studies
Akadeemia tee 21
12618 Tallinn
Estonia

Raquel F. Epand
McMaster University
Department of Biochemistry and
Biomedical Sciences
Health Science Centre
1200 Main Street West
Hamilton, Ontario, L8N 3Z5
Canada

Richard M. Epand
McMaster University
Department of Biochemistry and
Biomedical Sciences
Health Science Centre
1200 Main Street West
Hamilton, Ontario, L8N 3Z5
Canada

Roland Favier
Université Joseph Fourier
Bioénergétique Fondamentale et
Appliquée
INSERM U884
BP 53X
38041 Grenoble Cedex
France

Eric Fontaine
Université Joseph Fourier
Bioénergétique Fondamentale et
Appliquée
INSERM U884
BP 53X
38041 Grenoble Cedex
France

Frank Norbert Gellerich
KeyNeurotek AG
ZENIT – Technology Park
Magdeburg
Leipziger Strasse 44
39120 Magdeburg
Germany

Zemfira Gizatullina
KeyNeurotek AG
ZENIT – Technology Park
Magdeburg
Leipziger Strasse 44
39120 Magdeburg
Germany

Rita Guzun
Université Joseph Fourier
Bioénergétique Fondamentale et
Appliquée
INSERM U884
BP 53X
38041 Grenoble Cedex
France

D. Grahame Hardie
University of Dundee
College of Life Science
Division of Molecular Physiology
Sir James Black Centre
Dow Street
Dundee DD1 5EH
United Kingdom

Arend Heerschap
Radboud University Nijmegen
Medical Centre
Department of Radiology
P.O. Box 9101
6500 HB Nijmegen
The Netherlands

Robert J. Heine
VU University Medical Center
Department of Endocrinology
Institute for Cardiovascular Research
De Boelelaan 1117
1007 MB Amsterdam
The Netherlands

Thorsten Hornemann
University Hospital Zurich
Institute of Clinical Chemistry
Rämistrasse 100
8091 Zurich
Switzerland

René in 't Zandt
Radboud University Nijmegen
Medical Centre
Department of Radiology
P.O. Box 9101
6500 HB Nijmegen
The Netherlands

Hikari Jo
Kyoto University
Department of Physiology and
Biophysics
Graduate School of Medicine
Yoshida-Konoe, Sakyo-ku
Kyoto 606-8501
Japan

Caroline Jost
Radboud University Nijmegen
Medical Centre
Department of Cell Biology
NCMLS 283
P.O. Box 9101
6500 HB Nijmegen
The Netherlands

Klaas Krab
VU University
Faculty of Earth and Life Sciences
Department of Molecular Cell
Physiology
Institute for Molecular Cell
Biology
De Boelelaan 1085
1081 HV Amsterdam
The Netherlands

Jan Kuiper
Radboud University Nijmegen
Medical Centre
Department of Cell Biology
P.O. Box 9101
6500 HB Nijmegen
The Netherlands

Andrey V. Kuznetsov
Innsbruck Medical University
Department of General and Transplant
Surgery
Daniel-Swarovski Research Laboratory
Innrain 66
6020 Innsbruck
Austria

Masanori Kuzumoto
Discovery Research Laboratories
Shionogi & Co., Ltd.
12-4, Sagisu 5-chome
Fukushima-ku
Osaka 553-0002
Japan

Xavier M. Leverve
Université Joseph Fourier
Bioénergétique Fondamentale et
Appliquée
INSERM U884
BP 53X
38041 Grenoble Cedex
France

Satoshi Matsuoka
Kyoto University
Graduate School of Medicine
Department of Physiology and
Biophysics
Yoshida-Konoe, Sakyo-ku
Kyoto 606-8501
Japan

Sybille Mazurek
Institute of Biochemistry and
Endocrinology
Veterinary Faculty
University of Giessen
Frankfurter Strasse 100
35392 Giessen
Germany
and
ScheBo Biotech AG
Netanyastrasse 3
35394 Giessen
Germany

Claire Monge
Université Joseph Fourier
Bioénergétique Fondamentale et
Appliquée
INSERM U884
BP 53X
38041 Grenoble Cedex
France

Arnaud Mourier
Université de Bordeaux 2
Institut de Biochimie et de
Génétique
Cellulaires du CNRS
1 Rue Camille Saint Saëns
33077 Bordeaux Cedex
France

Dietbert Neumann
ETH Zurich
Hoenggerberg
Institute of Cell Biology
HPM-D24
Schafmattstrasse 18
8093 Zurich
Switzerland

Akinori Noma
Kyoto University
Graduate School Medicine
Department of Physiology and
Biophysics
Yoshida-Konoe, Sakyo-ku
Kyoto 606-8501
Japan

Frank Oerlemans
Radboud University Nijmegen
Medical Centre
Department of Cell Biology
NCMLS 283
P.O. Box 9101
6500 HB Nijmegen
The Netherlands

Brian O'Rourke
Johns Hopkins University
Institute of Molecular Cardiobiology
720 Rutland Avenue
1059 Ross Building
Baltimore, MD 21205
USA

Helma Pluk
Radboud University Nijmegen
Medical Centre
Department of Cell Biology
NCMLS 283
P.O. Box 9101
6500 HB Nijmegen
The Netherlands

Klaas Jan Renema
Radboud University Nijmegen
Medical Centre
Department of Radiology
P.O. Box 9101
6500 HB Nijmegen
The Netherlands

Mark H. Rider
Christian de Duve Institute of
Cellular Pathology
Hormone and Metabolic Research
Unit
Avenue Hippocrate 75
1200 Brussels
Belgium
and
University of Louvain
Medical School
1200 Brussels
Belgium

Michel Rigoulet
Université de Bordeaux 2
Institut de Biochimie et de
Génétique
Cellulaires du CNRS
1 Rue Camille Saint Saëns
33077 Bordeaux Cedex
France

Kent Sahlin
GIH
Swedish School of Sport and
Health Sciences
Department of Physiology and
Pharmacology
Box 5626
11486 Stockholm
Sweden
and
University of Southern Denmark
Institute of Physiology and
Clinical Biomechanics
5230 Odense
Denmark

Ryuta Saito
Discovery Technology Laboratory
Mitsubishi Pharma Corporation
1000, Kamoshida-cho, Aoba-ku
Yokohama 227-0033
Japan

Valdur Saks
Université Joseph Fourier
Bioénergétique Fondamentale et
Appliquée
INSERM U884
BP 53X
38041 Grenoble Cedex
France
and
National Institute of Chemical and
Biological Physics
Laboratory of Bioenergetics
Akadeemia tee 23
12618 Tallinn
Estonia

Uwe Schlattner
ETH Zurich
Hoenggerberg
Institute of Cell Biology
HPM-D24
8093 Zurich
Switzerland
and
Université Joseph Fourier
Laboratoire de Bioénergétique
Fondamentale et Appliquée
INSERM U884
BP 53X
38041 Grenoble Cedex
France

Enn Seppet
University of Tartu
Department of Pathophysiology
19 Ravila Street
50411 Tartu
Estonia

Femke Streijger
Radboud University Nijmegen
Medical Centre
Department of Cell Biology
NCMLS 283
P.O. Box 9101
6500 HB Nijmegen
The Netherlands

Frank Striggow
KeyNeurotek AG
ZENIT – Technology Park
Magdeburg
Leipziger Strasse 44
39120 Magdeburg
Germany

Ayako Takeuchi
Kyoto University
Graduate School of Medicine
Department of Physiology and
Biophysics
Yoshida-Konoe, Sakyo-ku
Kyoto 606-8501
Japan

Nellie Taleux
Université Joseph Fourier
Bioénergétique Fondamentale et
Appliquée
INSERM E-0221
BP 53X
38041 Grenoble Cedex
France

André Terzic
Mayo Clinic College of Medicine
Departments of Medicine,
Medicinal Genetics, Molecular
Pharmacology, and Experimental
Therapeutics
200 First Street SW
Rochester, MN 559095
USA

Malgorzata Tokarska-Schlattner
Université Joseph Fourier
Laboratoire de Bioénergétique
Fondamentale et Appliquée
INSERM U884
BP 53X
38041 Grenoble Cedex
France

Sonata Trumbeckaite
Kaunas University of Medicine
Institute for Biomedical Research
4 Eiveniu Street
50009 Kaunas
Lithuania

Ineke van der Zee
Radboud University Nijmegen
Medical Centre
Department of Cell Biology
NCMLS 283
P.O. Box 9101
6500 HB Nijmegen
The Netherlands

Marko Vendelin
Tallinn University of Technology
Institute of Cybernetics
Laboratory of Systems Biology
Centre for Nonlinear Studies
Akadeemia tee 21
12618 Tallinn
Estonia

Benoit Viollet
Université Paris 5
Department of Endocrinology,
Metabolism, and Cancer
Institut Cochin
INSERM U567, CNRS UMR 8104
24 Rue du Faubourg Saint-Jacques
75014 Paris
France

Theo Wallimann
ETH Zurich
Hoenggerberg
Institute of Cell Biology
HPM-D24
Schafmattstrasse 18
8093 Zurich
Switzerland

Hans V. Westerhoff
VU University
Faculty of Earth and Life Sciences
Department of Molecular Cell
Physiology
Institute for Molecular Cell
Biology
De Boelelaan 1085
1081 HV Amsterdam
The Netherlands
and
The University of Manchester
Manchester Centre for Integrative
Systems Biology, MIB
Sackville Street
Manchester M60 1QD
United Kingdom

Hans Rudolf Widmer
University of Bern
Department of Neurosurgery
Inselspital
Freiburgerstrasse 10
3010 Bern
Switzerland

Bé Wieringa
Radboud University Nijmegen
Medical Centre
Department of Cell Biology
NCMLS 283
P.O. Box 9101
6500 HB Nijmegen
The Netherlands

Stephan Zierz
Universität Halle/Wittenberg
Neurologische Klinik und Poliklinik
der Martin-Luther-Universität
06097 Halle/Saale
Germany

Introduction: From the Discovery of Biological Oxidation to Molecular System Bioenergetics

Valdur Saks

Energy is the capacity to perform work and produce heat, as the first law of thermodynamics quantitatively states. Bioenergetics as a science studies the energetic transformations within living cells, and as such it explains how cells obtain energy to live and perform work: mechanically (via motility and contraction), osmotically (via ion transport), or chemically (via biosynthesis) and how a cell maintains its specific structure and metabolism. Any aspect of cellular life involves, directly or indirectly, free energy conversion. More than 200 years ago it was discovered that respiration through oxidation of organic substrates is the main source of energy for living cells. During the years 1774 and 1775, Joseph Priestley in England, Carl Scheele in Sweden, and Antoine Laurent Lavoisier in France almost simultaneously discovered oxygen, and the priority of this discovery is sometimes still disputed [1–3]. However, it was Lavoisier who discovered later, in 1780–1783 in collaboration with Laplace, the process of biological oxidation – consumption of oxygen and production of carbon dioxide by living organisms during respiration [1–3]. Lavoisier and Laplace had developed the precise technique of calorimetry and used it to measure the quantity of heat produced by a guinea pig during respiration, along with the simultaneous analysis of the changes in the air composition within the calorimeter. They came to the remarkable conclusion that "respiration is a process similar to the burning of coal": this was the discovery of biological oxidation. At the same time, between 1771 and 1780, Priestley in collaboration with Ingen-Housz discovered the light-dependent capacity of plants to consume carbon dioxide and produce oxygen: photosynthesis [1, 2]. Thus, the birth dates of bioenergetics are known to be within the period 1771–1783. The 19th century gave the theory of thermodynamics, mostly through the work of Helmholtz, Mayer, Clausius, Gibbs, van't Hoff, Boltzmann, and other remarkable scientists. In particular, the connection between biological oxidation, muscular work, and heat production was observed and emphasized in the works of Mayer and Helmholtz [1]. At the beginning of the 20th century, the development of bioenergetics was related mostly to the study of muscle biochemistry and biochemical mechanisms of muscular contraction. Curiously, for some period of time, biological oxidation was left out of the focus of attention in many of these important studies. This happened in the 1920s, when Meyerhof, Embden, Parnas, and others performed fundamental studies of the mechanisms of glycolytic reactions of lactate production in muscle and alcoholic fermentation in yeast cells [4–6], while England's Archibald Hill carried out very precise measurements of heat production during muscular contraction. In close collaboration, the laboratories of two Nobel laureates, Meyerhof and Hill, reached the temporary conclusion that muscular contraction may be based on lactate production; thus, lacking other information, this biochemical hypothesis was put forward to explain the mechanism of muscular contraction even before the discovery of ATP (the controversial history of this period of research is well described in several reviews and books [1, 2, 7, 8]). Only Starling, Evans, and their coworkers in England followed in the footsteps of Lavoisier, assuming that the rate of oxygen consumption "is a measure of total energy set free in the heart during its activity" and finding out that this parameter increases with an increase in work performance of an isolated lung-heart preparation from a dog [9]. The controversial situation changed quickly when Lundsgaard, then a young investigator in Copenhagen, showed in 1930 that inhibition of lactate production by iodacetate did not stop contraction (he called this "alactacid" contraction) [7, 8, 10, 11]. Some years before this, in 1927, phosphocreatine (then called phosphagen) was discovered in muscle extracts by Eggleton and Eggleton and by Fiske and Subbarow, and in 1929 ATP was discovered by Lohmann and by Fiske and Subbarow, its molecular structure identified by Lohmann [1, 7, 8, 12]. In 1934, Karl Lohmann discovered the creatine kinase reaction, thus showing that these two phosphorus compounds, ATP and phosphocreatine, are interrelated [13]. Lundsgaard found that muscular contractile force always declined in parallel with a decrease in the phosphocreatine content of muscles [7, 8, 10], and this classical observation has been repeated over several decades in numerous laboratories on different muscles, including ischemic heart. From these works, it became immediately clear that lactate production is not the necessary basis of contraction and that other important (then unknown) processes are involved. In his famous paper "The Revolution in Muscle Physiology," Hill emphasized the importance of the "phosphagen" discovery and Lundsgaards's work, accepting the end of the lactate era in muscle bioenergetics by writing that "we have all been right sometimes and we have all been wrong often" [14]. Interestingly enough, the charm of lactate as an important compound in bioenergetics survived, and even now it is the focus of studies in many laboratories [15]. A breakthrough in revealing the nature of unknown processes related to muscle contraction came in 1930 when Engelhardt showed that in nucleated avian blood cells, oxygen consumption is coupled to the metabolism of phosphorus compounds, namely, to the production of ATP [11]; soon thereafter, Kalckar, Belitzer, and Tsybakova and Ochoa showed that during respiration, ATP (or phosphocreatine, if creatine is present, as in the work of Belizer and Tsybakova) is formed with a ratio of ATP (or PCr) to molecular oxygen O_2 close to 6 [1, 11, 16, 17]. In this way, oxidative phosphorylation was discovered, and the view of bioenergetics as biological oxidation and respiration returned [11, 17]. In 1939, Engelhardt and his wife Militza Ljubimova introduced another important discovery by showing that the major muscular protein myosin is an ATPase [18]. The role of ATP in the contraction process became fully evident in the works of the Szent-Gyorgyi and his collaborators [19]. A new picture of cellular bioenergetics based on the central role of ATP in energy metabolism emerged [11, 17–19].

During the same period, the pioneering works of Keilin, Wieland, Warburg, and many others led to the discovery of cytochromes and cytochrome c oxidase, involved in the biological oxidation and reduction of oxygen [20–22]. The rather dramatic and controversial history of the discoveries during this period is described in two recent excellent books by Pierre Vignais [1, 2]. After mitochondria were isolated and shown to be responsible for both respiration and ATP synthesis during 1948 and 1949, a new trend for bioenergetics – mitochondrial bioenergetics – was established. This has been the main subject of research in bioenergetics during the last five decades in many laboratories across the world (e.g., work by Lehninger, Green, Chance, Slater, Klingenberg, Mitchell, Skulachev, Racker, Pedersen, Boyer, Walker) [23]. Two Nobel prizes were awarded for deciphering the molecular mechanisms of free energy conversion in mitochondria and of the synthesis of ATP: to Peter Mitchell in 1978 and to Paul Boyer and John Walker in 1997. It is noteworthy that the radical new approach to oxidative phosphorylation introduced by Mitchell in 1961 as chemiosmotic coupling was and is still very modern [24]. Indeed, according to his theory, the structural (or supramolecular) features determine the vectorial nature of the energy transfer through a network of "enzyme-catalyzed group translocations" located in membranes. Such a mechanism differs fundamentally from the orthodox one previously developed [25, 26] in that (1) it depends absolutely on the supramolecular organization of the enzyme system concerned and (2) the driving force on a given chemical reaction is due to the specifically directed channeling of the chemical compounds or groups along a pathway specified in space by the physical organization of the system. Such a conception of metabolism is the basis of current systems biology.

The reductionist approach to bioenergetics described so far – excellent and successful studies of isolated mitochondria and their components, including structural studies on the atomic level of resolution – while glorious and important, is, however, only one aspect of bioenergetics, as it describes the production of ATP. Equally important is knowing how ATP is used for cellular work, how all these processes are integrated, and how these integrated processes are regulated. As a result of the enormous efforts of many laboratories over several decades, the muscular contractile cycle is now well understood. The discovery of Na/K-ATPase and of the CaATPase of the sarcoplasmic reticulum and its role in calcium metabolism opened the way to the study of membrane ATPases and ion pumps, which consume a significant portion of cellular energy, such as in brain cells. Both the structure and mechanism of action of these ion pumps are now known in great detail.

An increasingly important and challenging task now is to integrate this information in order to describe quantitatively the functioning and fine mechanisms of regulation of cellular energetics under *in vivo* conditions. An important aim of this strategy is to understand how they work in the cells as a whole and what may go wrong in pathology. This task has become of utmost importance in cellular energetics and biochemistry as well as in physiology, particularly in what concerns the rapid development of systems biology. The basic question now is how all the information collected in the studies of isolated parts of the cell helps us to understand the functioning of living cell, organs, and organisms. It has become evident that to solve this task one needs to account for

1. the complex structural organization of the cell and the nature of the intracellular medium where metabolic reactions occur,
2. specific multiple interactions between cellular components,
3. compartmentation of metabolic processes and energy conversion in cells, and
4. dynamic aspects of structural organization and compartmentation.

The pioneering work of Porter, Clegg, Srere, Welch, Ovadi, and many others has led the way to understanding the importance of a cell's structural organization in metabolic regulation [27, 28]. It is correct to say that modern cellular biochemistry, including bioenergetics, is the biochemistry of organized metabolic systems. Today a general consensus exists about the idea that integrated cellular metabolism is not a simple sum of reactions in homogeneous media; it appears to be much more than that, including specific interactions leading to novel-type regulatory mechanisms, as will be shown in several chapters of this book. The quantitative description of integrated metabolic systems cannot be reduced to the simple use of chemical or enzyme kinetics in diluted solutions. This is the philosophical background of systems biology [29–31].

It is important to take into account that the roots of the study of integrated metabolism go far back into history. The observation by Pasteur that fermentation of glucose is suppressed in aerobiosis, known thereafter as the Pasteur effect, was the first observation of metabolic regulation. One also can mention here the work of Claude Bernard and his discovery of homeostasis, the classical work of Starling on the heart's respiration as a function of workload, the discovery of the Krebs cycle, and many others [21]. Studies on heart energy metabolism have been an important chapter in cellular energetics, rendering enlightening results. The experimental investigation of cardiac metabolism was pioneered by Richard Bing, who was the first to use catheterization methods for studying metabolism in human hearts and established the major role played by free fatty acids in the energy supply for cardiac muscle contraction [32]. The intracellular mechanisms of regulation of these processes were significantly elucidated in fundamental work from the laboratories of Opie, Neely, Williamson, LaNoue, and many others, as will be analyzed in several chapters of this volume [33–35]. Important studies on energy metabolism have been carried out on skeletal muscle (reviewed in Chapter 14) and other tissues such as liver, brain, and smooth muscle.

Throughout the years, a controversial topic has concerned the mechanism of communication between different cellular compartments – as it relates to intracellular energy transfer and regulation of energy metabolism. Central to this problem is how energy is supplied to intracellular sites where needed and how energetic fluxes are regulated in response to changes in demand. These questions have been actively discussed since Lundsgaard's work and the discovery of creatine kinase and adenylate kinase isoenzymes and their compartmentation. This role of adenine nucleotides was assembled into the first concept of energy transfer through the "adenylate wire" by Lipmann [22], who noticed a remarkable analogy between the energy-carrying adenine nucleotide system and the electrical circuit in industrial energetic systems. At the same time, Kalckar presented similar ideas to Lipmann's ATP cycle, also suggesting that phosphorylation of creatine "very likely represents the transmission of the phosphorylation energy to the contractile system" [17]. However, the idea of the central role of creatine kinase and phosphocreatine in the intracellular energy supply was first clearly formulated by Samuel Bessman [36, 37]. The importance of these findings was underscored by pathological conditions such as hypoxia or ischemia, where the contractile function disappears in the presence of high amounts of ATP. Results obtained from studies on energy transfer show that this process is not a simple diffusion of ATP in the intracellular space but the outcome of phosphotransfer networks able to sustain high fluxes. These networks are based on protein–protein interactions, enzyme and metabolite compartmentation, metabolic channeling, functional coupling, etc., and all of these relate to the specific structural organization of the cell.

The networks of energy transfer assume a central role in the cell, connecting the processes of ATP production – mitochondria – with all sites of energy utilization and are therefore central to the integration and regulation of energy metabolism. For this reason, the majority of the chapters in this book deal with the description of systems of energy transfer and their regulatory role within the structurally organized intracellular milieu of cells. The central chapters are written by authors who have been actively working on the problem of energy transfer for decades and who have made major contributions to the field. The complex nature of the processes involved has required the intensive use of mathematical modeling for the quantitative analysis of integrated energy metabolism; these new approaches are also described in several chapters. The final chapters describe the most important applications of the methods and knowledge of cellular energetics for elucidating the mechanisms of pathogenesis arising from disease or stress – the metabolic syndrome.

Computational methods have been intensively developed and successfully applied for the description of cardiac electrophysiology by Denis Noble and whole-heart function within the Physiome project by Peter Hunter and others, as a basis of systems biology [29–31]. Another important system approach in analyzing dynamics and organization of cellular biochemical reactions was the development of network thermodynamics, where network topology and the way in which the elements are interconnected and communicate with each other are of crucial importance for optimal system function [38].

At present, systems biology is understood as a quantitative description of integrated systems at the cellular, organ, and body levels [29–31, 39, 40]. The meaning of "systems" itself varies from two macromolecules that interact to the whole organism, and the aim of systems biology is to understand the system-level properties of these complex multicomponent processes [39, 40]. System-level properties are those that are not predictable from the properties of the isolated components of the cell, and not even from gene expression, but depend upon the sum of interactions within the whole system [29, 39, 40]. Ironically, this may seem to be a return to the theories of "vitalism" of the 19th century [1, 27], but it is not. Systems biology takes into account the realities of intracellular life that were previously unknown [40]. The main focus of this book is the process of energy conversion at both the molecular and cellular levels, with special emphasis on the structure and function of energy transfer and regulatory networks, mechanisms of interaction between their components, and a quantitative description of these networks by computational models. This explains and justifies the title of the book, *Molecular System Bioenergetics*, as describing a new area of cellular bioenergetics in transition from the molecular to the system level. One of the authors of this book, Miguel Aon, has proposed the following definitions for this new direction of research. Molecular system bioenergetics is a broad research field accounting not only for metabolism as a reaction network but also for its spatial (organization) and temporal (dynamics) aspects. It considers intracellular spatial organization as dynamic and evolving, with a topology that carries information and takes into account that metabolic networks are not only mass–energy transforming but also information carrying. Finally, the field of molecular system bioenergetics emphasizes a combined theoretical-experimental approach to the study of mass–energy information cellular networks. All authors of this book agree with these definitions.

In this book it is shown that the molecular system approach to the study of energy conversion in cells allows us to fully explain many classical observations in the cellular physiology of respiration, such as the metabolic aspects of the Frank–Starling law of the heart and the regulation of substrate supply to the cell. This approach helps us to understand how a cell senses its energy status in adjusting its functional activity under stressful conditions or others aspects of its life. We have come a long way since biological oxidation was discovered 200 years ago, to now understanding how this fundamental process is regulated and how the cell handles energy to live.

References

1 Vignais P. (2001) La biologie des origines a nos jours, EDP Sciences, Les Ulis.

2 Vignais P. (2006) Science expérimentale et connaissance du vivant. La méthode et les concepts. EDP Sciences, Les Ulis.

3 Figurovsky N.A. (1969) History of chemistry from ancient times to 19th century, Nauka, Moscow.

4 Meyerhof O. (1935) Über die Intermediärvorgänge bei der biologischen Kohlenhydratspaltung *Ergebn. Enzymforsch.* 4:208.

5 Meyerhof O. (1935) Neuere Untersuchungen über die Reaktionskette der alkoholischen Gärung *Helv. Chim. Acta* 18:1030.

6 Parnas J.K. (1937) Der Mechanismus der Glycogenolyse im Muskel *Ergebn. Enzymforsch.* 6:57.

7 Ivanov I.I., Korovkin B.F., Pinaev G.P. (1997) Biochemistry of muscles, Meditsina, Moscow.

8 Mommaerts W.F. (1969) Energetics of muscular contraction. *Physiol Rev. 49*, 427–508.

9 Starling E.H., Visscher M.B. (1926). The regulation of the energy output of the heart. *J Physiol 62*, 243–261.

10 Lundsgaard E. (1930) Untersuhungen uber Muskelkontraktionen ohne Milchsaurebildung. *Biochem. Z.* 217, 162–177.

11 Engelhardt V.A. (1982) Life and Science. *Ann Rev. Biochem. 51*, 1–19.

12 Lohmann K. (1929) Über die Pyrophosphatfraktion im Muskel. *Naturwiss.* 17, 624.

13 Lohmann K. (1934) Uber die enzymatische Aufspaltung der Kreatinphosphorsaure; zugleich ein Beitrag zum Mechanismus der Muskelkontraktsion. *Biochem. Z.* 271, 264.

14 Hill A.V. (1932) The revolution in muscle physiology. *Physiol. Rev. 12*, 56–67.

15 Gladden L.B. Lactate metabolism: a new paradigm for the third millennium. *J. Physiol.* 558, 5–30.

16 Belitzer V.A., Tsybakova E.T. (1939) Sur le mécanisme des phosphorylations couplées avec la respiration. *Biochimia (Russian) 4*, 516–535.

17 Kalckar H. (1941). The nature of energetic coupling in biological syntheses. *Chem. Rev. 28*, 71–178.

18 Engelhardt V.A., Ljubimova M.N. (1939) Myosin and adenosinetriphosphatase. *Nature* 144, 668–669.

19 Szent-Gyorgyi A. (1953) Chemical physiology of contraction in body and heart muscle, Academic Press, New York.

20 Keilin D. (1938) Cytochrome and intracellular respiratory enzymes. *Ergebnisse Enzymforschung* 7:210.

21 Krebs H.A. (1953) Oxidative phosphorylation. *Biochem. J.* 54:107.

22 Lipmann E. (1941) Metabolic generation and utilization of phosphate bond energy. *Adv. Enzymol.* I, 99–162.

23 Nicholls D., Ferguson S.J. (2002) *Bioenergetics*, Academic Press, London – New York.

24 Mitchell P. (1961) Coupling of phosphorylation to electron transfer by a chemi-osmotic type of mechanism. *Nature*, 191, 144–148.

25 Slater E.C. (1953) Mechanism of phosphorylation in the respiratory chain. *Nature* 172, 975–978.

26 Suelter Ch., DeLuca M., Peter J.B., Boyer P.D. (1961) Detection of a possible intermediate in oxidative phosphorylation, *Nature*, 192, 43–47.

27 Srere P.A. (2000). Macromolecular interactions: tracing the roots. *Trends Biochem Sci* 25, 150–153.

28 Ovàdi, J. (1995) Cell architecture and metabolic channeling, R. G. Landes Co. Springer-Verlag, Austin. New York, Berlin, Heidelberg, London, Paris, Tokyo, Hong Kong, Barcelona, Budapest.

29 Kitano H. (2002). Systems biology: a brief overview. *Science* 295, 1662–1664.

30 Noble D. (2002). Modeling the heart – from genes to cells to the whole organ. *Science* 295, 1678–1682.

31 Hunter P., Nielsen P. (2005) A strategy for integrative computational physiology. *Physiology* 20, 316–325.

32 Bing R.J. (1965) Cardiac metabolism. *Physiol. Rev. 45*, 171–213.

33 Neely J.R., Morgan H.E. (1974) Relationship between carbohydrate and lipid metabolism and the energy balance of heart muscle. *Annu. Rev. Physiology* 63, 413–459.

34 Williamson J.R. (1979) Mitochondrial function in the heart. *Ann. Rev. Physiol.* 41, 485–506.

35 Opie L.H. (1998) The heart. Physiology, from cell to circulation, pp. 43–63. Lippincott-Raven Publishers, Philadelphia.

36 Bessman S.P. (1980) The origin of the creatine – creatine phosphate energy shuttle. In Heart creatine kinase. The integration of enzymes for energy distribution, Jacobus, W.E., Ingwall, J.S. (eds.), Williams & Wilkins, Baltimore, 75–79.

37 Bessman S.P. (1966) A molecular basis for the mechanism of insulin action. *Am. J. Med.* 40, 740–749.

38 Oster G.F., Perelson A.S., Katzir-Katchalsky A. (1971) Network thermodynamics. *Nature*, 234, 393–399.

39 Strange K. (2005) The end of "naive reductionism": rise of systems biology or renaissance of physiology. *Am. J. Physiol.* 288, 968–974.

40 Noble D. (2006) The music of life. Biology beyond the genome. Oxford University Press, Oxford, UK.

Part I
Molecular System Bioenergetics:
Basic Principles, Organization, and Dynamics
of Cellular Energetics

Molecular System Bioenergetics: Energy for Life. Edited by Valdur Saks
Copyright © 2007 WILEY-VCH Verlag GmbH & Co. KGaA, Weinheim
ISBN: 978-3-527-31787-5

1
Cellular Energy Metabolism and Integrated Oxidative Phosphorylation

Xavier M. Leverve, Nellie Taleux, Roland Favier, Cécile Batandier,
Dominique Detaille, Anne Devin, Eric Fontaine,
and Michel Rigoulet

Abstract

Energy metabolism in living organisms is supported by the oxidation of carbohydrates and lipids, which are metabolized with several similarities as well as major differences. Conversely to prokaryotes and inferior eukaryotes, a complete oxidation leading to CO_2 and H_2O formation is mandatory. Our knowledge about oxidative phosphorylation is mostly based on simplified *in vitro* models, i.e., isolated mitochondria where only those parameters included in the experimental systems can be appreciated. However, relationships between mitochondria and the host cell are of major importance in the regulation of the pathway of ATP synthesis and oxygen consumption by the respiratory chain. By determining the respective rate of glucose or fatty acid oxidation, cellular intermediary metabolism affects the ratio between NADH and $FADH_2$, upstream of the Krebs cycle, thus affecting the yield ATP synthesis. The mechanism for translocating reducing equivalents across the mitochondrial inner membrane (the malate–aspartate and glycerol-3-phosphate–dihydroxyacetone phosphate shuttles) also plays an important role. Indeed, because of such characteristics of electron supply to the respiratory chain, oxidative phosphorylation activity also participates in the determination of the ratio of NADH to $FADH_2$ oxidation, by modulation of the protonmotive force. Besides the role of thermodynamic and kinetic constraints applied to the oxidative phosphorylation pathway, it now appears that the supramolecular organization of oxidative phosphorylation and of cellular energy circuits introduces new regulatory factors of oxidative phosphorylation.

1.1
Introduction

Energy metabolism in living organisms is supported by the oxidation of two substrates: carbohydrates and lipids. Amino acids are also good oxidative substrates;

however, after deamination, they enter the pathway of carbohydrate oxidation. Interestingly, the specific metabolism of these two families of substrates exhibits several similarities, while some major differences explain the advantage of maintaining these two different pathways throughout evolution. In particular physiological or pathological situations, lipid and/or glucose oxidation has alternatively both advantages and drawbacks, and choosing the right substrate may confer a substantial advantage. In prokaryotes, as in inferior eukaryotes, a complete oxidation, i.e., involving a respiratory chain, is not mandatory because these living organisms can release an excess of reducing equivalents in the medium. By contrast, in mammals, as in all superior eukaryotes, full oxidation of energetic substrates is required, leading to CO_2 and H_2O formation. Actually, while some cells lacking mitochondria (such as red blood cells and a few other cells) can survive with glucose fermentation to lactate as a unique energetic pathway, the reducing compound lactate is further oxidized in other aerobic cells of the same organism in such a way that, as a whole, energy metabolism in these organisms is completely aerobic [1].

From a strictly bioenergetic point of view, carbohydrates and lipids exhibit major differences regarding rate and efficiency of oxidative phosphorylation, while they both contribute to the reduction of NAD^+ and FAD, although not in the same proportion. Because of the characteristics of the pathways of electron supply to the respiratory chain, oxidative phosphorylation activity participates in the determination of the ratio of NADH to $FADH_2$ oxidation, by the level of the generated steady-state protonmotive force. Reciprocally, the nature of the electron donors (NADH or $FADH_2$) regulates the rate and efficiency of oxidative phosphorylation as well as their relationship with the protonmotive force.

Besides such thermodynamic and kinetic constraints applied to the oxidative phosphorylation pathway, other important parameters are emerging, such as the supramolecular organization of oxidative phosphorylation and of cellular energy circuits. The highly dynamic characteristics of supramolecular organization introduce new regulatory factors of oxidative phosphorylation in addition to classical kinetic and thermodynamic parameters. Most of our knowledge about energy metabolism is based on simplified *in vitro* models, where the number of significant regulatory parameters is artificially limited. Hence, only those parameters included in the considered systems can be appreciated, and thus the others are ingnored. Furthermore, minor parameters are often overemphasized because of the characteristics of the considered experimental system. In fact, the difference between *in vitro* and *in vivo* situations regarding glucose or lipid as a preferred substrate for oxidation and ATP synthesis is a good example. When carbohydrates (glucose) and lipids (octanoate) are provided simultaneously to isolated cells (hepatocytes), lipid oxidation will be preferred and pyruvate oxidation powerfully inhibited. This is due to the negative feedback effect of β-oxidation provided by acetyl-CoA on pyruvate dehydrogenase [2, 3]. However, the same competition between lipids and glucose *in vivo* (in humans) results in preferred glucose oxidation, while lipids are stored as a consequence of a rise in insulin [4]. Hence, the competition between the two major metabolites as substrate for ATP synthesis results in opposite pictures *in vivo* and *in vitro*. Of course, this chapter will also suffer from these limitations; however, we will attempt to present the most integrative perspective.

1.2
Membrane Transport and Initial Activation

Plasma membrane transports followed by activation are the initial steps of glucose and fatty acid metabolism through glycolysis and β-oxidation, respectively. Glucose transport is allowed through a family of 13 carriers (GLUT), which differ in their kinetic characteristics [5]. Among these different carriers, GLUT4 is recognized as being regulated by insulin and other effectors and is connected to cellular energy status via AMP-activated protein kinase (AMPK) phosphorylation. GLUT4 carriers are stored in cytoplasmic vesicles and translocated to the plasma membrane in response to appropriate signaling [6]. Interestingly, similar events occur during fat transport across plasma membrane. FAT/CD36, the fatty acid carrier, is stored in the cytoplasm and translocated to the membrane upon appropriate signaling events. Furthermore, AMPK activation is also involved in the initiation of such translocation, reinforcing the similarity between both pathways [7]. Fatty acid activation by acyl-CoA synthetase is a cytosolic step requiring ATP and free CoA, while AMP and PP_i are released. Hence, fatty acid activation affects AMP levels and thus may participate in AMP kinase signaling processes (see Chapter 7). Glucose phosphorylation is permitted by a family of four enzymes (hexokinase) with different kinetic characteristics [8]. Interestingly, it has been shown that these enzymes are present as free compounds in the cytoplasm or bound to the outer mitochondrial membrane voltage-dependent anion channel (VDAC) (see Chapter 6). Interestingly, these enzymes, except glucokinase, are powerfully inhibited by the product glucose 6-phosphate (high elasticity) when present in the free form, while they are insensitive to this product when bound to the mitochondrial membrane [9]. This feature of hexokinase when bound to mitochondria is of major importance in explaining high glycolytic rates in some conditions, such as in pancreatic β-cells, in cancer cells, or the occurrence of cellular energy deficits.

1.3
Cytosolic Pathway

Numerous effectors regulate glycolysis, and it is not the purpose of this chapter to describe this as it has already been summarized in several good reviews [10–12]. However, we would like to focus on two important thermodynamic parameters: redox and phosphate potentials. Two successive steps work at near-equilibrium downstream of triose phosphate: glyceraldehyde-3-phosphate dehydrogenase (GAPdh) and phosphoglycerate kinase. Glycolysis is activated when the cytosolic

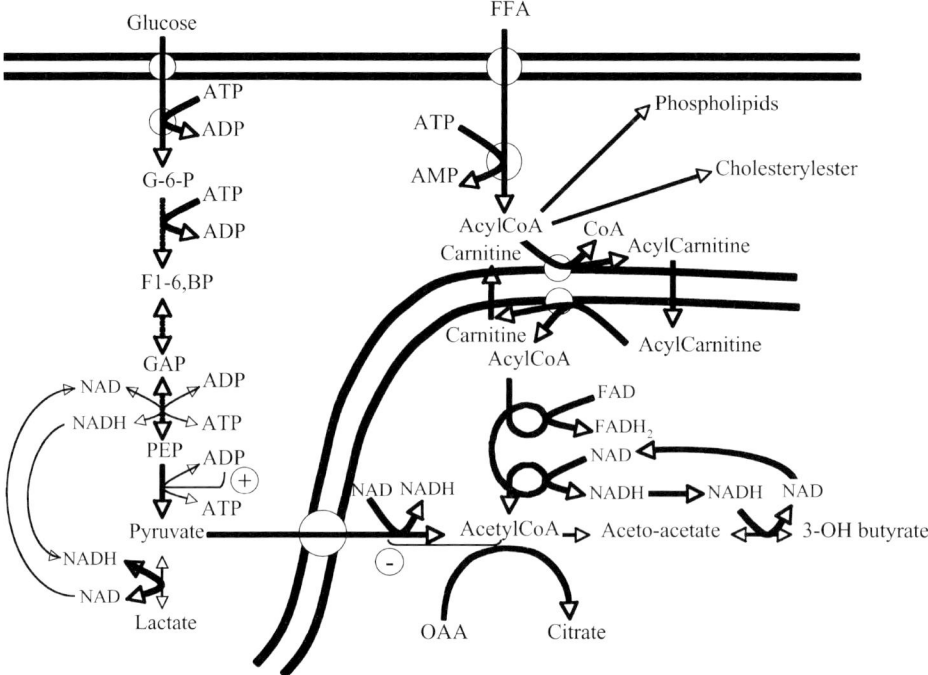

Fig. 1.1 Comparison between glycolysis and fatty acid oxidation. Glycolysis and fatty acid oxidation exhibit several similarities such as plasma membrane transport by a specific carrier (GLUT or Fat CD36), activation by an energy-dependent process (hexokinase or fatty acyl synthetase), followed by an oxidative pathway leading to a common product, acetyl-CoA. The glycolytic pathway is located in the cytoplasm, and its product, pyruvate, is translocated into the matrix by a carrier before oxidative decarboxylation to acetyl-CoA. By contrast, fatty acyl-CoA is first translocated into the matrix by the carnitine shuttle before β-oxidation, leading to acetyl-CoA, whose oxidation in the Krebs cycle is a pathway common to both carbohydrate and fatty acid complete oxidations. The high elasticity of pyruvate oxidative decarboxylation via pyruvate dehydrogenase by its product acetyl-CoA is responsible for a tight reciprocal control of pyruvate oxidation by the β-oxidation rate. Finally, while the redox balance of glycolysis can be canceled by lactate formation, ketone synthesis (3-hydroxybutyrate) can only partially compensate for the reducing equivalents generated by β-oxidation. This indicates that mitochondrial oxidation is mandatory for achieving fatty acid oxidation. G6,P: glucose 6-phosphate; F1-6,BP: fructose 1-6-bisphosphate; GAP: glyceraldehyde phosphate; PEP: phosphoenolpyruvate; FFA: free fatty acid.

compartment is oxidized and when phosphate potential is lowered. The last step of glycolysis, pyruvate kinase, is far from equilibrium and is allosterically activated by ADP (see Fig. 1.1). A defect in ATP supply by oxidative phosphorylation induces a decrease in phosphate potential, while the cytosolic and mitochondrial compartments are even more reduced, a situation that would induce opposite effects on glycolysis: a low phosphate potential favors high phosphoenolpyruvate concentration, while a high NADH:NAD$^+$ ratio favors high glyceraldehyde-3-phosphate concentration. However, the allosteric activation of pyruvate kinase by ADP [10] results in increasing the glycolytic flux, allowing a cellular release of reducing equivalents via pyruvate fermentation to lactate. Hence, it appears that the glycolytic rate is finely tuned by both the cytosolic redox state and phosphate potential and that oxidative phosphorylation is tightly connected to glycolytic rate. Three main pathways represent the cytosolic fate of fatty acyl-CoA metabolism: (1) mitochondrial transport and subsequent β-oxidation, (2) phospholipid synthesis, and (3) cholesteryl ester synthesis, plus two others in some specific tissues: triglyceride synthesis and peroxisomal metabolism (Fig. 1.1). Recent data indicate that cytosolic metabolism of acyl-CoA is, at least partly, dependent on channeling processes at the level of acyl-CoA synthetase [13].

1.4
Mitochondrial Transport and Metabolism

The next step is represented by the transport across the mitochondrial membrane of both pyruvate and acyl-CoA, which are ultimately oxidized in the matrix. Fatty acid translocation across the mitochondrial inner membrane has been extensively studied and represents a major controlling step of long-chain fatty acid oxidation [14, 15]. Non-activated medium-chain fatty acids (non-esterified medium-chain fatty acids) can cross the inner mitochondrial membrane; therefore, they can be oxidized in a carnitine-independent manner. However, such a process requires prior matricial activation by mitochondrial medium-chain acyl-CoA synthetase, which is present mostly in liver [16, 17]. Therefore, the carnitine-independent oxidation of medium-chain fatty acids occurs mainly in liver. Besides being the major controlling step of acyl-CoA translocation into the matrix, the mitochondrial NADH:NAD$^+$ ratio also plays a key role in the control of β-oxidation, mainly through the redox state of enzymes directly involved in this pathway [14]. As in the respiratory chain, the pathway of β-oxidation involves several electron carriers, and the requirement of a simultaneous oxidation of both NADH and FADH$_2$ in order to complete the entire pathway is a very important feature (see Fig. 1.2). Because NADH oxidation must occur at the complex I level (except for 3-hydroxybutyrate dehydrogenase, see below), this respiratory chain complex represents a major controlling step for β-oxidation, and NADH oxidation by complex I must parallel the rate of β-oxidation in all tissues, except liver. In liver mitochondria, NADH can also be substantially oxidized by reducing acetoacetate to β-hydroxybutyrate in the ketogenic pathway. It is therefore possible to compare glycolysis with lactate fermentation (anaerobic glycolysis) versus liver β-oxidation with ketogenesis: both pathways generate reducing equivalents (NADH production), which in turn negatively control the rate, while in both cases NADH can be oxidized in the last step (lactate dehydrogenase or β-hydroxybutyrate dehydrogenase), thus allowing maintenance of the flux (Fig. 1.1). However, a striking dif-

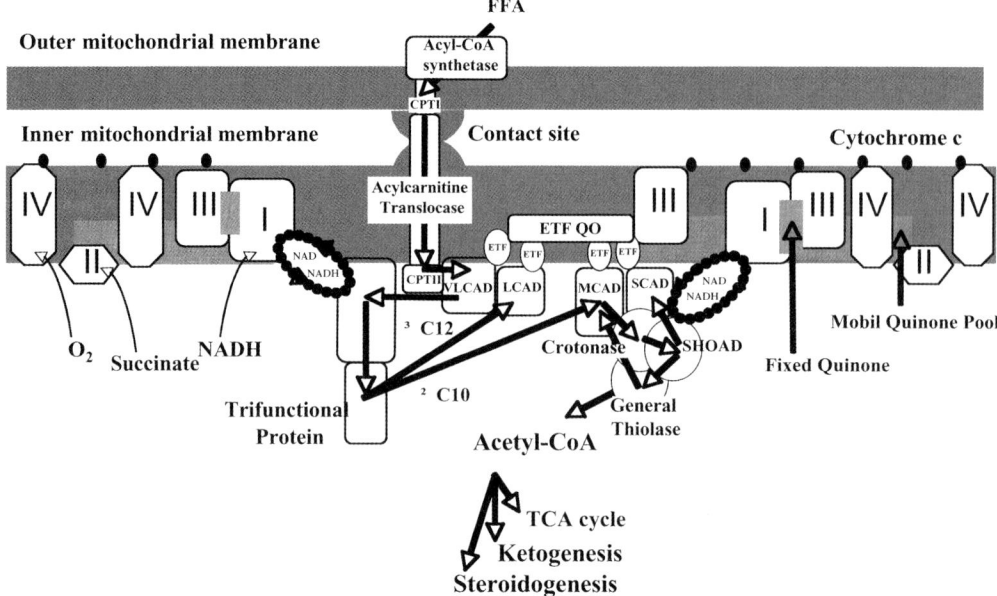

Fig. 1.2 Electron supply to the respiratory chain and β-oxidation. Following their entry into the mitochondrial matrix by means of the carnitine shuttle, fatty acids undergo β-oxidation, which is compartmentalized in different pathways according to chain length: very-long-chain (VLCFA), long-chain (LCFA), medium-chain (MCFA), or short-chain (SCFA) fatty acids. Electrons provided by the first step, FAD-dependent acyl-CoA dehydrogenase (VLCAD, LCAD, MCAD, and SCAD), are transferred to complex III via a specific carrier, electron transfer flavin (ETF) (second step); electrons provided from the third step, 3-hydroxyacyl-CoA dehydrogenase (trifunctional protein, short chain hydroxyacy-CoA dehydrogenase: SHOAD), are channeled to complex I (adapted from [14]).

ference remains between the two pathways: the net redox balance of glycolysis and fermentation is null, while there is a net production of reducing equivalents with β-oxidation, even when followed by ketogenesis. In conclusion, fatty acid β-oxidation requires mitochondrial respiratory chain activity.

Pyruvate entry into the matrix via the pyruvate carrier has long been studied [18]. This electroneutral transport involves one proton; therefore, pyruvate transport is affected by the difference in pH through the inner membrane. The next step is represented by the oxidative decarboxylation of pyruvate by pyruvate dehydrogenase, a step highly regulated by many effectors including two major forces related to oxidative phosphorylation: redox and phosphate potentials [3, 19]. It is important to note that the product of this step, acetyl-CoA, represents the ultimate and single common compound of both pathways. Indeed, the negative feedback by acetyl-CoA, provided by the β-oxidation, towards pyruvate oxidation represents the reciprocal metabolic control of β-oxidation on glucose oxidation (Fig. 1.1). However, as stated above, numerous effectors are involved in the regulation of both pathways, such as insulin, resulting in a much more complicated physio-logical response.

1.5
Respiratory Chain and Oxidative Phosphorylation

Respiratory chain activity has three main functions: (1) to oxidize reduced coen-zymes, (2) to lower cellular oxygen concentration, and (3) to maintain a high protonmotive force. In addition, a high protonmotive force allows several mito-chondrial enzymatic activities, including, of course, ATP synthesis, the net result being a chemiosmotic coupling between oxidation and phosphorylation. Because the cellular ATP requirement is not stoichiometrically linked to the need of reox-idation of reduced equivalent production, it is therefore of importance to finely adjust phosphorylation and oxidation separately, i.e., to modulate the ratio of ATP to O. There are three physiological ways to disjointedly tune oxidation and phosphorylation: (1) the site of electron supply to the respiratory chain; (2) the in-trinsic stoichiometry of respiratory chain proton pumps, and (3) the degree of the proton conductance of the inner mitochondrial membrane [20].

1.6
Electron Supply

Electron supply to the respiratory chain is provided either upstream (NADH) or downstream ($FADH_2$) of complex I. This difference has important consequences because in the former case there are three coupling sites (complexes I, III, and IV), while in the latter only two coupling sites are involved (complexes III and IV). Hence, the yield of ATP synthesis is lowered by approximately 40% when $FADH_2$ is oxidized as compared with NADH. The nature of the cellular sub-strates (i.e., fatty acids versus carbohydrate) affects the stoichiometry of oxidative phosphorylation by affecting the ratio between NADH and $FADH_2$. Conversely to carbohydrate metabolism, fatty-acid β-oxidation results in the formation of equi-molar amounts of NADH and $FADH_2$. Regarding the β-oxidation pathway, elec-trons are provided both to complex I from 3-hydroxyacyl-CoA dehydrogenase, via the bulk phase or by a channeling process (Fig. 1.2), and downstream of complex I to the quinone pool via the electron transfer flavin (ETF). Hence, the stoichiom-etry of ATP synthesis to oxygen consumption is lower when lipids rather than car-bohydrates are oxidized. In the case of acetyl-CoA oxidation by the Krebs cycle, which is common to both carbohydrate and lipid oxidation, reducing equivalents are provided simultaneously to complex I (3 NADH), either via the bulk phase or by channeling, and to the quinone pool via complex 2 (1 $FADH_2$). The net result is then 10 NADH:2 $FADH_2$ for the complete glucose oxidation and 12 NADH:6 $FADH_2$ for the complete oxidation of hexanoate, a six-carbon fatty acid, by β-oxidation.

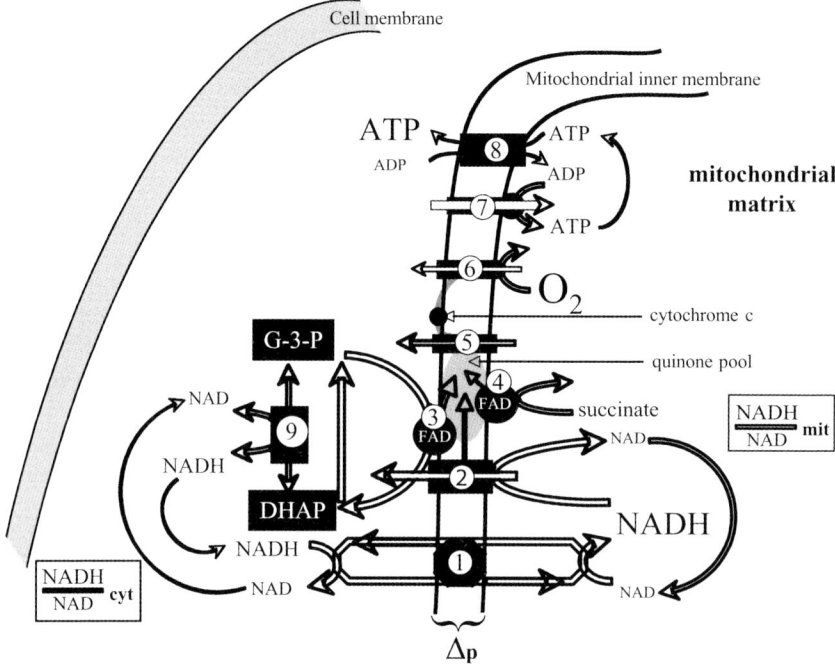

Fig. 1.3 Mitochondrial reducing power translocation and oxidative phosphorylation. The oxidative phosphorylation pathway consists of successive transductions of potentials from the chemical energy contained in the nutrient-to-phosphate potential (ATP:ADP·P_i), which is the energy source for the different biological functions. The chemical energy, supplied as reducing equivalent (NADH at complex I and $FADH_2$ at complex 2), is first converted in membrane potential (Δp) by the respiratory chain, which links redox reaction to proton extrusion from the matrix to the intermembrane space. The high electrochemical gradient (-180 mV) generated permits ATP synthesis from ADP and P_i, as well as other functions such as Ca^{2+} uptake and substrate transport. The inner membrane is impermeable to NADH/NAD^+ and to ATP/ADP; therefore, these compounds must be translocated by carrier systems: the malate–aspartate shuttle and adenine nucleotide translocase. These metabolite exchanges across the mitochondrial membrane are electrogenic and therefore depend on an electrochemical gradient. The gradient allows the entry of reducing equivalents in the matrix and the transport of ATP into cytosol. Hence, the net result is that the higher the electrochemical gradient is, the higher the matricial NADH:NAD^+ ratio will be, allowing a high concentration of respiratory substrate NADH. Similarly, the higher the gradient, the higher the export of ATP, which helps to maintain a low ATP:ADP·P_i ratio in the matrix (facilitating ATP synthesis) and a high ATP:ADP·P_i ratio in the cytosol (facilitating ATP hydrolysis and energy utilization). (1) malate–aspartate shuttle; (2) complex I; (3) mitochondrial glycerol-3-phosphate dehydrogenase; (4) respiratory chain complex 2 – succino-dehydrogenase; (5) complex III; (6) complex IV; (7) ATP synthetase; (8) adenine nucleotide translocator; (9) cytosolic glycerol-3-phosphate dehydrogenase.

1.7
Reducing Power Shuttling Across the Mitochondrial Membrane

Because the mitochondrial inner membrane is impermeable to NADH, shuttle systems are required to carry the reducing power into the mitochondrial matrix. Two shuttles are involved in this exchange: the malate–aspartate shuttle, which depends on protonmotive force (Δp), and the glycerol-3-phosphate–dihydroxyacetone phosphate shuttle, which does not (see Fig. 1.3). While the former system provides electrons to complex I (i.e., as NADH), the latter supplies electrons directly to the quinone pool from the mitochondrial glycerol-3-phosphate dehydrogenase ($FADH_2$). Thus, by adjusting the flux through these two shuttles, the yield of oxidative phosphorylation (i.e., the cellular metabolism of oxygen and ATP) can be regulated. One of the major effects of thyroid hormones on mitochondrial energy metabolism is achieved through this mechanism because these hormones affect transcription of the mitochondrial glycerol-3-phosphate dehydrogenase, which regulates the flux through the glycerol-3-phosphate–dihydroxyacetone phosphate shuttle [21–23].

1.8
Electron Transfer in the Respiratory Chain: Prominent Role of Complex I in the Regulation of the Nature of Substrate

Because complex I is the final common obligatory step for NADH oxidation, all pathways leading to NADH production are in competition at this level, which must be a location of tight control. When considering carbohydrate and lipid mitochondrial oxidation, the dehydrogenases of the specific pathways (pyruvate dehydrogenase and 3-hydroxyacyl-CoA dehydrogenase) as well as the Krebs cycle dehydrogenases (isocitrate dehydrogenase, α-ketoglutarate dehydrogenase, and malate dehydrogenase) compete. Furthermore, because malate dehydrogenase also represents the matricial NADH supplier of the malate–aspartate shuttle, this step represents a crossroad between (1) cytosolic and (2) mitochondrial redox state, (3) mitochondrial protonmotive force, (4) pyruvate, and (5) fatty acyl oxidation.

Considering this highly composite situation of multiple reciprocal regulations of different and interconnected pathways, two opposite and extreme pictures can be envisaged. First, the redox state of the bulk phase of the matricial compartment represents the common intermediate. In this situation the flux control of the different pathways depends on the capacity of complex I to oxidize NADH and on the specific elasticity of each dehydrogenase of the whole system towards the matricial NADH:NAD ratio. The second possibility is based on channeling of electron transfers between each dehydrogenase (or some of them) of the whole system and complex I. In this situation the supramolecular organization represents the main controlling factor (see below). Most likely, the actual situation is the result of a combination of these two extreme possibilities in a dynamic compromise, which is continuously adjusted and resettled. Nevertheless, the oxido-reduction status of complex I probably represents the key regulator of these complex and interconnected pathways.

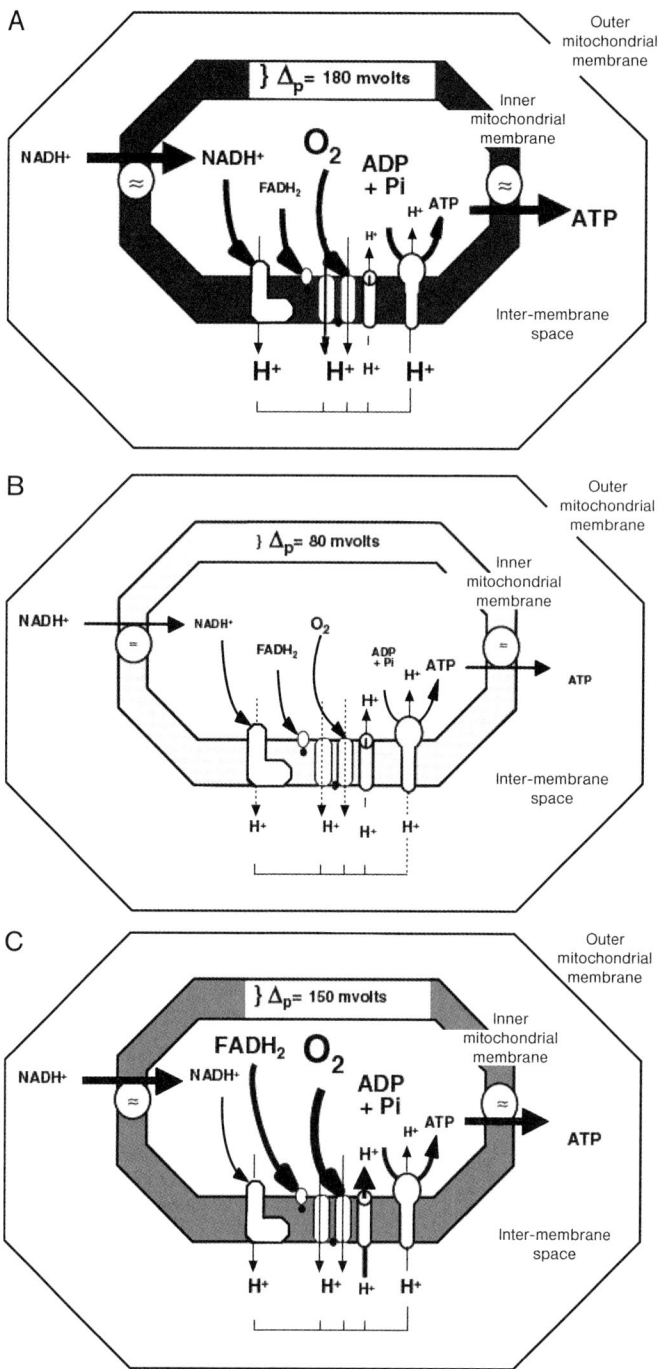

Fig. 1.4 (legend see p. 21)

Complex I is well known to work at near-equilibrium: it equilibrates the matricial redox state (NADH:NAD) and the protonmotive force [24]. Complex I generates a high protonmotive force while oxidizing NADH to NAD, i.e., when electron flux is transported in the forward direction. When electrons are transported in the reverse direction, it reduces NAD to NADH at the expense of the high protonmotive force generated at other coupling sites of the respiratory chain (complexes III and IV) and of electron supply downstream of complex I (i.e., FADH$_2$). Hence, thanks to this near-equilibrium between redox state and protonmotive force by complex I, mitochondrial protonmotive force appears to be the key regulatory factor in determining both the rate and the nature of the substrate used for fuelling the cell.

1.9
Modulation of Oxidative Phosphorylation by Respiratory Chain Slipping and Proton Leak

The occurrence of slipping processes between electron flux and proton transfer at the level of the respiratory chain is now well established [20, 25–28]. Recent data on the structure of cytochrome oxidase support the occurrence of slipping processes at this level [29–32]. In contrast to a proton leak, a slipping mechanism permits modulation of the rate of oxidation while the protonmotive force and the rate of ATP synthesis are not modified. Despite the low permeability of the inner membrane, proton leaks across the membrane do occur and result in the uncoupling of oxygen consumption from ATP synthesis, the energy being dissipated as heat (see Fig. 1.4). Uncoupling through proton leak permits dissociation

---

Fig. 1.4 Schematic view of oxidative phosphorylation and its uncoupling. (A) Coupled oxidative phosphorylation (see legend to Fig. 1.3). (B) Uncoupling with carbohydrates. By permitting the protons to freely reenter into the matrix, the uncoupling process (via uncoupling protein, for instance) creates a "futile cycling," dissipating energy into heat at the expense of oxygen consumption and water production. In the presence of carbohydrate as exogenous source of energy, reduced substrates are supplied to the respiratory chain as NADH, and energy-dependent import of NADH is required. Because of the uncoupling, the electro-chemical gradient collapses, impairing the active transport of NADH. Therefore, a sufficiently high reducing state in the matrix may not be sustained. In these conditions the net result is a collapse of Δp and the ATP/ADP·P$_i$ ratio, while oxygen consumption is low, despite the uncoupling state. (C) Uncoupling with fatty acids. In the presence of fatty acids, the metabolic effects of uncoupling are different. In this case the production of FADH$_2$ in the matrix by β-oxidation allows the supplying of substrates directly to complex II, even when the electrochemical gradient is collapsed. High levels of substrate supply to the respiratory chain lead to a strong activity, as evidenced by the very large increase in oxygen consumption. This high-level respiration activity permits maintenance, to some extent, of the electrochemical gradient, and therefore some ATP synthesis is maintained. In this case the main effect of uncoupling is an increase in oxygen consumption and heat production.

of the rate of oxidation from that of phosphorylation and thus a decrease in the yield of oxidative phosphorylation. This mechanism is similar to uncoupling through uncoupling proteins. The discovery of the physiological function of brown fat in mammals, related to the presence of uncoupling protein-1 (UCP1), has opened a new era in our understanding of the regulation of oxidative phosphorylation by describing a role for energy waste. Several other UCPs have recently been described [33, 34], and some of these (UCP2 and UCP3) have been found in most tissues, including white adipose tissue, muscle, macrophages, spleen, thymus, Kupffer cells, etc. [35, 36] . Whether proton leak occurs through these UCPs appears to be a legitimate question to ask.

In summary, the primary effects of slippage of proton pumps appear to modulate the rate of oxidation at a given level of protonmotive force, while proton leak primarily affects the level of protonmotive force. The secondary effects of slipping are related to its effect on redox state, and the physiological result is an increased reoxidation rate without a major effect on the nature of substrate involved (NADH or $FADH_2$). By contrast, the secondary effect of proton leak is related to the change in Δp with all the consequences related to it, thus including the effect mediated by complex I (i.e., a modulation of the nature of substrate supply [NADH versus $FADH_2$]).

1.10
The Nature of Cellular Substrates Interferes with the Metabolic Consequences of Uncoupling

Irrespective of the molecular mechanism(s), the metabolic consequences of a protonophoric leak (uncoupling) can be classified into three categories: (1) those related to the change in oxidation rate and redox state, (2) those related to the change in protonmotive force, and (3) those related to the change in ATP synthesis and phosphate potentials. In isolated mitochondria incubated in the presence of saturating concentrations of respiratory substrates, uncouplers invariably decrease Δp and redox and phosphate potentials and consequently increase respiratory rates. By contrast, in intact cells these forces are involved in a complex metabolic network that may significantly affect the outcome of uncoupling on the same parameters. On the one hand, when uncoupling is achieved without fatty acid, it results in a profound decrease in both Δp and cytosolic and mitochondrial ATP:ADP ratios, while the rate of respiration is not increased. This is due to a decline in the matricial reducing state linked to the collapsed protonmotive force. On the other hand, in the presence of octanoate, a large increase in respiration is associated with limited effects on Δp and ATP:ADP ratios because of the matricial supply of reducing equivalents downstream of complex I ($FADH_2$) [37–39]. Hence, the metabolic consequences of uncoupling in intact liver cells are variable and critically depend on the metabolic state of the cells. In the presence of a large supply of fatty acids and oxygen, the main effect of uncoupling is a dramatic increase in oxygen consumption as well as energy waste. The active mitochondrial β-oxidation permits the sustaining of a very high rate of mitochondrial respiration and a high membrane potential, while ATP synthesis can be at least partially maintained because of this high respiratory chain activity. When glycolysis is the unique pathway for substrate supply to the respiratory chain, the decreased mitochondrial membrane potential resulting from uncoupling strongly affects the mitochondrial redox potential, because the malate–aspartate shuttle, which depends on maintenance of Δp, is not able to sustain a highly reduced redox potential in the matrix. Hence, under these conditions, uncouplers do not significantly affect the respiratory rate, because the supply of reducing equivalents to complex I becomes controlling [40]. The main effect of uncoupling would be a striking decrease in Δp and ATP:ADP ratio, with an overall decrease in cell metabolic activity. It is not surprising that uncoupling by UCP1 results in a huge increase in the rate of fatty acid oxidation, oxygen consumption, and heat production in brown fat, where the storage of triglycerides is associated with a large number of mitochondria with high oxidative capacity. Thus, depending on substrate oxidation and heat production, uncoupling in intact cells may have very different effects on mitochondrial depolarization and on its consequences on cell energy status. Hence, on the one hand, uncoupling may be a very efficient way of decreasing oxygen concentration by reducing it to water; on the other hand, by decreasing mitochondrial membrane potential and the ATP:ADP ratio, uncoupling may affect all cellular pathways related to these potentials.

1.11
Dynamic Supramolecular Arrangement of Respiratory Chain and Regulation of Oxidative Phosphorylation

Considering the complex situation resulting from the numerous interactions of various parameters involved in many steps, either common or specific to these different pathways, the large number of common controlling steps may lead to excessive reciprocal dependence of these interconnected pathways. Hence, it seems important to maintain some degree of independence between these considered pathways, even if a high degree of coordination is mandatory. A biological response to this crucial question is given by a supramolecular organization of the pathway. Indeed, channeling in the glycolysis pathway has long been recognized. The cellular plasma membrane and cytoskeleton binding of the glycolytic enzyme lead to channeling of NADH to the respiratory chain in yeast, thanks to the presence of an external NADH dehydrogenase at the outer surface of the inner membrane [41, 42]. In mammals, including humans, such an organization has been shown to play a role in the compartmentation of the glycolytic supply of ATP for fuelling the sodium–potassium ATPase [43]. More recently the impact of channeling on fatty acid metabolism has been emphasized for both the cytosolic and mitochondrial parts of the pathway [13, 44]. Indeed, the fate of fatty acids appears to be determined by supramolecular organization immediately after cellular entry, i.e., activation and orientation towards the main pathways (mitochondrial oxidation, phospholipids synthesis, cholesterol esterification, etc.). In addition, similar processes of channeling are also involved in mitochondrion translocation and matricial β-oxidation. At the end of the pathway, electrons are probably channeled to the respiratory chain, and supramolecular organization of Krebs cycle enzymes has long been reported [45–48]. Interestingly, one of the Krebs cycle dehydrogenases, malate dehydrogenase, which is also involved in the reducing equivalent's translocation across the inner mitochondrial membrane, has been reported to preferentially provide NADH towards complex I [49]. Finally, the organization of the respiratory chain is another example of metabolic channeling between reduced coenzymes and oxygen. In the classical model of the respiratory chain arrangement, several multiproteic blocks are defined (complexes I, III, and IV), which are interconnected by small and mobile electron carriers (i.e., quinone and cytochrome *c*). In such a view, a "common" quinone pool interconnects complexes I and III, as complexes III and IV are connected by a "common" cytochrome *c* pool ("liquid-state" model). Such an organization has been challenged by two kinds of experimental data. First, experiments with mild detergents have permitted the obtaining of several types of supramolecular organizations of the respiratory chain with different fixed stoichiometry, including, for instance, complexes I, III, and IV associated with quinones and cytochrome *c* or complexes III and IV associated with quinone and cytochrome *c* (see [50] for review). Of course, such a "unit of electron transfer" is not fixed but represents a dynamic supramolecular organization that can be modulated depending on environmental conditions. This view of a "solid-state" model in which orderly sequences of redox compounds catalyze electron flux is also supported by kinetics analysis [51, 52]. Secondly, the origin of electron supply, i.e., from the different dehydrogenases, may or may not lead to competition. Hence, in yeast mitochondria, NADH supply by external NADH dehydrogenase inhibits all matricial dehydrogenases except succinate dehydrogenase, indicating a preferred channeling pathway [53, 54].

From these considerations, it appears that besides the long-recognized role of the various regulatory effectors involved in the tight reciprocal control of the two main substrates involved in cellular energy metabolism (carbohydrates and lipids), the global organization of the system is also a major parameter that, like metabolic effectors, is subject to continuous adaptation.

In view of several data sets, already published or not, it seems unlikely, at least for liver metabolism, that succinate is oxidized in the absence of complex I, because of the importance of the reverse electron flux from succinate to NADH [24, 55]. By contrast, the lack of evidence of such a reverse electron flux on complex I when fatty acid oxidation represents the electron source does not favor the presence of complex I in such a respiratory chain organization. Of course, NADH formation by 3-hydroxyacyl-CoA dehydrogenase in the β-oxidation must oxidized; however, in such a view, this could be achieved in a different respiratory chain organization.

Acknowledgments

This work was supported by INSERM, by the Ministère de l'Enseignement, de la Recherche et de la Technologie (MERT), and by GIP ANR (QuinoMitEAO).

References

1 Leverve XM. (1999). Energy metabolism in critically ill patients: lactate is a major oxidizable substrate. *Curr Opin Clin Nutr Metab Care.* **2**: 165–169.

2 Sumegi B, Batke J, Porpaczy Z. (1985). Substrate-induced structural changes of the pyruvate dehydrogenase multienzyme complex. *Arch Biochem Biophys.* **236**: 741–752.

3 Kerbey AL, Randle PJ, Cooper RH, Whitehouse S, Pask HT, Denton RM. (1976). Regulation of pyruvate dehydrogenase in rat heart. Mechanism of regulation of proportions of dephosphorylated and phosphorylated enzyme by oxidation of fatty acids and ketone bodies and of effects of diabetes: role of coenzyme A, acetyl-coenzyme A and reduced and oxidized nicotinamide-adenine dinucleotide. *Biochem J.* **154**: 327–348.

4 Cahill GF, Jr., Owen OE, Felig P. (1968). Insulin and fuel homeostasis. *Physiologist.* **11**: 97–102.

5 Joost HG, Thorens B. (2001). The extended GLUT-family of sugar/polyol transport facilitators: nomenclature, sequence characteristics, and potential function of its novel members (review). *Mol Membr Biol.* **18**: 247–256.

6 Ishiki M, Klip A. (2005). Minireview: recent developments in the regulation of glucose transporter-4 traffic: new signals, locations, and partners. *Endocrinology.* **146**: 5071–5078.

7 Koonen DP, Glatz JF, Bonen A, Luiken JJ. (2005). Long-chain fatty acid uptake and FAT/CD36 translocation in heart and skeletal muscle. *Biochim Biophys Acta.* **1736**: 163–180.

8 Wilson JE. (1995). Hexokinases. *Rev Physiol Biochem Pharmacol.* **126**: 65–198.

9 Gerbitz KD, Gempel K, Brdiczka D. (1996). Mitochondria and diabetes. Genetic, biochemical, and clinical implications of the cellular energy circuit. *Diabetes.* **45**: 113–126.

10 Hers HG, Hue L. (1983). Gluconeogenesis and related aspects of glycolysis. *Annu Rev Biochem.* **52**: 617–653.

11 Pilkis SJ, el-Maghrabi MR, Claus TH. (1988). Hormonal regulation of hepatic gluconeogenesis and glycolysis. *Annu Rev Biochem.* **57**: 755–783.

12 Van Schaftingen E. (1993). Glycolysis revisited. *Diabetologia.* **36**: 581–588.

13 Muoio DM, Lewin TM, Wiedmer P, Coleman RA. (2000). Acyl-CoAs are functionally channeled in liver: potential role of acyl-CoA synthetase. *Am J Physiol Endocrinol Metab.* **279**: E1366–1373.

14 Eaton S. (2002). Control of mitochondrial beta-oxidation flux. *Prog Lipid Res.* **41**: 197–239.

15 Eaton S, Bartlett K, Pourfarzam M. (1996). Mammalian mitochondrial beta-oxidation. *Biochem J.* **320 (Pt 2)**: 345–357.

16 Papamandjaris AA, MacDougall DE, Jones PJ. (1998). Medium chain fatty acid metabolism and energy expenditure: obesity treatment implications. *Life Sci.* **62**: 1203–1215.

17 Fujino T, Takei YA, Sone H, Ioka RX, Kamataki A, Magoori K, Takahashi S, Sakai J, Yamamoto TT. (2001). Molecular identification and characterization of two medium-chain acyl-CoA synthetases, MACS1 and the Sa gene product. *J Biol Chem.* **276**: 35961–35966.

18 Papa S, Paradies G. (1974). On the mechanism of translocation of pyruvate and other monocarboxylic acids in rat-liver mitochondria. *Eur J Biochem.* **49**: 265–274.

19 Wieland OH. (1983). The mammalian pyruvate dehydrogenase complex: structure and regulation. *Rev Physiol Biochem Pharmacol.* **96**: 123–170.

20 Rigoulet M, Leverve X, Fontaine E, Ouhabi R, Guerin B. (1998). Quantitative analysis of some mechanisms affecting the yield of oxidative phosphorylation: dependence upon both fluxes and forces. *Mol Cell Biochem*. **184**: 35–52.

21 Dummler K, Muller S, Seitz HJ. (1996). Regulation of adenine nucleotide translocase and glycerol 3-phosphate dehydrogenase expression by thyroid hormones in different rat tissues. *Biochem J*. **317**: 913–918.

22 Kalderon B, Hertz R, Bar Tana J. (1992). Effect of thyroid hormone treatment on redox and phosphate potentials in rat liver. *Endocrinology*. **131**: 400–407.

23 Muller S, Seitz HJ. (1994). Cloning of a cDNA for the FAD-linked glycerol-3-phosphate dehydrogenase from rat liver and its regulation by thyroid hormones. *Proc Natl Acad Sci U S A*. **91**: 10581–10585.

24 Grivennikova VG, Vinogradov AD. (2006). Generation of superoxide by the mitochondrial Complex I. *Biochim Biophys Acta*. **1757**: 553–561.

25 Azzone GF, Zoratti M, Petronilli V, Pietrobon D. (1985). The stoichiometry of H+ pumping in cytochrome oxidase and the mechanism of uncoupling. *J Inorg Biochem*. **23**: 349–356.

26 Piquet MA, Nogueira V, Devin A, Sibille B, Filippi C, Fontaine E, Roulet M, Rigoulet M, Leverve XM. (2000). Chronic ethanol ingestion increases efficiency of oxidative phosphorylation in rat liver mitochondria. *FEBS Lett*. **468**: 239–242.

27 Nogueira V, Piquet MA, Devin A, Fiore C, Fontaine E, Brandolin G, Rigoulet M, Leverve XM. (2001). Mitochondrial adaptation to *in vivo* polyunsaturated fatty acid deficiency: increase in phosphorylation efficiency. *J Bioenerg Biomembr*. **33**: 53–61.

28 Nogueira V, Rigoulet M, Piquet MA, Devin A, Fontaine E, Leverve XM. (2001). Mitochondrial respiratory chain adjustment to cellular energy demand. *J Biol Chem*. **276**: 46104–46110.

29 Capitanio N, Capitanio G, De Nitto E, Villani G, Papa S. (1991). H+/e− stoichiometry of mitochondrial cytochrome complexes reconstituted in liposomes. Rate-dependent changes of the stoichiometry in the cytochrome *c* oxidase vesicles. *FEBS Lett*. **288**: 179–182.

30 Frank V, Kadenbach B. (1996). Regulation of the H+/e− stoichiometry of cytochrome *c* oxidase from bovine heart by intramitochondrial ATP/ADP ratios. *FEBS Lett*. **382**: 121–124.

31 Rohdich F, Kadenbach B. (1993). Tissue-specific regulation of cytochrome *c* oxidase efficiency by nucleotides. *Biochemistry*. **32**: 8499–8503.

32 Sone N, Nicholls P. (1984). Effect of heat treatment on oxidase activity and proton-pumping capability of proteoliposome-incorporated beef heart cytochrome aa3. *Biochemistry*. **23**: 6550–6554.

33 Klingenberg M, Echtay KS. (2001). Uncoupling proteins: the issues from a biochemist point of view. *Biochim Biophys Acta*. **1504**: 128–143.

34 Klingenberg M, Winkler E, Echtay K. (2001). Uncoupling protein, H+ transport and regulation. *Biochem Soc Trans*. **29**: 806–811.

35 Bouillaud F, Couplan E, Pecqueur C, Ricquier D. (2001). Homologues of the uncoupling protein from brown adipose tissue (UCP1): UCP2, UCP3, BMCP1 and UCP4. *Biochim Biophys Acta*. **1504**: 107–119.

36 Ricquier D, Bouillaud F. (2000). The uncoupling protein homologues: UCP1, UCP2, UCP3, StUCP and AtUCP. *Biochem J*. **345 Pt 2**: 161–179.

37 Sibille B, Filippi C, Piquet MA, Leclercq P, Fontaine E, Ronot X, Rigoulet M, Leverve X. (2001). The mitochondrial consequences of uncoupling intact cells depend on the nature of the exogenous substrate. *Biochem J*. **355**: 231–235.

38 Sibille B, Keriel C, Fontaine E, Catelloni F, Rigoulet M, Leverve XM. (1995). Octanoate affects 2,4-dinitrophenol uncoupling in intact isolated rat hepatocytes. *Eur J Biochem*. **231**: 498–502.

39 Sibille B, Ronot X, Filippi C, Nogueira V, Keriel C, Leverve X. (1998). 2,4 Dinitrophenol-uncoupling effect on delta psi in living hepatocytes depends on reducing-equivalent supply. *Cytometry*. **32**: 102–108.

40 Leverve XM, Fontaine E. (2001). Role of substrates in the regulation of mitochondrial function *in situ*. *IUBMB Life*. **52**: 221–229.

41 Rigoulet M, Aguilaniu H, Averet N, Bunoust O, Camougrand N, Grandier-Vazeille X, Larsson C, Pahlman IL, Manon S, Gustafsson L. (2004). Organization and regulation of the cytosolic NADH metabolism in the yeast Saccharomyces cerevisiae. *Mol Cell Biochem.* **256–257**: 73–81.

42 Boubekeur S, Bunoust O, Camougrand N, Castroviejo M, Rigoulet M, Guerin B. (1999). A mitochondrial pyruvate dehydrogenase bypass in the yeast Saccharomyces cerevisiae. *J Biol Chem.* **274**: 21044–21048.

43 Novel-Chate V, Rey V, Chiolero R, Schneiter P, Leverve X, Jequier E, Tappy L. (2001). Role of Na^+/K^+-ATPase in insulin-induced lactate release by skeletal muscle. *Am J Physiol Endocrinol Metab.* **280**: E296–300.

44 Sumegi B, Srere PA. (1984). Binding of the enzymes of fatty acid beta-oxidation and some related enzymes to pig heart inner mitochondrial membrane. *J Biol Chem.* **259**: 8748–8752.

45 Haggie PM, Verkman AS. (2002). Diffusion of tricarboxylic acid cycle enzymes in the mitochondrial matrix *in vivo*. Evidence for restricted mobility of a multienzyme complex. *J Biol Chem.* **277**: 40782–40788.

46 Ovadi J, Srere PA. (2000). Macromolecular compartmentation and channeling. *Int Rev Cytol.* **192**: 255–280.

47 Velot C, Mixon MB, Teige M, Srere PA. (1997). Model of a quinary structure between Krebs TCA cycle enzymes: a model for the metabolon. *Biochemistry.* **36**: 14271–14276.

48 Velot C, Srere PA. (2000). Reversible transdominant inhibition of a metabolic pathway. *In vivo* evidence of interaction between two sequential tricarboxylic acid cycle enzymes in yeast. *J Biol Chem.* **275**: 12926–12933.

49 Fukushima T, Decker RV, Anderson WM, Spivey HO. (1989). Substrate channeling of NADH and binding of dehydrogenases to complex I. *J Biol Chem.* **264**: 16483–16488.

50 Schagger H. (2002). Respiratory chain supercomplexes of mitochondria and bacteria. *Biochim Biophys Acta.* **1555**: 154–159.

51 Boumans H, Berden JA, Grivell LA, van Dam K. (1998). Metabolic control analysis of the bc1 complex of Saccharomyces cerevisiae: effect on cytochrome *c* oxidase, respiration and growth rate. *Biochem J.* **331 (Pt 3)**: 877–883.

52 Boumans H, Grivell LA, Berden JA. (1998). The respiratory chain in yeast behaves as a single functional unit. *J Biol Chem.* **273**: 4872–4877.

53 Bunoust O, Devin A, Averet N, Camougrand N, Rigoulet M. (2005). Competition of electrons to enter the respiratory chain: a new regulatory mechanism of oxidative metabolism in Saccharomyces cerevisiae. *J Biol Chem.* **280**: 3407–3413.

54 Pahlman IL, Larsson C, Averet N, Bunoust O, Boubekeur S, Gustafsson L, Rigoulet M. (2002). Kinetic regulation of the mitochondrial glycerol-3-phosphate dehydrogenase by the external NADH dehydrogenase in Saccharomyces cerevisiae. *J Biol Chem.* **277**: 27991–27995.

55 Batandier C, Guigas B, Detaille D, El-Mir MY, Fontaine E, Rigoulet M, Leverve XM. (2006). The ROS production induced by a reverse-electron flux at respiratory-chain complex 1 is hampered by metformin. *J Bioenerg Biomembr.* **38**: 33–42.

2
Organization and Regulation of Mitochondrial Oxidative Phosphorylation

Michel Rigoulet, Arnaud Mourier, and Anne Devin

Abstract

Mitochondrial bioenergetics is a broad field. In this chapter we will focus on the state of the art of oxidative phosphorylation regulation. The main theoretical basis of energy transfer processes was laid by Peter Mitchell in the 1960s. Since this pioneering work, great progress has been made in the description of the systems involved at the molecular level. Moreover, the three-dimensional structure of most of the proton pumps has now been ascertained. Our understanding of oxidative phosphorylation kinetic regulations has also greatly progressed. However, numerous experimental data cannot be explained in the framework of non-integrative biology. We will summarize the structural and functional evidence showing a high degree of dynamic organization of the oxidative phosphorylation components in supramolecular states and discuss some of the functional consequences of such integrated systems.

2.1
Introduction

Non-photosynthetic organisms get their energy from nutriment oxidation (cf. Chapter 1). The useful form of energy for a cell is the phosphate potential (i.e., ΔG for the ATP synthesis reaction; cf. Chapter 3). This potential is maintained either through substrate phosphorylation or through oxidative phosphorylation. Substrate phosphorylation can occur both in the cytosol (two glycolysis enzymes catalyze this reaction: phosphoglycerate kinase and pyruvate kinase) and in the mitochondria (via the Krebs cycle enzyme α-ketoglutarate dehydrogenase). Besides this substrate phosphorylation, there is only one way (the main pathway) to maintain the phosphate potential, i.e., through oxidative phosphorylation. Although mitochondrial substrate-level phosphorylation might, under certain conditions, play a major role in the maintenance of the phosphate potential in lower eukaryotic organisms such as yeast [1], higher organisms use mostly oxidative phosphorylation for this process. In this review, we will consider mitochondrial oxidative phosphorylation, its organization, its functioning, and its regulation.

2.2
Oxidative Phosphorylation and the Chemiosmotic Theory

Oxidative phosphorylations take place in the mitochondrial inner membrane, which contains both the respiratory chain and the ADP-phosphorylating systems (Fig. 2.1). In higher eukaryotes, the mitochondrial respiratory chain is divided into four complexes: complexes I–IV. Complex I (NADH-ubiquinone oxidoreductase) is a huge multi-subunit complex that contains 40–50 different polypeptides. While the structure of complex I is not yet completely resolved, those of complexes II, III, and IV are known in great detail due to X-ray crystallography analyses (for a review, see [2]). Complex II is the succinate oxidase–ubiquinone reductase, which oxidizes succinate (into fumarate) and reduces the ubiquinone pool. Complex III, also called the bc1 complex, is the ubiquinol–cytochrome c oxidoreductase, which transfers electrons from ubiquinol to cytochrome c. Complex IV is the cytochrome c oxidase; it oxidizes cytochrome c and reduces molecular oxygen into water. The ADP-phosphorylating systems are composed of two carriers, the phosphate and the ATP/ADP carriers and of the multi-enzymatic ATP synthase complex.

In the chemiosmotic theory, [3] the energy transduction is due to the redox and ATP synthase proton pumps indirectly coupled through the energetic intermediate $\Delta\tilde{\mu}_{H^+}$, the difference in electrochemical proton potential across the mitochondrial inner membrane. Indeed, during the electron transfer through the respiratory chain, complexes I, III, and IV couple electron transfer to proton extrusion, thanks to a sufficient redox span that allows proton extrusion against their transmembranal electrochemical potential difference. The energetic intermediate $\Delta\tilde{\mu}_{H^+}$ is used for multiple functions (such as substrate transports and ionic gradient maintenance) and, for the most part, ATP synthesis. ATP synthase is a reversible proton pump that is able to transduce osmotic energy ($\Delta\tilde{\mu}_{H^+}$) into ATP for-

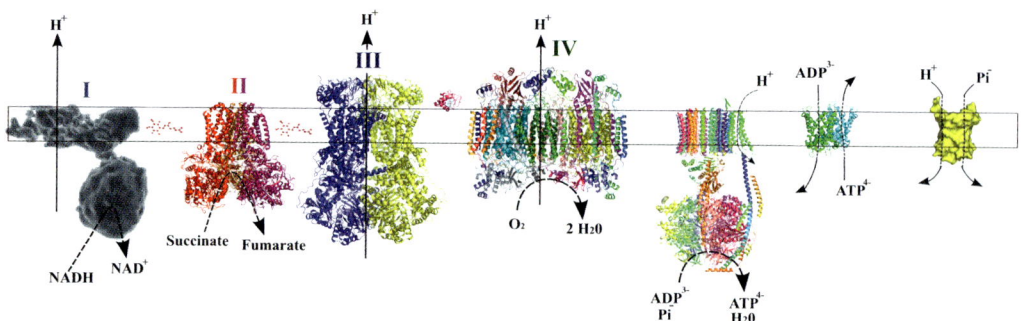

Fig. 2.1 Mammalian respiratory chain composition.

mation. The phosphate carrier catalyzes an electroneutral co-transport P_i/H^+, while the ADP/ATP carrier catalyzes the electrogenic exchange of ATP^{4-} against ADP^{3-}.

In the classical view of oxidative phosphorylation systems, (1) constant stoichiometries are needed, i.e., the amount of protons extruded by the respiratory chain or necessary for ATP synthesis are constant; and (2) from NADH oxidation to cytochrome c reduction, the respiratory chain is in thermodynamic near-equilibrium with $\Delta\tilde{\mu}_{H^+}$ and the mitochondrial phosphate potential is in near-equilibrium with $\Delta\tilde{\mu}_{H^+}$. In such a model, cytochrome oxidase plays a key role in the control of oxidative phosphorylation flux by being the only enzyme operating far from thermodynamic equilibrium and having its activity regulated by many effectors [4–10].

This view was rapidly reconsidered because the respiratory chain up to cytochrome c is in near-equilibrium only when the fluxes are very low [11, 12]. Moreover, numerous data have shown the following.

1. Kinetic control of the respiratory chain is shared among the different complexes and is not only localized at the cytochrome c oxidase level [13–15] and the ATP/ADP carrier is also a controlling step during the oxidative phosphorylation process [13]. Electron flux through the respiratory chain can also be regulated by complex I phosphorylation [16].

2. The proton pump $H^+/2e^-$ and ATP synthase H^+/ATP stoichiometries that are determined on isolated mitochondria can vary according to the fluxes and forces [17]. It has also been shown that the overall yield of the oxidative phosphorylation, i.e., the amount of ATP synthetized per oxygen consumed (ATP:O ratio), can vary [18]. This can be explained either by an increase in energy waste or by changes in the proton pumps' stoichiometries. It should be stressed that changes in oxidative phosphorylation yield have a physiological purpose. Indeed, reduced equivalents (NADH, $FADH_2$) have to be oxidized for the proper functioning of metabolic pathways. The rate at which these molecules have to be oxidized cannot be strictly correlated to cellular ATP turnover. Depending on the cell priority (i.e., oxidation of reduced equivalent, phosphate potential maintenance, or both), the mechanism allowing a change in the oxidative phosphorylation yield can vary.

2.3
The Various Mechanisms of Energy Waste

2.3.1
Passive Leak

The first mechanism that allows a decrease in the coupling efficiency of oxidative phosphorylation, the proton leak, is a direct consequence of the nature of the energetic intermediary, the protonmotive force (see Fig. 2.2). Indeed, biological membranes always present some proton conductance (L_H). It is generally ac-

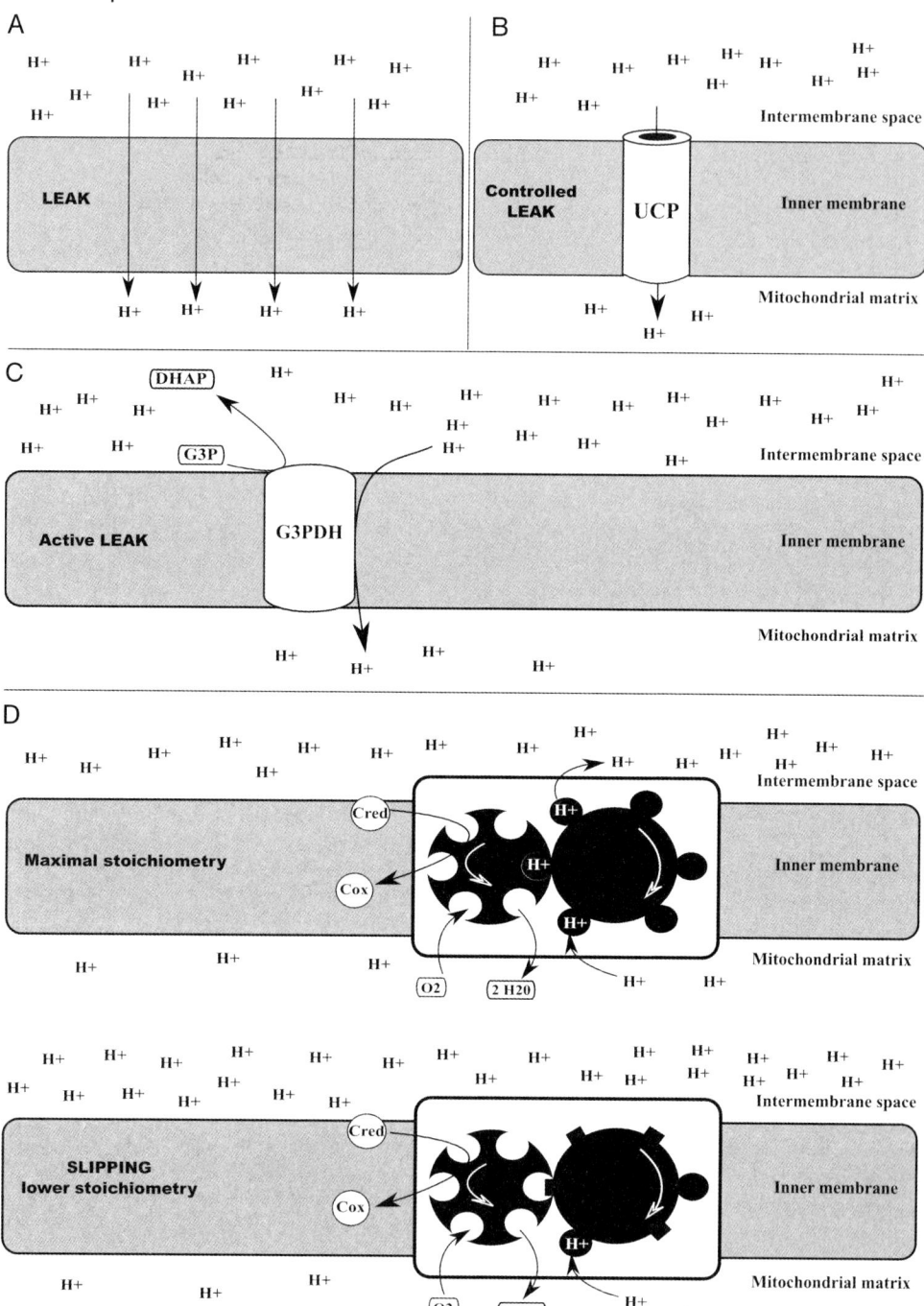

Fig. 2.2 Wastage mechanisms; (A) passive leak, (B) controlled leak, (C) active leak, (D) slipping.

cepted that the resulting proton influx is strictly dependent upon protonmotive force ($J_H = L_H \times \Delta p$). This membrane conductance is a specific property of the membrane itself but is not entirely independent of the protonmotive force: at high values of this force, the membrane proton conductance increases. The consequence of this is a non-ohmic relationship between passive proton flux (proton leak) and protonmotive force, which has been observed in many kinds of mitochondria [3, 19–25], including yeast mitochondria [26]. Obviously, the size of this proton leak may modulate the yield of oxidative phosphorylation (ATP:O), and it has been clearly shown that a large group of chemical molecules, called protonophores, uncouples oxidative phosphorylation by increasing membrane proton conductance [27–32]. In the classical definition of passive proton leak, it depends on only two factors: (1) the nature of the membrane and (2) the size of the protonmotive force. In this uncoupling process, the first event is a dissipation of the protonmotive force, which leads to an increase in respiratory rate and a decrease in ATP synthesis rate, thus inducing a decrease in ATP:O ratio.

2.3.2
Leak Catalyzed by Uncoupling Proteins

From a physiological point of view, the only well-known example of mitochondrial inner membrane uncoupling is the UCP1 protein [33, 34]. It was first isolated from brown adipose tissue, in which its main function is to produce heat [35, 36]. The activity of this protein generates a proton influx and thus uncouples oxidation from phosphorylation. However, it is very different from passive leak in that (1) the uncoupling has to go through a carrier and (2) it is kinetically regulated by numerous effectors. Contrary to passive leak, which is permanent, this proton permeability is controlled by the level of expression of the protein as well as by its activity in such a way that when it is active, there is a complete change in the cell physiology [37]. It has been shown that there is a family of UCPs. However, their respective physiological role and catalytic activity are still under scrutiny. The hypothesis that such a protein family is involved in slight and highly regulated proton permeability in order to limit reactive oxygen species (ROS) production by the respiratory chain is very attractive [38–40].

2.3.3
The Active Leak

On isolated mitochondria, under non-phosphorylating conditions, the relationship between respiratory rate and protonmotive force is known to be non-ohmic (see [41] for review). However, the value of the respiratory rate depends on the number of coupling sites (i.e., 3 for NADH and 2 for $FADH_2$) as well as the yield of the proton pumps. This yield has been shown to vary only when the protonmotive force reaches its highest value, which means that when this yield varies, the respiratory rate increases without any change in the protonmotive force. These modifications result mostly from changes in the efficiency of the proton pumps

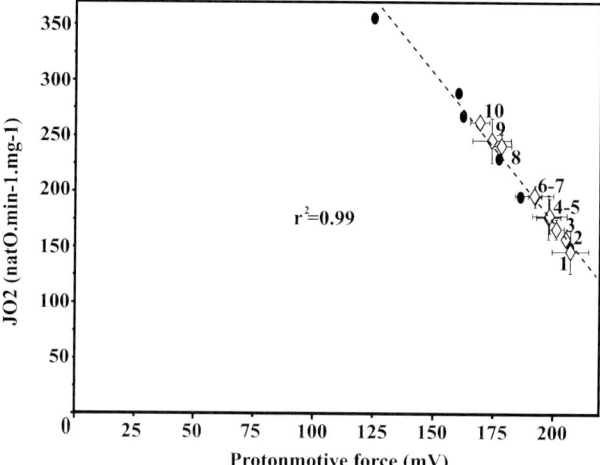

Fig. 2.3 Non-phosphorylating respiratory rates versus the protonmotive force (◇). In the presence of a single substrate: (1) ethanol, (2) glycerol-3-phosphate, (3) succinate, (4) NADH. With paired substrates: (5) NADH–ethanol, (6) NADH– succinate, (7) succinate–ethanol, (8) glycerol-3-phosphate–ethanol, (9) glycerol-3-phosphate–succinate, (10) glycerol-3-phosphate–NADH. Also represented are the non-phosphorylating ethanol conditions with increasing concentrations of CCCP (●).

through slip mechanisms that will be discussed later on. On the other hand, whichever substrate is used, if one is to assess an increase in the membrane leak, both protonmotive force and respiratory rate would vary. This is what is observed with slight concentrations of uncoupling agents (CCCP) or by activating UCP1 [42]. On isolated yeast mitochondria, which do not have a coupling site at the complex I level, the different dehydrogenases (external and internal) give their electrons to the quinone pool. It has been shown that when multiple substrates for the distinct dehydrogenases are used (e.g., glycerol-3-phosphate and NADH), there is a competition for electron entrance in the respiratory chain in such a way that NADH oxidation is favored [43]. In these conditions the relationship between respiratory rate and protonmotive force (Fig. 2.3) is clearly comparable to the one observed in the presence of increasing concentrations of protonophoric agent (CCCP). Indeed, when mitochondria oxidize NADH, the addition of substrates for the other dehydrogenases (glycerol-3-phosphate, ethanol) induces an uncoupling process. It should be stressed that this is an active leak process, because under conditions where the redox proton pumps are completely inhibited and electrons are accepted with an artificial acceptor, the protonmotive force maintained by mitochondrial ATPase activity decreases when the dehydrogenases are activated (unpublished results). There is a major difference between passive and active leak. Passive leak depends only on the membranal properties, although it is also related to the amount of proteins embedded in the membrane. Therefore, we propose the term *active leak* for the leak induced by the activity of membranal enzymes, which are not proton pumps. Furthermore, this leak is directly related to the level of activity of the enzyme involved (unpublished results). One should point out that a major physiological advantage of such a leak is that even if multiple substrates are at high concentrations in the cell together with a high phosphate potential, the reduced equivalent can be oxidized at a higher rate than the one obtained with a unique substrate.

2.3.4
The Slipping Mechanism

Even though much experimental work has established strong evidence that some protonophoric action can quantitatively account for the uncoupling of oxidative phosphorylation [44], no definite proof that the decrease in oxidative phosphorylation yield is exclusively and quantitatively due to an increase in cation membranal conductance has been obtained. In fact, from a growing amount of data it is evident that the question of the multiplicity of energy waste mechanisms is still largely open. Among them, two types of experimental evidence can be noted. First, it has been shown that some uncoupling effects are not linked to a significant decrease in protonmotive force, which indicates that the decrease in oxidative phosphorylation efficiency is not, in this case, the consequence of an increase in membranal proton conductance [17, 45–50]. Secondly, direct or indirect estimations of the coupled flow through different proton pumps indicate that their intrinsic stoichiometry, i.e., the $H^+/2e^-$ stoichiometry of the respiratory chain and the H^+/ATP stoichiometry of the ATP synthase, may vary as a function of many physical parameters or some drug addition [17, 18, 26, 48, 49, 51–54]. This led Azzone's team [55, 56] to propose another mechanism that induces a loss of oxidative phosphorylation yield. Such a new possibility, called "slip," is a decrease in the efficiency of a proton pump due to the partial and variable coupling efficiency of chemical reaction and proton transport, i.e., a decrease in the H^+/O stoichiometry of the respiratory chain or an increase in the H^+/ATP stoichiometry of the ATP synthase. A kinetic model for proton pump functioning, using a Hill diagram, has been proposed by Pietrobon and Caplan [57] (see also Fig. 2.2). The whole reaction is divided into two parts: the catalytic pathway of the chemical reaction and the pathway for the protons. In the absence of slip, chemical reaction and proton transport are closely coupled. Slip results from the possibility of chemical reaction without concomitant proton movement or *vice versa*. Consequently, this uncoupling generates two kinds of slipping: (1) chemical slipping in which the chemical reaction is not associated to proton transfer and (2) proton slipping in which the proton transfer through the proton pump is not associated to a chemical reaction. In the latter case and experimentally speaking, it seems impossible to distinguish proton slip from an active leak. Depending on the number of proton pumps intrinsically coupled or uncoupled, for a given steady state, one can define a degree of slipping leading to a peculiar oxidative phosphorylation yield. On the other hand, the chemical slipping that has been studied mainly on the respiratory chain, and in this case called redox slipping, confers original oxidative phosphorylation regulation properties. In this model, the electron flux should vary largely without any change in protonmotive force.

In conclusion, at the level of the mitochondrial inner membrane, there are four mechanisms that can modulate the coupling between oxidation and phosphorylation. Obviously, these are not exclusive mechanisms.

1. The classical leak, or passive leak, depends only on the protonmotive force and the membrane's nature. Even if the entrance proton flux is localized essentially at the interface lipids/proteins, this leak is independent of the membranal enzymes' activities.

2. The leak catalyzed by proton carriers (e.g., uncoupling proteins) depends mostly on the amount of carrier; the protonmotive force is highly regulated by numerous effectors activating or inhibiting the carrier.

3. The active leak is strictly dependent on the activity of some membranal proteins that are not proton pumps. In such a case, the proton permeability seems to be proportional to the activity of the enzymes. Whatever the mechanism for the leak process, any increase in the leak is linked to a decrease in the protonmotive force.

4. The redox slipping is linked to the activity of membranal redox proton pumps. In this mechanism, the electron transfer flux increases when the pump coupling state decreases, independently of the protonmotive force. It is obvious that the protonmotive force somehow controls the slipping process. However, this mechanism, which decreases the coupling efficiency, should not alter the protonmotive force.

On isolated mitochondria, one can manipulate fluxes and forces in such a way that these mechanisms can be assessed separately. However, in an integrated system such as the cell, one has to consider that all these mechanisms can operate simultaneously. The main consequence of these wastage processes is that the energy transformation processes at the mitochondrial level have a yield that is quite a bit lower than what is classically described in biochemical textbooks, while it allows a great flexibility of energy waste regulation. Furthermore, these processes are necessary for the adjustment of the coenzymes' reoxidation and ATP synthesis fluxes in accordance with the cell physiology.

2.4
Mechanisms of Coupling in Proton Pumps

In conceptual models of proton translocation, the proton chemical reaction coupling may be direct or indirect. Direct models for redox proton pumps initially proposed by Mitchell [3, 22, 58, 59] stipulated that protons were directly carried by redox elements such as quinones or flavins, which are disposed in the membrane in such a manner that the redox dependence on protonation/deprotonation results in a translocation from the inside to the outside of the mitochondria (redox loop). This model predicts a fixed stoichiometry H^+/e^- of 1 at each coupling site. It is accepted that the example of such an arrangement is the quinone cycle mechanism for the cytochrome bc1 and b6f complexes [60, 61]. The consequence of this direct mechanism is that the protonmotive force value, in the absence of enzymatic H^+-gradient–consuming pathways, depends exclusively on the redox span and the membrane proton permeability coefficient. The most recent data concerning ATP synthase or cytochrome oxidase mechanism studies support the indirect coupling view. For instance, a model for energy coupling by ATP synthase is based on the following assumptions [62–64]:

1. Proton translocation is needed to promote ATP release from a high affinity site where it forms spontaneously.
2. Substrate binding is also associated with the energization step.
3. Both steps occur simultaneously at separate but interacting sites.
4. The binding change is due to the rotation of the γ-subunit that modifies the structure of the catalytic sites.

From this model, a mechanistic H^+/ATP stoichiometry change should be related to the energy required to induce a rotation of the subunit, which involves the number and nature of the different bonds existing between the subunits concerned. It is not difficult to admit that this energy can vary with a modification of the conformational state of the enzyme or with a change in its surroundings. Moreover, this indirect conformational coupling between ATP synthesis and proton flow can lead to slippage if there is not a perfect concordance between the high- and low-affinity conformational state transition of the catalytic site in the presence of ATP.

Cytochrome c oxidase and ubiquinol oxidase are now considered proton pumps [65, 66]. Protons are required for two different purposes. First, "chemical" protons are consumed upon reduction of dioxygen by four electrons to form two water molecules; the H^+/e^- stoichiometry of 1 is fixed by the chemical reaction and cannot vary. Second, the "pumped" protons are translocated from the inside to the outside of the mitochondria, and the stoichiometry H^+/e^- must depend on the energy made available by the redox reaction and on the way in which electron and proton flux are coupled. An important observation is that the substitution of asparagine for aspartate-135 in subunit I of the ubiquinol oxidase of $E.$ $coli$ abolishes proton pumping with little change in electron transfer activities [67, 68]. This observation led to two predictions: (1) two separate proton pathways may exist, one for "chemical" and the other for "pumped" protons, and (2) the H^+/e^- stoichiometry of pumped protons may vary from 0 to its maximal value. This prediction was supported by the structural determination of the cytochrome oxidase in both $Paracoccus$ $denitrificans$ [69] and heart bovine mitochondria [70]. The notion of stoichiometry changes according to the conditions applied to the system was significantly supported by reports on reconstituted cytochrome c oxidase, where stoichiometry varies according to the electron flux, the ΔpH value [54], and the medium composition in adenylic nucleotides [71]. Combining their results with the structural information available, Papa et al. proposed that the actual H^+/e^- in oxidase is determined by the relative contribution of two electron transfer pathways, one coupled to proton pumping and the other not [54, 65]. The relative contributions of these pathways should be determined by kinetic and thermodynamic factors. At constant protonmotive force, ATP:O decreased when the electron flux rose (in part due to a decrease in the cytochrome c oxidase $H^+{:}e^-$ ratio) to reach a minimal value, and a further increase in respiration rate at constant ATP:O was accompanied by a decrease in protonmotive force (see also Ref. [26]). Another model for redox slippage is that the rate of the conformational transition of the protein residue(s) involved in moving the proton-binding site from the input to the output states is not completely synchronized with the electron flux through the oxidase. Whatever the mechanism, slippage should depend on both electron rate and protonmotive force, the latter of which acts as a resistance to proton efflux. Contrary to a redox loop (see above), the maximal value of the protonmotive force established by a proton pump, in the absence of a consuming system, should depend not only on redox span and membrane H^+ conductance but also on slippage. This means that oxidase alone should maintain a lower Δp than the bc1 complex does. In such an explanation, it is interesting to note that if optimal H^+/e^- is lower in the bc1 complex than in the cytochrome oxidase, this complex maintains a lower Δp value under steady state and at the same electron flux.

2.5
Oxidative Phosphorylation Control and Regulation

2.5.1
Metabolic Control Analysis

Ever since the work of Chance and Williams [72, 73], numerous studies have tried to determine which parameters control the activity of oxidative phosphorylation. The observation that a purely thermodynamic interpretation does not allow the understanding of the origin of this control led to the development of a kinetic approach. However, there was a bias in that people concentrated on the study of the limiting step, leading to unproductive disputes over this very step. A major step forward occurred with the application of the control theory to mitochondrial metabolism. The reader can find detailed and useful information about the use of this theory in biology in numerous reviews and books [74, 75]. This theory led to quantification of control elements through some key parameters. First, the flux control coefficient of an enzyme over a flux represents the relative variation of flux in response to the relative variation of the enzyme-specific activity. This parameter is a systemic property because it depends on a defined network. Second, the elasticity coefficient that is a local parameter represents the relative variation of local properties (e.g., the activity of an enzyme towards another local parameter such as the concentration of some effectors). This theory has been applied successfully by Tager's team to mitochondrial oxidative phosphorylation [14]. This pioneering work and many others have shown that the kinetic control is shared among several steps and that the control is dynamic in such a way that the nature of the controlling step and the value of the control coefficient depends on the oxidative phosphorylation steady state. This theory conciliated previous studies aimed at determining the limiting step. In the field of oxidative phosphorylation, this theory has been successfully applied to assess the targets of drugs and to better understand the threshold effects of mitochondrial diseases linked to a deficiency in some oxidative phosphorylation complexes [76–78]. It also allowed the questioning of mitochondrial respiratory chain organization; metabolic control analysis put into question the hypothesis of mitochondrial respiratory chain complexes as free diffusible complexes and led to the hypothesis of a supramolecular organization of the respiratory chain [79, 80]. Even though the applicability of this theory has been extended by theoretical developments, it appears that its applicability to biology is becoming more and more limited due to the increased complexity of biological systems.

2.5.2
Regulations

2.5.2.1 Kinetic Regulation of Mitochondrial Oxidative Phosphorylation:
Complex I Covalent cAMP-dependent Phosphorylation

In C6 glioma cells, the relative contribution of mitochondrial oxidative phosphorylation to ATP synthesis is between 70% and 85% [81]. It has been shown that in these cells, cellular respiratory rate decreased when the growth rate decreased, without any change in mitochondrial enzymatic equipment [81, 82]. However, there is a good positive correlation between the respiratory rate and the specific activity of complex I, indicating that the cellular respiratory rate is controlled mainly by the activity of this complex [82]. Furthermore, the specific activity of this complex is directly related to the phosphorylation degree of one of its subunits. This phosphorylated protein is likely to be the subunit ESSS, which has been identified in bovine heart mitochondria as one of two complex I phosphorylation sites, the other being MWFE, a 10-kDa protein whose expression is supposed to be linked to cAMP signaling [16]. Further experiments demonstrated that this phosphorylation is at least under the control of the cAMP signaling pathway [82]. Thus, in C6 glioma cells, the regulation of the phosphorylation level of a subunit of complex I is a key mechanism of oxidative phosphorylation adaptation to energy demand during these cancer cells' multiplication.

2.5.2.2 Cytochrome Oxidase: An Example of Coordinate Regulation

Cytochrome c oxidase (COX), which catalyzes the last and irreversible step of the electron transfer pathway, has been shown to be a controlling step in rat liver [83], rat heart [84], and yeast [85] mitochondria. Numerous studies have shown that the stoichiometry of its proton pumping is variable and dependent on many parameters [86–93], including allosteric regulations [94]. The rate of electron transfer through cytochrome oxidase is controlled by the associated thermodynamic forces: (1) the span in redox potential between oxygen and cytochrome c (i.e., ΔEh) and (2) the proton electrochemical potential difference. However, enzyme turnover has been shown to be more sensitive to ΔpH (per millivolt) than to $\Delta \Psi$ due to the effect of matrix pH on the enzyme [95, 96]. Moreover, the electron flux through the cytochrome oxidase also depends on various effectors as well as on kinetic regulation.

Action of Adenylic Nucleotides on COX It has long been suspected that the final products of oxidative phosphorylation (i.e., $ATP/ADP \times P_i$) have a regulatory effect on electron transfer at the cytochrome c oxidase level. Indeed, specific binding sites for ATP and ADP have been identified on cytochrome c; this binding may in turn decrease the electron transfer from cytochrome c to the oxidase [97, 98]. In mammalian cytochrome oxidase, high-affinity binding sites for ATP and ADP have been identified, and because ATP and ADP can compete for the same binding sites, this indicates that the ATP/ADP ratio, rather than the ATP and ADP concentrations, is recognized by the enzyme. The number of binding sites depends on the origin of the COX and is still a matter of debate [99–101].

Kinetics studies carried out on reconstituted COX from different origins showed that extraliposomal ATP and ADP increase the apparent K_m of this enzyme for cytochrome c [100, 102, 103] without any change in its maximal turnover [100]. By contrast, intraliposomal ATP increased and ADP decreased the K_m for cytochrome c of bovine COX [102, 103] but had no effect on the bacterial enzyme of *Paracoccus denitrificans* [103]. Intraliposomal ATP and ADP were able to stimulate the maximal turnover of COX obtained from bovine heart but not from liver [104] (see Fig. 2.4). By comparing measurements performed under coupling and uncoupling conditions, it was shown that intramatrical ADP, in contrast to ATP, increases the respiratory control ratio, thereby suggesting that H^+ pumping is affected by intraliposomal adenylic nucleotides [104]. Furthermore, a decrease in the H^+/e^- stoichiometry at high intraliposomal ATP:ADP ratios was finally observed by direct measurements of H^+ translocation [94, 105]. It is worth noting that this ATP-induced decrease in H^+ pumping stoichiometry seems to be specific to subunit VIa-H expression, because it was not observed in bovine liver, in turkey liver, or in turkey heart [94, 105].

Allosteric Inhibition of COX by High Internal ATP:ADP Ratios Further experiments carried out with outside-out submitochondrial particles demonstrated that high ATP:ADP ratios induce allosteric inhibition from the inner matrix side of COX but not from the cytosolic side. It has also been shown that cardiolipin is involved in the cooperative interaction of the two cytochrome c binding sites of the dimeric enzyme complex [106, 107].

In addition, the allosteric properties of bovine heart COX have been shown to be short-term controlled by signaling pathways mediated by thyroid hormone interactions as well as by cAMP-dependent phosphorylations [108]. Moreover, the allosteric ATP inhibition of Tween 20–solubilized COX is reversibly switched on

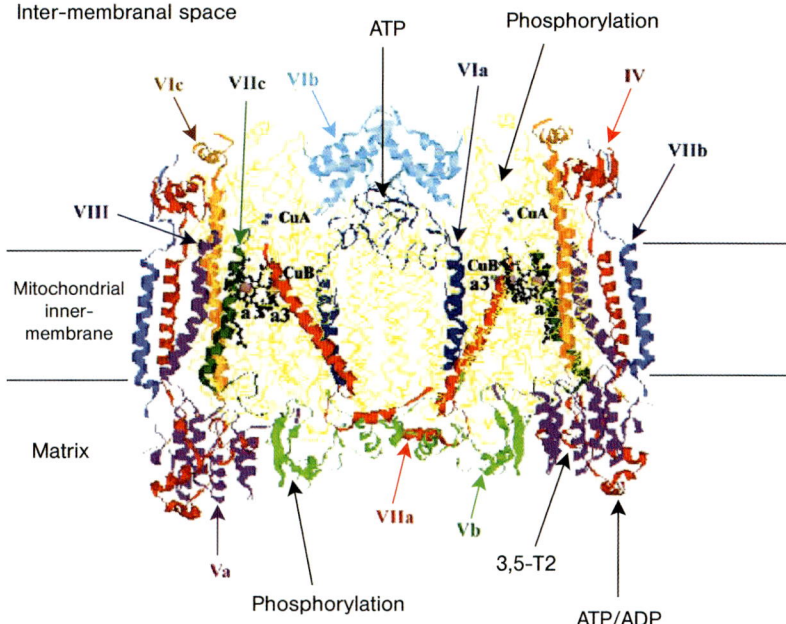

Fig. 2.4 Crystal structure of the dimeric cytochrome oxidase from bovine heart [168] located in the mitochondrial inner membrane. The crystallographic data were obtained from Protein Data Brookhaven (cyto oxidase.pdb) and were processed by the RasMol 2.6 program. Marked in the backbone are the mitochondrially encoded proteins: subunits I, II, and III. The 10 nuclear-encoded subunits are shown as ribbons. The hemes a and a_3, iron atoms, and copper atoms are indicated. The regulatory sites for effectors (phosphorylations, ATP, ADP, and 3,5-diiodothyronine [3,5-T2]) are indicated.

by cAMP-dependent phosphorylation and switched off by subsequent treatment with protein phosphatase [109]. These kinetic effects have been correlated with phosphorylation of subunits II (and/or III) and Vb [109]. Furthermore, the matrix-oriented subunit Vb of COX was shown to interact with RIα, the regulatory subunit of the cAMP-dependent protein kinase A [110]. On the other hand, cAMP-independent phosphorylation of subunit IV has been described [111].

Modifications of Flow–Force Relationships by External ATP in Yeast Mitochondria
To extrapolate these data to the *in vivo* situation, near physiological conditions have to be tested (e.g., respiratory flux, ATP synthesis, $\Delta\tilde{\mu}_{H^+}$ size, cytochrome *c*, and a + a_3 reduction level) in which the oxidative phosphorylation activity may be more or less controlled by effectors acting on COX. Several studies performed on isolated yeast mitochondria have given original insights into the physiological relevance of the regulation of COX by adenylic nucleotides. The key term [112] that thermodynamically controls the respiratory rate is $2\Delta Eh$-$n\Delta p$, where ΔEh is the span of redox potential between electron carriers over a segment of the entire

respiratory chain that translocates n protons (and/or charges) upon the transfer of two electrons. Under conditions where the redox span may be unchanged, the relationship between the respiratory flux and Δp is not unique and depends on the way in which the force varies (e.g., ATP synthesis stimulation, proton leak increase) [113–115]. One of the proposed explanations is the possible allosteric link between the respiratory chain and ATP synthase [116, 117], which leads to a change in the response of the oxygen uptake to Δp. In mitochondria isolated from yeast, [118] we observed that, as in rat liver mitochondria [114, 115], there is more than one relationship between the respiratory rate and the size of Δp, when the respiratory rate is modified by either ADP plus P_i or by an uncoupler. Added $ADP + P_i$ stimulates respiration more than uncoupler at the same Δp value. Moreover, at the same respiratory rate, cytochrome $a + a_3$ is more reduced by uncoupler than by $ADP + P_i$ additions. In contrast, the unique relationship between, on the one hand, the respiratory rate and the reduction level of COX and, on the other hand, the respiratory rate and Δp is obtained under ATP synthesis or uncoupling in the presence of 1 mM external ATP. The kinetic control analysis provides support for regulation of the respiratory chain by ATP addition at the COX level. Indeed, control of respiration exerted by COX is low when the respiratory rate is varied by ATP synthesis, whereas it increases as a function of the respiratory rate with uncoupler in the absence of ATP. However, ATP addition decreases the control coefficient under uncoupling conditions to reach the same value as measured during ATP synthesis [118]. Taken together, these results indicate a regulatory effect of external ATP on COX: as ATP increases the enzymatic capacity of COX without stimulation of respiratory rate, it decreases the control exerted by this complex on respiration, leading to a shift in the control distribution to one or many other steps. Moreover, the fact that ATP addition enhances Δp in the presence of the same amount of uncoupler and at the same respiratory rate strongly suggests an increase in H^+/e^- stoichiometry at the COX level. It is worth noting that in isolated rat liver mitochondria, the relationship between respiratory rate and Δp is also not unique and depends on how the flux is modified. However, the differences observed in either uncoupling or phosphorylating conditions are not linked to a regulatory effect of ATP on COX during ATP synthesis. Considering the tissue specificity of the subunit expression described previously, this result does not mean that such an ATP modulation of flow–force relationships is not present in other rat tissues, such as heart or muscle.

In conclusion, the kinetic studies indicate that there should be a balance between activation and inhibition with respect to the ATP concentration (or the ATP:ADP ratio) in the cytosol and matrix. Therefore, the question becomes, what is the physiological meaning of such multiple binding and regulating sites for adenylic nucleotides in the control and regulation of cellular energy metabolism? Extrapolation of previous findings to the *in vivo* situation is complicated because the cellular respiration is a function of (1) the electron supply to the respiratory chain, sustained by dehydrogenase activity and intermediate metabolism; (2) ATP turnover; (3) the magnitude of the proton leak and the stoichiometry of the different proton pumps; and (4) signal transduction pathways, e.g., phosphorylation of respiratory chain complexes [119, 120]. Indeed, during oxidative phosphorylation, intracellular adenine nucleotides could play a key role at two levels, i.e., as components of the free energy of the ATP synthesis (i.e., phosphate potential) and as kinetic effectors of the COX. In fact, in the mitochondrial matrix, a synergistic effect of the matrix phosphate potential and nucleotide effectors might occur that would amplify the increase in respiration as a consequence of a decrease in the cellular energy status. In contrast, on the cytosolic side, there could be an antagonist effect of the cytosolic phosphate potential and the nucleotidic effectors, which might minimize the effect of a workload increase on cellular respiration. The *in vivo* balance between these two hypothetical scenarios would depend on the sensitivity of the COX to the ATP:ADP ratio maintained in the matrix and the cytosol, respectively [9].

2.6
Supramolecular Organization of the Respiratory Chain

Numerous studies show that most of the enzymes involved in mitochondrial metabolism can be organized in supramolecular structures. There are now numerous experimental data clearly showing that some metabolic pathways are organized on a supramolecular level. For instance, part of the citric acid cycle is clearly channeled [121–123]. Channeling involves an organization of the metabolic network in such a way that metabolites are conveyed from one enzyme to the other without dilution in the bulk phase. In the citric acid cycle, an "assembly mutation," i.e., a mutation that causes a tricarboxylic acid cycle deficiency without affecting the citrate synthase activity, has been isolated in yeast. Moreover, a 15-amino-acid peptide from wild-type Cit1p encompassing the mutation point inhibits the tricarboxylic acid cycle in a dominant manner, and the inhibitory phenotype is overcome by a co-overexpression of Mdh1p, the mitochondrial malate dehydrogenase. These data provide direct *in vivo* evidence of an interaction between two sequential tricarboxylic acid cycle enzymes, Cit1p and Mdh1p, and indicate that the characterization of assembly mutations by the reversible transdominant inhibition method may be a powerful way to study multi-enzyme complexes in their physiological context [121]. Furthermore, it has been shown in hepatocytes that the urea cycle (a pathway that is partly in the cytosol and partly in the mitochondria, both portion being connected through a carrier) is completely channeled [124, 125]. It has also been shown that the following exist: (1) NADH channeling from malate dehydrogenase to mitochondrial complex I [126], (2) associations between malate dehydrogenase and complex I, and (3) associations between malate dehydrogenase and 3-L-hydroxyacyl-CoA dehydrogenase [127]. It has also been shown in yeast spheroplasts that NADH was channeled from alcohol dehydrogenase or glyceraldehyde-3-phosphate dehydrogenase to the external NADH dehydrogenase [128, 129]. Taken together, all these data clearly show that in cells there is a clear restriction for metabolite diffusion and that metabolic pathways are often organized. In this review, we will particularly focus our attention on the supramolecular organization of the complexes involved in oxidative phosphorylation. For more information relating to organization as energetic units in the cardiac cell, see Chapter 11.

2.6.1
Structural Data

2.6.1.1   ATP Synthase Organization

The F_1F_0-ATP synthase is composed of two different sectors: the membrane-embedded F_0-part, which is responsible for proton translocation across the membrane, and the hydrophilic F_1-part, which contains the catalytic sites for ATP synthesis. In the yeast *Saccharomyces cerevisiae*, the F_1F_0-ATP synthase is a 600-kDa complex constituted of 17 distinct subunits and related in composition to the mammalian ATP synthase. Three proteins (subunits 6, 8, and 9) are encoded by the mitochondrial genome, while all the other subunits are nuclear encoded. A ring of subunit 9 is associated with one subunit 6, one subunit 8, and the hydrophobic domain of subunit 4. These components make up the basic portion of the proton translocating F_0-sector. Subunits OSCP, 4, h, d, and f constitute a peripheral stalk, known as the "second stalk." The F_1-portion consists of an $(\alpha\beta)3$ barrel that surrounds a coiled-coil helical domain of the γ-subunit [130] and a central stalk formed by subunits δ, var epsilon, and the remainder of the γ-subunit [131]. Each of the 13 subunits mentioned above has been shown to be indispensable for enzyme activity [132].

Four supernumerary proteins (subunits i, e, g, and k) belonging to the F_0-sector and that are not essential for ATP synthase activity were identified recently [133, 134]. Subunit e was found in bovine heart, rat liver, and yeast mitochondrial ATP synthases [135–137]. Cross-linking experiments have shown that subunit e is in the vicinity of subunit g and close to another subunit e [138]. By using the Blue Native polyacrylamide gel electrophoresis (BN-PAGE) technique, Arnold et al. [133] identified a dimeric state of the ATP synthase. Subunits e, g, and k were detected only in the dimeric form of the enzyme. The ATP synthase dimers were absent in extracts from strains devoid of subunits e or g, indicating that these two proteins are involved in the dimerization of the ATP synthase complex. By using a similar approach and cross-linking experiments performed on a strain with a mutated version of subunit i, it has been shown that ATP synthase dimers exist in the inner membrane. Recent results suggest that, in addition to subunits e and g, the first membrane-spanning segment of subunit 4, bTM1, is also required for ATP synthase dimerization or for a higher state of oligomerization of the ATP synthase complex.

Furthermore, mitochondria are highly compartmentalized organelles. High-resolution scanning electron microscopy and electron tomography have recently provided a new mitochondrion model in which the inner mitochondrial membrane forms a structured network of cristae [139, 140]. Electron microscopic studies performed on yeast cells lacking either subunit e, subunit g, or bTM1 have shown that in the absence of these specific dimerization/oligomerization components, the inner mitochondrial membrane is completely disorganized and does not form cristae [141, 142]. These results indicate that a link exists between yeast F_1F_0-ATP synthase oligomerization and cristae genesis [141, 143].

The terminal steps involved in making ATP in mitochondria require the ATP synthase, a phosphate carrier, and an adenine nucleotide carrier. Under mild conditions, these entities subfractionate as an ATP synthase–phosphate carrier–adenine nucleotide carrier complex or "ATP synthasome" [144]. As a first step toward obtaining three-dimensional information about this large complex, or "metabolon," and the locations of the phosphate carrier and adenine nucleotide carrier therein, ATP synthasomes have been dispersed into single complexes, and negatively stained images clearly showing the classical headpiece, central stalk, and base piece have been visualized by electron microscopy (EM). Parallel immuno-EM studies revealed the presence of the phosphate carrier and adenine nucleotide carrier located non-centrally in the base piece, and other studies implicated an ATP synthase/phosphate carrier/adenine nucleotide carrier stoichiometry near 1:1:1. Docking studies with known structures together with the immuno-EM studies suggest that the phosphate carrier or adenine nucleotide carrier may be located in the smaller domain, whereas the other transporter resides nearby in the larger domain. Collectively, these findings support a mechanism in which the entry of the substrates ADP and P_i into mitochondria, the synthesis of ATP on F_1, and the release and exit of ATP are very localized and highly coordinated events [145].

2.6.1.2 Respiratory Chain Supramolecular Organization

Two alternative models for the arrangement of respiratory chain complexes in the membrane have been proposed (Fig. 2.5). According to the currently favored random collision model [146], all components of the respiratory chain diffuse individually in the membrane, and electron transfer depends on the random, transient encounter of the individual protein complexes with the smaller electron carriers. In the solid-state model [147] proposed 50 years ago, the substrate is channeled directly from one enzyme to the next. Recently isolated stoichiometric assemblies, so-called supercomplexes, support this model. Respiratory supercomplexes of different compositions have been described in bacteria (e.g., *Paracoccus denitrificans* [148]), in mitochondria from *Saccharomyces cerevisiae* [149, 150], in other fungi [151], in higher plants [152–154], and in mammals [149, 155, 156] by means of BN-PAGE, gel filtration, and immunoprecipitation. In these studies, supercomplexes of various stoichiometries have been detected, such as assemblies of monomeric complex I (I1), dimeric complex III (III2), and complex IV in various copy numbers (IVx). A crucial function of these respiratory supercomplexes may be the stabilization of the individual complexes [148, 151, 156, 157]. Kinetic evidence by inhibitor–titration studies in bovine heart mitochondria is consistent with both models for the arrangement of the respiratory chain complexes and suggests their coexistence [158].

Recently, an electron projection map of a plant supercomplex consisting of only complex I and dimeric complex III (III2) was presented for *Arabidopsis thaliana*

Liquid state model

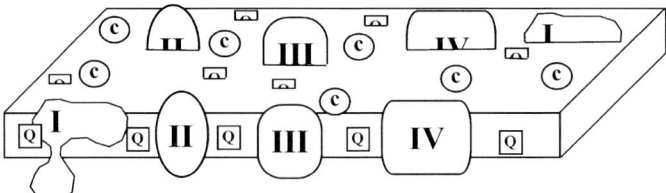

Solid state model

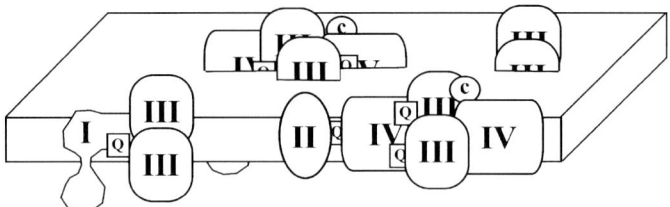

Fig. 2.5 The two models of respiratory chain organization. In the liquid state, the respiratory complexes are randomly positioned in the inner mitochondrial membrane, and the mobile elements such as quinone and cytochrome *c* can freely diffuse from one complex to another. In the solid-state model, the respiratory complexes are organized into a respirasome structure, and neither quinone nor cytochrome *c* can diffuse from one supercomplex to another.

[159]. But so far, the supramolecular architecture of a stoichiometrically defined supercomplex comprising all three complexes (I, III, and IV) has not been determined. According to BN-PAGE analysis, the most abundant supercomplexes in bovine heart mitochondria are one supercomplex consisting of complex I and dimeric III and another supercomplex containing complexes I and dimeric III and IV with molecular masses of ~1500 kDa and 1700 kDa, respectively [149, 155].

All these data argue in favor of a supramolecular organization of the respiratory chain depending on distinct modalities. In response to stimuli or effectors, these organizations could change, meaning that they would be dynamically modulated. The question raised by these results is the physiological significance and the plasticity of such an organization. It has been shown by Schägger's group that supramolecular organization of the respiratory chain in yeast mitochondria can vary depending on the carbon source in the culture medium [149]. This reinforces the idea that supramolecular organization could answer kinetic and thermodynamic constraints due to the nature of substrates available in the cell.

2.6.2
Functional Data

Through studies relative to the functioning of the respiratory chain in yeast mitochondria, some authors were led to propose an organization of the respiratory chain as functional units [79, 80]. These authors have conducted two kinds of studies. First, by studying the response of the respiratory flux to distinct inhibitors of the respiratory chain, the results obtained could only be interpreted in a scheme of the respiratory chain behaving as a single unit [80]. Second, by studying a collection of mutants of the bc1 complex, they showed that there is a very good correlation between the bc1 complex and cytochrome *c* oxidase activity, which strongly suggests that both complexes are somehow related [79]. Furthermore, considering various yeast strains growing on distinct non-fermentable media, we have clearly shown that there is a linear relationship between growth rate and respiratory rate as well as between respiratory rate and each and every one of the cytochromes. This clearly indicates that the respiratory chain in yeast behaves as a whole and is a functional argument in favor of a supramolecular organization [160]. This work also indicates that one of the advantages of such an organization is that different strains can maintain a constant growth yield.

Another argument in favor of a supramolecular organization of the respiratory chain in yeast comes from the study of oxidative phosphorylation yield in yeast mitochondria. Indeed, in yeast mitochondria the ATP:O ratios obtained by classical procedures with respiratory chain inhibitor titration are in agreement with those found in mammalian mitochondria: 0.5 and 1 for sites 2 and 3, respectively [161] (this classical procedure consists of a saturating amount of substrate and addition of an inhibitor of the respiratory chain). In contrast, when the electron flux through each respiratory chain unit is decreased by means of substrate availability, the ATP:O ratio increases. Thus, the value of ATP:O measured under state 3 is not the highest one. Moreover, the highest value of the ATP:O ratio is linked to a nearly total kinetic control of oxidative activity upstream of the respiratory chain. The nature of the mechanism involved here is very likely to be slipping at the level of cytochrome oxidase, which would decrease with the flux. Indeed, many studies have shown that respiratory chain slipping is due to an increase in protonmotive force (see above), but, as predicted from the redox slipping model [57], we found that it may also depend on the flux. At the level of respiratory chain organization, the only model that could explain such a result would be an organization as single units.

In yeast mitochondria, which lack a proton-pumping complex I, there are two external NADH dehydrogenases (Nde1p and Nde2p) and one internal NADH dehydrogenase (Ndip). Both external dehydrogenases share the same function, leaving one to wonder about the role of this redundancy. Previous studies [162] have shown that Nde1p is the main enzyme necessary for cytosolic NADH reoxidation during growth on a non-fermentable carbon source. However, NDE2 gene expression is induced after the diauxic shift [163], i.e., for growth on ethanol. Thus, under these conditions, both enzymes were present. We have shown that mitochondria bearing only the Nde1p or the Nde2p enzyme have a comparable oxygen consumption rate on NADH. Even though both enzymes share the same function, they do not have the same role in regulating oxygen consumption fluxes. We have shown that both Gut2p and Ndip were inhibited when functioning in conjunction with Nde1p, whereas Nde2p did not have such an effect. When it was functionally isolated, Gut2p was not inhibited by Nde1p, ruling out an enzyme interaction hypothesis. Thus, the only remaining hypothesis that can explain our results is a competition between dehydrogenases for the electron supply to the respiratory chain [43, 164]. We have shown that electrons coming from Nde1p have the right of way over those coming from either Gut2p or Ndip. Finally, this shows that even though both Nde1p and Nde2p share the same function, they do not have the same role in providing electrons to the respiratory chain and/or reoxidizing NADH. That electrons coming from Nde1p have the right of way over electrons coming from Gut2p and Ndip supports the specific organization of the dehydrogenases and the respiratory chain. Our hypothesis is that the supercomplexes can interact specifically either with complex II, to form a respiratory chain using succinate, or with the other supercomplex containing most of the dehydrogenases, i.e., Nde1p, Nde2p, Ndip, and Gut2p. This priority implies that the dehydrogenases involved belong to the same supramolecular complex. This is reinforced by the fact that the dehydrogenases themselves are not able to inhibit each other. Second (and this is a different situation), if one considers the case of succinate, the succinodehydrogenase can belong to a different supramolecular complex, and the electron flux is then shared between the different dehydrogenases involved.

Whichever mechanism is involved, the fact that electrons coming from certain dehydrogenases have priority over electrons coming from other dehydrogenases must have physiological consequences. Indeed, it is generally accepted that the yeast mitochondrial respiratory chain is able to oxidize matrical or cytosolic NADH in a comparable manner. Moreover, if all the dehydrogenases are able to oxidize their substrates at the same time, the redox potential, i.e., the $NADH:NAD^+$ ratio in either the cytosol or the mitochondria, should be the same, in contrast to the situation observed in mammalian cells. In these cells, it is well known that the transfer of reduced equivalents from the cytosol to the mitochondria by the malate–aspartate shuttle is dependent on the protonmotive force and leads to an important difference between the redox potentials, i.e., the mitochondria are more reduced than the cytosol. The results shown in this study scrutinize the validity of our thinking concerning yeast mitochondrial and cytosolic redox potentials. Indeed, we have shown previously [128] that in permeabilized spheroplasts, the NADH produced by cytosolic dehydrogenases, i.e., alcohol dehydrogenase and glyceraldehyde-3-phosphate dehydrogenase, is channeled through the porin to external mitochondrial NADH dehydrogenases. The main point of this study is that this very NADH is then oxidized by Nde1p and thus has the right of way over other substrates such as the matrical NADH. This tight organization, which favors the reoxidation of cytosolic NADH, has important consequences on the functioning of the Krebs cycle and possibly on the redox status of the mitochondrial matrix. Indeed, when cytosolic NADH concentration is high, the functioning of the Krebs cycle might be altered (consequent to the inability of matrical NADH reoxidation), and it has been shown that in this condition the Krebs cycle does not operate as a cycle but rather as two branches, one oxidative and one reducing [165]. Furthermore, when cytosolic NADH is lowered, the Krebs cycle returns to functioning as a cycle [165]. Thus, cytosolic NADH concentrations, through Nde1p, can regulate mitochondrial oxidative metabolism. Under physiological conditions, the kinetic constraints for electron transfer at the level of the respiratory chain should lead to a higher redox potential in mitochondria than in the cytosol.

2.7
Conclusions

It is noteworthy that the radical new approach to oxidative phosphorylation introduced by Peter Mitchell in 1961 as chemiosmotic coupling is still up to date. Indeed, in his theory, the structural (or supramolecular) features determine the vectorial nature of the energy transfer through a network of "enzyme-catalyzed group translocations" located in membranes. Such a mechanism differed fundamentally from the orthodox one previously developed [166, 167] because (1) it depends absolutely on the supramolecular organization of the enzyme system concerned, and (2) the driving force of a given chemical reaction is due to the specifically directed channeling of the diffusion of the chemical compounds or groups along a pathway specified in space by the physical organization of the system. The fact that Mitchell defined the energetic intermediary in oxidative phosphorylation as being the Δp between the bulk phases led researchers to focus on the relationships between forces (considered as bulk phase elements) and fluxes (oxygen consumption, ATP synthesis). This led them to consider oxidative phosphorylation complexes as "black boxes." Recent studies on the determination of the structures of the different complexes and the attempt to understand the functioning and kinetic regulation of oxidative phosphorylation led us to put back into focus the basis of Mitchell's theory.

Acknowledgments

Work presented here and done in our laboratory was supported by ANR-05-BLAN-0339-01.

The authors wish to thank M.F. Giraud for her incommensurable help in the elaboration of Fig. 2.1.

References

1 Rigoulet, M., Velours, J., Guerin, B. (1985) Substrate-level phosphorylation in isolated yeast mitochondria, *Eur J Biochem 153*, 601–607.

2 Rich, P. R. (2003) The molecular machinery of Keilin's respiratory chain, *Biochem Soc Trans 31*, 1095–1105.

3 Mitchell, P. (1961) Coupling of phosphorylation to electron and hydrogen transfer by a chemi-osmotic type of mechanism, *Nature 191*, 144–148.

4 Brunori, M., Sarti, P., Colosimo, A., Antonini, G., Malatesta, F., Jones, M. G., Wilson, M. T. (1985) Mechanism of control of cytochrome oxidase activity by the electrochemical-potential gradient, *EMBO J 4*, 2365–2368.

5 Mason, M. G., Nicholls, P., Wilson, M. T., Cooper, C. E. (2006) Nitric oxide inhibition of respiration involves both competitive (heme) and noncompetitive (copper) binding to cytochrome *c* oxidase, *Proc Natl Acad Sci USA 103*, 708–713.

6 Nicholls, P., Sharpe, M., Torres, J., Wilson, M. T., Cooper, C. E. (1998) Nitric oxide as an effector and a substrate for cytochrome *c* oxidase, *Biochem Soc Trans 26*, S323.

7 Beauvoit, B., Rigoulet, M., Guerin, B. (1989) Thermodynamic and kinetic control of ATP synthesis in yeast mitochondria: role of delta pH, *FEBS Lett 244*, 255–258.

8 Beauvoit, B., Bunoust, O., Guérin, B., Rigoulet, M. (1999) ATP-regulation of cytochrome oxidase in yeast mitochondria: role of subunit VIa, *Eur J Biochem 263*, 118–127.

9 Beauvoit, B., Rigoulet, M. (2001) Regulation of cytochrome *c* oxidase by adenylic nucleotides. Is oxidative phosphorylation feedback regulated by its end-products?, *IUBMB Life 52*, 143–152.

10 Kadenbach, B., Napiwotzki, J., Frank, V., Arnold, S., Exner, S., Hüttemann, M. (1998) Regulation of energy transduction and electron transfer in cytochrome *c* oxidase by adenine nucleotides, *J Bioenerg Biomembr 30*, 25–33.

11 Forman, N. G., Wilson, D. F. (1982) Energetics and stoichiometry of oxidative phosphorylation from NADH to cytochrome *c* in isolated rat liver mitochondria, *J Biol Chem 257*, 12908–12915.

12 Küster, U., Letko, G., Kunz, W., Duszyńsky, J., Bogucka, K., Wojtczak, L. (1981) Influence of different energy drains on the interrelationship between the rate of respiration, proton-motive force and adenine nucleotide patterns in isolated mitochondria, *Biochim Biophys Acta 636*, 32–38.

13 Duszynski, J., Groen, A. K., Wanders, R. J., Vervoorn, R. C., Tager, J. M. (1982) Quantification of the role of the adenine nucleotide translocator in the control of mitochondrial respiration in isolated rat-liver cells, *FEBS Lett 146*, 262–266.

14 Groen, A. K., Wanders, R. J., Westerhoff, H. V., van der Meer, R., Tager, J. M. (1982) Quantification of the contribution of various steps to the control of mitochondrial respiration, *J Biol Chem 257*, 2754–2757.

15 Tager, J. M., Wanders, R. J., Groen, A. K., Kunz, W., Bohnensack, R., Küster, U., Letko, G., Böhme, G., Duszynski, J., Wojtczak, L. (1983) Control of mitochondrial respiration, *FEBS Lett 151*, 1–9.

16 Chen, R., Fearnley, I. M., Peak-Chew, S. Y., Walker, J. E. (2004) The phosphorylation of subunits of complex I from bovine heart mitochondria, *J Biol Chem 279*, 26036–26045.

17 Rigoulet, M., Ouhabi, R., Leverve, X., Putod-Paramelle, F., Guérin, B. (1989) Almitrine, a new kind of energy-transduction inhibitor acting on mitochondrial ATP synthase, *Biochim Biophys Acta 975*, 325–329.

18 Fitton, V., Rigoulet, M., Ouhabi, R., Guérin, B. (1994) Mechanistic stoichiometry of yeast mitochondrial oxidative phosphorylation, *Biochemistry 33*, 9692–9698.

19 Brown, G. C. (1989) The relative proton stoichiometries of the mitochondrial proton pumps are independent of the proton motive force, *J Biol Chem 264*, 14704–14709.

20 Brown, G. C., Brand, M. D. (1986) Changes in permeability to protons and other cations at high proton motive force in rat liver mitochondria, *Biochem J 234*, 75–81.

21 Krishnamoorthy, G., Hinkle, P. C. (1984) Non-ohmic proton conductance of mitochondria and liposomes, *Biochemistry 23*, 1640–1645.

22 Mitchell, P. (1961) Coupling of phosphorylation to electron and hydrogen transfer by a chemi-osmotic type of mechanism, *Nature 191*, 144–148.

23 Nicholls, D. G. (1974) The influence of respiration and ATP hydrolysis on the proton-electrochemical gradient across the inner membrane of rat-liver mitochondria as determined by ion distribution, *Eur J Biochem 50*, 305–315.

24 O'Shea, P. S., Petrone, G., Casey, R. P., Azzi, A. (1984) The current-voltage relationships of liposomes and mitochondria, *Biochem J 219*, 719–726.

25 Zoratti, M., Favaron, M., Pietrobon, D., and Azzone, G. F. (1986) Intrinsic uncoupling of mitochondrial proton pumps. 1. Non-ohmic conductance cannot account for the nonlinear dependence of static head respiration on delta microH, *Biochemistry 25*, 760–767.

26 Ouhabi, R., Rigoulet, M., Lavie, J. L., and Guérin, B. (1991) Respiration in non-phosphorylating yeast mitochondria. Roles of non-ohmic proton conductance and intrinsic uncoupling, *Biochim Biophys Acta 1060*, 293–298.

27 Mitchell, P., Moyle, J. (1967) Acid-base titration across the membrane system of rat-liver mitochondria. Catalysis by uncouplers, *Biochem J 104*, 588–600.

28 Mitchell, P., Moyle, J. (1967) Respiration-driven proton translocation in rat liver mitochondria, *Biochem J 105*, 1147–1162.

29 Bielawski, J., Thompson, T. E., Lehninger, A. L. (1966) The effect of 2,4-dinitrophenol on the electrical resistance of phospholipid bilayer membranes, *Biochem Biophys Res Commun 24*, 948–954.

30 Hopfer, U., Lehninger, A. L., Thompson, T. E. (1968) Protonic conductance across phospholipid bilayer membranes induced by uncoupling agents for oxidative phosphorylation, *Proc Natl Acad Sci USA 59*, 484–490.

31 Liberman, E. A., Topaly, V. P. (1968) Selective transport of ions through bimolecular phospholipid membranes, *Biochim Biophys Acta 163*, 125–136.

32 McLaughlin, S. (1972) The mechanism of action of DNP on phospholipid bilayer membranes, *J Membr Biol 9*, 361–372.

33 Ricquier, D. (1998) Neonatal brown adipose tissue, UCP1 and the novel uncoupling proteins, *Biochem Soc Trans 26*, 120–123.

34 Del Mar Gonzalez-Barroso, M., Ricquier, D., Cassard-Doulcier, A. M. (2000) The human uncoupling protein-1 gene (UCP1): present status and perspectives in obesity research, *Obes Rev 1*, 61–72.

35 Nicholls, D. G., Locke, R. M. (1984) Thermogenic mechanisms in brown fat, *Physiol Rev 64*, 1–64.

36 Klingenberg, M., Huang, S. G. (1999) Structure and function of the uncoupling protein from brown adipose tissue, *Biochim Biophys Acta 1415*, 271–296.

37 Nicholls, D. G., Rial, E. (1999) A history of the first uncoupling protein, UCP1, *J Bioenerg Biomembr 31*, 399–406.

38 Duval, C., Nègre-Salvayre, A., Dogilo, A., Salvayre, R., Pénicaud, L., Casteilla, L. (2002) Increased reactive oxygen species production with antisense oligonucleotides directed against uncoupling protein-2 in murine endothelial cells, *Biochem Cell Biol 80*, 757–764.

39 Abe, T., Mujahid, A., Sato, K., Akiba, Y., Toyomizu, M. (2006) Possible role of avian uncoupling protein in down-regulating mitochondrial superoxide production in skeletal muscle of fasted chickens, *FEBS Lett 580*, 4815–4822.

40 Vidal-Puig, A. J., Grujic, D., Zhang, C. Y., Hagen, T., Boss, O., Ido, Y., Szczepanik, A., Wade, J., Mootha, V., Cortright, R., Muoio, D. M., Lowell, B. B. (2000) Energy metabolism in uncoupling protein 3 gene knockout mice, *J Biol Chem 275*, 16258–16266.

41 Nicholls, D. G. (1997) The non-Ohmic proton leak – 25 years on, *Biosci Rep 17*, 251–257.

42 Kadenbach, B. (2003) Intrinsic and extrinsic uncoupling of oxidative phosphorylation, *Biochim Biophys Acta 1604*, 77–94.

43 Bunoust, O., Devin, A., Avéret, N., Camougrand, N., Rigoulet, M. (2005) Competition of electrons to enter the respiratory chain: a new regulatory mechanism of oxidative metabolism in Saccharomyces cerevisiae, *J Biol Chem 280*, 3407–3413.

44 Hanstein, W. G. (1976) Uncoupling of oxidative phosphorylation, *Biochim Biophys Acta 456*, 129–148.

45 Rottenberg, H. (1983) Uncoupling of oxidative phosphorylation in rat liver mitochondria by general anesthetics, *Proc Natl Acad Sci USA 80*, 3313–3317.

46 Rottenberg, H., Hashimoto, K. (1986) Fatty acid uncoupling of oxidative phosphorylation in rat liver mitochondria, *Biochemistry 25*, 1747–1755.

47 Pick, U., Weiss, M., Rottenberg, H. (1987) Anomalous uncoupling of photophosphorylation by palmitic acid and by gramicidin D, *Biochemistry 26*, 8295–8302.

48 Luvisetto, S., Pietrobon, D., Azzone, G. F. (1987) Uncoupling of oxidative phosphorylation. 1. Protonophoric effects account only partially for uncoupling, *Biochemistry 26*, 7332–7338.

49 Luvisetto, S., Azzone, G. F. (1989) Nature of proton cycling during gramicidin uncoupling of oxidative phosphorylation, *Biochemistry 28*, 1100–1108.

50 Rottenberg, H., Koeppe, R. E. (1989) Mechanism of uncoupling of oxidative phosphorylation by gramicidin, *Biochemistry 28*, 4355–4360.

51 Canton, M., Gennari, F., Luvisetto, S., Azzone, G. F. (1996) The nature of uncoupling by n-hexane, 1-hexanethiol and 1-hexanol in rat liver mitochondria, *Biochim Biophys Acta 1274*, 39–47.

52 Luvisetto, S., Conti, E., Buso, M., Azzone, G. F. (1991) Flux ratios and pump stoichiometries at sites II and III in liver mitochondria. Effect of slips and leaks, *J Biol Chem 266*, 1034–1042.

53 Capitanio, N., Capitanio, G., De Nitto, E., Villani, G., Papa, S. (1991) H+/e– stoichiometry of mitochondrial cytochrome complexes reconstituted in liposomes. Rate-dependent changes of the stoichiometry in the cytochrome *c* oxidase vesicles, *FEBS Lett 288*, 179–182.

54 Capitanio, N., Capitanio, G., Demarinis, D. A., De Nitto, E., Massari, S., Papa, S. (1996) Factors affecting the H+/e– stoichiometry in mitochondrial cytochrome *c* oxidase: influence of the rate of electron flow and transmembrane delta pH, *Biochemistry 35*, 10800–10806.

55 Pietrobon, D., Azzone, G. F., Walz, D. (1981) Effect of funiculosin and antimycin A on the redox-driven H+-pumps in mitochondria: on the nature of ''leaks', *Eur J Biochem 117*, 389–394.

56 Pietrobon, D., Zoratti, M., Azzone, G. F. (1983) Molecular slipping in redox and ATPase H+ pumps, *Biochim Biophys Acta 723*, 317–321.

57 Pietrobon, D., Caplan, S. R. (1985) Flow-force relationships for a six-state proton pump model: intrinsic uncoupling, kinetic equivalence of input and output forces, and domain of approximate linearity, *Biochemistry 24*, 5764–5776.

58 Mitchell, P. (1976) Vectorial chemistry and the molecular mechanics of chemiosmotic coupling: power transmission by proticity, *Biochem Soc Trans 4*, 399–430.

59 Mitchell, P. (1976) Possible molecular mechanisms of the protonmotive function of cytochrome systems, *J Theor Biol 62*, 327–367.

60 Mitchell, P. (1975) The protonmotive Q cycle: a general formulation, *FEBS Lett 59*, 137–139.

61 Gennis, R. B., Barquera, B., Hacker, B., Van Doren, S. R., Arnaud, S., Crofts, A. R., Davidson, E., Gray, K. A., Daldal, F. (1993) The bc1 complexes of Rhodobacter sphaeroides and Rhodobacter capsulatus, *J Bioenerg Biomembr 25*, 195–209.

62 Boyer, P. D. (1993) The binding change mechanism for ATP synthase – some probabilities and possibilities, *Biochim Biophys Acta 1140*, 215–250.

63 Duncan, T. M., Bulygin, V. V., Zhou, Y., Hutcheon, M. L., Cross, R. L. (1995) Rotation of subunits during catalysis by Escherichia coli F_1-ATPase, *Proc Natl Acad Sci USA 92*, 10964–10968.

64 Zhou, Y., Duncan, T. M., Bulygin, V. V., Hutcheon, M. L., Cross, R. L. (1996) ATP hydrolysis by membrane-bound Escherichia coli F0F1 causes rotation of the gamma subunit relative to the beta subunits, *Biochim Biophys Acta 1275*, 96–100.

65 Papa, S., Lorusso, M., Capitanio, N. (1994) Mechanistic and phenomenological features of proton pumps in the respiratory chain of mitochondria, *J Bioenerg Biomembr 26*, 609–618.

66 Wikström, M. (1989) Identification of the electron transfers in cytochrome oxidase that are coupled to proton-pumping, *Nature 338*, 776–778.

67 Thomas, J. W., Puustinen, A., Alben, J. O., Gennis, R. B., Wikström, M. (1993) Substitution of asparagine for aspartate-135 in subunit I of the cytochrome bo ubiquinol oxidase of Escherichia coli eliminates proton-pumping activity, *Biochemistry 32*, 10923–10928.

68 Garcia-Horsman, J. A., Puustinen, A., Gennis, R. B., Wikström, M. (1995) Proton transfer in cytochrome bo3 ubiquinol oxidase of Escherichia coli: second-site mutations in subunit I that restore proton pumping in the mutant Asp135 → Asn, *Biochemistry 34*, 4428–4433.

69 Iwata, S., Ostermeier, C., Ludwig, B., Michel, H. (1995) Structure at 2.8 A resolution of cytochrome *c* oxidase from Paracoccus denitrificans, *Nature 376*, 660–669.

70 Tsukihara, T., Aoyama, H., Yamashita, E., Tomizaki, T., Yamaguchi, H., Shinzawa-Itoh, K., Nakashima, R., Yaono, R., Yoshikawa, S. (1996) The whole structure of the 13-subunit oxidized cytochrome *c* oxidase at 2.8 A, *Science 272*, 1136–1144.

71 Frank, V., Kadenbach, B. (1996) Regulation of the H+/e− stoichiometry of cytochrome *c* oxidase from bovine heart by intramitochondrial ATP/ADP ratios, *FEBS Lett 382*, 121–124.

72 Chance, B., Williams, G. R. (1956) The respiratory chain and oxidative phosphorylation, *Adv Enzymol Relat Subj Biochem 17*, 65–134.

73 Chance, B., Williams, G. R. (1956) Respiratory enzymes in oxidative phosphorylation. VI. The effects of adenosine diphosphate on azide-treated mitochondria, *J Biol Chem 221*, 477–489.

74 Qian, H., Beard, D. A. (2005) Thermodynamics of stoichiometric biochemical networks in living systems far from equilibrium, *Biophys Chem 114*, 213–220.

75 Fell, D. (1997) *Understanding the Control of Metabolism*, Portland Press, London.

76 Salter, M., Knowles, R. G., Pogson, C. I. (1994) in *Essays in biochemistry, Metabolic control*, pp 1–12, Portland Press, London.

77 Rossignol, R., Letellier, T., Malgat, M., Rocher, C., Mazat, J. P. (2000) Tissue variation in the control of oxidative phosphorylation: implication for mitochondrial diseases, *Biochem J 347 Pt 1*, 45–53.

78 Rigoulet, M., Averet, N., Mazat, J. P., Guerin, B., Cohadon, F. (1988) Redistribution of the flux-control coefficients in mitochondrial oxidative phosphorylations in the course of brain edema, *Biochim Biophys Acta 932*, 116–123.

79 Boumans, H., Berden, J. A., Grivell, L. A., van Dam, K. (1998) Metabolic control analysis of the bc1 complex of Saccharomyces cerevisiae: effect on cytochrome *c* oxidase, respiration and growth rate, *Biochem J 331 (Pt 3)*, 877–883.

80 Boumans, H., Grivell, L. A., Berden, J. A. (1998) The respiratory chain in yeast behaves as a single functional unit, *J Biol Chem 273*, 4872–4877.

81 Martin, M., Beauvoit, B., Voisin, P. J., Canioni, P., Guérin, B., Rigoulet, M. (1998) Energetic and morphological plasticity of C6 glioma cells grown on 3-D support; effect of transient glutamine deprivation, *J Bioenerg Biomembr 30*, 565–578.

82 Pasdois, P., Deveaud, C., Voisin, P., Bouchaud, V., Rigoulet, M., Beauvoit, B. (2003) Contribution of the phosphoryl-able complex I in the growth phase-dependent respiration of C6 glioma cells *in vitro*, *J Bioenerg Biomembr 35*, 439–450.

83 Groen, A. K., Wanders, R. J., Westerhoff, H. V., van der Meer, R., Tager, J. M. (1982) Quantification of the contribution of various steps to the control of mitochondrial respiration, *J Biol Chem 257*, 2754–2757.

84 Doussiere, J., Ligeti, E., Brandolin, G., Vignais, P. V. (1984) Control of oxidative phosphorylation in rat heart mitochondria. The role of the adenine nucleotide carrier, *Biochim Biophys Acta 766*, 492–500.

85 Mazat, J., Jean-Bart, E., Rigoulet, M., Guerin, B. (1986) Control of oxidative phosphorylations in yeast mitochondria. Role of the phosphate carrier, *Biochimica*

et Biophysica Acta (BBA) – Bioenergetics 849, 7–15.

86 Wikstrom, M., Morgan, J. E., Verkhovsky, M. I. (1997) Proton and electrical charge translocation by cytochrome-*c* oxidase, *Biochimica et Biophysica Acta (BBA) – Bioenergetics 1318*, 299–306.

87 Sarti, P., Jones, M. G., Antonini, G., Malatesta, F., Colosimo, A., Wilson, M. T., Brunori, M. (1985) Kinetics of redox-linked proton pumping activity of native and subunit III-depleted cytochrome *c* oxidase: a stopped-flow investigation, *Proc Natl Acad Sci USA 82*, 4876–4880.

88 Steverding, D., Kadenbach, B., Capitanio, N., Papa, S. (1990) Effect of chemical modification of lysine amino groups on redox and protonmotive activity of bovine heart cytochrome *c* oxidase reconstituted in phospholipid membranes, *Biochemistry 29*, 2945–2950.

89 Papa, S., Capitanio, N., De Nitto, E. (1987) Characteristics of the redox-linked proton ejection in beef-heart cytochrome *c* oxidase reconstituted in liposomes, *Eur J Biochem 164*, 507–516.

90 Papa, S., Capitanio, N., Capitanio, G., De Nitto, E., Minuto, M. (1991) The cytochrome chain of mitochondria exhibits variable H+/e− stoichiometry, *FEBS Lett 288*, 183–186.

91 Capitanio, N., Capitanio, G., De Nitto, E., Villani, G., Papa, S. (1991) H+/e− stoichiometry of mitochondrial cytochrome complexes reconstituted in liposomes. Rate-dependent changes of the stoichiometry in the cytochrome *c* oxidase vesicles, *FEBS Lett 288*, 179–182.

92 Fitton, V., Rigoulet, M., Ouhabi, R., Guerin, B. (1994) Mechanistic stoichiometry of yeast mitochondrial oxidative phosphorylation, *Biochemistry 33*, 9692–9698.

93 Nogueira, V., Rigoulet, M., Piquet, M. A., Devin, A., Fontaine, E., Leverve, X. M. (2001) Mitochondrial respiratory chain adjustment to cellular energy demand, *J Biol Chem 276*, 46104–46110.

94 Frank, V., Kadenbach, B. (1996) Regulation of the H+/e− stoichiometry

of cytochrome *c* oxidase from bovine heart by intramitochondrial ATP/ADP ratios, *FEBS Lett 382*, 121–124.

95 Gregory, L., Ferguson-Miller, S. (1989) Independent control of respiration in cytochrome *c* oxidase vesicles by pH and electrical gradients, *Biochemistry 28*, 2655–2662.

96 Capitanio, N., De Nitto, E., Villani, G., Capitanio, G., Papa, S. (1990) Protonmotive activity of cytochrome *c* oxidase: control of oxidoreduction of the heme centers by the protonmotive force in the reconstituted beef heart enzyme, *Biochemistry 29*, 2939–2945.

97 Craig, D. B., Wallace, C. J. (1991) The specificity and Kd at physiological ionic strength of an ATP-binding site on cytochrome *c* suit it to a regulatory role, *Biochem J 279 (Pt 3)*, 781–786.

98 Lin, J., Wu, S., Lau, W. T., Chan, S. I. (1995) 8-Azido-ATP modification of cytochrome *c*: retardation of its electron-transfer activity to cytochrome *c* oxidase, *Biochemistry 34*, 2678–2685.

99 Napiwotzki, J., Shinzawa-Itoh, K., Yoshikawa, S., Kadenbach, B. (1997) ATP and ADP bind to cytochrome *c* oxidase and regulate its activity, *Biol Chem 378*, 1013–1021.

100 Napiwotzki, J., Kadenbach, B. (1998) Extramitochondrial ATP/ADP-ratios regulate cytochrome *c* oxidase activity via binding to the cytosolic domain of subunit IV, *Biol Chem 379*, 335–339.

101 Rieger, T., Napiwotzki, J., Huther, F. J., Kadenbach, B. (1995) The number of nucleotide binding sites in cytochrome C oxidase, *Biochem Biophys Res Commun 217*, 34–40.

102 Huther, F. J., Kadenbach, B. (1987) ADP increases the affinity for cytochrome *c* by interaction with the matrix side of bovine heart cytochrome *c* oxidase, *Biochem Biophys Res Commun 147*, 1268–1275.

103 Huther, F. J., Kadenbach, B. (1988) Intraliposomal nucleotides change the kinetics of reconstituted cytochrome *c* oxidase from bovine heart but not from Paracoccus denitrificans, *Biochem Biophys Res Commun 153*, 525–534.

104 Rohdich, F., Kadenbach, B. (1993) Tissue-specific regulation of cytochrome

c oxidase efficiency by nucleotides, *Biochemistry 32*, 8499–8503.

105 Huttemann, M., Arnold, S., Lee, I., Muhlenbein, N., Linder, D., Lottspeich, F., Kadenbach, B. (2000) Turkey cytochrome *c* oxidase contains subunit VIa of the liver type associated with low efficiency of energy transduction, *Eur J Biochem 267*, 2098–2104.

106 Arnold, S., Kadenbach, B. (1997) Cell respiration is controlled by ATP, an allosteric inhibitor of cytochrome-*c* oxidase, *Eur J Biochem 249*, 350–354.

107 Palmieri, F., Indiveri, C., Bisaccia, F., Kramer, R. (1993) Functional properties of purified and reconstituted mitochondrial metabolite carriers, *J Bioenerg Biomembr 25*, 525–535.

108 Arnold, S., Goglia, F., Kadenbach, B. (1998) 3,5-Diiodothyronine binds to subunit Va of cytochrome-*c* oxidase and abolishes the allosteric inhibition of respiration by ATP, *Eur J Biochem 252*, 325–330.

109 Bender, E., Kadenbach, B. (2000) The allosteric ATP-inhibition of cytochrome *c* oxidase activity is reversibly switched on by cAMP-dependent phosphorylation, *FEBS Lett 466*, 130–134.

110 Yang, W. L., Iacono, L., Tang, W. M., Chin, K. V. (1998) Novel function of the regulatory subunit of protein kinase A: regulation of cytochrome *c* oxidase activity and cytochrome *c* release, *Biochemistry 37*, 14175–14180.

111 Steenaart, N. A., Shore, G. C. (1997) Mitochondrial cytochrome *c* oxidase subunit IV is phosphorylated by an endogenous kinase, *FEBS Lett 415*, 294–298.

112 Ferguson, S. J., Sorgato, M. C. (1982) Proton electrochemical gradients and energy-transduction processes, *Annu Rev Biochem 51*, 185–217.

113 Woelders, H., Putters, J., van Dam, K. (1986) Flow-force relationships in mitochondrial oxidative phosphorylation, *FEBS Lett 204*, 17–21.

114 Azzone, G. F., Pozzan, T., Massari, S., Bragadin, M. (1978) Proton electrochemical gradient and rate of controlled respiration in mitochondria, *Biochim Biophys Acta 501*, 296–306.

115 Jumelle-Laclau, M., Rigoulet, M., Guérin, B. (1994) Age-dependent change in the redox proton pump intrinsic uncoupling, *Biothermokinetics*, 129–140.

116 Padan, E., Rottenberg, H. (1973) Respiratory control and the proton electrochemical gradient in mitochondria, *Eur J Biochem 40*, 431–437.

117 Azzone, G. F., Massari, S., Pozzan, T. (1977) The generation of the proton electrochemical potential and its role in energy transduction, *Mol Cell Biochem 17*, 101–112.

118 Rigoulet, M., Guerin, B., Denis, M. (1987) Modification of flow-force relationships by external ATP in yeast mitochondria, *Eur J Biochem 168*, 275–279.

119 Papa, S., Sardanelli, A. M., Scacco, S., Technikova-Dobrova, Z. (1999) cAMP-dependent protein kinase and phosphoproteins in mammalian mitochondria. An extension of the cAMP-mediated intracellular signal transduction, *FEBS Lett 444*, 245–249.

120 Scacco, S., Vergari, R., Scarpulla, R. C., Technikova-Dobrova, Z., Sardanelli, A., Lambo, R., Lorusso, V., and Papa, S. (2000) cAMP-dependent phosphorylation of the nuclear encoded 18-kDa (IP) subunit of respiratory complex I and activation of the complex in serum-starved mouse fibroblast cultures, *J Biol Chem 275*, 17578–17582.

121 Vélot, C., Srere, P. A. (2000) Reversible transdominant inhibition of a metabolic pathway. *In vivo* evidence of interaction between two sequential tricarboxylic acid cycle enzymes in yeast, *J Biol Chem 275*, 12926–12933.

122 Srere, P. A., Sumegi, B., Sherry, A. D. (1987) Organizational aspects of the citric acid cycle, *Biochem Soc Symp 54*, 173–178.

123 Robinson, J. B., Inman, L., Sumegi, B., Srere, P. A. (1987) Further characterization of the Krebs tricarboxylic acid cycle metabolon, *J Biol Chem 262*, 1786–1790.

124 Cheung, C. W., Cohen, N. S., Raijman, L. (1989) Channeling of urea cycle intermediates *in situ* in permeabilized hepatocytes, *J Biol Chem 264*, 4038–4044.

125 Watford, M. (1989) Channeling in the urea cycle: a metabolon spanning two

compartments, *Trends Biochem Sci 14*, 313–314.

126 Ushiroyama, T., Fukushima, T., Styre, J. D., Spivey, H. O. (1992) Substrate channeling of NADH in mitochondrial redox processes, *Curr Top Cell Regul 33*, 291–307.

127 Fukushima, T., Decker, R. V., Anderson, W. M., Spivey, H. O. (1989) Substrate channeling of NADH and binding of dehydrogenases to complex I, *J Biol Chem 264*, 16483–16488.

128 Avéret, N., Aguilaniu, H., Bunoust, O., Gustafsson, L., Rigoulet, M. (2002) NADH is specifically channeled through the mitochondrial porin channel in Saccharomyces cerevisiae, *J Bioenerg Biomembr 34*, 499–506.

129 Avéret, N., Fitton, V., Bunoust, O., Rigoulet, M., Guérin, B. (1998) Yeast mitochondrial metabolism: from *in vitro* to *in situ* quantitative study, *Mol Cell Biochem 184*, 67–79.

130 Abrahams, J. P., Leslie, A. G., Lutter, R., Walker, J. E. (1994) Structure at 2.8 A resolution of F_1-ATPase from bovine heart mitochondria, *Nature 370*, 621–628.

131 Gibbons, C., Montgomery, M. G., Leslie, A. G., Walker, J. E. (2000) The structure of the central stalk in bovine F_1-ATPase at 2.4 A resolution, *Nat Struct Biol 7*, 1055–1061.

132 Velours, J., Arselin, G. (2000) The Saccharomyces cerevisiae ATP synthase, *J Bioenerg Biomembr 32*, 383–390.

133 Arnold, I., Pfeiffer, K., Neupert, W., Stuart, R. A., Schägger, H. (1998) Yeast mitochondrial F1F0-ATP synthase exists as a dimer: identification of three dimer-specific subunits, *EMBO J 17*, 7170–7178.

134 Vaillier, J., Arselin, G., Graves, P. V., Camougrand, N., Velours, J. (1999) Isolation of supernumerary yeast ATP synthase subunits e and i. Characterization of subunit i and disruption of its structural gene ATP18, *J Biol Chem 274*, 543–548.

135 Walker, J. E., Lutter, R., Dupuis, A., Runswick, M. J. (1991) Identification of the subunits of F_1F_0-ATPase from bovine heart mitochondria, *Biochemistry 30*, 5369–5378.

136 Higuti, T., Yoshihara, Y., Kuroiwa, K., Kawamura, Y., Toda, H., Sakiyama, F. (1992) A simple, rapid method for purification of epsilon-subunit, coupling factor 6, subunit d, and subunit e from rat liver H(+)-ATP synthase and determination of the complete amino acid sequence of epsilon-subunit, *J Biol Chem 267*, 22658–22661.

137 Arnold, I., Bauer, M. F., Brunner, M., Neupert, W., Stuart, R. A. (1997) Yeast mitochondrial F_1F_0-ATPase: the novel subunit e is identical to Tim11, *FEBS Lett 411*, 195–200.

138 Belogrudov, G. I., Tomich, J. M., Hatefi, Y. (1996) Membrane topography and near-neighbor relationships of the mitochondrial ATP synthase subunits e, f, and g, *J Biol Chem 271*, 20340–20345.

139 Perkins, G. A., Frey, T. G. (2000) Recent structural insight into mitochondria gained by microscopy, *Micron 31*, 97–111.

140 Frey, T. G., Mannella, C. A. (2000) The internal structure of mitochondria, *Trends Biochem Sci 25*, 319–324.

141 Paumard, P., Vaillier, J., Coulary, B., Schaeffer, J., Soubannier, V., Mueller, D. M., Brèthes, D., di Rago, J. P., Velours, J. (2002) The ATP synthase is involved in generating mitochondrial cristae morphology, *EMBO J 21*, 221–230.

142 Soubannier, V., Vaillier, J., Paumard, P., Coulary, B., Schaeffer, J., Velours, J. (2002) In the absence of the first membrane-spanning segment of subunit 4(b), the yeast ATP synthase is functional but does not dimerize or oligomerize, *J Biol Chem 277*, 10739–10745.

143 Giraud, M. F., Paumard, P., Soubannier, V., Vaillier, J., Arselin, G., Salin, B., Schaeffer, J., Brèthes, D., di Rago, J. P., Velours, J. (2002) Is there a relationship between the supramolecular organization of the mitochondrial ATP synthase and the formation of cristae?, *Biochim Biophys Acta 1555*, 174–180.

144 Ko, Y. H., Delannoy, M., Hullihen, J., Chiu, W., Pedersen, P. L. (2003) Mitochondrial ATP synthasome. Cristae-enriched membranes and a multiwell detergent screening assay yield

dispersed single complexes containing the ATP synthase and carriers for Pi and ADP/ATP, *J Biol Chem 278*, 12305–12309.

145 Chen, C., Ko, Y., Delannoy, M., Ludtke, S. J., Chiu, W., Pedersen, P. L. (2004) Mitochondrial ATP synthasome: three-dimensional structure by electron microscopy of the ATP synthase in complex formation with carriers for Pi and ADP/ATP, *J Biol Chem 279*, 31761–31768.

146 Hackenbrock, C. R., Chazotte, B., Gupte, S. S. (1986) The random collision model and a critical assessment of diffusion and collision in mitochondrial electron transport, *J Bioenerg Biomembr 18*, 331–368.

147 Chance, B., Williams, G. R. (1955) A method for the localization of sites for oxidative phosphorylation, *Nature 176*, 250–254.

148 Stroh, A., Anderka, O., Pfeiffer, K., Yagi, T., Finel, M., Ludwig, B., Schägger, H. (2004) Assembly of respiratory complexes I, III, and IV into NADH oxidase supercomplex stabilizes complex I in Paracoccus denitrificans, *J Biol Chem 279*, 5000–5007.

149 Schägger, H., Pfeiffer, K. (2000) Supercomplexes in the respiratory chains of yeast and mammalian mitochondria, *EMBO J 19*, 1777–1783.

150 Cruciat, C. M., Brunner, S., Baumann, F., Neupert, W., Stuart, R. A. (2000) The cytochrome bc1 and cytochrome *c* oxidase complexes associate to form a single supracomplex in yeast mitochondria, *J Biol Chem 275*, 18093–18098.

151 Krause, F., Scheckhuber, C. Q., Werner, A., Rexroth, S., Reifschneider, N. H., Dencher, N. A., Osiewacz, H. D. (2004) Supramolecular organization of cytochrome *c* oxidase- and alternative oxidase-dependent respiratory chains in the filamentous fungus Podospora anserina, *J Biol Chem 279*, 26453–26461.

152 Eubel, H., Jänsch, L., Braun, H. P. (2003) New insights into the respiratory chain of plant mitochondria. Super-complexes and a unique composition of complex II, *Plant Physiol 133*, 274–286.

153 Eubel, H., Heinemeyer, J., Braun, H. P. (2004) Identification and characteriza-

tion of respirasomes in potato mitochondria, *Plant Physiol 134*, 1450–1459.

154 Krause, F., Reifschneider, N. H., Vocke, D., Seelert, H., Rexroth, S., Dencher, N. A. (2004) "Respirasome"-like supercomplexes in green leaf mitochondria of spinach, *J Biol Chem 279*, 48369–48375.

155 Schägger, H., Pfeiffer, K. (2001) The ratio of oxidative phosphorylation complexes I–V in bovine heart mitochondria and the composition of respiratory chain supercomplexes, *J Biol Chem 276*, 37861–37867.

156 Schägger, H., de Coo, R., Bauer, M. F., Hofmann, S., Godinot, C., Brandt, U. (2004) Significance of respirasomes for the assembly/stability of human respiratory chain complex I, *J Biol Chem 279*, 36349–36353.

157 Schägger, H. (2002) Respiratory chain supercomplexes of mitochondria and bacteria, *Biochim Biophys Acta 1555*, 154–159.

158 Bianchi, C., Genova, M. L., Parenti Castelli, G., Lenaz, G. (2004) The mitochondrial respiratory chain is partially organized in a supercomplex assembly: kinetic evidence using flux control analysis, *J Biol Chem 279*, 36562–36569.

159 Dudkina, N. V., Eubel, H., Keegstra, W., Boekema, E. J., Braun, H. P. (2005) Structure of a mitochondrial super-complex formed by respiratory-chain complexes I and III, *Proc Natl Acad Sci USA 102*, 3225–3229.

160 Devin, A., Dejean, L., Beauvoit, B., Chevtzoff, C., Avéret, N., Bunoust, O., Rigoulet, M. (2006) Growth yield homeostasis in respiring yeast is due to a strict mitochondrial content adjustment, *J Biol Chem 281*, 26779–26784.

161 Ouhabi, R., Rigoulet, M., Guerin, B. (1989) Flux-yield dependence of oxidative phosphorylation at constant $\Delta\mu$H, *FEBS Letters 256*, 245–245.

162 Small, W. C., McAlister-Henn, L. (1998) Identification of a cytosolically directed NADH dehydrogenase in mitochondria of Saccharomyces cerevisiae, *J Bacteriol 180*, 4051–4055.

163 DeRisi, J. L., Iyer, V. R., Brown, P. O. (1997) Exploring the metabolic and genetic control of gene expression on a genomic scale, *Science 278*, 680–686.

164 Påhlman, I. L., Larsson, C., Averét, N., Bunoust, O., Boubekeur, S., Gustafsson, L., Rigoulet, M. (2002) Kinetic regulation of the mitochondrial glycerol-3-phosphate dehydrogenase by the external NADH dehydrogenase in Saccharomyces cerevisiae, *J Biol Chem 277*, 27991–27995.

165 Gombert, A. K., Moreira dos Santos, M., Christensen, B., Nielsen, J. (2001) Network identification and flux quantification in the central metabolism of Saccharomyces cerevisiae under different conditions of glucose repression, *J Bacteriol 183*, 1441–1451.

166 Slater, E. C. (1953) Mechanism of phosphorylation in the respiratory chain, *Nature 172*, 975–978.

167 Boyer, P. D. (1965) *Oxidases and related redox systems*, pp 994–1008, Wiley, New York.

168 Tsukihara, T., Aoyama, H., Yamashita, E., Tomizaki, T., Yamaguchi, H., Shinzawa-Itoh, K., Nakashima, R., Yaono, R., Yoshikawa, S. (1996) The whole structure of the 13-subunit oxidized cytochrome *c* oxidase at 2.8 A, *Science 272*, 1136–1144.

3

Integrated and Organized Cellular Energetic Systems: Theories of Cell Energetics, Compartmentation, and Metabolic Channeling

Valdur Saks, Claire Monge, Tiia Anmann, and Petras P. Dzeja

> *The essential thing in metabolism is that the organism succeeds in freeing itself from all entropy it cannot help producing while alive.*
>
> Erwin Schrödinger, *What Is Life?* (1944)

Abstract

Bioenergetics as a part of biophysical chemistry and biophysics is a quantitative science that is based on several fundamental theories. The definition of energy itself is given by the first law of thermodynamics, and application of the second law of thermodynamics shows that living cells can function as open systems only where the internal order (low entropy state) is maintained due to increasing the entropy in the surrounding medium (Schrödinger's principle of negentropy). For this, the exchange of mass is needed, which gives rise to metabolism as the sum of catabolism and anabolism, and the free energy changes during catabolic reactions supply energy for all cellular work – osmotic, mechanical, and biochemical. The free energy changes during metabolic reactions obey the rules of chemical thermodynamics, which deals with Gibbs free energy of chemical reactions and with electrochemical potentials. For application of these theories to the integrated systems *in vivo*, complex cellular organization should be accounted for: macromolecular crowding; metabolic channeling and functional coupling mechanisms due to close and tight protein–protein interactions; compartmentation of the enzymes due to their attachment to the subcellular membranes or connection to the cytoskeleton; and both macro- and microcompartmentation of substrates and metabolites in the cells. Because of this, almost all processes important for cell life are localized within small areas of the cell, and for their integration effective systems of communication, including compartmentalized energy transfer systems, are required. These are represented in muscle cells by the phosphotransfer networks, mostly by the creatine kinase and adenylate kinase systems, whose alterations under ischemic conditions contribute significantly to acute ischemic contractile failure of the heart.

Molecular System Bioenergetics: Energy for Life. Edited by Valdur Saks
Copyright © 2007 WILEY-VCH Verlag GmbH & Co. KGaA, Weinheim
ISBN: 978-3-527-31787-5

3.1
Introduction

Living cells belong to the group of open thermodynamic systems because they exchange both energy and mass with their environment [1–6]. This is necessary to maintain cell structure and function and to avoid decay into equilibrium, which would lead to death. This thermodynamic principle as a physical basis of life was first discovered and described some 60 years ago by Erwin Schrödinger in his famous book *What Is Life?* [7]. Mass and energy exchange are the functions of metabolism [7], including both catabolism and anabolism (see below), and these constitute a lawful way for cells to live in a thermodynamic sense. At the same time, metabolism is the basis of biological energy transformations in the cells, as any chemical or biochemical reaction is inevitably linked to free energy changes according to the laws of chemical thermodynamics. In the first part of this chapter, we describe briefly and in a general manner these theoretical bases of free energy conversion and metabolism. In the second part, we analyze the energy transfer networks in muscle and brain cells. Because of structural organization and macromolecular crowding, almost all processes important for cell life are localized within small areas of the cell, and for their integration effective systems of communication, including compartmentalized energy transfer systems, are required. These are represented in muscle cells by the phosphotransfer networks, mostly by the creatine kinase and adenylate kinase systems, whose alterations under ischemic conditions contribute significantly to acute ischemic contractile failure of the heart.

3.2
Theoretical Basis of Cellular Metabolism and Bioenergetics

3.2.1
Thermodynamic Laws, Energy Metabolism, and Cellular Organization

The first law of thermodynamics, formulated by Mayer in 1842, is the law of energy conservation for isolated systems. It is also useful for defining of energy itself as the capacity to perform any kind of work and/or to produce heat. This law says that a change in the internal energy of a system (dE)

$$dE = E_f - E_0 = \delta q + \delta w \tag{1}$$

is the result of heat exchanged and taken up by system δq and work done by the surrounding medium on the system δw (E_f and E_0 are the final and initial energy levels, respectively). By convention, if the system performs work, the sign of δw is negative.

The second law of thermodynamics describes a direction in which all processes can occur spontaneously. It tells us that irreversible processes move in the direction of the increase of entropy. The entropy function S was introduced by Clausius as $dS = \delta q/T$ (for reversible processes) to describe the Carnot cycle, and Boltzmann gave its statistical explanation in terms of molecular kinetic theory:

$$S = k_B \log W \qquad (2)$$

Here, k_B is Boltzmann's constant and W is the thermodynamic probability function associated with the number of possible arrangements of molecules (each arrangement is called a "microstate") in the given "macrostate" of a system [1, 4]. Thus, the less organized a system is, the larger the number of possible arrangements of molecules in it (or number of "microstates") and the higher the entropy of the system will be. The second law of thermodynamics tells us that under irreversible conditions all processes in all systems move toward a state of a greater probability or greater disorder.

The two thermodynamic laws taken together describe the relationship between changes in the functions of the thermodynamic state of a system – Gibbs free energy G (which is the capacity to do useful work), enthalpy H (which is the maximal amount of exchangeable heat a system can produce), and entropy S (related to the amount of heat that cannot be used to perform useful work but related to the order exhibited by the system):

$$dG = dH - TdS = dE + pdV - TdS \leq 0 \qquad (3)$$

According to these laws, all irreversible processes can proceed spontaneously only when the free energy decreases in the direction of equilibrium, at which it reaches a minimum and does not change anymore ($dG = 0$). The decrease in free energy in any spontaneous process is due mainly to an increase in entropy. If the system is under non-equilibrium steady-state conditions, the free energy is continuously dissipated (see below) [1, 2, 4–6].

Thus, the universal laws of thermodynamics described by Eq. (3) allow all processes and reactions to proceed only in the direction in which the free energy decreases and entropy increases and thus in which disorder increases. This seems to be in contradiction with what is known about cell life. In cells, biochemical catalysis needs protein catalysts with precise conformational structure; transmission of genetic information needs the permanence of the structure of DNA and RNA; and metabolic pathways are organized into supramolecular complexes, metabolons, etc. Schrödinger showed that the apparent contradiction is overcome if we consider that living cells are open systems that, by means of metabolism, decrease or maintain their low entropy state by increasing the entropy of the medium [7]. This is shown by the general scheme in Fig. 3.1 describing the energetics of the cell as an open system exchanging both energy and masses with surrounding medium. Catabolic reactions through coupling to anabolic reactions (biosynthesis) maintain cell structure and organization as an expression of the decrease in entropy. They are also the source of metabolic energy for the performance of any kind of cellular work.

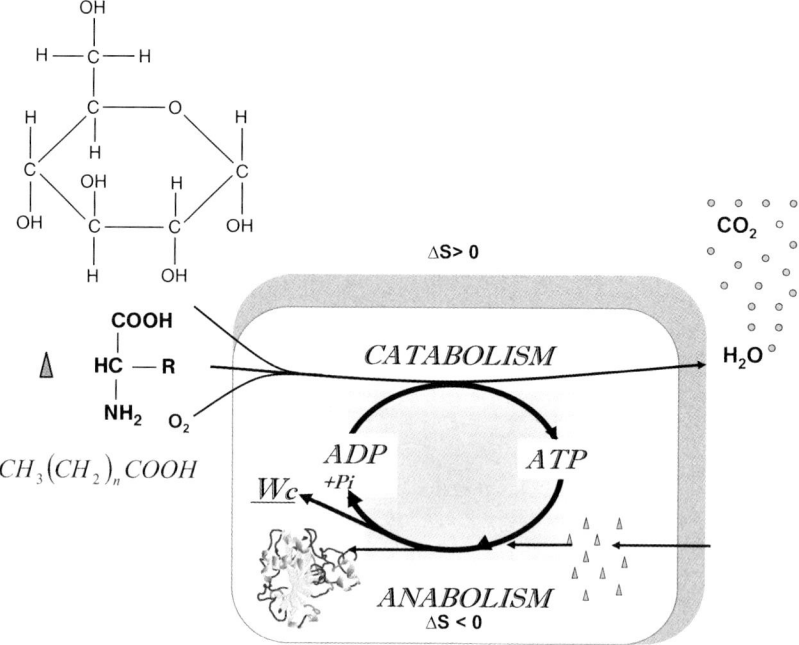

Fig. 3.1 General scheme of cellular metabolism. A cell is a thermodynamically open system. In accordance with Schrödinger's principle of negentropy, extraction (increasing the entropy in the extracellular medium and decreasing it in the cell) via metabolism is necessary for maintenance of the structural organization of both biopolymers as proteins, DNA and RNA, and for maintenance of the fine structural organization of the cell for effectively running compartmentalized metabolic processes. In this way, the cell can live in agreement with thermodynamic laws (see the text). W_c means cellular work.

3.2.2

Chemical and Electrochemical Potentials: Energy of Transmembrane Transport and Metabolic Reactions

In general terms, the capacity to do useful work, quantified by the Gibbs free energy, is a function of the pressure p, temperature T, and chemical composition [1], and its full differential can be written as:

$$dG = \left(\frac{\partial G}{\partial p}\right)_{T, n_i} \cdot dp + \left(\frac{\partial G}{\partial T}\right)_{p, n_i} \cdot dT + \sum_i \left(\frac{\partial G}{\partial n_i}\right)_{p, T} \cdot dn_i \qquad (4)$$

where $\dfrac{\partial G}{\partial n_i} = \mu_i$ is a chemical potential of a component i.

Usually biochemical reactions occur at constant temperature and pressure, and thus $dp = 0$ and $dT = 0$. Under these conditions, the free energy changes are caused by changes in the chemical composition, i.e., the free energy changes in the system are directly related to the chemical or biochemical reactions [1, 2]:

$$dG = \sum_i \left(\frac{\partial G}{\partial n_i}\right)_{T,p} \cdot dn_i = \sum_i \mu_i \cdot dn_i \qquad (5)$$

or in the integral form:

$$G = \sum_i \mu_i \cdot n_i \qquad (6)$$

The first derivative of the Gibbs free energy with respect to the molar concentration of component n_i, its chemical potential $\mu_i = (\partial G/\partial n_i)_{T,p}$ (see Eq. 4), is a function of the molar concentration C_i of component n_i. For species bearing electric charges z (cations or anions) in the presence of electrical potential Ψ, it becomes an electrochemical potential [1–3]:

$$\tilde{\mu}_i = \mu_i^0 + RT \ln C_i + zF\Psi \qquad (7)$$

$F = 0.0965$ kJ mol^{-1} mV^{-1} is the Faraday constant [3], and $\mu^0{}_i$ is a standard chemical potential, corresponding to the standard reference state of the system, in which the concentration is $C^0 = 1$ M. To remind one that all concentrations should be given in reference to this state, the equation can be rewritten as:

$$\tilde{\mu}_i = \mu_i^0 + RT \ln \frac{C_i}{C^0} + zF\Psi \qquad (8)$$

Equations (8) and (9) are the basic equations of bioenergetics and describe the energy conversion in biophysical or biochemical processes [1–6]. Thus, according to Mitchell's chemiosmotic theory (see Chapter 2), the protons translocated by the respiratory chain (see Chapter 2) result in a transmembrane gradient of H^+ or ΔpH and other ions (e.g., K^+, Na^+, Ca^{2+}) indirectly coupled to the H^+ gradient or $\Delta\Psi$, which together represent an electrochemical potential of ions or protonmotive force:

$$\Delta p = \frac{\Delta\tilde{\mu}_{H^+}}{F} = \Delta\Psi - \frac{2.3RT}{F} \cdot \Delta pH \qquad (9)$$

The free energy changes in the metabolic reactions are in a similar way explained by changes of chemical potentials of reaction components, substrates and products. Thus, free energy change in the reaction of the conversion of substrate A into product P,

$$A \underset{k_{-1}}{\overset{k_1}{\rightleftharpoons}} P$$

is given by the difference of chemical potentials of P and A:

$$\Delta G = \mu_P - \mu_A = \left(\mu_P^0 + RT \ln \frac{[P]}{C^0} \right) - \left(\mu_A^0 + RT \ln \frac{[A]}{C^0} \right)$$

$$= (\mu_P^0 - \mu_A^0) + RT \ln \frac{[P]}{[A]} \qquad (10)$$

where $\Delta G^0 = (\mu_P^0 - \mu_A^0)$ is the standard free energy change and $\Gamma = \frac{[P]}{[A]}$ is the mass action ratio of the reaction (for multi-reactant reactions, this ratio is defined as $\Gamma = \frac{\prod_i [P]_i}{\prod_i [A]_i}$), and therefore:

$$\Delta G = \Delta G^0 + RT \ln \Gamma \qquad (11)$$

At equilibrium,

$$\Delta G = \Delta G^0 + RT \ln \Gamma = 0 \qquad (12)$$

and Γ corresponds to the equilibrium constant K_{eq}:

$$\Gamma_{eq} = K_{eq} = \frac{[P]_{eq}}{[A]_{eq}} = \frac{k_1}{k_{-1}} = e^{-(\mu_P^0 - \mu_A^0)/RT} \qquad (13)$$

That is typically written as

$$\Delta G^0 = -RT \ln K_{eq} \qquad (14)$$

This is the famous equation relating the equilibrium constant to standard free energy change, first described by van't Hoff. This gives the final equation for the free energy change in any reaction:

$$\Delta G = -RT \ln K_{eq} + RT \ln \Gamma \qquad (15)$$

This equation can be rearranged in the following way:

$$\Delta G = RT \ln \frac{\Gamma}{K_{eq}} \qquad (16)$$

Figure 3.2A shows the plot of the free energy G of a system calculated on the basis of Eq. (6) as functions of $\log \Gamma / K_{eq}$ (the ratio Γ / K_{eq} is shown in logarithmic scale), and Fig. 3.2B shows the corresponding ΔG. In Fig. 3.2, one can see a minimum of free energy G in the equilibrium position, where $\Delta G = 0$ [8]. If the reaction is shifted from equilibrium to the right in the direction of an increase in

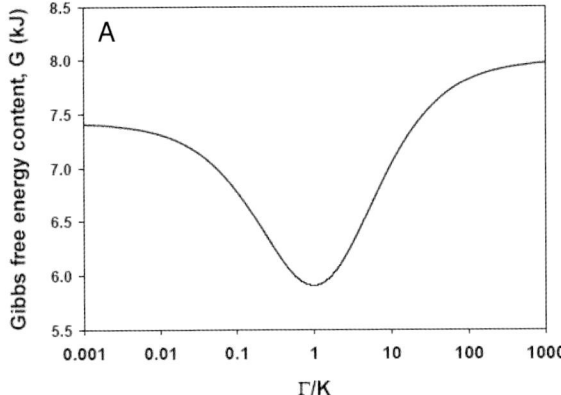

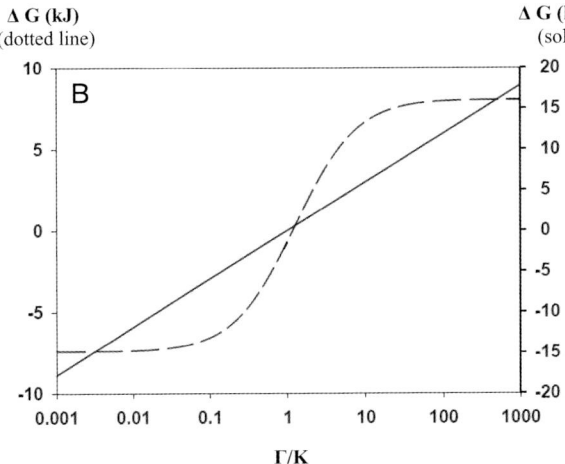

Fig. 3.2 (A) The free energy of a system as a function of the log Γ/K_{eq} (Γ/K_{eq} on the abscissa axis is given in the logarithmic scale). Γ is shown for the direction of conversion of A into P. Free energy of the system was calculated according to Eq. (6): $G = n_A\mu_A + n_P\mu_P$. (B) ΔG as a function of log Γ/K for two cases. The solid line shows ΔG calculated by using Eq. (16) and expressed in kilojoules per mole. This line shows the amount of the free energy, which could be obtained (in exergonic reactions) or used (in endergonic reactions) for conversion of one mole of substrate under given conditions. The dotted line shows the ΔG available in the "real" system according to our calculations, taking into account an amount of substrate present, $\Delta G = n_P\mu_P - n_A\mu_A$, in kilojoules. The calculations were made by Marc Jamin (Joseph Fourier University, France) using the following arbitrary conditions: $\mu_A^o = 7.4$ kJ mol^{-1}, $\mu_P^o = 8$ kJ mol^{-1}, $K = 0.8$, $[A]_0 = 1$ M, $[P]_0 = 0$ M. (The figure is adapted from [8] and is courtesy of Marc Jamin).

Γ/K_{eq}, free energy and ΔG increase (Fig. 3.2). Free energy of a system is also higher left of the equilibrium position of the reaction (Fig. 3.2A), but in this case ΔG is negative (Fig. 3.2B). This means that left of equilibrium, the reaction in the direction $A \rightarrow P$ is exergonic (spontaneous) and right of equilibrium it is endergonic (non-spontaneous). The opposite is true for the reaction $P \rightarrow A$, which is endergonic left of and exergonic right of equilibrium. Fig. 3.2B shows two ways of expression of ΔG: the solid line shows this parameter in kilojoules per mole of a substrate, which is a linear function of log Γ/K in accordance with Eq. (16). This is the free energy change during conversion of one mole of substrate into product under given conditions (given value of Γ). The dotted line shows ΔG available in the actual system by taking into account the number of moles present in the system, $\Delta G = n_P \mu_P - n_A \mu_A$, where n_A and n_P are the number of moles of A and P, respectively. This latter value of ΔG is directly related to "real" changes in G shown in Fig. 3.2A.

These relationships mean that any biochemical reaction in the cell that is away from equilibrium is a potential source of free energy.

3.2.3
Non-equilibrium, Steady-state Conditions

As has recently been shown by Qian and Beard, in the case of biochemical networks in living systems, when the metabolic processes are in the steady state far from equilibrium, Eq. (16) can be modified to include fluxes or reaction rates [5, 6]. For processes that are in the non-equilibrium steady state, for example, the reaction

$$\rightarrow A \underset{k_{-1}}{\overset{k_1}{\rightleftarrows}} P \rightarrow$$

the forward and backward fluxes are $J_+ = k_1[A]$, $J_- = k_{-1}[P]$, respectively, and by definition $J = J_+ - J_- \neq 0$ (only in equilibrium does $J = 0$). The equilibrium constant is $K_{eq} = k_1/k_{-1}$, and by replacing K_{eq} in Eq. (16) with this expression, for chemical potential differences in steady state of this reaction in the system, we obtain

$$\Delta \mu = RT \ln \frac{J_-}{J_+} \tag{17}$$

This represents the chemical driving force under steady-state conditions. In an open system, Eq. (17) also shows the amount of energy to be applied to the system in order to maintain the reactions in the steady state (constant concentrations of A and P), and in the absence of work done this is the energy dissipated as heat and transformed into entropy of the surroundings [5, 6]. This is explained by non-equilibrium thermodynamics [2, 9–11], according to which for open systems the entropy may be split into two contributing parts [9, 10]:

$$dS = d_i S + d_e S \tag{18}$$

where d_iS is the entropy production from irreversible catabolic processes inside the system, which always increases, and d_eS is the entropy flow from the surroundings, whose contribution is null in an isolated system since entropy always increases irreversibly in such a system. However, in systems open to the exchange of matter and energy, such as biological systems, d_eS may be negative, so that the system may decrease its entropy because statistical fluctuations may favor the organized state. The stabilization of an organized state after statistical fluctuations or instabilities (order through fluctuations) may occur because of the existence of nonlinear kinetic laws. These states are maintained by energy and matter dissipation; therefore, they are also known as *dissipative structures* [9]. This clarifies the answer given by Schrödinger to the question: How do living organisms avoid decay? Or, otherwise stated, how do they avoid contradicting the second law of thermodynamics? The basic answer given by Schrödinger is that organisms do not disobey the second law because as first formulated it applies to closed rather than open systems; more explicitly, although a living organism continually increases entropy (positive entropy or d_iS) and tends to approach the dangerous state of maximum entropy (or death), it keeps away from it by continually drawing negative entropy (d_eS) from its environment [7], or, more rigorously, free energy to keep its organization (see [11] for an excellent discussion).

3.2.4
Free Energy Changes and the Problem of Intracellular Organization of Metabolism

Coupling of catabolism with anabolism (the metabolism) is the way through which "negentropy" or free energy is extracted from the medium. Evolution has selected the adenine nucleotides to fulfill this important task of coupling catabolism and anabolism (Fig. 3.1). A possible explanation is the rather high standard free energy change ($\Delta G° = 31.5$ kJ mol^{-1}) and high affinity of ATP and ADP for many enzymes and carriers [12]. Cellular energetics is thus based on the reactions of ATP synthesis and utilization: $ADP + P_i \Leftrightarrow ATP + H_2O$. Taking water content as a constant and in excess (not changing in the reaction), the mass action ratio of the reaction of ATP synthesis is usually written as:

$$\Gamma = \frac{[ATP]}{[ADP][P_i]} \tag{19}$$

By maintaining a high mass action ratio for the reaction of ATP synthesis, catabolic reactions also supply free energy for cellular work, W_c. The free energy available in the cellular system is, according to Eqs. (14)–(16), a function of the ratio of Γ to the equilibrium constant of the ATP synthesis, or:

$$\Delta G_{ATP} = \Delta G^0_{ATP} + RT \ln \frac{[ATP]}{[ADP][P_i]} \tag{20}$$

This function is usually called the phosphorylation potential [13, 14]. The principal purpose of free energy transformation associated with catabolic reactions is to maintain the phosphorylation potential at a high value, which is achieved mostly through mitochondrial oxidative phosphorylation or photosynthesis in autotrophic organisms. The theory of phosphorylation potential for the analysis of cellular life was first used by Veech et al. [13] and Kammermeier et al. [14].

However, now it has become clear that applying Eq. (20) as well as any quantitative theory of physical chemistry to the real intracellular medium is not a simple task, because of the complex organization of cell structure and metabolism. It has become clear that it is not the global ATP content that is important; rather, the ATP and the free energy available in micro- and macrocompartments have to be accounted for, as will be described below.

3.2.5
Macromolecular Crowding, Heterogeneity of Diffusion, Compartmentation, and Vectorial Metabolism

The first phenomenon to be taken into account in all cells is macromolecular crowding: the high concentrations of macromolecules in the cells [15–22] decrease the volume available for free diffusion of substrates and accordingly make it difficult to use correctly the enzyme kinetics and equations usually worked out in enzymatic studies carried out in diluted solutions [20, 22, 23]. Cytoplasmic protein concentration may be as high as 200–300 mg mL^{-1}, which corresponds to a volume fraction of about 20–30% of intracellular medium occupied by these proteins [15, 16]. In the mitochondria the high density of enzymes and other proteins constitutes more than 60% of the matrix volume [17].

At first sight, this macromolecular crowding should cause real chaos by making intracellular communication by diffusion of reaction intermediates difficult. This is similar to the situation that occurs in any megalopolis at rush hour if the traffic lights are turned off. This chaos and the related problems are well described by Denis Noble in his recent book [24]. In reality, however, macromolecular crowding gives rise to new mechanisms based on specific protein–protein interactions – microcompartmentation, metabolic channeling, and functional coupling – and in this way leads to a fine organization and regulation of the metabolism. The traffic lights are on, and movements are perfectly organized. The origin of this order as well as the program and source of these specific interactions and cellular organizations are still a mystery hidden in undiscovered genetic laws, including feedback control of gene expression, and in rules governing system-level properties of cellular systems [24], and their elucidation is the main challenge for cell sciences in the post-genomic era [24–27].

3.2.5.1   Heterogeneity of Intracellular Diffusion and Metabolic Channeling
Many new experimental techniques have been developed to study the molecular networks formed by protein–protein interactions [28]. In the cells, the high protein density predominantly determines the major characteristics of the cellular environment, such as diffusion in heterogeneous compartments [23, 29–31]. There are distinct barriers to diffusion of solutes within the cells, i.e., binding

and crowding. Whereas molecular crowding and sieving restrict the mobility of very large solutes, binding can severely restrict the mobility of smaller solutes (see Ref. [30] and references therein). This also explains the heterogeneity of the diffusion behavior of ATP in cells. Studies using pulsed-gradient ^{31}P-NMR showed that the diffusion of ATP and phosphocreatine is anisotropic in muscle cells [31, 32]. Recent mathematical modeling of the decreased affinity of mitochondria for exogenous ADP *in situ* in permeabilized cardiac cells also showed that ADP or ATP diffusion in cells is heterogeneous and that the apparent diffusion coefficient for ADP (and ATP) may be locally decreased (diffusion locally restricted) by an order, or even several orders, of magnitude [33]. A similar limited diffusion of ATP in the subsarcolemmal area in cardiac cells was proposed by Terzic and Dzeja's group [34, 35]. There is also firm experimental evidence for compartmentalization of ATP in cardiomyocytes (see next section).

Because of molecular crowding and hindered diffusion, cells need to compartmentalize metabolic pathways in order to overcome diffusive barriers. Biochemical reactions can successfully proceed and even be facilitated by metabolic channeling of intermediates due to structural organization of enzyme systems into organized multi-enzyme complexes. Metabolite channeling directly transfers the intermediate from one enzyme to an adjacent enzyme without the need of free aqueous-phase diffusion [15, 18, 36–38]. This property was suggested to be a unique catalytic behavior of enzyme complexes due to their specific structural organization [37]. Gaertner [37] reported that "physically associated multi-enzyme systems (enzyme clusters) have the potential of expressing unique catalytic properties in contrast to their non-associated counterparts." It is quite clear that the enhanced probability for intermediates to be transferred from one active site to the other by sequential enzymes requires stable or transient interactions between the relevant enzymes. Enzymes are able to associate physically in non-dissociable, static multi-enzyme complexes, which are not random associations but an assembly of sequentially related enzymes.

3.2.5.2 Compartmentation Phenomenon and Vectorial Metabolism

The principal mechanisms of organization of cell metabolism are macro- and microcompartmentation, metabolic channeling, and functional coupling. By definition, the term compartmentation is usually related to the existence of intracellular *macrocompartments* – subcellular regions that are large relative to the molecular dimension – and *microcompartments* that are on the order of the size of metabolites. A compartment is a "subcellular region of biochemical reactions kinetically isolated from the rest of cellular processes" [39]. Macrocompartments are easy to understand, and they can be visualized by electron or confocal microscopy. They are compartments inside the organelles, such as mitochondria or lysosomes, that are insulated from the cytoplasm by membranes. The concept of microcompartmentation is more mysterious, because these microcompartments are not usually visible by electron microscopy. However, now it is becoming clear that microcompartments related to multi-enzyme complexes and metabolic channeling, as described above, are the principal basis of organization and compartmentation of cellular metabolism. They are formed by specific protein–protein interactions within multi-enzyme complexes that are due to macromolecular crowding, anchoring of glycolytic [40–43] and other enzymes to the cytoskeleton, or membrane channels and transporters [43–48]. Investigations carried out in Clegg and Deutscher's laboratories have shown that mammalian cells behave as highly organized macromolecular assemblies dependent on the cytoskeleton [47, 48]. Multi-enzyme complexes may be of different size and may even include whole metabolic pathways; these are then called metabolons, according to the terminology introduced by Paul Srere in 1985 [49]. Thus, there are glycolytic metabolons [41, 45], Krebs cycle metabolons [50], and others [51, 52]. New techniques such as FRET have been developed to study and visualize microcompartmentation, e.g., the study of microcompartmentation of cyclic AMP [53, 54]. Microcompartmentation is sometimes taken to be synonymous with metabolic channeling [16, 39]. However, metabolic channeling of the reaction intermediate between two enzymes (or a transporter and an enzyme) may occur via microcompartment or by direct transfer [55, 56]. In both cases, it results in functional coupling (see below). Importantly, microcompartmentation may be of a dynamic nature, and this may result in the coexistence of a whole set of organized metabolic networks [55].

Interestingly enough, there is an exciting hypothesis that these phenomena, in particular metabolic channeling, are even older than life itself and are related to its origin. Edwards and others (see [57] for review) have put forward the hypothesis that on prebiotic Earth, when no enzyme or metabolic complexes were initially present, archetypal catalytic complexes were formed at the mineral surface (e.g., iron sulfide minerals) and biomolecules and catalysts were formed at specific sites relative to these complexes [57]. The evolution of metabolic pathways in this case would have been dictated by the relative positions of substrates and catalysts, where only closely juxtaposed species would have been allowed to interact [57]. Thus, the cell as a ''bag of enzymes'' probably never existed. It is now clear that living cells are much more complicated and better organized than previously thought [58, 59].

In the case of metabolic channeling of species via microcompartments with only a small number of molecules, the validity of introducing chemical potentials is questionable [6], whereas for macrocompartments and larger microdomains, this classical theory remains useful. In the first case, application of thermodynamic activities instead of concentrations [18–20] and stochastic kinetics based on probabilities of different states of enzymes are essential [60, 61].

An important consequence of the organization of enzymes into multi-enzyme complexes is vectorial metabolism and ligand conduction, a general principle proposed by Peter Mitchell after extensive enzymological studies and detailed characterization of mitochondrial proteins. One important example is the chemiosmotic coupling of energetic processes through the ''protonic current'' [62, 63]. It brought together ''transport and metabolism into one and the same chemiosmotic molecular level – biochemical process catalyzed by group-conducting or conformationally mobile group-translocating enzyme system'' [63]. In his reviews, Mitchell encouraged the wider use of the chemiosmotic principle and the biochemical concept of specific ligand conduction to explain the organization and operation of metabolic and transport processes within the cell [63]. Today, this idea is receiving increased attention and is certainly another important insight by Mitchell into the understanding of cellular energy conversion processes.

3.3
Compartmentalized Energy Transfer and Metabolic Sensing

3.3.1
Compartmentation of Adenine Nucleotides in Cardiac Cells

The application to cells of Eq. (20) in its general form is complicated by the compartmentation of ATP and adenine nucleotides making the use of the easily measurable total ATP content very questionable and practically useless. For cardiac cells, macrocompartmentation of adenine nucleotides was demonstrated in Ruth Altschuld's laboratory by using a controlled permeabilization technique [64].

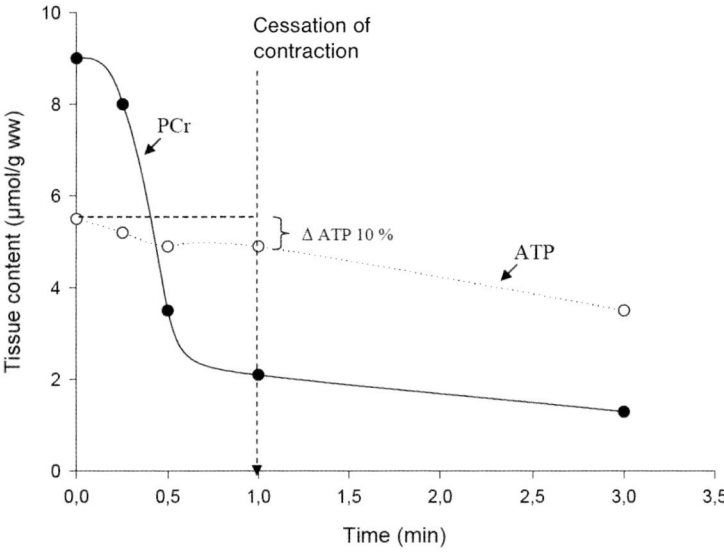

Fig. 3.3 Origin of the problem of the compartmentalization of adenine nucleotides and metabolic energy sensing in cardiac cells. The data show metabolic changes in totally ischemic dog hearts [65]. Within 1 min, PCr content falls by 80% and contraction stops, but 90% of cellular ATP is still intact. The general problems to explain are why the contraction stops when most of the ATP is not used up, and which mechanism allows sensing of the decrease in PCr level as the main source of cellular energy. For explanation, see the text and Fig. 3.5. (Data are redrawn from [65]; only the mean values of metabolites are shown).

They found that 74% of total nucleotides are localized in the cytoplasm, about 20% in mitochondria, and the remaining 6% in cellular structures. A similar distribution was found in other cells [64]. Before this, however, important data on the possible compartmentation of ATP were obtained in studies of the metabolism of ischemic heart by Gudbjarnason [65] and Neely [66]: they always observed very rapid decreases in contractile force in parallel with a decrease in PCr concentration (exactly as Lundsgaard detected in skeletal muscle; see the Introduction to this volume) and a complete interruption of contraction in the presence of about 80–90% intact ATP (see Fig. 3.3). These basic, classical observations show that there is a very precise energy-sensing mechanism present in the cells, responding

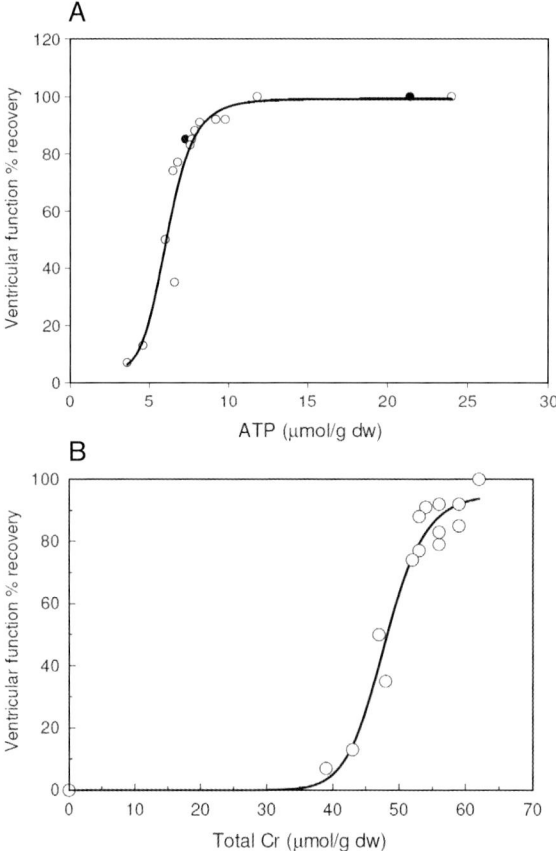

Fig. 3.4 (A) Absence of any correlation between total ATP content in the cells and contractile function of heart muscle. (Data are redrawn from [69–71]). ATP was changed by periods of hypoxic perfusion followed by reperfusion with normal solutions [69] (open circles) or by treatment of the hearts with 2-deoxyglucose [70, 71]. (B) Permanence of the total creatine in the experiments shown in Fig. 3.4A [69–70].

mostly to changes in PCr content (this mechanism will be discussed below). The interpretation of these experiments posed the following problem: Why does the heart stop contracting in the absence of oxygen supply so quickly if there is still plenty of ATP in the cells? Many of the explanations proposed – such as pH changes, changes in the sensitivity of troponin to calcium, etc. – are not sufficient because these changes occur at longer time intervals [67, 68]. The explanation proposed by Gudbjarnason highlights the importance of ATP compartmentation, only about 10% of which is in the functionally important pool(s) [65]. While the experiments described above showed almost constant ATP during a decrease in contraction force, further experiments with perfusion of heart during different hypoxic periods or perfusion with 2-deoxyglucose (a phosphate trap that decreases ATP content via adenylate kinase and 5′-nucleotidase reactions) [69–71] showed that the total content of ATP can be decreased by 70% (corresponding to the amount present in cytoplasm) without a significant effect on contraction (Fig. 3.4A), provided that total creatine content is not changed (Fig. 3.4B). Thus, the total content of ATP is dissociated from contractile force, which seems to depend only upon a fraction (not more than 20–30%) of ATP. At the same time, inactivation of the creatine kinase system by inhibition of creatine kinase or replacing creatine with less active analogues significantly decreases the maximal work capacity of the muscle [72, 73]. ATP compartmentation was confirmed in multiple indirect experiments performed on ischemic hearts, showing the importance of so-called glycolytic ATP, and in studies of energy supply to membrane ion pumps [74–77]. Compartmentation of ATP has been demonstrated recently by imaging techniques in several cell types [78].

Thirty-five years ago, the experiments described above initiated very intensive investigations that led to the discovery and description of phosphotransfer networks and metabolic signaling connecting different functional pools of ATP. Now, we have an answer to the question raised by Gudbjarnason and Neely concerning the complex phenomenon of ischemic heart failure. The answer is described below in this and other chapters of this book.

3.3.2
Unitary (Modular) Organization of Energy Metabolism and Compartmentalized Energy Transfer in Cardiac Cells

Cardiac cells present a highly organized structure in which mitochondria are localized at the A-band level within the sarcomere limits [79–82]. Intermyofibrillar mitochondria are arranged in a highly ordered crystal-like pattern in a muscle-specific manner with relatively small deviations in the distances separating neighboring mitochondria [81, 82]. Contrary to many other cells with less developed intracellular structure, dynamic changes in mitochondrial position due to their fission and fusion [83–85] are not found in adult and healthy cardiac and skeletal muscle cells because of their rigid intracellular structural organization, and the mitochondria in these cells are morphologically heterogeneous (see Chapter 5). In this structurally organized medium, energy transfer between different subcellular micro- and macrocompartments (called compartmentalized energy transfer) is of central importance. The existence of these rather complicated networks of energy transfer and signaling is a direct consequence of the compartmentalization of adenine nucleotides in the cells [33–35, 62, 65, 71, 78, 86]. This is due to the significant heterogeneity and local restrictions in the diffusion of adenine nucleotides in cells and to the necessity of rapid removal of ADP from the vicinity of Mg-ATPases to avoid their inhibition by the accumulating product MgADP. As is described in many chapters of this book, not only is ATP delivered by diffusion but also intracellular energy transfer is facilitated via networks consisting of phosphoryl group–transferring enzymes such as creatine kinase (CK), adenylate kinase (AK), and glycolytic phosphoryl-transferring enzymes [39, 87–105]. Most important among them is the creatine kinase system. CK catalyzes the reversible reaction of adenine nucleotide transphosphorylation, the forward reaction of phosphocreatine (PCr) and MgADP synthesis, and the reverse reaction of creatine (Cr) and MgATP production:

$$MgATP^{2-} + Cr \leftrightarrow MgADP^- + PCr^{2-} + H^+ \tag{21}$$

Four CK isoforms, each with compartmentalized cellular location, exist in mammals. Specific mitochondrial CK isoenzymes (MtCK), called ubiquitous (uMtCK) and sarcomeric (sMtCK), are functionally coupled to oxidative phosphorylation and produce PCr from mitochondrial ATP. PCr in turn is used for local regeneration of ATP by the muscle cytoplasmic isoform of CK (M-CK), driving myosin ATPases or ion pump ATPases [87–105]. Recent studies in CK-deficient transgenic animals indicate that energy transfer and communication between ATP-generating and ATP-utilizing sites within a muscle cell do not rely exclusively upon the activity of CK but rather may include a number of additional intracellular phosphotransfer systems such as AK and glycolysis [106–111]. AK-catalyzed reversible phosphotransfer between ADP, ATP, and AMP molecules may process cellular signals associated with ATP production and utilization [104, 105]. Cluster organization and the high rate of unidirectional phosphoryl exchange in phosphotransfer systems promote ligand conduction and signal communication at cellular distances, providing enhanced thermodynamic efficiency.

Remarkably, in the heart the intracellular energy transfer networks are structurally organized in the intracellular medium, where macromolecules and organelles surrounding a regular mitochondrial lattice are involved in multiple structural and functional interactions [112–114]. Figure 3.5 summarizes the available information about such an organized and compartmentalized energy metabolism in cardiac cells. This scheme also illustrates the view that mitochondria in muscle cells are structurally organized into functional complexes with myofibrils and sarcoplasmic reticulum [112–114]. These complexes are called "intracellular energetic units" (ICEUs) and are taken to represent the basic pattern of organization of muscle energy metabolism [112]. There are no physical barriers between ICEUs; each mitochondrion (or several adjacent mitochondria) can be taken to be in the center of its own ICEU. This concept is consistent with the very regular,

Intracellular Energy Unit (ICEU)

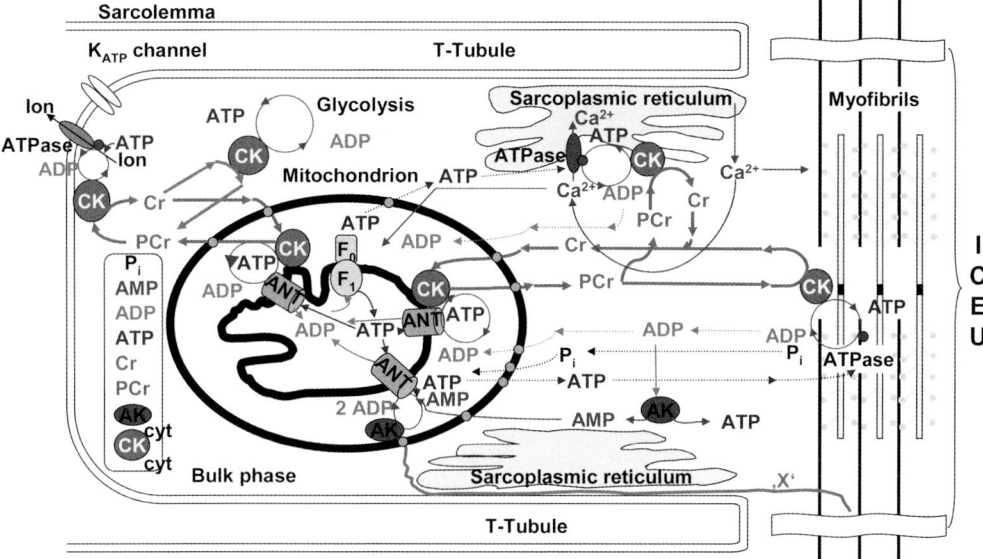

Fig. 3.5 Organization of compartmentalized energy transfer and metabolism in cardiac cells by intracellular energetic units (ICEUs). The scheme shows the structural organization of the energy transfer networks of coupled CK and AK reactions within an ICEU. By interaction with cytoskeletal elements, the mitochondria and sarcoplasmic reticulum (SR) are precisely fixed with respect to the structure of the sarcomere of myofibrils between two Z-lines and correspondingly between two t-tubules. Calcium is released from the SR into the space in the ICEU in the vicinity of mitochondria and sarcomeres to activate contraction and mitochondrial dehydrogenases. Adenine nucleotides within ICEUs do not equilibrate rapidly with adenine nucleotides in the bulk water phase. The mitochondria, SR, and Mg-ATPase of myofibrils and the ATP-sensitive systems in the sarcolemma are interconnected by metabolic channeling of reaction intermediates and energy transfer within ICEUs by the creatine kinase–phosphocreatine and myokinase systems. The protein factors (still unknown and marked as "X"), most likely connected to the cytoskeleton, fix the position of mitochondria and possibly also control the permeability of the VDAC channels for ADP and ATP. Adenine nucleotides within the ICEU and bulk water phase may be connected by more rapidly diffusing metabolites such as Cr–PCr. (Reproduced from [112] with permission).

crystal-like arrangement of mitochondria in cardiac cells [79–82] and describes the organized functional connections of mitochondria with their neighbors. ICEUs are analogous to calcium release units (CRUs), structurally organized sites of Ca^{2+} microdomains (Ca^{2+} sparks) that form a discrete, stochastic system of intracellular calcium signaling in cardiac cells [115, 116]. The structural organization of ICEUs results in local confinement of adenine nucleotides and Cr–PCr couples in discrete dynamic energetic circuits between actomyosin ATPases and mitochondrial ATPsynthases [90–106]. Similar discrete microdomains in cardiomyocytes have been shown for cAMP in the range of approximately 1 μm exhibiting high local concentrations [53, 116–118].

The concept of ICEUs is a useful basis for the mathematical modeling of compartmentalized energy metabolism (see Chapter 11). It allows dividing the complex problem of regulation of mitochondrial functions into two aspects: metabolic regulation of respiration inside these units and synchronization of the mitochondrial activities in the cell in all ICEUs. The first problem is analyzed in Chapter 11, the second in Chapter 4. Inside the ICEUs, the phosphotransfer pathways of energy channeling and metabolic signaling are based on (1) the mechanisms of functional coupling of compartmentalized CK isoenzymes with ANT in mitochondria and (2) Mg-ATPases in myofibrils and subcellular membranes. These will be analyzed below in more detail.

3.3.3
Functional Coupling of Mitochondrial Creatine Kinase and Adenine Nucleotide Translocase

The central mechanism in compartmentalized energy transfer is given by the functional coupling between the mitochondrial ATP/ADP translocase (ANT) and mitochondrial creatine kinase (MtCK). The structural basis of this coupling is described in Chapter 7; here, we will analyze some functional peculiarities.

MtCK was discovered by Klingenberg's group in 1964 [119]. This CK isoenzyme is localized on the outer surface of the mitochondrial inner membrane, in close vicinity to the ANT [119–123]. The structure of the ANT was recently resolved at 2.2 Å [124]. The translocation of both ATP and ADP in the Mg^{2+}-free forms is related to the conformational changes in pore-forming monomers [125]. This conformational change ("gated pore") mechanism leads in its simplest version to the ping-pong transport mechanism [126–139], but the kinetics of ATP–ADP exchange conforms to the sequential mechanism with simultaneous binding of nucleotides on both sides [130]. The structural data and the kinetics of ATP–ADP exchange by ANT fit well together by assuming that the dimers with alternatively activated monomers function in a coordinated manner in the tetramers, where the export of ATP from mitochondria by one monomer in a dimer occurs simultaneously with import of ADP by another monomer in another dimer [132, 133], or both monomers in the dimmer may be active and transport nucleotides simultaneously in opposite directions [125, 131].

In the mitochondrial matrix side, ANT forms a supercomplex, a synthesome with ATP synthase F_0F_1 and P_i carrier (PIC) [134]. ANT in the inner mitochondrial membrane forms tight complexes with negatively charged cardiolipin (1:6 ratio) [135–137]. It has been shown that positively charged MtCK is fixed to this cluster by electrostatic forces through three C-terminal lysines that strongly interact with the negatively charged cardiolipin in complex with ANT at the outer surface of the inner mitochondrial membrane [122, 123]. The peculiarity of the MtCK, in contrast with other dimeric CK isoenzymes (MM and BB), is that it forms octamers [90, 98, 100]. Thus, in the heart, brain, skeletal and smooth muscle, and some other cells, both ANT and MtCK function within a real supercomplex that connects mitochondrial ATP production with the cytoplasmic reactions of energy utilization via MtCK and VDAC (see also Chapter 7) [138, 139].

3.3.3.1 Kinetic Evidence of Functional Coupling

There is a good kinetic method for identifying and quantifying functional coupling between MtCK and ANT [140–145], the principles of which are shown in Fig. 3.6.

The creatine kinase reaction mechanism follows a Bi-Bi quasi-equilibrium random-type mechanism, according to the Cleland's classification, and is characterized by two dissociation constants for each substrate – from the binary and ternary (central) complexes of CK [146, 147] (see the scheme in Table 3.1). The values of these constants are easily determined by measuring reaction rates in series at different concentrations of both substrates. First, the concentration of one substrate is fixed and that of the other is changed. This procedure is repeated for several other fixed concentrations of the first substrate. In Fig. 3.6B, creatine is

---

Fig. 3.6 Kinetic evidence for the functional coupling mechanism in mitochondria: CK kinetics in isolated cardiac mitochondria without and with oxidative phosphorylation. Experiments and analyses were carried out according to the protocols and equations described in [143].

(A) Scheme of the protocols to study the kinetics of the creatine kinase reaction. MIM: mitochondrial inner membrane; MOM: mitochondrial outer membrane; RC: respiratory chain; ANT: adenine nucleotide translocase; MtCK: mitochondrial creatine kinase: Cr: creatine; PCr: phosphocreatine; I: with oxidative phosphorylation; II: without oxidative phosphorylation; PEP: phosphoenol pyruvate; PK: pyruvate kinase; LDH: lactate dehydrogenase.

(B) Initial normalized rates of the MtCK reaction in isolated heart mitochondria as functions of MgATP concentration for different fixed Cr concentrations without oxidative phosphorylation (dotted lines) and with oxidative phosphorylation (solid lines). All rates were normalized by V_{max}, $v^n = v/V_{max}$.

(C) Double reciprocal plot of data shown in Fig. 3.6B.

(D) Secondary analysis of MtCK kinetic data from (C) for different fixed creatine and changing ATP concentrations. Values of intercepts of ordinate, i_o, in (C) are plotted as a function of 1/[Cr]; the abscissa intercepts

correspond to reciprocals of a dissociation constant K_b of creatine from ternary complex [143]. No effect of oxidative phosphorylation is seen.

(E) Secondary analysis of the kinetic data when the reaction rates were measured for different fixed MgATP concentrations as functions of creatine concentration; from double reciprocal plots, the intercepts of ordinate were found as functions of 1/[ATP] (in these experiments, ATP means MgATP). Abscissa intercepts show the reciprocals of K_a, a dissociation constant for MgATP from ternary complex into medium [143]. This constant is decreased by oxidative phosphorylation by an order of magnitude (see Table 3.1), showing direct transfer of ATP from ANT to MtCK and recycling of ADP and ATP in the coupled reactions.

(F) Complete kinetic analysis of the forward MtCK reaction in heart mitochondria. The temperature dependencies of the kinetic constants (see scheme in Table 3.1) are shown at a semi-logarithmic scale in the presence (+) and absence (−) of oxidative phosphorylation. The dissociation constants were expressed in mM. Only K_a (see E) is changed by oxidative phosphorylation by an order of magnitude; smaller changes are seen for K_{ia}, but practically no changes are seen for the dissociation constant of creatine. (Reproduced from [96] with permission).

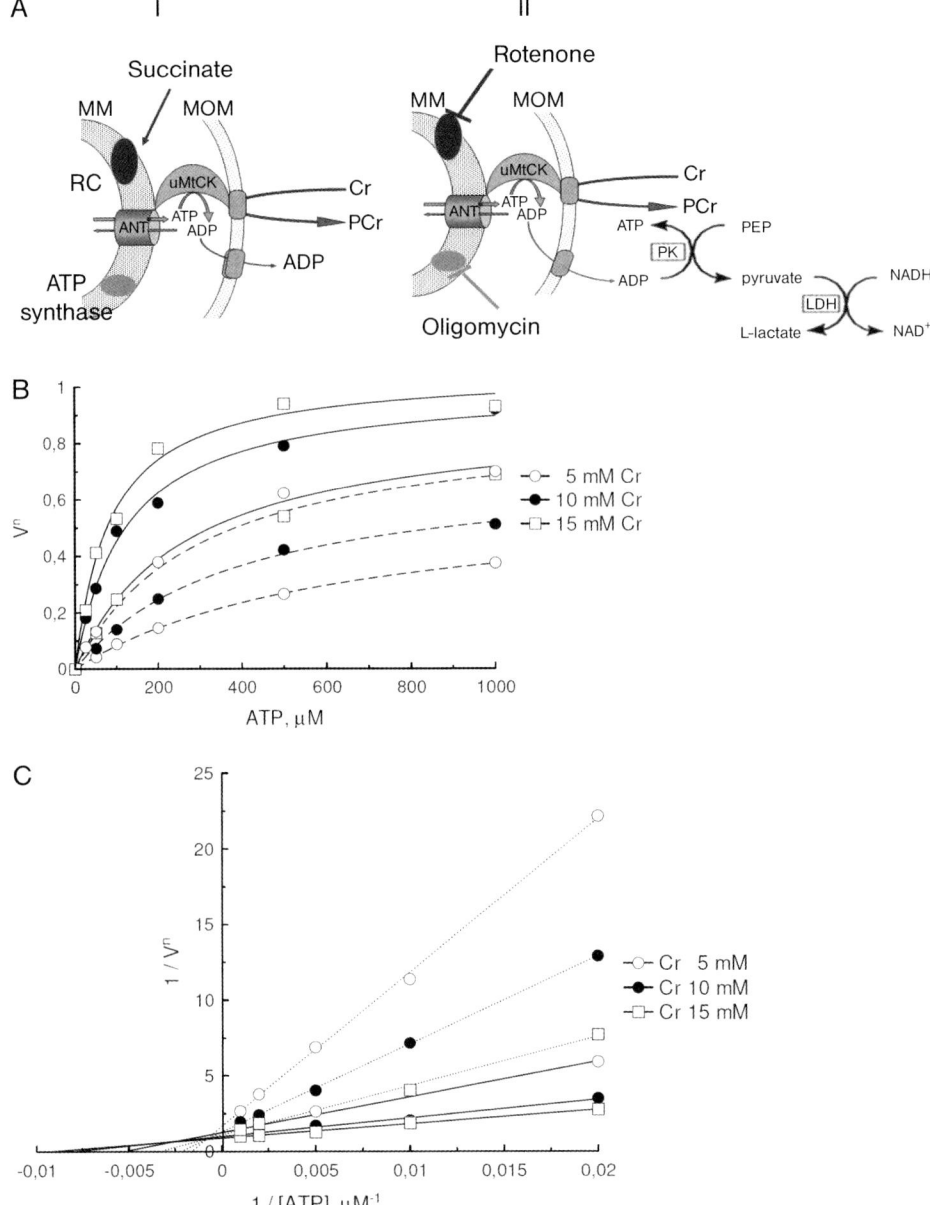

Fig. 3.6 (A–C) (legend see p. 77)

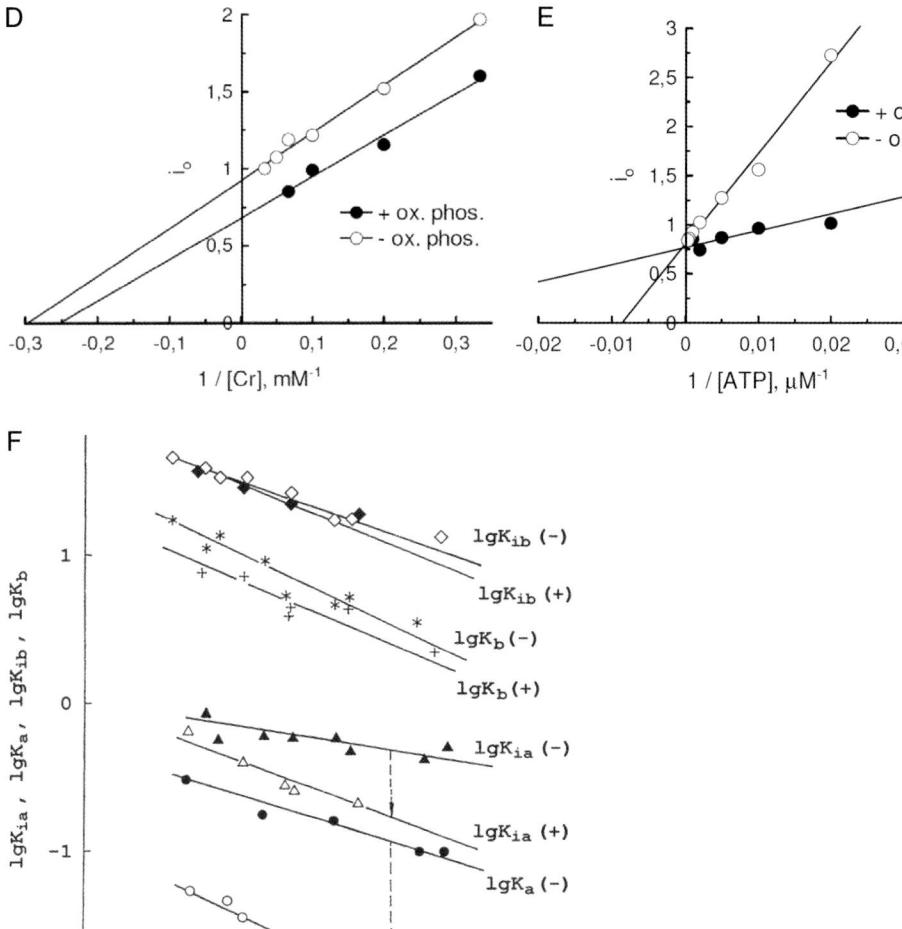

Fig. 3.6 (D–F) (legend see p. 77)

Table 3.1 The kinetic scheme of the Bi-Bi random type quasi-equilibrium creatine kinase reaction and dissociation constants for the forward MtCK reaction in the absence and presence of oxidative phosphorylation.

Dissociation constant	Without oxidative phosphorylation	With oxidative phosphorylation	Reference
K_{ia} (mM)	0.75 ± 0.06	0.29 ± 0.04	[143]
	0.64 ± 0.04	0.20 ± 0.03	[144]
	1.1 ± 0.1	0.3 ± 0.06	[145]
	0.92 ± 0.09	0.44 ± 0.08	[1]
K_a (mM)	0.15 ± 0.01	0.014 ± 0.005	[143]
	0.14 ± 0.02	0.016 ± 0.001	[144]
	0.22 ± 0.03	0.02 ± 0.004	[144]
	0.12 ± 0.23	0.023 ± 0.003	[1]
K_{ib} (mM)	28.8 ± 8.45	29.4 ± 12	[143]
	30 ± 4.5	30 ± 4	[144]
	51.1 ± 11	30 ± 1	[145]
	28.67 ± 7.15	24.54 ± 5.32	[1]
K_b (mM)	5.2 ± 0.3	5.2 ± 2.3	[143]
	9.5 ± 4.5	3.5 ± 1	[144]
	9 ± 1	5 ± 1.2	[145]
	3.37 ± 0.41	3.96 ± 0.64	[1]

[1] Determined by Tiia Anmann for this publication (see Fig. 3.6).

the substrate used at different fixed concentrations, and MgATP is varied. When oxidative phosphorylation is not activated, the mitochondrial creatine kinase reaction does not differ kinetically and thermodynamically from that catalyzed by other creatine kinase isoenzymes: the reaction favors ATP production, ADP and phosphocreatine binding is more effective due to higher affinities than that of ATP or creatine, respectively, and the kinetic constants obey the Haldane relationship [142–146]. However, under conditions of oxidative phosphorylation, the mitochondrial creatine kinase reaction is strongly shifted in the direction of

PCr synthesis and may use all ATP produced in mitochondria for PCr production [96, 143–145]. This is clearly seen in Fig. 3.6B,C: for any creatine concentration (5, 10, and 15 mM), the rates of phosphocreatine production in the presence of active oxidative phosphorylation (solid lines) were much higher than in its absence (dotted lines). Secondary analysis of the data in double reciprocal plots (Fig. 3.6C–E) directly yields the dissociation constants. While the kinetic constants for creatine were not changed and were the same in both conditions, the oxidative phosphorylation had a specific effect on the kinetic parameters for adenine nucleotides [96, 143–145]. These changes in the MtCK kinetics induced by oxidative phosphorylation are very clearly illustrated in Fig. 3.6E,F. Complete kinetic analysis of the MtCK reaction under conditions of oxidative phosphorylation (Fig. 3.6F) revealed that both constants for MgATP dissociation from the MtCK–MgATP and MtCK–Cr–MgATP complexes, Kia and Ka, respectively, were decreased, most significantly Ka, which was dramatically decreased by one order of magnitude [96, 143–145] (see Fig. 3.6 and Table 3.1). The explanation proposed was the direct transfer of ATP from ANT to MtCK due to their spatial proximity, which also results in an increased uptake of ADP by ANT from MtCK (reversed direct transfer). As a result, the turnover of adenine nucleotides is increased manifold at low external concentrations of MgATP, thus maintaining high rates of oxidative phosphorylation and coupled phosphocreatine production in the presence of creatine. This was confirmed in direct experiments in which MtCK was detached from the inner mitochondrial membrane into the intermembrane space by isotonic KCl solutions: in this case the effect of oxidative phosphorylation on the MtCK kinetics was lost [145]. It is important to emphasize that the structural and functional coupling of the MtCK–ANT system does not prevent its participants from working independently under some conditions. For example, it is well known that in a medium with only ADP, mitochondria can carry out the oxidative phosphorylation reactions without any limitations, despite the structural associations of ANT with MtCK; on the other hand, the inhibition of oxidative phosphorylation does not result in inhibition of MtCK but only alters its apparent kinetic behavior [142, 148]. These facts clearly indicate that the structural associations of ANT with MtCK are flexible and do not result in formation of a completely isolated space between complexes. The metabolite molecules can leave this space, but they also can be arrested in it to realize functional coupling between the partners ANT and MtCK, if all substrates including creatine are present. Quantitative analysis of the experimental data on coupled MtCK and ANT by a complicated mathematical model, based on the analysis of the energy profile of the reaction (see Chapter 12), gave evidence for the direct transfer of ATP from ANT to MtCK, with significant changes in the energy level of complexes of ATP bound to ANT [149]. This interesting possibility merits further experimental studies.

The direct transfer of mitochondrial ATP to MtCK and the increased mitochondrial turnover of adenine nucleotides in the presence of creatine were confirmed by the isotopic method [150], by the thermodynamic approach [144, 151, 152], and by showing that inhibition of MtCK by monoclonal antibodies also inhibited

ADP–ADP exchange due to the proximity of MtCK to ANT [153]. An effective competitive enzyme method for studying the functional coupling phenomenon – namely, the pathway of ADP movement from MtCK back to ANT [154] – is to use the phosphoenolpyruvate (PEP)–pyruvate kinase (PK) to trap ADP and thus to compete with ANT for this substrate [155]. The use of this method is demonstrated in Chapter 11.

3.3.3.2 Thermodynamic Evidence of Functional Coupling

Because of its importance, we reproduce this evidence again in Fig. 3.7. The creatine kinase reaction equilibrium favours the direction of ATP production, the apparent $K'_{eq} = K_{eq} \times [H^+]$ in this direction:

$$K'_{eq} = ([ATP] \times [Cr])/([PCr] \times [ADP])$$

having the following values: pH 7.0, 38 °C, free $[Mg^{2+}] = 1$ mM equal to ~170, and the standard free energy change of the creatine kinase reaction in the direction of ATP synthesis $\Delta G° = -13.4$ kJ mol^{-1} [156–158]. In the reverse direction, the apparent K'_{eq} is equal to 6×10^{-3} (see dotted line in Fig. 3.7B). In these experiments, reproduced from Ref. [144], the reaction conditions are such that $\Gamma = ([PCr][ADP])/([Cr][ATP])$ in the medium is higher than K'_{eq}; therefore, according to the laws of chemical thermodynamics, the reaction of phosphocreatine production should have proceeded under these conditions in the direction of lowering Γ to K'_{eq} by decreasing phosphocreatine concentration, as could be

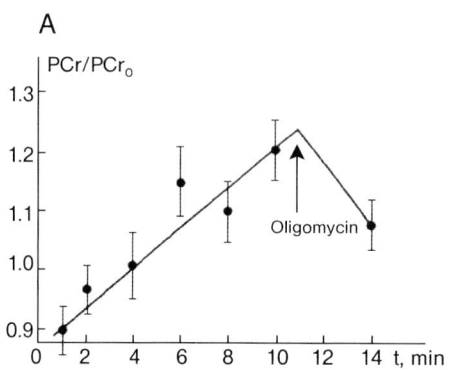
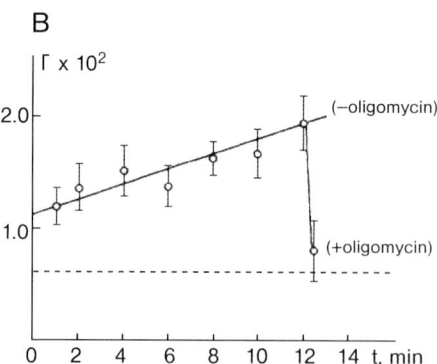
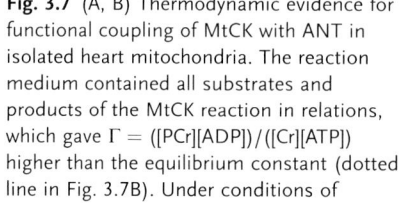

Fig. 3.7 (A, B) Thermodynamic evidence for functional coupling of MtCK with ANT in isolated heart mitochondria. The reaction medium contained all substrates and products of the MtCK reaction in relations, which gave $\Gamma = ([PCr][ADP])/([Cr][ATP])$ higher than the equilibrium constant (dotted line in Fig. 3.7B). Under conditions of oxidative phosphorylation, PCr/PCr_0 and Γ increased. Thus, Γ moved away from the equilibrium position, uphill on the curve shown in Fig. 3.2. Inhibition of oxidative phosphorylation by oligomycin reversed the direction of the reaction. (Data are reproduced from [144] with permission).

predicted from Fig. 3.2. Figure 3.7 shows, however, that in the presence of oxidative phosphorylation the concentration of PCr increases and Γ moves uphill (along the curve in Fig. 3.2) away from equilibrium. There should be a strong force and a significant amount of free energy change underlying this uphill movement, according to Fig. 3.2, meaning that oxidative phosphorylation is controlling the MtCK reaction. Most likely, the control occurs by direct supply of mitochondrial ATP to the active center of MtCK by ANT and by rapid removal of ADP by the latter – the functional coupling mechanism. When the oxidative phosphorylation reaction is inhibited by oligomycin, the MtCK can fully obey the thermodynamic rules and Γ decreases to the value of K'_{eq} (Fig. 3.7B).

It is interesting to recall again some facts from history. In 1939 Belitzer and Tsybakova observed in well-washed skeletal muscle homogenates a strong stimulation of respiration by Cr without addition of exogenous adenine nucleotides [159]. This was the earliest indication of the functional coupling of MtCK with oxidative phosphorylation. Much later, Kim and Lee showed the same effect for isolated pig heart mitochondria [160] and Dolder et al. for liver mitochondria from transgenic mice expressing active MtCK [161]. All these data show rapid recycling of endogenous ADP and ATP already present in mitochondria, due to the functional coupling of MtCK with ANT.

3.3.4
Heterogeneity of ADP Diffusion in Permeabilized Cells: Importance of Structural Organization

Studies on the regulation of mitochondrial respiration in permeabilized muscle cells and fibers have rendered important information about the structure–function relationships of cellular energetic systems, the role of creatine kinases, and calcium effects on respiration. In these experiments, the important role of the mitochondrial outer membrane in strengthening the functional coupling between MtCK and ANT (as predicted by Gellerich; see [155]) became evident.

Figure 3.8 illustrates this information. As mentioned above, in intact cardiac cells mitochondria are arranged very regularly at the level of A-bands of sarcomeres (Fig. 3.8A) and are functionally coupled to multiple intracellular ATP-consuming processes by energy transfer and metabolic feedback signaling networks within highly organized energetic units (see above). However, Claycomb and coworkers recently described the cultured (continuously dividing) HL-1 cell line with a differentiated cardiac phenotype [162, 163]. Pelloux et al. found that growing HL-1 cells with a different serum (GIBCO fetal bovine serum, batch 1147078) for five weeks (four passages) led to cells devoid of beating properties, which we classified as NB HL-1 cells [164]. Thus, the NB HL-1 cells represent an original phenotype displaying cardiac characteristics. Most remarkably, these cells are devoid of sarcomeric structures and possess randomly organized filamentous dynamic mitochondria [164, 165]. Unlike in adult rat cardiomyocytes, rapid movement, fission, and fusion of mitochondria (Y. Usson, personal communication) characterize filamentous mitochondria in NB HL-1 cells (Fig. 3.8B)

A

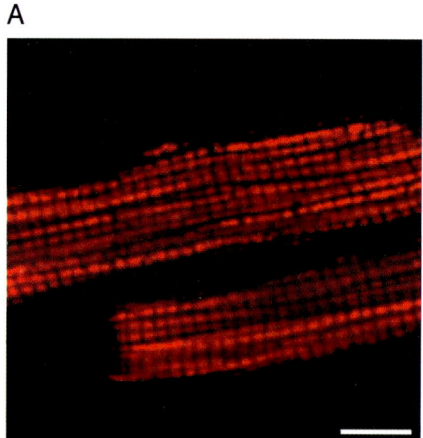

B

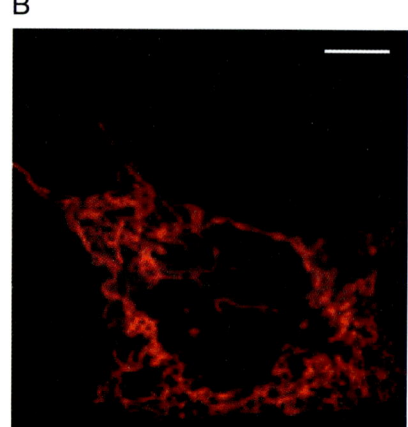

Fig. 3.8 Confocal fluorescent images of mitochondria in adult isolated cardiomyocytes (A) and cultured non-beating HL-1 cells (B) (scale bars: 10 μm). Mitochondrial arrangement was visualized by MitoTracker® Deep Red as described in [178]. (C, D) Different kinetics of regulation of the respiration in permeabilized adult cardiomyocytes and NB HL-1 cells. (C) ADP kinetic protocols and representative respiration traces for permeabilized cardiomyocytes recorded using a two-channel, high-resolution respirometer (Oroboros Oxygraph-2k, Oroboros, Innsbruck, Austria) in respirometry medium Mitomed [178]. Mitochondrial respiration increases with stepwise increasing concentrations of ADP in the range of 0.05 to 2 mM for cardiomyocytes to reach the saturated rate of respiration (V_{max}). At the end of the measure, cytochrome c (Cyt c, 8 μM) addition does not change the respiration, indicating that the outer membrane is intact. Atractyloside (Atr, 30 μM) results in a decrease in respiration back to V_o due to inhibition of adenine-dinucleotide translocase. (Reproduced from [178] with permission). (D) Comparison of respiration kinetics as normalized respiration rates versus [ADP] in permeabilized adult cardiomyocytes and in permeabilized HL-1 NB cells determined in experiments described in (C). Values of apparent K_m for exogenous ADP were equal to 360 ± 51 μM for isolated and permeabilized cardiomyocytes and 25 ± 4 μM for HL-1 NB cells [178]. For normalization, the respiration rates were expressed as fractions of the maximal rates, V_{max}, found by analysis of experimental data in double reciprocal plots. (E) Comparison of respiration regulation kinetics in isolated heart mitochondria and in permeabilized cardiomyocytes without and with creatine (20 mM). (Taken from [179, 180]).

[164]. Also, striking differences in the kinetics of respiration regulation by exogenous ADP are exhibited by these cells (Fig. 3.8C–E). In permeabilized adult cardiomyocytes, the apparent K_m for exogenous ADP is very high (360 ± 51 μM). This has been previously reported [166–176] and explained by local restrictions of ADP diffusion in the cells, including limitation of its diffusion across the mitochondrial outer membrane [33, 177]. In permeabilized cardiomyocytes, treatment with trypsin resulted in dramatic changes in intracellular structure that were associated with a threefold decrease in apparent K_m for ADP in the regulation of respiration [176, 178]. In contrast to permeabilized cardiomyocytes, in NB HL-1 cells the apparent K_m for exogenous ADP (25 ± 4 μM) was 14 times lower

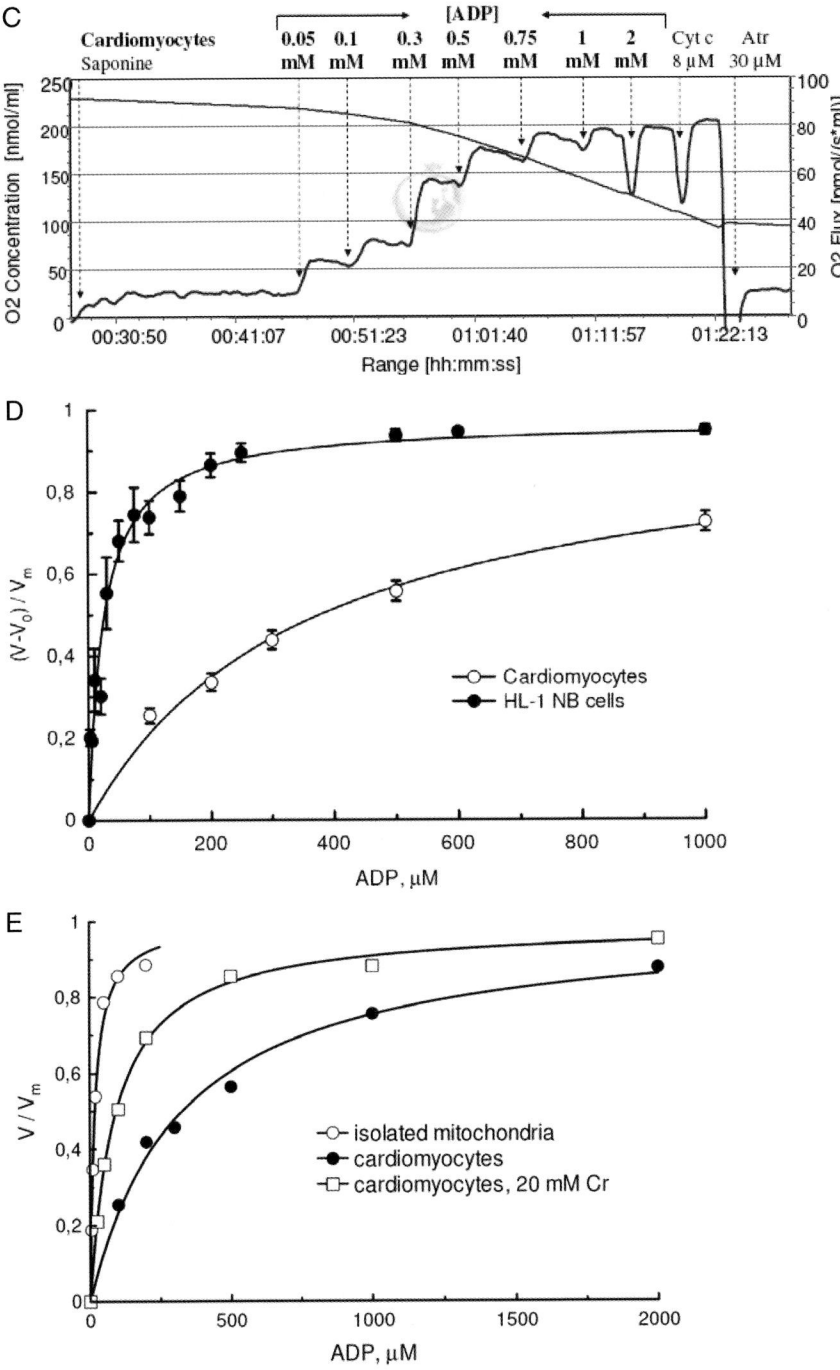

Fig. 3.8 (C–E)

(Fig. 3.8C–E). The regulation of respiration by exogenous ADP in NB HL-1 cells is very close to that in isolated mitochondria (Fig. 3.8D,E). While in normal adult cardiomyocytes creatine significantly activates respiration and, as in the experiments described in Fig. 3.8E, decreases the apparent K_m for ADP (see also Chapter 11), in HL-1 cells it exerts very little influence, if any, due to the downregulation of creatine kinase [178]. These results show that in normal adult cardiomyocytes intracellular local restrictions of diffusion of adenine nucleotides and metabolic feedback regulation of respiration via phosphotransfer networks are related to the complex structural organization of these cells.

The high K_m for exogenous ADP has been shown to reflect the local restriction of diffusion of ADP and ATP, which in cardiac cells are bypassed by the creatine kinase system [177]. Most likely, these diffusion limitations are localized at or close to the outer mitochondrial membrane due to close connections to some cytoskeletal elements [33, 177, 179, 180]. In fact, simple logical analysis shows that the limited permeability of the mitochondrial outer membrane for ADP in cardiac cells *in vivo* should be expected. In the resting heart, or in the diastolic phase of contracting heart, when cytoplasmic creatine kinase is in the equilibrium state (see Chapter 11), the cytoplasmic ADP concentration calculated from the CK equilibrium is in the range of 50–100 µM [181]. However, in isolated mitochondria, the apparent K_m of ANT for ADP is 10–20 µM [182]. If mitochondria behave in the cells *in vivo* as they do *in vitro*, the mitochondrial respiration rate should be maximal in resting heart or during the diastolic phase of the contraction cycle due to the saturation of ANT with ADP that is never observed. This means that the apparent K_m probably also is high for endogenous ADP, as was found in experiments with exogenous ADP (see Fig. 3.8), due to the decreased permeability of VDAC channels in the outer mitochondrial membrane for this substrate. This conclusion is further confirmed in experiments described in Chapter 11, and that makes functionally coupled MtCK even more necessary for energy transfer and respiration regulation at high workloads.

Interestingly, the apparent K_m for exogenous ADP (which is very easy to measure in biopsy samples), by reflecting the complexity of intracellular interactions and organization in oxidative muscle cells, has become an important diagnostic parameter in studies of skeletal muscle bioenergetics and exercise physiology (see also Chapter 14). In both experimental and clinical studies, endurance training has been shown to significantly increase (2–3 times) the apparent K_m for exogenous ADP in animal and human skeletal muscle fibers, and it correspondingly increases the stimulatory effect of creatine on respiration and thus the role of the creatine kinase system in intracellular communication [183–187]. Thus, this parameter has a clear practical and diagnostic value for muscle performance [182–188]. Identification of cell components responsible for specific organization of energy transfer systems and intracellular diffusion merits further investigation. For elucidation of the nature of cytoskeletal components responsible for specific organization of both mitochondrial arrangement and complex metabolic signaling and intracellular energy transfer pathways [178], further investigations on the proteome are needed.

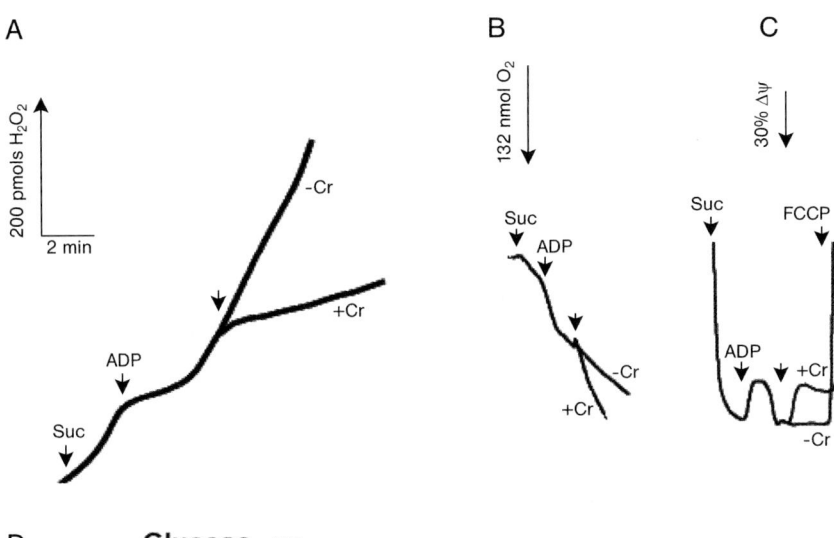

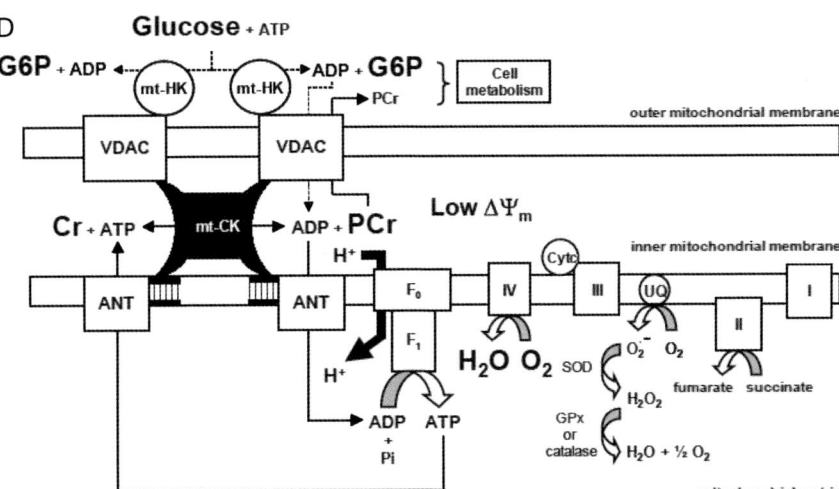

Fig. 3.9 H_2O_2 production (A), oxygen consumption (B), and membrane potential (C) in rat brain mitochondria are regulated by Cr phosphorylation by MtCK. Succinate was added to 10 mM; ADP, 0.15 mM; Cr, 10 mM; and FCCP 5 μM [194]. (D) Scheme of the reactions by which MtCK and hexokinase (Mt-HK) bound to the outer mitochondrial membrane regulate oxygen consumption and membrane potential, $\Delta\Psi m$, and inhibit ROS formation in the respiratory chain. SOD: superoxide dismutase; GPx: glutathione peroxidase; UQ; ubiquinone; cyt. C: cytochrome *c*. (Reproduced from [194] with permission).

3.3.5
Evidence for the Importance of Functional Coupling for Cell Life

The functional coupling mechanism is supported by the structural and functional studies of MtCK performed by Wallimann's group (see Chapter 7). *In vivo*, the functional coupling between MtCK and ANT was verified recently in studies of heart energy metabolism performed *in vivo* by ^{31}P-NMR inversion transfer. This approach showed that in intact heart cells mitochondrial creatine kinase is strongly shifted in the direction of phosphocreatine production under aerobic conditions [189, 190]. It was also shown in mice with knockout of MtCK that, as predicted by the theory described above, these hearts had lower levels of PCr and reduced post-ischemic recovery [191, 192]. In human heart, decreased PCr/ATP ratio is a diagnostic index of increased mortality [193].

A new, important role for MtCK's control over ANT is to hinder the opening of the mitochondrial permeability transition pore [161] and cell death. Important new data were published recently [194] showing that due to the active functional coupling between MtCK and ANT, MtCK-induced ADP recycling strongly decreases the production of reactive oxygen species (ROS) in brain mitochondria. Some of these important results are reproduced in Fig. 3.9. Taking into account that ROS production is now considered a major cause of many age-related diseases and aging itself [195, 196], the importance of these results is difficult to overestimate. It may soon become clear that our good health and longevity significantly depend upon the degree of functional coupling between MtCK and ANT in heart, skeletal and smooth muscles, brain, and many other tissues in which the creatine kinase network is important.

3.3.6
Myofibrillar Creatine Kinase

The myofibrillar end of the creatine–phosphocreatine cycle is represented by the MM isozyme of creatine kinase localized in the sarcomere and functionally coupled to the actomyosin Mg-ATPase [90, 93, 94, 197, 198]. Within the contraction cycle, ADP release is a necessary step for new binding of MgATP, dissociation of actomyosin cross-bridges, and muscle relaxation in order to start the contraction cycle anew [199–206]. This step is often found to be the slowest in the contraction cycle and therefore is rate controlling. MgADP may compete with MgATP for the substrate site on myosin and inhibit cross-bridge detachment by MgATP, the inhibition constant K_i being approximately 200 μM both in the Mg-ATPase reaction and in the sliding of fluorescent actin on myosin [200, 203–207]. Thus, MgADP should be rapidly removed from actomyosin, and the high local value of the MgATP:MgADP ratio along with the local phosphorylation potential should be maintained. The MM-CK is bound specifically to the M-line, and a significant portion of this isozyme is found in the I – band of sarcomeres [208–210]. An increasing amount of evidence points out that this MM creatine kinase is intimately involved in the contraction cycle at the level of ADP-release and

ATP-rebinding steps by the mechanism of functional coupling, likely including myofibrillar microcompartments of adenine nucleotides [93, 207, 211–213]. Ogut and Brozovich studied the kinetics of force development in skinned trabeculae from mice hearts and found that despite the presence of 5 mM MgATP, the rate of force development depended on the concentration of PCr. These authors concluded that there is a direct functional link between the creatine kinase reaction and the actomyosin contraction cycle at the step of ADP release in myofibrils [214]. This effective interaction probably occurs via small microcompartments of adenine nucleotides in myofibrils and is facilitated by anisotropic diffusion. When PCr is exhausted, the rate of local regeneration of ATP in myofibrils decreases, slowing the contraction cycle irrespective of high cytoplasmic ATP concentrations (see Fig. 3.3) [86, 212–214]. In agreement with this interpretation, it was recently shown that inhibition or knockout of the creatine kinase system by genetic manipulations results in a significant loss of work capacity [215, 216] (see Chapter 11). The rigor state known as ischemic contracture appears only after complete exhaustion of all ATP in the cell [217] and can be significantly delayed by local production of ATP by the myofibrillar glycolytic system or accelerated by inhibition of this process (the so-called glycolytic ATP effect [74, 75, 86]). Under these conditions, the rate of glycolytic ATP production and its diffusion from the cytoplasm are too slow to sustain contraction.

3.3.7
Membrane-bound Creatine Kinases and Membrane Energy Sensing

Other ATP-consuming systems are found in both the membranes of the sarcoplasmic reticulum (SR) and in the plasmalemma (sarcolemma). Their function is to maintain the ionic homeostasis and, particularly, the regulation of the calcium cycle. The role of the MM-CK connected to the membrane of the SR and functionally coupled to the Ca/MgATP-dependent ATPase (SERCA) is to rapidly rephosphorylate local MgADP produced in the Ca/Mg-ATPase reaction, thus maintaining a high local value of phosphorylation potential to keep a high efficiency of calcium pumping and avoid its reversal or inhibition due to accumulation of ADP [218–222].

In the control of the excitation–contraction coupling in the heart, a principal step is the sarcolemmal membrane metabolic sensor complex [34, 35]. Its main component is the sarcolemmal ATP-sensitive K^+ (K_{ATP}) channel that acts as an alarm system to adjust cell electrical activity to the metabolic state of the cell [34, 223–231]. The sarcolemmal MM-CK rephosphorylates the local ADP, maintaining a high ATP/ADP level in these microcompartments for coordination of membrane electrical activity with cellular metabolic status, notably with PCr levels. The K_{ATP} channel was discovered by Noma [227], and it was found that this channel has a high affinity for ATP, about 100 μM [227, 228]. Nevertheless, the channel is opened in the presence of millimolar ATP, as seen from rapid membrane repolarization and shortening of action potential in ischemic and hypoxic hearts or as shown directly in experiments with internally perfused cardiomyo-

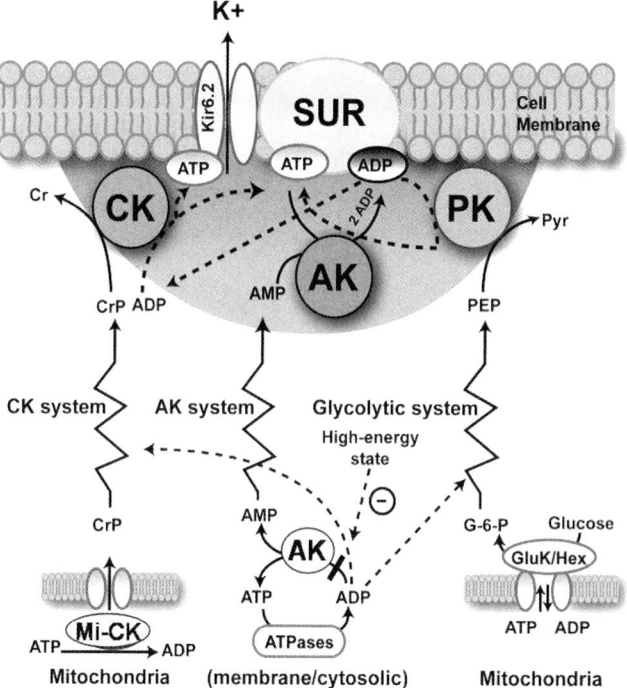

Fig. 3.10 Paradigm of phosphotransfer-mediated energetic signaling: coupling of cellular metabolic and electrical activities. Dynamic interaction between CK, AK, and glycolytic (represented by PK) phosphotransfer relays determines the behavior of a prototypic metabolic sensor – the K_{ATP} channel – and subsequent cellular responses, such as excitability, hormone secretion, intracellular calcium homeostasis, and vascular tone. The shadowed area represents a metabolic sensor "sensing zone," where intimate local changes in nucleotide ratios are sensed and transduced into an appropriate cellular response. Phosphotransfer circuits connect the "sensing zone" with cellular processes. Dashed lines indicate pathways signaling the high-energy state, while solid lines represent low-energy state signal transmission. Kir6.2: potassium channel subunit; SUR: sulfonylurea receptor; GluK/Hex: glucokinase and hexokinase. (Reproduced from [102] with permission).

cytes [34, 35]. This is explained by strong diffusional restriction, and thus ATP compartmentation, in the subsarcolemma and is linked to the cellular pool of PCr via CK reactions (Fig. 3.10). Pool exhaustion in the first minutes of ischemia certainly contributes to the cessation of contraction due to opening of the K_{ATP} channel and decreased calcium entry (see also Chapter 11). These energy transfer and control functions are shared by the whole system, including the creatine kinase, adenylate kinase, and glycolytic systems, as was seen in experiments involving genetic manipulation [35, 225]. Similar cell membrane metabolic sensors also may be important in brain cells. The basic principles of phosphotransfer-mediated metabolic signaling to metabolic sensors at the sarcolemma are presented in Fig. 3.10. The premembrane area, where diffusion of molecules is restricted due to the unstirred layer of structured water and molecular crowding [34, 35, 177, 229], represents the "sensing zone" of a metabolic sensor, where intimate changes in nucleotide ratios are sensed and transduced into appropriate cellular responses [35, 104–106]. Because cellular and even submembrane ATP concentrations are higher than required for half-maximal channel inhibition (between 10 μM and 100 μM) [34, 35, 226, 227], production of ADP by AK or membrane ATPases and ADP scavenging by CK and PK reactions are critical for channel activity. It may be assumed that dynamic interaction between CK, AK, and glycolytic enzyme (represented as PK) phosphotransfer relays determines the behavior of a prototypical metabolic sensor – the K_{ATP} channel – and subsequent cellular responses, such as excitability, hormone secretion, intracellular calcium homeostasis, and vascular tone [34, 35, 104, 225, 228–232]. These phosphotransfer relays communicate metabolic signals that originate in mitochondria or at cellular ATPases to metabolic sensors, conveying information about "high" or "low" cellular energy, oxygen supply, or hormonal states [34, 35, 104–106]. CK and AK deficiencies or their altered activity ratio compromises metabolic signaling to K_{ATP} channels [34, 225]. Decreases in CK activity, such as in CK knockouts or failing hearts, can be partially compensated by the increase in glycolytic phosphotransfer to alleviate cardiomyocyte electrical instability [106, 109].

3.3.8
The Mechanism of Acute Contractile Failure of the Ischemic Heart

Taken together, the results described in this section help us to explain why the heart stops contracting in the first minutes of total ischemia, when oxidative phosphorylation and thus PCr production in mitochondria are inhibited but the total ATP level is still very high. Clearly, the major contributors are the changes happening in the compartmentalized energy transfer. A rapid decrease in PCr concentration in the cytoplasm results in rapid exhaustion of the local ATP pools in myofibrils, which slows down the contraction cycle [175, 212–214], favors the opening of the sarcolemmal K_{ATP} channels, affects membrane repolarization, decreases the entry of calcium via L-type calcium channels, and inhibits Ca^{2+}-induced Ca^{2+} release (CICR) in the sarcoplasmic reticulum. As a result, contraction stops because of lack of fuel, but myofibrils stay relaxed (due to glycolytic ATP production in myofibrils) and also slow diffusion of ATP from the cytoplasm. The latter reveals itself as a protective mechanism that allows the cells to preserve a significant portion of the ATP and thus to survive an energetic crisis. Only total exhaustion of ATP results in development of ischemic contracture and irreversible damage of the heart [217]. Thus, the compartmentalized energy transfer mechanism permits an accurate explanation of the fundamental observations about the ischemic heart shown in Figs. 3.3 and 3.4. All these changes occur very rapidly, and there are no compensatory mechanisms available, in contrast to hypertrophy or in experimental animals with knockout of CK genes (see Chapter 10). The initial proposal by Gudbjarnason was proved to be correct in general, but in the detailed investigations described above, several functionally very

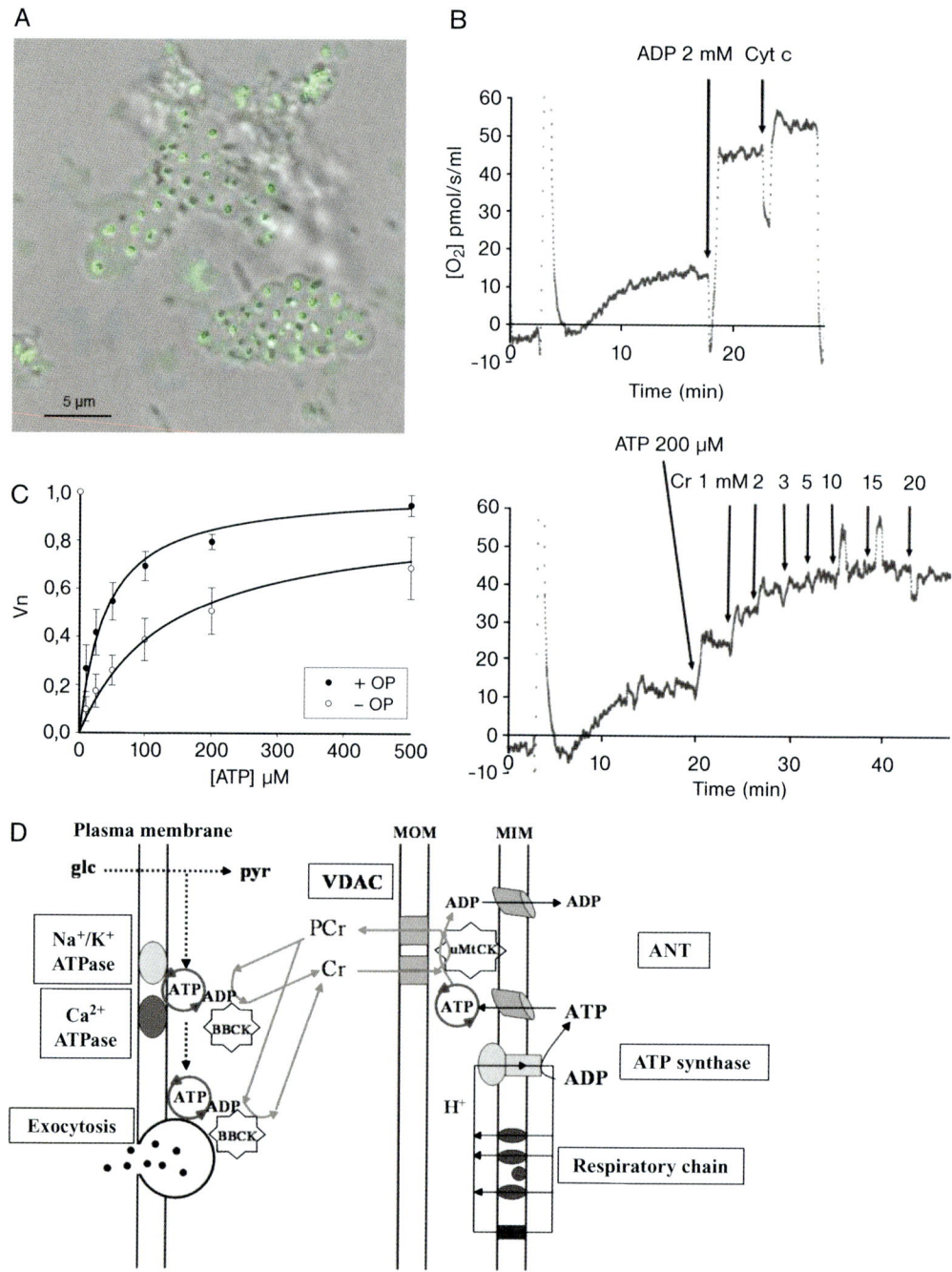

Fig. 3.11 (legend see p. 93)

important compartments of ATP were identified: in myofibrils and close to the subcellular membranes – the sarcoplasmic reticulum and the sarcolemma. All these functionally important pools of ATP are connected to the cytoplasmic phosphocreatine via compartmentalized CK isoenzymes (see Fig. 3.5).

In chronic heart failure the changes in the energy transfer system are of a structural and functional character and represent severe pathogenic mechanisms of contractile failure in both heart and skeletal muscle, as shown in recent reviews from Ventura-Clapier's group [175, 188, 233] and by group of Neubauer [191–193].

3.3.9
Creatine Kinase System in Brain Cells

Similar to cardiac cells, brain cells exhibit high levels of CK activity, represented in the mitochondria by the ubiquitous isoenzyme uMtCK and in the cytosol by BB-CK [234–238]. However, the role of the CK system in brain cells has been investigated to a much lesser extent than in cardiac cells [103, 105, 236–238], likely due to the high degree of tissue heterogeneity. In our recent studies we adopted the method of Clark for purification of rat brain synaptosomes and synaptosomal mitochondria. These purified preparations of synaptosomes contain a large amount of mitochondria (Fig. 3.11A) with high respiratory activity, which in the presence of ATP is well controlled by creatine due to the MtCK reaction (Fig. 3.11B). Figure 3.11C shows that the rate of CK for PCr production is enhanced with oxidative phosphorylation in brain: mitochondria as in heart (Fig. 3.6). This shows that in the brain synaptosomes, the MtCK reaction is also tightly coupled to oxidative phosphorylation via ANT and the phosphocreatine–creatine kinase pathway is actively operative (Fig. 3.11D).

◀──

Fig. 3.11 Creatine kinase in brain synaptosomes. (A) Confocal imaging of mitochondria in rat brain synaptosomes. Mitochondria were visualized by the mitochondria-specific probe MitoTracker Green. Merging of transmission and fluorescent confocal images is shown. (B, C) Recordings of the rate of respiration of isolated rat brain mitochondria activated by exogenous ADP (B) and creatine at different concentrations in the presence of ATP (right panel in C). Respiration rates were recorded using a two-channel, high-resolution respirometer (Oroboros Oxygraph-2k, Oroboros, Innsbruck, Austria). (C, left panel) Comparison of the kinetics of regulation of the rate of CK reaction in isolated mitochondria of rat brain by exogenous ATP in the presence of 10 mM Cr in the presence or absence of oxidative phosphorylation (OP). (D) Schematic presentation of energy transfer in brain synaptosomes. Mitochondrial coupled reactions are very close to those in heart mitochondria (see explanation in the text). The energy-consuming reactions are related to the ion transport across the outer membrane and exocytose of glutamate at the expense of ATP, produced locally by BB-CK from PCr. (Reproduced from [238] with permission).

3.3.10
Maxwell's Demon and Organized Cellular Metabolism

Vectorial metabolism, metabolic channeling, and functional coupling may well be explained and illustrated by Maxwell's demon, a general and important philosophical concept in the history of physical sciences. In 1994 Azzone and Mae-Wan Ho [239, 240] discussed the idea that Maxwell's demon is behind Schrödinger's principle of negentropy in living cells. This metaphor is useful for explaining general mechanisms that help the cell to keep entropy low through a very precise structural organization of all metabolic networks [101, 238].

In 1871 James Clerk Maxwell analyzed, in his book *Theory of Heat*, the nature of the second law of thermodynamics and described the following imaginary situation. In the state of thermodynamic equilibrium, all parameters of the system, such as temperature and pressure, have constant values and no work is possible. This is due to the constant average value of the kinetic energy. However, this average value is of a statistical nature due to the large number of molecules with different rates that distribute according to the Boltzmann function. Maxwell proposed consideration of the following situation: the homogenous system is divided into two parts separated by a small hole that can be closed or opened by a hypothetical being of intelligence but of molecular order. This hypothetical being, which was later nicknamed a "demon" by William Thomson, permits the molecules with a rate higher than the average to traverse the hole but closes it for the molecules with rates lower than the average. In this way Maxwell's demon disturbs the equilibrium by increasing the order, thus creating a temperature difference between the two compartments and making work possible without using an external energy supply. This imaginary experiment immediately initiated vivid philosophical discussions, which have lasted until today, has been particularly useful in information theory, and is often used for analysis of biological systems. It has been discussed by Szilard and Brillouin that the energy demanded for obtaining the information about molecules would be greater than that gained, and this could make Maxwell's demon's actions ineffective [241]. However, the information needed for the demon may be given by the proper system's organization [239]. In this case, its aim is not to break the second law of thermodynamics but to save the energy and help to avoid increasing the entropy. Most importantly, however, the Maxwell's demon metaphor is not in disagreement with either kinetic or thermodynamic aspects of metabolic channeling in organized multi-enzyme complexes and networks.

Equilibrium of an enzymatic reaction proceeding in a homogenous medium implies that the average concentrations of the substrates and products involved in the reaction are constant all over the cellular cytoplasm, i.e., determined by the equilibrium constant value and thus the value of the standard free energy change ΔG° of the reaction. It may be assumed that any given enzyme molecule catalyzes on average an equal number of direct and reverse reactions in a time unit (principle of microscopic reversibility) in a random manner.

At non-equilibrium steady states, the net flow through the reaction is steady but non-zero. Such a situation is kept through a continuous supply of substrate and removal of product. The multiple components of an enzymatic reaction system leave Maxwell's demon a much larger choice of parameters to play with than it had in the classical situation of Maxwell's time. The most interesting and important game could be to look at each enzyme molecule and decide in which direction it will catalyze the reaction, simply by giving it a necessary substrate and removing the product from it at the same time. If the demon wishes, it can constantly keep any given enzyme molecule working irreversibly in one direction, as in endergonic, energy-dependent reactions. Other enzyme molecules can be kept working in the reverse direction, to keep the metabolic system in the overall permanent steady state. In fact, it is the precise structural organization of the metabolic system that creates Maxwell's demon. This principle of intelligence, the concept of Maxwell's demon, is well realized in compartmentalized energy transfer pathways, such as creatine kinase systems with structurally fixed (bound) creatine kinase isoenzymes interacting with the adjacent ATP-producing, -transporting, or -consuming systems, as illustrated in Fig. 3.12. In this figure the Maxwell's demon metaphor is used to describe the functional coupling between MtCK and ANT, which is analyzed in detail in this chapter. ANT, which supplies the substrate ATP for MtCK and removes the product ADP from local microcompartments, fulfils the intelligent role of the demon directing the MtCK reaction uphill from equilibrium, as described in Figs. 3.2 and 3.7. At the same time, by using ATP and supplying ADP for ANT, MtCK fulfils the same role of the demon for ANT. For regulation of mitochondrial respiration in many cells with high-energy fluxes, this is the key mechanism (see also Chapter 11). The demons shown in Fig. 3.12 are very good ones: by recycling ADP in mitochondria they help to avoid ROS production and protect the cells from unpleasant outcomes. Similar schemes may be used for description of the roles of other CK and AK isoenzymes (see Fig. 3.10) in other subcellular compartments and for the phenomena of metabolic channeling in general.

Thus, the intelligence of Maxwell's demon is realized in the proper structural organization of cellular metabolic systems. By controlling the direct supply of substrate to the enzyme and removing the product, the intelligence of Maxwell's demon helps to avoid an unnecessary increase in entropy and thus helps to increase the efficiency of energy transduction. In Schrödinger's days, the fine structural organization of metabolism was unknown, but it clearly contributes to the foundations of the basic law that he discovered [7].

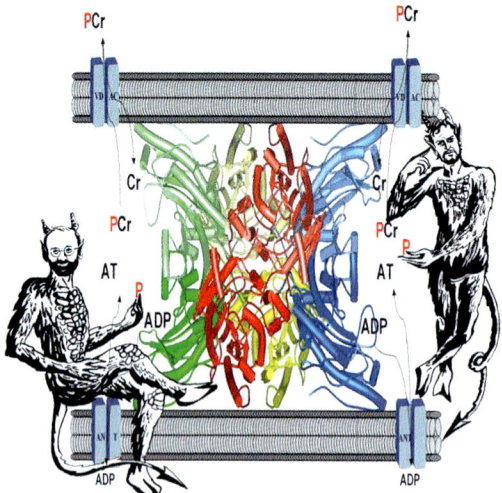

Fig. 3.12 Principle of Maxwell's demon as a metaphor for explaining and illustrating the functional coupling and metabolic channeling phenomena. The coupled MtCK reaction and ANT are shown. The hypothetical being of molecular size (Maxwell's demon) takes ATP molecules from ANT and puts them into the active site of creatine kinase for transfer of the terminal phosphate of ATP to creatine. In exchange, it takes ADP from CK and puts it back into the ANT. The intelligence of Maxwell's demon is realized by the precise structural arrangement of MtCK and ANT. For further explanation, see text. The structural image of the octameric form of the MtCK is courtesy of Theo Wallimann (Zurich, Switzerland, right) and Uwe Schlattner (Grenoble, France, left). (This figure was drawn by Tatiana Samoilova, Grenoble, France, and published with her permission).

Acknowledgments

This work was supported by INSERM, France, and by grants from the Estonian Science Foundation (Nos. 7117 and 6142 to V.S.), the National Institutes of Health (HL64822, HL07111), the Marriott Program for Heart Disease Research, the Marriott Foundation, and the Miami Heart Research Institute (P.D). The authors thank Prof. Agu Laisk, Tartu, Estonia, for stimulating discussions and Rita Guzun, Grenoble, France, for excellent technical assistance. The authors are very grateful to Miguel Aon, the John Hopkins University, Institute of Molecular Cardiobiology, Baltimore, MD, USA, and Marc Jamin, The Joseph Fourier University, Grenoble, France, for constructive discussions, critical reviewing of the manuscript, and editing of its English. Due to space limitations, as well as restriction of cited references, we apologize to all those colleagues and researchers in the field whose work is not directly cited here, although they significantly contributed to this synopsis through their work and discussions.

References

1 Marshall, A. G. (1978) *Biophysical Chemistry. Principles, techniques and applications*, John Wiley & Sons, New York.

2 Caplan, S. R., Essig, A. (1983) *Bioenergetics and linear nonequilibrium thermodynamics*, Harvard University Press, Cambridge, Massachusetts, London.

3 Nicholls, D., Ferguson, S. J. (2002) *Bioenergetics 3*, Academic Press, London–New York.

4 Westerhoff, H. V., Van Dam, K. (1987) *Thermodynamics and Control of Biological Free-Energy Transduction*, Elsevier, Amsterdam-New York-Oxford.

5 Qian, H., Beard, D., Liang, S. (2003) Stoichiometric network theory for nonequilibrium biochemical systems. *Eur. J. Biochem. 270*, 415–421.

6 Qian, H., Beard, D. (2005) Thermodynamics of stoichiometric biochemical networks in living systems far from equilibrium. *Biophys. Chem. 114*, 213–220.

7 Schrödinger, E. (1944) *What is life? The Physical Acpect of Living cell*, Cambridge University Press.

8 Cornish-Bowden, A., Jamin, M., Saks, V. A. (2005) *Cinétique Enzymatique*, EDP Sciences, Les Ulis, France, 1–462.

9 Nicolis, G. and Prigogine, I. (1977) *Self-organization in non-equilibrium systems*. Wiley-Interscience, London.

10 Aon, M. A. and Cortassa, S. (1997) *Dynamic biological organization*. Fundamentals as applied to cellular systems. Chapman and Hall, London.

11 Schneider, E. D. and Sagan, D. (2005) *Into the cool. Energy flow, thermodynamics and life*. The University of Chicago Press, Chicago.

12 Rosing, J., Slater, E. S. (1972) The value of G^0 for the hydrolysis of ATP. *Biochim. Biophys. Acta 267*, 275–290.

13 Veech, R. L., Lawson, J. W. R., Cornell, N. W., Krebs, H. A. (1979) Cytosolic phosphorylation potential. *J. Biol. Chem. 254*, 6538–6547.

14 Kammermeier, H., Schmidt, P., Jungling, E. (1982) Free energy change of ATP hydrolysis: a causal factor of early hypoxic failure of the myocardium? *J. Mol. Cell. Cardiol. 14*, 267–277.

15 Fulton, A. B. (1982) How crowded is the cytoplasm? *Cell 30*, 345–347.

16 Srere, P. A. (2000) Macromolecular interactions: tracing the roots. *Trends Biochem. Sci. 25*, 150–153.

17 Scalettar, B. A., Abney, J. R., Hackenbrock, C. R. (1991) Dynamics, structure, and function are coupled in the mitochondrial matrix. *Proc. Natl. Acad. Sci. U.S.A. 88*, 8057–8061.

18 Ovàdi, J. (1995) *Cell Architecture and Metabolic Channeling*, R. G. Landes Co., Springer-Verlag, Austin, New York, Berlin, Heidelberg, London, Paris, Tokyo, Hong Kong, Barcelona, Budapest.

19 Minton, A. P. (2001) The influence of macromolecular crowding and macromolecular confinement on biochemical reactions in biological media. *J. Biol. Chem. 276*, 10577–10580.

20 Hall, D., Minton, A. P. (2003) Macromolecular crowding: qualitative and semiquantitative successes, quantitative challenges Biochim. *Biophys. Acta 1649*, 127–139.

21 Schliwa, M. (2002) The evolving complexity of cytoplasmic structure. *Nat. Rev. Mol. Cell Boil. 3*, 1–6.

22 Schnell, S., Turner, T. E. (2004) Reaction kinetics in intracellular environments with macromolecular crowding: simulations and rate laws Progr. *Biophys. Mol. Biol. 85*, 235–260.

23 Agius, L., Sherratt, H. S. A. (eds.) (1996) In *Channelling in Intermediary Metabolism*, Portland Press, London, 237–268.

24 Noble, D. (2006) *The music of life. Biology beyond the genome*, Oxford University Press, Oxford, UK.

25 Kitano, H. (2002) Systems biology: a brief overview *Science 295*, 1662–1664.

26 Noble, D. (2002) Modeling the heart – from genes to cells to the whole organ. *Science 295*, 1678–1682.

27 Hunter, P., Nielsen, P. (2005) A strategy for integrative computational physiology. *Physiology 20*, 316–325.

28 Xia, Y., Yu, H., Jansen, R., Seringhaus, M., Baxter, S., Greenbaum, D., Zhao, H., Gerstein, M. (2004) Analyzing cellular biochemistry in terms of molecular networks. *Annu. Rev. Biochem. 73*, 1051–1087.

29 Ovàdi, J., Saks, V. (2004) On the origin of intracellular compartmentation and organized metabolic systems. *Mol. Cell. Biochem. 256/257*, 5–12.

30 Verkman, A. S. (2002) Solute and macromolecule diffusion in cellular aqueous compartments. *Trends Biochem. Sci. 27*, 27–33.

31 de Graaf, R. A., Van Kranenburg, A., Nicolay, K. (2000) *In vivo* 31P-NMR spectroscopy of ATP and phospho-creatine in rat skeletal muscle. *Biophys. J. 78*, 1657–1664.

32 Kinsey, S. T., Locke, B. R., Benke, B., Moerland, T. S. (1999) Diffusional anisotropy is induced by subcellular barriers in skeletal muscle. *NMR Biomed. 12*, 1–7.

33 Vendelin, M., Eimre, M., Seppet, E., Peet, N., Andrienko, T., Lemba, M., Engelbrecht, J., Seppet, E. K., Saks, V. A. (2004) Intracellular diffusion of adenosine phosphates is locally restricted in cardiac muscle. *Mol. Cell. Biochem. 256/257*, 229–241.

34 Abraham, M. R., Selivanov, V. A., Hodgson, D. M., Pucar, D., Zingman, L. V., Wieringa, B., Dzeja, P., Alekseev, A. E., Terzic, A. (2002) Coupling of cell energetics with membrane metabolic sensing. Integrative signaling through creatine kinase phosphotransfer disrupted by M-CK gene knock-out. *J. Biol. Chem. 277*, 24427–24434.

35 Selivanov, V. A., Alekseev, A. E., Hodgson, D. M., Dzeja, P. P., Terzic, A. (2004) Nucleotide-gated K_{ATP} channels integrated with creatine and adenylate kinases: Amplification, tuning and sensing of energetics signals in the compartmentalized cellular environment. *Mol. Cell. Biol. 256/257*, 243–256.

36 Welch, G. R., Clegg, J. S. (1986) In The organization of cell metabolism (Welch,

G. R. and Clegg, J. S., eds.), Plenum Press, New York, 57–74.

37 Gaertner, F. H. (1978) Unique catalytic properties of enzyme clusters. *Trends Biochem. Sci. 3*, 63–65.

38 Ovàdi, J., Srere, A. P. (2000) Macro-molecular compartmentation and channeling. *Intern. Rev. Cytol. 192*, 255–280.

39 Saks, V. A., Khuchua, Z. A., Vasilyeva, E. V., Belikova, Yu, O., Kuznetsov, A. (1994) Metabolic compartmentation and substrate channeling in muscle cells. Role of coupled creatine kinases *in vivo* regulation of cellular respiration – a synthesis. *Mol. Cell. Biochem. 133/134*, 155–192.

40 Ottaway, J. H., Mowbray, J. (1977) The role of compartmentation in the control of glycolysis. *Curr. Top. Cell Regul. 12*, 107–208.

41 Maughan, D. W., Henkin, J. A., Vigoreaux, J. O. (2005) Concentrations of glycolytic enzymes and other glycolytic proteins in the diffusible fraction of a vertebrate muscle proteome. *Mol. Cell. Proteomics 4*, 1541–1549.

42 Waingeh, V. F., Gustafson, C. D., Kozliak, E. I., Lowe, S. L., Knull, H. R., Thomasson, K. A. (2006) Glycolytic enzyme interactions with yeast and skeletal muscle F-actin. *Biophys. J. 90*, 1371–1384.

43 Penman, S. (1995) Rethinking cell structure. *Proc. Natl. Acad. Sci. U.S.A. 92*, 5251–5257.

44 Arnold, H., Pette, D. (1968) Binding of glycolytic enzymes to structure proteins in muscle. *Eur. J. Biochem. 6*, 163–171.

45 Kurganov, B. I., Sugrobova, N. P., Mil'man, L. S. (1985) Supramolecular organization of glycolytic enzymes. *J. Theor. Biol. 116*, 509–526.

46 Saier, M. H. (2000) Vectorial Metabolism and the Evolution of Transport Systems. *J. Bacteriol. 182*, 5029–5035.

47 Clegg, J. S. (1984) Properties and metabolism of the aqueous cytoplasm and its boundaries. *Am. J. Physiol. 246*, R133–R151.

48 Hudder, A., Nathanson, L., Deutscher, M. P. (2003) Organization of mammalian cytoplasm. *Mol. Cell. Biol. 23*, 9318–9326.

49 Srere, P. A. (1985) The metabolon. *Trends Biochem. Sci. 10*, 109–110.

50 Vélot, C., Mixon, M. B., Teige, M., Srere, P. (1997) Model of a quinary structure between Krebs TCA cycle enzymes: a model for metabolon. *Biochemistry 36*, 14271–14276.

51 Haggie, P. M., Verkman, A. S. (2002) Diffusion of tricarboxylic acid cycle enzymes in the mitochondrial matrix *in vivo. J. Biol. Chem. 277*, 40782–40788.

52 Eaton, S., Bursby, T., Middleton, B., Pourfarzam, M., Mills, K., Johnson, A. W., Bartlett, K. (2000) The mitochondrial trifunctional protein: centre of a β–oxidation metabolon? *Biochem. Soc. Transactions 28*, 177–182.

53 Zaccolo, M., Pozzan, T. (2002) Discrete microdomains with high concentration of cAMP in stimulated rat neonatal cardiac myocytes. *Science 295*, 1711–1715.

54 Jurevicius, J., Fischmeister, R. (1996) cAMP compartmentation is responsible for a local activation of cardiac Ca channels by β–adrenergic agonists. *Proc. Natl. Acad. Sci. U.S.A. 93*, 295–299.

55 Friedrich, P. (1985) Dynamic compartmentation in soluble multienzyme system. In: *Organized multienzyme systems. Catalytic properties*, Welch, G. R. (Ed.), Academic Press, New – York – London, 141–176.

56 Huang, X., Holden, H. M., Raushel, F. M. (2001) Channeling of substrates and intermediates in enzyme-catalyzed reactions. *Annu. Rev. Biochem. 70*, 149–180.

57 Edwards, M. R. (1996) Metabolite channeling in the origin of life. *J. Theor. Biol. 179*, 313–322.

58 Alberts, B. (1998) The cell as a collection of protein machines: preparing the next generation of molecular biologists. *Cell 92*, 291–294.

59 Clegg, J. S. (1992) Cellular infrastructure and metabolic organization. *Curr. Top. Cell. Regul. 33*, 3–14.

60 Qian, H., Elson, E. L. (2002) Single-molecule enzymology: stochastic Michaelis-Menten kinetics. *Biophys. Chem. 101/102*, 565–576.

61 Aliev, M. K., Saks, V. A. (1993) Quantitative analysis of the "phosphocreatine shuttle". I. A probability approach to the description of phosphocreatine production in the coupled creatine kinase – ATP/ADP translocase – oxidative phosphorylation reaction in heart mitochondria. *Biochim. Biophys. Acta 1143*, 291–300.

62 Mitchell, P. (1961) Coupling of phosphorylation to electron transfer by a chemi-osmotic type of mechanism. *Nature 191*, 144–148.

63 Mitchell, P. (1979) The Ninth Sir Hans Krebs Lecture. Compartmentation and communication in living systems. Ligand conduction: a general catalytic principle in chemical, osmotic and chemiosmotic reaction systems. *Eur. J. Biochem. 95*, 1–20.

64 Geisbuhler, T., Altschuld, R. A., Trewyn, R. W., Ansel, A. Z., Lamka, K., Brierley, G. P. (1984) Adenine nucleotide metabolism and compartmentalization in isolated adult rat heart cells. *Circ. Res. 54*, 536–546.

65 Gudbjarnason, S., Mathes, P., Raven, K. G. (1970) Functional compartmentation of ATP and creatine phosphate in heart muscle. *J. Mol. Cell. Cardiol. 1*, 325–339.

66 Neely, J. R., Rovetto, M. J., Whitmer, J. T., Morgan, H. (1973) Effects of ischemia on function and mtabolism of the isolated working rat heart. *Am. J. Physiol. 225*, 651–658.

67 Opie, L. H. (1998) The Heart. Physiology, from cell to circulation, Lippincott-Raven Publishers, Philadelphia, 43–63.

68 Goldhaber, J. I. (1997) Metabolism in normal and ischemic myocardium. In *The myocardium*, Langer, G. (ed.), Academic Press, San – Diego – New – York, 325–393.

69 Neely, J. R., Grotyohann, L. W. (1984) Role of glycolytic products in damage to ischemic myocardium. Dissociation of adenosine triphosphate levels and recovery of function of reperfused ischemic myocardium. *Circ. Res. 55*, 816–824.

70 Kupriyanov, V. V., Lakomkin, V. L., Kapelko, V. I., Steinschneider, A. Ya., Ruuge, E. K., Saks, V. A. (1987) Dissociation of adonosine diphosphate levels and contractile function of in isovolumic hearts perfused with 2-deoxyglycose. *J. Mol. Cell. Cardiol.* *19*, 729–740.

71 Kupriyanov, V. V., Lakomkin, V. L., Korchazhkina, O. V., Steinschneider, A. Ya., Kapelko, V. I., Saks, V. A. (1991) Control of cardiac energy turnover by cytoplasmic phosphates: 31P-NMR study. *Am. J. Physiol. 261*, 45–53.

72 Gerken, G., Schlette, U. (1968) Metabolite status of the heart in acute insufficiency due to 1-fluoro-2,4-dinitrobenzene. *Experientia 24*, 17–19.

73 Kapelko, V. I., Kupriyanov, V. V., Novikova, N. A., Lakomkin, V. L., Steinschneider, A., Severina, M., Veksler, V. I., Saks, V. A. (1988) The cardiac contractile failure induced by chronic creatine and phosphocreatine deficiency. *J. Mol. Cell. Cardiol. 20*, 465–479.

74 Weiss, J., Hiltbrand, B. (1985) Functional compartmentation of glycolytic versus oxidative metabolism in isolated rabbit heart. *J. Clin. Invest. 75*, 436–447.

75 Bricknell, O. L., Opie, L. H. (1981) A relationship between adenosine triphosphate, glycolysis and ischemic contracture in the isolated rat heart. *J. Mol. Cell. Card. 13*, 941–945.

76 Mercer, R. W., Dunham, P. B., (1981) Membrane-bound ATP fuels the Na/K pump. Studies on membrane-bound glycolytic enzymes on inside-out vesicles from human red cell membranes. *J. Gen. Physiol. 78*, 547–568.

77 Weiss, J. N., Lamp, S. T. (1987) Glycolysis preferentially inhibits ATP-sensitive K-channels in isolated guinea-pig cardiac myocytes. *Science 238*, 67–69.

78 Kennedy, H. J., Pouli, A. E., Ainscow, E. K., Jouaville, L. S., Rizzuto, R., Rutter, G. A. (1999) Glucose generates sub-plasma membrane ATP microdomains in single islet-cells. *J. Biol. Chem. 274*, 13291–13291.

79 Fawcett, D. W., McNutt, N. S. (1969) The ultrastructure of the cat myocardium. I Ventricular papillary muscle. *J. Cell Biol. 42*, 1–45.

80 Sommer, J. R., Jennings, R. B. (1986) Ultrastructure of cardiac muscle. In

The Heart and Catdiovascular System, Fozzard, H. A., Haber, E., Jennings, R. B., Katz, A. M., Morgan, H. E. (eds.), Raven Press, New York, 61–100.

81 Aon, M., Cortassa, S., O'Rourke, B. (2004) Percolation and criticality in a mitochondrial network. *Proc. Natl. Acad. Sci. 101*, 4447–4452.

82 Vendelin, M., Beraud, N., Guerrero, K., Andrienko, T., Kuznetsov, A. V., Olivares, J., Kay, L., Saks, V. A. (2005) Mitochondrial regular arrangement in muscle cells: a "crystal-like" pattern. *Am. J. Physiol. Cell Physiol. 288*, C757–767.

83 Bereiter-Hahn, J., Voth, M. (1994) Dynamics of mitochondria in living cells: shape changes, dislocations, fusion and fission of mitochondria. *Microsci. Res. Tech. 27*, 198–219.

84 Yi, M., Weaver, D., Hajnocsky, G. (2004) Control of mitochondrial motility and distribution by the calcium signal: a homeostatic circuit. *J. Cell Biol. 167*, 661–672.

85 Rube, D. A., van den Bliek, A. M. (2004) Mitochondrial morphology is dynamic and varied. *Mol. Cell. Biochem. 256/257*, 331–339.

86 McLellan, G., Weisberg, A., Winegrad, S. (1983) Energy transport from mitochondria to myofibril by a creatine phosphate shuttle in cardiac cells. *Am. J. Physiol. 254*, C423–C427.

87 Bessman, S. P., Carpenter, C. L. (1985) The creatine-creatine phosphate energy shuttle. *Annu. Rev. Biochem. 54*, 831–862.

88 Bessman, S. P., Geiger, P. J. (1981) Transport of energy in muscle: the phosphorylcreatine shuttle. *Science 211*, 448–452.

89 Saks, V. A., Rosenshtraukh, L. V., Undrovinas, A. I., Smirnov, V. N., Chazov, E. I. (1976) Studies of energy transport in heart cells. Intracellular creatine content as a regulatory factor of frog heart energetics and force of contraction. *Biochem. Med. 16*, 21–36.

90 Wallimann, T., Wyss, M., Brdiczka, D., Nicolay, K., Eppenberger, H. M. (1992) Intracellular compartmentation, structure and function of creatine kinase isoenzymes in tissues with high and fluctuating energy demands: the 'phosphocreatine circuit' for cellular energy homeostasis. *Biochem. J. 281*, 21–40.

91 Saks, V. A., Ventura-Clapier, R. (eds.) (1994) Cellular Bioenergetics. *Role ocoupled creatine kinase*, Kluver Academic Publishers, Dordrecht-Boston, 1–348.

92 Saks, V. A., Ventura-Clapier, R., Aliev, M. K. (1996) Metabolic control and metabolic capacity: two aspects of creatine kinase functioning in the cells. *Biochim. Biophys. Acta 1274*, 81–92.

93 Ventura-Clapier, R., Veksler, V., Hoerter, J. A. (1994) Myofibrillar creatine kinase and cardiac contraction. *Mol. Cell. Biochem. 133*, 125–144.

94 Ventura-Clapier, R., Kuznetsov, A., Veksler, V., Boehm, E., Anflous, K. (1998) Functional coupling of creatine kinases in muscles: species and tissue specificity. *Mol. Cell. Biochem. 184*, 231–247.

95 Wyss, M., Smeitink, J., Wevers, R. A., Wallimann, T. (1992) Mitochondrial creatine kinase: a key enzyme of aerobic energy metabolism. *Biochim. Biophys. Acta 1102*, 119–166.

96 Saks, V. A., Dos Santos, P., Gellerich, F. N., Diolez, P. (1998) Quantitative studies of enzyme – substrate compartmentation, functional coupling and metabolic channelling in muscle cells. *Mol. Cell Biochem. 184*, 291–307.

97 Wyss, M., Kaddurah-Daouk, R. (2000) Creatine and creatinine metabolism. *Physiol. Rev. 80*, 1107–1213.

98 Schlattner, U., Wallimann, T. (2004) Metabolite channeling: creatine kinase microcompartments. In *Encyclopedia of Biological Chemistry*, Lennarz, W. J., Lane, M. D. (eds.), Academic Press, New York, USA, 646–651.

99 Saks, V. A., Kuznetsov, A. V., Vendelin, M., Guerrero, K., Kay, L., Seppet, E. K. (2004) Functional coupling as a basic mechanism of feedback regulation of cardiac energy metabolism. *Mol. Cell. Biochem. 256–257*, 185–199.

100 Schlattner, U., Wallimann, T. (2006) Molecular structure and function of mitochondrial creatine kinases. In *Creatine kinase – biochemistry, physiology,* *structure and function*, V. N. Uversky (ed.), Nova Science Publishers, New York, pp. 123–170.

101 Saks, V., Guerrero, K., Vendelin, M., Engelbrecht, J., Seppet, E. (2006) The creatine kinase isoenzymes in organized metabolic networks and regulation of cellular respiration: a new role for Maxwell's demon. In *Creatine Kinase*, Vial, C. (ed.), In series: *Molecular Anatomy and Physiology of proteins*, V. N. Uversky (series ed.), NovaScience Publisher, New York, pp. 223–267.

102 Saks, V., Dzeja, P., Schlattner, U., Vendelin, M., Terzic, A., Wallimann, T. (2006) Cardiac system bioenergetics: metabolic basis of Frank-Starling law. *J. Physiol. 571*, 253–273.

103 Saks, V., Vendelin, M., Aliev, M. K., Kekelidze, T., Engelbrecht, J. (2007) Mechanisms and modeling of energy transfer between and among intracellular compartments. In *Handbook of Neurochemistry & Molecular Neurobiology: Brain Energetics*, Gibson, G., Dienel, G. (eds.), Springer Science, New York, pp. 815–860.

104 Dzeja, P. P., Terzic, A. (2003) Phosphotransfer networks and cellular energetics. *J. Exp. Biol. 206*, 2039–2047.

105 Dzeja, P. P., Terzic, A. (2005) Mitochondrial-nucleus energetic communication: role of phosphotransfer networks in processing cellular information. In *Handbook of Neurochemistry & Molecular Neurobiology: Brain Energetics*, Gibson, G., Dienel, G. (eds.), Springer Science, New York, pp. 614–666.

106 Dzeja, P. P., Zelenznikar, R. J., Goldberg, N. D. (1998) Adenylate kinase: kinetic behaviour in intact cells indicates it is integral to multiple cellular processes. *Mol. Cell. Biochim. 184*, 169–182.

107 Steeghs, K., Benders, A., Oerlemans, F., de Haan, A., Heerschap, A., Ruitenbeek, W., Jost, C., van Deursen, J., Perryman, B., Pette, D., Bruckwilder, M., Koudijs, J., Jap, P., Veerkamp, J., Wieringa, B. (1997) Altered Ca^{2+} responses in muscles with combined mitochondrial and cytosolic creatine kinase deficiencies. *Cell 89*, 93–103.

108 de Groof, A. J., Oerlemans, F. T., Jost, C. R., Wieringa, B. (2001) Changes in glycolytic network and mitochondrial design in creatine kinase-deficient muscles. *Muscle Nerve 24*, 1188–1196.

109 Dzeja, P. P., Vitkevicius, K. T., Redfield, M. M., Burnett, J. C., Terzic, A. (1999) Adenylate kinase-catalyzed phosphotransfer in the myocardium: increased contribution in heart failure. *Circ. Res. 84*, 1137–1143.

110 Dzeja, P. P., Zeleznikar, R. J., Goldberg, N. D. (1998) Adenylate kinase: kinetic behavior in intact cells indicates it is integral to multiple cellular processes. *Mol. Cell. Biochem. 184*, 169–182.

111 Dzeja, P. P., Terzic, A., Wieringa, B. (2004) Phosphotransfer dynamics in skeletal muscle from creatine kinase gene-deleted mice. *Mol. Cell. Biochem. 256–257*, 13–27.

112 Saks, V. A., Kaambre, T., Sikk, P., Eimre, M., Orlova, E., Paju, K., Piirsoo, A., Appaix, F. Kay, L., Regiz-Zagrosek, V., Fleck, E., Seppet, E. (2001) Intracellular energetic units in red muscle cells. *Biochem. J. 356*, 643–657.

113 Seppet, E. K., Kaambre, T., Sikk, P., Tiivel, T., Vija, H., Tonkonogi, M., Sahlin, K., Kay, L., Appaix, F., Braun, U., Eimre, M., Saks, V. A. (2001) Functional complexes of mitochondria with Ca,MgATPases of myofibrils and sarcoplasmic reticulum in muscle cells. *Biochim. Biophys. Acta 1504*, 379–395.

114 Kaasik, A., Veksler, V., Boehm, E., Novotova, M., Minajeva, A., Ventura-Clapier, R. (2001) Energetic crosstalk between organelles. Architectural integration of energy production and utilization. *Circ. Res. 89*, 153–159.

115 Wang, S. Q., Wei, C., Zhao, G., Brochet, D., Shen, J., Song, L. S., Wang, W., Yang, D., Cheng, H., (2004) Imaging microdomain Ca^{2+} in muscle cell. *Circ. Res. 94*, 1011–1022.

116 Rizzuto, R., Pozzan, T. (2006) Microdomains of intracellular Ca2+: molecular determinants and functional consequences. *Physiol. Rev. 86*, 369–408.

117 Nicolaev, V. O., Lohse, M. J. (2005) Monitoring of cAMP synthesis and degradation in living cells. *Physiology 21*, 86–92.

118 Karpen, J. W., Rich, T. C. (2004) Resolution of cAMP signals in three-dimensional microdomains using novel, real-time sensors. *Proc. West. Pharmacol. Soc. 47*, 1–5.

119 Jacobs, H., Heldt, H. W., Klingenberg, M. (1964) High activity of creatine kinase in mitochondria from muscle and brain and evidence for a separate mitochondrial isoenzyme of creatine kinase. *Biochem. Biophys. Res. Commun. 16*, 516–521.

120 Scholte, H. R. (1973) On the triple localization of creatine kinase in heart and skeletal muscle cells of the rat: evidence for the existence of myofibrillar and mitochondrial isoenzymes. *Biochim. Biophys. Acta 305*, 413–427.

121 Scholte, H. R. (1973) The separtation and enzymatic characterization of inner and outer membranes of rat heart mitochondria. *Biochim. Biophys. Acta 330*, 283–293.

122 Schlattner, U., Gehring, F., Vernoux, N., Tokarska-Schlattner, M., Neumann, D., Marcillat, O., Vial, C., Wallimann, T. (2004) C-terminal lysines determine phospholipid interaction of sarcomeric mitochondrial creatine kinase. *J. Biol. Chem. 279*, 24334–24342.

123 Muller, M., Moser, R., Cheneval, D., Carafoli, E., (1985) Cardiolipin is the membrane receptor for mitochondrial creatine phosphokinase. *J. Biol. Chem. 260*, 3839–3843.

124 Pebay-Peyroula, E., Dahout-Gonzalez, C., Trézéguet, V., Lauquin, G., Brandolin, G. (2003) Structure of mitochondrial ADP/ATP carrier in complex with carboxyatractyloside. *Nature 426*, 39–44.

125 Nury, H., Dahout-Gonzalez, C., Trezeguet, V., Lauquin, G. J., Brandolin, G., Pebay-Peyroula, E. (2006) Relations between structure and function of the mitochondrial ADP/ATP carrier. *Annu. Rev. Biochem. 75*, 713–41.

126 Huang, S.-G., Oday, S., Klingenberg, M. (2001) Chimers of two fused ADP/ATP carrier monomers indicate a single channel for ADP/ATP transport. *Arch. Biochem. Biophys. 394*, 67–75.

127 Huber, T., Klingenberg, M., Beyer, K. (1999) Binding of nucleotides by mitochondrial ADP/ATP carrier as studied by 1H nuclear magnetic resonance spectroscopy. *Biochemistry 38*, 762–769.

128 Gropp, T., Brustovetsky, N., Klingenberg, M., Müller, V., Fendler, K., Bamberg, E. (1999) Kinetics of electrogenic transport by the ADP/ATP Carrier. *Biophys. J. 77*, 714–726.

129 Klingenberg, M. (1985) The ADP/ATP carrier in mitochondrial membranes. In *The enzymes of biological membranes: bioenergetics of electron and proton transport, volume 4*, Martonosi, A. N. (ed.), Plenum Press, New York, London, 511–553.

130 Duyckaerts, C., Sluse-Coffart, C. M., Fux, J. P., Sluse, F. E., Liebecq, C. (1980) Kinetic mechanism of the exchanges catalysed by the adenine nucleotide carrier. *Eur. J. Biochem. 106*, 1–6.

131 Metelkin, E., Goryanin, I. and Demin, O. (2006) Mathematical modeling of mitochondrial adenine nucleotide translocase. *Biophys. J. 90*, 423–432.

132 Palmieri, F. (2004) The mitochondrial transporter family (SLC25): physiological and pathological implications. *Pflugers Arch. 447*, 689–709.

133 Kramer, R., Palmieri, F. (1992) Metabolic carriers in mitochondria. In *Molecular Mechanisms in Bioenergetics*, Ernster, L. (ed.), Elsevier Science Publishers, 359–384.

134 Chen, C., Ko, Y., Delannoy, M., Ludtke, S. J., Chiu, W., Pedersen, P. L. (2004) Mitochondrial ATP synthasome: three-dimensional structure by electron microscopy of the ATP synthase in complex formation with carriers for Pi and ADP/ATP. *J. Biol. Chem. 279*, 31761–31768.

135 Beyer, K., Klingenberg, M. (1985) ADP/ATP carrier protein from beef heart mitochondria has high amounts of tightly bound cardiolipin, as revealed by ^{31}P nuclear magnetic resonance. *Biochemistry 24*, 3821–3826.

136 Beyer, K., Nuscher, B. (1996) Specific cardiolipin binding interferes with labeling of sulfhydryl residues in the adenosine diphosphate/adenosine triphosphate carrier protein from beef heart mitochondria. *Biochemistry 35*, 15784–90.

137 Nury, H., Dahout-Gonzalez, C., Trezeguet, V., Lauquin, G., Brandolin, G., Pebay-Peyroula, E. (2005) Structural basis for lipid-mediated interactions between mitochondrial ADP/ATP carrier monomers. *FEBS Lett. 579*, 6031–6036.

138 Schlattner, U., Dolder, M., Wallimann, T., Tokarska-Schlattner, M. (2001) Mitochondrial creatine kinase and mitochondrial outer membrane porin show a direct interaction that is modulated by calcium. *J. Biol. Chem. 276*, 48027–48030.

139 Schlattner, U., Forstner, M., Eder, M., Stachowiak, O., Fritz-Wolf, K., Wallimann, T. (1998) Functional aspects of the X-ray structure of mitochondrial creatine kinase: a molecular physiology approach. *Mol. Cell. Biochem. 184*, 125–140.

140 Jacobus, W. E., Lehninger, A. L. (1973) Creatine kinase of rat heart mitochondria. Coupling of creatine phosphorylation to electron transport. *J. Biol. Chem. 248*, 4803–4810.

141 Saks, V. A., Chernousova, G. B., Voronkov, U. I., Smirnov, V. N., Chazov, E. I. (1974) Study of energy transport mechanism in myocardial cells. *Circ. Res. 35*, 138–149.

142 Saks, V. A., Chernousova, G. B., Gukovsky, D. E., Smirnov, V. N., Chazov, E. I. (1975) Studies of energy transport in heart cells. Mitochondrial isoenzyme of creatine phosphokinase: kinetic properties and regulatory action of Mg^{2+} ions. *Eur. J. Biochem. 57*, 273–290.

143 Jacobus, W. E., Saks, V. A. (1982) Creatine kinase of heart mitochondria: changes in its kinetic properties induced by coupling to oxidative phosphorylation. *Arch. Biochem. Biophys. 219*, 167–178.

144 Saks, V. A., Kuznetsov, A. V., Kupriyanov, V. V., Miceli, M. V., Jacobus, W. J. (1985) Creatine kinase of rat heart mitochondria. The demonstration of functional coupling to oxidative phosphorylation in an inner membrane-matrix preparation. *J. Biol. Chem. 260*, 7757-7764.

145 Kuznetsov, A. V., Khuchua, Z. A., Vassil'eva, E. V., Medved'eva, N. A., Saks, V. A. (1989) Heart mitochondrial creatine kinase revisited: the outer

mitochondrial membrane is not important for coupling of phospho-creatine production to oxidative phosphorylation. *Arch. Biochem. Biophys. 268*, 176–190.

146 Kenyon, G. L., Reed, G. H. (1983) Creatine kinase: structure-activity relationships. *Adv. Enzymol. 54*, 367–426.

147 McLeish, M. J., Kenyon, G. L. (2005) Relating structure to mechanism in creatine kinase. *Crit. Rev. Biochem. Mol. Biol. 40*, 1–20.

148 Jacobus, W. E., Moreadith, R. W., Vandegaer, K. M. (1982) Mitochondrial respiratory control. Evidence against the regulation of respiration by extramitochondrial phosphorylation potentials or by [ATP]/[ADP] ratios. *J. Biol. Chem. 257*, 2397–2402.

149 Vendelin, M., Lemba, M., Saks, V. A. (2004) Analysis of functional coupling: mitochondrial creatine kinase and adenine nucleotide translocase. *Biophys. J. 87*, 696–713.

150 Barbour, R. L., Ribaudo, J., Chan, S. H. (1984) Effect of creatine kinase activity on mitochondrial ADP/ATP transport. Evidence for a functional interaction. *J. Biol. Chem. 259*, 8246–8251.

151 DeFuria, R. A., Ingwall, J. S., Fossel, E. T., Dygert, M. K. (1980) Micro-compartmentation of the mitochondrial creatine kinase reaction. In *Heart Creatine Kinase. The Integration of Enzymes for Energy Distribution*, Jacobus, W. E., Ingwall, J. S. (eds.), Williams & Wilkins, Baltimore, 135–141.

152 Soboll, S., Gonrad, A., Hebish, S. (1994) Influence of mitochondrial creatine kinase on the mitochondrial/extramitochondrial distribution of high energy phosphates in muscle tissue: evidence for the leak in the creatine shuttle. *Mol. Cell. Biochem. 133/134*, 105–115.

153 Saks, V. A., Khuchua, Z. A., Kuznetsov, A. V. (1987) Specific inhibition of ATP-ADP translocase in cardiac mitoplasts by antibodies against mitochondrial creatine kinase. *Biochim. Biophys. Acta. 891*, 138–144.

154 Moreadith, R. W., Jacobus, W. E. (1982) Creatine kinase of heart mitochondria. Functional coupling of ADP transfer to the adenine nucleotide translocase. *J. Biol. Chem. 257*, 899–905.

155 Gellerich, F. N., Schlame, M., Bohnensack, R., Kunz, W. (1987) Dynamic compartmentation of adenine nucleotides in the mitochondrial intermembrane space of rat-heart mitochondria. *Biochim. Biophys. Acta 890*, 117–126.

156 Teague, W. E., Dobson, G. P. (1992) Effect of temperature on the creatine kinase equilibrium. *J. Biol. Chem. 267*, 14084–14093.

157 Lawson, J. W. R., Veech, R. L. (1979) Effect of pH and free Mg2+ on the K_{eq} of the creatine kinase reaction and other phosphate hydrolysis and phosphate transfer reactions. *J. Biol. Chem. 254*, 6528–6537.

158 Teague, W. E., Golding, E. M., Dobson, G. P. (1996) Adjustment of K′ for the creatine kinase, adenylate kinase and ATP hydrolysis equilibria to varying temperaure and ionicf strength. *J. Exp. Biol. 199*, 509–512.

159 Belitzer, V. A., Tsybakova, E. T. (1939) Sur le mécanisme des phosphorylations couplées avec la respiration. *Biochimia (Russian) 4*, 516–535.

160 Kim, I. H., Lee, H. J. (1987) Oxidative phosphorylation of creatine by respiring pig heart mitochondria in the absence of added adenine nucleotides. *Biochem. Internatl. 14*, 103–110.

161 Dolder, M., Walzel, B., Speer, O., Schlattner, U., Wallimann, T. (2003) Inhibition of the mitochondrial permeability transition by creatine kinase substrates. Requirement for microcompartmentation. *J. Biol. Chem. 278*, 17760–17766.

162 Claycomb, W. C., Lanson, N. A. Jr., Stallworth, B. S., Egeland, D. B., Delcarpio, J. B., Bahinski, A., Izzo, N. J. Jr. (1998) HL-1 cells: a cardiac muscle cell line that contracts and retains phenotypic characteristics of the adult cardiomyocyte. *Proc. Natl. Acad. Sci. U.S.A. 95*, 2979–2984.

163 White, S. M., Constantin, P. E., Claycomb, W. C. (2004) Cardiac physiology at the cellular level: use of cultured HL-1 cardiomyocytes for studies of cardiac muscle cell structure

and function. *Am. J. Physiol. Heart Circ. Physiol.* 286, H823–829.

164 Pelloux, S., Robillard, J., Ferrera, R., Bilbaut, A., Ojeda, C., Saks, V., Ovize, M., Tourneur, Y. (2006) Non-beating HL-1 cells for confocal microscopy: Application to mitochondrial functions during cardiac preconditioning. *Prog. Biophys. Mol. Biol.* 90, 270–298.

165 Kuznetsov, A. V., Troppmair, J., Sucher, R., Hermann, M., Saks, V., Margreiter, R. (2006) Mitochondrial subpopulations and heterogeneity revealed by confocal imaging; possible physiological role? *Biochim. Biophys. Acta 1757*, 686–691.

166 Kummel, L. (1988) Ca,MgATPase activity of permeabilized rat heart cells and its functional coupling to oxidative phosphorylation in the cells. *Cardiovasc. Res. 22*, 359–367.

167 Saks, V. A., Kapelko, V. I., Kupriyanov, V. V., Kuznetsov, A. V., Lakomkin, V. L., Veksler, V. I., Sharov, V. G., Javadov, S. A., Seppet, E. K., Kairane, C. (1989) Quantitative evaluation of relationship between cardiac energy metabolism and post-ischemic recovery of contractile function. *J. Mol. Cell Cardiol. 21*, 67–78.

168 Saks, V. A., Belikova, Yu. O., Kuznetsov, A. V. (1991) *In vivo* regulation of mitochondrial respiration in cardiomyocytes: Specific restrictions for intracellular diffusion of ADP. *Biochim. Biophys. Acta 1074*, 302–311.

169 Veksler, V. I., Kuznetsov, A. V., Anflous, K., Mateo, P., van Deursen, J., Wieringa, B., Ventura-Clapier, R. (1995) Muscle creatine-kinase deficient mice. II Cardiac and skeletal muscles exhibit tissue-specific adaptation of the mitochondrial function. *J. Biol. Chem. 270*, 19921–19929.

170 Saks, V. A., Veksler, V. I., Kuznetsov, A. V., Kay, L., Sikk, P., Tiivel, T., Tranqui, L., Olivares, J., Winkler, K., Wiedemann, F., Kunz, W. S. (1998) Permeabilized cell and skinned fiber techniques in studies of mitochondrial function *in vivo*. *Mol. Cell. Biochem. 184*, 81–100.

171 Anflous, K., Armstrong, D. D., Craigen, W. J. (2001) Altered mitochondrial sensitivity for ADP and maintenance of creatine-stimulated respiration in oxidative striated muscles from VDAC1-deficient mice. *J. Biol. Chem. 276*, 1954–1960.

172 Boudina, S., Laclau, M. N., Tariosse, L., Daret, D., Gouverneur, G., Boron-Adele, S., Saks, V. A., Dos Santos, P. (2002) Alteration of mitochondrial function in a model of chronic ischemia *in vivo* in rat heart. *Am. J. Physiol. 282*, H821–H831.

173 Burelle, Y., Hochachka, P. W. (2002) Endurance training induces muscle-specific changes in mitochondrial function in skinned muscle fibers. *J. Appl. Physiol. 92*, 2429–2438.

174 Liobikas, J., Kopustinskiene, D. M., Toleikis, A. (2001) What controls the outer mitochondrial membrane permeability for ADP: facts for and against the oncotic pressure. *Biochim. Biophys. Acta 1505*, 220–225.

175 Zoll, J., Ponsot, E., Doutreleau, S., Mettauer, B., Piquard, F., Mazzucotelli, J. P., Diemunsch, P., Geny, B. (2005) Acute myocardial ischaemia induces specific alterations of ventricular mitochondrial function in experimental pigs. *Acta Physiol. Scand. 185*, 25–32.

176 Kuznetsov, A. V., Tiivel, T., Sikk, P., Käämbre, T., Kay, L., Daneshrad, Z., Rossi, A., Kadaja, L., Peet, N., Seppet, E., Saks, V. (1996) Striking difference between slow and fast twitch muscles in the kinetics of regulation of respiration by ADP in the cells *in vivo*. *Eur. J. Biochem. 241*, 909–915.

177 Saks, V., Kuznetsov, A. V., Andrienko, T., Usson, Y., Appaix, F., Guerrero, K., Kaambre, T., Sikk, P., Lemba, M., Vendelin, M. (2003) Heterogeneity of ADP diffusion and regulation of respiration in cardiac cells. *Biophys. J. 84*, 3436–3456.

178 Anmann, T., Guzun, R., Beraud, N., Pelloux, S., Kuznetsov, A. V., Kogerman, L., Kaambre, T., Sikk, P., Paju, K., Peet, N., Seppet, E., Ojeda, C., Tourneur, Y., Saks, V. (2006) Different kinetics of the regulation of respiration in permeabilized cardiomyocytes and HL-1 cells. Importance of cell structure/organization for respiration regulation. *Biochim. Biophys. Acta 1757*, 1597–1606.

179 Saks, V. A., Vasilyeva, E., Belikova, Yu. O., Kuznetsov, A. V., Lyapina, S., Petrova, L., Perov, N. A. (1993) Retarded diffusion of ADP in cardiomyocytes: possible role of mitochondrial outer membrane and creatine kinase in cellular regulation of oxidative phosphorylation. *Biochim. Biophys. Acta 1144*, 134–148.

180 Saks, V. A., Kuznetsov, A. V., Khuchua, Z. A., Vasilyeva, E. V., Belikova, J. O., Kesvatera, T., Tiivel, T. (1995) Control of cellular respiration *in vivo* by mitochondrial outer membrane and by creatine kinase. A new speculative hypothesis: possible involvement of mitochondrial-cytoskeleton interactions. *J. Mol. Cell Cardiol. 27*, 625–645.

181 Ingwall, J. S. (2002) *ATP and the heart*, Kluwer Academic Publishers, Dordrecht-Boston-London. 1–244.

182 Vignais, P. V. (1976) Molecular and physiological aspects of adenine nucleotide transport in mitochondria. *Biochim. Biophys. Acta 456*, 1–38.

183 Zoll, J., Koulmann, N., Bahi, L., Ventura-Clapier, R., Bigard, A. X. (2002) Quantitative and qualitative adaptation of skeletal muscle mitochondria to increased physical activity. *J. Cell Physiol. 194*, 186–193.

184 Zoll, J., Sanchez, H., N'Guessan, B., Ribera, F., Lampert, E., Bigard, X., Serrurier, B., Fortin, D., Geny, B., Veksler, V., Ventura-Clapier, R., Mettauer, B. (2002) Physical activity changes the regulation of mitochondrial respiration in human skeletal muscle. *J. Physiol. 543*, 191–200.

185 Zoll, J., N'Guessan, B., Ribera, F., Lampert, E., Fortin, D., Veksler, V., Bigard, X., Geny, B., Lonsdorfer, J., Ventura-Clapier, R., Mettauer, B. (2003) Preserved response of mitochondrial function to short-term endurance training in skeletal muscle of heart transplant recipients. *J. Am. Coll. Cardiol. 42*, 126–32.

186 Guerrero, K., Wuyam, B., Mezin, P., Vivodtzev, I., Vendelin, M., Bore, J. C., Hacini, R., Chavanon, O., Imbeaud, S., Saks, V., Pison, C. (2005) Functional coupling of adenine nucleotide translocase and mitochondrial creatine kinase is enhanced after exercise training in lung transplant skeletal muscle. *Am. J. Physiol. 289*, R1144–R1154.

187 Ponsot, E., Dufour, S. P., Zoll, J., Doutrelau, S., N'Guessan, B., Geny, B., Hoppelerr, H., Lampert, E., Mettauer, B., Ventura-Clapier, R., Ruddy, R. (2006) Exercise training in normobaric hypoxia in endurance runners. II. Improvement of mitochondrial properties in skeletal muscle. *J. Appl. Physiol. 100*, 1249–1257.

188 Mettauer, B., Zoll, J., Garnier, A., Ventura-Clapier, R. (2006) Heart failure: a model of cardiac and skeletal muscle energetic failure. *Pfugers Arch. 452*, 653–666.

189 Joubert, F., Hoerter, J. A., Mazet, J. L. (2001) Discrimination of cardiac subcellular creatine kinase fluxes by NMR spectroscopy: a new method of analysis. *Biophys. J. 81*, 2995–3004.

190 Joubert, F., Mateo, P., Gillet, B., Beloeil, J. C., Mazet, J. L., Hoerter, J. A. (2004) CK flux or direct ATP transfer: versatility of energy transfer pathways evidenced by NMR in the perfused heart. *Mol. Cell. Biochem. 256–257*, 43–58.

191 Spindler, M., Niebler, R., Remkes, H., Horn, M., Lanz, T., Neubauer, S. (2002) Mitochondrial creatine kinase is critically necessary for normal myocardial high-energy phosphate metabolism. *Am. J. Physiol. Heart Circ. Physiol. 283*, H680–H687.

192 Spindler, M., Meyer, K., Stromer, H., Leupold, A., Boehm, E., Wagner, H., Neubauer, S. (2004) Creatine kinase-deficient hearts exhibit increased susceptibility to ischemia-reperfusion injury and impaired calcium homeostasis. *Am. J. Physiol. Heart Circ. Physiol. 287*, H1039–H1045.

193 Neubauer, S. (2007) The failing heart – an engine out of fuel. *N. Engl. J. Med. 356*, 1140–1151.

194 Meyer, L. E., Machado, L. B., Santiago, A. P. S. A., da-Silva, S., De Felice, F. G., Holub, O., Oliviera, M., Galina, A. (2006) Mitochondrial creatine kinase activity prevents reactive oxygen species generation: Antioxidant role of mitochondrial kinases-dependent ADP re-cycling activity. *J. Biol. Chem. 281*, 37361–37371.

195 Jezek, P., Hlavata, L. (2005) Mitochondria in homeostasis of reactive oxygen species in cell, tissues and organism. *Intl. J. Biochem. Cell Biol. 37*, 2478–2503.

196 Zorov, D. B., Filburn, C. R., Klotz, L. O., Zweier, J. L., Sollott, S. J. (2000) Reactive oxygen species (ROS)-induced ROS release: a new phenomenon accompanying induction of the mitochondrial permeability transition in cardiac myocytes. *J. Exp. Med. 192*, 1001–1014.

197 Wallimann, T., Schlosser, T., Eppenberger, H. (1984) Function of M-line-bound creatine kinase as intramyofibrillar ATP regenerator at the receiving end of the phosphorylcreatine shuttle in muscle. *J. Biol. Chem. 259*, 5238–5246.

198 Yamashita, K., Yoshioka, T. (1991) Profiles of creatine kinase isoenzyme compositions in single muscle fibres of different types. *J. Muscle. Res. Cell. Motil. 12*, 37–44.

199 Korge, P. (1995) Factors limiting adenosine triphosphase function during high intensity exercise. Thermodynamic and regulatory considerations. *Sports Med. 20*, 215–225.

200 Cook, R., Pate, E. (1985) The effects of ADP and phosphate on the contraction of muscle fibers. *Biophys. J. 48*, 789–798.

201 Goldman, Y. (1987) Kinetics of the actomyosin ATPase in muscle fibers. *Annu. Rev. Physiol. 49*, 637–654.

202 Rayment, I., Holden, H. M., Whittaker, M., Yohn, C. B., Lorenz, M., Holmes, K. C., Milligan, R. A. (1993) Structure of the actin-myosin complex and its implications for muscle contraction. *Science 261*, 58–65.

203 Karatzaferi, C., Myburgh, K. H., Chinn, M. K., Franks-Skiba, K., Cook, R. (2003) Effect of an ADP analog on isometric force and ATPase activity of active muscle fibers. *Am. J. Physiol. 284*, C816–825.

204 Yamashita, H., Sata, M., Sugiura, S., Monomura, S. I., Serizawa, T., Masahiko, I. (1994) ADP inhibits the sliding velocity of fluorescent actin filaments on cardiac and skeletal myosins. *Circ. Res. 74*, 1027–1033.

205 Fukuda, N., Fujita, H., Fujita, T., Ishiwata, S. (1998) Regulatory roles of MgADP and calcium in tension development of skinned cardiac muscle. *J. Muscle Res. Cell. Motil. 19*, 909–921.

206 Fukuda, N., Kajiwara, H., Ishiwata, S., Kurihara, S. (2000) Effects of MgADP on length dependence of tension generation in skinned rat cardiac muscle. *Circ. Res. 86*, E1-6.

207 Sata, M., Sugiura, S., Yamashita, H., Momomura, S. I., Serizawa, T. (1996) Coupling between myosin ATPase cycle and creatine kinase cycle facilitates cardiac actomyosin sliding *in vitro*: a clue to mechanical dysfunction during myocardial ischemia. *Circulation 93*, 310–317.

208 Wegmann, G., Zanolla, E., Eppenberger, H. M., Wallimann, T. (1992) *In situ* compartmentation of creatine kinase in intact sarcomeric muscle: the actomyosin overlap zone as a molecular sieve. *J. Muscle Res. Cell Motil. 13*, 420–435.

209 Hornemann, T., Stolz, M., Wallimann, T. (2000) Isoenzyme-specific interaction of muscle-type creatine kinase with the sarcomeric M-line is mediated by NH(2)-terminal lysine charge-clamps. *Eur. J. Cell Biol. 149*, 1225–1234.

210 Hornemann, T., Kempa, S., Himmel, M., Hayess, K., Furst, D. O., Wallimann, T. (2003) Muscle-type creatine kinase interacts with central domains of the M-band. proteins myomesin and M-protein. *J. Mol. Biol. 332*, 877–87.

211 Krause, S. M., Jacobus, W. E. (1992) Specific enhancement of the cardiac myofibrillar ATPase activity by bound creatine kinase. *J. Biol. Chem. 267*, 2480–2486.

212 Bessman, S. P., Yang, W. C., Geiger, P. J., Erickson-Viitanen, S. (1980) Intimate coupling of creatine phosphokinase and myofibrillar adenosinetriphosphatase. *Biochem. Biophys. Res. Commun. 96*, 1414–1420.

213 Ventura-Clapier, R., Mekhi, H., Vassort, G. (1987) Role of creatine kinase in force development in chemically skinned rat cardiac muscle. *J. Gen. Physiol. 89*, 815–837.

214 Ogut, O., Brozovich, F. V. (2003) Creatine phosphate consumption and the actomyosin crossbridge cycle in cardiac muscles. *Circ. Res. 93*, 54–60.

215 Momken, I., Lechene, P., Koulmann, N., Fortin, D., Mateo, P., Hoerter, J., Doan, B. T., Bigard, X., Veksler, V., Ventura-Clapier, R. F. (2005) Impaired voluntary running capacity in CK deficient mice. *J. Physiol. 565*, 951–964

216 Kinding, C. A., Howlett, R. A., Stary, C. M., Walsh, B., Hogan, M. C. (2004) Effects of acute creatine kinase inhibition on metabolism and tension development in isolated single myocytes. *J. Appl. Physiol. 98*, 541–549.

217 Koretsune, Y., Marban, E. (1990) Mechanism of ischemic contracture in ferret hearts: relative roles of [Ca2+]i elevation and ATP depletion. *Am. J. Physiol. 258*, H9–H16.

218 Rossi, A. M., Eppenberger, H. M., Volpe, P., Cotrufo, R., Wallimann, T. (1990) Muscle-type MM creatine kinase is specifically bound to sarcoplasmic reticulum and can support Ca^{2+} uptake and regulate local ATP/ADP ratios. *J. Biol. Chem. 265*, 5258–5266.

219 Minajeva, A., Ventura-Clapier, R., Veksler, V. (1996) Ca^{2+} uptake by cardiac sarcoplasmic reticulum ATPase *in situ* strongly depends on bound creatine kinase. *Pflugers Arch. 432*, 904–912.

220 Korge, P., Byrd, S. K., Campbell, K. B. (1993) Functional coupling between sarcoplasmic-reticulum-bound creatine kinase and Ca^{2+} ATPase. *Eur. J. Biochem. 213*, 973–980.

221 Korge, P., Campbell, K. B. (1994) Local ATP regeneration is important for sarcoplasmic reticulum Ca^{2+} pump function. *Am. J. Physiol. 267*, C357–366.

222 Yang, Z., Steele, D. S. (2002) Effects of phosphocreatine on SR regulation in isolated saponin-permeabilised rat cardiac myocytes. *J. Physiol. 539*, 767–777.

223 Lederer, W. J., Nichols, C. G. (1989) Nucleotide modulation of the activity of rat heart ATP-sensitive K^+ channels in isolated membrane patches. *J. Physiol. 419*, 193–211.

224 Lorenz, E., Terzic, A. (1999) Physical association between recombinant cardiac ATP-sensitive K^+ subunits Kir6 and

SUR2A. *J. Mol. Cell. Cardiol. 31*, 425–434.

225 Crawford, R. M., Ranki, H. J., Botting, C. H., Budas, G. R., Jovanovic, A. (2002) Creatine kinase is physically associated with the cardiac ATP-sensitive K+ channel *in vivo*. *FASEB Journal 16*, 102–104.

226 Carrasco, A. J., Dzeja, P. P., Alekseev, A. E., Pucar, D., Zingman, L. V., Abraham, M. R., Hodgson, D., Bienengraeber, M., Puceat, M., Janssen, E., Wieringa, B., Terzic, A. (2001) Adenylate kinase phosphotransfer communicates cellular energetic signals to ATP-sensitive potassium channels. *Proc. Natl. Acad. Sci. U.S.A. 98*, 7623–7628.

227 Noma, A. (1983) ATP-regulated K^+ channel in cardiac muscle. *Nature 305*, 147–148.

228 Noma, A., Shibasaki, T. (1985) Membrane current through adenosine-triphosphate regulated potassium channels in guinea-pig ventricular cells. *J. Physiol. 363*, 463–480.

229 Carmeliet, E. (1992) A fuzzy sub-sarcolemmal space for intracellular Na^+ in cardiac cells? *Cardiovasc. Res. 26*, 433–442.

230 Saks, V. A., Lipina, N. V., Sharov, V. G., Smirnov, V. N., Chazov, E. I., Grosse, R. (1977) The localization of the MM isoenzyme of creatine phosphokinase on the surface membrane of myocardial cells and its functional coupling to oubain-inhibited (Na+, K+) ATPase. *Biochem. Biophys. Acta 465*, 550–558.

231 Sasaki, N., Sato, T., Marban, E., O'Rourke, B. (2001) ATP consumption by uncoupled mitochondria activates sarcolemmal K_{ATP} channels in cardiac myocytes. *Am. J. Physiol. 280*, H1882–H1888.

232 Alekseev, A. E., Hodgso, D. M., Karger, A. B., Park, S., Zingman, L. V., Terzic, A. (2005) ATP-sensitive K+ channel channel/enzyme multimer: metabolic gating in the heart. *J. Mol. Cell. Cardiol. 38*, 895–905.

233 Ventura-Clapier, R., Garnier, A., Veksler, V. (2004) Energy metabolism in heart failure. *J. Physiol. 555*, 1–13.

234 Kottke, M., Wallimann, T., Brdiczka, D. (1994) Dual electron microscopic localization of mitochondrial creatine kinase in brain mitochondria. *Biochem. Med. Metab. Biol. 51*, 105–117.

235 Ames, A. III. (2000) CNS energy metabolism as related to function. *Brain Res. Rev. 34*, 42–68.

236 Burklen, T. S., Schlattner, U., Homayouni, R., Gough, K., Rak, M., Szeghalmi, A., Wallimann, T. (2006) The creatine kinase/creatine connection to Alzheimer's disease: CK-inactivation, APP-CK complexes and focal creatine deposits. *J. Biomed. Biotechnol. 2006*, 1–11.

237 Peuffer, J., Tkac, I., Gruetter, R. (2000) Extracellular – intracellular distribution of glucose and lactate in the rat brain assessed noninvasively by diffusion – weighted 1H – nuclear magnetic resonance spectroscopy *in vivo*. *J. Cereb. Blood Flow Metab. 20*, 736–746.

238 Saks, V., Aliev, M., Guzun, R., Beraud, N., Monge, C., Anmann, T., Kuznetsov, A. V., Seppet, E. (2006) Biophysics of the organized metabolic networks in muscle and brain cells. In *Recent Research Developments in Biophysics 5*, 269–318; Transworld Research Networks, *37*, 612(2), Kerala, India.

239 Ho, M.-W. (1994) What is (Schrodinger's) Negentropy? In *Modern Trends in BioThermoKinetics, Volume 3*, Gnaiger, E., Gellerich, F., Wyss, M. (eds.), Innsbruck University Press, Innsbruck, Austria, 50–61.

240 Azzone, G. (1994) Negentropy and the historical arrow of time. Thermo-dynamical and informational aspects of the Darwinian revolution. In *Modern Trends in BioThermoKinetics, Volume 3*, Gnaiger, E., Gellerich, F., Wyss, M. (eds.), Innsbruck University Press, Innsbruck, Austria, 38–49.

241 Brillouin, L. (1962) *Science and Information Theory*, 2nd Edition, Academic Press, New York.

4

On the Network Properties of Mitochondria

Miguel A. Aon, Sonia Cortassa, and Brian O'Rourke

Abstract

In this chapter, we introduce a network view of mitochondrial function that includes consideration of how the spatial and temporal organization of mitochondria is crucial for understanding the behavior of living cells. The network is characterized by applying advanced experimental techniques – including optical mapping, patch-clamp techniques, and two-photon microscopy – combined with mathematical methods, including computational modeling, power spectral analysis, and relative dispersion analysis for studying complex nonlinear interactions with the potential for self-organized behavior. Remarkably, the mitochondrial network is shown to have scale-free fractal characteristics in both time and space that govern the timing and extent of functional synchronization between individual organelles. This new framework is applied not only to better understand the established role of mitochondria in bioenergetics but also to reveal a novel physiological signaling system centered on reactive oxygen species (ROS) and how this system can undergo a transition to a critical pathophysiological state under metabolic stress. The major insight from these studies is that mitochondria act as a network of coupled oscillators with the potential for active intracellular ROS signaling through a frequency- and amplitude-encoded process. These newly identified spatiotemporal features of mitochondria extend the traditional view as mere ATP generators to explain their impact on signaling, cell injury, cell death, and whole-organ dysfunction associated with aging and a variety of diseases.

4.1
Introduction

It has been argued that respiration and protonmotive gradients represent powerful primordial evolutionary forces that contributed to a rapid increase in the complex organization of eukaryotic cells and organisms subsequent to the endosymbiosis of the protomitochondria [1]. Over the past century since their first de-

Molecular System Bioenergetics: Energy for Life. Edited by Valdur Saks
Copyright © 2007 WILEY-VCH Verlag GmbH & Co. KGaA, Weinheim
ISBN: 978-3-527-31787-5

scription, mitochondria have been found to be vital for the integration of cell function, fulfilling multiple roles as providers of energy for mechanical work, macromolecular synthesis, and ion homeostasis; transmitters of genetic information; signaling organelles that modulate Ca^{2+} and ROS-dependent pathways; and cellular executioners. Moreover, alterations in mitochondrial function and ROS production are thought to be major determinants of age-associated disease and the human lifespan [1–3].

Poised at the convergence of most anabolic and catabolic biochemical pathways, mitochondria appear as "hubs" in the intracellular metabolic network (Fig. 4.1). Understanding of the extent and connectivity of the cytoplasmic and mitochondrial components of the biochemical network has been the subject of recent developments in network theory [4] and is crucial to comprehending the web of linked reactions underlying the global metabolic status of an organism.

Organizational complexity also extends to the morphology of the mitochondrial network. As noted in early studies of mitochondria in living cells, rapid fission, fusion, and translocation events lead to continuous remodeling of the mitochondrial network, and intense recent interest has focused on the proteins involved in the dynamic reorganization of mitochondria, particularly with respect to disease, aging, and development (reviewed in [5, 6]). Mitochondrial plasticity contributes to mitochondrial distribution patterns that vary markedly in different cell types and under different environmental conditions, raising questions about how the spatial organization of the network influences function. While the links between morphology and function are still unclear, provocative new findings show that altering the structure of the mitochondrial network, and/or the topology of the mitochondrial cristae, leads to changes in metabolic fluxes. Moreover, mitochondrial membranes appear to be structural organizing centers for a variety of intracellular signaling pathways (reviewed in [7, 8]).

Morphologically, mitochondria can appear as a regular lattice-like network, as in heart cells [9–11], or as an irregular, filamentous structure, as in neurons or cancer cells (Fig. 4.2) [12, 13]. Interconversion of the elongated filamentous network to dispersed punctate mitochondria may occur under metabolic stress, including intense neuronal activity, changes in nutrient availability, oxidative stress, or apoptotic stimuli [14]. Furthermore, several studies have even attributed functional specialization to mitochondria in different cellular locations [12, 15] or, at a minimum, noted functional heterogeneity within the mitochondrial network [16]. These observations have given rise to the idea that mitochondria may operate as networks and have spurred investigation into how they communicate with each other. Early studies suggested that mitochondria formed an extended reticular structure [17], and, based on experiments employing laser fluorescence photobleaching in living cells, it was postulated that mitochondria were organized as clusters and might be electrically coupled through cable properties [18]. In contrast, other investigators have demonstrated that individual mitochondria can respond independently of the network as a whole [9, 19–21].

The mitochondrial network view extends beyond the complex connectivity of the biochemical circuit diagrams and the morphological dynamics of the mitochon-

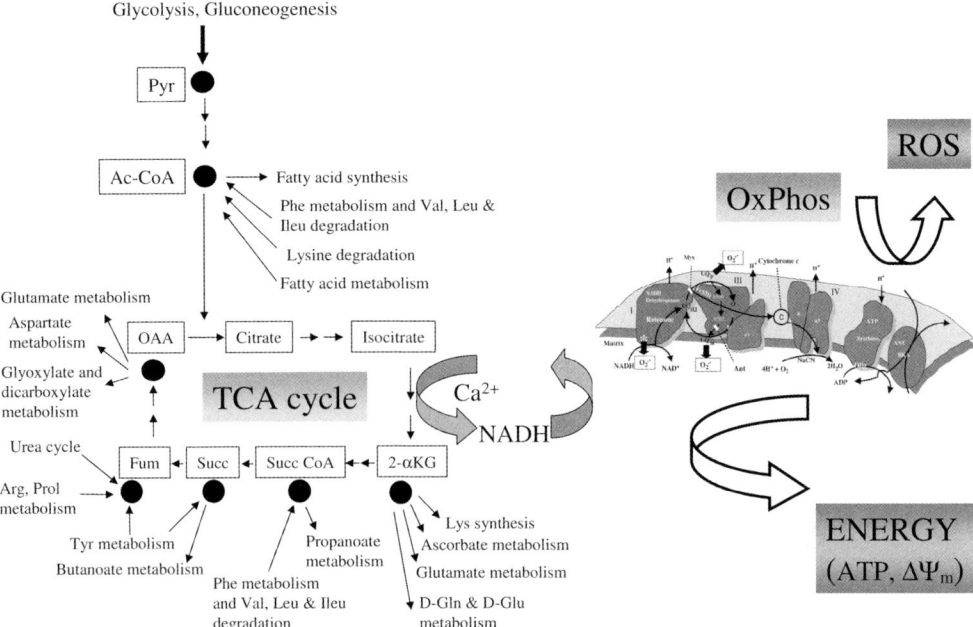

Fig. 4.1 Mitochondria as metabolic hubs. Through the tricarboxylic acid cycle (TCA cycle), mitochondria represent a hub within the cellular network due to their multiple links to other pathways as either an input (source) or an output (sink) [126, 127]. Several of the TCA cycle intermediates, such as oxaloacetate (OAA) and alpha-ketoglutarate (2-αKG), feed several amino acid production pathways, while others such as succinyl-CoA (Succ coA) or fumarate (Fum) are fed by the phenylalanine (Phe), valine (Val), leucine (Leu), isoleucine (Ileu), or tyrosine (Tyr) degradation pathways. Two of the main dehydrogenases catalyzing the conversion of isocitrate and 2-αKG and responsible for producing the electron donor NADH are activated by calcium along with pyruvate dehydrogenase. Reactive oxygen species (ROS) produced in the respiratory chain – such as superoxide anion, $O_2^{\bullet -}$, at the level of complex I or complex III, and its product hydrogen peroxide, H_2O_2 – have a signaling role under normal physiological conditions. Through ROS production, mitochondria expand their role as the cell's powerhouse to a signaling task by contributing to the intracellular redox environment. By changing the thiol redox status of cysteine residues in kinases and phosphatases associated with mitochondria, ROS contribute to modifying signaling phosphorylation/dephosphorylation events in proteins (e.g., transcriptional factors) that activate or inactivate gene expression [85, 88, 89].

drial network. Complex nonlinear control mechanisms governing mitochondrial oxidative phosphorylation and emergent macroscopic responses of the mitochondrial network highlight the inseparability of the spatial and temporal aspects of the network under both physiological and pathophysiological conditions. The role of spatial organization on mitochondrial dynamic responses has been recognized previously [22–25]. However, only recently have we begun to understand the mechanisms underlying stable, unstable, and oscillatory modes of mitochondrial

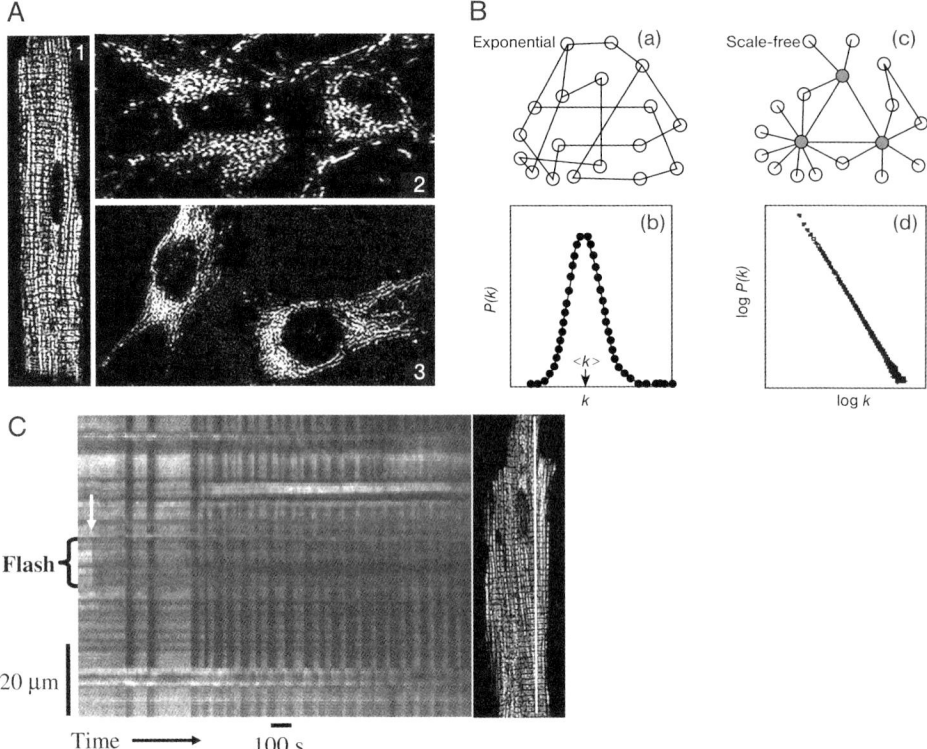

Fig. 4.2 Architectural, topological, and dynamic views of networks.

(A) The *archi-tectural* approach is based on microscopic imaging as applied to fixed specimens or living cells. Its main emphasis is the visual morphological appearance of intracellular structures, e.g., mitochondria, as can be judged by, e.g., fluorescent probes. In panels 1–3, the different architecture of mitochondrial networks is shown after loading of cardiomyocytes (1), cortical neurons (2), and HeLa (3) cells, with the cationic potentiometric dye tetramethylrhodamine ethyl ester (TMRE) used to monitor $\Delta\Psi$m.

(B) The *topological* approach conceives networks as a collection of nodes and edges. Its main emphasis is to describe network connectivity, which can be characterized by the probability, $P(k)$, that a node has k links. For a random network $P(k)$ peaks strongly at $k = \langle k \rangle$ and decays exponentially for a large k. For a scale-free network most nodes have only a few links, but a few nodes, called hubs (dark gray), have a very large number of links (reproduced from [42]).

(C) The *dynamic* view emphasizes the autonomous self-organization of cellular (patho)physiological processes in time. In the example, a self-organized transition visualized as a cell-wide $\Delta\Psi_m$ depolarization (dark gray bands) as visualized by TMRE, a potentiometric $\Delta\Psi_m$ sensor, after a laser flash is shown. This synchronized $\Delta\Psi_m$ depolarization followed by sustained oscillations is triggered after the ongoing autonomous activity of ROS production in the cardiac mitochondrial lattice reaches a threshold level. The situation apparent under oxidative stress is defined as mitochondrial criticality and has been shown to correspond to a cluster of mitochondria with a threshold level of ROS that spans the whole cell [10, 69]. For visualization of the spatiotemporal responses of TMRE presented in panel C, a two- to three-pixel wide line was drawn along the length of the myocyte (as shown in the cell to the right of panel C), and the average fluorescence profile along the line was determined for the entire time series of 2D images for a given experiment. A new image was then created, showing the line fluorescence as a function of time (time-line image).

bioenergetics. Moreover, the fundamental importance of these complex spatio-temporal aspects of mitochondria to physiological signaling [26] and the scaling of network instability to whole-cell and whole-organ function is just beginning to be explored [10, 27–29].

4.1.1
Conceptual Issues About Networks

As applied to mitochondria, the concept of a network must include several inter-pretations of the term. In this chapter, network can refer to *spatial* (structural and topological) as well as *temporal* (different dynamics) aspects of mitochondrial function. From this point of view, networks have been analyzed as regular in structure and topology (e.g., lattice) and dynamics (e.g., each node in the network exhibiting stable fixed points or limit-cycle oscillations) or irregular in structure and topology (e.g., random or scale-free networks) and dynamics (e.g., each node in the network exhibiting chaotic attractors) [30]. Network as used herein mainly refers to the collective dynamics exhibited by mitochondria resulting in emergent self-organized spatiotemporal behavior under certain conditions. This distinction attempts to avoid confusion with the definition of a network as a collection of nodes and edges as conceived in graph theory, a branch of mathematics [31] that emphasizes the topological aspects of network connectivity.

The highly concerted action that is described for mitochondrial networks must fulfill the requirements of robustness (i.e., be able to continue to operate under variable conditions such as changes in substrates, etc.), together with properties of both constancy and flexibility. Constancy is required to provide a steady supply of ATP to fuel the energy demands of the cell (e.g., for muscle contraction, pro-tein synthesis, ion homeostasis), while flexibility is required to adapt the rate of energy production to meet the changing metabolic demand as workload varies [32, 33]. In the particular case of heart cells, this organization supports the rapid and robust response of mitochondrial energetics to large increases or decreases in workload, without fail, for the life of the individual. This network view of mito-chondria is based on the function of mitochondria as powerhouses. However, this perspective has been dramatically expanded in the last two decades to novel functions involving intracellular signaling [8, 20]. Existing and emerging theoret-ical and experimental evidence suggests that the global response of cardiomyo-cytes depends on the coordinated action of thousands of mitochondria arranged in a network of coupled oscillators.

4.2
The Study of (Sub)Cellular Networks and the Emerging View of Cells as Dynamic Mass Energy Information Networks

Three main approaches have characterized the study of cellular networks (Fig. 4.2): (1) architectural (structural morphology), (2) topological (connectivity proper-ties), and (3) dynamical.

The *architecture* of cellular networks based on morphology corresponds to the more classical view of structural organization as visualized mainly by electron microscopy [34], fluorescence microscopy (Fig. 4.2A), or, more recently, electron tomography [35, 36]. In the same framework, a more quantitative approach has been introduced by fractal geometry, which is able to deal with disparate and detail-rich geometries exhibited by living systems in general. Morphological changes induced by environmental or genetic challenges in normal as well as pathological situations have been analyzed using fractal techniques [28, 37, 38]. The fractal view of cellular architecture proposes that the geometric self-similarity exhibited by the complex intracellular spatial patterns [22] may be derived from the recursive application of inherently simple and invariant patterns, spanning several length scales [28].

The *topological* view is based on network theory and the classical work of Erdös and Rényi [39] based on random graphs, further elaborated by Watts and Strogatz [40] adding the two properties of short paths and high clustering. These features confer high-speed communication channels between distant parts of the network, favoring global coordination as shown for natural, social, and technological networks [30]. More recently, these initial efforts have been extended to accommodate other characteristics exhibited by real networks (Fig. 4.2B). A profuse literature has emerged in recent years on this topic, which is still growing (see, e.g., [31]) and has been applied to various sorts of networks, including metabolic ones [41–44].

The original ideas on networks by Erdös and Rényi were aimed at describing random connectivity, whereas the new developments have introduced the view of "scale-free" networks based on concepts of scaling and inverse power laws. According to this view, most of the nodes in a network will have only a few links that are held together by a small number of nodes exhibiting high connectivity, rather than most of the nodes having the same number of links as in "random" networks. The few nodes having many links represent the network's "hubs" and explain, at least in part, the existence of the inverse power law distribution of the connectivity in scale-free networks. Thus, the statistical distribution of the connectivity degree of scale-free networks (given by the number of links exhibited by the nodes of a network) exhibits a continuous hierarchy of nodes, spanning from rare hubs to numerous tiny nodes, rather than having a single, characteristic scale (Fig. 4.2B) [45]. The scale-free topology exhibited by networks can be explained by growth and "preferential attachment" among nodes with a higher probability of expanding their links, concepts derived largely from studies on network topology of the Internet [45]. These two concepts introduced a more dynamic view of the emergence of topology of network connectivity, explaining the origin of hubs and power laws.

In the cellular realm, *dynamic organization* encompasses both the architectural and the topological views of network analysis, accounting for the autonomous dynamics exhibited by their components (nodes) and their defined interactions (connectivity) based on kinetic and thermodynamic principles (Fig. 4.2C) [22]. Dynamic organization obeys two foundational and distinctive characteristics of living systems: "operational closure" and self-organization. More than 30 years ago Maturana and Varela [46] introduced the concept of operational closure to describe the capacity of living systems for self-determination or self-making of their organization (autopoietic networks). This key concept together with that of "dissipative structures" through self-organization by continuous exchange of matter and energy [47] define living networks as continually self-producing, i.e., they are produced by their components and in turn they produce those components (see [48] for an excellent synthesis).

It is interesting to note the conceptual parallels existing between the three different approaches utilized to study networks and the emerging view of the cell and living systems in general [46, 48]:

1. Architecture or structural morphology = structure (the physical embodiment of the system's pattern of organization).
2. Topology or connectivity = pattern of organization (the configuration of relationships).
3. Dynamic organization = process (the ongoing activity from which the embodiment of the pattern of organization in the physical structure emerges).

Thus, in the view of cells as networks, mass, energy, and information continually flow through them, but the system maintains a stable form autonomously through self-organization, i.e., dynamic organization [22, 49, 50].

4.3
Mitochondrial Morphodynamics

The classical work of Hackenbrock [51] demonstrated that mitochondria undergo ultrastructural changes from a "condensed" to an "orthodox" conformation from high (state 3) to low (state 4) respiration rates (Fig. 4.3). In this way, morphological changes within mitochondria were linked to its physiology. This anticipatory finding has received experimental support over the years, notably most recently by electron tomography [52] or during respiratory oscillations in yeast [53].

More recent work shows evidence of the potential importance of mitochondrial morphodynamics for cell (patho)physiology by utilizing different approaches: (1) affecting proteins that mediate fusion and fission processes in cells; (2) influencing cellular energetics or Ca^{2+} levels and observing changes in mitochondrial morphology [54, 55]; and (3) changing mitochondrial mobility by disrupting cytoskeletal architecture [56].

Recent work suggests that specific proteins mediate the dynamics of mitochondrial morphology (or morphodynamics). These proteins modulate the processes of fusion and fission that produce the remodeling of mitochondrial networks [57, 58] or mitochondrial inner membranes [6, 52, 59] in a dynamic way. Dynamin-related protein (DRP1), mitofusin-1 and mitofusin-2 (Mnf1, Mfn2), and OPA1 have been shown to be associated with mitochondrial morphody-

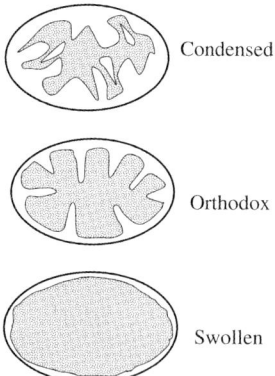

Condensed

Orthodox

Swollen

Fig. 4.3 Changes in mitochondria ultra-structure according to respiratory activity and swelling. The scheme shows the transition between condensed and orthodox ultrastructural morphologies of mitochondria as a result of water uptake. During this transition the matrix becomes less electron dense while its volume increases [51]. In this way, Hackenbrock [51] in his classical work reported the first results on the relationship between mitochondrial physiology and morphodynamics (see Section 4.3). This early report established that mitochondria undergo a transition from condensed to orthodox conformation during the transition from state 3 (high respiration) to state 4 (low respiration). Paraphrasing Hackenbrock, this transition is effected by the expansion of an ultrastructurally organized protein reticular network rather than by random dispersion of proteins due to water dilution [51]. Osmotic swelling, either passive or active (the swollen state), results in expansion of the matrix and the outer mitochondrial membrane, eventually leading to its rupture [128, 129].

namics [60–62]. Mfn2 and Opa1 were shown to be involved in pathologies such as dominant optic atrophy and Charcot–Marie–Tooth disease [63].

Taken together, all these approaches show that continuous morphodynamic processes contribute not only to mitochondrial overall morphology but also to the cell's energetic capacity, intracellular signaling pathways [8], pathology [63], and cell apoptosis [59, 62].

4.4
The Key Role of Inner and Outer Membrane Ion Channels on Mitochondrial Network Dynamics

The controlled flux of ions through the inner and outer mitochondrial membranes is crucial for mitochondrial physiology. A decisive contribution of the chemiosmotic hypothesis [64] was to show that impermeability of the inner membrane to protons was a key property for energy transduction through the protonmotive force and its dissipation by the F_1F_0-ATPase to synthesize ATP. These findings from the golden era of mitochondrial energetics belong to the now classical view of mitochondria as the powerhouse of the cell.

Other ions (e.g., Ca^{2+}, K^+, Na^+) indirectly coupled to the proton gradient have been shown to be transported through a variety of ion channels recently described to be present on the inner and outer mitochondrial membranes [65–68]. The opening of these channels changes the degree of leakiness of the membrane and thus the degree of coupling of oxidative phosphorylation. More recent compelling evidence shows that following the effect of different stressors (due to age, illness, etc.), mitochondrial ion channels also may elicit the collapse of $\Delta\Psi_m$.

Several new features of the multifaceted role of mitochondria in cells have become increasingly clear in recent years. Overall, these new roles revolve around the ionic conduction properties through channels present in the inner and outer mitochondrial membranes. These channels can be divided according to their pro-life and pro-death effects on cells following different types of metabolic stress (see [29, 69] for recent reviews). Among them we have

- mitoK_{ATP}, an inner-membrane, ATP-sensitive K^+ channel activated by ischemia or K^+ channel openers that is thought to play a central role in protection against ischemic injury;
- mitoK_{Ca}, a toxin-sensitive (charybdotoxin, iberiotoxin) mitochondrial Ca^{2+}-activated K^+ channel that has been detected in the cardiac mitochondrial inner membrane and has properties similar to K_{Ca} channels found in the plasmalemma of other cell types;
- the Ca^{2+} uniporter, considered to be the major Ca^{2+} uptake pathway of the mitochondria, thus playing a role in the stimulation of NADH production and oxidative phosphorylation when workload increases;
- the permeability transition pore (PTP), a large, nonselective ion channel responsible for the catastrophic increase in inner membrane permeability induced by Ca^{2+} and/or ROS;
- the inner membrane anion channel (IMAC), which apparently is involved in energy dissipation and ROS transport under normal or oxidative stress conditions; and
- VDAC (or porin), an outer membrane channel whose changes in permeability have been implicated in cell death processes and activation of PTP, along with modulation of metabolite transport [29].

In addition to the rapid electrophoretic movement of cations and anions through mitochondrial ion channels (uniporters), a variety of metabolite transporters are present on the inner membrane and can mediate ion movements [70]. Some, such as the adenine nucleotide transporter (ANT), are electrogenic, while others are proton symports or antiports, thus contributing to energy dissipation across the membrane. In addition, several possible sources of proton leak, including that through uncoupling proteins (UCPs), must be taken into consideration.

From the point of view of new functions bestowed on mitochondria and cells by these channels, there are at least three unifying themes that culminate in the induction of slight, severe, or total collapse of $\Delta\Psi_m$:

1. Redox regulation mediated by ion channel opening that may involve changes in mitochondrial ROS production (e.g., mitoK$_{ATP}$-induced preconditioning).
2. Permeability transition induced by metabolic stressors (Ca^{2+} overload or severe energy depletion or oxidative stress) leading to cytochrome c release and the triggering of apoptosis.
3. Intracellular signaling elicited by calcium- and redox-dependent signaling mechanisms influencing transcription, translation, (de)phosphorylation cascades, and programmed cell death.

We will concentrate on the Ca^{2+}- and ROS-dependent signaling functions of the mitochondrial network that are dependent on some of the channels mentioned above.

4.5
Mitochondrial Network Behavior Associated with Intracellular Signaling

A common theme exhibited by network-associated mitochondrial functions concerns signaling. Mitochondria function as a source of signaling molecules, such as ROS, or as a target of second messengers such as Ca^{2+} or intermediary metabolites such as ADP [71–73]. All of these molecules function as intracellular messengers. Mitochondrial physiology is influenced by the dynamics of sequestering and exchanging (Ca^{2+} and ADP) or producing (ROS) these molecules.

The mitochondrial network properties are related to precise spatial and temporal delivery or reception of intracellular signals [74], such as with Ca^{2+} during EC coupling in heart cells [75]. Frequency- rather than amplitude-encoded signaling has been suggested for Ca^{2+} oscillations because frequency is more responsive to hormonal stimulation [76–78]. In theoretical simulations, Ca^{2+} sequestering by mitochondria influences the shape as well as the amplitude constancy of the oscillations [79]. Frequency-modulated Ca^{2+} signaling occurs in many cells (e.g., hepatocytes, endothelial and smooth muscle cells) and may regulate differential gene transcription and mitochondrial redox state [80, 81], among other intracellular processes [76].

Mitochondria represent the main site of ROS production in the cell [82]. Oxidative stress is a major pathophysiological route to the collapse of $\Delta\Psi_m$, but, when kept under control, ROS serve as important signaling molecules [83–88]. ROS-mediated signaling of gene expression is highly dependent on the ability of cells to maintain a reductive intracellular environment to counterbalance the highly oxidizing extracellular environment [83, 84]. By bringing about post-translational modification of key target proteins, or through the activation of redox-sensitive transcription factors, ROS influence the expression of multiple genes [86, 88, 89]. Numerous transcription factors are especially sensitive to ROS in biological responses such as inflammatory reactions, growth control, apoptosis, and cardiac disease [85, 87, 90–95]. Moreover, nuclear transcription factors can be activated either indirectly, via the abovementioned signal cascades, or directly by ROS [87], and recent evidence also suggests that mitochondrial ROS production may be important for the activation of hypoxia-inducible factor (HIF) [96].

ROS signaling can occur through a number of ROS-sensitive kinases [97, 98]. ROS can enhance the activation of FGF-, NGF-, and PDGF-β receptor tyrosine kinases and inhibit protein tyrosine phosphatase activity, leading to a net increase in protein tyrosine phosphorylation [84]. Other cytoplasmic signaling kinases such as mitogen-activated MAP kinases (e.g., p38, JNK, ERK), apoptosis-regulating kinase (ASK1), and several isoforms of protein kinase C (e.g., PKC-α), can be activated by ROS or by a pro-oxidative shift in the antioxidant capacity [99–101].

4.5.1
Mitochondria as Intracellular Timekeepers

Timing exerted by oscillatory mechanisms are found throughout the biological world, and their periods range from milliseconds, as in the action potential of neurons and the heart, to the slow evolutionary changes that require thousands of generations [50, 102]. In effect, the living world has been keeping track of time since the rise of the cyanobacterium more than 3 billion years ago [103].

One of the main interests of oscillations resides in the fact that they have been harnessed to provide rhythms, and where these rhythms are embedded so as to become heritable and provide anticipatory advantages, they turned into biological clocks [50, 102, 104, 105]. In virtually all light-sensitive organisms from cyanobacteria to humans, biological clocks adapt cyclic physiology to geophysical time with time-keeping properties in the circadian (24 h), ultradian (<24 h), and infradian (>24 h) domains [106]. Within this framework, it becomes clear why it is important to understand the synchronization of a population of coupled oscillators as an important problem for the dynamics of physiology in living systems.

Through frequency and amplitude modulation, oscillatory dynamics may function as a temporal-encoding signaling mechanism. So far, mitochondria have not been seriously considered as an organelle with the potential for active temporal encoding beyond the role of being a target of second messengers such as Ca^{2+}. This situation is changing, because mitochondria recently were shown experimentally and theoretically to be autonomous oscillators [9, 107], and physiological production of ROS may be required to sustain hormone-linked Ca^{2+} oscillations [108].

The idea that mitochondria may function as a coordinated network of oscillators emerged from studies in living cardiomyocytes subjected to metabolic stress [9, 24]. The network behavior of mitochondria depends on local as well as global coordination in the cell. ROS-induced ROS release [109, 110] is a mechanism that was recently shown in the cardiac mitochondrial network to exert both local and cell-wide influence during synchronized mitochondrial oscillations of cardiomyocytes under oxidative stress [9, 10].

The most relevant aspect of synchronization among coupled oscillators concerns temporally encoded signaling. The possibility of high-frequency, low-amplitude oscillations in mitochondrial ROS and $\Delta\Psi_m$ was predicted from simulations using our computational model of the mitochondrial oscillator (Fig. 4.4;

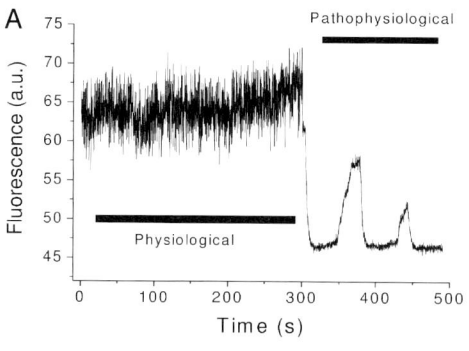

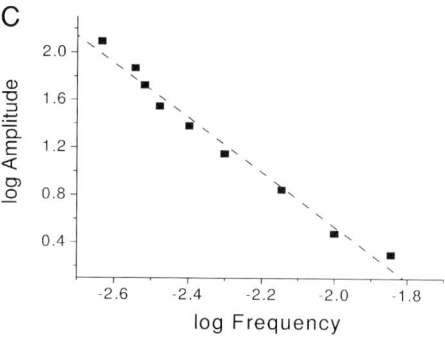

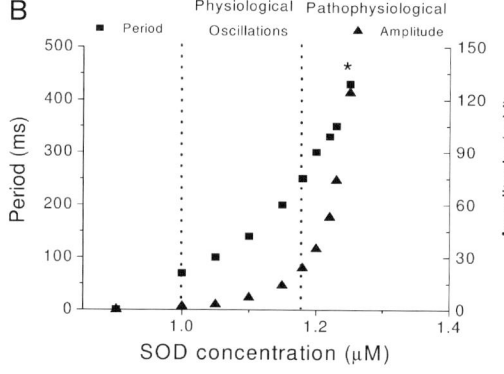

Fig. 4.4 Cardiac mitochondrial network oscillatory dynamics.

(A) Freshly isolated ventricular cardiomyocytes were loaded with 100 nM TMRM and imaged by two-photon microscopy (150-ms time resolution). The results obtained from a stack of 3720 images are shown. Before the mitochondrial network reaches criticality [10, 29, 69], the $\Delta\Psi_m$ (as measured by TMRM) oscillates at high frequencies and small amplitudes. After criticality, the network behavior evolves into "pathophysiological" behavior characterized by low-frequency, high-amplitude oscillations [9, 107].

(B) The modulation of the period and amplitude of oscillations through changes in the rate of ROS scavenging by superoxide dismutase (SOD) is shown. The simulations for shunt = 0.1, and SOD concentrations as shown in panel B were performed with the following set of parameters (see [32, 107] for detailed parameter descriptions):

concentration of respiratory chain carriers, $\rho^{REN} = 2.50 \times 10^{-6}$ mM; concentration of F_1F_0-ATPase, $\rho^{F1} = 2.03 \times 10^{-3}$ mM; $[Ca^{2+}]_i = 0.1$ μM; $K_{cc} = 0.01$ mM; $k_{SOD}^1 = 2.4 \times 10^6$ mM^{-1}s^{-1}; $k_{CAT}^1 = 1.7 \times 10^4$ mM^{-1}s^{-1}; $G_T = 0.5$ mM; maximal rate of the adenine nucleotide translocase, $V_{maxANT} = 5$ mM s^{-1}; and maximal rate of the mitochondrial Na-Ca exchanger, $V_{max}^{NaCa} = 0.015$ mM s^{-1}.

(C) The inverse power law behavior of the amplitude-versus-frequency relationship exhibited by the mitochondrial oscillator is shown. Oscillations were simulated with our computational model of the mitochondrial oscillator [107]. The double-log graph of the amplitude versus frequency (1/period) was plotted from $\Delta\Psi_m$ oscillations with amplitudes in the range of 2–124 mV and periods ranging from 70 ms to 430 ms. The simulations were performed as described in panel B.

see also next section). Recent experimental findings provide support for the hypothesis that mitochondrial oscillatory behavior may function as a frequency- and/or amplitude-encoded signaling mechanism under physiological conditions. Theoretical simulations indicate that the mitochondrial oscillator's period can be modulated over a wide range of time scales [26, 107], suggesting that it may play a role as an intracellular timekeeper (Fig. 4.4C). Although the frequency distribution is broad under normal conditions, the long-term temporal correlations of the mitochondrial network could theoretically allow a change in one time scale to be felt across the frequency range.

Winfree [111], Kuramoto [112], and Strogatz [113], among others (see [114] for a review), have addressed the problem of how hundreds or thousands of coupled oscillators achieve synchrony. A main finding arising from those studies is that synchronization occurs cooperatively from an initial nucleus where a few oscillators happen to sync and then recruit other oscillators, making the initial nucleus even larger and amplifying its signal [114]. After the initial nucleus achieves a threshold given by a critical mass of oscillators in phase, the population spontaneously self-synchronizes as in a phase transition [113]. In these pioneering studies, idealized mathematical structures took the place of the oscillators. It is now time that the oscillators become cells, genes [114], or, as in this work, mitochondria.

4.6
Mitochondria as a Network of Coupled Oscillators

A surprising recent finding is that cardiac mitochondria behave as a network of coupled oscillators not only in the pathophysiological but also in the physiological domain [26]. Under "physiological" conditions, mitochondrial $\Delta\Psi_m$ fluctuates at a high frequency within a restricted amplitude range, implying depolarizations of only microvolts to a few millivolts. Oxidative stress may elicit low-frequency, high-amplitude oscillations that characterize the "pathophysiological" response (Fig. 4.4A). The distribution of frequencies and amplitudes of oscillation can change at the junction between physiology and pathophysiology from an apparently "noisy" appearance to a highly correlated limit cycle–type oscillation, respectively.

The self-sustained and highly coordinated oscillations observed under pathophysiological conditions of oxidative stress [9] could be reproduced in a computational model of ROS-induced ROS release [107]. The model represents mitochondrial energetics [32] production, transport, and scavenging [107] and is mathematically described by a system of 15 ordinary differential equations (ODEs). This ODE system is able to exhibit Hopf bifurcations, a signature of the presence of stable limit-cycle (oscillatory) dynamics, under appropriate conditions [69, 107]. We became interested in determining whether the physiological domain also shows temporal correlations after model simulations revealed the potential for high-frequency, low-amplitude oscillations in mitochondria (Fig. 4.5).

A

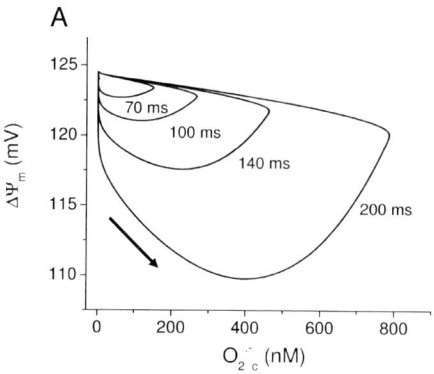

B

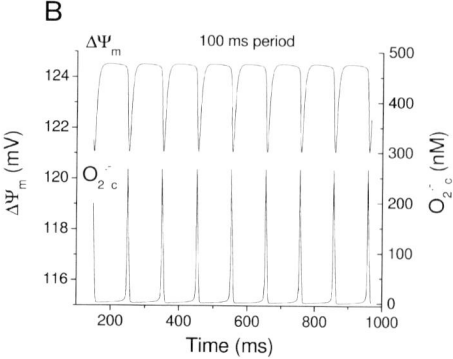

Fig. 4.5 Behavior of the mitochondrial oscillator in the high-frequency domain. Phase plane plot showing the relationship between $\Delta\Psi$m and the superoxide anion released to the periplasmic mitochondrial space, $O_2{}^{\bullet-}{}_c$, after each depolarization. (A) Oscillatory periods ranging from 70 ms to 200 ms showing the increasing ROS concentrations associated with the increase in amplitude as the frequency (1/period) decreases. (B) A time series of $\Delta\Psi$m and $O_2{}^{\bullet-}{}_c$ shown as an example. The simulations were performed as described in the caption of Fig. 4B.

By using relative dispersional analysis (RDA) and power spectral analysis (PSA) of time series from heart cells loaded with TMRM, reporting $\Delta\Psi_m$, and subjected to two-photon laser scanning microscopy, we found the following results.

- Under physiological conditions, the mitochondrial network exhibits a high temporal correlation for $\Delta\Psi_m$ that is different from random behavior (Fig. 4.6A). This is indicative of long-term memory.

- The power spectrum of such a distribution scales according to an inverse power law spanning at least three orders of magnitude (from milliseconds to a few minutes) and obeys a homogeneous power law $(1/f^\beta)$ with $\beta = 1.74$ (Fig. 4.6B) in the mid-range between pink $(\beta = 1.0)$ and brown $(\beta = 2.0)$ noise.

- Decreasing mitochondrial ROS production at the level of the respiratory chain, or blocking the ROS-induced ROS release mechanism by inhibiting the mitochondrial benzodiazepine receptor in the physiological domain, consistently diminished the extent of correlated behavior of the mitochondrial network in the high-frequency domain [115].

The inverse power law behavior in the mitochondrial network dynamics obeys a precise mechanism. In fact, according to the computational model, the double-log plot of amplitude versus oscillatory frequency exhibits an inverse relationship (Fig. 4.4C). Mechanistically, two key factors contribute to this dependence – the superoxide dismutase (SOD) activity and the balance between the rate of ROS production and scavenging.

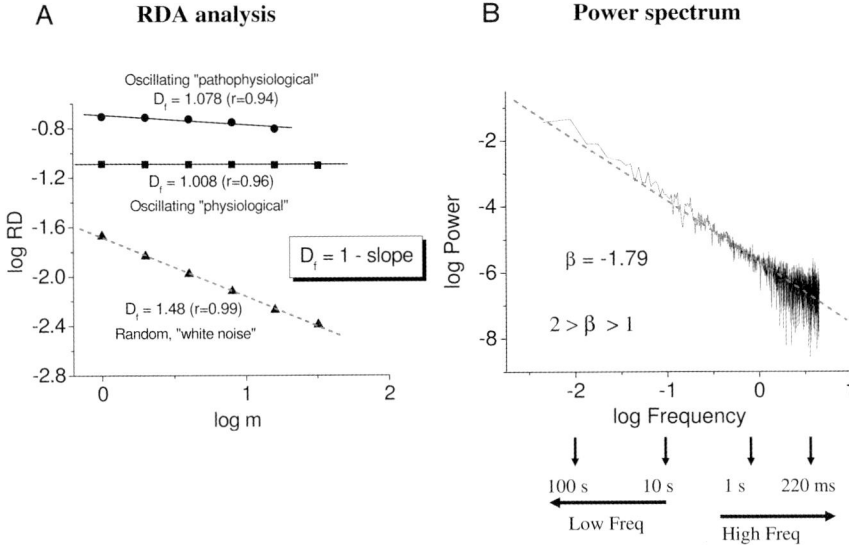

Fig. 4.6 Statistical analysis of $\Delta\Psi$m time series of the mitochondrial network of living cardiomyocytes loaded with TMRM showing that it functions as a highly correlated network of oscillators.
(A) RDA as a function of the aggregation parameter, m, gives a fractal dimension, D_f, close to 1.0 for myocytes showing large oscillations in $\Delta\Psi_m$ or those under physiological conditions, indicating a deterministic control process, while D_f is 1.5 for a completely random process.
(B) PSA of $\Delta\Psi$m time series after fast Fourier transform (FFT) also reveals a broad spectrum of oscillation in normally polarized mitochondria with a spectral exponent, $\beta = 1.79$, while a random process gives $\beta = 0$, meaning that there is no relationship between the amplitude and the frequency in a random signal.

The inverse power law behavior revealed by RDA and PSA also indicates a self-similar fractal process (Fig. 4.6). This corresponds to a broad range of frequencies and amplitudes spanning at least three orders of magnitude in time (from milliseconds to a few minutes). Scale-free architecture has been shown in network connectivity but not in dynamics. In fact, the mitochondrial network of cardiac cells appears as a regular lattice (in 2D and 3D), with each mitochondrion (node) exhibiting limit-cycle oscillations under certain conditions.

The scale-free behavior exhibited by mitochondrial network dynamics also allows simultaneous modulation of intracellular timekeeping in several time scales. Thus, two main features of mitochondria as metabolic "hubs" and as producers of ROS as signaling molecules with scale-free dynamics are in agreement with (1) the cardiac mitochondrial network having properties of both constancy and flexibility, i.e., providing a steady supply of ATP to fuel contraction, and adapting the rate of energy production to meet changing metabolic demands [72, 73]; and (2) achieving the right ROS balance in both amount and dynamics, compatible with intracellular signaling [1].

4.6.1
Spatial and Temporal Aspects of Synchronization

The finding that cardiac mitochondria lock to a dominant frequency and high-amplitude $\Delta\Psi_m$ oscillation under pathological conditions (e.g., ischemia–reperfusion) as a self-organized phase transition raises parallels between the mitochondrial network and many other physical, chemical, and engineered systems. Systems of disparate nature, when subjected to excessive loads, approach a *critical* state at which they become extremely sensitive to perturbations that can be efficiently propagated under these conditions. Because of the intrinsic nonlinear properties of mitochondria, new emergent macroscopic behavior appears, including spatiotemporal synchronization visualized as oscillations in energetics and waves of $\Delta\Psi_m$ depolarization.

As Winfree and Kuramoto discussed (reviewed in [30, 114]), we also observed the analogue of a phase transition at the turning point between the physiological and pathophysiological regimes in the mitochondrial network (Fig. 4.4A) [10, 69]. Experimentally, this global phase transition (visualized as a cell-wide mitochondrial depolarization) is attained when a critical density of mitochondria accumulates ROS above a threshold to form an extended spanning cluster. In fact, the spanning cluster in our work may be considered analogous to the nucleus of synchronized oscillators as described in the work of Winfree and Kuramoto. However, an important distinction arising from our studies is that in the percolation cluster any member of the network may initiate the $\Delta\Psi_m$ depolarization wave, instead of always arising from the same point as in Ca^{2+} waves.

We coined the term "mitochondrial criticality" to refer to the state of the system just before network depolarization [10, 69]. At this point, mitochondrial dynamics becomes rapidly unstable and pronounced transitions occur in all of the energetic state variables (i.e., $\Delta\Psi_m$, redox potential, ATP:ADP ratio, VO_2, etc.). This behavior involves a coordinated response of at least 60% of the mitochondrial population. These results are in good agreement with the quantitative predictions derived from percolation theory, especially concerning the percolation threshold, the fractal (spatial) organization exhibited by percolation processes at the threshold, and the critical exponents [10, 69]. Furthermore, we showed that this phase transition is associated with a depolarization wave traveling ~20 μm s^{-1} all throughout the mitochondrial network of the cell [10].

4.6.2
Mitochondrial Network Collapse and Scaling: A Cascade of Failures from the Bottom Up

When the normal function of the heart cell becomes severely compromised under pathophysiological conditions (e.g., ischemic injury) or metabolic stress (e.g., oxidative stress, substrate deprivation), the inherent vulnerability of the mitochondrial network is revealed. Several examples reported in the literature illustrate this point [116–120]. As pointed out above, we have shown that inverse power laws characterize the normal, physiological, and pathophysiological behavior of the mitochondrial network. A profound implication of the power law behavior exhibited by the mitochondrial network is that beyond a critical level, the loss of a single element may cause the entire network to fail because of their multiplicative interdependency [45, 121, 122]. The loss of mitochondrial function through collapse of $\Delta\Psi_m$ following metabolic stress directly affects the sarcolemmal K_{ATP} channel and alters the cellular action potential (AP) [123]. The rapid uncoupling of oxidative phosphorylation during depolarization of $\Delta\Psi_m$ is closely linked to the activation of sarcolemmal K_{ATP} currents, consequently shortening the cellular AP and rendering the myocyte electrically inexcitable during the nadir of $\Delta\Psi_m$ oscillation [9]. This mechanism recently was shown to contribute to the destabilization of AP repolarization during reperfusion after ischemic injury in the whole heart, leading to arrhythmias [27]. This confirms the prediction derived from the inherent organization of mitochondria as a network, i.e., that failures can *scale* to higher levels of organization [29, 69] and lead to fatal arrhythmias.

4.7
Discussion

The present work emphasizes the view of mitochondria as a network whose organization in spatial and dynamic terms is crucial for understanding its function in cellular and tissue contexts. This framework brings about a novel and, hopefully, fruitful exploration path of mitochondrial function at the cellular level during normal and pathophysiological conditions.

Our exploration of cardiac mitochondrial network behavior and organization was performed with the following tools:

- state-of-the-art two-photon scanning laser microscopy as applied to the study of mitochondrial function in living cells;
- nonlinear dynamics as applied to the computational modeling and experimental study of spatiotemporal self-organization (e.g., waves, oscillations);
- statistical analysis (relative dispersional analysis) and signal processing of time series (fast Fourier transform and power spectral analysis) in the fast temporal domain from physiological variables of relevance in mitochondrial and cellular function (e.g., $\Delta\Psi m$); and
- fractal analysis of spatial patterns emerging from network physiology in the framework of physical theories such as two-dimensional percolation to interpret critical phenomena exhibited by mitochondria.

4.7.1
What Is the Contribution of the Network View of Mitochondrial Function?

By employing all the techniques mentioned above in hypothesis-driven research, we found that cardiac mitochondria behave as a network of coupled oscillators. In cardiac mitochondrial and cell physiology, this finding is relevant for understanding the transition between physiological and pathophysiological behaviors in the mitochondrial network. Moreover, and in contrast to the results obtained from previous models of synchronized oscillators, our finding of an inverse power law by RDA and PSA in the physiological regime indicates that despite their weak coupling, the oscillators' population does not behave incoherently or randomly, as predicted by previous models [111, 114]. If that were the case, we should have found pure (or close to pure) white noise in the physiological regime (Fig. 4.4A). Instead, we found a broad band of frequencies proceeding simultaneously at different time scales, tied by an inverse power law different from random behavior (Fig. 4.6B). This result is physiologically meaningful in at least two ways.

1. The interconnectedness inferred from the power law behavior exhibited by the mitochondrial network explains why beyond a critical level (ROS), the loss of a single element ($\Delta\Psi$m depolarization of a mitochondrion) may cause the entire (mitochondrial) network to fail because of their multiplicative interdependency (ROS-induced ROS release in the mitochondrial spanning cluster). Such an inherent organization of mitochondria predicts that failures can scale to higher levels of organization [29, 69], as has been shown to be the case of mitochondrial criticality in heart cells under oxidative stress [9, 10] or the whole organ after ischemia–reperfusion [27].

2. It is possible that the mitochondrial network functions as a frequency- and amplitude-modulated signaling system, sensitive to relevant physiological variables such as those related to ROS scavenging (Figs. 4.4B and 4.5). This result may be relevant for understanding why the correct balance between ROS and antioxidants should be in place for transcription factors to be sensitive to changes in the cellular redox status [1].

It has been proposed that the basic idea of the scaling hypothesis is that the long-range correlations in the fluctuations of a variable such as $\Delta\Psi$m are due to the existence of a critical point that induces local instabilities in the system dynamics while retaining global stability [121]. In the case of the cardiac mitochondrial network, the local instability is produced through the basic mechanism of the mitochondrial oscillator, involving ROS production and scavenging mechanisms as bifurcation parameters. Under physiological conditions, the network retains global stability while exhibiting high-frequency, low-amplitude oscillations. With metabolic stress, a spanning mitochondrial cluster develops and, at criticality, loses global stability, resulting in a transition between physiological and pathophysiological behaviors, i.e., a prelude to cell death and fatal arrhythmias.

4.7.2
Why a Network View of Mitochondria May Contribute to Understanding Aging and Illness

As pointed out in this review, the signaling properties of mitochondria, through ROS production and its relationship with Ca^{2+} [108, 124], may be intimately related to their network (and oscillatory) behavior. We propose that mitochondrial network dynamics contribute to a sophisticated signaling modality able to serve as a timekeeper or a finely tuned graded-response mechanism accomplished by modulating the amplitude, frequency, and synchronization of the intrinsic oscillatory response (Fig. 4.4B,C). In this respect, recognizing the dynamic component of mitochondrial ROS signaling is necessary to advance the so-called mitochondrial theory of aging [1, 125]. Instead of a monotonic accumulation of ROS with age, culminating in an increasing incidence of mutations and molecular damage [125], intracellular ROS bursting may occur abruptly when mitochondrial criticality is reached, as a result of either scavenger depletion or increased ROS production. It is therefore imperative to understand how mitochondrial network organization contributes to the crucial transition to necrotic or apoptotic stages during aging and disease.

The incidence of cardiovascular disease, cancer, and neurodegenerative (e.g., Alzheimer's, Parkinson's) diseases increases exponentially within the period of life between 45 and 65 years [2, 89]. It has been shown that the same mutations in nuclear genes cause the same age-related (neurodegenerative) diseases in different species [3]. In these cases, the rate of illness progression appears to be strongly correlated with the rate of ROS production by mitochondria (i.e., in the species that produce ROS quickly, the disease sets in early and progresses more quickly) [3].

4.8
Concluding Remarks

The study of mitochondrial physiology has progressed from exploring their established role as the cell's powerhouse to uncovering emerging functions in the delivery of signals crucial for cell survival or death. Based on the network view of mitochondrial function, we present a new framework for better understanding this evolving field that emphasizes the spatial and temporal organization of mitochondria in cells. A main proposal arising from this view is that mitochondrial network dynamics may constitute a sophisticated intracellular timekeeper through amplitude- and frequency-encoded signaling of ROS. However, the intrinsic organization of the mitochondria as a network of coupled oscillators can be the basis for instability and catastrophic failure under stress. This viewpoint is relevant for understanding the dynamic component of mitochondrial ROS production and signaling as a crucial link to disease and aging.

References

1 Lane, N. (2005) *Power, sex, suicide. Mitochondria and the meaning of life*, Oxford University Press, Oxford.

2 Finkel, T. (2005) Opinion: Radical medicine: treating ageing to cure disease. *Nat Rev Mol Cell Biol* 6, 971–6.

3 Wright, A. F., Jacobson, S. G., Cideciyan, A. V., Roman, A. J., Shu, X., Vlachantoni, D., McInnes, R. R., Riemersma, R. A. (2004) Lifespan and mitochondrial control of neurode-generation. *Nat Genet 36*, 1153–8.

4 Barabasi, A. L., Oltvai, Z. N. (2004) Network biology: understanding the cell's functional organization. *Nat Rev Genet 5*, 101–13.

5 Chan, D. C. (2006) Mitochondria: dynamic organelles in disease, aging, and development. *Cell 125*, 1241–52.

6 Mannella, C. A. (2006) The relevance of mitochondrial membrane topology to mitochondrial function. *Biochim Biophys Acta 1762*, 140–7.

7 Dimmer, K. S., Scorrano, L. (2006) (De)constructing mitochondria: what for? *Physiology (Bethesda) 21*, 233–41.

8 McBride, H. M., Neuspiel, M., Wasiak, S. (2006) Mitochondria: more than just a powerhouse. *Curr Biol 16*, R551–60.

9 Aon, M. A., Cortassa, S., Marban, E., O'Rourke, B. (2003) Synchronized whole cell oscillations in mitochondrial metabolism triggered by a local release of reactive oxygen species in cardiac myocytes. *J Biol Chem 278*, 44735–44.

10 Aon, M. A., Cortassa, S., O'Rourke, B. (2004) Percolation and criticality in a mitochondrial network. *Proc Natl Acad Sci USA 101*, 4447–52.

11 Vendelin, M., Beraud, N., Guerrero, K., Andrienko, T., Kuznetsov, A. V., Olivares, J., Kay, L., Saks, V. A. (2005) Mitochondrial regular arrangement in muscle cells: a "crystal-like" pattern. *Am J Physiol Cell Physiol 288*, C757–67.

12 Collins, T. J., Berridge, M. J., Lipp, P., Bootman, M. D. (2002) Mitochondria are morphologically and functionally heterogeneous within cells. *Embo J 21*, 1616–27.

13 Dedov, V. N., Roufogalis, B. D. (1999) Organisation of mitochondria in living sensory neurons. *FEBS Lett 456*, 171–4.

14 Skulachev, V. P., Bakeeva, L. E., Chernyak, B. V., Domnina, L. V., Minin, A. A., Pletjushkina, O. Y., Saprunova, V. B., Skulachev, I. V., Tsyplenkova, V. G., Vasiliev, J. M., Yaguzhinsky, L. S., Zorov, D. B. (2004) Thread-grain transition of mitochondrial reticulum as a step of

mitoptosis and apoptosis. *Mol Cell Biochem 256–257*, 341–58.

15 Palmer, J. W., Tandler, B., Hoppel, C. L. (1977) Biochemical properties of subsarcolemmal and interfibrillar mitochondria isolated from rat cardiac muscle. *J Biol Chem 252*, 8731–9.

16 Kuznetsov, A. V., Troppmair, J., Sucher, R., Hermann, M., Saks, V., Margreiter, R. (2006) Mitochondrial subpopulations and heterogeneity revealed by confocal imaging: possible physiological role? *Biochim Biophys Acta 1757*, 686–691.

17 Bakeeva, L. E., Chentsov Yu, S., Skula-chev, V. P. (1983) Intermitochondrial contacts in myocardiocytes. *J Mol Cell Cardiol 15*, 413–20.

18 Amchenkova, A. A., Bakeeva, L. E., Chentsov, Y. S., Skulachev, V. P., Zorov, D. B. (1988) Coupling membranes as energy-transmitting cables. I. Filamentous mitochondria in fibroblasts and mitochondrial clusters in cardiomyocytes. *J Cell Biol 107*, 481–95.

19 Duchen, M. R., Leyssens, A., Crompton, M. (1998) Transient mitochondrial depolarizations reflect focal sarcoplasmic reticular calcium release in single rat cardiomyocytes. *J Cell Biol 142*, 975–88.

20 Juhaszova, M., Zorov, D. B., Kim, S. H., Pepe, S., Fu, Q., Fishbein, K. W., Ziman, B. D., Wang, S., Ytrehus, K., Antos, C. L., Olson, E. N., Sollott, S. J. (2004) Glycogen synthase kinase-3beta mediates convergence of protection signaling to inhibit the mitochondrial permeability transition pore. *J Clin Invest 113*, 1535–49.

21 Loew, L. M., Tuft, R. A., Carrington, W., Fay, F. S. (1993) Imaging in five dimensions: time-dependent membrane potentials in individual mitochondria. *Biophys J 65*, 2396–407.

22 Aon, M. A., Cortassa, S. (1997) *Dynamic biological organization. Fundamentals as applied to cellular systems*, Chapman & Hall, London.

23 Cortassa, S., Aon, M. A. (1994) Spatio-temporal regulation of glycolysis and oxidative phosphorylation *in vivo* in tumor and yeast cells. *Cell Biol Int 18*, 687–713.

24 Romashko, D. N., Marban, E., O'Rourke, B. (1998) Subcellular metabolic transients and mitochondrial redox waves in heart cells. *Proc Natl Acad Sci USA 95*, 1618–23.

25 Saks, V. A., Kaambre, T., Sikk, P., Eimre, M., Orlova, E., Paju, K., Piirsoo, A., Appaix, F., Kay, L., Regitz-Zagrosek, V., Fleck, E., Seppet, E. (2001) Intracellular energetic units in red muscle cells. *Biochem J 356*, 643–57.

26 Aon, M. A., Cortassa, S. C., O'Rourke, B. (2006). The fundamental organization of cardiac mitochondria as a network of coupled oscillators. *Biophys J 91*, 4317–4327.

27 Akar, F. G., Aon, M. A., Tomaselli, G. F., O'Rourke, B. (2005) The mitochondrial origin of postischemic arrhythmias. *J Clin Invest 115*, 3527–35.

28 Aon, M. A., O'Rourke, B., Cortassa, S. (2004) The fractal architecture of cytoplasmic organization: scaling, kinetics and emergence in metabolic networks. *Mol Cell Biochem 256–257*, 169–84.

29 O'Rourke, B., Cortassa, S., Aon, M. A. (2005) Mitochondrial ion channels: Gatekeepers of life and death. *Physiology 20*, 303–315.

30 Strogatz, S. H. (2001) Exploring complex networks. *Nature 410*, 268–76.

31 Albert, R. (2005) Scale-free networks in cell biology. *J Cell Sci 118*, 4947–57.

32 Cortassa, S., Aon, M. A., Marban, E., Winslow, R. L., O'Rourke, B. (2003) An integrated model of cardiac mitochondrial energy metabolism and calcium dynamics. *Biophys J 84*, 2734–55.

33 Ingwall, J. S., Weiss, R. G. (2004) Is the failing heart energy starved? On using chemical energy to support cardiac function. *Circ Res 95*, 135–45.

34 Porter, K. R., Palade, G. E. (1957) Studies on the endoplasmic reticulum. III. Its form and distribution in striated muscle cells. *J Biophys Biochem Cytol 3*, 269–300.

35 Baumeister, W. (2002) Electron tomography: towards visualizing the molecular organization of the cytoplasm. *Curr Opin Struct Biol 12*, 679–84.

36 Frank, J., Wagenknecht, T., McEwen, B. F., Marko, M., Hsieh, C. E., Mannella, C. A. (2002) Three-dimensional imaging of biological complexity. *J Struct Biol 138*, 85–91.

37 Crosta, G. F., Urani, C., Fumarola, L. (2006) Classifying structural alterations of the cytoskeleton by spectrum enhancement and descriptor fusion. *J Biomed Opt 11*, 024020.

38 Dey, P. (2005) Basic principles and applications of fractal geometry in pathology: a review. *Anal Quant Cytol Histol 27*, 284–90.

39 Erdos, P., Renyi, A. (1960) On the evolution of random graphs. *Publ. Math. Inst. Hung. Acad. Sci. 5*, 17–61.

40 Watts, D. J., Strogatz, S. H. (1998) Collective dynamics of 'small-world' networks. *Nature 393*, 440–2.

41 Almaas, E., Kovacs, B., Vicsek, T., Oltvai, Z. N., Barabasi, A. L. (2004) Global organization of metabolic fluxes in the bacterium Escherichia coli. *Nature 427*, 839–43.

42 Jeong, H., Tombor, B., Albert, R., Oltvai, Z. N., Barabasi, A. L. (2000) The large-scale organization of metabolic networks. *Nature 407*, 651–4.

43 Papin, J. A., Hunter, T., Palsson, B. O., Subramaniam, S. (2005) Reconstruction of cellular signalling networks and analysis of their properties. *Nat Rev Mol Cell Biol 6*, 99–111.

44 Wagner, A., Fell, D. A. (2001) The small world inside large metabolic networks. *Proc Biol Sci 268*, 1803–10.

45 Barabasi, A. L. (2003) *Linked*, Plume, New York.

46 Varela, F., Maturana, H., Uribe, R. (1974) Autopoiesis: the organization of living sytems, its characterization and a model. *Biosystems 5*, 187–196.

47 Nicolis, G., Prigogine, I. (1977) *Self-organization in nonequilibrium systems: from dissipative structures to order through fluctuations*, Wiley, New York.

48 Capra, F. (1996) *The web of life*, Anchor books Doubleday, New York.

49 Aon, M. A., Cortassa, S. (2006) *Metabolic dynamics in cells viewed as multilayered, distributed, mass-energy-information networks*, Vol. Part 3. Proteomics. 3.8 Systems Biology, Wiley Interscience.

50 Lloyd, D., Aon, M. A., Cortassa, S. (2001) Why homeodynamics, not

homeostasis? *ScientificWorldJournal 1*, 133–45.

51 Hackenbrock, C. R. (1968) Chemical and physical fixation of isolated mitochondria in low-energy and high-energy states. *Proc Natl Acad Sci USA 61*, 598–605.

52 Mannella, C. A., Pfeiffer, D. R., Bradshaw, P. C., Moraru, II, Slepchenko, B., Loew, L. M., Hsieh, C. E., Buttle, K., Marko, M. (2001) Topology of the mitochondrial inner membrane: dynamics and bioenergetic implications. *IUBMB Life 52*, 93–100.

53 Lloyd, D., Salgado, L. E., Turner, M. P., Suller, M. T., Murray, D. (2002) Cycles of mitochondrial energization driven by the ultradian clock in a continuous culture of Saccharomyces cerevisiae. *Microbiology 148*, 3715–24.

54 Rossignol, R., Gilkerson, R., Aggeler, R., Yamagata, K., Remington, S. J., Capaldi, R. A. (2004) Energy substrate modulates mitochondrial structure and oxidative capacity in cancer cells. *Cancer Res 64*, 985–93.

55 Yi, M., Weaver, D., Hajnoczky, G. (2004) Control of mitochondrial motility and distribution by the calcium signal: a homeostatic circuit. *J Cell Biol 167*, 661–72.

56 Anesti, V., Scorrano, L. (2006) The relationship between mitochondrial shape and function and the cytoskeleton. *Biochim Biophys Acta 1757*, 692–9.

57 Chen, H., Chan, D. C. (2006) Critical dependence of neurons on mitochondrial dynamics. *Curr Opin Cell Biol 18*, 453–9.

58 Twig, G., Graf, S. A., Wikstrom, J. D., Mohamed, H., Haigh, S. E., Elorza, A., Deutsch, M., Zurgil, N., Reynolds, N., Shirihai, O. S. (2006) Tagging and tracking individual networks within a complex mitochondrial web with photoactivatable GFP. *Am J Physiol Cell Physiol 291*, C176–84.

59 Frezza, C., Cipolat, S., Martins de Brito, O., Micaroni, M., Beznoussenko, G. V., Rudka, T., Bartoli, D., Polishuck, R. S., Danial, N. N., De Strooper, B., Scorrano, L. (2006) OPA1 controls apoptotic cristae remodeling independently from mitochondrial fusion. *Cell 126*, 177–89.

60 Cipolat, S., Martins de Brito, O., Dal Zilio, B., Scorrano, L. (2004) OPA1 requires mitofusin 1 to promote mitochondrial fusion. *Proc Natl Acad Sci USA 101*, 15927–32.

61 Pich, S., Bach, D., Briones, P., Liesa, M., Camps, M., Testar, X., Palacin, M., Zorzano, A. (2005) The Charcot-Marie-Tooth type 2A gene product, Mfn2, up-regulates fuel oxidation through expression of OXPHOS system. *Hum Mol Genet 14*, 1405–15.

62 Youle, R. J., Karbowski, M. (2005) Mitochondrial fission in apoptosis. *Nat Rev Mol Cell Biol 6*, 657–63.

63 Bach, D., Naon, D., Pich, S., Soriano, F. X., Vega, N., Rieusset, J., Laville, M., Guillet, C., Boirie, Y., Wallberg-Henriksson, H., Manco, M., Calvani, M., Castagneto, M., Palacin, M., Mingrone, G., Zierath, J. R., Vidal, H., Zorzano, A. (2005) Expression of Mfn2, the Charcot-Marie-Tooth neuropathy type 2A gene, in human skeletal muscle: effects of type 2 diabetes, obesity, weight loss, and the regulatory role of tumor necrosis factor alpha and interleukin-6. *Diabetes 54*, 2685–93.

64 Mitchell, P. (1961) Coupling of phosphorylation to electron and hydrogen transfer by a chemi-osmotic type of mechanism. *Nature 191*, 144–8.

65 Gunter, T. E., Pfeiffer, D. R. (1990) Mechanisms by which mitochondria transport calcium. *Am J Physiol 258*, C755–86.

66 O'Rourke, B. (2000) Pathophysiological and protective roles of mitochondrial ion channels. *J Physiol 529 Pt 1*, 23–36.

67 O'Rourke, B. (2004) Evidence for mitochondrial K+ channels and their role in cardioprotection. *Circ Res 94*, 420–32.

68 O'Rourke, B. (2006) Mitochondrial Ion Channels. *Annu Rev Physiol.*

69 Aon, M. A., Cortassa, S., Akar, F. G., O'Rourke, B. (2006) Mitochondrial criticality: A new concept at the turning point of life or death. *Biochim Biophys Acta 1762*, 232–40.

70 Palmieri, F. (2004) The mitochondrial transporter family (SLC25): physiological and pathological implications. *Pflugers Arch 447*, 689–709.

71 Balaban, R. S. (2002) Cardiac energy metabolism homeostasis: role of cytosolic calcium. *J Mol Cell Cardiol 34*, 1259–71.

72 Cortassa, S. C., Aon, M. A., O'Rourke, B., Jacques, R., Tseng, H. J., Marban, E., Winslow, R. L. (2006) A computational model integrating electrophysiology, contraction and mitochondrial bioenergetics in the ventricular myocyte. *Biophys J. 91*, 1564–89.

73 Saks, V., Dzeja, P., Schlattner, U., Vendelin, M., Terzic, A., Wallimann, T. (2006) Cardiac system bioenergetics: metabolic basis of the Frank-Starling law. *J Physiol 571*, 253–73.

74 Babcock, D. F., Hille, B. (1998) Mitochondrial oversight of cellular Ca2+ signaling. *Curr Opin Neurobiol 8*, 398–404.

75 Mackenzie, L., Roderick, H. L., Berridge, M. J., Conway, S. J., Bootman, M. D. (2004) The spatial pattern of atrial cardiomyocyte calcium signalling modulates contraction. *J Cell Sci 117*, 6327–37.

76 Berridge, M. J., Bootman, M. D., Roderick, H. L. (2003) Calcium signalling: dynamics, homeostasis and remodelling. *Nat Rev Mol Cell Biol 4*, 517–29.

77 Meyer, T., Stryer, L. (1988) Molecular model for receptor-stimulated calcium spiking. *Proc Natl Acad Sci USA 85*, 5051–5.

78 Schuster, S., Marhl, M., Hofer, T. (2002) Modelling of simple and complex calcium oscillations. From single-cell responses to intercellular signalling. *Eur J Biochem 269*, 1333–55.

79 Marhl, M., Schuster, S., Brumen, M. (1998) Mitochondria as an important factor in the maintenance of constant amplitudes of cytosolic calcium oscillations. *Biophys Chem 2*, 125–32.

80 Hajnoczky, G., Robb-Gaspers, L. D., Seitz, M. B., Thomas, A. P. (1995) Decoding of cytosolic calcium oscillations in the mitochondria. *Cell 82*, 415–24.

81 Maack, C., Cortassa, S., Aon, M. A., Ganesan, A. N., Liu, T., O'Rourke, B. (2006) Elevated cytosolic Na+ decreases mitochondrial Ca2+ uptake during excitation-contraction coupling and impairs energetic adaptation in cardiac myocytes. *Circ Res. 99*, 172–182.

82 Turrens, J. F. (2003) Mitochondrial formation of reactive oxygen species. *J Physiol 552*, 335–44.

83 Cadenas, E. (2004) Mitochondrial free radical production and cell signaling. *Mol Aspects Med 25*, 17–26.

84 Droge, W. (2002) Free radicals in the physiological control of cell function. *Physiol Rev 82*, 47–95.

85 Haddad, J. J. (2004) Oxygen sensing and oxidant/redox-related pathways. *Biochem Biophys Res Commun 316*, 969–77.

86 Marin-Garcia, J., Goldenthal, M. J. (2004) Heart mitochondria signaling pathways: appraisal of an emerging field. *J Mol Med 82*, 565–78.

87 Morel, Y., Barouki, R. (1999) Repression of gene expression by oxidative stress. *Biochem J 342 Pt 3*, 481–96.

88 Storz, P. (2006) Reactive oxygen species-mediated mitochondria-to-nucleus signaling: a key to aging and radical-caused diseases. *Sci STKE 2006*, re3.

89 Finkel, T., Holbrook, N. J. (2000) Oxidants, oxidative stress and the biology of ageing. *Nature 408*, 239–47.

90 Bao, J., Sato, K., Li, M., Gao, Y., Abid, R., Aird, W., Simons, M., Post, M. J. (2001) PR-39 and PR-11 peptides inhibit ischemia-reperfusion injury by blocking proteasome-mediated I kappa B alpha degradation. *Am J Physiol Heart Circ Physiol 281*, H2612–8.

91 Kis, A., Yellon, D. M., Baxter, G. F. (2003) Role of nuclear factor-kappa B activation in acute ischaemia-reperfusion injury in myocardium. *Br J Pharmacol 138*, 894–900.

92 Maulik, N., Goswami, S., Galang, N., Das, D. K. (1999) Regulation of cardiomyocyte apoptosis by redox-sensitive transcription factors. *FEBS Lett 443*, 331–6.

93 Misra, A., Haudek, S. B., Knuefermann, P., Vallejo, J. G., Chen, Z. J., Michael, L. H., Sivasubramanian, N., Olson, E. N., Entman, M. L., Mann, D. L. (2003) Nuclear factor-kappaB protects the adult cardiac myocyte against ischemia-induced apoptosis in a murine model of acute myocardial infarction. *Circulation 108*, 3075–8.

94 Sawa, Y., Morishita, R., Suzuki, K., Kagisaki, K., Kaneda, Y., Maeda, K., Kadoba, K., Matsuda, H. (1997) A novel strategy for myocardial protection using *in vivo* transfection of cis element 'decoy' against NFkappaB binding site: evidence for a role of NFkappaB in ischemia-reperfusion injury. *Circulation* 96, II-280-4; discussion II-285.

95 Waris, G., Ahsan, H. (2006) Reactive oxygen species: role in the development of cancer and various chronic conditions. *J Carcinog 5*, 14.

96 Schroedl, C., McClintock, D. S., Budinger, G. R., Chandel, N. S. (2002) Hypoxic but not anoxic stabilization of HIF-1alpha requires mitochondrial reactive oxygen species. *Am J Physiol Lung Cell Mol Physiol 283*, L922–31.

97 Storz, P., Doppler, H., Toker, A. (2005) Protein kinase D mediates mitochondrion-to-nucleus signaling and detoxification from mitochondrial reactive oxygen species. *Mol Cell Biol 25*, 8520–30.

98 Storz, P., Toker, A. (2003) Protein kinase D mediates a stress-induced NF-kappaB activation and survival pathway. *Embo J 22*, 109–20.

99 Gopalakrishna, R., Anderson, W. B. (1989) Ca^{2+}- and phospholipid-independent activation of protein kinase C by selective oxidative modification of the regulatory domain. *Proc Natl Acad Sci USA 86*, 6758–62.

100 Ono, Y., Fujii, T., Igarashi, K., Kuno, T., Tanaka, C., Kikkawa, U., Nishizuka, Y. (1989) Phorbol ester binding to protein kinase C requires a cysteine-rich zinc-finger-like sequence. *Proc Natl Acad Sci USA 86*, 4868–71.

101 Quest, A. F., Bardes, E. S., Bell, R. M. (1994) A phorbol ester binding domain of protein kinase C gamma. Deletion analysis of the Cys2 domain defines a minimal 43-amino acid peptide. *J Biol Chem 269*, 2961–70.

102 Lloyd, D. (1998) Circadian and ultradian clock-controlled rhythms in unicellular microorganisms. *Adv Microb Physiol 39*, 291–338.

103 Editorial, B. (2000) Biological timing mechanisms: a thematic special issue for the new millennium. *BioEssays 22*, 1–2.

104 Edmunds, L. N., Jr. (1988) *Cellular and molecular basis of biological clocks. Models and mechanisms for circadian timekeeping*, Springer-Verlag, New York.

105 Sweeney, B. M., Hastings, J. W. (1960) Effects of temperature upon diurnal rhythms. *Cold Spring Harb Symp Quant Biol 25*, 87–104.

106 Schibler, U., Naef, F. (2005) Cellular oscillators: rhythmic gene expression and metabolism. *Curr Opin Cell Biol 17*, 223–9.

107 Cortassa, S., Aon, M. A., Winslow, R. L., O'Rourke, B. (2004) A mitochondrial oscillator dependent on reactive oxygen species. *Biophys J 87*, 2060–73.

108 Camello-Almaraz, M. C., Pozo, M. J., Murphy, M. P., Camello, P. J. (2006) Mitochondrial production of oxidants is necessary for physiological calcium oscillations. *J Cell Physiol 206*, 487–94.

109 Zorov, D. B., Filburn, C. R., Klotz, L. O., Zweier, J. L., Sollott, S. J. (2000) Reactive oxygen species (ROS)-induced ROS release: a new phenomenon accompanying induction of the mitochondrial permeability transition in cardiac myocytes. *J Exp Med 192*, 1001–14.

110 Zorov, D. B., Juhaszova, M., Sollott, S. J. (2006) Mitochondrial ROS-induced ROS release: An update and review. *Biochim Biophys Acta 1757*, 509–17.

111 Winfree, A. T. (1967) Biological rhythms and the behavior of populations of coupled oscillators. *J Theor Biol 16*, 15–42.

112 Kuramoto, Y. (1984) *Chemical oscillations, waves, and turbulence*, Springer-Verlag, Berlin.

113 Strogatz, S. H. (2000) From Kuramoto to Crawford: exploring the onset of synchronization in population of coupled oscillators. *Physica D 143*, 1–20.

114 Strogatz, S. H. (2003) *Sync. The emerging science of spontaneous order*, Hyperion books, New York.

115 Aon, M. A., Cortassa, S., O'Rourke, B. (2006) The fundamental organization of cardiac mitochondria as a network of coupled oscillators. *Biophys J 91*, 4317–27.

116 Crompton, M. (1999) The mitochondrial permeability transition pore and its role in cell death. *Biochem J 341 (Pt 2)*, 233–49.

117 Duchen, M. R. (1999) Contributions of mitochondria to animal physiology: from homeostatic sensor to calcium signalling and cell death. *J Physiol 516 (Pt 1)*, 1–17.

118 Huser, J., Blatter, L. A. (1999) Fluctuations in mitochondrial membrane potential caused by repetitive gating of the permeability transition pore. *Biochem J 343 Pt 2*, 311–7.

119 Ichas, F., Jouaville, L. S., Mazat, J. P. (1997) Mitochondria are excitable organelles capable of generating and conveying electrical and calcium signals. *Cell 89*, 1145–53.

120 Siemens, A., Walter, R., Liaw, L. H., Berns, M. W. (1982) Laser-stimulated fluorescence of submicrometer regions within single mitochondria of rhodamine-treated myocardial cells in culture. *Proc Natl Acad Sci USA 79*, 466–70.

121 West, B. J. (1999) *Physiology, promiscuity and prophecy at The Millennium: A tale of tails*, Vol. 7, World Scientific, Singapore.

122 West, B. J., Deering, B. (1995) *The Lure of Modern Science. Fractal Thinking*, Vol. 3, World Scientific, Singapore.

123 O'Rourke, B., Ramza, B. M., Marban, E. (1994) Oscillations of membrane current and excitability driven by metabolic oscillations in heart cells. *Science 265*, 962–6.

124 Brookes, P. S., Yoon, Y., Robotham, J. L., Anders, M. W., Sheu, S. S. (2004) Calcium, ATP, and ROS: a mitochondrial love-hate triangle. *Am J Physiol Cell Physiol 287*, C817–33.

125 Harman, D. (1956) Aging: a theory based on free radical and radiation chemistry. *J Gerontol 11*, 298–300.

126 Mathews, C. K., van Holde, K. E., Ahern, K. G. (2000) *Biochemistry*, Addison Wesley Longman, Inc., San Francisco, CA.

127 Nelson, D. L., Cox, M. M. (2005) *Lehninger principles of biochemistry*, Fourth ed., W.H. Freeman, New York.

128 Hunter, G. R., Kamishima, Y., Brierley, G. P. (1969) Ion transport by heart mitochondria. XV. Morphological changes associated with the penetration of solutes into isolated heart mitochondria. *Biochim Biophys Acta 180*, 81–97.

129 Stoner, C. D., Sirak, H. D. (1969) Osmotically-induced alterations in volume and ultrastructure of mitochondria isolated from rat liver and bovine heart. *J Cell Biol 43*, 521–38.

5

Structural Organization and Dynamics of Mitochondria in the Cells *in Vivo*

Andrey V. Kuznetsov

Abstract

As is observed in mature cardiac and skeletal muscle cells, mitochondrial localization can be strongly fixed, which creates a very regular arrangement and promotes numerous and tight interactions with other intracellular structures such as the ER and cytoskeleton. These interactions may involve the association of various proteins (e.g., desmin, tubulin, actin, dynein family) with porin (VDAC) in the outer mitochondrial membrane. Moreover, the detailed analysis of the natural interactions of mitochondria together with the analysis of coupled creatine kinase and other energy-transferring systems have led to the discovery of mitochondrial functional complexes, termed the intracellular energetic units (ICEU) in cardiac muscle. In many other cell types, mitochondria can be highly dynamic, undergoing continual fission and fusion. Mitochondrial fragmentation occurs under various stressful conditions, representing an early event in apoptosis. Specific mitochondrial distribution can be achieved by movement along the cytoskeleton and attachment to the cytoskeleton using specific motor and connector proteins. Mitochondrial movement can be highly dynamic in cells such as neurons, and it serves to direct mitochondria to cellular regions of high ATP demands or to transport damaged or old mitochondria destined for elimination. Furthermore, this movement can be strongly regulated by metabolic signals (e.g., by local cytosolic Ca^{2+} concentrations). A growing body of evidence demonstrates that the proper regulation of mitochondrial dynamics is absolutely crucial for the cell. Many diseases, in particular neurological diseases, are associated with mutations in proteins that control mitochondrial dynamics and morphology. Defects in mitochondrial dynamics may have serious deleterious effects on normal cell function. Moreover, mitochondrial dynamics and interactions of mitochondria with cytoskeletal components may significantly change during development and aging of the cell, or in response to changing environmental conditions. Taken together, recently reported data suggest very important physiological roles and various further implications of highly dynamic mitochondrial behavior and complex, cell/

Molecular System Bioenergetics: Energy for Life. Edited by Valdur Saks
Copyright © 2007 WILEY-VCH Verlag GmbH & Co. KGaA, Weinheim
ISBN: 978-3-527-31787-5

tissue-specific patterns of their intracellular organization and interactions with other cellular structures.

5.1
Introduction

The recognition of the central role of mitochondria in apoptosis and calcium regulation has greatly renewed interest in mitochondria. In the past few years, significant progress has also been made in the study of the mechanisms controlling morphology and intracellular dynamics of mitochondria. Recent evidence demonstrates that specific cellular organization and numerous interactions of mitochondria play a fundamental role in their regulation *in vivo* and may be strongly cell/tissue specific. In cardiac and skeletal muscles, mitochondria are firmly fixed, creating a very regular arrangement and interacting with other intracellular structures, such as the ER and elements of the cytoskeleton. Conversely, in cell types such as neurons or cultured HL-1 cells, mitochondria can be highly dynamic, undergoing continual fission and fusion events modulated by mitochondria-associated GTP-binding proteins. A number of proteins that modulate mitochondrial shape, division, and fusion have been recently discovered. Importantly, fission and fusion processes that regulate mitochondrial morphology may occur under normal and pathological conditions and can also be involved in the regulation of mitochondrial metabolism. Thread-grain transitions may be transient, may occur under various stressful conditions (hypoxia-reoxygenation, oxidative stress, etc.), and may represent an early event in apoptosis. Mitochondrial fragmentation can either enhance or reduce mitochondrial outer membrane permeabilization and subsequent cell death depending on the nature of the death-inducing signal transduction pathway. We know that specific mitochondrial distributions in the cell can be achieved by the movement of these organelles along the cytoskeleton and attachment to the cytoskeleton using specific motor and connector proteins. However, not all cells use the same cytoskeletal elements and motor proteins for mitochondrial movement and attachment. Mitochondrial movement can be very dynamic in cells such as neurons, and it serves to direct mitochondria to cellular regions of locally high ATP demand to provide energy and Ca^{2+} buffer capacity or to transport damaged or old mitochondria destined for elimination. Furthermore, this movement can be strongly regulated by local cytosolic Ca^{2+} concentrations in the physiological range. Importantly, fragmentation of the mitochondrial network (mitochondrial fission) may facilitate mitochondrial movement in the cell, suggesting a link between mitochondrial motility and morphology and thus demonstrating the physiological significance of both aspects of mitochondrial dynamics. Here, we review the proteins involved in the morphology and movement of mitochondria and some regulatory aspects of mitochondrial intracellular organization and dynamics that play an important role in mitochondrial function and pathology.

5.2
Intracellular Organization of Mitochondria

5.2.1
Multiple Functions of Mitochondria in the Cells *in Vivo*

Mitochondria have been recognized in recent years not only as the main intracellular source of energy (in the form of ATP) necessary for normal cell function and viability but also as a major controller in a variety of other important cellular functions (Fig. 5.1). These organelles may synthesize different metabolites; regulate cellular redox potential; and play very important roles in programmed cell death, thermogenesis, and different ionic regulations (in particular, in calcium homeostasis). Mitochondria participate in Ca^{2+} signaling as a result of their close apposition to Ca^{2+} release (endoplasmic reticulum [ER]) and entry sites (plasma membrane), where microdomains with high local Ca^{2+} concentrations are formed. Moreover, mitochondria are directly involved in pathophysiological mechanisms of ischemia–reperfusion injury, oxidative stress, inherited diseases, toxicological injury, and side effects of pharmacological treatments. Damaged mitochondria also cause organ injury by several mechanisms, including diminished cellular energy status, production of reactive oxygen species, disturbance of ionic balance, cytochrome *c* release, and induction of apoptosis.

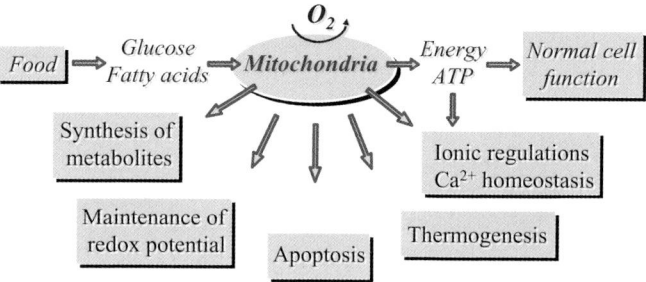

Fig. 5.1 Key roles of mitochondria in normal cell function and viability.

5.2.2
Analysis of Mitochondrial Dynamics and Intracellular Organization *in Vivo*: Confocal Fluorescent Microscopy

Mitochondrial analyses *in vitro* using mitochondrial or cellular suspensions usually fail to resolve functional, morphological differences and responses of mitochondrial subsets localized in different regions of the cell. In contrast, imaging approaches are well suited for studying complex mitochondrial dynamics critical for understanding how mitochondria actually work in the cell. This method clearly indicates that mitochondria may be highly motile organelles that show a well-organized distribution pattern. Fluorescence measurements performed with perfused tissues, cell or mitochondrial suspensions can be used for the sensitive assessment of mitochondrial function and dysfunction. These studies, however, cannot discriminate between the fluorescence of mitochondria localized in different intracellular regions. Alternatively, confocal fluorescence imaging distinguishes the signals from different mitochondrial subpopulations, from small clusters, or even from single mitochondrion in different cell regions, therefore providing quantitative measurement of their region-specific function and dynamics [1–4]. Fluorescent imaging techniques are intensively used for the quantitative digital imaging of several important functional and morphological characteristics of mitochondria in living cells *in vivo* [5–10]. Imaging approaches provide both spatial and temporal analysis and can be used to study mitochondrial morphology and networks, complex patterns in mitochondrial dynamics, and various functional defects, representing the most direct way to visualize intracellular heterogeneity of mitochondria and complex mitochondrial interactions with other intracellular structures. Several mitochondria-specific markers or green fluorescent proteins specifically targeted to mitochondria can be used in combination with high-resolution imaging techniques. Recently developed methods include (1) fluorescence resonance energy transfer (FRET) for imaging protein–protein interactions and conformational changes; (2) recording the image data in three spatial dimensions over time (i.e., 4D imaging); and (3) novel 4Pi and STED microscopy [11]. All these imaging approaches allow a comprehensive analysis of the role of the cytoskeleton in complex regulations of mitochondrial function, organization, and dynamics.

5.2.3
Mitochondrial Clusters and Subpopulations, Mitochondrial Heterogeneity, and their Physiological and Pathophysiological Roles

Confocal imaging of mitochondria *in situ* in permeabilized, non-fixed preparations of tissue or in living cells revealed that mitochondria are either clustered or arranged in a highly organized, tissue-specific manner. Moreover, mitochondria in the cell may be very heterogeneous in many structural and functional aspects under normal [1, 7, 12] and pathological [2, 13] conditions. Mitochondrial clustering has been reported as one specific organization in various cell types that may be associated with specific cellular demands. For example, mitochondrial clusters surrounding the nuclei may serve to drive mitochondrial metabolism to generate ATP close to the nucleus, as has been shown recently for parotid acinar cells [14]. This observation is in line with the concepts of an integrated phosphotransfer network and energetic channeling between mitochondria and nuclei as suggested by Dzeja et al. [15]. Such a clustering may play an important physiological role in the mechanisms for nuclear import, as well as for regulating a variety of other nuclear functions. Structural and functional interactions between mitochondrial and endoplasmic reticulum domains were found to be necessary for Ca^{2+} signaling [16–18], and intermitochondrial interactions may participate in the generation and propagation of apoptotic signals [19]. The closeness of mitochondrial clusters to the plasma membrane may be important for functional coupling to ATP-driven ion pumps (e.g., Ca^{2+} entry) [20]. Other subpopulations of mitochondria with cluster organization are the subsarcolemmal mitochondria in skeletal muscle [1, 9]. This physical (spatial) compartmentalization of mitochondria is frequently accompanied by a functional heterogeneity with respect to mitochondrial membrane potential, Ca^{2+} sequestration, and sensitivity to permeability transition activation. Monitoring of flavoprotein autofluorescence by confocal fluorescent imaging of intact and permeabilized cells or by fluorescence-activated cell sorting (FACS) analysis indicated that mitochondria may be highly heterogeneous with respect to their oxidative state. In mice skeletal muscle, a much higher oxidized state of subsarcolemmal mitochondria as compared with intermyofibrillar mitochondria has been demonstrated [1, 9]. Using the confocal imaging technique, we have presented similar phenomena for rat soleus and gastrocnemius muscles, where a higher oxidative state of mitochondrial flavoproteins correlates with elevated mitochondrial calcium. In addition, these mitochondria show less regular arrangement than intermyofibrillar or cardiac mitochondria [9]. Interestingly, in rat soleus, confocal imaging of mitochondria located near the outer cell membrane demonstrates numerous contacts and interconnections, showing a clear mitochondrial network in this muscle. The metabolic differences between subsarcolemmal and intermyofibrillar mitochondria may have important functional and physiological consequences. Indeed, subsarcolemmal mitochondria are located close to the cell periphery and therefore are exposed to higher oxygen levels than other mitochondria inside the cell. Such localization close to the source of oxygen may explain the more oxidized state of this mitochondrial subset and potentially indicates a more active mitochondrial respiration. Subsarcolemmal mitochondria may serve as a "protection barrier" maintaining permissive levels of oxygen in the cell. Therefore, this subsarcolemmal population of mitochondria may defend intracellular structures against high oxygen concentration outside the cell and thus provide an important shielding mechanism against oxidative stress inside the cell. Static and dynamic heterogeneity of mitochondria has been reported for a number of cells, including hepatocytes, HUVEC, astrocytes, and various human carcinoma cells [7, 12]. Heterogeneity of mitochondrial calcium has been studied in cardiac cells under pathological conditions [21]. Interestingly, cells with targeted null mutations in fusion proteins Mfn1 or Mfn2 (Table 5.1) and lacking mitochondrial fusion ability show a high degree of mitochondrial functional heterogeneity [22]. However, the metabolic consequences and causes of mitochondrial heterogeneity, as well as its physiological and pathophysiological significance, are still not clear and require further investigation. Taken together, recent findings suggest that the structural organization of the cell, its local energy, and other cellular demands may determine features of mitochondrial region-specific behavior, implying a strong relationship between mitochondrial function and cell organization. Thus, different mitochon-

Table 5.1 Proteins involved in mitochondrial dynamics: fusion–fission and shape.

Protein	Localization/Function/Organism	Protein characteristics	References
Drp1/Dnm1 Dynamin-related protein-1	OMM/Fission/Human, *C. elegans*, yeast	GTPase from the dynamin family; Overexpression sensitizes to staurosporine, but inhibits apoptosis induced by ceramide, H_2O_2, and serum deprivation	[47, 48, 61]
Fis1/hFis1/Mdv1, Mdv2	OMM/Fission/Yeast, fly, human	OMM integral protein, recruits Dnm1 to fission foci; Overexpression induces fragmentation, cytochrome *c* release, and apoptosis, which can be blocked by Drp1	[62]
Rab32	OMM/Fission/Human	Small GTPase, member of the Rab family, colocalizes with mitochondria	[43, 110]
Caf4	Cytoplasm, OMM/Peripheral/Fission/Yeast	Fis1p-dependent localization to the outer mitochondrial membrane, recruits Dnm1p into mitochondria	[67]
Fzo1p/Mfn1, Mfn2 (Mitofusin)	OMM/Fusion/Yeast, fly, human, mouse	Large GTPase containing C-terminal coiled-coil domain. Overexpression inhibits Bax translocation to OMM, Bax/Bak oligomerization in OMM, and OMM permeabilization	[63–65]
Ugo1	OMM/Fusion/Yeast	Integral protein of the OMM	[43, 108, 109]
Mgm1/OPA1	OMM, IMM/Fusion/Yeast, human, mouse	GTPase from the dynamin family; regulation of the cristae morphology; downregulation induces fragmentation, remodeling of cristae and apoptosis	[45]
Mmm1p, Mmm2p; Mdm10p, Mdm12p	OMM/Tubulation/Yeast	Mitochondrial tubulation via interactions with cytoskeletal elements or scaffolding structures	[51, 53, 54]

Table 5.1 (continued)

Protein	Localization/Function/ Organism	Protein characteristics	References
Mdm31p, Mdm32p	IMM integrated/ Tubulation/Yeast	IMM remodeling; may link IMM and OMM	[52]
Mdm33p	IMM/Yeast	May be involved in IMM fission	[111]

OMM: outer mitochondrial membrane; IMM: inner mitochondrial membrane

drial subpopulations, clusters, and even single mitochondrion may carry out diverse processes within distinct regions of a cell.

5.2.4
Mitochondria as Discrete Units in Highly Differentiated Cells Versus the Mitochondrial Interconnected Network

In some cell types mitochondria exist singly and are randomly dispersed. In other cells mitochondria may also exist as dynamic networks (resembling the structure of the ER) that often change shape and subcellular distribution [23–25]. For example, mitochondria in actively growing yeast cells show a dynamic reticulum of branched tubules. In yeast and several other lower eukaryotic organisms, electron microscopy using serial sectioning has demonstrated the presence of one or a few giant, highly branched mitochondria per cell [26]. Mitochondrial networks (*reticulum mitochondriale*) were found first in diaphragm muscle and in various neuronal and non-neuronal cells types and were shown to interact with other organelles such as the ER. Mitochondrial imaging also revealed important roles of specific interactions of mitochondria with components of the cytoskeleton, which were responsible for both the arrangement and the intracellular repositioning of mitochondria. It has been proposed that the existence of mitochondrial networks may be vital for mitochondrial synchronization and that electrical connections across a mitochondrial reticulum may be crucial for cellular physiology [23, 24]. For example, such an electrical continuity would transfer the potential across the inner membrane of all mitochondria, whether they are exposed to oxygen or not. It is assumed that oxygen and mitochondrial substrates from a capillary are first consumed by a subsarcolemmal mitochondrial cluster. H^+ gradients created by these mitochondria may then be transferred to intermyofibrillar mitochondria through connecting mitochondrial filaments (a cable-like energy-transporting system) [24]. In addition, a continuous and coupled mitochondrial network may provide a mechanism for Ca^{2+} tunneling similar to the ER, leading to Ca^{2+} wave propagation to an area of the cell distant from the site of the original signal.

It is known that mitochondria play a key role in the induction of apoptosis by release of proapoptotic factors such as cytochrome *c*. Normal mitochondrial function is inhibited by permeability transition and cytochrome *c* depletion. On the other hand, apoptosis requires ATP, which is produced mainly in mitochondrial oxidative phosphorylation. Availability of ATP is a prerequisite for apoptosis, while energy deprivation shifts the cell towards necrosis [27]. Therefore, a non-continuous mitochondrial network would allow one mitochondrial population to be actively involved in apoptotic signaling, while another could continue to provide the ATP necessary for apoptosis [27]. However, the physiological roles of complex patterns of mitochondrial morphology, networks, and, in particular, regulatory mechanisms still require further investigation.

5.2.5
Role of Mitochondrial Interactions with Other Intracellular Structures for Regulating Mitochondrial and Cellular Function

It is well known that the cytoskeleton is very important for mitochondrial and cell morphology and motility, intracellular traffic (see below), and mitosis [28]. Notably, mitochondrial dynamics and interactions can be very important not only for control of their morphology and organization but also for their functioning. The complex cytoskeletal network and specific cytoskeleton-associated proteins interact with mitochondria to regulate mitochondrial respiratory function and to control the permeability of the mitochondrial outer membrane to ADP (for details, see Chapter 3). Mitochondria are associated with the three major cytoskeletal structures: microfilaments, microtubules, and intermediate filaments [29, 30]. For example, in frog neurons, cross-bridging between the mitochondrial outer membrane and microtubules has been demonstrated [31]. The microtubules are composed mostly of tubulin, and their assembly and function are regulated by microtubule-associated proteins, the major classes of which are the motor proteins (kinesin, dynein, etc.). Figure 5.2 shows immunochemical staining of the tubulin network (red) in adult rat cardiomyocytes tightly associated with cardiac mitochondria (mitochondrial flavoprotein autofluorescence, green). In many other cells and tissues, mitochondria typically display a subcellular distribution corresponding to that of the microtubular network [32], and many chemical agents that depolymerize microtubules significantly modify intracellular distribution of mitochondria [29, 32]. Similar effects on mitochondrial localization and motility are found with actin-encoding gene mutations [33], demonstrating an important role of the actin cytoskeleton. Mitochondria could be either anchored on cytoskeletal filaments or shaped by the forces generated by actin. Annexins may be connected to the phospholipid membrane in a calcium-dependent manner, thus interfering with cytoskeleton-dependent functions of mitochondria [34]. Experiments performed with desmin-null cardiac muscle have indicated that the respiratory function of mitochondria *in vivo* can be significantly changed [35]. This includes loss of functional coupling between the mitochondrial creatine

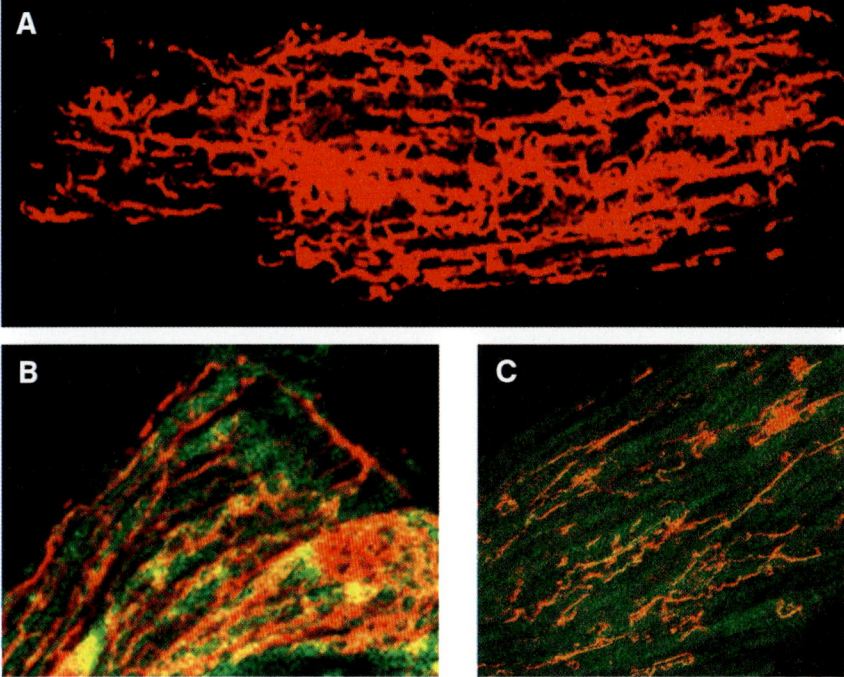

Fig. 5.2 Confocal fluorescence imaging of a tubulin network and mitochondria in isolated adult rat cardiomyocytes. (A) Immunochemical staining of tubulin. (B) Immunochemical staining of tubulin (red) is combined with imaging of mitochondrial flavoprotein autofluorescence (green). (C) Disintegration of the tubulin network after treatment of cardiomyocytes with colchicine (1 μM).

kinase and the ATP–ADP translocase. Remarkably, a decrease in the maximal rate of ADP-stimulated state 3 respiration as well as in the apparent K_m for ADP in desmin-null cardiac and soleus muscles is also reported, although these last two characteristics were not changed in desmin-null glycolytic gastrocnemius muscle [35]. Moreover, ultrastructural analysis of desmin-null cardiomyocytes has demonstrated widespread proliferation of mitochondria that is dramatically visible after work overload. Taken together, all these findings show a crucial role of components of the cytoskeleton in the regulation of mitochondrial function, dynamics, and morphology.

5.2.6
Tissue and Cell-type Specificity of Mitochondrial Intracellular Organization and Dynamics

Analysis of mitochondrial motility, using specially designed quantitative approaches, provides strong evidence that mitochondria are firmly fixed and do not

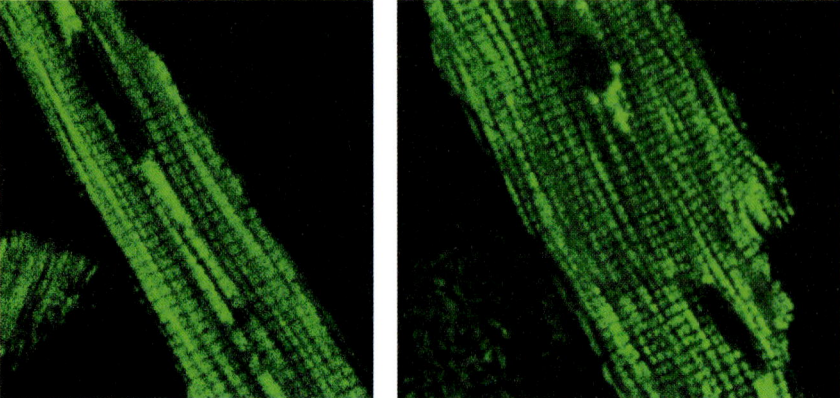

Fig. 5.3 Regular intracellular organization of mitochondria in adult rat cardiomyocytes. Regular mitochondrial arrangement was visualized by confocal imaging of flavoprotein autofluorescence.

move in cells such as rat heart cardiomyocytes (Y. Usson, unpublished data) or skeletal muscle. Cardiac mitochondria are organized in a "lattice" of parallel rows surrounding the contractile myofilaments, creating a very regular intracellular arrangement (Fig. 5.3 and [36]). These mitochondria form functional complexes with other intracellular structures – such as the ER, sarcomere, and cytoskeleton – termed the intracellular energetic units (ICEUs) [37, 38], to recruit proteins such as desmin, tubulin, actin, dynein family, etc., which may be associated with porin (voltage-dependent anion channel [VDAC]) in the outer mitochondrial membrane. Importantly, mitochondrial interactions with elements of the cytoskeleton effectively modulate mitochondrial respiration and ADP kinetics [39, 40] and play a fundamental role in the metabolic regulation of mitochondrial function *in vivo* in living cells, and probably also in the regulation of calcium homeostasis. Controversially, HL-1 cells with a differentiated adult cardiac phenotype do not exhibit the strictly regular ("crystal-like") mitochondrial distribution typical of rat cardiac cells (Fig. 5.4). In these cells, mitochondria can be highly dynamic and undergo continual displacements. Mitochondrial short- and long-range movements along microtubules have also been detected in H9c2 myoblasts [41]. Wide variations in mitochondrial shape and morphology can be observed in HL-1 cells, including small spheres or short rod-like shapes, together with long filamentous spaghetti-like mitochondria, irrespective of a particular cellular region (Fig. 5.4; see also [5, 9]). Figure 5.4 demonstrates heterogeneity in mitochondrial membrane potential and reactive oxygen species (ROS) production. Thus, all recently obtained experimental observations point to a strong cell and tissue specificity of mitochondrial dynamics and organization. This tissue/cell mitochondrial specificity may have high physiological significance.

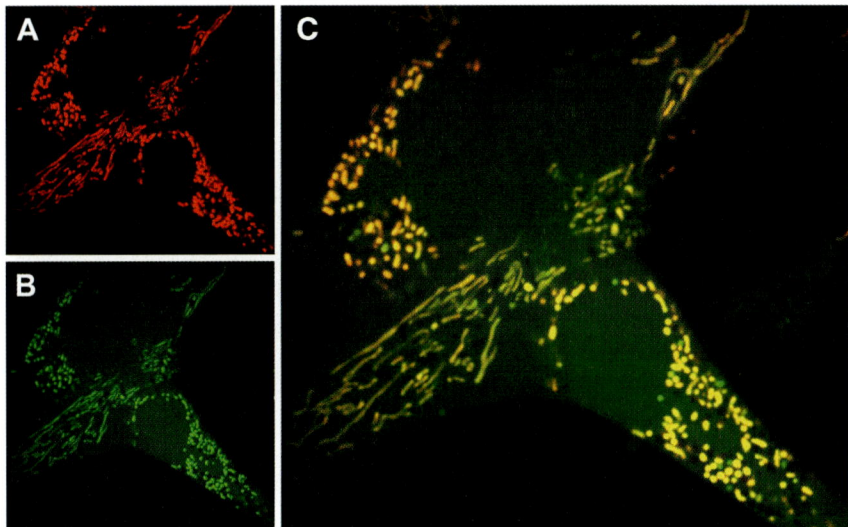

Fig. 5.4 Confocal fluorescent imaging of mitochondria in cultured HL-1 cells demonstrates the heterogeneity of mitochondrial morphology, membrane potential, and ROS. Simultaneous confocal imaging of (A) red fluorescence of mitochondrial membrane potential–sensitive probe TMRM (100 nM) and (B) green fluorescence of ROS-sensitive probe DCF (20 μM). (C) Fluorescence of TMRM and DCF is shown as a merge image.

Mitochondria in cultured HL-1 cells show clearly different morphology: long thread and small isolated grain mitochondria. The homogeneous yellow-brown color of mitochondria is seen as a result of spatial colocalization of mitochondria and ROS. However, some mitochondria with only green fluorescence indicate partially depolarized mitochondria with a lower TMRM signal.

5.3
Mitochondrial Dynamics: Regulation of Mitochondrial Morphology

Depending on cell type, mitochondria may be highly dynamic [42], undergoing frequent cycles of fusion and fission (Fig. 5.6), as well as remarkable intracellular movement in some cells. The development of specific probes and GFP fusions has allowed visualization of mitochondrial dynamics in cells *in vivo*. Fusion and fission continuously change mitochondrial shape under physiological (e.g., cellular division) and pathophysiological (apoptotic) conditions [43, 44]. Defects in fusion frequently lead to fragmentation of the mitochondrial network [45, 46], whereas the result of fission inability is a formation of the network of excessively elongated and interconnected organelles [47]. The precise balance between these two processes might play a key role in mitochondrial and cellular function. The very different mitochondrial shapes (small spheres; short, rod-like shapes or, at the same time, spaghetti-like shapes; and long, branched tubules) visible in HL-

1 cells (Fig. 5.4) may reflect continuous fusion and fission of mitochondria in these cells.

5.3.1
Mitochondrial Shape Proteins

In some cells mitochondria are able to change intracellular location and shape, which is regulated by specific shape proteins and by complex interactions with the cytoskeleton. Normal mitochondrial shape usually needs external attachment of mitochondria to the cytoskeleton (an internal scaffold). Several potential shape-forming proteins have been recently identified in yeast mutants with alterations in mitochondrial morphology or intracellular organization [48–50]. Some proteins may control shape by mediating attachment of mitochondria to the cytoskeleton, while other proteins may be involved in the creation of connections between the mitochondrial outer and inner membranes. Formation of the normal tubular shape of mitochondria requires proteins of the mitochondrial outer membrane (Table 5.1), such as Mmm1p, Mmm2p, Mdm10p, and Mdm12p, and proteins in the inner membrane, such Mdm31p and Mdm32p [49, 51–54]. In mutants lacking any one of these proteins, mitochondrial tubulation, the elongated and branched shapes of tubules disappear and mitochondria organize large spherical clusters. It has been demonstrated that Mmm1p–Mdm10p–Mdm12p form an MMM complex responsible for mitochondrial tubular shape in cooperation with Mmm2p, Mdm31p, and Mdm32p. This complex participates in the attachment of mitochondria to the actin cytoskeleton (Fig. 5.5A), also interacting with other cytoskeletal elements or scaffolding structures. Mitochondrial morphology also requires Mdm33p and Gem1p proteins, which are possibly involved in inner membrane remodeling [55]. Furthermore, recent findings have shown the involvement of mitofilin [56] and ATP synthase [57] as critical factors in the modeling of the cristae morphology.

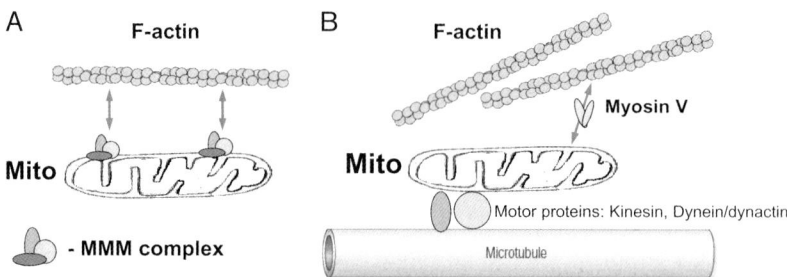

Fig. 5.5 Interactions of mitochondria with the actin cytoskeleton. (A) The attachment of mitochondria to the actin cytoskeleton via the MMM complex regulates mitochondrial shape. (B) The interaction of mitochondria with F-actin (via myosin V) is required for short mitochondrial dislocations in axons.

5.3.2
Mitochondrial Fusion and Fission

The two opposing fission and fusion processes (Fig. 5.6) may be modulated by organelle-associated proteins [58] or by energy substrates [59]. Thread-grain transitions may be transient and may occur under various stressful conditions, also representing an early event in apoptosis (see below). Under physiological conditions, however, mitochondrial fission is counteracted by fusion, causing mitochondrial network stability [60]. In mutants lacking fusion or fission, small fragmented organelles or large, branched mitochondrial networks can be observed. Fusion is a complexly regulated process in which the tips of mitochondrial tubules fuse. The process requires the coordinated joining of outer membranes with other outer membranes and of inner membranes with other inner membranes. This GTP- and membrane potential–dependent process maintains the integrity of mitochondrial compartments (intermembrane space and matrix). It has been shown that dissimilar mechanisms of mitochondrial fusion can be involved in mammals and yeast. Similarly, fission is a GTP-dependent process by which a mitochondrial tubule is fragmented in a way that prevents the loss of contents of mitochondrial intermembrane space or matrix.

Key proteins in the mitochondrial fusion, fission, and tubulation pathways are shown in Table 5.1. Fission and fusion processes are controlled by the dynamin family GTPases. Dynamin-related protein 1 (Drp1 or DLP1) in mammals and Dnm1 in yeast together with other proteins (Fis1 and Mdv1 in yeast, hFis1 in humans) participate in mitochondrial fission [47, 48, 61, 62], whereas mitofusin-1 (Mfn1), mitofusin-2 (Mfn2), and optic atrophy-1 (OPA1) in mammals and Mgm1 and Fzo1 in budding yeast regulate mitochondrial fusion [46, 63–65] (Fig. 5.6 and Table 5.1). Importantly, both cytoskeletal microfilaments and microtubules can contribute to the recruitment of Drp-1 to mitochondria [66]. Some proteins controlling yeast mitochondrial fusion and fission are evolutionarily conserved in flies, worms, plants, mice, and humans, which points to common

(Mfn1,2 Fzo1 OPA1 Mgm1 Ugo1)
Fusion

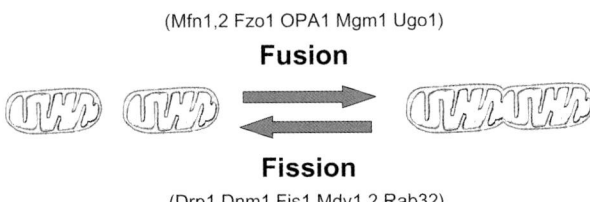

Fission
(Drp1 Dnm1 Fis1 Mdv1,2 Rab32)

Fig. 5.6 Mitochondrial fusion and fission. The balance of mitochondrial fusion and fission can be changed in response to various stimuli (e.g., fusion can serve in the transmission of energy and Ca^{2+} signals, mitochondrial DNA inheritance, cell development, etc., whereas fission can be associated with apoptosis or cellular division).

mechanisms involved in mitochondrial morphology and dynamics. *In vitro* experiments have indicated the requirement of high GTP levels for inner membrane fusion. At low GTP levels, only the outer mitochondrial membranes, not the inner membranes, may fuse. Addition of a GTP-generating system causes a fusion of the inner mitochondrial membrane. In yeast, GTPase Dnm1 creates a fission complex with proteins that modulate GTPase activity, such as Fis1, an integral protein of the outer mitochondrial membrane, and Mdv1 and Caf4 (binding partner adapter proteins) [67]. Interestingly, mutations in mitofusin-2 and OPA1 may lead to neuropathy, which demonstrates a key role of the proper regulation of mitochondrial dynamics in neurons [68, 69]. Specific mitochondrial uncouplers (e.g., CCCP) may block fusion and cause mitochondrial fragmentation [70]. Thus, mitochondrial fusion may be dependent on the mitochondrial membrane potential, and this effect is specific for fusion of the inner mitochondrial membrane. Moreover, such an effect is reversible, and the mitochondrial networks can be restored upon withdrawal of the uncoupler [70]. Possible mechanisms may include membrane conformational changes that are potential dependent or structural modifications that affect the fusion machinery.

5.3.3
Mitochondrial Fragmentation During Apoptosis

In apoptosis, during the early stages of cell death, mitochondria may undergo remarkable structural changes (extensive fragmentation of the mitochondrial reticulum and remodeling of the cristae) that can be associated with progression of the apoptotic cascade. Fission and cristae remodeling can be associated with a higher degree of cytochrome *c* release. Mitochondrial fragmentation after activation of apoptosis [43] may occur before caspase activation and is not inhibited by caspase inhibitors, but is linked to the activation of the mitochondrial fission machinery [71, 72]. On the other hand, inhibition of proteins involved in mitochondrial fission, such as Drp1 (with siRNA or in DN mutants) or Fis1 (with siRNA), prevents fragmentation, reduces release of proapoptotic factors from mitochondria, and thus diminishes apoptosis [71–73]. Additionally, inhibition of mitochondrial fusion is a general phenomenon during apoptosis, occurring almost simultaneously with the coalescence of Bax and Bak on the mitochondrial membrane and cytochrome *c* release [74], whereas this fragmentation can be suppressed by the overexpression of proteins (e.g., mitofusins) involved in mitochondrial fusion. Typical fragmentation of the mitochondrial network in HL-1 cells during oxidative stress can be seen in Fig. 5.7 after addition of pro-oxidant to the culture medium. Very similar mitochondrial fragmentation also can be observed during photooxidative stress, or very early after induction of apoptosis by various pro-apoptotic inducers, such as in staurosporine (STS)- or etoposide-treated Cos-7 cells [43, 71], and this fragmentation cannot be prevented by overexpression of the antiapoptotic factor Bcl-2.

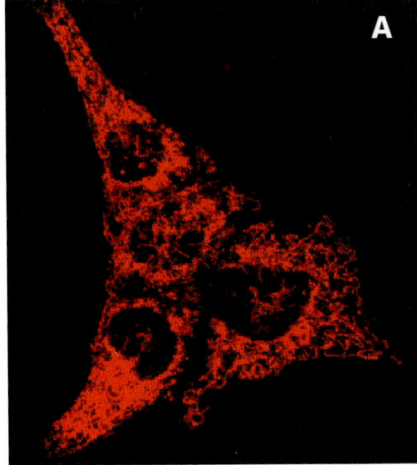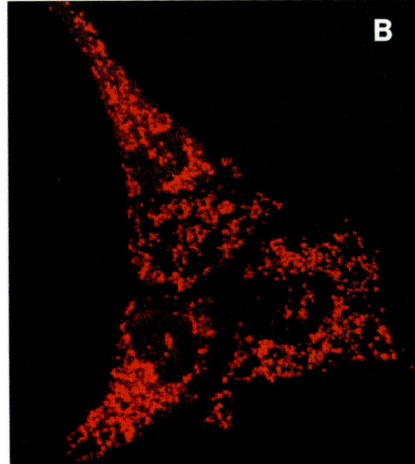

Fig. 5.7 Fragmentation of the mitochondrial network in cultured HL-1 cells during oxidative stress. HL-1 cells were stained with the mitochondria membrane potential–sensitive probe TMRM (100 nM) and imaged over time using confocal microscopy.

Disintegration of the mitochondrial network and formation of punctate; fragmented mitochondria occur early during oxidative stress induced by the variety of the stimuli. (A) Control HL-1 cells, (B) HL-1 cells treated with pro-oxidant (t-BHP) for 20 min.

5.3.4
Apoptotic Control and Mitochondrial Dynamics

Importantly, proteins that control mitochondrial morphology (e.g., proteins of the mitochondrial fusion machinery such as OPA1) also may participate in mitochondrial steps of apoptosis (Table 5.1). In the same way, proteins associated with the regulation of apoptosis (both pro- and antiapoptotic) may contribute to morphological changes in mitochondria (Table 5.2). For example, the proapoptotic protein Bax may colocalize with Drp1 and Mfn2 in apoptotic cells at sites of mitochondrial division [75]. Both Bax and Bak in their apoptotic conformation may interact with Mfn2 and cause Drp1-dependent fission. Bax also may heterodimerize with another fission protein, endophilin B1, on mitochondria, leading to Bax activation and induction of apoptosis [76]. Paradoxically, in normal non-apoptotic cells, Bax seems to be required for mitochondrial fusion by activating Mfn2 assembly [77]. This fusion can be inhibited in cells lacking Bax and Bak. Therefore, fusion and fission proteins can interfere with regulation pathways of apoptotic cell death (although fragmentation of mitochondria may take place without apoptosis). Overexpression of pro-fusion proteins or inhibition of pro-fission molecules may reduce apoptosis. Conversely, inhibition of mitochondrial fusion may result in mitochondrial fragmentation, leading to apoptosis induction. For example, OPA1 depletion may produce mitochondrial fragmentation, permeability

Table 5.2 Pro- and antiapoptotic proteins potentially involved in morphological alterations of mitochondria.

Protein	Function localization	Characteristics	References
PB1-F2	Proapoptotic (MITO)	Viral product of influenza A PB1; induces mitochondrial fragmentation	[112]
Bax/Bak	Proapoptotic (OMM)	Overexpression induces fragmentation of mitochondria; upon induction of apoptosis, coalesce into mitochondrial scission foci, colocalize with Drp1 and Mfn2. Overexpression inhibits mitochondrial fusion, causes fragmentation and OMM permeabilization	[44, 71, 75, 77, 113, 114]
Bid	Proapoptotic (OMM)	Upon induction of apoptosis translocates to mitochondria and activates Bax and Bak; induces fusion of mitochondrial cristae; possesses lipid translocase activity	[115, 116]
Bcl-2	Antiapoptotic (OMM)	Overexpression causes an increase in the size and complexity of mitochondria	[117]
vMIA	Antiapoptotic (MITO)	Viral mitochondria-localized inhibitor of apoptosis. Induces mitochondrial fragmentation	[118]

transition, Bax translocation, and cytochrome *c* release [73, 78], which is protected by Bcl-2. In addition, a deficiency of the fission protein Drp1 in dominant-negative mutants suppresses mitochondrial fission and inhibits cell death, suggesting that the interconnected mitochondrial network may be advantageous for cell survival. OPA1 mutations may sensitize cells to ROS-induced apoptosis. Interestingly, mitochondria–cytoskeleton interactions also seem to be involved in the apoptotic pathways [79]. An increase in actin-mitochondria associations has been implicated in the apoptotic induction triggered by various stimuli [80]. The associations of the actin cytoskeleton with mitochondria actively participate in the mitochondrial recruitment of the cytosolic proapoptotic proteins.

5.3.5
The Importance of Spatial Mitochondrial Organization for Mediating Lethal Ca^{2+} Waves

In contrast to the proapoptotic role of mitochondrial fission in the programmed cell death pathways, mitochondrial fragmentation can inhibit Ca^{2+}-mediated apoptosis induced by ceramide, pro-oxidant (H_2O_2) treatment, or serum deprivation. Apoptosis is frequently associated with Ca^{2+} release from the ER and subsequent mitochondrial Ca^{2+} overload. This leads to the permeability transition being transmitted as a wave along the mitochondrial network, causing propagation of the apoptotic signal [81]. Therefore, interconnected mitochondria may play an important role in the propagation of Ca^{2+} and apoptotic waves. On the other hand, mitochondrial Ca^{2+} overload may induce activation of Bax and Bak, their coalescence with Drp1, and mitochondrial fission. In this case, mitochondrial fragmentation can interrupt the Ca^{2+} waves required for apoptotic signaling. Indeed, Drp1-induced fission blocks Ca^{2+} waves in Drp1-overexpressing HeLa cells, thus protecting against apoptosis [82].

5.4
Mitochondrial Dynamics: Mitochondrial Movement (Motility) in the Cell

Another type of mitochondrial dynamics is mitochondria's ability for fast, continuous movements within cells of certain types, such as HL-1 cells, neurons, and human pancreatic cells. Confocal imaging demonstrates that mitochondria tubular networks can be highly dynamic, frequently changing their intracellular location and moving long distances on cytoskeletal tracks. Similar to morphological alterations, this dynamic motion can be tightly linked to mitochondrial energy metabolism. For example, intracellular transport of mitochondria may be significantly inhibited under conditions of mitochondrial membrane depolarization or suppressed mitochondrial ATP synthesis. The usual types of mitochondrial movements include (1) small oscillatory repositioning; (2) larger movements, filament enlargements, and branching; and (3) intracellular transfer movement and relocation of distinct mitochondria. The remarkable movement of mitochondria was first described in embryonic heart and intestine of chick and later in cells such as neurons (transport along axons and dendrites), the budding yeast *Saccharomyces cerevisiae*, epithelial cells, and superficial cortical fiber cells of bovine lenses [83–86]. Mitochondria need to be localized at subcellular sites for both providing ATP and participating in intracellular signaling [87]. This specific mitochondrial distribution involves mitochondrial intracellular transfer by a cytoskeleton-based transportation system and is highly coordinated in response to various cellular demands.

5.4.1
Key Role of the Cytoskeleton and Specific Motor and Connector Proteins for Mitochondrial Movement

Mitochondrial movement is based on interactions with a variety of cytoskeletal proteins (microtubule dependent) that may regulate mitochondrial function, shape, and distribution at sites of high ATP utilization. Molecular motors (kinesin family of mitochondria-bound proteins) are responsible for the transport of organelles along the cytoskeletal fibers [88, 89]. In animal cells – such as embryonic mouse cells, human kidney and pancreatic cells, extruded axoplasm of squid, and axons of mammalian cells and *Drosophila* – motor proteins play a major role in mitochondrial movement along microtubules [83, 90–92]. A growing body of evidence suggests that in many other cells, the actin cytoskeleton is mainly responsible for mitochondrial distribution and motility. For example, in *S. cerevisiae* [84, 93, 94], *Aspergillus nidulans* [95], photoreceptor cells [96], and cultured tobacco cells [97], mitochondrial movement is largely actin based (along actin filaments) [79]. Also in neurons, where cytoskeletal microtubules are considered to be elements for long-distance mitochondrial intracellular traffic, the actin cytoskeleton is certainly necessary for short mitochondrial dislocations (Fig. 5.5B) [98]. It is suggested that the actin cytoskeleton is specifically important for redistribution of mitochondria during inheritance. Notably, the dynein–dynactin complex may contribute to the mitochondrial targeting of Drp1 that promotes mitochondrial fission and facilitates mitochondrial transfer [99], suggesting a coupling between mitochondrial motility and morphology. Mitochondrial motility may be controlled by messengers such as Ca^{2+} at its physiological concentration, implicating the interactions with mitochondrial motor proteins. In H9c2 myoblasts, it has been shown that while mitochondrial movement is active at resting cytosolic Ca^{2+} concentrations, the movement of mitochondria can be completely blocked at 1–2 μM local $[Ca^{2+}]$ [41]. This arrest in movement is reversible, can be restored by removing excess Ca^{2+} to its resting levels, and can be repeated without any desensitization. The diminished mitochondrial movement in the region of high cytosolic Ca^{2+} [41] may result in increased local Ca^{2+} buffering capacity and rates of ATP synthesis, providing a feedback mechanism in metabolic and calcium signaling and thus representing a dynamic control of local mitochondrial function.

5.4.2
Motility of Mitochondria in Neurons

Mitochondrial oxidative phosphorylation is the main source of energy in the form of ATP for neurons. In these cells, movement of mitochondria can be highly dynamic (with velocities of 0.3–2.0 μm sec^{-1}) to transfer mitochondria to regions of increased demand for mitochondrial function or to remove defective or old mitochondria. Neurons are cells with tremendous dimensions, irregularity, and, thus, high heterogeneity of their energetic demands over long intracellular distances, and translocation of mitochondria along axons is necessary for normal functioning of these cells and is tightly regulated by the metabolic signals [83]. Interesting, in younger cells mitochondria are more mobile [100], allowing faster ATP supply (e.g., for synaptogenesis). Mitochondrial transport along microtubules is based on specific motor proteins such as kinesin families and dynein to move organelles along microtubules (Fig. 5.5B). In addition, myosin V may contribute to transport along actin tracks (Fig. 5.5B). It has been suggested that the protein kinases and phosphorylation of microtubule-associated proteins may be involved in the regulation of organelle transport directly or by interactions with the mitochondrial membrane. In axons, PI 3-kinase signaling pathways were found to be important mechanisms controlling mitochondrial motility [83, 101]. The role of changes in the mitochondrial volume via opening of mitochondrial K–ATP channels and decreased mitochondrial membrane potential also has been proposed in the control of mitochondrial distribution to improve ATP supply [83, 102, 103]. Interestingly, mitochondrial network fragmentation in neurons may facilitate mitochondrial transport to cellular regions of high ATP demands, thus demonstrating the physiological significance of both aspects of mitochondrial dynamics: changes in their morphology and motility [104]. Defects in mitochondrial morphology and dynamics may have serious deleterious effects on normal cell function in many diseases. Because of the existence of huge distances between sites of mitochondrial biogenesis and function, neurons seem to be especially susceptible to mitochondrial injury. It has been shown that defects in the normal dynamic behavior of mitochondria in these cells may be associated with neurodegenerative diseases, such as amyotrophic sclerosis and Alzheimer's, Huntington's, and Parkinson's diseases [105]. Moreover, mutations in the two mitochondrial fusion genes OPA1 and Mfn2 are associated with neurodegenerative human diseases, such as optic atrophy, leading to progressive blindness and Charcot–Marie–Tooth peripheral neuropathy with a loss of motor neurons [68, 69, 106, 107].

5.5
Concluding Remarks

Growing evidence demonstrates that dynamic mitochondrial function may be tightly regulated by mitochondria's specific intracellular organization, which integrates mitochondria in multiple cell signaling cascades. Many structural and functional interactions have been found to be involved in the integration of mitochondria with other cellular structures (such as the cytoskeleton and ER), connecting mitochondrial dynamics with extremely complex regulation of the entire cell physiology and programmed cell death. Recent studies have discovered numerous specific proteins that regulate mitochondrial respiration, morphology, fission–fusion events, and motility. Mitochondrial movement within the cell provides a local ATP supply at sites of high ATP demand. Mitochondrial fusion and fission also actively participate in the regulation of early steps of mitochondria-dependent apoptosis, thus suggesting new concepts for modulating cell death in cancer. Furthermore, recent findings have demonstrated that the proper regulation of mitochondrial dynamics is vital for various cellular pathways and that defective mitochondrial behavior affects human health, leading to disorders such as diabetes, obesity, and various mitochondrial myopathies. Many diseases, in particular neurological diseases, are associated with mutations in proteins that control mitochondrial dynamics and morphology (e.g., the fusion proteins OPA1 and Mfn2).

Acknowledgments

This work was supported in part by a research grant from the Austrian Cancer Society/Tyrol. The skillful assistance of Karen Guerrero, Florence Appaix, and Martin Hermann is gratefully acknowledged.

References

1 Kuznetsov, A.V., Mayboroda, O., Kunz, D., Winkler, K., Schubert, W., Kunz, W.S. (1998) Functional imaging of mitochondria in saponin-permeabilized mice muscle fibers. *J. Cell. Biol. 140*, 1091–1099.

2 Romashko, D.N., Marban, E., O'Rourke, B. (1998) Subcellular metabolic transients and mitochondrial redox waves in heart cells. *Proc. Natl. Acad. Sci. USA 95*, 1618–1623.

3 Appaix, F., Kuznetsov, A.V., Usson, Y., et al. (2003) Possible role of cytoskeleton in intracellular arrangement and regulation of mitochondria. *Exp. Physiol. 88*, 175–190.

4 Andrienko, T., Kuznetsov, A.V., Kaambre, T., Usson, Y., Orosco, A., Appaix, F., Tiivel, T., Sikk, P., Vendelin, M., Margreiter, R., Saks, V.A. (2003) Metabolic consequences of functional complexes of mitochondria, myofibrils and sarcoplasmic reticulum in muscle cells. *J. Exp. Biol. 206*, 2059–2072.

5 Anmann, T., Guzun, R., Beraud, N., Pelloux, S., Kuznetsov, A.V., Kogerman, L., Kaambre, T., Sikk, P., Paju, K., Peet, N., Seppet, E., Ojeda, C., Tourneur, Y., Saks, V. (2006) Different kinetics of the regulation of respiration in permeabilized cardiomyocytes and in HL-1 cardiac cells: Importance of cell structure/organization for respiration regulation. *Biochim. Biophys. Acta 1757*, 1597–1606.

6 Huang, S., Heikal, A.A., Webb, W.W. (2002) Two-photon fluorescence spectroscopy and microscopy of NAD(P)H and flavoprotein. *Biophys. J. 82*, 2811–2825.

7 Collins, T.J., Berridge, M.J., Lipp, P., Bootman, M.D. (2002) Mitochondria are morphologically and functionally heterogeneous within cells. *EMBO J. 21*, 1616–1627.

8 Zorov, D.B., Filburn, C.R., Klotz, L.O., Zweier, J.L., Sollott, S.J. (2000) Reactive oxygen species (ROS)-induced ROS release: a new phenomenon accompanying induction of the mitochondrial permeability transition in cardiac myocytes. *J. Exp. Med. 192*, 1001–1014.

9 Kuznetsov, A.V., Troppmair, J., Sucher, R., Hermann, M., Saks, V., Margreiter, R. (2006) Mitochondrial subpopulations and heterogeneity revealed by confocal imaging: possible physiological role? *Biochim. Biophys. Acta 1757*, 686–691.

10 Anmann, T., Eimre, M., Kuznetsov, A.V., Andrienko, T., Kaambre, T., Sikk, P., Seppet, E., Tiivel, T., Vendelin, M., Seppet, E., Saks, V.A. (2005) Calcium–induced contraction of sarcomeres changes regulation of mitochondrial respiration in permeabilized cardiac cells. *FEBS Journal 272*, 3145–3161.

11 Egner, A., Jakobs, S., Hell, S.W. (2002) Fast 100-nm resolution three-dimensional microscope reveals structural plasticity of mitochondria in live yeast. *Proc Natl Acad Sci USA 99*, 3370–3375.

12 Kuznetsov, A.V., Usson, Y., Leverve, X., Margreiter, R. (2004) Subcellular heterogeneity of mitochondrial function and dysfunction: evidence by confocal imaging. *Mol. Cel. Biochem. 256/257*, 359–365.

13 Kuznetsov, A.V., Schneeberger, S., Renz, O., Meusburger, H., Saks, V., Usson, Y., Margreiter, R. (2004) Functional heterogeneity of mitochondria after cardiac cold ischemia and reperfusion revealed by confocal imaging. *Transplantation 77*, 754–756.

14 Bruce, J.I.E., Giovannucci, D.R., Blinder, G., Shuttleworth, T.J., Yule, D.I. (2004) Modulation of $[Ca2+]i$ signaling dynamics and metabolism by perinuclear mitochondria in mouse parotid acinar cells. *J. Biol. Chem. 279*, 12909–12917.

15 Dzeja, P.P., Bortolon, R., Perez-Terzic, C., Holmuhamedov, E.L., Terzic, A. (2002) Energetic communication between mitochondria and nucleus directed by catalyzed phosphotransfer. *Proc. Natl. Acad. Sci. USA 99*, 10156–10161.

16 Rizzuto, R., Brini, M., Murgia, M., Pozzan, T. (1993) Microdomains with high Ca2+ close to IP3-sensitive channels that are sensed by neighboring mitochondria. *Science 262*, 744–747.

17 Rizzuto, R., Pinton, P., Carrington, W., Fay, F.S., Fogarty, K.E., Lifshitz, L.M., Tuft, R.A., Pozzan, T. (1998) Close contacts with the endoplasmic reticulum as determinants of mitochondrial Ca2+ responses. *Science 280*, 1763–1766.

18 Csordas, G., Thomas, A.P., Hajnoczky, G. (1999) Quasi-synaptic calcium signal transmission between endoplasmic reticulum and mitochondria. *EMBO J. 18*, 96–108.

19 Pacher, P., Hajnoczky, G. (2001) Propagation of the apoptotic signal by mitochondrial waves. *EMBO J. 20*, 4107–4121.

20 Lawrie, A.M., Rizzuto, R., Pozzan, T., Simpson, A.W. (1996) A role for calcium influx in the regulation of mitochondrial calcium in endothelial cells. *J. Biol. Chem. 271*, 10753–10759.

21 Bowser, D.N., Minamikawa, T., Nagley, P., Williams, D.A. (1998) Role of mitochondria in calcium regulation of spontaneously contracting cardiac muscle cells. *Biophys. J. 75*, 2004–2014.

22 Chen, H., Chomyn, A., Chan, D.C. (2005) Disruption of fusion results in mitochondrial heterogeneity and dysfunction. *J. Biol. Chem. 280*, 26185–26192.

23 Amchenkova, A.A., Bakeeva, L.E., Chentsov, Y.S., Skulachev, V.P., Zorov, D.B. (1988) Coupling membranes as energy-transmitting cables. I. Filamentous mitochondria in fibroblasts

and mitochondrial clusters in cardiomyocytes. *J. Cell. Biol. 107*, 481–495.

24 Skulachev, V.P. (2001) Mitochondrial filaments and clusters as intracellular power-transmitting cables. *Trends Biochem. Sci. 26*, 23–29.

25 Knowles, M.K., Guenza, M.G., Capaldi, R.A., Marcus, A.H. (2002) Cytoskeletal-assisted dynamics of the mitochondrial reticulum in living cells. *Proc. Natl. Acad. Sci. USA 99*, 14772–14777.

26 Davidson, M.T., Garland, P.B. (1975) Mitochondrial structure studied by high voltage electron microscopy of thick sections of Candida utilis. *J. Gen. Microbiol. 91*, 127–138.

27 Leist, M., Single, B., Castoldi, A.F., Kuhnle, S., Nicotera, P. (1997) Intracellular adenosine triphosphate (ATP) concentration: a switch in the decision between apoptosis and necrosis. *J. Exp. Med. 185*, 1481–1486.

28 Anesti, V., Scorrano, L. (2006) The relationship between mitochondrial shape and function and the cytoskeleton. *Biochim. Biophys. Acta 1757*, 692–699.

29 Ball, E.H., Singer, S.J. (1982) Mitochondria are associated with microtubules and not with intermediate filaments in cultured fibroblasts. *Proc. Natl. Acad. Sci. USA 79*, 123–126.

30 Mose-Larsen, P., Bravo, R., Fey, S.J., Small, J.V., Celis, J.E. (1982) Putative association of mitochondria with a subpopulation of intermediate-sized filaments in cultured human skin fibroblasts. *Cell 31*, 681–692.

31 Hirokawa, N. (1982) Cross-linker system between neurofilaments, microtubules and membranous organelles in frog axons revealed by quick-freeze, deep etching method. *J. Cell. Biol. 94*, 129–142.

32 Heggeness, M.H., Simon, M., Singer, S.J. (1978) Association of mitochondria with microtubules in cultured cells. *Proc. Natl. Acad. Sci. USA 75*, 3863–3866.

33 Smith, M.G., Simon, V.R., O'Sullivan, H., Pon, L.A. (1995) Organelle-cytoskeletal interactions: Actin mutations inhibit meiosis-dependent mitochondrial rearrangement in the budding yeast. *Saccharomyces cerevisiae Mol. Biol. Cell.* 6, 1381–1396.

34 Sun, J., Bird, C.H., Salem, H.H., Bird, P. (1993) Association of annexin V with mitochondria. *FEBS Lett.* 329, 79–83.

35 Milner, D.J., Mavroidis, M., Weisleder, N., Capetanaki, Y. (2000) Desmin cytoskeleton linked to muscle mitochondrial distribution and respiratory function. *J. Cell. Biol.* 150, 1283–1298.

36 Vendelin, M., Beraud, N., Guerrero, K., Andrienko, T., Kuznetsov, A.V., Olivares, J., Kay, L., Saks, V.A. (2005) Mitochondrial regular arrangement in muscle cells: a "crystal-like" pattern. *Am. J. Physiol. Cell. Physiol.* 288, C757–C767.

37 Seppet, E., Kaambre, T., Sikk, P., Tiivel, T., Vija, H., Kay, L., Appaix, F., Tonkonogi, M., Sahlin, K., Saks, V.A. (2001) Functional complexes of mitochondria with MgATPases of myofibrils and sarcoplasmic reticulum in muscle cells. *Biochim. Biophys. Acta* 1504, 379–395.

38 Kaasik, A., Veksler, V., Boehm, E., Novotova, M., Minajeva, A., Ventura-Clapier, R. (2001) Energetic crosstalk between organelles. Architectural integration of energy production and utilization. *Circ. Res.* 89, 153–159.

39 Kay, L., Li, Z., Mericskay, M., Olivares, J., Tranqui, L., Fontaine, E., Tiivel, T., Sikk, P., Kaambre, T., Samuel, J.L., Rappaport, L., Usson, Y., Leverve, X., Paulin, D., Saks, V.A. (1997) Study of regulation of mitochondrial respiration *in vivo*. An analysis of influence of ADP diffusion and possible role of cytoskeleton. *Biochim. Biophys. Acta.* 1322, 41–59.

40 Saks, V., Kuznetsov, A.V., Andrienko, T., Usson, Y., Appaix, F., Guerrero, K., Kaambre, T., Sikk, P., Lemba, M., Vendelin, M. (2003) Heterogeneity of ADP diffusion and regulation of respiration in cardiac cells. *Biophys. J.* 84, 3436–3456.

41 Yi, M., Weaver, D., Hajnoczky, G. (2004) Control of mitochondrial motility and distribution by the calcium signal: a homeostatic circuit. *J. Cell. Biol.* 167, 661–72.

42 Yaffe, M.P. (1999) Dynamic mitochondria. *Nat. Cell. Biol.* 1, E149–E150.

43 Karbowski, M., Youle, R.J. (2003) Dynamics of mitochondrial morphology in healthy cells and during apoptosis. *Cell Death and Differentiation* 10, 870–880.

44 Perfettini, J.L., Roumier, T., Kroemer, G. (2005) Mitochondrial fusion and fission in the control of apoptosis. *TRENDS in Cell Biology* 15, 179–183.

45 Griparic, L., van der Wel, N.N., Orozco, I.J., Peters, P.J., van der Bliek, A.M. (2004) Loss of the intermembrane space protein Mgm1/OPA1 induces swelling and localized constrictions along the lengths of mitochondria. *J. Biol. Chem.* 279, 18792–18798.

46 Chen, H., Chan, D.C. (2005) Emerging functions of mammalian mitochondrial fusion and fission. *Hum. Mol. Genet.* 14, R283–R289.

47 Smirnova, E., Griparic, L., Shurland, D.L., van der Bliek, A.M. (2001) Dynamin-related protein Drp1 is required for mitochondrial division in mammalian cells. *Mol. Biol. Cell.* 12, 2245–2256.

48 Sesaki, H., Jensen, R.E. (1999) Division versus fusion: Dnm1p and Fzo1p antagonistically regulate mitochondrial shape. *J. Cell. Biol.* 147, 699–706.

49 Burgess, S.M., Delannoy, M., Jensen, R.E. (1994) MMM1 encodes a mitochondrial outer membrane protein essential for establishing and maintaining the structure of yeast mitochondria. *J. Cell. Biol.* 126, 1375–1391.

50 Jensen, R.E. (2005) Control of mitochondrial shape. *Curr. Opin. Cell. Biol.* 17, 384–388.

51 Berger, K.H., Sogo, L.F., Yaffe, M.P. (1997) Mdm12p, a component required for mitochondrial inheritance that is conserved between budding and fission yeast. *J. Cell. Biol.* 136, 545–553.

52 Dimmer, K.S., Jakobs, S., Vogel, F., Altmann, K., Westermann, B. (2005) Mdm31 and Mdm32 are inner membrane proteins required for maintenance of mitochondrial shape and stability of mitochondrial DNA nucleoids in yeast. *J. Cell. Biol. 168*, 103–115.

53 Sogo, L.F., Yaffe, M.P. (1994) Regulation of mitochondrial morphology and inheritance by Mdm10p, a protein of the mitochondrial outer membrane. *J. Cell. Biol. 126*, 1361–1373.

54 Youngman, M.J., Hobbs, A.E., Burgess, S.M., Srinivasan, M., Jensen, R.E. (2004) Mmm2p, a mitochondrial outer membrane protein required for yeast mitochondrial shape and maintenance of mtDNA nucleoids. *J. Cell. Biol. 164*, 677–688.

55 Frederick, R.L., McCaffery, J.M., Cunningham, K.W., Okamoto, K., Shaw, J.M. (2004) Yeast Miro GTPase, Gem1p, regulates mitochondrial morphology via a novel pathway. *J. Cell. Biol. 167*, 87–98.

56 John, G.B., Shang, Y., Li, L., Renken, C., Mannella, C.A., Selker, J.M., Rangell, L., Bennett, M.J., Zha, J. (2005) The mitochondrial inner membrane protein mitofilin controls cristae morphology. *Mol. Biol. Cell. 16*, 1543–1554.

57 Paumard, P., Vaillier, J., Coulary, B., Schaeffer, J., Soubannier, V., Mueller, D.M., Brethes, D., di Rago, J.P., Velours, J. (2002) The ATP synthase is involved in generating mitochondrial cristae morphology. *EMBO J. 21*, 221–230.

58 Thomson, M. (2002) The regulation of mitochondrial physiology by organelle-associated GTP-binding proteins. *Cell. Biochem. Funct. 20*, 273–278.

59 Rossignol, R., Gilkerson, R., Aggeler, R., Yamagata, K., Remington, S.J., Capaldi, R.A. (2004) Energy substrate modulates mitochondrial structure and oxidative capacity in cancer cells. *Cancer Res. 64*, 985–993.

60 Mozdy, A.D., Shaw, J.M. (2003) A fuzzy mitochondrial fusion apparatus comes into focus. *Nat. Rev. Mol. Cell. Biol. 4*, 468–78.

61 Bleazard, W., McCaffery, J.M., King, E.J., Bale, S., Mozdy, A., Tieu, Q., Nunnari, J., Shaw, J.M. (1999) The dynamin-related GTPase Dnm1 regulates mitochondrial fission in yeast. *Nat. Cell. Biol. 1*, 298–304.

62 James, D.I., Parone, P.A., Mattenberger, Y., Martinou, J.C. (2003) hFis1, a novel component of the mammalian

mitochondrial fission machinery. *J. Biol. Chem. 278*, 36373–36379.

63 Chen, H., Detmer, S.A., Ewald, A.J., Griffin, E.E., Fraser, S.E., Chan, D.C. (2003) Mitofusins Mfn1 and Mfn2 coordinately regulate mitochondrial fusion and are essential for embryonic development. *J. Cell. Biol. 160*, 189–200.

64 Cipolat, S., Martins de Brito, O., Dal Zilio, B., Scorrano, L. (2004) OPA1 requires mitofusin 1 to promote mitochondrial fusion. *Proc. Natl. Acad. Sci. USA 101*, 15927–15932.

65 Hermann, G.J., Thatcher, J.W., Mills, J.P., Hales, K.G., Fuller, M.T., Nunnari, J., Shaw, J.M. (1998) Mitochondrial fusion in yeast requires the transmembrane GTPase Fzo1p. *J. Cell. Biol. 143*, 359–373.

66 Cereghetti, G.M., Scorrano, L. (2006) The many shapes of mitochondrial death. *Oncogene 25*, 4717–4724.

67 Griffin, E.E., Graumann, J., Chan, D.C. (2005) The WD40 protein Caf4p is a component of the mitochondrial fission machinery and recruits Dnm1p to mitochondria. *J. Cell Biol. 170*, 237–248.

68 Zuchner, S., Mersiyanova, I.V., Muglia, M., Bissar-Tadmouri, N., Rochelle, J., Dadali, E.L., Zappia, M., Nelis, E., et al. (2004) Mutations in the mitochondrial GTPase mitofusin 2 cause Charcot–Marie–Tooth neuropathy type 2A. *Nat. Genet. 36*, 449–451.

69 Alexander, C., Votruba, M., Pesch, U.E., Thiselton, D.L., Mayer, S., Moore, A., Rodriguez, M., Kellner, U., et al. (2000) OPA1, encoding a dynamin-related GTPase, is mutated in autosomal dominant optic atrophy linked to chromosome 3q28. *Nat. Genet. 26*, 211–215.

70 Ishihara, N., Jofuku, A., Eura, Y., Mihara, K. (2003) Mitochondrial fusion is completely inhibited by protonophores that dissipate the inner membrane potential. *Biochem. Biophys. Res. Commun. 301*, 891–898.

71 Frank, S., Gaume, B., Bergmann-Leitner, E.S., Leitner, W.W., Robert, E.G., Catez, F., Smith, C.L., Youle, R.J. (2001) The role of dynamin-related protein 1, a mediator of mitochondrial fission, in apoptosis. *Dev. Cell. 1*, 515–525.

72 Breckenridge, D.G., Stojanovic, M., Marcellus, R.C., Shore, G.C. (2003) Caspase cleavage product of BAP31 induces mitochondrial fission through endoplasmic reticulum calcium signals, enhancing cytochrome *c* release to the cytosol. *J. Cell. Biol.* 160, 1115–1127.

73 Lee, Y.J., Jeong, S.Y., Karbowski, M., Smith, C.L., Youle, R.J. (2004) Roles of the mammalian mitochondrial fission and fusion mediators Fis1, Drp1, and Opa1 in apoptosis. *Mol. Biol. Cell.* 15, 5001–5011.

74 Youle, R.J., Karbowski, M. (2005) Mitochondrial fission in apoptosis. *Nat. Rev. Mol. Cell. Biol.* 6, 657–663.

75 Karbowski, M., Lee, Y.-J., Gaume, B., Jeong, S.-Y., Frank, S., Nechushtan, A., Santel, A., Fuller, M., Smith, C.L., Youle, R.J. (2002) Spatial and temporal association of Bax with mitochondrial fission sites, Drp1, and Mfn2 during apoptosis. *J. Cell. Biol.* 159, 931–938.

76 Takahashi, Y., Karbowski, M., Yamaguchi, H., Kazi, A., Wu, J., Sebti, S.M., Youle, R.J., Wang, H.G. (2005) Loss of Bif-1 suppresses Bax/Bak conformational change and mito-chondrial apoptosis. *Mol. Cell. Biol.* 25, 9369–9382.

77 Karbowski, M., Norris, K.L., Cleland, M.M., Jeong, S.Y., Youle, R.J. (2006) Role of Bax and Bak in mitochondrial morphogenesis. *Nature* 443, 658–662.

78 Olichon, A., Baricault, L., Gas, N., Guillou, E., Valette, A., Belenguer, P., Lenaers, G. (2002) Loss of OPA1 perturbates the mitochondrial inner membrane structure and integrity, leading to cytochrome *c* release and apoptosis. *J. Biol. Chem.* 278, 7743–7746.

79 Boldogh, I.R., Pon, L.A. (2006) Interactions of mitochondria with the actin cytoskeleton. *Biochim. Biophys. Acta 1763*, 450–462.

80 Tang, H.L., Le, A.H., Lung, H.L. (2006) The increase in mitochondrial association with actin precedes Bax translocation in apoptosis. *Biochem. J.* 396, 1–5.

81 Pacher, P., Hajnoczky, G. (2001) Propagation of the apoptotic signal by mitochondrial waves. *EMBO J. 20*, 4107–4121.

82 Szabadkai, G., Simoni, A.M., Chami, M., Wieckowski, M.R., Youle, R.J., Rizzuto, R. (2004) Drp-1-dependent division of the mitochondrial network blocks intraorganellar Ca2+ waves and protects against Ca2+-mediated apoptosis. *Mol. Cell. 16*, 59–68.

83 Hollenbeck, P.J., Saxton, W.M. (2005) The axonal transport of mitochondria. *J. Cell. Sci. 118*, 5411–5419.

84 Simon, V.R., Swayne, T.C., Pon, L.A. (1995) Actin-dependent mitochondrial motility in mitotic yeast and cell-free systems: identification of a motor activity on the mitochondrial surface. *J. Cell. Biol. 130*, 345–354.

85 Fehrenbacher, K.L., Yang, H.C., Gay, A.C., Huckaba, T.M., Pon, L.A. (2004) Live cell imaging of mitochondrial movement along actin cables in budding yeast. *Curr. Biol. 14*, 1996–2004.

86 Bantseev, V., Sivak, J.G. (2005) Confocal laser scanning microscopy imaging of dynamic TMRE movement in the mitochondria of epithelial and superficial cortical fiber cells of bovine lenses. *Mol. Vis. 11*, 518–523.

87 Park, M.K., Ashby, M.C., Erdemli, G., Petersen, O.H., Tepikin, A.V. (2001) Perinuclear, perigranular and sub-plasmalemmal mitochondria have distinct functions in the regulation of cellular calcium transport. *EMBO J. 20*, 1863–1874.

88 Vale, R.D., Funatsu, T., Pierce, D.W., Romberg, L., Harada, Y., Yanagida, T. (1996) Direct observation of single kinesin molecules moving along microtubules. *Nature 380*, 451–453.

89 Vale, R.D. (2003) The molecular motor toolbox for intracellular transport. *Cell* 112, 467–480.

90 Brady, S.T., Lasek, R.J., Allen, R.D. (1982) Fast axonal transport in extruded axoplasm from squid giant axon. *Science* 218, 1129–1131.

91 Pereira, A.J., Dalby, B., Stewart, R.J., Doxsey, S.J., Goldstein, L.S. (1997) Mitochondrial association of a plus end-directed microtubule motor expressed during mitosis in *Drosophila*. *J. Cell Biol. 136*, 1081–1090.

92 Stowers, R.S., Megeath, L.J., Gorska-Andrzejak, J., Meinertzhagen, I.A.,

Schwarz, T.L. (2002) Axonal transport of mitochondria to synapse depends on milton, a novel *Drosophila* protein. *Neuron 36*, 1063–1077.

93 Boldogh, I.R., Yang, H.C., Nowakowski, W.D., Karmon, S.L., Hays, L.G., Yates, 3rd J.R., Pon, L.A. (2001) Arp2/3 complex and actin dynamics are required for actin-based mitochondrial motility in yeast. *Proc. Natl. Acad. Sci. USA 98*, 3162–3167.

94 Smith, M.G., Simon, V.R., O'Sullivan, H., Pon, L.A. (1995) Organelle-cytoskeletal interactions: actin mutations inhibit meiosis-dependent mitochondrial rearrangement in the budding yeast. *Saccharomyces cerevisiae Mol. Biol. Cell. 6*, 1381–1396.

95 Suelmann, R., Fischer, R. (2000) Mitochondrial movement and morphology depend on an intact actin cytoskeleton in *Aspergillus nidulans. Cell. Motil. Cytoskeleton 45*, 42–50.

96 Sturmer, K., Baumann, O., Walz, B. (1995) Actin-dependent light-induced translocation of mitochondria and ER cisternae in the photoreceptor cells of the locust *Schistocerca gregaria. J. Cell. Sci. 108*, 2273–2283.

97 Van Gestel, K., Kohler, R.H., Verbelen, J.P. (2002) Plant mitochondria move on F-actin, but their positioning in the cortical cytoplasm depends on both F-actin and microtubules. *J. Exp. Bot. 53*, 659–667.

98 Morris, R.L., Hollenbeck, P.J. (1995) Axonal transport of mitochondria along microtubules and F-actin in living vertebrate neurons. *J. Cell. Biol. 131*, 1315–1326.

99 Varadi, A., Johnson-Cadwell, L.I., Cirulli, V., Yoon, Y., Allan, V.J., Rutter, G.A. (2004) Cytoplasmic dynein regulates the subcellular distribution of mitochondria by controlling the recruitment of the fission factor dynamin-related protein-1. *J. Cell. Sci. 117*, 4389–4400.

100 Chang, D.T., Reynolds, I.J. (2006) Differences in mitochondrial movement and morphology in young and mature primary cortical neurons in culture. *Neuroscience 141*, 727–736.

101 Chada, S.R., Hollenbeck, P.J. (2003) Mitochondrial movement and positioning in axons: the role of growth factor signaling. *J. Exp. Biol. 206*, 1985–1992.

102 Miller, K.E., Sheetz, M.P. (2004) Axonal mitochondrial transport and potential are correlated. *J. Cell. Sci. 117*, 2791–2804.

103 Rintoul, G.L., Filiano, A.J., Brocard, J.B., Kress, G.J., Reynolds, I.J. (2003) Glutamate decreases mitochondrial size and movement in primary forebrain neurons. *J. Neurosci. 23*, 7881–7888.

104 Li, Y.C., Zhai, X.Y., Ohsato, K., Futamata, H., Shimada, O., Atsumi, S. (2004) Mitochondrial accumulation in the distal part of the initial segment of chicken spinal motoneurons. *Brain Res. 1026*, 235–243.

105 Trimmer, P.A., Swerdlow, R.H., Parks, J.K., Keeney, P., Bennett, J.P., Miller, S.W., Davis, R.E., Parker, W.D. (2000) Abnormal mitochondrial morphology in sporadic Parkinson's and Alzheimer's disease cybrid cell lines. *Exp. Neurol. 162*, 37–50.

106 Delettre, C., Lenaers, G., Griffoin, J.M., Gigarel, N., Lorenzo, C., Belenguer, P., Pelloquin, L., Grosgeorge, J., Turc-Carel, C., Perret, E., Astarie-Dequeker, C., Lasquellec, L., Arnaud, B., Ducommun, B., Kaplan, J., Hamel, C.P. (2000) Nuclear gene OPA1, encoding a mitochondrial dynamin-related protein, is mutated in dominant optic atrophy. *Nat. Genet. 26*, 207–210.

107 Olichon, A., Guillou, E., Delettre, C., Landes, T., Arnaune-Pelloquin, L., Emorine, L.J., Mils, V., Daloyau, M., Hamel, C., Amati-Bonneau, P., Bonneau, D., Reynier, P., Lenaers, G., Belenguer, P. (2006) Mitochondrial dynamics and disease. *Biochim. Biophys. Acta 1763*, 500–509.

108 Sesaki, H., Jensen, R.E. (2001) UGO1 encodes an outer membrane protein required for mitochondrial fusion. *J. Cell. Biol. 152*, 1123–1134.

109 Sesaki, H., Jensen, R.E. (2004) Ugo1p links the Fzo1p and Mgm1p GTPases for mitochondrial fusion. *J. Biol. Chem. 279*, 28298–28303.

110 Alto, N.M., Soderling, J., Scott, J.D. (2002) Rab32 is an A-kinase anchoring protein and participates in

mitochondrial dynamics. *J. Cell Biol.* *158*, 659–668.

111 Messerschmitt, M., Jakobs, S., Vogel, F., Fritz, S., Dimmer, K.S., Neupert, W., Westermann, B. (2003) The inner membrane protein Mdm33 controls mitochondrial morphology in yeast. *J. Cell Biol. 160*, 553–564.

112 Chen, W., Calvo, P.A., Malide, D., Gibbs, J., Schubert, U., Bacik, I., Basta, S., O'Neill, R., Schickli, J., Palese, P., Henklein, P., Bennink, J.R., Yewdell, J.W. (2001) A novel influenza A virus mitochondrial protein that induces cell death. *Nat. Med. 7*, 1306–1312.

113 Nechushtan, A., Smith, C.L., Lamensdorf, I., Yoon, S.-H., Youle, R.J. (2001) Bax and Bak coalesce into novel mitochondria-associated clusters during apoptosis. *J. Cell. Biol. 153*, 1265–1276.

114 Wolter, K.G., Hsu, Y.T., Smith, C.L., Nechushtan, A., Xi, X.G., Youle, R.J. (1997) Movement of Bax from the cytosol to mitochondria during apoptosis. *J. Cell. Biol. 139*, 1281–1292.

115 Scorrano, L., Ashiya, M., Buttle, K., Weiller, S., Oakes, S.A., Mannella, C.A., Korsmeyer, S.J. (2002) A distinct pathway remodels mitochondrial cristae and mobilizes cytochrome *c* during apoptosis. *Dev. Cell. 2*, 55–67.

116 Esposti, M.D., Erler, J.T., Hickman, J.A., Dive, C. (2001) Bid, a widely expressed proapoptotic protein of the Bcl-2 family, displays lipid transfer activity. *Mol. Cell. Biol. 21*, 7268–7276.

117 Kowaltowski, A.J., Cosso, R.G., Campos, C.B., Fiskum, G. (2002) Effect of Bcl-2 overexpression on mitochondrial structure and function. *J. Biol. Chem. 277*, 42802–42807.

118 Goldmacher, V.S., Bartle, L.M., Skaletskaya, A., Dionne, C.A., Kedersha, N.L., Vater, C.A., Han, J.-w., Lutz, R.J., Watanabe, S., Cahir McFarland, E.D., Kieff, E.D., Mocarski, E.S., Chittenden, T. (1999) A cytomegalovirus-encoded mitochondrialocalized inhibitor of apoptosis structurally unrelated to Bcl-2. *Proc. Natl. Acad. Sci. USA 96*, 12536–12541.

Part II
Energy Transfer Networks, Metabolic Feedback Regulation, and Modeling of Cellular Energetics

Molecular System Bioenergetics: Energy for Life. Edited by Valdur Saks
Copyright © 2007 WILEY-VCH Verlag GmbH & Co. KGaA, Weinheim
ISBN: 978-3-527-31787-5

6
Mitochondrial VDAC and Its Complexes

Dieter Brdiczka

Abstract

Research over the last decade has extended the current view of mitochondria to include functions well beyond their role in bioenergetics of supplying ATP. It is now recognized that mitochondria play a crucial role in cell signaling events, inter-organelle communication, aging, many diseases, cell proliferation, and cell death. Apoptotic signal transmission to the mitochondria results in the efflux of a number of potential apoptotic regulators to the cytosol that trigger caspase activation and lead to cell destruction. Accumulating evidence indicates that the voltage-dependent anion channel (VDAC) is involved substantially as a receptor in this release of proteins via the outer mitochondrial membrane. VDAC is in a key position in the outer mitochondrial membrane, forming the main interface between the mitochondrial and the cellular metabolisms. It has been recognized that VDAC interacts on the one hand with several cytosolic proteins of the pro- and antiapoptotic Bcl-2 protein family and, due to its receptor function, causes release of proapoptotic proteins located in the intermembrane space. On the other hand, VDAC interacts with proteins in the intermembrane space and the main regulator of oxidative phosphorylation, the adenine nucleotide translocator (ANT). The interaction with ANT in the inner membrane leads to formation of contact sites. This chapter focuses on the structure and function of protein complexes in mitochondrial contact sites and their regulatory role in cellular bioenergetics, intra and extramitochondrial Ca^{2+} levels, and the release of apoptosis-inducing factors. I will only shortly point out the complex of mitochondrial creatine kinase (mCK) with VDAC, as Chapter 7 discusses this in detail. The text will provide insights into the central role of the transient VDAC ANT complex in cell life and death. It will be shown that the complex indicates the functional state of the mitochondria and is also a target of hormonal regulation of mitochondrial function through protein kinases.

Molecular System Bioenergetics: Energy for Life. Edited by Valdur Saks
Copyright © 2007 WILEY-VCH Verlag GmbH & Co. KGaA, Weinheim
ISBN: 978-3-527-31787-5

6.1
The Role of VDAC in Controlling the Interaction of Mitochondria with the Cytosol

For a thorough understanding of the function of VDAC, it is informative to recapitulate the evolutionary origin of the mitochondrial permeability transition pore
(MPT pore). The original ATP-providing system of the cell was anaerobic glycolysis ending with lactate production. According to the endosymbiont hypothesis [1],
a second ATP energy–providing system was acquired: oxygen-dependent phosphorylation. In order to integrate the latter into the energy metabolism and convert the previous bacteria into mitochondria, the cell had to overcome difficult
tasks. It had to keep the two compartments, the cytosol and the inner membrane
matrix, separated but also had to allow metabolite exchange. The metabolic parti-
tion is necessary because ATP synthesis in the two compartments has opposite
prerequisites: firstly, while ATP synthesis through glycolysis depends on a high
NAD:NADH quotient in the cytosol (700), the opposite is true for oxidative
phosphorylation (the NAD:NADH quotient is 7 in mitochondria, see Fig. 6.1);
[2] secondly, energy-consuming systems, for example, ion pumps and actomyosin
ATPase, depend on high free energy ΔG of the phosphoryl group transfer from
ATP. To provide a ΔG of 59 KJ per mol, the ATP:ADP quotient has to be approximately 800 in the cytosol, whereas it is about 5 in the mitochondria in order to
support ATP synthesis [3, 4] (see Fig. 6.1). The hydrogen from cytosolic NADH
is shuttled to the rotenone-sensitive NADH oxidase in the inner membrane as
glycerol phosphate or malate through the glutamate–aspartate shuttle. However,
a direct electron transfer exists to cytochrome oxidase from rotenone-insensitive
outer membrane NADH oxidase, via cytochrome b_5, involving cytochrome c and
bypassing complex III [5]. The direct electron transfer by cytochrome c depends
on the existence of contact sites [6]. As shown in Fig. 6.1 a chain of cytochrome c
molecules might be involved, transferring the electrons to cytochrome oxidase in
the cristae membranes.

For ATP exchange the brilliant mechanism is not to export the ATP but
simply to transfer the phosphoryl group from mitochondrial ATP to either glucose or creatine. This phosphoryl group transfer is performed by conversion of
the ANT from an unspecific uniporter pore to an antiporter specific for ATP and
ADP and coupling of either hexokinase or creatine kinase to it (see Fig. 6.2) [7].
The transfer of the phosphoryl group to creatine serves to conserve high ΔG as
phosphocreatine.

VDAC [8] is the major transport protein in the outer mitochondrial membrane. As depicted in Fig. 6.1, the same isotype I of VDAC present in the outer
mitochondrial membrane is also a component of the cell membrane [9]. According to the aforementioned endosymbiont hypothesis, the incorporated bacteria
(mitochondria) can still be considered extracellular, with the outer mitochondrial
membrane providing a barrier between the mitochondrial inner membrane matrix and the cell cytosol, and in this membrane VDAC mediates the multifaceted
interactions between mitochondria and other parts of the cell by transporting
anions, cations, ATP, Ca^{2+}, and metabolites. Thus, VDAC plays an important

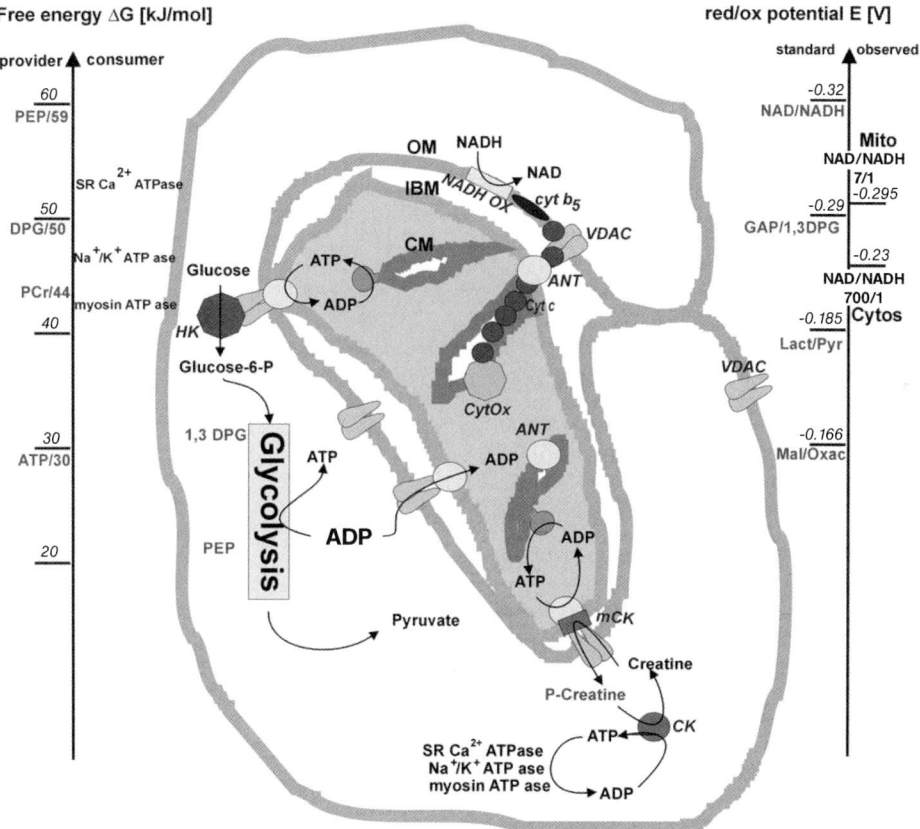

Free energy ΔG [kJ/mol]

red/ox potential E [V]

Fig. 6.1. Biochemical consequences of the mitochondrial origin as an endosymbiont. The mitochondrial outer membrane (OM) originates from the cell membrane and contains the same VDAC isotype I. The peripheral inner membrane (inner boundary membrane, IBM) has the specific function of interacting with the outer membrane, while the cristae membrane (CM) has no contact with the OM. VDAC regulates the integration of glycolysis and oxidative phosphorylation by forming complexes with the adenine nucleotide translocator (ANT), hexokinase (HK), and mitochondrial creatine kinase (mCK). The HK–VDAC–ANT complex has the function of adapting glucose phosphorylation and the rate of glycolysis to intramitochondrial ATP synthesis. The VDAC–mCK–ANT complex has the function of increasing the high free energy ΔG of the phosphoryl group transfer from ATP. On the left side, the standard ΔG (in kilojoules per mol) is shown for phosphor-enol pyruvate (PEP; 59 kJ mol^{-1}), diphosphoglycerate (DPG; 50 kJ mol^{-1}), and phosphocreatine (PCr; 44 kJ mol^{-1}). The standard ΔG is compared with the energy consumption of the SR Ca^{2+} pump, 52 kJ mol^{-1}; Na^+/K^+-ATPase, 48 kJ mol^{-1}; and actomyosin ATPase, 43 kJ mol^{-1}. On the right side, the direct NADH oxidation pathway is shown, involving outer membrane rotenone-insensitive NADH oxidase, cytochrome b_5, cytochrome c transferring electrons to cytochrome oxidase, and bypassing complex III. The noncalibrated scale on the right side lists standard redox potentials of NAD, glyceraldehyde phosphate, lactate, and malate and compares this with the physiologically observed redox potentials of NAD/NADH in the cytosol and the mitochondria, which differ significantly.

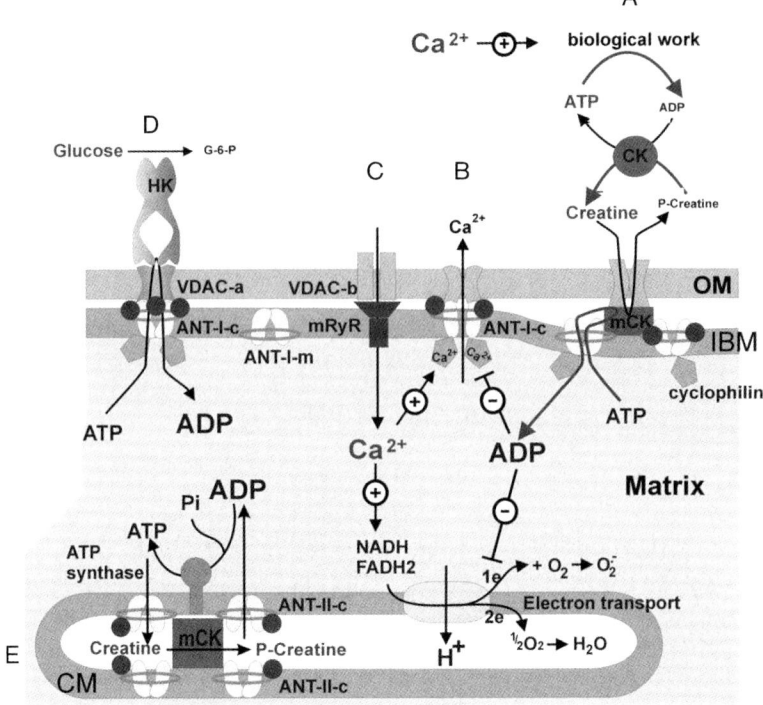

Fig. 6.2. Metabolite circuits involving the VDAC–ANT–kinase complexes.

(A) Biological work (e.g., contraction) triggers cytosolic ADP–ATP turnover and leads to an increase in cytosolic creatine through the activity of cytosolic creatine kinase (CK). Creatine causes an increase in intramitochondrial ADP by the activity of mitochondrial creatine kinase (mCK), which is linked to the adenine nucleotide translocator (ANT) through cardiolipin (dark rings). In a complex with mCK, the ANT acts as anti-porter in the inner membrane.

(B) In the absence of ADP and the presence of high Ca^{2+}, ANT can form an unspecific permeability transition pore (MPT pore) that releases Ca^{2+}.

(C) The mitochondrial ryanodine receptor (mRyR) may be liked to VDAC. Ca^{2+} activates substrate supply to the electron transport chain. This may lead to increased ROS formation (e.g., superoxide) if it is not accompanied by ADP increase.

(D) In the hexokinase complex, the enzyme forms a tetramer that is linked to two dimers

of VDAC and ANT. In brain and heart mitochondria, the ANT isotype 1 is in this complex. ANT-1 is in its c-conformation (ANT-1-c, induced by atractyloside), where it can operate either as antiporter to exchange ATP for ADP or as uniporter to form an unspecific pore. ANT in the m-conformation (ANT-m, induced by bongkrekate) does not form complexes with VDAC and cannot form the MPT pore. The oligomerization of hexokinase leads to activation of the enzyme. VDAC obtains a different structure by interaction with ANT in the c-conformation, which has higher capacity to bind hexokinase. Both kinase complexes contain cytochrome *c* (represented by black dots). It is bound to cardiolipin (dark ring).

(E) ANT isotype 2 was found in the cristae together with the octamer of mitochondrial creatine kinase. This enzyme attaches specifically to the cardiolipin that is surrounding ANT-2. OM: outer mitochondrial membrane, IBM: inner boundary membrane, CM: cristae membrane.

role in coordinating communication between mitochondria and the cytosol. An essential aspect of this management is the transient formation of complexes of VDAC with other proteins (see below). The complexes of VDAC with either ANT or mitochondrial creatine kinase have been found to generate contact sites between the mitochondrial outer membrane and the peripheral inner membrane (IBM) (Fig. 6.1) [10–13]. The bridging between membranes by intermembrane proteins in the contact sites provides the basis for phospholipid exchange such as cardiolipin [14]. Because energy metabolism is decisive especially in excitable tissues, it appears plausible that structures such as contacts present the preferred targets for either apoptosis or growth regulation [7].

6.2
Molecular Structure and Membrane Topology of VDAC

The assumptions about VDAC structure are, in the absence of a crystal structure, derived from the known structure of bacterial porins [15] to which VDAC is related. Mammalian VDAC is a 31-kDa protein with stretches of alternating hydrophobic and hydrophilic amino acids that form a transmembrane β-barrel composed of a single α-helix at the N-terminus and several transmembrane amphipathic β-strands. According to different models, the α-helix at the N-terminus protrudes outside, [16] crosses the membrane [17], or interacts with the membrane or membrane proteins [9]. Computer modeling of VDAC's primary amino acid sequence has led to the development of models showing the transmembrane organization, consisting of a single α-helix and 13-, 16- or 19-transbilayer β-strands. These β-strands are connected by several peptide loops of different sizes on both sides of the membrane that serve as potential protein interacting sites [18–20]. It has been proposed that monomeric VDAC reconstituted in artificial membranes can serve as the functional channel [21]. However, evidence consistent with the oligomerization of purified VDAC from rat liver or *Neurospora crassa*, and in yeast outer membrane suggests that VDAC forms dimers [22, 23] and possibly tetramers. In brain mitochondria [24] or recombinant human VDAC [25], the existence of VDAC dimers and tetramers in mitochondria has been reported. Thus, it was recently proposed that VDAC oligomerization functions in mitochondria-mediated apoptosis [24, 25].

6.3
VDAC Conductance, Voltage Dependence, and Ion Selectivity

VDAC has been purified by using various procedures and detergents and was characterized by reconstitution into a planar lipid bilayer. Bilayer-reconstituted VDAC forms a voltage-dependent channel and possesses multiple conformational states, resulting in varying selectivity and permeability [15, 26–28]. Depending on the substate, the VDAC pore diameter has been estimated to be between 1.8 and

3 nm [15, 29]. VDAC shows a symmetrical voltage-dependent conductance [30], depending on the transmembrane potential. The channel has its highest conductance states at low potentials (4 nS at 1 M KCl), with a preference for anions such as phosphate, chloride, adenine nucleotides, and many anionic metabolites. At higher positive or negative potentials (>30–60 mV), the pore size is reduced to 2 nm and VDAC is at subconductance states, with 50–60% reduced conductance; the selectivity shifts to small cations and the channel becomes almost impermeable to ATP and ADP [31–33]. As a voltage-gated channel, VDAC has a voltage sensor that responds to changes in transmembrane voltage. It is believed that VDAC channels have two separate gating processes: one at positive transmembrane potentials and the other at negative potentials, where VDAC closure is associated with motion of the positively charged voltage sensor [19, 34]. It has been postulated that a positively charged loop moves out of the channel [34], but it is also possible that negative charges move into the mouth of the channel, as has been observed for bacterial porins [35]. Thus, the molecular nature of the VDAC gating mechanism is not yet resolved.

6.4
The Physiological Significance of VDAC Voltage Gating

As discussed above, at high positive or negative membrane potentials, VDAC is in the low-conductance, cation-selective state that is not permeable to ATP, ADP, or other negatively charged small molecules. Thus, voltage changes can control the fluxes of ions and metabolites in and out of the mitochondria. The physiological significance of the voltage dependence of VDAC is not clear and has been quite controversial. For years, it was assumed that the outer mitochondrial membrane was freely permeable to various ions and metabolites due to VDAC and thus no membrane potential could be maintained. If there is no significant membrane potential across the outer membrane, one would not expect modulation of VDAC conductance and selectivity by voltage changes. However, it was found that VDAC is a controlled channel and that the outer membrane limits transport in and out of heart and red muscle mitochondria [36]. A variety of ways to generate a membrane potential have been proposed. It is known that an intrinsic membrane potential exists across the outer membrane generated by charge difference as a natural consequence of charged membrane–integrated protein molecules and the different phospholipid composition of the inner and outer membrane leaflet [37]. Another proposal takes into account the motion of charged substrates associated with mitochondrial metabolism [38] and differential permeability of VDAC to metabolites that also could result in a transmembrane potential. It has been postulated that VDAC is voltage regulated in the contact sites where the potential across the inner membrane would somehow spill over by capacitive coupling to the outer membrane due to the proximity of the two membranes [31]. While this specific proposal is rather speculative, metabolic cycling associated with ATP, ADP, and P_i translocation could couple potential generation in the inner membrane to the formation of a potential in the outer membrane (for details, see Ref. [38]). The demonstration of a membrane potential across the outer membrane and its physiological significance await additional studies.

6.5
VDAC Isoforms and Functions

In human, rat, and mouse, three different VDAC genes encoding distinctly expressed isoforms have been reported [39, 40]. VDAC isoforms appear to have specialized functions, because the more complex the multicellular organism is, the more VDAC isoforms it possesses. *Neurospora crassa* have only one form of VDAC, *Saccharomyces cerevisiae* have two, and mice and humans have three [40, 41]. In yeast, only VDAC1 (not VDAC2) forms channels when reconstituted into phospholipid membranes, and it seems to provide the major pathway for NADH transport across the outer membrane [40, 42, 43]. Moreover, when type I VDAC is deleted, type II can not replace it; instead, TOM 40, a pore for mitochondrial precursor peptides, is overexpressed [44]. Human and mouse VDACs have more than 85% similarity at the level of amino acids, but they have less than 30% sequence identity with yeast VDACs, although they share similar channel characteristics. Human and mouse VDAC1 and VDAC2 isoforms show channel-forming activity *in vitro* and can complement the yeast VDAC1 deficiency [45], while VDAC3 can only partially complement this defect [46].

The three genes that encode distinct isoforms of mouse VDAC, VDACs 1–3, have been characterized [39]. Each isoform is 65–70% identical to the other isoforms [46]. Phylogenetic analysis indicates that VDAC3 is the more primordial of the vertebrate VDAC genes, suggesting that if the multiple isoforms arose from a gene duplication and divergence event, VDAC3 diverged from the primordial VDAC first, with VDAC1 and VDAC2 arising more recently [47]. Mammalian VDAC isoforms are expressed in a wide variety of tissues and are present at high levels in heart, kidney, brain, and skeletal muscle. VDAC1 was the predominant form observed in the tissues that have been tested [48]. Both VDAC1 and VDAC2 are ubiquitously expressed, while VDAC3 has a more restricted organ distribution [39, 49, 50]. However, the tissue-specific distributions and intracellular localization of the different VDAC isoforms are not well established. VDAC isoforms appear to display different physiological functions. Some differences in voltage-dependent gating and other properties of the VDAC isoforms have been reported [46], and binding of hexokinase (HK) appears to be a specific feature of VDAC1 [45]. However, the possible relationship between such differences and distinct physiological roles for these isoforms remains to be investigated. One difference in function that has been found is that VDAC1 is connected with apoptosis through its interaction with different proteins and factors, while VDAC2 has been associated with antiapoptotic activity [51].

6.6
Mitochondria and Apoptosis

Apoptosis, or programmed cell death, is a genetically regulated process that plays an important role in tissue homeostasis and development in multicellular metazoans. Defects in apoptosis regulation are often associated with diseases such as neuronal degenerative diseases, tumor genesis, autoimmune disorders, and viral infections. Mitochondria play an important role in the regulation of programmed cell death via release of proapoptotic agents and/or disruption of cellular energy metabolism [52]. The proapoptotic signal can be generated inside the mitochondria, for example, by increased Ca^{2+} and/or ROS concentrations (see Fig. 6.2), but it can also be initiated from outside by proapoptotic proteins such as t-Bid and Bax [53, 54].

Following the idea of the bacterial origin of mitochondria discussed above (Fig. 6.1), we may define specific characteristics of internal and external initiation of apoptosis. These include first the permeability transition disrupting the separation of the two compartments, which is fundamental for mitochondrial integration into energy metabolism, and second the access of proapoptotic proteins to cardiolipin at the mitochondrial surface as a target that is a specific component of the inner membrane.

6.7
VDAC and ANT May Form the Mitochondrial Permeability Transition Pore

ANT is a member of the superfamily of membrane transporters. Six membrane-spanning domains of 18–22 residues are separated by hydrophilic domains facing either the mitochondrial matrix or the intermembrane space. According to this model the C- and N-termini protrude into the intermembrane space [55]. Biochemical isolations and reconstitutions in liposomes of the permeability transition pore suggested that the protein complex consists of ANT in the inner membrane and VDAC in the outer membrane. These aggregates display a striking voltage dependence with gating effects that are consistent with the voltage dependence in intact mitochondria [56]. Moreover, reconstituted ANT facilitated malate transport that was inhibited by the same reagents as the MPT pore [57]. As mentioned above, the use of the immunosuppressant cyclosporin A (CsA) marked an important advance in the field in that it showed that the MPT pore can be inhibited by specific ligands. Cyclophilin D (CypD) has since been shown to be the target of CsA. It interacts directly with ANT [58, 59], and a CypD affinity matrix isolated ANT and VDAC and the reconstituted complexes mimicked an MPT pore [58]. CypD is a soluble protein in the mitochondrial matrix. Its crystal structure shows that it shares a common protein structure with other cyclophilins as expected based on the similarity of their primary sequences [60]. A β-barrel is formed by 10 antiparallel β-sheets, which are connected by loops of varying length. This β-barrel is flanked by two α-helices.

6.8
Accessory MPT Pore Subunits

In addition to the classical components of the MPT pore enumerated above, which were supposed to constitute the core of the MPT pore, a number of accessory proteins associated with the MPT pore – such as hexokinase (HK), shown in Fig. 6.2 – have been detected. This protein interacts directly with VDAC at the outer membrane of mitochondria. Its displacement from VDAC is necessary for Bax binding and cell death induction. Hence, Bax as well as other Bcl-2 family members can, depending on the physiological state of the cell, interact with MPT pore components in the outer mitochondrial membrane. This is also the locale of the peripheral benzodiazepine receptor (PBR), an evolutionarily conserved, 18-kDa protein with five membrane-spanning domains that is characterized by its high-affinity binding of a number of pharmacologically important substances. Also, mitochondrial creatine kinase (mCK) is found in the intermembrane space of mitochondria and associates with ANT-1 (MPT pore) at the contact sites as well as with ANT-2 in the cristae (Fig. 6.2) [61]. New data also indicate that cytochrome c, a pivotal apoptosis inducer, is associated with the MPT pore. Moreover, specific lipids interact with components of the MPT pore and thereby modulate the activity of this protein aggregate.

6.9
ANT Knockout Studies

In one study, the two ANT genes in mice were knocked out in hepatocytes [62]. This led to a number of secondary responses such as the upregulation of cytochrome c, the downregulation of UCP2, the stimulation of the respiratory rate, and an increased mitochondrial membrane potential. Experiments on the MPT pore revealed that its sensitivity to Ca^{2+} was reduced by a factor of three. Also, ANT ligands such as atractyloside and ADP no longer had a modifying effect on the permeability transition pore. However, the fact that the permeability transition could still be induced led the authors of the study to suggest that ANT is nonessential for the MPT pore [62].

Since this study was published, another isoform of ANT (ANT-4) has been discovered in the mouse genome [63]. So far its expression has been detected only in stem cells and testis, but it might substitute for the inactivated genes in other tissues as well.

Nevertheless, the fact that the deletion of the two mouse ANT forms led to only a minor effect on the MPT pore was surprising and caused several different inter-pretations, ranging from doubt that the ANT-1/2-deficient tissues are indeed devoid of ANT activity, to speculations about residual ANT activity (an argument supported by the recent discovery of ANT-4, see above), to the notion that other carriers of this gene family might substitute for ANT [64]. These issues as well as the above-described adaptive responses of the cells put into perspective the conclusion that ANT causes a permeability transition regulated by ANT ligands but is dispensable for permeability transition and can be replaced by other membrane components.

6.10
The Role of VDAC in Organizing Kinases at the Mitochondrial Surface

Hexokinase [65, 66] and the octamer of mitochondrial creatine kinase [67] bind specifically to VDAC. The coupling of hexokinase to mitochondrial ATP by the VDAC–ANT complex has the function of regulating the rate of glycolysis by glucose phosphorylation according to the activity of mitochondrial oxidative phosphorylation (Fig. 6.1). While in neurons this coupling is permanent through hexokinase I, it varies depending on the cytosolic Ca^{2+} level in muscle because hexokinase II is missing the hydrophobic anchor [68] and therefore does not permanently interact with the MPT pore. The coupling of VDAC to mitochondrial creatine kinase and ANT has the function of providing a high ΔG by transferring the ATP phosphoryl group to creatine (Fig. 6.1, left scale).

Kinetic analysis of hexokinase isotype I binding to VDAC reconstituted in liposomes, isolated outer mitochondrial membrane, and functionally intact mitochondria showed the different binding behaviors of hexokinase. The capacity and the binding kinetics differed depending on the presence of contact sites with the mitochondrial inner membrane [69]. An increase in mitochondrial contact sites by dextran led to higher binding capacity and cooperative binding behavior in contrast to isolated outer membranes or mitochondria with depressed contact sites by glycerol or DNP [70]. Moreover, induction of contact sites by dextran resulted in a functional coupling of the bound hexokinase with oxidative phosphorylation. The more ATP was produced by the respiratory chain, the more active were hexokinase and the enzymatic reactions of glycolysis. This was observed by following the stimulation of mitochondrial respiration by ADP produced through bound hexokinase. The respiration was studied in the presence of extramitochondrial ADP scavenging by pyruvate kinase and phospho-enol phosphate. The ADP scavenging system was less effective after induction of contact sites by dextran [71]. Xie and Wilson [72] found that hexokinase I forms tetramers by binding to the mitochondrial surface, and Hashimoto and Wilson [73] analyzed the surface of bound hexokinase by using surface domain–specific antibodies. The authors observed that the bound hexokinase changed its structure according to the function of the ANT and oxidative phosphorylation. This suggested a tight coupling between hexokinase and ANT transmitted by VDAC. This finding emphasizes the idea that the bound hexokinase has the function of adapting the rate of glycolysis to the rate of oxidative phosphorylation. In tumor cells this kind of regulation is missing (see below).

6.11
Hexokinase–VDAC–ANT Complexes in Tumor Cells

Cancer cells are characterized by a high rate of glycolysis, which serves as their primary energy-generating pathway [74]. The molecular basis of this high rate of glycolysis involves a number of genetic and biochemical events, including overexpression of the mitochondria-bound isoforms HK-I and HK-II [74]. It has been shown that cancer cells possess high levels of HK activity and that drugs that detach HK from mitochondrial membranes decrease cell viability [75].

Specific to tumor cells is the glucose-induced inhibition of oxidative phosphorylation, known as the Crabtree effect [76]. It has been suggested that mitochondrial oxidative phosphorylation lacks ADP because it is consumed by the high activity of pyruvate kinase outside the mitochondria. This behavior is comparable to the ADP scavenging system described above that was artificially established to study the functional coupling of hexokinase to the ANT. Thus, the Crabtree effect in tumor cells indicates that the functional coupling of hexokinase to the ANT in the contact sites is missing. Indeed, by applying freeze-fracture analysis it was observed that contact sites were absent in HT 29 tumor cells that had a high glycolytic rate and displayed a typical Crabtree effect [77]. However, contact sites reappeared and the Crabtree effect disappeared when the same cells were converted to normal by growing in the absence of glucose. In this case oxidative phosphorylation was the only source of ATP, and VDAC–ANT complexes formed to facilitate ATP/ADP exchange.

6.12
Hexokinase as a Marker Enzyme of Contact Sites

Contact sites between the outer mitochondrial and peripheral inner membrane were analyzed by freeze-fracturing mitochondria and were found to be transient structures that formed depending on the activity of the oxidative phosphorylation of mitochondria [70]. ADP and atractyloside induced the contacts, while glycerol and DNP depressed them [78].

Hexokinase I could be localized by immune electron microscopy in the contacts between the inner and outer boundary membrane. The enzyme was therefore used as a specific marker during isolation of contact sites. Mitochondrial membranes were fragmented by osmotic shock. Centrifugation of the membrane fragments on a sucrose density gradient resulted in a fraction with high hexokinase activity and intermediate density that was distinct from outer (monoamine oxidase) and inner membrane (succinate dehydrogenase) marker enzymes. The contacts contained VDAC and ANT [10, 11] as determined by specific antibodies.

6.13
VDAC–ANT Complexes

The investigations described above suggested that complexes between VDAC and ANT were forming contact sites. This assumption was supported by the observation that a change in ANT structure resulted in an increase or decrease in contact sites [13]. Kidney mitochondria were pretreated with either atractyloside or bongkrekate and subjected to osmotic shock and contact site isolation by sucrose density centrifugation. Hexokinase activity as a marker for contact sites disappeared in the contact site fraction in the presence of bongkrekate, while in the presence of atractyloside it increased to a higher level compared with untreated mitochondria. The results supported the assumption stated above that VDAC–ANT complexes form contact sites to which hexokinase binds specifically. The contact site marker hexokinase was subsequently used to isolate VDAC and ANT complexes from a 1% Triton extract. Hexokinase in the Triton extract was bound to an anion exchanger column and was eluted by a KCl gradient. The hexokinase activity peak contained a tetramer of hexokinase, VDAC, ANT, and cyclophilin D [79, 80]. As observed with hexokinase, cyclophilin D bound to immobilized antibodies was used to isolate the VDAC–ANT complex from a Chaps extract of heart muscle mitochondria [58]. The hexokinase–VDAC–ANT complex as well as the cyclophilin–VDAC–ANT complex were reconstituted in liposomes and resembled the Ca^{2+}-dependent, CsA-sensitive permeability transition pore.

6.14
VDAC–ANT Complexes Contain Cytochrome *c*

Release of cytochrome *c* from mitochondria is considered to be a key step in the apoptotic process, although the precise mechanisms regulating cytochrome *c* release remain unknown [81]. To date, all of the mitochondrial components known to translocate to the cytoplasm following an apoptotic stimulus reside in the mitochondrial intermembrane space. Therefore, only the outer membrane needs to be modified. Indeed, some models suggest that the release exclusively involves an increase in outer membrane permeability that is due to the formation of a channel large enough to account for the release of proteins such as cytochrome *c*, while others consider its release the result of a disruption of the outer membrane [25, 82, 83]. However, it has been observed that cytochrome *c* can be released without rupture of the outer membrane and that contact sites are involved [83]. Approximately 20% of the total cytochrome *c* was found in the contacts site fraction in isolated kidney and brain mitochondria. This amount increased upon induction of contact sites by pre-incubation with atractyloside or by addition of dextran, whereas it decreased after depression of contact sites by pre-incubation with bongkrekate or treatment with glycerol or DNP [13]. It has already been suggested above that tetramers of hexokinase preferentially bind to contact sites formed by VDAC–ANT interaction. Thus, for stoichiometric reasons it may be assumed that tetramers of VDAC also are present in the cytochrome *c*–containing contact sites (see Fig. 6.2, one HK–VDAC–ANT complex in front, another behind).

6.15
VDAC Oligomerization and Cytochrome *c* Binding

Both soluble, purified VDAC and membrane-embedded VDAC can assemble into dimers, trimers, and tetramers, as demonstrated by cross-linking with five cross-linking reagents bearing different spacer lengths (3–16 Å) and membrane permeability [25]. The mechanisms of how cytochrome *c* is organized in the contact sites remain elusive. The involvement of VDAC oligomerization in the release of cytochrome *c* from mitochondria was recently suggested [25]. It was also demonstrated that VDAC oligomerization is induced by the presence of cytochrome *c*. In VDAC-reconstituted liposomes containing encapsulated cytochrome *c*, but not in VDAC liposomes free of cytochrome *c*, VDAC dimers, trimers, and tetramers were observed even without chemical cross-linking [25]. This suggests that cytochrome *c* interacts preferentially with VDAC oligomers present in the contact sites where the peripheral inner membrane contributes cardiolipin, which is known as a specific binding site of cytochrome *c*.

6.16
Possible Function of Cytochrome *c* in the Contact Sites

The nature of cytochrome *c* binding in the contact sites is not yet known. It can only be speculated that the negatively charged phospholipid cardiolipin that is associated with ANT [84] may be the cytochrome *c* (which is a polycation at physiological pH) docking site (Fig. 6.2). The function of cytochrome *c* in the VDAC–ANT complex remains unknown. Bernardi and Azzone [5] described an electron transport pathway between the outer membrane and the peripheral inner membrane in which cytochrome *c* was involved. This pathway channels electrons from external NADH through rotenone-insensitive oxidase via cytochrome b_5 and cytochrome *c* to cytochrome oxidase (see Fig. 6.1). Marzulli et al. [6] observed that oxidation of external NADH was stimulated by induction of contact sites by atractyloside or ADP, whereas it was suppressed by glycerol, which is known to decrease the number of contact sites [70]. This electron transport pathway may involve the cytochrome *c* fraction located in contact sites as described here (Figs. 6.1 and 6.3). The cytochrome *c* bound in the VDAC–ANT complex may act as a sensor of the NAD/NADH redox state in the cytosol (Fig. 6.1) and also may provide feedback of the redox state in the electron transport chain.

A

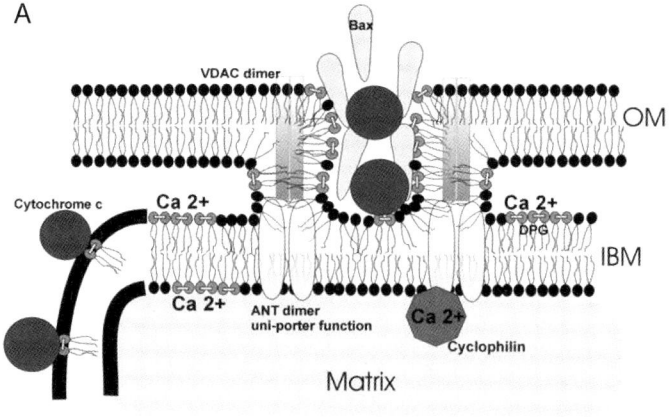

B

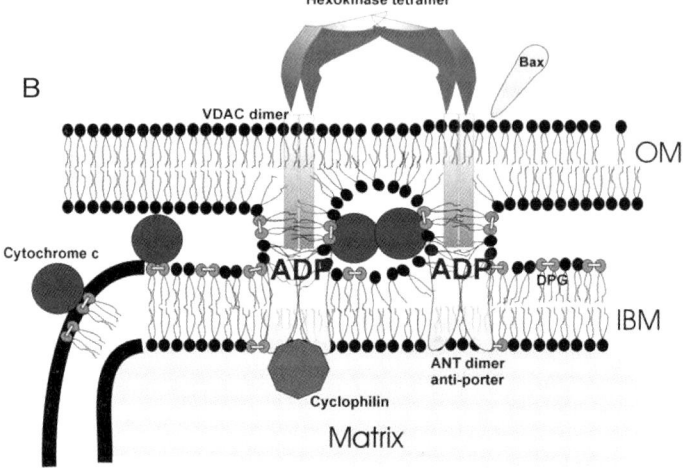

C

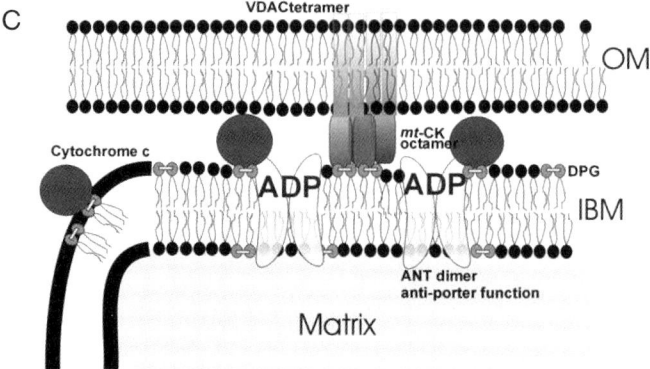

Fig. 6.3 (legend see p. 179)

6.17
Cholesterol and Cardiolipin Influence VDAC Structure and Function

Sterols have been found to be associated with purified mitochondrial VDAC, such as ergosterol in *N. crassa* [85] and cholesterol in VDAC from bovine heart [18]. Sterols were observed to be essential for proper insertion of VDAC into bilayer membranes, and cholesterol increases VDAC conductance [86]. It was assumed that sterols function to improve the hydrophobic surface of the VDAC β-barrel exposed to the phospholipid matrix of the membrane. The outer membrane contains 10% cholesterol, while almost none is found in the inner membrane [87]. The cholesterol in the outer membrane is nonrandomly distributed [88], and VDAC is localized inside or outside the cholesterol-containing lipid raft-like areas [89]. This might explain the observed varying properties and structure of VDAC in different protein complexes. The hexokinase–VDAC–ANT complex has low cholesterol but high cardiolipin content compared with mean membrane values [90]. Interestingly, in certain tumor cells, a threefold higher cholesterol concentration was found and, as described above, contact sites were absent. The observed Crabtree effect suggested that VDAC–ANT complexes were missing [77], and this may explain suppression of apoptosis in tumor cells. Treatment of tumor cells with inhibitors of cholesterol synthesis made them again susceptible to apoptosis [91], thus emphasizing the functional importance of cholesterol content in the outer membrane.

Cardiolipin is a specific component of the inner membrane, where it is tightly associated with several inner membrane proteins, controlling their activities. It may also indirectly regulate VDAC functions. VDAC forms contact sites with ANT containing tightly bound cardiolipin [86]. Thus, when Ca^{2+} affects cardiolipin bound to ANT, it also may affect the interaction of VDAC with ANT. Because cytochrome *c* displays high affinity for cardiolipin, it is as well concentrated in the contact sites [13]. A representation of the possible structure of the VDAC–ANT complex is shown in Fig. 6.3. According to the model of Van Venetie and Verkleij

---

Fig. 6.3. Structure of VDAC–ANT and VDAC–mCK–ANT complexes in the contact site. Two dimers of VDAC and ANT are connected by non-bilayer cardiolipin arranged as inverted micelles according to [92] (B). The non-bilayer arrangement also can result in opening to the cytosol (A). This arrangement would explain how inner membrane cardiolipin (diphosphatidylglycerol [DPG]) could come to the mitochondrial surface and how cytochrome *c* could be detached from its binding sites in the contact sites by external Bax without forming pores. Hexokinase binds to the four VDACs as a tetramer, hindering access of Bax to the cytochrome *c* binding sites.

(A) ANT in the presence of Ca^{2+} and the absence of ADP changes to the uniporter structure, forming an unspecific permeability transition pore.
(B) ANT in the absence of Ca^{2+} and the presence of ADP generated by hexokinase obtains the antiporter structure specific for ATP and ADP exchange.
(C) VDAC and ANT are not connected when the octamer of the mitochondrial creatine kinase binds to cardiolipin surrounding ANT and on the other side to VDAC. A non-bilayer arrangement of cardiolipin as shown in (A) and (B) is not possible in the creatine kinase contact site.

[93], cardiolipin forms inverted micelles or arranges into other non-bilayer structures. These authors postulated formation of contact sites by a pure non-bilayer phospholipid arrangement induced by Ca^{2+}. We postulate that cardiolipin may arrange along the hydrophobic surface of the VDAC–ANT complex, which after interaction of the two proteins, penetrates the two boundary membranes. As depicted in Figs. 6.2 and 6.3, cytochrome c can arrange in the VDAC–ANT complex in a way that is accessible to external Bax and t-Bid. It has recently been observed that proteins bridging the peripheral inner membrane and the outer membrane, such as mitochondrial creatine kinase, can cause exchange of lipids between the membranes [14].

There is evidence that cardiolipin is important in apoptosis induction [93] as a specific binding site for cytochrome c [94]. The MPT pore, meaning the VDAC–ANT complex, also appears to be a target of the proapoptotic proteins Bax [95] and t-Bid [96, 97]. Bax may replace cytochrome c from cardiolipin in the contacts and get incorporated into the VDAC–ANT complex as shown in Fig. 6.3. Cardiolipin is also required for targeting t-Bid to mitochondria [97], where it appears to interact especially in the contact sites [96]. As a second step, t-Bid changes the structure of the inner membrane and thereby mobilizes cytochrome c from the cristae surface to the peripheral intermembrane space to facilitate its release [98].

6.18
The Importance of VDAC Complexes in Regulation of Energy Metabolism and Apoptosis

Apoptosis, or programmed cell death, is a genetically regulated process that plays an important role in tissue homeostasis and development. Thus, a balance between apoptosis and cell growth has to be maintained, meaning that regulation between apoptosis and stabilization of energy metabolism is needed, as cell growth depends on energy supply. In fact, many investigations suggest that the two kinases interacting with VDAC play an important role in this regulation. The hexokinase and creatine kinase contact sites are important domains where regulation operates through association and dissociation of the VDAC kinase complexes.

The proapoptotic signal can be generated inside the mitochondria, (e.g., by increased Ca^{2+} and/or ROS concentrations; see Fig. 6.2), but it also can be initiated from outside by proapoptotic proteins such as t-Bid and Bax [53, 54]. The internal initiation of apoptosis is performed by permeability transition resulting not only in loss of mitochondrial membrane potential and ATP synthesis but also in ADP increase in the cytosol. This decrease in free energy ΔG of the phosphoryl group transfer from ATP is critical in nerve and heart cells as it affects ion pumps such as K^+/Na^+- and Ca^{2+}-ATPases (see Fig. 6.1) with the consequence of retarded repolarization of cell membrane potential and increased cytosolic Ca^{2+} levels. The external initiation of apoptosis is achieved by binding of proapoptotic Bax and t-Bid to cardiolipin exposed in the contact sites followed by liberation of cytochrome c from the VDAC–ANT complexes (see Fig. 6.3).

Release of cytochrome c from mitochondria proceeds in two steps. The first step involves the release of a small fraction of cytochrome c from the compartment between the outer and peripheral inner membranes and the contact sites (Fig. 6.2). It is induced by low concentrations of Bax [99] in the early phase [100]. The second step includes the release of additional fractions of cytochrome c bound to the surface of the cristae membranes; it results from t-Bid perturbation of the cristae membranes [96] and opening of the permeability transition pore, followed by mitochondrial swelling and membrane disruption [83]. Opening of the permeability transition pore is induced by high Ca^{2+} levels (see below) that also may detach cytochrome c from its cardiolipin-binding sites.

6.19
Suppression of Bax-dependent Cytochrome *c* Release and Permeability Transition by Hexokinase

As shown in Fig. 6.4, the activity of mitochondrial bound hexokinase is increased by the action of growth hormone or insulin. It was found that the two hormones act via protein kinase B [101]. The activation of protein kinase B/Akt leads to activation of cell membrane glucose transport and glycogen synthesis as well as binding of hexokinase to the mitochondrial surface. The bound hexokinase was found to be important for protein kinase B–linked suppression of cytochrome c release and apoptosis [102, 103]. The experiments suggested that following the withdrawal of growth hormone, Bax liberated a small fraction of cytochrome c from the contact sites. In complete agreement with this, Bax-dependent release of cytochrome c from VDAC–ANT complexes was inhibited by hexokinase and enforced binding of the enzyme in the presence of glucose and ATP [13, 104]. This was explained on the basis of the observation by Pastorino et al. [105] that Bax and hexokinase compete for the same binding site at VDAC. Besides this effect on Bax-dependent cytochrome c release, activation of hexokinase inhibited Ca^{2+}-dependent opening of the permeability transition pore by ADP production [79, 106].

A second protein kinase involved in apoptosis regulation through hexokinase binding to the mitochondrial surface is AMP kinase (Fig. 6.4) [107]. The enzyme is activated by increased AMP levels (e.g., under chronic stimulation of white muscle fibers), induces enzymes of the citric acid cycle and fatty acid oxidation as well as the glucose transporter GLUT4 in the cell membrane, and increases hexokinase activity at the mitochondrial surface [108, 109]. It was found that activator ICAR-induced stimulation of AMP kinase suppresses apoptosis [110].

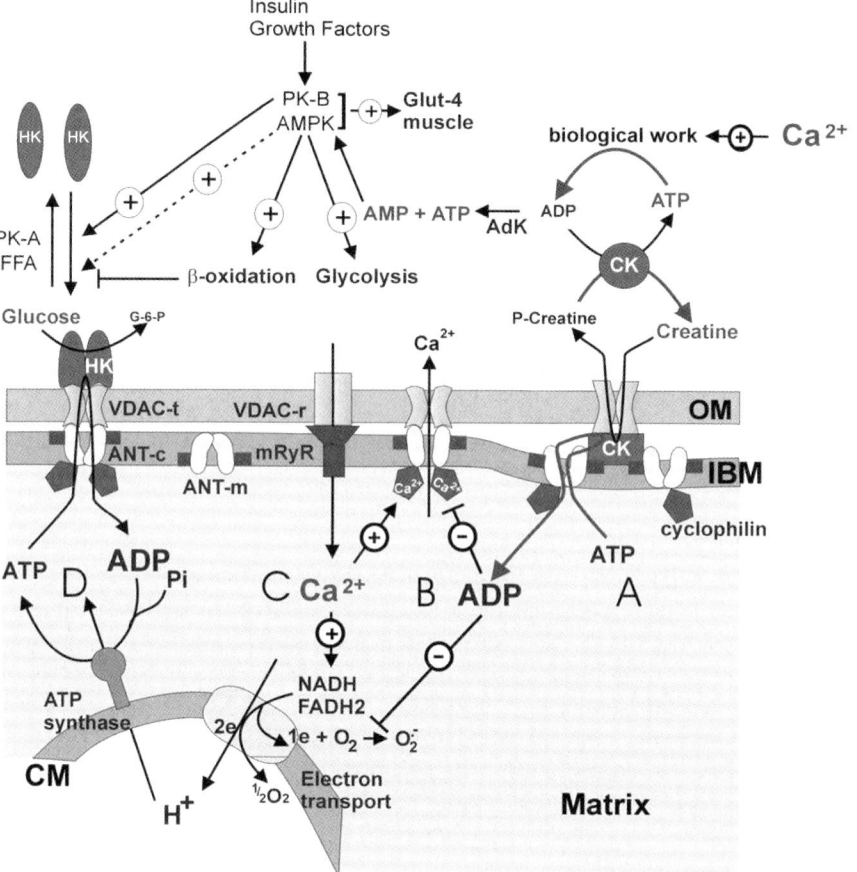

Fig. 6.4. Hormonal regulation involving VDAC–ANT–kinase complexes.

(A) Biological work (e.g., contraction) triggers cytosolic ADP–ATP turnover and leads to increased cytosolic creatine through the activity of cytosolic creatine kinase (CK). Creatine causes an increase in intramitochondrial ADP supply by the activity of mitochondrial creatine kinase (mCK), which is linked to the adenine nucleotide translocator (ANT) through cardiolipin (green squares). In a complex with mCK, the ANT acts as antiporter in the inner membrane.

(B) In the absence of ADP and the presence of high Ca^{2+}, the ANT can form an unspecific permeability transition pore (PTP) that releases Ca^{2+}.

(C) The mRyR may be linked to VDAC. Ca^{2+} activates substrate supply to the electron transport chain. This may lead to increased ROS formation if it is not accompanied by an ADP increase, which in turn decreases ROS production through activation of ATP synthesis.

(D) In the hexokinase complex, the enzyme forms a tetramer that is linked to two dimers of VDAC and ANT (not shown, but see Fig. 6.3). The ANT in this complex is in its c-conformation (ANT-c, induced by atractyloside), where it can operate either as antiporter to exchange ATP for ADP or as uniporter to form an unspecific pore. The ANT in the m-conformation (ANT-m, induced by bongkrekate) does not form complexes with VDAC and cannot form the MPT pore. The oligomerization of hexokinase leads to activation of the enzyme. VDAC obtains a different (VDAC-a) structure by interaction

6.20
Suppression of Permeability Transition and Cytochrome *c* Release by Mitochondrial Creatine Kinase

Mitochondrial creatine kinase interacts with VDAC exclusively in the octameric state, while the dimer has only weak affinity to the pore [67, 111]. Thus, the association–dissociation equilibrium between octamer and dimer determines on one hand the formation of octamer–VDAC complexes and on the other hand the induction of VDAC–ANT complexes. Because VDAC–ANT complexes can open as a permeability transition pore, octamer dissociation is crucial in the regulation of energy metabolism and apoptosis (Figs. 6.3 and 6.4). Besides acting as a lock between VDAC and ANT to prevent permeability transition, the octamer's presence has an influence on the number and stability of contact sites. Liver mitochondria from transgenic mice expressing ubiquitous mitochondrial creatine kinase in this organ showed a threefold increase in contact sites and increased resistance against detergent-induced lysis compared with controls [112]. These transgenic livers also became largely tolerant against tumor necrosis factor TNF-α–induced apoptosis [113].

Free radicals can shift the octamer–dimer equilibrium of the mitochondrial creatine kinase towards dimerization [114]. Sarcoplasmic as well as ubiquitous mitochondrial creatine kinases (present in brain and kidney) are especially highly susceptible to peroxynitrite treatment *in vitro* [115], well within the physiological range [116]. Because mitochondrial creatine kinase inactivation occurs at peroxynitrite concentrations far below those affecting the respiratory chain itself, this isoenzyme is indeed a prime target of peroxynitrite-induced modification and inactivation [115, 117]. Peroxynitrite also leads to a decrease in octamer formation from dimeric mitochondrial creatine kinase and a dose-dependent dissociation of octamers into dimers [115, 117]. This might explain why ligation of the coronary artery in guinea pigs led to a decrease in octamer from 85% to 60% in the infracted area of the heart [118], because it is known that peroxynitrite may play a pathologic role in ischemia–reperfusion injury. As a consequence of creatine kinase octamer dissociation, the probability of a direct VDAC–ANT interaction is increased. The formation of the latter complex has two outputs: (1) cytochrome *c* is rearranged (compare Fig. 6.2A, B) and becomes a target for Bax and (2) the VDAC–ANT complex provides the conditions for a Ca^{2+}-induced permeability

---

Fig. 6.4 (cont.)
with ANT in the c-conformation, which has a higher capacity to bind hexokinase. AMP protein kinase (AMPK) and protein kinase B (PK-B, Akt) stimulate the binding of hexokinase to the VDAC–ANT complex, while protein kinase A (PK-A) and free fatty acids (FFA) support the desorption of hexokinase from the mitochondrial membrane. OM: outer mitochondrial membrane, IBM: inner boundary membrane, CM: cristae membrane, AdK: adenylate kinase, GLUT4: glucose transporter. The figure was modified according to that published in [7].

transition (Fig. 6.2B), followed by mitochondrial swelling, membrane disruption, cytochrome *c* release, and apoptosis.

6.21
The General Importance of the Creatine Kinase System in Heart Performance

As mentioned above, energy-consuming systems, e.g., ion pumps and actomyosin ATPase, depend on high free energy ΔG of the phosphoryl group transfer from ATP. To provide a ΔG of 59 kJ per mol, the ATP:ADP quotient has to be approximately 800 in the cytosol, while it is about 5 in the mitochondria in order to support ATP synthesis [3, 4] (see Fig. 6.1). The transfer of the phosphoryl group to creatine through mitochondrial creatine kinase in the VDAC–mCK–ANT complex serves to conserve a high ΔG as phosphocreatine. This explains the importance of creatine and creatine kinase in heart function.

Several studies with perfused heart suggest that global creatine kinase inactivation directly affects heart accomplishment [119–121]. Interestingly, one consequence of ischemic preconditioning, as a cardioprotective tool [122], is a higher flux through the creatine kinase system and preservation of functional coupling between mitochondrial creatine kinase and ANT [123]. Decreased activity of kinase at sites of energy production and utilization may contribute to alterations in energy fluxes and calcium homeostasis in congestive heart failure [124]. The study using double knockout of cytosolic and mitochondrial creatine kinase in mouse muscle demonstrated that below a certain level of creatine kinase activity, the energy expenditure upon cardiac work become less efficient [125]. In animal models of dilated cardiomyopathy, hypertrophy, and heart failure, a decrease in the creatine kinase–creatine phosphate system preceded the development of contractile dysfunction and led to decreased energy reserve [126–128].

6.22
The Central Regulatory Role of ANT

ANT is a member of the superfamily of mitochondrial membrane transporters [129]. Among these substrate exchangers, ANT is the most abundant. Of the monomer, six membrane-spanning domains of 18–22 residues are separated by hydrophilic domains facing either the mitochondrial matrix or the intermembrane space. According to this model, the C- and N-termini protrude into the intermembrane space [130, 131]. The active ANT is a dimer. It is essential for these types of transport proteins that they can be switched between the antiporter state specific to substrate exchange such as ATP and ADP and the uniporter state, where they become unspecific pores that can be assigned for permeability transition [132]. This change between two functional ANT states (shown in Figs. 6.3 and 6.4) provides the basis for adjusting the mitochondrial function to the actual energy demand and regulatory signals such as Ca^{2+} and the hormonal control.

An increase in cytosolic Ca^{2+} upon excitation in muscle and nerve cells leads to uptake into the mitochondria by the mitochondrial Ca^{2+} transporters (including a mitochondrial ryanodine receptor [133]) that are functionally dependent on the mitochondrial membrane potential. High intramitochondrial Ca^{2+} concentration and depolarization of mitochondrial membrane potential could result in Ca^{2+}-induced Ca^{2+} release [134]. It has been suggested that transient opening of the permeability transition pore might be involved in this process [135, 136]. We hypothesize that the VDAC–ANT complex performs the Ca^{2+} release when ANT is in its uniporter state. The change between the two states of ANT is regulated by ATP turnover. The interplay of peripheral and mitochondrial creatine kinases (Fig. 6.2A) translates the ATP turnover into varying creatine levels. Thus, high ATP turnover through creatine kinase suppresses Ca^{2+}-dependent Ca^{2+} release although the intramitochondrial Ca^{2+} level is high. In addition, high ATP turnover and stimulation of oxidative phosphorylation by mitochondrial creatine kinase decreases ROS generation [137], which is a well-known factor for regulating the permeability transition pore [138]. High intramitochondrial ATP turnover (supported by the mitochondrial respiration rate) was found to be directly related to creatine concentration [139]. This and the activity of creatine kinases provide the basis for the energy reserve in the heart [119, 140, 141] (the ability of the heart to increase cardiac performance on demand), through increasing the number of complexes between mitochondrial creatine kinase and ANT also inside the mitochondria along the cristae [123].

By interacting with the regulatory proteins, ANT thus becomes an integral point over substrate supply through hexokinase, ATP turnover through creatine kinase, and Ca^{2+} level through cyclophilin. As schematically depicted in Fig. 6.4, hormonal control engages the VDAC–ANT complex to adjust the energy supply. This has an impact on apoptosis regulation as well. Thus, growth factors through protein kinase B and increased hexokinase binding suppress apoptosis, whereas epinephrine through c-AMP protein kinase A and decreased hexokinase binding promote apoptosis. Similarly, activated AMP kinase by inducing fatty acid oxidation (in the late phase) and glycolysis through increased hexokinase binding (in the early phase) in its turn suppresses apoptosis (see Fig. 6.4).

The same regulatory mechanisms apply for Bax-dependent release of mitochondrial cytochrome *c*. The latter is organized in the VDAC–ANT complex (Fig. 6.2A,B,D), which is a binding target that hexokinase and Bax are competing for. Thus, regulators that support hexokinase binding to the complex, such as PKB and AMPK, will suppress cytochrome *c* release, while systems inducing Bax binding, such as p53, will facilitate cytochrome *c* release.

In disorders in which defects in mitochondrial bioenergetics are detected, a protective effect of creatine supplementation has been observed (e.g., in Duchenne muscular dystrophy [142, 143], mitochondrial cytopathies [144], animal models of amyotrophic lateral sclerosis [145, 146], Huntington's disease [147, 148], Parkinson's disease [149], and brain ischemia [150, 151]). However, this field is still controversial, and there are some reported cases where creatine supplementation had no benefit for the patients [152]. So far, clinical investigations studying creatine supplementation in a large number of patients are missing, and it is therefore too early to draw final conclusions about the clinical application of creatine. However, the crucial role of mitochondrial contact sites in regulating cellular bioenergetics, Ca^{2+} homeostasis, ROS generation, and apoptosis will definitely draw more future work to this exciting research topic.

References

1 Martin, W., Hoffmeister, M., Rotte, C., Henze, K. (2001) An overview of endosymbiotic models for the origins of eukaryotes, their ATP-producing organelles (mitochondria and hydrogenosomes), and their heterotrophic lifestyle. *Biol. Chem 382*, 1521–1539.

2 Gumaa, K. A., McLean, P., Greenbaum, A. L. (1971) Compartmentation in relation to metabolic control in liver. *Essays Biochem 7*, 39–86.

3 Wallimann, T., Wyss, M., Brdiczka, D., Nicolay, K., Eppenberger, H. M. (1992) Intracellular compartmentation, structure and function of creatine kinase isoenzymes in tissues with high and fluctuating energy demands: the 'phosphocreatine circuit' for cellular energy homeostasis. *Biochem J 281*, 21–40.

4 Saks, V., Dzeja, P., Schlattner, U., Vendelin, M., Terzic, A., Wallimann, T. (2006) Cardiac system bioenergetics: metabolic basis of the Frank-Starling law. *J. Physiol. 571*, 253–273.

5 Bernardi, P., Azzone, G. F. (1981) Cytochrome *c* as an electron shuttle between the outer and inner mitochondrial membranes. *J. Biol. Chem. 256*, 7187–7192.

6 Marzulli, D., La Piana, G., Franseve, E., Lofrumento, N. E. (1999) Modulation of cytochrome *c*-mediated extramitochondrial NADH oxidation by contact site structure. *Biochem. Biophys. Res. Commun 259*, 325–330.

7 Brdiczka, D., Zorov, D. B., Sheu, S. S. (2006) Mitochondrial contact sites: their role in energy metabolism and apoptosis. *Biochim Biophys Acta 1762*, 148–63.

8 Schein, S. J., Colombini, M., Finkelstein, A. (1976) Reconstitution in planar lipid bilayers of a voltage-dependent anion-selective channel obtained from paramecium mitochondria. *J Membr Biol 30*, 99–120.

9 Reymann, S et al. (1995) Further evidence for multitopological localization of mammalian porin (VDAC) in the plasmalemma forming part of a chloride channel complex affected in cystic fibrosis and encephalomyopathy. *Biochem Mol Med 54*, 75–87.

10 Ohlendieck, K., Riesinger, I., Adams, V., Krause, J., Brdiczka, D. (1986) Enrichment and biochemical characterization of boundary membrane contact sites from rat-liver mitochondria. *Biochim Biophys Acta 860*, 672–89.

11 Adams, V., Bosch, W., Schlegel, J., Wallimann, T., Brdiczka, D. (1989) Further characterization of contact sites from mitochondria of different tissues: toplogy of peripheral kinases. *Biochim. Biophys. Acta 981*, 213–225.

12 Kottke, M., Adams, V., Wallimann, T., Nalam, V. K., Brdiczka, D. (1991) Location and regulation of octameric mitochondrial creatine kinase in the contact sites. *Biochim Biophys Acta 1061*, 215–25.

13 Vyssokikh, M. Y., Zorova, L., Zorov, D., Heimlich, G., Jürgensmeier, J. M., Schreiner, D., Brdiczka, D. (2004) The intramitochondrial cytochrome *c* distribution varies correlated to the formation of a complex between VDAC and the adenine nucleotide translocase: this affects Bax-dependent cytochrome *c* release. *Biochim. Biophys. Acta 1644*, 27–36.

14 Epand, R. F., Schlattner, U., Wallimann, T., Lacombe, M.-L., Epand, R. M. (2006) Novel lipid transfer property of two mitochondrial proteins that bridge the

inner and outer membranes. *Biophys. J.* *106*, 1529.

15 Benz, R. (1994) Permeation of hydrophilic solutes through mitochondrial outer membranes: Review on mitochondrial porins. *Biochim. Biophys. Acta* *1197*, 167–196.

16 De Pinto, V., Prezioso, G., Thinnes, F., Link, T. A., Palmieri, F. (1991) Peptide-specific antibodies and proteases as probes of the transmembrane topology of the bovine heart mitochondrial porin. *Biochemistry 30*, 10191–200.

17 Colombini, M. (2004) VDAC: the channel at the interface between mitochondria and the cytosol. *Mol Cell Biochem 256–257*, 107–15.

18 DePinto, V., Benz, R., Palmieri, F. (1989) Interaction of non-classical detergents with the mitochondrial porin. A new purification procedure and characterization of the pore-forming unit. *Eur J Biochem 183*, 179–187.

19 Thomas, L., Blachly-Dyson, E., Colombini, M., Forte, M. (1993) Mapping of residues forming the voltage sensor of the voltage-dependent anion-selective channel. *Proc Natl Acad Sci USA 90*, 5446–9.

20 Song, J., Midson, C., Blachly-Dyson, E., Forte, M., Colombini, M. (1998) The topology of VDAC as probed by biotin modification. *J Biol Chem 273*, 24406–13

21 Rostovtseva, T. K., Liu, T. T., Colombini, M., Parsegian, V. A., Bezrukov, S. M. (2000) Positive cooperativity without domains or subunits in a monomeric membrane channel. *Proc Natl Acad Sci USA 97*, 7819–22.

22 Krause, J., Hay, R., Kowollik, C., Brdiczka, D. (1986) Cross-linking analysis of yeast mitochondrial outer membrane. *Biochim. Biophys. Acta 860*, 690–698.

23 Linden, M., Gellerfors, P. (1983) Hydrodynamic properties of porin isolated from outer membranes of rat liver mitochondria. *Biochim Biophys Acta 736*, 125–9.

24 Shoshan-Barmatz, V., Zalk, R., Gincel, D., Vardi, N. (2004) Subcellular localization of VDAC in mitochondria and ER in the cerebellum. *Biochim Biophys Acta 1657*, 105–14.

25 Zalk, R., Israelson, A., Garty, E. S., Azoulay-Zohar, H., Shoshan-Barmatz, V. (2005) Oligomeric states of the voltage-dependent anion channel and cytochrome *c* release from mitochondria. *Biochem. J. 386*, 73–83.

26 De Pinto, V. e. a. (2003) New functions of an old protein: the eukaryotic porin or voltage dependent anion selective channel (VDAC). *Ital J Biochem 52*, 17–24.

27 Shoshan-Barmatz, V., Gincel, D. (2003) The voltage-dependent anion channel: characterization, modulation, and role in mitochondrial function in cell life and death. *Cell Biochem Biophys 39*, 279–92.

28 Colombini, M. (2004) VDAC: the channel at the interface between mitochondria and the cytosol. *Mol Cell Biochem 256–257*, 107–15.

29 Krasilnikov, O. V., Carneiro, C. M., Yuldasheva, L. N., Campos-de-Carvalho, A. C., Nogueira, R. A. (1996) Diameter of the mammalian porin channel in open and closed states: direct measurement at the single channel level in planar lipid bilayer. *Braz J Med Biol Res 29*, 1691–7.

30 Colombini, M. (1989) Voltage gating in the mitochondrial channel, VDAC. *J Membr Biol*, 103–11.

31 Benz, R., Kottke, M., Brdiczka, D. (1990) The cationically selective state of the mitochondrial outer membrane pore: a study with intact mitochondria and reconstituted mitochondrial porin. *Biochim. Biophys. Acta 1022*, 311–318.

32 Benz, R., Brdiczka, D. (1992) The cation-selective substate of the mitochondrial outer membrane pore: single-channel conductance and influence on intermembrane and peripheral kinases. *J Bioenerg Biomembr 24*, 33–9.

33 Rostovtseva, T., Colombini, M. (1997) VDAC channels mediate and gate the flow of ATP: implications for the regulation of mitochondrial function. *Biophys. J. 72*, 1954–1962.

34 Song, J., Midson, C., Blachly-Dyson, E., Forte, M., Colombini, M. (1998) The sensor regions of VDAC are translocated from within the membrane to the surface during the gating processes. *Biophys. J. 74*, 2926–2944.

35 Welte, W., Nestel, U., Wacker, T., Diederichs, K. (1995) Structure and function of the porin channel. *Kidney Int 48*, 930–40.

36 Saks, V., Kuznetsov, A., Khuchua, Z., Vasilyeva, E., Belikova, J., Kesvatera, T., Tiivel, T. (1995) Control of cellular respiration *in vivo* by mitochondrial outer membrane and by creatine kinase. A new speculative hypothesis: possible involvement of mitochondrial-cytoskeleton interactions. *J. Mol. Cell. Cardiol. 27*, 625–645.

37 Wojtczak, L., Adams, V., Brdiczka, D. (1988) Effect of oleate on the apparent Km of monoamine oxidase and the amount of membrane-bound hexokinase in isolated rat hepatocytes. *Mol. Cell. Biochem. 79*, 25–3.

38 Lemeshko, V. V. (2002) Model of the outer membrane potential generation by the inner membrane of mitochondria. *Biophys. J. 82*, 684–692.

39 Sampson, M. J., Lovell, R. S., Craigen, W. J. (1997) The murine voltage-dependent anion channel gene family. Conserved structure and function. *J Biol Chem 272*, 18966–73.

40 Blachly-Dyson, E., Song, J., Wolfgang, W. J., Colombini, M., Forte, M. (1997) Multicopy suppressors of phenotypes resulting from the absence of yeast VDAC encode a VDAC-like protein. *Mol Cell Biol 17*, 5727–38.

41 Sampson, M. J., Decker, W. K., Beaudet, A. L., Ruitenbeek, W., Armstrong, D., Hicks, M. J., Craigen, W. J. (2001) Immotile sperm and infertility in mice lacking mitochondrial voltage-dependent anion channel type 3. *J Biol Chem 276*, 39206–12.

42 Averet, N., Fitton, V., Bunoust, O., Rigoulet, M., Guerin, B. (1998) Yeast mitochondrial metabolism: from *in vitro* to *in situ* quantitative study. *Mol Cell Biochem 184*, 67–79.

43 Averet, N., Aguilaniu, H., Bunoust, O., Gustafsson, L., Rigoulet, M. (2002) NADH is specifically channeled through the mitochondrial porin channel in Saccharomyces cerevisiae. *J Bioenerg Biomembr 34*, 499–506.

44 Kmita, H., Budzinska, M. (2000) Involvement of the TOM complex in external NADH transport into yeast mitochondria depleted of mitochondrial porin1. *Biochim Biophys Acta 1509*, 86–94.

45 Blachly-Dyson, E., Zambronicz, E. B., Yu, W. H., Adams, V., McCabe, E. R., Adelman, J., Colombini, M., Forte, M. (1993) Cloning and functional expression in yeast of two human isoforms of the outer mitochondrial membrane channel, the voltage-dependent anion channel. *J Biol Chem 268*, 1835–41.

46 Xu, X., Decker, W., Sampson, M. J., Craigen, W. J., Colombini, M. (1999) Mouse VDAC isoforms expressed in yeast: channel properties and their roles in mitochondrial outer membrane permeability. *J Membr Biol 170*, 89–102.

47 Sampson, M. J., Lovell, R. S., Davison, D. B., Craigen, W. J. (1996) A novel mouse mitochondrial voltage-dependent anion channel gene localizes to chromosome 8. *Genomics 36*, 192–6.

48 Huizing, M., Ruitenbeek, W., van den Heuvel, L. P., Dolce, V., Iacobazzi, V., Smeitink, J. A., Palmieri, F., Trijbels, J. M. (1998) Human mitochondrial transmembrane metabolite carriers: tissue distribution and its implication for mitochondrial disorders. *J Bioenerg Biomembr 30*, 277–84.

49 Shinohara, Y., Ishida, T., Hino, M., Yamazaki, N., Baba, Y., Terada, H. (2000) Characterization of porin isoforms expressed in tumor cells. *Eur J Biochem 267*, 6067–73.

50 Yu, W. H., Wolfgang, W., Forte, M. (1995) Subcellular localization of human voltage-dependent anion channel isoforms. *J Biol Chem 270*, 13998–4006.

51 Cheng, E. H., Sheiko, T. V., Fisher, J. K., Craigen, W. J., Korsmeyer, S. J. (2003) VDAC2 inhibits BAK activation and mitochondrial apoptosis. *Science 301*, 513–7.

52 Gottlieb, R. A. (2000) Role of mitochondria in apoptosis. *Crit Rev Eukaryot Gene Expr*, 231–9.

53 Wang, G. Q. e. a. (2001) A role for mitochondrial Bak in apoptotic response to anticancer drugs. *J Biol Chem 276*, 34307–17.

54 Crompton, M. (1999) The mitochondrial permeability transition pore and its role in cell death. *Biochem J 341*, 233–49.

55 Halestrap, A. P., Brenner, C. (2003) The adenine nucleotide translocase: a central component of the mitochondrial permeability transition pore and key player in cell death. *Curr Med Chem 10*, 1507–25.

56 Brustovetsky, N., Klingenberg, M. (1996) Mitochondrial ADP/ATP carrier can be reversibly converted into a large channel by Ca^{2+}. *Biochemistry 35*, 8483–8.

57 Ruck, A., Dolder, M., Wallimann, T., Brdiczka, D. (1998) Reconstituted adenine nucleotide translocase forms a channel for small molecules comparable to the mitochondrial permeability transition pore. *FEBS Let. 426*, 97–101.

58 Crompton, M., Virji, S., Ward, J. M. (1998) Cyclophilin-D binds strongly to complexes of the voltage-dependent anion channel and the adenine nucleotide translocase to form the permeability transition pore. *Eur J Biochem 258*, 729–35.

59 Woodfield, K., Ruck, A., Brdiczka, D., Halestrap, A. P. (1998) Direct demonstration of a specific interaction between cyclophilin-D and the adenine nucleotide translocase confirms their role in the mitochondrial permeability transition. *Biochem J 336*, 287–90.

60 Schlatter, D., Thoma, R., Kung, E., Stihle, M., Muller, F., Borroni, E., Cesura, A., Hennig, M. (2005) Crystal engineering yields crystals of cyclophilin D diffracting to 1.7 A resolution. *Acta Crystallogr D Biol Crystallogr 61*, 513–9.

61 Vyssokikh, M. Y., Katz, A., Rück, A., Wuensch, C., Dörner, A., Zorov, D. B., Brdiczka, D. (2001) Adenine nucleotide translocator isoforms 1 and 2 are differently distributed in the mitochondrial inner membrane and have distinct affinities to cyclophilin D. *Biochem. J. 358*, 349–358.

62 Kokoszka, J. E., Waymire, K. G., Levy, S. E., Sligh, J. E., Cai, J., Jones, D. P., MacGregor, G. R., Wallace, D. C. (2004) The ADP/ATP translocator is not essential for the mitochondrial permeability transition pore. *Nature 427*, 461–5.

63 Rodic, N., Oka, M., Hamazaki, T., Murawski, M. R., Jorgensen, M., Maatouk, D. M., Resnick, J. L., Li, E., Terada, N. (2005) DNA methylation is required for silencing of ant4, an adenine nucleotide translocase selectively expressed in mouse embryonic stem cells and germ cells. *Stem Cells 23*, 1314–1323.

64 Halestrap, A. P. (2004) Mitochondrial permeability: dual role for the ADP/ATP translocator? *Nature 430*, 1 p following 983.

65 Fiek, C., Benz, R., Roos, N., Brdiczka, D. (1982) Evidence for identity between the hexokinase-binding protein and the mitochondrial porin in the outer membrane of rat liver mitochondria. *Biochim Biophys Acta 688*, 429–40.

66 Lindén, M., Gellerfors, P., Nelson, B. D. (1982) Pore protein and the hexokinase-binding protein from the outer membrane of rat liver mitochondria are identical. *FEBS-Letters 141*, 189–192.

67 Brdiczka, D., Kaldis, P., Wallimann, T. (1994) *In vitro* complex formation between octamer of creatine kinase and porin. *J. Biol. Chem. 269*, 27640–27644.

68 Wilson, J. E. (1978) Ambiquitous enzymes: variation in intracellular distribution as a regulatory mechanism. *Trends. Biochem. Sci 3*, 124–125.

69 Wicker, U., Bücheler, K., Gellerich, F. N., Wagner, M., Kapischke, M., Brdiczka, D. (1993) Effect of macromolecules on the structure of the mitochondrial intermembrane space and the regulation of hexokinase. *Biochim. Biophys. Acta 1142*, 228–239.

70 Knoll, G., Brdiczka, D. (1983) Changes in freeze-fracture mitochondrial membranes correlated to their energetic state. *Biochim. Biophys. Acta 733*, 102–110.

71 Laterveer, F. D., Gellerich, F. N., Nicolay, K. (1995) Macromolecules increase the channeling of ADP from externally associated hexokinase to the matrix of mitochondria. *Eur J Biochem. 232*, 569–77.

72 Xie, G., Wilson, J. E. (1990) Tetrameric structure of mitochondrially bound rat brain hexokinase: A crosslinking study. *Arch. Biochem. Biophys 276*, 285–293.

73 Hashimoto, M., Wilson, J. E. (2000) Membrane potential-dependent conformational changes in mitochondrially bound hexokinase in brain. *Arch. Biochem. Biophys. 884*, 163–173.

74 Pedersen, P. L., Mathupala, S., Rempel, A., Geschwind, J. F., Ko, Y. H. (2002) Mitochondrial bound type II hexokinase: a key player in the growth and survival of many cancers and an ideal prospect for therapeutic intervention. *Biochim Biophys Acta 1555*, 14–20.

75 Penso, J., Beitner, R. (1998) Clotrimazole and bifonazole detach hexokinase from mitochondria of melanoma cells. *Eur J Pharmacol 342*, 113–7.

76 Ibsen, H. K. (1961) The Crabtree effect: a review. *Cancer Res 21*, 829–841.

77 Denis-Pouxviel, C., Riesinger, I., Bühler, C., Brdiczka, D., Murat, J.-C. (1987) Regulation of mitochondrial hexokinase in cultured HT 29 human cancer cells. *Biochim. Biophys. Acta 902*, 335–348.

78 Bücheler, K., Adams, V., Brdiczka, D. (1991) Localization of the ATP/ADP translocator in the inner membrane and regulation of contact sites between mitochondrial envelope membranes by ADP. A study on freeze fractured isolated liver mitochondria. *Biochim Biophys Acta 1061*, 215–225.

79 Beutner, G., Rück, A., Riede, B., Welte, W., Brdiczka, D. (1996) Complexes between kinases, mitochondrial porin and adenylate translocator in rat brain resemble the permeability transition pore. *FEBS Lett 396*, 189–195.

80 Beutner, G., Ruck, A., Riede, B., Brdiczka, D. (1997) Complexes between hexokinase, mitochondrial porin and adenylate translocator in brain: regulation of hexokinase, oxidative phosphorylation and permeability transition pore. *Biochem Soc Trans 25*, 151–7.

81 Liu, X., Kim, C. N., Yang, J., Jemmerson, R., Wang, X. (1996) Induction of apoptotic program in cell-free extracts: requirement for dATP and cytochrome c. *Cell 86*, 147–157.

82 Martinou, J. C., Desagher, S., Antonsson, B. (2000) Cytochrome c release from mitochondria: all or nothing. *Nat Cell Biol 2*, E41–3.

83 Doran, E., Halestrap, A. P. (2000) Cytochrome c release from isolated rat liver mitochondria can occur independently of outer-membrane rupture: possible role of contact sites. *Biochem. J.*, 343–350.

84 Beyer, K., Klingenberg, M. (1985) ADP/ATP carrier protein from beef heart mitochondria has high amounts of tightly bound cardiolipin, as revealed by 31P nuclear magnetic resonance. *Biochemistry 24*, 3821–3826.

85 Freitag, H., Neupert, W., Benz, R. (1982) Purification and characterization of a pore protein of the outer mitochondrial membrane from Neurospora crassa. *Eur J Biochem 162*, 629–636.

86 Popp, B., Schmid, A., Benz, R. (1995) Role of sterols in the functional reconstitution of water-soluble mitochondrial porins from different organisms. *Biochemistry 34*, 3352–3361.

87 Daum, G. (1985) Lipids of mitochondria. *Biochim Biophys Acta 822*, 1–42.

88 Ardail, D., Privat, J.-P., Egret-Charlier, M., Levrat, C., Lerme, F., Louisot, P. (1990) Mitochondrial contact sites, lipid composition and dynamics. *J. Biol. Chem. 265*, 18797–18802.

89 Simons, K., Ikonen, E. (1997) Functional rafts in cell membranes. *Nature 387*, 569–572.

90 Vyssokikh, M., Brdiczka, D. (2004) Function of the Outer Mitochondrial Membrane Pore (Voltage-dependent Anion Channel) in Intracellular Signaling, in *Bacterial and Eukaryotic Porins Structure, Function, Mechanism* (Benz, R., Ed.) pp 339–358, Wiley-VCH, Weinheim Germany.

91 Dimitroulakos, J., Nohynek, D., Backway, K. L., Hedley, D. W., Yeger, H., Freedman, M. H., Minden, M. D., Penn, L. Z. (1999) Increased Sensitivity of Acute Myeloid Leukemias to Lovastatin-Induced Apoptosis: A Potential Therapeutic Approach. *Blood 93*, 1308–1318.

92 Van Venetie, R., Verkleij, A. J. (1982) Possible role of non-bilayer lipids in the structure of mitochondria A freeze-fracture electron microscopy study. *Biochim Biophys Acta 692*, 397–405.

93 Newmeyer, D. D., Ferguson-Miller, S. (2003) Mitochondria: releasing power for life and unleashing the machineries of death. *Cell 112*, 481–90.

94 Iverson, S. L., Orrenius, S. (2004) The cardiolipin–cytochrome *c* interaction and the mitochondrial regulation of apoptosis. *Archives of Biochemistry and Biophysics 423*, 37–46.

95 De Giorgi, F., Lartigue, L., Bauer, M. K. A., Schubert, A., Grimm, S., Hanson, G. T., Remington, J., Youle, R. J. I., Ichas, F. (2002) The permeability transition pore signals apoptosis by directing Bax translocation and multimerization. *FASEB Journal 16*, 607–609.

96 Kim, T. H., Zhao, Y., Ding, W. X., Shin, J. N., He, X., Seo, Y. W., Chen, J., Rabinowich, H., Amoscato, A. A., Yin, X. M. (2004) Bid–cardiolipin interaction at mitochondrial contact site contributes to mitochondrial cristae reorganization and cytochrome *c* release. *Molecular Biology of the Cell 15*, 3061–3072.

97 Lutter, M., Fang, M., Luo, X., Nishijima, M., Xie, X., Wang, X. (2000) Cardiolipin provides specificity for targeting of tBid to mitochondria. *Nat Cell Biol 2*, 754–61.

98 Scorrano, L., Ashiya, M., Buttle, K., Weiler, S., Oakes, S. A., Mannella, C. A., Korsmeyer, S. J. (2002) A distinct pathway remodels mitochondrial cristae and mobilizes cytochrome *c* during apoptosis. *Developmental Cell, 2*, 55–67.

99 Pastorino, J. G., Tafani, M., Rothman, R. J., Marcineviciute, A., Hoek, J. B., Farber, J. L. (1999) Functional consequences of sustained or transient activation by Bax of the mitochondrial permeability transition pore. *J. Biol. Chem. 274*, 31734–31739.

100 Capano, M., Crompton, M. (2002) Biphasic translocation of BAX to mitochondria. *Biochem J 367*, 169–178.

101 AlessiI, D. R., Jelcovic, M., Caudwell, B., Cron, P., Morrice, N., Cohen, P., A., H. B. (1996) Mechanism of Activation of Protein Kinase B by Insulin and IGF-1. *EMBO journal 15*, 6541–6551.

102 Gottlob, K., Majewski, N., Kennedy, S., Kandel, E., Robey, R., Hay, N. (2001) Inhibition of early apoptotic events by Akt/PKB is dependent on the first committed step of glycolysis and mitochondrial hexokinase. *Genes Dev. 15*, 1406–1418.

103 Majewski, N., Nogueira, V., Bhaskar, P., Coy, P. E., Skeen, J. E., Gottlob, K., Chandel, N. S., Thompson, C. B., Robey, R. B., Hay, N. (2004) Hexokinase-mitochondria interaction mediated by Akt is required to inhibit apoptosis in the presence or absence of Bax and Bak. *Mol. Cell 16*, 819–830.

104 Vyssokikh, M. Y., Zorova, L., Zorov, D., Heimlich, G., Jürgensmeier, J. M., Brdiczka, D. (2002) Bax releases cytochrome *c* preferentially from a complex between porin and adenine nucleotide translocator. Hexokinase activity suppresses this effect. *Mol. Biol. Rep. 29*, 93–96.

105 Pastorino, J. G., Shulga, N., Hoek, J. B. (2002) Mitochondrial binding of hexokinase II inhibits bax-induced cytochrome *c* release and apoptosis. *J. Biol. Chem. 277*, 7610–7618.

106 Beutner, G., Rück, A., Riede, B., Brdiczka, D. (1998) Complexes between porin, hexokinase, mitochondrial creatine kinase and adenylate translocator display properties of the permeability transition pore. Implication for regulation of permeability transition by the kinases. *Biochim. Biophys. Acta 1368*, 7–18.

107 Rutter, G. A., Da Silva Xavier, G., Leclerc, I. (2003) Roles of 5′-AMP-activated protein kinase (AMPK) in mammalian glucose homeostasis. *Biochem. J. 375*, 1–16.

108 Putman, C. T., Kiricsi, M., Pearcey, J., MacLean, I. M., Bamford, J. A., Murdoch, G. K., Dixon, W. T., Pette, D. (2003) AMPK activation increases uncoupling protein-3 expression and mitochondrial enzyme activities in rat muscle without fibre type transitions. *J. Physiol. 551*, 169–178.

109 Parra, J., Brdiczka, D., Cusso, R., Pette, D. (1997) Enhanced catalytic activity of hexokinase by work-induced mitochondrial binding in fast-twitch muscle of rat. *FEBS Lett 403*, 279–282.

110 Blasquez, C., Geelen, M. J., Velasco, G., Guzman, M. (2001) The AMP-activated protein kinase prevents ceramide synthesis *de novo* and apoptosis in astrocytes. *FEBS Lett 489*, 149–153.

111 Schlattner, U., Dolder, M., Wallimann, T., Tokarska-Schlattner, M. (2001) Mitochondrial creatine kinase and mitochondrial outer membrane porin show direct interaction that is modulated by calcium. *J. Biol. Chem. 276*, 48027–48030.

112 Speer, O., Back, N., Buerklen, T., Brdiczka, D., Koretsky, A., Wallimann, T., Eriksson, O. (2005) Octameric mitochondrial creatine kinase induces and stabilizes contact sites between the inner and outer membrane. *Biochem J 385*, 445–450.

113 Hatano, E., Tanaka, A., Kanazawa, A., Tsuyuki, S., Tsunekawa, S., Iwata, S., Takahashi, R., Chance, B., Yamaoka, Y. (2004) Inhibition of tumor necrosis factor-induced apoptosis in transgenic mouse liver expressing creatine kinase. *Liver Int. 24*, 384–93.

114 Koufen, P., Rück, A., Brdiczka, D., Wendt, S., Wallimann, T., Stark, G. (1999) Free radical induced inactivation of creatine kinase: influence on the octameric and dimeric states of the mitochondrial enzyme (Mib-CK). *Biochem J. 344*, 413–417.

115 Stachowiak, O., Dolder, M., Wallimann, T., Richter, C. (1998) Mitochondrial creatine kinase is a prime target of peroxynitrite-induced modification and inactivation. *J Biol Chem 273*, 16694–9.

116 Darley-Usmar, V., Wiseman, H., Halliwell, B. (1995) Nitric oxide and oxygen radicals: a question of balance. *FEBS Lett 369*, 131–5.

117 Wendt, S., Schlattner, U., Wallimann, T. (2003) Differential effects of peroxynitrite on human mitochondrial creatine kinase isoenzymes Inactivation, octamer destabilization, and identification of involved residues. *J Biol Chem 278*, 1125–30.

118 Soboll, S., Brdiczka, D., Jahnke, D., Schmidt, A., Schlattner, U., Wendt, S., Wyss, M., Wallimann, T. (1999) Octamer-dimer transitions of mitochondrial creatine kinase in heart disease. *J. Mol. Cell. Cardiol. 31*, 857–866.

119 Hamman, B. L., Bittl, J. A., Jacobus, W. E., Allen, P. D., Spencer, R. S., Tian, R., Ingwall, J. S. (1995) Inhibition of the creatine kinase reaction decreases the contractile reserve of isolated rat hearts. *Am J Physiol 269*, H1030–6.

120 Spindler, M., Niebler, R., Remkes, H., Horn, M., Lanz, T., Neubauer, S. (2002) Mitochondrial creatine kinase is critically necessary for normal myocardial high-energy phosphate metabolism. *Am J Physiol Heart Circ Physiol 283*, H680–7.

121 Spindler, M., Meyer, K., Stromer, H., Leupold, A., Boehm, E., Wagner, H., Neubauer, S. (2004) Creatine kinase-deficient hearts exhibit increased susceptibility to ischemia-reperfusion injury and impaired calcium homeostasis. *Am J Physiol Heart Circ Physiol 287*, H1039–45.

122 Zaugg, M., Schaub, M. C. (2003) Signaling and cellular mechanisms in cardiac protection by ischemic and pharmacological preconditioning. *J Muscle Res Cell Motil 24*, 219–49.

123 Laclau, M. N., Boudina, S., Thambo, J. B., Tariosse, L., Gouverneur, G., Bonoron-Adele, S., Saks, V. A., Garlid, K. D., Dos Santos, P. (2001) Cardio-protection by ischemic preconditioning preserves mitochondrial function and functional coupling between adenine nucleotide translocase and creatine kinase. *J Mol Cell Cardiol 33*, 947–56.

124 Ventura-Clapier, R., Garnier, A., Veksler, V. (2004) Energy metabolism in heart failure. *J Physiol 555*, 1–13.

125 Saupe, K. W., Spindler, M., Tian, R., Ingwall, J. S. (1998) Impaired cardiac energetics in mice lacking muscle-specific isoenzymes of creatine kinase. *Circ Res. 82*, 898–907.

126 Liao, R., Nascimben, L., Friedrich, J., Gwathmey, J. K., Ingwall, J. S. (1996) Decreased energy reserve in an animal model of dilated cardiomyopathy. Relationship to contractile performance. *Circ Res. 78*, 893–902.

127 Spindler, M., Engelhardt, S., Niebler, R., Wagner, H., Hein, L., Lohse, M. J., Neubauer, S. (2003) Alterations in the myocardial creatine kinase system precede the development of contractile dysfunction in beta[1]-adrenergic receptor transgenic mice. *J Mol Cell Cardiol 35*, 389–97.

128 Cha, Y. M., Dzeja, P. P., Shen, W. K., Jahangir, A., Hart, C. Y., Terzic, A., Redfield, M. M. (2003) Failing atrial myocardium: energetic deficits accompany structural remodeling and electrical instability. *Am J Physiol Heart Circ Physiol 284*, H1313–20.

129 Krämer, R., Palmieri, F. (1992) Metabolite carriers in mitochondria, in *Molecular Mechanisms in Bioenergetics* (Ernster, L., Ed.) pp 359–384.

130 Klingenberg, M. (1992) Structure-function of the ADP/ATP carrier *Biochem Soc Trans 20*, 547–50.

131 Pebay-Peyroula, E., Dahout-Gonzalez, C., Kahn, R., Trézéguet, V., Lauquin, G., Brandolin, G. (2003) Structure of mitochondrial ADP/ATP carrier in complex with carboxyatractyloside *Nature 426*, 39–44.

132 Dierks, T., Salentin, A., Heberger, C., Krämer, R. (1990) The mitochondrial aspartate/glutamate and ADP/ATP carrier switch from obligate counterexchange to unidirectional transport after modification by SH-reagents. *Biochim. Biophys. Acta 1028*, 268–280.

133 Beutner, G., Sharma, V. K., Giovannucci, D. R., Yule, D. I., Sheu, S. S. (2001) Identification of a ryanodine receptor in rat heart mitochondria. *J. Biol. Chem. 276*, 21482–21488.

134 Haworth, R. A., Hunter, P. R. (1980) Allosteric inhibition of the Ca^{2+}-activated hydrophilic channel of the mitochondrial inner membrane by nucleotides. *J. Membr. Biol. 57*, 231–236.

135 Ichas, F., Jouaville, L. S., Mazat, J. P. (1997) Mitochondria are excitable organelles capable of generating and conveying electrical and calcium signals. *Cell 89*, 1145–1153.

136 Bernardi, P., Petronilli, V. (1996) The permeability transition pore as a mitochondrial calcium release channel: a critical appraisal. *J Bioenerg Biomembr 28*, 131–8.

137 Korshunov, S. S., Skulachev, V. P., Starkov, A. A. (1997) High protonic potential actuates a mechanism of production of reactive oxygen species in mitochondria. *FEBS Lett 416*, 15–18.

138 Jacobson, J., Duchen, M. R. (2004) Interplay between mitochondria and cellular calcium signalling. *Mol. Cell. Biochem. 256/257*, 209–218.

139 Dolder, M., Walzel, B., Speer, O., Schlattner, U., Wallimann, T. (2003) Inhibition of the mitochondrial permeability transition by creatine kinase substrates. Requirement for micro-compartmentation. *J Biol Chem 278*, 17760–17766.

140 Ingwall, J. S., Weiss, R. G. (2004) Is the Failing Heart Energy Starved? On Using Chemical Energy to Support Cardiac Function. *Circ Res. 95*, 35–145.

141 Tian, R., Ingwall, J. S. (1996) Energetic basis for reduced contractile reserve in isolated rat hearts. *Am. J. Physiol. 270*, H1207–H1216.

142 Walter, M. C., Lochmuller, H., Reilich, P., Klopstock, T., Huber, R., Hartard, M., Hennig, M., Pongratz, D., Muller-Felber, W. (2000) Creatine monohydrate in muscular dystrophies: A double-blind, placebo-controlled clinical study. *Neurology 54*, 1848–50.

143 Felber, S., Skladal, D., Wyss, M., Kremser, C., Koller, A., Sperl, W. (2000) Oral creatine supplementation in Duchenne muscular dystrophy: a clinical and 31P magnetic resonance spectroscopy study. *Neurol Res 22*, 145–50.

144 Tarnopolsky, M. A., Roy, B. D., MacDonald, J. R. (1997) A randomized, controlled trial of creatine monohydrate in patients with mitochondrial cytopathies. *Muscle Nerve 20*, 1502–9.

145 Klivenyi, P., Ferrante, R. J., Matthews, R. T., Bogdanov, M. B., Klein, A. M., Andreassen, O. A., Mueller, G., Wermer, M., Kaddurah-Daouk, R., Beal, M. F. (1999) Neuroprotective effects of creatine in a transgenic animal model of amyotrophic lateral sclerosis. *Nat Med 5*, 347–50.

146 Klivenyi, P., Kiaei, M., Gardian, G., Calingasan, N. Y., Beal, M. F. (2004) Additive neuroprotective effects of creatine and cyclooxygenase 2 inhibitors

in a transgenic mouse model of
amyotrophic lateral sclerosis.
J Neurochem 88, 576–82.

147 Andreassen, O. A., Dedeoglu, A.,
Ferrante, R. J., Jenkins, B. G., Ferrante,
K. L., Thomas, M., Friedlich, A.,
Browne, S. E., Schilling, G., Borchelt,
D. R., Hersch, S. M., Ross, C. A., Beal,
M. F. (2001) Creatine increase survival
and delays motor symptoms in a
transgenic animal model of
Huntington's disease. *Neurobiol Dis 8*,
479–91.

148 Dedeoglu, A., Kubilus, J. K., Yang, L.,
Ferrante, K. L., Hersch, S. M., Beal,
M. F., Ferrante, R. J. (2003) Creatine
therapy provides neuroprotection after
onset of clinical symptoms in
Huntington's disease transgenic mice.
J Neurochem 85, 1359–67.

149 Klivenyi, P., Gardian, G., Calingasan,
N. Y., Yang, L., Beal, M. F. (2003)

Additive neuroprotective effects of
creatine and a cyclooxygenase 2
inhibitor against dopamine depletion
in the 1-methyl-4-phenyl-1,2,3,6-
tetrahydropyridine (MPTP) mouse
model of Parkinson's disease. *J Mol
Neurosci 21*, 191–8.

150 Adcock, K. H., Nedelcu, J., Loenneker,
T., Martin, E., Wallimann, T., Wagner,
B. P. (2002) Neuroprotection of creatine
supplementation in neonatal rats with
transient cerebral hypoxia-ischemia *Dev
Neurosci 24*, 382–8.

151 Matthews, R. T., Ferrante, R. J., Kliveny,
P., Yang, L., Klein, A. M., Mueller, G.,
Kaddurah-Daouk, R., Beal, M. F. (1999)
Creatine and cyclocreatine attenuate
MPTP neurotoxicity. *Exp Neurol.*, 142–
149.

152 Brudnak, M. A. (2004) Creatine: are the
benefits worth the risk? *Toxicol Lett 150*,
123–130.

7

The Phosphocreatine Circuit: Molecular and Cellular Physiology of Creatine Kinases, Sensitivity to Free Radicals, and Enhancement by Creatine Supplementation*

Theo Wallimann, Malgorzata Tokarska-Schlattner, Dietbert Neumann, Richard M. Epand, Raquel F. Epand, Robert H. Andres, Hans Rudolf Widmer, Thorsten Hornemann, Valdur Saks, Irina Agarkova, and Uwe Schlattner

Abstract

Evidence for the important physiological role of the creatine kinase (CK)–phosphocreatine (PCr) system in energy homeostasis in sarcomeric muscle, brain, and other organs of high and fluctuating energy requirements is presented in the context of defined subcellular compartments of CK isoenzymes with processes of energy production, i.e., glycolysis and mitochondrial oxidative phosphorylation, and those of energy utilization, i.e., ATPases, ATP-gated ion channels, and ATP-dependent signaling events. The concepts of functional coupling and metabolite channeling are discussed, and the CK subcompartments that have been characterized in great detail are described. Mitochondrial CK (MtCK) plays a central role (1) in high-energy phosphoryl transfer and channeling, (2) in metabolic feedback regulation of mitochondrial respiration *in vivo*, (3) in stabilizing the contact sites of inner and outer mitochondrial membranes, (4) in delaying and preventing mitochondrial permeability pore opening, an early event in apoptosis, and (5) in preventing, by efficient ADP recycling inside mitochondria and by optimal coupling of respiration with ATP synthesis, excessive free oxygen radical (ROS) formation. Additional new and exciting findings indicate that octameric MtCK, a highly symmetrical cube-like molecule that is able to cross-link two membranes, is also involved in lipid transfer between mitochondrial membranes and in clustering of cardiolipin (CL) and the formation of membrane patches. Because of the high reactivity of the active-site cysteine of CK isoenzymes, this enzyme is highly susceptible to oxidation that leads to inactivation. Oxidation of other residues at the dimer–dimer interface of MtCK induces dimerization of the MtCK octamer, the functional entity in the mitochondrial intermembrane space, and dissociation of MtCK from the mitochondrial membranes.

---

* A list of abbreviations used can be found at the end of this chapter.

Under oxidative stress, as observed in mitochondrial myopathies and neurodegenerative diseases, these negative effects on MtCK function lead to further deterioration of cellular energy status, chronic calcium overload, and exacerbation of oxidative stress by generation of more ROS, events that eventually lead into apoptosis. On the other hand, the very same exquisite sensitivity of CK to free radical damage renders this enzyme vulnerable to anthracyclines. Sarcomeric sMtCK in the heart is especially susceptible to ROS damage caused by anthracyclines and more sensitive than ubiquitous uMtCK this fact can explain at least in part the specific cardiotoxicity of this potent class of anticancer drugs. After the importance of the CK–PCr system was recognized, transgenic mice lacking expression of the various CK isoforms were created. The most severe phenotype is seen after knockout of both brain CK isoforms (BB-CK and uMtCK), and this is fully in line with the recently discovered creatine deficiency syndrome in mentally retarded patients. Based on genetic deficiencies either in endogenous creatine (Cr) synthesis or in the Cr transporter (CRT) gene, these patients are basically devoid of Cr in their brains. These data support the immanent physiological importance of the CK–PCr system for normal function of brain, muscle, and other cells and organs with high and fluctuating energy demands. Alternatively, enhancing Cr content by Cr supplementation, now widely and successfully used by athletes, also revealed astonishing neuroprotective effects in cases of acute traumatic brain and spinal cord injury as well as in brain ischemia. In addition, as supported by clinical trials, Cr supplementation positively affects skeletomuscular health, i.e., muscle strength and bone mineral density. Cr also improves mental performance and short-term memory. Cr supplementation may become an adjuvant therapeutic measure for chronic neuromuscular and neurodegenerative diseases, such as muscular dystrophy, ALS, MS, and Huntington's, Parkinson's, and Alzheimer's diseases.

7.1
Phosphotransfer Enzymes: The Creatine Kinase System

7.1.1
Microcompartments: A Principle of Life

Life most likely originated autotrophically *de novo* in metabolic complexes organized on FeS_2 (pyrite) mineral surfaces, the earliest form of microcompartments [1]. Thus, because subcellular compartments were already formed at the origin of life, it is hard to understand why the notion was conceived, and still persists, that a living cell, which evolved by this principle, should behave as a well-mixed bag filled with enzymes, substrates, and products [2]. For example, using ^{31}P-NMR magnetization transfer measurements with muscle *in vivo* on a global level, involving a volume of approximately 1 cm^3 of tissue, it was assumed that creatine kinase (CK) ($PCr^{2-} + MgADP^- + H^+ \leftarrow CK \rightarrow MgATP^{2-} + Cr$), as well as the adenylate kinase (AdK) ($2MgADP^- \leftarrow AdK \rightarrow MgATP^{2-} + MgAMP$) were in chemical equilibrium [3]. For this, it was concluded (i) that in muscle cells, CK, with its cytosolic and mitochondrial isoforms already known at that time [4], were in equilibrium, (ii) that the CK metabolites ATP and ADP, as well as phosphoc-reatine (PCr) and creatine (Cr), were freely mixing in these cells as in solution, and thus (iii) that the CK system would behave according to solution kinetics [2, 5]. However, based on our data showing subcellular compartmentation of CK isoenzymes, this equilibrium concept must be seriously questioned.

7.1.2
Subcellular Compartments and Microcompartments of CK

In fact, recent developments in modern cell biology show that even so-called soluble cytosolic enzymes, such as the cytosolic isoforms of CK, are organized by specific subcellular localizations into metabolically coupled subcompartments, microcompartments, or specialized "metabolons," where they may work far from chemical equilibrium, in the direction of either entirely ATP regeneration or PCr production [4, 6–9] (see also Chapters 3 and 11). In parallel, compart-mentalized AdK [10] is at work and, depending on the workload, operates more or less independently or in tandem with the CK system [11–13]. Microcompart-ments and other subcellular compartments, consisting of multi-enzyme com-plexes that may be embedded within the cellular "ground substance," associated with the cytoskeleton, or situated in an unstirred layer along cellular membranes, operate according to exclusion principles and favor preferred pathways of inter-mediates [14]. Such an association between two or more sequential enzyme or transport reactions in a microcompartment, forming a distinct functional pool of intermediates, is also called "functional coupling" (for a detailed explanation, see Chapter 11). This greatly facilitates metabolite channeling in such a way that those substrates and products of sequential reactions do not mix with the bulk cytosolic phase [15]. These general mechanisms are thought to increase the effi-ciency of sequential reactions in a metabolic pathway by helping to avoid a poten-tially chaotic situation due to macromolecular crowding (see also Chapter 3) [16]. Such mechanisms also may play a significant role in cell signaling where, upon cell activation or stress, a host of different signaling cascades are activated simultaneously, proceeding either independently, or restricted to subcellular microcompartments, or jointly by crosstalk with each other. If these mecha-nisms were random in a cell, precise spatial and temporal signaling would be impossible.

This idea is supported by the fact that, as recently discovered, even small metabolites – such as calcium [17] or cyclic AMP [18], which work as second messengers – are diffusion restricted in living cells and thus form subcellular mi-crodomains that can be visualized, for example, as calcium sparks or propagating calcium waves [19]. Thus, they do not behave as if they were freely diffusible in solution [9, 20]. This also holds true for the rather bulky and charged adeno-sine nucleotides ATP, ADP, and AMP. As a matter of fact, there are abundant data available in the literature showing compartmentation of adenine nucleotides, particularly in muscle (see Chapters 3 and 11). Such a concept is attractive especially for AMP, which in resting cells is present at submicromolar con-centrations. Upon activation, [AMP] may rise by several orders of magnitude to stimulate AMP-stimulated protein kinase (AMPK), a metabolic master switch that is also compartmentalized (see Chapter 9).

Thus, enzymes and their corresponding metabolites, substrates, and products, e.g., working as part of a metabolic or cellular signaling cascade, may form structurally, functionally, and temporally defined four-dimensional microcompartments and other subcellular compartments [21], either via strong, static interactions or via fickle, dynamic interactions with other enzymes, proteins, or subcellular structures, the cytoskeleton representing a principle basis for such interactions.

There are interesting data concerning compartmentation of ATP coming from NMR studies *in vivo* [22–24]. For example, dipolar coupling effects, due to anisotropic motional averaging, have been observed for ^{1}H proton resonances of creatine (Cr), taurine, and lactate in muscle [25]. The most likely explanation for these effects is that even small compounds such as Cr and PCr, together with their hydration spheres, are large enough to be hindered by isotropic tumbling in the

---

Fig. 7.1 The CK–phosphocreatine circuit. Creatine (Cr), either synthesized in the body or taken up from alimentary sources, e.g., meat and fish, is transported into muscle and other cells with high and fluctuating energy requirements by a specific creatine transporter (CRT). Imported Cr is charged to the high-energy compound phosphocreatine (PCr) by the action of strictly soluble cytosolic CK (CK$_c$, **3**), by CK coupled to glycolysis (CK$_g$, **2**), or by mitochondrial CK coupled to oxidative phosphorylation (MtCK, **1**). According to the CK equilibrium reaction, in a resting cell, this results at equilibrium in approximately two-thirds [PCr] and one-third [Cr] and in a very high ATP:ADP ratio (≥100:1). Isoenzymes of CK not only are cytosolic (CK$_c$, **3**) but also are associated (CK$_a$, **4**) in compartments to ATP-consuming processes (ATPase), such as the myofibrillar actomyosin ATPase, the SR Ca^{2+}-ATPase, the plasma membrane Na$^+$/K$^+$-ATPase, the ATP-gated K$^+$ channel, or ATP-requiring cell signaling (**4**), where CK regenerates *in situ* the ATP utilized by these processes, drawing from the large PCr pool. This represents the ATP-consuming end of the CK–PCr–Cr circuit (**4**). In addition, CK is also associated to glycolytic enzyme complexes (CK$_g$, **2**), where glycolytically generated ATP is transphosphorylated into PCr that is fed into the large PCr pool. This represents the first of the two ATP-producing sites of the CK–PCr–Cr circuit (**2**). Mitochondrial MtCK is specifically located in the intermembrane space of mitochondria (MtCK [1]), and by functional coupling to the adenine nucleotide translocator (ANT) of the mitochondrial inner membrane, this MtCK accepts mitochondrially generated ATP and transphosphorylates it into PCr, which then leaves the mitochondria. This represents the second ATP-producing site of the CK–PCr–Cr circuit (**1**). A large cytosolic PCr pool of up to 30 mM is built up by CK using ATP from oxidative phosphorylation (**1**), as in the heart, or from glycolysis (**2**), as in fast-twitch glycolytic skeletal muscle. PCr is then used to buffer global cytosolic (**3**) and local (**4**) ATP:ADP ratios. This represents the temporal buffer function of the system. In cells that are polarized and/or have very high or localized ATP consumption (**4**), the differentially localized CK isoenzymes, together with easily diffusible PCr and Cr, can also maintain a high-energy PCr circuit or shuttle between ATP-providing (**1**, **2**) and ATP-consuming processes (**3**, **4**). Thus, the energy-producing and -consuming terminals of the shuttle are connected via PCr and Cr, with no need for ATP to diffuse, e.g., from mitochondria (**1**) to the sites of ATPases (**4**). Metabolite channeling occurs where CK is associated with ATP-providing (**1**, **2**) or ATP-consuming processes (**4**), such as ATPases, ATP-gated ion channels, ion pumps and transporters, metabolic enzymes, and protein kinases for cell signaling.

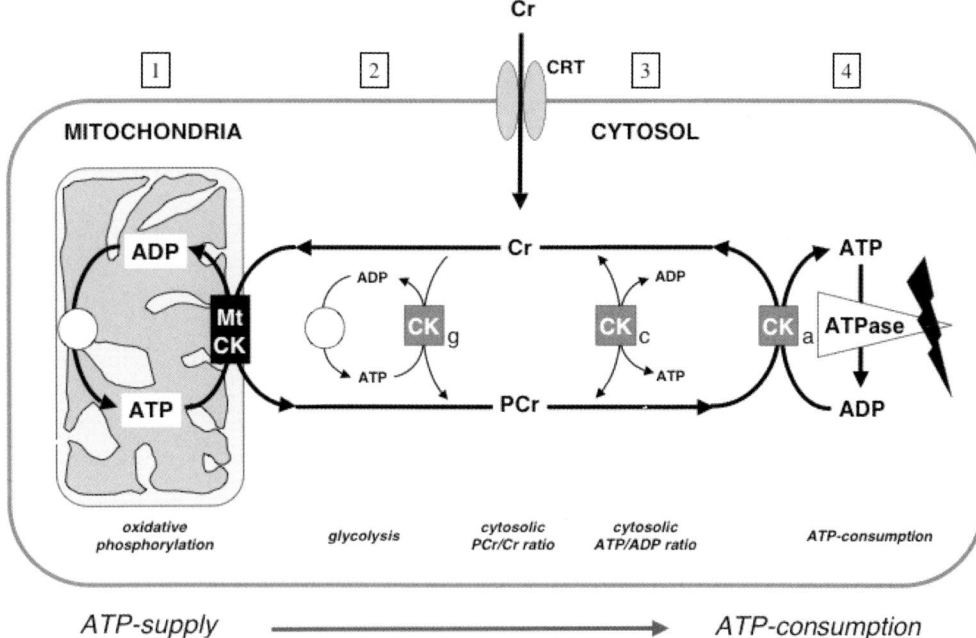

Fig. 7.1 (legend see p. 198)

elongated spaces between the thick and thin filaments in muscle, which in the actomyosin overlap zone are extremely crowded [26] and thus lead to aniso-tropic ordering of Cr and PCr along the myofilament axis [27]. This orientation-dependent dipolar splitting of the methyl group protons of Cr and PCr in muscle is no longer observable two hours postmortem [28]. Thus, PCr and Cr, and even more so the adenosine nucleotides ATP and ADP, may form distinct subcellular pools that are more or less movable or movement restricted, and some but not others, due to additional subcellular compartmentation, may be immediately available for CK-mediated phosphoryl transfer reactions [29]. These data are fully in line with considerations made by the late Peter Hochachka, who surmised that intracellular ordering of metabolic pathways, together with intracellular circula-tion or convection, is a fundamental principle for homeostasis and metabolic reg-ulation that may explain the cellular energy stability paradox [6, 30–32]. These considerations are also in line with the philosophical negentropy principle of Schrödinger, according to which cellular metabolism is always linked to a high degree of intracellular organization (see Chapters 3 and 11).

7.1.3
Tissue-specific Expression of Creatine Kinase Isoenzymes

The existence of tissue- and compartment-specific isoenzymes of creatine kinase (CK) is an important property of this long-known enzyme [33] and is a key to its functions in cellular energy metabolism. Most vertebrate tissues express two CK isoenzyme combinations: either dimeric, cytosolic, muscle-type MM-CK together with mostly octameric muscle-type mitochondrial sMtCK or, alternatively, ubiquitous brain-type BB-CK together with ubiquitous uMtCK [4]. The first CK isoenzyme combination, MM-CK with sarcomeric sMtCK, is expressed in differentiated sarcomeric muscle, cardiac muscle [34], and skeletal muscle [26], whereas the second combination, BB-CK with uMtCK, is prominently expressed in brain [35], neuronal cells [36], retina photoreceptor cells [37, 38], hair bundles of the inner ear [39], smooth muscle [40], kidney [41], endothelial cells [42], spermatozoa [43], and skin [44], while CK expression levels are generally very low or absent in liver [45]. Hybrid cytosolic MB-CK, on the other hand, is expressed only transiently during muscle differentiation but persists in adult cardiac muscle (for reviews, see [4, 46, 47]).

Octameric MtCK is localized in the cristae as well as in the intermembrane space of mitochondria, preferentially at the contact sites between the inner and outer mitochondrial membrane (Figs. 7.1 and 7.4) [7, 8, 38, 47–49]. Mitochondrial and cytosolic CK diverged at least 670 million years ago [50], suggesting that compartmentalized CK isoenzymes evolved very early during evolution in the context of functional coupling between MtCK and oxidative phosphorylation [51–53] and metabolite channeling [7, 8, 54].

7.1.4
Temporal and Spatial Buffering Functions of Creatine Kinases: The CK–Phosphocreatine Circuit

Life is based on the conversion of free energy by cells that are open thermodynamic systems, exchanging both mass and energy with the surrounding medium (see Chapter 3). Cellular energy demand and supply are balanced and tightly regulated for economy and efficiency of energy use. Tissues and cells with high and fluctuating energy requirements may increase the rate of ATP hydrolysis within seconds by several orders of magnitude, but intracellular ATP levels remain amazingly constant. This stability paradox [30, 31] can be explained by the action of immediately available, fast, and efficiently working energy-supporting and -backup systems, such as CK and AdK that connect sites of energy consumption with those of energy production via phosphoryl transfer networks [4, 6, 11, 12]. CK is a major enzyme of higher eukaryotes that copes with high and fluctuating energy demands to maintain cellular energy homeostasis in general and to guarantee stable, locally buffered ATP:ADP ratios in particular [4, 47, 53, 55–59].

"Cytosolic" CK isoenzymes are not strictly soluble, but a certain variable fraction thereof is bound permanently to or associated transiently with subcellular structures (e.g., the myofibrillar M-band [60], the myofibrillar I-band in conjunction with a glycolytic enzyme metabolon [61]) or cellular membranes, such as the plasma membrane, the sarcoplasmic reticulum (SR), or the Golgi apparatus, where bound CK is functionally coupled to the Na^+/K^+-ATPase [41, 62], the Ca^{2+}-ATPase [63], or the Golgi matrix protein GM-130 [64]. On the other hand, MtCK is exclusively found in mitochondria [48]. This compartmentation of CK involves direct or indirect association of CK with ATP-providing (mitochondria and glycolysis) and ATP-consuming processes (ATPases and ATP-dependent cell functions), forming distinct compartments that facilitate a direct exchange of ADP and ATP between associated CK and its substrates (PCr and Cr) and the respective association partners of CK, without mixing with the bulk cytosol (Fig. 7.1). The interplay between cytosolic and mitochondrial CK isoenzymes fulfils multiple roles in cellular energy homeostasis (see also Chapters 3 and 11) [4, 6–8, 29, 65, 66].

First, both isoenzymes contribute to the buildup of a large intracellular pool of phosphocreatine (Fig. 7.1) that represents an efficient temporal energy buffer and prevents a rapid decrease in global ATP concentrations upon cell activation or sudden stress conditions (Fig. 7.1, cytosolic CK_c equilibrates the cytosolic overall ATP:ADP ratio) [56]. This buffering function of CK also could be demonstrated by an inverse genetics approach. Introducing the genes for phosphagen kinases, either arginine kinase (AK) or CK, into bacteria or yeast cells leads to functional expression of CK or AK and, together with creatine or arginine supplementation, renders these cells resistant to transient metabolic stresses by stabilizing intracellular ATP concentrations [67, 68]. A clear advantage of the PCr–Cr system is that the compounds are "metabolically inert." Unlike the adenosine nucleotides, which interact with and regulate a plethora of cellular proteins or enzymes, respectively, no other enzyme seems to exist that binds and metabolizes PCr or Cr except for CK. This allows an accumulation of this guanidino compound in the cell basically without interfering with basic metabolism.

Secondly, a very important additional function of CK is based on the concepts of subcellular compartmentation of CK isoenzymes and diffusion limitations of ATP and especially ADP. The so-called PCr–Cr circuit or shuttle theory postulates that not ATP and ADP, but rather PCr and Cr, diffuse from subcellular sites of ATP generation (mitochondria or glycolysis) to distant ATPases, e.g., the myofibrillar calcium-activated MgATPase, the SR Ca^{2+} pump ATPase, the Na^+/K^+ pump ATPase, or ATP-gated processes such as the K^+_{ATP} channel or ATP-dependent cell signaling (Fig. 7.1) [4, 47, 55, 57, 59, 65, 66, 69, 70]. In fact, the advantage of PCr and Cr as metabolically inert "energy transport compounds" over ATP and especially ADP has been experimentally confirmed in an *in vivo* model system with spermatozoa. These are elongated, highly polar cells, where energy production by mitochondria in the midpiece, near the sperm head, is distant from energy utilization in the long sperm tail [43, 71]. However, this principle is also likely to hold true for skeletal and cardiac muscle as well as for brain and retina, where cells are operated by a very high-energy flux that puts enormous strain on subcellular energy delivery and regulation of these processes (see also Chapters 3 and 11) [6].

Third, subcellularly compartmented CK isoenzymes allow maintenance of high local ATP:ADP ratios in the vicinity of cellular ATPases. Thus, they maintain a maximal change in the Gibbs free energy of ATP hydrolysis, $\Delta G_{ATP} = \Delta G°_{obs} - RT \times \ln([ATP]/[ADP] \times [P_i])$, which is proportional to the [ATP]:[ADP] ratio $\Delta G°_{obs}$ corresponds to the standard tree energy change at 25°C and pH 7.0). This energetic aspect is especially important for thermodynamically unfavorable reactions, i.e., for ATPases that need a high free energy change for optimal function, such as the sarcoplasmic reticulum Ca^{2+} pump (see below). On the other hand, localization of MtCK at the outer surface of the mitochondrial inner membrane in conjunction with the adenine nucleotide transporter (ANT) allows, by a direct transfer of substrates, for immediate transphosphorylation of matrix-generated ATP into PCr and also for a rapid recycling of ADP into the mitochondrial matrix [72]. This results in a strong stimulation of oxidative phosphorylation by MtCK (for a detailed explanation, see Chapters 3 and 11 and Figs. 7.1 and 7.4) [51]. Because of the specific localization of mitochondrial and cytosolic CK isoenzymes, the much faster diffusion rate of PCr as compared to ATP (the latter being decreased in some subcellular compartments by orders of magnitude [70, 73]), and the very much faster diffusion rate of Cr compared with ADP [43], the CK–PCr system provides for a spatial "energy shuttle" or "energy circuit" (Fig. 7.1) that bridges sites of ATP generation with sites of ATP consumption. Thus, MtCK and the CK–PCr system contribute to an intricate metabolic energy transfer network in the cell, connecting mitochondria with myofibrils, the sarcoplasmic reticulum, and nuclei [11, 24, 65, 74–76].

A tight functional coupling of CK to ATPases has the advantage (1) that product inhibition of the ATPase by ADP and H^+ is avoided, since the latter are both substrates of the CK reaction that are continuously removed if the enzyme is active (PCr + ADP + H^+ ↔ Cr + ATP) and (2) that the high Gibbs free energy change of ATP hydrolysis (ΔG_{ATP}) at sites of ATP consumption is preserved by keeping very high local ATP:ADP ratios due to coupling of CK with said ATPases *in situ* and thus preventing energy dissipation caused by transport of ATP and avoiding mixing it with the adenine nucleotides of the surrounding bulk solution [4, 11, 46].

Another important function of CK is the release of P_i due to PCr hydrolysis by the CK reaction with its manifold regulatory consequences for energy metabolism [4]. Finally, via the CK reaction, the inhibitory actions of ADP on ATPases are minimized by keeping [ADP] very low as long as possible. In this way, the buildup of AMP and the subsequent degradation of AMP and IMP into inosine, which would ultimately leave the cell, are delayed, and thus a loss of valuable nucleotides is largely prevented (see Chapters 3, 8 and 9) [4].

To summarize, convincing evidence for the importance of the compartmentalized CK system and the *in vivo* operation of a CK-mediated PCr–Cr circuit as well as the existence of phosphotransfer networks in tissues and cells with high and fluctuating energy requirements comes from

1. *in situ* immuno-localization studies of CK isoenzymes in tissues [4, 46, 66];
2. kinetic and thermodynamic analysis of the MtCK reaction coupled to oxidative

phosphorylation via ANT in both isolated mitochondria and permeabilized cells (see also Chapters 3, 11, 12, and 15) [48, 75, 77];

3. careful analysis of the phenotype of CK transgenic knockout mice, e.g., showing obvious deficiency in skeletal and cardiac muscle Ca^{2+} handling [78–80]; and

4. direct *in vivo* ^{18}O labeling of phosphoryl moieties in intact muscle (see also Chapters 3, 8, and 11) [12, 81].

Fig. 7.2 The function of muscle-type MM-CK at the sarcomeric M-band. This model combines biochemical data on the molecular interactions of the myomesin with the thick filament, titin, and MM-CK with the information provided by immunoelectron microscopy.

(A) Model of myomesin arrangement in the M-band (modified from [98]). Myosin and titin filaments are represented extremely schematically (A, B). The myomesin molecule is drawn according to structural predictions. The unique N-terminal domain My1 (A, **1**), which is predicted to have a disordered conformation and possibly to wrap around the thick filament to provide the essential strength for a myomesin–myosin interface [88], is followed by 12 structural modules with strong homology to either fibronectin (Fn) or Ig-like domains (shown as ellipses). Myomesin molecules, forming antiparallel dimers via their last C-terminal domain My13 (A, **2**) [98], are shown to associate with their My4–6 domains with the m4 titin domain (A, **3**, black ellipse) and run along the antiparallel titin molecules that overlap with their C-terminal ends (titin N and C). (B) Superimposed onto the basic structural scaffold (A), binding of MM-CK dimers [437] to myomesin domains My7–My8 is shown [103] (B). Note that the interaction site of MM-CK with myomesin corresponds to the symmetrically off-center M4 and M4' m-bridge positions (indicated by black M4' and M4 boxes in A and B), which is fully in line with immunoelectron microscopic labeling by anti-MM-CK antibodies [95]. Thus, the presence of MM-CK together with myomesin at these locations explains the appearance of the electron-dense M4' and M4 lines on EM pictures of the sarcomere. For reasons of simplicity, and because the exact position of the M-protein is not known, it is omitted from the scheme. However, there is good reason to believe that M-protein, also likely to form antiparallel dimers, is largely responsible for the electron density of the central M1 m-bridges [91], the positions of which are indicated in A (black M1 box on top). Because MM-CK was also shown to specifically interact with M-protein at domains M6–8 [103], MM-CK binding to M-protein thus also may contribute to the electron density of this M1 line [102]. The MM-CK integrated into the M-band scaffold regenerates ATP that is hydrolyzed in the actomyosin overlap zone of the sarcomere on both sides of the central M-band by tapping from the large pool of more or less freely diffusible PCr. The microcompartmentation and functional coupling of the actomyosin ATPase cycle *in situ* with the CK reaction ($MgADP^- + H^+ \leftarrow CK \rightarrow Mg\ ATP^{2-} + Cr$) (symbolized by a central ellipsoid) lead to maintenance of locally high ATP:ADP ratios by CK and efficient local ATP regeneration on both sides of the central M-band, with adenosine nucleotides cycling along the filament axes within the crowded myofibrils [26, 108]. Because ADP and protons are substrates of the CK reaction, this prevents inhibition by protons and ADP of the actomyosin kinetic cycle [4, 105, 108]. Convection by the action of myosin cross-bridges [30, 31] and piston-like movement of the actin filament during contraction and relaxation would direct a forward flow of ADP towards the M-band for regeneration by CK and PCr and redirect a backward flow of ATP towards the actomyosin overlap regions, respectively [26].

The energy-shuttling function of the system has been experimentally demonstrated in a convincing fashion by the *in vivo* analysis of activated sea urchin spermatozoa by ^{31}P-NMR CK flux measurements in combination with oxygraphy [43, 82]. Recently, evidence of the importance of the CK–PCr circuit for organ function, as well as of the existence of distinct non-equilibrating pools of ATP, has come from ^{31}P- and ^{1}H-NMR studies with intact *ex vivo* perfused heart [23, 24, 79, 83–85]. The outcomes of these experimental approaches have also been substantiated by results obtained from computer modeling approaches (see Chapters 12 and 13) [72, 86, 87].

7.1.5
A Closer Look at CK (Micro)Compartments and High-energy Phosphate Channeling

In the following sections, we will discuss subcellular CK compartments in more detail (see Sections 7.1.6 to 7.1.13), because they represent important puzzle pieces for understanding the architecture and functioning of the CK–PCr circuit (Fig. 7.1). A significant fraction of "cytosolic" CK is structurally and functionally associated or co-localized with different structurally bound ATPases: (1) different ion pumps in the plasma membrane, (2) the sarcomeric M-band and I-band of

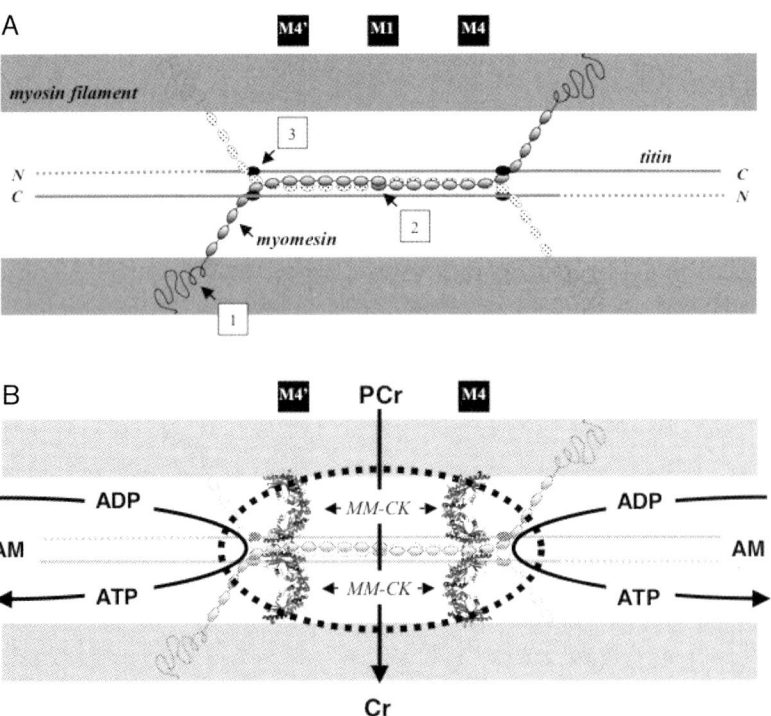

Fig. 7.2 (legend see p. 203)

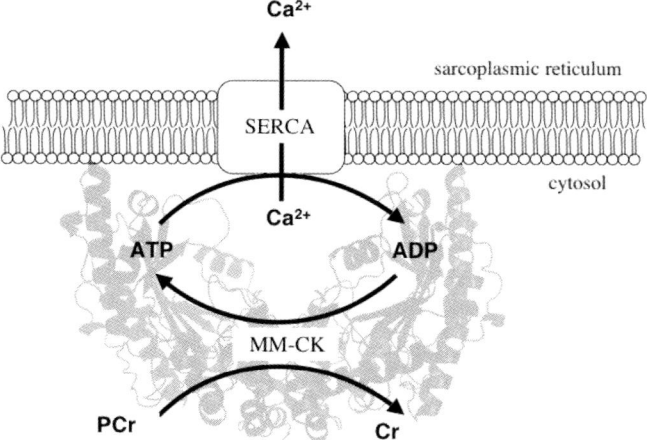

Fig. 7.3 The function of muscle-type MM-CK at the sarcoplasmic reticulum. The propinquity of MM-CK with the SR Ca^{2+} pump, the most abundant SR protein, allows for the formation of a shielded subcompartment between the two entities and for functional coupling between CK and the Ca^{2+}-ATPase. Thus, ATP generated *in situ* by the CK reaction at the expense of PCr has preferential access to the pump and facilitates maximal Ca^{2+} uptake into isolated SR vesicles [63]. This is corroborated by the fact that neither externally added ATP-regenerating systems (PEP and pyruvate kinase) nor ATP traps [128] significantly affect CK-mediated Ca^{2+} uptake [127], indicating that CK, due to its high affinity for ADP (K_m ADP of MM-CK = 15–35 μM [437]), is able to maintain a very high local ATP:ADP ratio in the vicinity of the SR Ca^{2+} pump and to channel the ATP directly to the Ca^{2+}-ATPase, thus maintaining the thermodynamic driving force of this energetically demanding ion pump [4, 132, 134, 438]. This illustration may be taken as representative of other subcellular CK compartments with ion pumps, ATP-gated ion channels, membrane proteins, or membrane-associated signaling by protein kinases, such the Na^+/K^+ pump ATPase [41], the ATP-gated K^+ channel [146], the Golgi matrix protein GM130 [64], and insulin [155] or thrombin [156] signaling events, respectively, where CK was shown to be associated and functionally coupled with these processes.

the myofibrils in muscle, and (3) the calcium pump of the muscular sarcoplasmic reticulum. In all these cases, PCr is used for local regeneration of ATP, which is directly channeled from CK to the consuming ATPase without major dilution by the surrounding bulk solution. On the ATP-generating side, part of the cytosolic CK is associated with glycolytic enzymes, and, more importantly, the mitochondrial proteolipid complexes containing MtCK are coupled to oxidative ATP production. In particular, three well-characterized functionally coupled subcellular CK compartments will be presented in more detail. Two are at the receiving end of the CK–PCr circuit at ATP-utilizing sites: muscle-type MM-CK at the sarcomeric M-band of myofibrils, regenerating ATP for the actomyosin ATPase (Fig. 7.2; see Section 7.1.6), and at the sarcoplasmic reticulum (SR), regenerating ATP for the Ca^{2+} pump ATPase (Fig. 7.3; see Section 7.1.7). The third well-characterized example is concerned with high-energy phosphate channeling by

MtCK in mitochondria, constituting an energy-producing terminus of the circuit (Fig. 7.4; see Section 7.1.13).

Prerequisites for a functionally coupled CK compartment at ATP-utilizing sites include (1) a co-localization and association of CK to such processes, (2) a confirmation that PCr would preferentially drive such ATPases via associated CK, and

Fig. 7.4 Functions of mitochondrial MtCK compartments and metabolite channeling by MtCK. After the import of nascent MtCK over the mitochondrial outer membrane and cleavage of the N-terminal targeting sequence, MtCK first assembles into dimers. Dimers rapidly associate into octamers (not shown); although this reaction is reversible, octamer formation is strongly favored by the high MtCK concentration in the intermembrane space. The symmetrical and cube-like MtCK octamers (top left) then directly bind to acidic phospholipids of the outer mitochondrial membrane (OM) and preferentially to cardiolipin (CL) of the inner mitochondrial membrane (IM). MtCK is found in two locations: in the so-called mitochondrial contact sites associated with ANT and VDAC (shown here) and in the cristae associated with ANT only (not shown). In contact sites, MtCK simultaneously binds to the IM, as well as to the OM, due to the identical top and bottom faces of the MtCK octamer. The binding partner in the IM is the twofold negatively charged CL (marked in dark grey), which allows a functional interaction with the adenine nucleotide translocator (ANT) that is situated in CL membrane patches (marked in dark grey). In the OM, MtCK interacts with other acidic phospholipids and, in a calcium-dependent manner, directly with VDAC. The main substrate and product fluxes are indicated by dark arrows. Minor or alternative fluxes are shown in light gray. It should be noted that for simplification and guidance of the reader, the main fluxes are shown to flow through two different VDAC and ANT molecules, although ANT is a true antiporter that stoichiometrically exchanges one ATP for one ADP. In this scheme, ATP generated by oxidative phosphorylation via the F_1-ATPase (**1**) is transported through the IM by ANT (**2**) in exchange for ADP. This ATP may either leave the mitochondrion directly via outer membrane VDAC or is preferentially accepted and transphosphorylated into PCr by octameric MtCK in the intermembrane space (**3**). PCr then preferentially leaves the mitochondrion via VDAC and feeds into the large cytosolic PCr pool (**4**). ADP generated from the MtCK transphosphorylation reaction is accepted by ANT and immediately transported back into the matrix to be recharged (**5**). In contact sites, this substrate channeling allows for a constant supply of substrates and removal of products at the active sites of MtCK. In cristae, only ATP/ADP exchange is facilitated through direct channeling to the MtCK active site, while Cr and PCr have to diffuse along the cristae space to reach the VDAC (not shown; for details, see Refs. [7] and [8]). The tight functional coupling of MtCK to the ANT leads to saturation of the ANT on the outer side of the IM with ADP, which is transported back into the matrix to be recharged by the F_1-ATPase, thus efficiently coupling electron transport to ATP generation and thus lowering the production of free oxygen radicals (ROS) [192]. On the other hand, the tight functional coupling of ANT to MtCK leads to a saturation of MtCK with ANT-delivered ATP and a locally high ATP:ADP ratio in the vicinity of MtCK, and in combination with cytosolic Cr, entering the intermembrane space via VDAC (**6**), it drives the synthesis by MtCK of PCr from ATP without a loss of its energy content, thus maintaining maximal thermodynamic efficiency for high-energy phosphate synthesis in the form of PCr (**4**), which then is exported into the cytosol. This represents an instructive example of functional coupling and metabolite channeling. (Adapted from [4, 192, 294]).

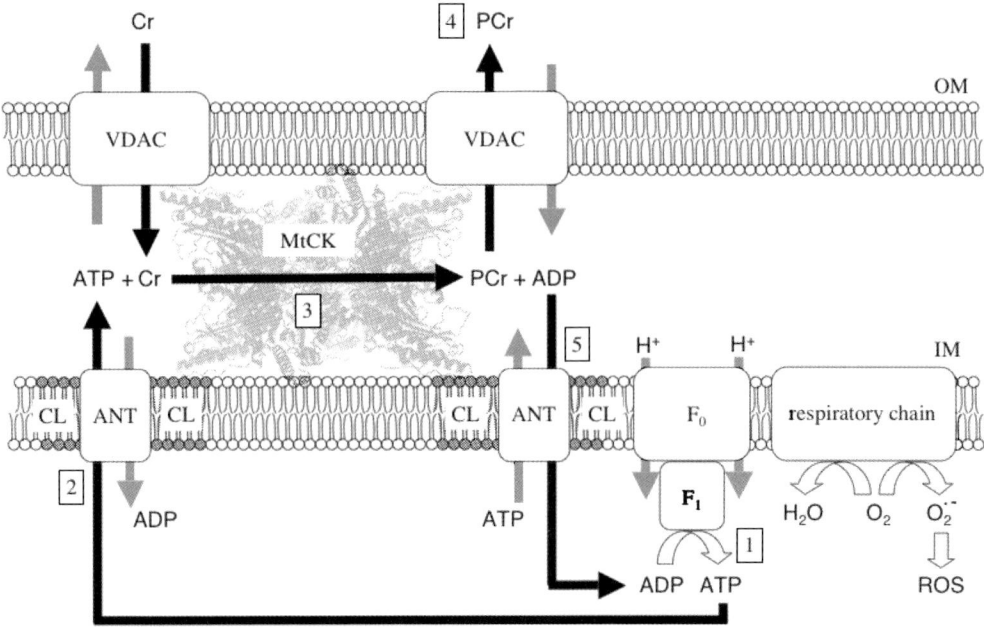

Fig. 7.4 (legend see p. 206)

(3) experimental evidence that PCr-driven ATPase function is not significantly affected by externally added ATP or soluble ATP-regenerating systems, such as phosphoenolpyruvate (PEP) and pyruvate kinase, nor by added ATP traps. If these three conditions were fulfilled, this would indicate that CK-generated ATP (from PCr) has preferential access to the ATPase due to the presence of a shielded microcompartment. As a consequence of such functional coupling of CK with the ATPase, the apparent K_m for ATP of the ATPase is lowered. Thus, for maintaining full ATPase activity, a lower adenine nucleotide concentration is necessary in the presence of PCr compared with that for ATP alone. Such data are available both for M-band–associated and SR-associated CK, as discussed in some detail below (see also Figs. 7.2 and 7.3), and the same principles apply for MtCK-generated ADP that is preferentially recycled in mitochondria to stimulate oxidative phosphorylation (see Fig. 7.4).

7.1.6
Subcellular Compartmentation of CK at the Myofibrillar M-band: ATP Regeneration for Muscle Contraction

The M-band of the striated muscle sarcomere represents the most thoroughly characterized CK microcompartment. The M-band is visible as a substriated electron-dense structure at the central region of individual sarcomeres (for review, see [59, 88]). There, the so-called m-bridges, running perpendicular to the thick filament axis, connect individual thick filaments at their bare zones that are free of myosin heads [89]. So far, besides the giant titin molecules that extend from both sides of the adjacent sarcomeric Z-disks into the M-band [90] and overlap there with their C-termini [91] (Fig. 7.2A), three additional major constituents of the M-band have been identified: the 165-kDa M-protein [92, 93], the 185-kDa myomesin [94], and the 84-kDa dimeric muscle-type MM-CK [59, 60, 95] (for review, see [4]) (Fig. 7.2B). Additionally, other minor proteins such as enolase, AMPK, calpain, and DRAL/FHL-2 have been identified at this location [96, 97], suggesting a general role of the M-band as a targeting site for metabolic enzymes. Myomesin and M-protein are highly modular proteins composed mainly from the conserved sequence of Ig- and Fn-like domains. The antiparallel dimers of myomesin molecules [98] might form the elastic connection between the neighboring thick filaments (Fig. 7.2A). This cross-link was suggested to play an important mechanical role and promote sarcomere stability in the contracting muscle [88, 99]. We have found that some 5–10% of total cytosolic MM-CK is located in an isoenzyme-specific manner at this sarcomeric structure [60, 100, 101]. Based on the fact that a significant proportion of the electron density of the M-band is lost upon incubation of muscle fibers with an excess of specific, high-affinity monovalent Fab antibodies against MM-CK, it was concluded that the MM-CK is responsible for the electron-dense appearance of the M-band on electron micrographs [102]. Accordingly, direct immunolabeling with anti–MM-CK IgG antibodies resulted in a double-striped pattern corresponding to the off-center m-bridge stripes [95]. This pattern indicated that MM-CK bound to the M-band mainly follows the location of the so-called M4 m-bridges that connect thick filaments in the bare zone (Fig. 7.2B). Localization at the M-band is specific for the MM-CK isoform and is not found for the highly homologous BB-CK. It results from two MM-CK–specific lysine "charge clamps" at positions K8 and K24 and K104 and K115 [100]. Binding of MM-CK via these lysine charge clamps leads to a strong but pH-dependent interaction of this CK isoform with the two central fibronectin-like Fn7-8 domains of myomesin, as well as with the homologous Fn6-8 domains of M-protein [103]. According to the current model, the MM-CK interacting Fn7-8 domains of myomesin are localized at the level of M4/M4′ bridges (see Fig. 7.2B), in agreement with the higher electron density found at the M4 m-bridge level. Because M-protein is likely to be located in the vicinity of the central M1-bridge [91], it is most likely that MM-CK attached to the M-protein is also responsible for the electron density of this structure. This may explain the correlation between the M1 line appearance and M-protein expression [89]. Another exciting fact results from reengineering the above lysine charge clamps of MM-CK into the corresponding amino acid positions of BB-CK. This results in a gain of function of the mutant BB-CK, which is now able to bind specifically to the myofibrillar M-band [100] and to interact with myomesin, just like MM-CK [103]. Thus, the specific interaction sites of MM-CK with the two M-band proteins, myomesin and M-protein are now well defined.

The amount of MM-CK bound to the M-band is sufficient to regenerate the ATP hydrolyzed by the actomyosin ATPase, running at maximal velocity, in isolated myofibrils [57, 104, 105]. The apparent K_m of ATP for the actomyosin ATPase was reduced by endogenous MM-CK because of its specific location at the M-band, an effect that could not be mimicked by external addition of an alternative ATP regeneration system [105, 106]. This is consistent with tight functional coupling between CK and the actomyosin ATPase. This *in situ* role of MM-CK was corroborated by contraction and rigor experiments with chemically skinned intact muscle fibers [107, 108]. The functional development of this role of CK could be followed during perinatal maturation of the heart [109]. Ablation of the MM-CK gene leads to a distinct but not very severe phenotype affecting muscle contraction, insofar as muscles of these mice lack burst activity [110] and show less peak muscle force that cannot be maintained [111]. Alterations in myofibrillar functions were obvious as well [112]. Later on, it became clear that in transgenic knockout mice, lacking either cytosolic MM-CK or mitochondrial MtCK, these isoforms can to some extent compensate for each other [113] and, together with a host of metabolic and structural adaptations that take place in the muscles of these transgenic animals [114], lead to a relatively mild muscle phenotype [115]. This muscle phenotype, however, becomes more severe if both MM-CK and MtCK genes are ablated simultaneously [80, 115]. This complex situation argues for the importance of the CK system in muscle but at the same time reveals that muscle cells, with their potential for plasticity, can undergo remarkable structural and functional adaptations. For example, adenylate kinase or glycolysis as an ATP provider may take over parts of the functions that are otherwise guaranteed by CK flux reactions [12, 116–118]. In addition, amazing structural remodeling of cell architecture to compensate for the loss of PCr shuttling can be observed. These adaptations can be considered safeguard mechanisms to sustain important functions of energy provision in muscle. For example, in the double-knockout mouse lacking both CK isoforms (MM-CK and sMtCK) in muscle [80], the mitochondrial density and location are increased drastically in muscle, and the organelles are placed in rows between individual myofibrils [116–118]. By this adaptive strategy, the diffusion distance for ATP from mitochondria to myofibrils is shortened because PCr shuttling is no longer possible [119]. These compensatory adaptations in CK double-knockout mice seem to be even more prominent in the heart muscle, to safeguard the performance of this organ that is essential for life [120]. The most clear-cut phenotype of these double-knockout CK animals was again a muscle phenotype affecting contraction [121] and leading to a significant disturbance of Ca^{2+} homeostasis (see below) [80, 117].

7.1.7
Subcellular Localization of CK at the Sarcoplasmic Reticulum and Plasma Membrane

CK activity in crude sarcoplasmic reticulum (SR) preparations was first measured in 1970 [122]. This finding was extended by functional assays demonstrating that the fraction of CK, which is specifically bound to highly purified SR membrane vesicles, is able to fully support, at the expense of PCr, Ca^{2+} uptake into the SR vesicles by the SR Ca^{2+}-ATPase pump [63, 123]. Inhibition of CK bound to SR vesicles by *N*-nitroso-glutathione (a nitric oxide donor) or DNFB (an inhibitor) leads to a significant decrease in [124] or complete abolition of Ca^{2+} uptake [63] if ATP derived from PCr and catalyzed by SR-bound CK drives the reaction. CK has been localized along SR membranes *in situ* by immunogold labeling of ultra-thin cryosections of muscle [26, 63], as well as on isolated SR vesicles *in vitro* [63]. Functional coupling of this membrane-bound CK with the Ca^{2+}-ATPase pump, representing the most prominent polypeptide in isolated SR preparations, also can be shown directly [125]. By overlay blot binding assays, a direct pH-dependent association of MM-CK to a number of SR proteins was demonstrated, and two domains on MM-CK could be identified that are responsible for this interaction [126]. Binding of CK to the SR Ca^{2+}-ATPase results in a close propinquity of the two molecules (Fig. 7.3). Functional coupling of SR-bound CK to the Ca^{2+}-ATPase also has been shown *in situ* with permeabilized intact muscle fibers, where it could be demonstrated that ATP generated by SR-bound CK is more effective than externally added ATP at similar concentrations [127]. Similarly, ATP generated by a soluble ATP-regenerating system, such as PEP and pyruvate kinase [127], or by addition of an ATP trap to the system did not significantly interfere with PCr-supported Ca^{2+} uptake by the SR [128]. This indicates that the compartment is shielded against outside influences and that ATP generated by CK is preferentially channeled to the Ca^{2+}-ATPase. Thus, coupling of CK to the Ca^{2+}-ATPase results in preferential delivery of CK–PCr-derived ATP to this ion pump (Fig. 7.3) as well as in promotion of a disengagement of the MgADP-bound state of the Ca^{2+}-ATPase. Both lead to a significant decrease in the apparent K_m for ATP of the Ca^{2+}-ATPase. The importance of CK for Ca^{2+} handling has been corroborated with skinned fibers from normal muscle [129] and with muscle from CK double-deficient transgenic mice [130], as well as with intact single cardiomyocytes [131]. The SR Ca^{2+} uptake operates with a $\Delta G_{Ca2+\text{-transport}}$ of approximately $+51$ kJ mol^{-1}, and the Gibbs free energy change of ATP hydrolysis (ΔG_{ATP}) at physiological concentrations of ATP (5–8 mM), free ADP (10–50 µM), and P_i (5–10 mM) in resting muscle may be estimated to be approximately -55 kJ mol^{-1} (for review, see Ref. [46]). The latter is only slightly higher than the Gibbs free energy change ($\Delta G_{Ca2+\text{-transport}}$) needed to support the thermodynamically unfavorable Ca^{2+} uptake and indicates that the SR Ca^{2+}-ATPase is working close to thermodynamic equilibrium and therefore greatly depends on a high local ATP:ADP ratio for efficient sequestration of Ca^{2+} into the SR lumen [46, 63, 132]. Inhibition of SR-bound CK or CK associated with other ion pumps (e.g., by oxidative damage, as seen in many neuromuscular and neurodegenerative diseases, including mitochondrial myopathies [133]) therefore would decrease ΔG_{ATP}, limit the thermodynamic driving force of the SR Ca^{2+} pump or other ion pumps, and thus cause a decrease in contractile reserve [134]. This lowered cellular energy status, under chronic conditions, would lead to pathological Ca^{2+} overload and generation of even more ROS damage, which is typical for such diseases [135]. Because CK, including MtCK, is extremely susceptible to oxidative damage (see Section 7.4), it is obvious that compromising this enzyme by ROS damage under pathological conditions will hamper the function of the Ca^{2+} pump, thus acerbating any preexisting Ca^{2+} overload situation and eventually leading to cell death.

Muscle MM-CK and sMtCK double-knockout mice revealed that CK failure profoundly affects the ability of these mice to engage in chronic bouts of endurance running exercise and that this decrease in performance is associated with muscle wasting [136]. As a main phenotype, these mice display significant difficulties with intracellular Ca^{2+} handling and muscle relaxation [80], again emphasizing the physiological importance of the CK system for the energetics of intracellular Ca^{2+} homeostasis, the delivery of ATP to the energetically demanding Ca^{2+} pump, and specifically for optimal refilling of the SR Ca^{2+} store [78]. It has been suggested that Ca^{2+} uptake into the SR is supported not only by CK but also by glycolysis, via glycolytic enzymes bound to the SR triads and SR vesicles [26, 137, 138]. Indeed, in the above CK double-knockout mice, the absence of MM-CK at the SR could be compensated at least in part by glycolysis, which took over, as an adaptational measure, some energetic support by delivering glycolytic ATP to drive calcium uptake into the SR [116].

7.1.8
Subcellular Compartments of CK at the Plasma Membrane and Functional Coupling with Na+/K+-ATPase

A certain fraction of total CK is associated with plasma membranes [139], where the enzyme is functionally coupled to the Na+/K+-ATPase or sodium pump [41, 140, 141]. As a nice example to support this notion, significant amounts of CK are associated to postsynaptic membranes of the *Torpedo* electric organ [142]. After discharge of this organ, the levels of PCr fall rapidly due to the almost exclusive utilization by CK of PCr for fueling the fully activated Na+/K+-ATPase to recharge the electrocytes. Indeed, an intimate functional coupling between CK and this Na+/K+ pump could directly be demonstrated by *in vivo* saturation transfer ^{31}P-NMR measurement [143].

7.1.9
Structural and Functional Coupling of Cytosolic CK with ATP-sensitive K+ Channels and KCC2

Sarcolemmal ATP-sensitive K+ channels (K^+_{ATP}) belong to the group of intracellular energy sensors, coupling the metabolic status of the cell with membrane excitability. These channels are selectively permeable to K+ ions and are closed at normal millimolar concentrations of intracellular ATP. In other words, if this channel saw normal bulk cytosolic [ATP], it would stay closed all the time, because bulk [ATP] is kept homeostatically constant in a cell as long as possible [31]. However, active membrane ATPases, such as Na+/K+-ATPase, reduce the local ATP concentrations, thereby setting the ATP:ADP ratio near the plasma membrane distinct from the bulk cytosol [20, 70] and allowing opening of the K^+_{ATP} even if bulk cytosolic [ATP] remains high. The direct physical association of cytosolic CK with the K^+_{ATP} channel [144], as well as functional coupling of CK with vicinal Na^+/K^+ pumps (see above), together with the dynamics of high-energy phosphoryl transfer through the cytoplasm via the CK reaction, permit a high-fidelity transmission of energetic signals into the sub-membrane compartment, thus synchronizing cell metabolism with K^+_{ATP} electrical activity [145, 146]. In transgenic mice carrying a germ line deletion of the MM-CK gene, regulation of the K^+_{ATP} by PCr and concomitant signal delivery to K^+_{ATP} channels is disrupted, and a cellular phenotype with increased electrical vulnerability of the cardiomyocytes is generated [145]. In addition, CK can functionally couple with K^+_{ATP} through direct CK-dependent regulation of the ATPase catalytic cycle within SUR, the K^+_{ATP} regulatory subunit, promoting disengagement of the MgADP-bound state and thus channel closure [147]. In pancreatic β-cells, K^+_{ATP} helps to control insulin secretion, and thus functional coupling of the CK system to this channel is clinically relevant. Strategies of structural compartmentation of CK and functional coupling similar to those shown to take place with the K^+_{ATP} channel [145] could also apply to other ATP-gated processes in the cell. In fact, a specific association has been shown between CK and the neuron-specific K^+-Cl^- co-transporter (KCC2) that may be ATP induced [148]. Furthermore, it could be shown that CK activates KCC2 in its subcellular microenvironment at the plasma membrane [149]. This complex membrane sensor of the cellular energy state is responsible for decreasing the entry of Ca^{2+} into the cells under conditions of energy deficiency as manifested by a decrease in PCr content. Disturbance of the CK system and loss of functional coupling of CK to the KCC2 ion transporter could contribute to acute ischemic contractile failure of the heart, a condition in which CK is impaired (see Chapter 11).

7.1.10
CK Interaction with the Golgi GM130 Protein

Recently, brain-type BB-CK was shown to transiently associate specifically and co-localize with the *cis*-Golgi matrix protein (GM130) during early prophase of mitosis [64]. This is exactly the time point when energy is needed by signaling pathways to initiate the fragmentation of the Golgi apparatus [150–152]. Because high-energy phosphates are needed for the various phosphorylation processes, it is likely that the interaction between GM130 and BB-CK facilitates GM130 phosphorylation by ATP-requiring protein kinases. In support of this notion, recent results of a proteomic approach revealed a co-purification of BB-CK in a complex with Golgi casein kinase as well as with GM130 [153]. Taken together, these data strongly suggest that BB-CK is structurally and functionally coupled in a dynamic and cell cycle–dependent manner to the Golgi apparatus. There, BB-CK is most likely a member of a large protein complex that also comprises several protein kinases. Thus, as a novel function, CK family members may be linked to signaling cascades regulating the integrity of the Golgi apparatus and thus may be involved in the control of the cell cycle [64]. Interestingly, BB-CK was recently shown to be in a complex together with the chloride intracellular channel (CLIC), dynamin I, α-tubulin and β-actin, and isoforms of the 14-3-3-protein [154]. This may implicate BB-CK in further cellular processes different from those mentioned above.

7.1.11
Specific Compartments of Cytosolic CK with Insulin and Thrombin Signaling Pathways

Insulin signaling requires autophosphorylation of the insulin receptor kinase domain, and this process is strongly inhibited by ADP. Thus, CK associated at the plasma membrane locally provides the ATP needed for autophosphorylation and, at the same time, effectively removes ADP from this microcompartment and thus relieves the inhibition mentioned [155]. This also holds true for other signaling pathways, e.g., thrombin signaling, where a structural and functional interaction of cytosolic CK with the protease-activated receptor-1 (PAR-1) has been convincingly shown *in vitro* as well as *in vivo*. CK bound to PAR-1 is poised to provide bursts of site-specific, high-energy phosphate necessary for efficient thrombin receptor signal transduction during cytoskeleton reorganization [156]. Hence, we consider it likely that CK is also involved in similar functionally coupled compartments with a multitude of other receptors and protein kinases belonging to other signaling pathways. A connection between CK and AMPK has been suggested (see Chapter 9).

7.1.12
Subcellular Compartments of CK with Glycolytic Enzymes and Targeting of Glycolytic Multi-enzyme Complexes

In muscle, glycolytic enzymes are targeted to the actin-containing thin filaments at the sarcomeric I-band region, where they form highly complex glycolytic metabolons. By elegant experiments with transgenic *Drosophila*, which express the glycerol-3-phosphate dehydrogenase isoform (GDPH-1) but not the GDPH-3 isoform in their flight muscle, it could be shown that neither glyceraldehyde-3-phosphate dehydrogenase (GAPDH) nor aldolase would co-localize at the I-band in flight muscle of transgenic flies where GDPH-1 had been replaced by GDPH-3. Even though the full complement of active glycolytic enzymes was present in these transgenic flight muscles, with the exception of an isoenzyme switch from GDPH-1 to GDPH-3, the failure of the glycolytic enzymes to co-localize in the sarcomere resulted in the inability to fly [157]. Thus, formation of functionally coupled and correctly targeted multi-enzyme complexes with substrate/product channeling seems to be paramount for proper functioning of glycolysis and ultimately for correct muscle function.

Interestingly, CK also seems to participate in the glycolytic metabolon, because in muscle MM-CK was also localized specifically in the sarcomeric I-band together with other glycolytic enzymes, such as aldolase and phosphofructokinase (PFK) [26, 61]. Close functional coupling between CK and pyruvate kinase [158] and glycogen phosphorylase [132], both forming so-called diazyme complexes with CK, has been shown and discussed in theoretical terms, respectively. Recently, structural coupling of cytosolic CK to the key regulatory enzyme of glycolysis, PFK-1, which itself is strongly regulated by ATP, has also been demonstrated [61]. It thus is entirely reasonable to assume that CK co-recruited together with PFK and aldolase [61] into the glycolytic metabolon could be directly coupled to glycolysis and transfer the glycolytically produced ATP immediately into PCr (Fig. 7.1, CK_g). By this strategy, glycolytic ATP would not accumulate, which otherwise would result in inhibition of PFK and shutting off of glycolysis, especially during high-intensity performance of fast-twitch glycolytic muscle [61]. Thus, CK seems not only to be involved in removing glycolytically generated, metabolically highly active ATP into the PCr pool but also, by adjusting ATP:ADP ratios *in situ* within the glycolytic metabolon and by tight functional coupling with PFK-1, to regulate glycolytic flux [61] and glycogenolysis [132]. Finally, functional coupling of glycolysis and PCr utilization has been directly shown by ^{31}P-NMR experiments *in vivo* with anoxic muscle of a fish species that can survive under completely anaerobic conditions, deriving its total energy from glycogenolysis [159] (for review, see [4, 46, 61]).

7.1.13
High-energy Phosphate Channeling by MtCK in Energy-transducing Mitochondrial Compartments

Mitochondrial CK (MtCK) forms mainly large, cube-like octamers that are present (1) between the outer and inner mitochondrial membrane (the so-called *intermembrane space* of mitochondria) (Fig. 7.4) and preferentially localized at the so-called *mitochondrial contact sites* [38, 160] as well as (2) in the *cristae space* (for details, see [7, 8, 38]). The kinase catalyzes the direct transphosphorylation of ATP, produced inside mitochondria, using Cr from the cytosol to give PCr and ADP. Subsequently, the ADP thus generated enters the matrix space to stimulate oxidative phosphorylation, giving rise to mitochondrial recycling of a specific pool of ATP and ADP, while PCr is the primary "high-energy" phosphoryl compound that leaves the mitochondria and moves into the cytosol [48]. The molecular basis for such directed metabolite flux is channeling between the large, cube-like MtCK octamer [49, 161, 162] and two transmembrane proteins: the adenine translocator (ANT) and the mitochondrial porin or voltage-dependent anion channel (VDAC). ANT is an obligatory antiporter for ATP/ADP exchange across the inner mitochondrial membrane [163, 164], while VDAC is a nonspecific, potential-dependent pore in the outer mitochondrial membrane [165]. MtCK-linked metabolite channeling is based on co-localization, direct interactions, and diffusion barriers. MtCK tightly binds to cardiolipin that is specific for the mitochondrial inner membrane (see below). Because ANT is situated in a cardiolipin patch (see [166] and Section 7.3), this leads to co-localization and metabolite channeling between both proteins, MtCK and ANT, in the cristae and intermembrane space

[160]. MtCK in the intermembrane space further interacts with outer membrane phospholipids and, in a Ca^{2+}-dependent manner, with VDAC, thus virtually cross-linking the inner and outer membrane and contributing to the mitochondrial contact sites (Fig. 7.4). Increasing external Ca^{2+} concentrations strengthen the interaction of MtCK with VDAC [167], which may improve high energy phosphate channeling under cytosolic calcium overload that occurs at low cellular energy states. Some studies show that only the membrane-bound, octameric form of MtCK is able to maintain the metabolite channeling described above [168]. Finally, the limited permeability of VDAC and thus of the entire outer mitochondrial membrane in the cells *in vivo* creates a dynamic compartmentation of metabolites in the intermembrane space [169–171] that contributes to MtCK-linked channeling and separate mitochondrial ATP and ADP pools [77] (see Chapter 6 and Fig. 7.4). Similar to MtCK, hexokinase is able to use intramitochondrially produced ATP by binding to VDAC from the cytosolic mitochondrial surface at contact sites containing only ANT and VDAC [160]. The direct functional coupling of MtCK to oxidative phosphorylation can be demonstrated with respirometric oxygraphy on skinned muscle fibers from normal and transgenic mice lacking MtCK [51]. Oxidative damage of MtCK, induced by free reactive oxygen (ROS) and nitrogen species (RNS), such as peroxynitrite (PN), generated under cellular stress situations (e.g., in infarcted heart or under chemotherapeutic intervention by anthracyclines) leads to inactivation and dimerization of MtCK octamers as well as to dissociation of the enzyme from the mitochondrial inner membrane [172], thus negatively interfering with important prerequisites for efficient channeling of high-energy phosphates by MtCK (for details, see Section 7.4). These events contribute to cardiac energy failure [173] and specific cardiotoxicity [172], respectively.

7.2
Creatine Kinases and Cell Pathology

7.2.1
Mitochondrial MtCK, the Mitochondrial Permeability Transition Pore, and Apoptosis

Proteolipid complexes containing MtCK and ANT exhibit a direct protective effect on mitochondrial permeability transition (MPT). This reversible and cyclosporine-sensitive process is caused by a large pore in the inner mitochondrial membrane and is triggered by multiple signals, including Ca^{2+} and ROS [174]. MPT leads to dissipation of the membrane potential, mitochondrial swelling, and permeabilization of the outer membrane. The latter may trigger apoptosis by the release of proapoptotic proteins such as cytochrome *c* [175] or lead to necrotic cell death due to energy depletion [176]. Both *in vitro* and *in vivo*, contact site complexes containing ANT and VDAC exhibit many properties of the MPT pore [160, 177], and the ANT has become widely accepted as the putative pore-forming channel [175, 178]. However, ANT does not seem to be the only pore-forming protein of the inner membrane, since mitochondria from ANT-knockout mice still show MPT, although this occurs only at higher Ca^{2+} concentrations and the MPT is insensitive to cyclosporine [179, 180]. An involvement of MtCK in MPT and mitochondrial ultrastructure was demonstrated with liver mitochondria that were isolated either from transgenic mice expressing ubiquitous uMtCK in their liver [181] or from control animals lacking liver MtCK [182–184]. These experiments clearly show that MtCK, in complexes with ANT and together with its substrates creatine and ADP, is able to delay or even prevent Ca^{2+}-induced MPT pore opening [182, 183]. It is not the presence of MtCK *per se* that inhibits MPT but more precisely its enzymatic activity and its correct localization in mitochondrial complexes. No effect on MPT is observed if the enzyme is not supplied with its appropriate substrate, Cr, or if MtCK is only added externally to liver mitochondria that lack endogenous MtCK [182]. This mechanism may be explained by functional coupling of octameric MtCK to ANT. If MtCK is provided with substrate, it will maintain a high ADP concentration in the mitochondrial matrix (Fig. 7.4), which in turn is known to effectively inhibit MPT pore opening. MtCK does not merely act as a "plug" between porin and ANT to prevent MPT, as speculated earlier; rather, its presence has a clear influence on the number and stability of contact sites. Liver mitochondria from transgenic mice, expressing uMtCK in this organ, showed a threefold increase in the number of recognizable contact sites by electron microscopy and an increased resistance to detergent-induced lysis as compared with controls [185]. Remarkably, these transgenic CK livers, including those that express cytosolic BB-CK, become largely tolerant to liver toxins as well as to tumor necrosis factor α (TNF-α)-induced apoptosis [186]. The decisive involvement of MtCK, together with its substrate Cr, in the regulation of the mitochondrial permeability transition pore [182, 184] may explain at least in part the remarkable cell- and neuroprotective effects of creatine that have been reported lately (see Section 7.5). Thus, the CK system and its substrates seem to exert additional effects that are not necessarily directly coupled to an improved cellular energy status.

7.2.2
Mitochondrial MtCK and Intramitochondrial Inclusions in Mitochondrial Myopathy

Interestingly, MtCK expression levels are an indicator of cellular low-energy stress, that is, the expression of this enzyme is highly upregulated in patients with mitochondrial myopathies in the so called "ragged-red" skeletal muscle fibers, where mitochondrial volume and size are markedly increased and where characteristic intramitochondrial "railway-track inclusions" are observed as a hallmark of pathology. The latter have been shown to basically consist of crystalline sheets of MtCK (see Fig. 7.4) [133]. Similar MtCK inclusions can also be induced in animals by chronic Cr depletion, leading to cellular low-energy stress [187]. These data indicate that cellular low-energy stress, be it by chronic endurance training, fasting [188], Cr depletion, or pathologies in ATP generation (e.g., mitochondrial dysfunction, as seen in patients with mitochondrial cytopathies), induces a coordinated induction of the expression of genes related to energy metabolism, the most prominent among them being represented by MtCK [189]. Presumably, this is to compensate for a lack of energy supply and transport, which leads to overexpression and crystallization of MtCK under pathological conditions [133]. Interestingly, Cr supplementation of a patient with a novel cytochrome *b* mutation resulted in attenuation of free radical production and concomitant disappearance of the intramitochondrial MtCK inclusions in his muscles [190]. This could indicate that the para-crystalline MtCK inclusions form as a result of oxidative damage to MtCK.

Interestingly enough, sMtCK is also significantly upregulated in aged mice, but this can be entirely inhibited by caloric restriction [188]. Such a nutritional regime delays the aging process by increasing protein turnover, reducing oxidative stress, and thus decreasing macromolecular damage [191]. The compensatory upregulation of MtCK expression would be protective, since it was shown recently that MtCK can reduce mitochondrial free radical generation, based on its efficient recycling of ADP in mitochondria. If MtCK is fully functional, respiration is optimally coupled to ATP synthesis, thus reducing mitochondrial damage by ROS and RNS [192]. This is important also for MtCK function itself, as the enzyme is extremely susceptible to oxidation (see Section 7.4).

7.2.3
Overexpression of MtCK in Certain Malignancies

In normal liver, only very low levels of mostly cytosolic BB-CK are found, the origin of which may be vascular smooth muscle and endothelial cells. Normal hepatocytes do not seem to express CK at all. However, MtCK expression has been observed in regenerating and malignant liver [193], specifically in hepatocellular carcinomas [45], and in mammary carcinoma cell lines [194], where strong induction of BB-CK and uMtCK, respectively, was noted. Generally, it was found that in those tumors that start to overexpress MtCK, mostly adenomas and lymphomas [195], the mortality rate was significantly higher compared with patients with similar malignomas that did not express MtCK [196]. It also seems that leakage of MtCK into the bloodstream of patients can be used as a tumor-associated serum marker: 85% of a patient group with metastatic or infiltrating malignancy, mostly hepatic and gastric carcinomas, showed MtCK in the serum [197]. Considering the fact that MtCK stabilizes mitochondrial membranes and protects these organelles from undergoing permeability transition, the resistance of MtCK-expressing cancer cells to apoptosis-inducing chemotherapy may at least in part be due to the presence of elevated levels of MtCK in these cancer cells (see above).

7.3
Novel Membrane-related Functions of MtCK

Two additional and novel functions of the symmetrical, cube-like octameric MtCK protein [161, 162] were discovered very recently. They deal with the basic functions of lipid transfer from one mitochondrial membrane to another [198] as well as with MtCK-induced clustering of cardiolipin [199].

7.3.1
Transfer of Lipids by Proteins That Bridge the Inner and Outer Mitochondrial Membrane

Two mitochondrial proteins localized in contact sites, i.e., mitochondrial creatine kinase (MtCK) [38] and the mitochondrial isoform of nucleoside diphosphate kinase (NDPK-D), are both basic peripheral membrane proteins [200]. An important component of the interaction of MtCK and NDPK-D with membranes is through the cationic segments in these proteins that preferentially interact with anionic lipids to form cardiolipin-rich domains [199, 201]. In addition to forming membrane domains, these proteins can cross-bridge bilayers because of their symmetrical homo-oligomeric structure [202]. This feature gives them the ability to facilitate the transfer of lipid from one bilayer to another (Fig. 7.5) [198].

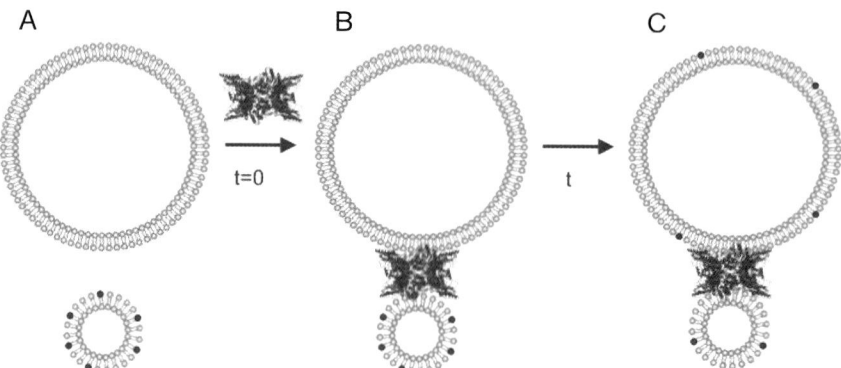

Fig. 7.5 Octameric MtCK cross-links two membrane vesicles and facilitates lipid transfer between these vesicles.
(A) A population of large (200 nm in diameter) unilamellar DOPC:DOPE:CL (1:1:1) phospholipid vesicles (large acceptor vesicles) were mixed with a population of smaller vesicles (50 nm in diameter) having the same lipid composition and also containing some fluorescent probe (small donor vesicle with a fluorescent lipid having a black-dotted head group).
(B) At $t = 0$, octameric MtCK (actual side view of the X-ray structure of octameric MtCK [161, 162] is shown) is added to the vesicles, and within minutes MtCK binds to and cross-links two vesicles based on its symmetrical top and bottom faces exposing membrane-interaction domains [54].
(C) MtCK sandwiched in between these vesicles facilitates, also within minutes, the transfer of fluorescently labeled phospholipid from the small donor to the large acceptor vesicle, as measured in a fluorometer cuvette by dequenching of the fluorescent probe and visualized here as black-dotted fluorescent phospholipid molecules present in the large acceptor vesicle population. Protein bridging of vesicles by MtCK causes probe transfer as observed by probe dequenching, with no fusion or leakage taking place (measured independently; for experimental details and original data, see [198]).

Mitochondrial Creatine Kinase   Both uMtCK and sMtCK have several C-terminal lysine residues that are important for binding to anionic lipids [54]. It is known that uMtCK binds only to liposomes containing anionic lipids [203]. One of the important anionic lipids for binding of MtCK to the mitochondrial membrane is cardiolipin (CL) [202, 204–206]. MtCK undergoes a concentration-dependent, reversible oligomerization between the octameric and dimeric forms [207], with octameric uMtCK dissociating 23–24 times slower compared with sMtCK [208]. The crystal structures of both octameric MtCKs are known [161, 162]. The octameric form binds rapidly to anionic phospholipids, while the dimeric forms binds much more slowly, possibly requiring prior octamerization [209]. This may simply be a thermodynamic consequence of the equilibration of the dimeric and octameric forms in solution, with the small amount of octameric form that is present, even in dilute solution, preferentially binding to the membrane.

Mitochondrial Nucleoside Diphosphate Kinase   The nucleoside diphosphate kinase (NDPK) is a large family of hexameric isoenzymes that are encoded by different nm23 genes that exert multiple functions in cellular energetics, signaling, proliferation, and differentiation. Although NDPK activity was reported in mitochondria with different localization depending on the species or organ [210], only NDPK-D has a mitochondrial targeting sequence and has been unambiguously localized as bound to the inner mitochondrial membrane (IM), in particular to contact sites [211, 212]. NDPK-D plays an important role in mitochondrial iron homeostasis [213]. NDPK-D also shows properties similar to uMtCK, insofar as it also binds to anionic phospholipids such as CL. This protein has a cationic cluster that is thought to be responsible for binding to anionic lipids. NDPK-D is also capable of cross-linking two lipid bilayers, probably due to the symmetry of its hexameric structure ([198] and Tokarska-Schlattner, Schlattner, and Lacombe, unpublished data). However, unlike uMtCK, NDPK-D remains hexameric at any protein concentration. There is particular interest in the products of the nm23 gene family because they are involved in tumor progression and metastasis [214]. NDPK-D was found to be overexpressed in a majority of gastric and colon cancers [215], possibly linking NDPK-D to the development of tumors.

7.3.2
Lipid Transfer at Contact Sites

Connections between the outer and peripheral inner membranes of mitochondria, termed contact sites, are revealed by morphological analysis (e.g., in chemically fixed mitochondria where the intermembrane space is enlarged [216]), as jumps in the fracture planes of freeze-fractured mitochondria [217], or in three-dimensional electron tomography [218]. At contact sites it is suggested that uMtCK binds to CL on the inner mitochondrial membrane (IM), stabilizing a CL-rich domain and binding to the voltage-dependent anion channel (VDAC) on the outer mitochondrial membrane (OM) to bridge the IM and OM [219, 220]. VDAC is part of the permeability transition pore and the sustained opening of this pore is important for necrosis [221], having important implications for energy flux processes as well as for apoptosis [6–8, 170, 182]. The function of contact sites is modulated by the presence of CL [166, 222, 223]. In general, mitochondrial lipids play an important and specific biological role [224].

Contact sites are considered a domain within the mitochondrial membrane [225]. They are regulated, dynamic structures, as their number depends on the metabolic activity of the cell [226]. Because of the close approach of the two membranes, these contact sites could favor interbilayer transfer of lipids. The transfer of phospholipids from one membrane to another is in general a very slow process. Therefore, specific pathways have to be postulated for this process to occur in apoptotic mitochondria. Likely candidates that could facilitate this process are the proteins MtCK and NDPK-D, which are able to bridge the IM and OM [29, 49, 212, 227] and, in the case of MtCK, interfere with apoptotic signaling [7, 8, 182].

7.3.3
Cardiolipin Transfer and Apoptosis

Cardiolipin is synthesized on the IM [228, 229], and the major fraction of CL in the cells is present on the inner leaflet of the IM [230]. CL moves to the surface of mitochondria early in apoptosis [231–233]. Monolysocardiolipin (MCL) is an important CL metabolite, naturally present in mitochondria, that is involved as an intermediate in the remodeling of the acyl-chain composition of CL [234]. Like CL, MCL also binds to truncated Bid (t-Bid), and its concentration in the mitochondria increases during FAS-induced apoptosis [235]. Thus, the movement and domain formation of MCL, as well as CL itself, are important to mitochondrial function and apoptosis. The process of movement of CL or MCL from the IM to the exterior of mitochondria involves both interbilayer transfer from the IM to OM and flip-flop across the OM membrane. Several proteins are involved in this lipid movement, including the phospholipid transferase PLS3 [236] and an unidentified flipase that promotes transfer of phospholipids into the mitochondria [237]. In addition, MtCK and NDPK-D also facilitate the transfer of lipids between these membranes [198]. Transfer of lipids between bilayers is more commonly facilitated by specific lipid transfer proteins [238–240]. There is a report indicating that t-Bid also has lipid transfer activity [241], but this may be a result of its structural similarity to other lipid transfer proteins [242]. Thus, the observation that uMtCK and NDPK-D have such lipid transfer activity is novel. The biological importance of this lipid transfer is shown by the fact that a promyelocytic leukemia cell line that exhibits less exposed CL is resistant to apoptosis [232]. In contrast, a lipid transfer protein from maize produces proapoptotic effects on mammalian mitochondria [243]. In addition, tumor cells that are poorly differentiated have fewer contact sites [244]. Such cells are also more resistant to apoptosis [245], again suggesting a relationship between apoptosis and the presence of contact sites where these proteins are known to be located. NDPK-D and MtCK also differ in their expression levels in different cancer cell lines. This may have consequences on the susceptibility of these cells to chemotherapeutic agents, since a major drawback of such therapies is the resistance of tumor cells to apoptosis. It is known that the Bcl-2 proteins t-Bid and Bax bind more readily to membranes with exposed CL [246–249]. Hence, there is a clear association between apoptosis and the exposure of CL on the surface of mitochondria.

7.3.4
Cardiolipin Domains and Mitochondrial Kinases

It would be expected that the lateral distribution of lipids in the membrane would affect their rate of interbilayer transfer, as well as affecting the stability of the protein bridges between the IM and OM. As shown recently, uMtCK promotes the segregation of anionic lipids (Fig. 7.6) [199]. The formation of CL-rich domains by the mitochondrial protein uMtCK was initially assessed by differential scanning calorimetry (DSC) using mixtures of tetraoleoyl cardiolipin (CL) and phosphatidylethanolamine (PE). These are two of the major lipid components of mitochondrial contact sites [201]. The mixture has a sufficiently small number of components that it will exhibit cooperative phase transitions detectable by DSC. One of two different species of PE was used: either 1-palmitoyl-2-oleoyl PE

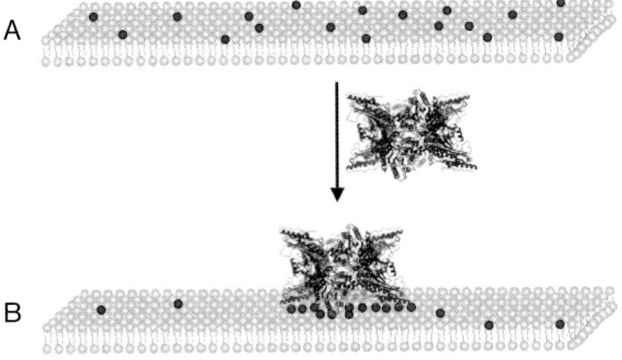

Fig. 7.6 MtCK binding to cardiolipin-containing phospholipid membranes leads to clustering of cardiolipin around the MtCK octamer. Cardiolipin (CL) clustering was measured by differential scanning calorimetry (DSC). Multilamellar vesicles composed of POPE:CL (0.75:0.25) are heated in a DSC cell, yielding a broad phase transition with no differentiated peaks for the mixture. When heating is performed in the presence of uMtCK, a partial separation of the components of the lipid mixture is observed, with sharpening of the peaks, as a result of the preferential association of the protein with CL (for original data, see [199]). This preferential association of octameric MtCK with CL and the clustering of CL molecules is shown here schematically in a diagram depicting a membrane leaflet containing a mixture of POPE (gray) and CL (with black head groups), with the CL uniformly distributed before MtCK protein is added (A). After the addition of MtCK (an MtCK octamer shown as an X-ray structure from its side view is depicted [161, 162]), the protein binds to the membrane by its symmetrical top or bottom face, and subsequently CL molecules (black head groups) begin to preferentially interact with and cluster around the MtCK octamer (B).

(POPE) or dipalmitoyl PE (DPPE). POPE is similar to forms of PE found in biological membranes, but mixtures of CL with DPPE accentuate the effects of the proteins, allowing a comparison of the relative potency of different proteins relative to domain formation in the absence of proteins. The CL does not exhibit any phase transition above 0 °C, while POPE and DPPE have gel-to-liquid crystalline transitions at 25 and 65 °C, respectively. In a lipid mixture of CL with one of the forms of PE, a broad phase transition can be observed over a range of temperatures between that of the two pure-lipid components. If a protein preferentially binds to CL, forming a domain of this lipid, the remainder of the membrane becomes enriched in the PE component. As a result, the phase transition in the presence of such a protein will exhibit a component with higher transition temperature and higher transition enthalpy, approaching the characteristics of a pure PE domain. In addition, the potency of the dimeric and octameric forms of uMtCK to induce the formation of domains was compared. This was done at the same concentrations of uMtCK for both forms, since a freshly diluted solution of the octameric form dissociates slowly, over a period of hours, to the dimeric form. Replacing CL with a structurally related anionic lipid (phosphatidylglycerol) abolishes protein-induced domain formation, pointing to the specificity of this protein–lipid interaction. Our results demonstrate that the octameric form of uMtCK is effective in preferentially interacting with CL, leaving a PE-enriched component that exhibits characteristic phase transition behavior [199]. This property is specific and is not seen with either the dimeric form of uMtCK or this protein after it has undergone irreversible thermal denaturation. Furthermore, the ability of octameric uMtCK to form CL-rich domains is much greater than that of other cationic proteins found in the mitochondria, and the formation of these domains is expected to affect the rates of movement of CL between monolayers in a bilayer. The presence of exposed CL-rich domains at contact sites in mitochondria is important for the binding of the proapoptotic protein t-Bid, which results in the promotion of apoptosis [250, 251].

7.3.5
Perspectives: Relationship Between the Surface Exposure of Cardiolipin and Expression Levels of Mitochondrial Kinases

Exposure of CL on the surface of mitochondria entails two processes. Because in mitochondria most of the CL is on the IM, CL must be transferred from the IM to the OM and undergo transbilayer diffusion (flip-flop) from one side of the OM to the other [252]. The rates of both types of lipid transfer determine how quickly CL becomes exposed on the surface of mitochondria, resulting in apoptosis.

One might expect that the interaction of CL with a protein at the membrane interface would inhibit its translocation across the bilayer. It is known that in many cancers, particularly those with poor prognosis, there is an overexpression of uMtCK [196, 197]. The cells in these tumors are also resistant to apoptosis. On the basis of the observation that uMtCK promotes interbilayer transfer, one would expect these tumors to be more sensitive to apoptosis. It is possible that the tendency of uMtCK to interact with CL and form membrane domains actually slows the rate of flip-flop. As a result, although there would be transfer of CL from the IM to the OM, there would be no exposure of this lipid on the surface of the mitochondria. Thus, one may suggest that the relationship between MtCK and NDPK-D expression and CL exposure on the mitochondrial surface is biphasic. At very low levels of expression of the kinases, there are fewer contact sites and hence few domains with a closely spaced IM and OM where interbilayer lipid transfer could occur. At very high expression levels, there is rapid transfer from the IM to the OM. However, CL domain formation could reduce the movement of CL across the OM. Thus, there would be less CL exposed on the surface of the mitochondria, despite the increased rate of interbilayer transfer. Only inter-mediate levels of expression lead to interbilayer CL transfer, but because of lim-ited CL domain formation, flip-flop could occur more rapidly, resulting in expo-sure of CL and apoptosis. This hypothesis provides an *ad hoc* explanation for the apparent anomaly of both low and high levels of MtCK expression resulting in resistance to apoptosis. The entire scheme of apoptosis and its control are more complex, but CL transfer to the outside of the mitochondria is likely an important contributing factor.

7.4
Exquisite Sensitivity of the Creatine Kinase System to Oxidative Damage

7.4.1
Molecular Damage of Creatine Kinase by Oxidative and Nitrosative Stress

Oxidative and nitrosative stress have been implicated in loss of cellular function in a number of pathological states. These include ischemia–reperfusion injury in heart or brain, acute and chronic heart failure (in particular, the heart failure induced by anthracyclins, a prominent class of anticancer drugs) [253–257] as well as inflammation [258], neurodegenerative and neuromuscular diseases [259, 260], and aging [261]. Elevated levels of reactive oxygen and nitrogen spe-cies (ROS and RNS, respectively), together with decreased anti-oxidative defense mechanisms, are at the origin of harmful modifications of various biomolecules, including proteins, lipids, and nucleic acids.

All CK isoenzymes are particularly susceptible targets of ROS and RNS (as superoxide anions, hydrogen peroxide, hydroxyl radicals, nitric oxide, and peroxy-nitrite [PN]) as well as of anthracyclines, which are known to generate ROS and RNS especially in the presence of iron (see [8, 257, 262]). Oxidative modifications affecting CK functions are thought to play a critical role during pathologies in-volving oxidative stress. Oxidative inhibition of the CK system is a hallmark of cardiac ischemia–reperfusion injury [263]. In permeabilized rat heart fibers, cyto-solic MM-CK has been shown to be the main target of ROS in the myofibrillar compartment. In a study with hydrogen peroxide or using a xanthine oxidase–xanthine system that generated superoxide anions, neither myosin ATPase nor other myofibrillar regulatory proteins were affected [264]. Specific oxidation of CK, correlating with loss of muscular performance, was observed in skeletal muscles of patients with chronic obstructive pulmonary disease [265]. PN-mediated nitration of myofibrillar creatine kinase has been shown to impair myocardial contractility in different models of experimental heart failure as well as in heart failure induced by anthracyclines [266]. BB-CK was identified, together with glutamine synthase and ubiquitin carboxy-terminal hydrolase L-1, as one of the major proteins specifically oxidized in Alzheimer's disease brain [267] as well as in aged brain [261, 268]. Oxidative damage to the CK system was shown to contribute to impaired energy metabolism in amyotrophic lateral sclerosis (ALS) [269]. More recently, an important reduction in MtCK function was related to oxidative stress caused by endotoxin challenge in sepsis [270].

ROS and RNS induce not only inactivation, which is often irreversible and occurs with all CK isoenzymes, but also specific damage to the mitochondrial CK isoform, namely, interference with its oligomeric state and membrane-binding capacity. In animal models of acute and chronic cardiac ischemia, impairment of heart function and mitochondrial energy metabolism was associated with a significant decrease in sMtCK octamer:dimer ratio and enzymatic activity, which were related to PN-induced damage [173]. MtCK has been recognized as a prime mitochondrial target of PN. Its inactivation in isolated heart mitochondria (after direct addition of PN) or in heart of anthracycline-treated mice (where PN is generated via redox cycling of the drugs) is observed before dysfunction of enzymes in oxidative phosphorylation becomes apparent [271, 272]. Finally, MtCK is a prime target of ROS and RNS not only because of high sensitivity but also due to its mitochondrial localization. Most of the reactive species originate directly or indirectly from the mitochondrial respiratory chain.

7.4.2
Molecular Basis of Creatine Kinase Damage

It is known that proteins that are highly sensitive to oxidative modifications have sulfur-containing amino acid residues, with the thiol (-SH) moiety on the side chain of the cysteine being particularly sensitive to redox reactions [273]. In CK isoforms, the active-site cysteine (Cys-283 for cytosolic CK and Cys-278 for MtCK) was identified as the prime residue responsible for oxidative inactivation [8, 274]. This cysteine is not only highly reactive and important for full enzymatic activity but also is easily accessible to modifying agents. Further modifications found at the dimer–dimer interface of MtCK involve oxidation of Met267 and nitration of Trp268 and/or Trp264, which is the most critical single residue for octamer formation. Finally, the C-terminal cysteine Cys358 was also found to be oxidized, which in turn can contribute to a reduced membrane affinity of the C-terminal phospholipid-binding motif [274].

The impact of oxidative modifications on protein function and the reversibility of these modifications depend on the character and concentration of the oxidizing molecule. During conditions of moderate oxidative stress, CK inactivation observed with ROS or anthracyclines is partially reversible by reducing agents (e.g.,

β-mercaptoethanol, dithiothreitol [DTT]) [172, 275]. However, at higher ROS concentrations or if caused by the very potent oxidant PN, CK damage is largely irreversible and less specific, involving other amino acid residues in addition to the active-site cysteine [172, 271, 274]. Recent studies on MtCK using site-directed mutagenesis and mass spectrometry showed unambiguously that PN induces reversible nitrosylation and irreversible oxidation of the active-site Cys278, the latter explaining the entire loss of enzymatic activity of the oxidized enzyme.

CK thiols have been shown to undergo reversible oxidative modifications, such as S-thiolation, S-nitrosylation [124], and formation of inter- or intramolecular disulfide bonds [172]. These modifications may have a potential regulatory significance in situations where pro-oxidizing conditions arise in a cell [273]. S-thiolation of CK, consisting of formation of mixed disulfides between CK thiols and the low-molecular-mass thiol glutathione, has been observed, and the active-site cysteine (Cys283) was identified as the site of thiolation [276]. CK S-thiolation, resulting in enzyme inactivation, can be reversed by reduced glutathione or DTT. It is not clear whether S-thiolation of CK would play a protective rather than a deleterious role *in vivo* [262]. Currently, protein S-glutathiolation is considered an adaptive response that protects critical regulatory molecules from permanent loss of function as a result of oxidative or nitrosative

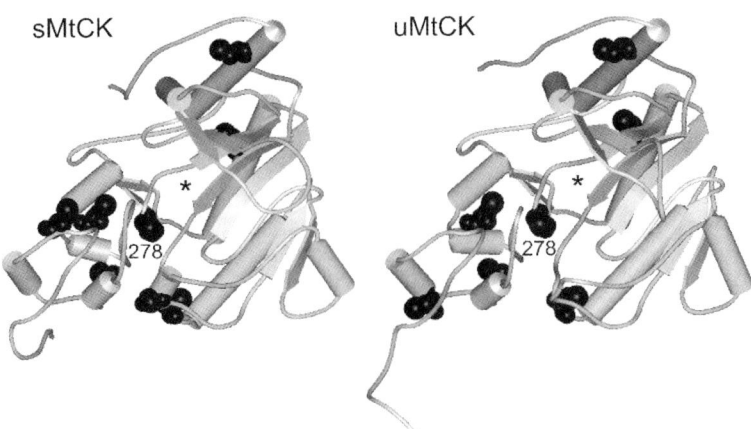

Fig. 7.7 The catalytic site cysteine residue of CK is exquisitely sensitive to oxidation. Cysteine residues (space-fill representation, black balls) in chicken sarcomeric muscle sMtCK (left) [162] and human ubiquitous uMtCK (right) [161] monomers. Locations of the active site (*) of MtCK and the highly reactive cysteine (Cys278) within the active-site pocket of MtCK are indicated. The active-site cysteine in all CK isoenzymes is exquisitely sensitive to oxidative damage, and its oxidation leads to almost complete inactivation of CK catalytic activity. The larger number of cysteines and their concentration around the active site in sMtCK (black balls, left) may contribute to the higher sensitivity of sMtCK to oxidative and nitrosative damage compared with uMtCK. Note that this may partly explain the specific cardiotoxic side effects of anthracyclines, a potent class of anticancer drugs, which inactivate sMtCK significantly more strongly than uMtCK (for details, see text).

insult [277]. In this case, the maintenance of appropriate intracellular levels of reduced glutathione, GSH, would be critical for regeneration of active CK from S-glutathionylated protein. Inactivation of purified MtCK by the anthracycline drug doxorubicin was accompanied by formation of cross-linked aggregates due to reversible intermolecular disulfide bridge formation between MtCK monomers [172]. The active-site cysteine does not seem to be essential for this process, because a C278G mutant CK protein shows doxorubicin-induced cross-link products quite similar to those of wild-type protein.

An interesting outcome of our studies with purified MtCK treated with PN or anthracyclines is the significantly higher sensitivity to oxidative damage found with sarcomeric sMtCK in comparison to ubiquitous uMtCK [172, 274], indicating that the deleterious effects of these compounds may be more pronounced in heart and skeletal muscles, both of which express sMtCK. This finding could contribute to the cardiac-specific side effects of anthracyclines. The different susceptibility of the two MtCK isoenzymes could be related to some differences in their molecular structures. Both isoenzymes share a number of conserved cysteine residues, including active-site Cys278 and Cys358, but sMtCK shows a distinct distribution, with more cysteines near the active site and one additional cysteine in total compared with uMtCK (Fig. 7.7).

7.4.3
Functional Consequences of Oxidative Damage in Creatine Kinase

Inevitably, ROS- and RNS-induced CK damage is at the origin of numerous deleterious processes that promote cellular dysfunction. Compromised MtCK functions, including inactivation, dimerization, and inhibition of its binding to cardiolipin, not only impair the energy channeling and signaling between mitochondria and cytosol [6–8] but also affect mitochondrial respiration. This molecular damage also contributes to destabilization of the mitochondrial contact sites consisting of ANT, MtCK, and VDAC [169, 185]. MtCK defects sensitize cells to mitochondrial permeability transition [182] and abolish MtCK function in prevention of radical generation, which was recently demonstrated [192]. In the cytosol, where MM-CK is functionally coupled to the Ca^{2+} pump of the sarcoplasmic reticulum, inhibition of this isoenzyme results in Ca^{2+} imbalance, which in turn interferes with muscle contraction and relaxation [4, 80] and could lead to apoptotic and/or necrotic cell death via chronic Ca^{2+} overload.

7.5
Enhancement of Brain Functions and Neuroprotection by Creatine Supplementation

7.5.1
Creatine Metabolism and Brain Energetics

The importance of the CK system for normal function *in vivo* of skeletal muscle [6, 80, 136] as well as heart muscle [79, 85, 278–280] has been established most convincingly by transgenic approaches. In addition, the ergogenic effects of oral

Cr supplementation on muscle performance and muscle power in humans, especially for repetitive, high-intensity tasks, are obvious [281–283], and the first positive effects of Cr supplementation on the skeletomuscular health status of elderly patients [284] as well as for those with muscle, neuromuscular, and heart diseases have been reported [190, 285–293].

As a recent development in the field, however, a plethora of new data provide evidence of the importance of the CK system and of Cr metabolism for normal function of the brain. Further, oral Cr supplementation shows remarkable positive effects on brain function and neuroprotective properties for acute and chronic neurological diseases (for review, see [262, 294, 295]).

As for skeletal and cardiac muscle, the specific functional properties of neuronal tissue require very high cellular energy resources. The brain may spend up to 20% of the body's energy resources. A very high turnover of ATP is necessary to maintain electrical membrane potentials, as well as signaling activities of the central nervous system (CNS) and peripheral neuronal system. Rapid changes in ATP demands occur during physiological function of neurons, while cellular energy reserves are small. Due to widely distributed cellular processes and sites of high-energy consumption that are localized at remote locations from the neuronal cell body, i.e., synapses, an effective coupling of ATP-generating and ATP-consuming processes is needed to maintain a sufficiently high energy transfer [296].

The CK–PCr circuit has been described as playing a key role in the energy metabolism of the brain and neurons [36, 46, 295, 297, 298]. Hence, this system is thought to play a pivotal role in neuronal ATP metabolism [298, 299]; consequently, Cr depletion in brain is associated with disruption of neuronal function, changes in morphology, and clinical pathology [300].

7.5.2
Disturbance of the CK System or Creatine Metabolism in the Brain

Interestingly, ablation of the cytosolic brain-type BB-CK or the ubiquitous mitochondrial uMtCK genes, individually or combined, leads to a significant neurological and brain-related phenotype in such transgenic mice. For example, mice with a gene knockout of cytosolic BB-CK show diminished open-field habituation, learn slower in the water maze, present with a delayed development of pentylenetetrazole-induced epileptic seizures, and anatomically show a loss in hippocampal mossy fiber connections [301]. Double-knockout transgenic mice, lacking both BB-CK and uMtCK, present with undetectable PCr and Cr levels in the brain [302] and display a strong phenotype that includes reduced body weight, severely impaired spatial learning, lower nest-building activity, and diminished acoustic startle reflex, as well as having lower brain weight and reduced hippocampal size [303].

On the other hand, if the CK substrate creatine (Cr) is depleted in the brain by feeding animals with the Cr analogue beta-guanidino propionic acid (βGPA) – a competitive inhibitor of the creatine transporter (CRT) – strong muscle and neurological phenotypes are also observed (for review, see [300]). Most importantly, in humans a new creatine-deficiency syndrome that affects either endogenous Cr synthesis or transport has been discovered recently (for review, see [304]). Patients with this syndrome present with an almost complete lack of Cr in the brain and show a severe neurological phenotype (for details, see below). Thus, either ablating the CK isoenzymes or drastically reducing the substrate Cr in the brain leads to similar and rather severe phenotypes. This represents strong evidence in favor of the physiological significance of the CK–PCr system for normal brain function. Because Cr obviously is crucial for normal brain function, better knowledge about Cr metabolism in the human body and specifically in the brain is needed (for review, see [262]).

7.5.3
Body and Brain Creatine Metabolism

In omnivores, a certain fraction of Cr (1.5–2.0 g of Cr per day, representing approximately 50% of daily Cr requirement in humans) is endogenously synthesized by a two-step synthesis involving the enzymes arginine:glycine amidinotransferase (AGAT) (Fig. 7.8, **1a**) and *S*-adenosyl-L-methionine:*N*-guanidinoacetate methyltransferase (GAMT) (Fig. 7.8, **2a**). Whereas AGAT, which produces guanidino acetate (GAA) as an intermediate, is expressed preferentially in kidney and pancreas (Fig. 7.8, **1a**), GAMT is found to some extent in the pancreas but mostly in the liver, the main organ of the final step of endogenous Cr synthesis (Fig. 7.8, **2a**). Cr, which leaves the liver by some unknown mechanism, is then transported through the bloodstream and is actively taken up, via a specific creatine transporter (CRT) (Fig. 7.8, **3**), by cells with high energy requirements, such as muscle (Fig. 7.8, **3**) or brain cells, where Cr is charged up by CK to high-energy PCr (Fig. 7.8, **4a,b**) (for review, see Ref. [262]). To get into the brain, Cr has to first pass the blood–brain barrier, where a CRT is localized in endothelial cells of microcapillaries (Fig. 7.8, **5**) but not in astrocytes lining these microcapillaries [305, 306]. From there, Cr is taken up by those brain cells that express the CRT, mainly represented by neurons and oligodendrocytes but not astrocytes, which seem to lack the CRT [305–307]. A similar situation has been found at the blood–retina barrier, where CRT is expressed at the luminal and abluminal membranes of retinal capillary endothelial cells, explaining the transcellular transport of Cr through the endothelium [308]. On the other hand, alimentary Cr, which is present in fresh fish and meat, is taken up by an intestinal CRT (Fig. 7.8, **8**) [309] and transported into the bloodstream, where it mixes with endogenously synthesized Cr.

  In embryonic brain (and to a limited extent also in adult brain), AGAT and GAMT can be detected in all of the brain cells mentioned above (Fig. 7.8, **1b,2b**), such that there seems to be a limited potential for endogenous Cr synthesis in the adult brain as well [305]. As a consequence, due to the presence of CRT in neurons and oligodendrocytes but not in astrocytes, trafficking of Cr between astrocytes and neurons or oligodendrocytes has been suggested (i.e., Cr synthesized by astrocytes could be released and taken up by neurons; Fig. 7.8, **6**) [305, 310].

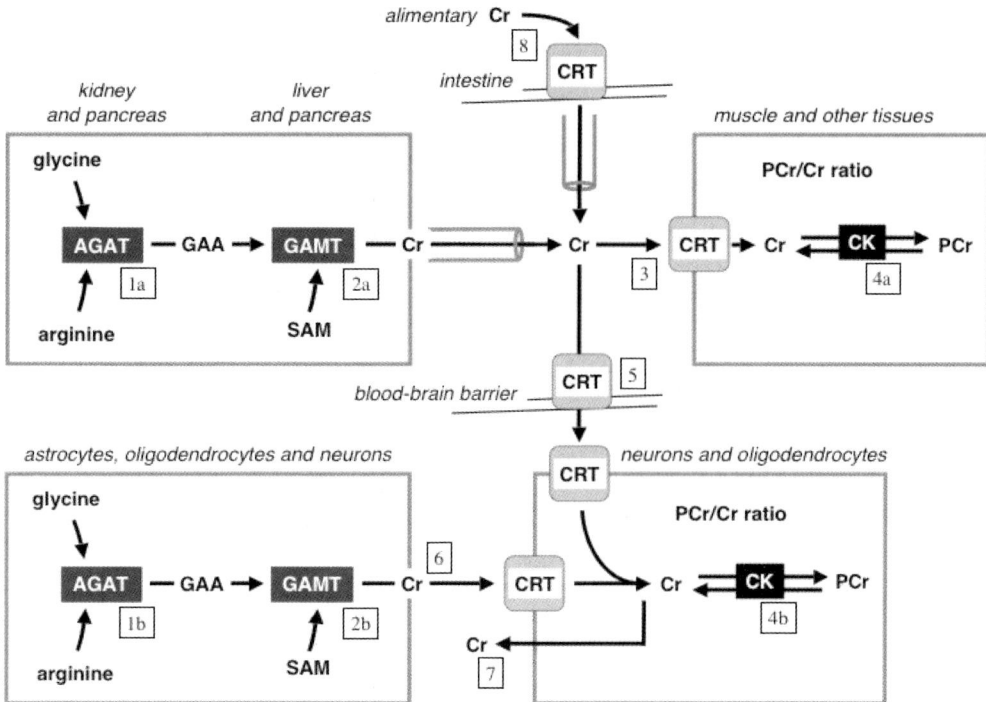

Fig. 7.8 Simplified scheme of whole-body creatine metabolism with emphasis on the brain. A certain fraction of Cr is endogenously synthesized by a two-step synthesis involving AGAT (**1a,b**) and GAMT (**2a,b**). AGAT, which produces guanidino acetate (GAA) as an intermediate, is expressed preferentially in the kidney and pancreas (**1a**). GAMT is found to some extent in the pancreas but mostly in liver (**2a**), the main organ of the final step of endogenous Cr synthesis [262]. Cr leaving the liver is transported through the bloodstream and is actively taken up via a specific creatine transporter (CRT) by cells with high energy requirements, such as muscle (**3**) or brain cells (**4**), where Cr is charged up by CK to high-energy PCr (**4a,b**). To get into the brain, Cr needs to pass the blood–brain barrier by the CRT, which is localized in the endothelial cells of microcapillaries (**5**). From there, Cr is taken up by those brain cells that express the CRT (**4**), represented mainly by neurons and oligodendrocytes but not by astrocytes, which seem to lack the CRT [305–307]. On the other hand, alimentary Cr, present in fresh fish and meat, is taken up by an intestinal CRT (**8**) [309] and transported into the bloodstream, where it mixes with endogenously synthesized Cr. Trafficking of Cr between astrocytes and neurons or oligodendrocytes (**6**) and exocytotic release of Cr from neurons (**7**) [312] are also shown.

A similar situation has been found in retina, where CRT is not expressed in Mueller cells but in retinal photoreceptor cell inner segments, suggesting that, as in brain, neurons are independent of these glial cells in accumulating Cr because they can take it up via CRT [311]. A new finding shows that upon electrical stimulation of superfused brain slices, endogenous Cr is specifically released (Fig. 7.8, **7**). This exocytotic release can be prevented by omitting Ca^{2+} from the medium or by adding the Na^+ channel blocker tetrodotoxin (TTX) [312]. These *in situ* data indicate that Cr not only is synthesized and taken up by central neurons but also is released in an action potential–dependent manner, providing strong evidence for its role as a neuromodulator in the brain (Fig. 7.8, **7**). Several important questions concerning the details of Cr metabolism, such as the exact mechanisms and regulation of transcellular Cr transport, interorgan Cr exchange, release of synthesized Cr from the liver, uptake of Cr into the brain, and intracellular trafficking and exocytotic release of Cr inside the brain, remain to be answered.

7.5.4
Effects of Creatine on Memory, Mental Performance, and Complex Tasks

High expression of CK isoenzymes has been found in hippocampal pyramidal cells, which are involved in learning and memory [35]. This finding indicates that the CK–PCr system plays an important role in the energy metabolism of these cells and that Cr supplementation may lead to improved functions of these systems. Indeed, positive effects of orally administered Cr on mental performance have been reported in healthy volunteers in a controlled double-blind study [313]. Using infrared spectroscopy, the authors found correspondingly increased blood oxygenation in the Cr-treated group. Moreover, a double-blind study investigating Cr supplementation for six weeks on healthy vegetarians, who have a reduced nutritional Cr supply, revealed significantly better results in intelligence test and working memory performance in the Cr-treated subjects compared with control individuals [314]. In a recent test, Cr supplementation (4×5 g Cr per day for 7 days), prior to 18–36 hours of sleep deprivation, was shown to significantly improve the performance of complex central executive tasks [315, 316]. Thus, it can be concluded that Cr supplementation enhances brain function under normal and stress conditions. This may be relevant for promoting Cr supplementation as a brain performance–enhancing nutritional supplement for humans.

7.5.5
Creatine Supplementation and Neurodegenerative Diseases

Neurodegenerative disorders are characterized by a progressive loss of cells from one or multiple regions of the nervous system. Despite intensive research, the etiology of neuronal cell death in most neurodegenerative diseases still remains enigmatic. However, there are a number of similarities in the fundamental biochemical processes involved in the pathogenesis and progression of these otherwise different pathological states. The concepts of energy depletion, oxidative stress by ROS and RNS, excitotoxicity, and mitochondrial dysfunction have been implicated in neurodegenerative disorders [317, 318]. Although these processes may be directly or indirectly involved in the pathogenesis of a given disease, they converge in final common pathways of either necrosis or apoptosis. Substantial evidence indicates that energy dysfunction plays either a primary or secondary role in cell death in neurodegenerative disorders, and even in normal aging. Mitochondria are critical organelles in the regulation of cellular energy status. Mitochondrial dysfunction results in ATP depletion, which may contribute to neuronal cell death. Moreover, these organelles are also involved in excitotoxicity, generation of free radicals, calcium buffering, and apoptotic pathways. Mitochondrial mutations, particularly at complexes I and III, can lead to generation of ROS, and accumulation of mitochondrial DNA mutations in aging and Alzheimer's disease has been shown to be linked to oxidative stress [319, 320]. These processes provide potential targets for the therapy of neurodegenerative diseases.

Parkinson's Disease   Parkinson's disease (PD) is a common neurodegenerative disorder affecting more than 1% of all individuals over the age of 50 years [321]. Clinical symptoms include resting tremor, bradykinesia, rigidity, and postural imbalance [322, 323]. PD is characterized by a progressive deterioration of dopaminergic neurons in the substantia nigra, leading to a profound loss of dopaminergic input into the striatum. An impaired function of the mitochondrial electron transport system, in particular complex I, is thought to be involved in the pathogenesis of PD [324–326], suggesting a mitochondrial basis for the disease. Recent research has focused on neuroprotective strategies for PD [327]. Using an experimental *in vitro* paradigm of PD, we observed neuroprotective properties of Cr against toxic insults induced by 6-hydroxydopamine (6-OHDA) [328] or 1-methyl-4-phenyl pyridinium (MPP+) [329] exposure in ventral mesencephalic rat cultures. Beneficial effects of Cr have also been demonstrated in an animal model of PD [330]. In clinical pilot studies, Cr supplementation resulted in improved patient mood and, in smaller doses, increased the effects of dopaminergic therapy [331] and reduced the progression of the disease [332]. These results make Cr a qualifier for clinical phase III trials with PD patients.

Huntington's Disease and Other Trinucleotide Repeat Disorders   Huntington's disease (HD) is an autosomal, dominantly inherited, neurodegenerative disorder that clinically presents with progressive choreoathetotic movements in combination with severe cognitive and emotional dysfunction, finally leading to death [333]. The main pathologic finding is a loss of striatal GABA-ergic projection neurons. A defect in energy metabolism has been proposed as one of the potential pathogenetic mechanisms leading to neuronal death [334]. Studies on cerebral metabolism using ^{18}F-fluorodeoxyglucose positron emission tomography (PET) showed typical patterns of diminished cerebral metabolic rates in the basal ganglia as well as in the frontal and parietal regions of HD patients, correlating with the severity of the disease [335]. Recently, evidence of impaired energy metabolism in HD due to reduced mitochondrial complex II and complex III activity has been reported [336, 337], resulting in increased cerebral lactate levels and a reduced PCr:P$_i$ ratio in muscle. Corresponding mitochondrial defects have been described in the brains of patients suffering from HD, particularly in the basal ganglia [338]. Further evidence of mitochondrial respiratory chain dysfunction has been provided by studies of transgenic mouse models of HD [339]. Increasing cellular PCr levels and thereby improving the impaired energy metabolism by exogenous Cr supplementation may therefore offer a feasible approach for reducing neuronal deterioration in the context of this severe disorder. Using an experimental *in vitro* model of HD, we detected that Cr supplementation provided significant neuroprotection against glucose and serum deprivation and against 3-nitropropionic acid (3-NP)-induced toxicity in striatal cultures [340]. Cr directly injected intraperitoneally at 12 mg kg^{-1} body weight was shown to protect experimental animals against convulsive behavior and lactate production elicited by intrastriatal injection of methylmalonate [341]. In addition, Cr administration increased survival, delayed motor symptoms, and significantly reduced brain lesion size in a transgenic animal model of HD [342, 343] and in 3-NP–exposed rats [344]. In recent clinical trials, it was reported that Cr is well tolerated and safe in HD patients [345]. Brain glutamate levels were significantly reduced after a Cr-enhanced diet [346], and serum 8-hydroxy-2′-deoxyguanosine (8OH2′dG) levels, an indicator of oxidative injury to DNA that is markedly elevated in HD, were reduced by Cr treatment [347], indicating some efficacy of Cr treatment for this devastating neurodegenerative disease. Finally, a Cr-supplemented diet significantly extends Purkinje cell survival in a spinocerebellar ataxia type 1 transgenic mouse model but does not prevent the ataxic phenotype [348].

Alzheimer's Disease   Alzheimer's disease (AD) is a neurodegenerative disease of the brain leading to progressive dementia. Mutations in the amyloid precursor protein (APP) result in abnormal processing of APP and accumulation of beta-amyloid peptide, the main constituent of amyloid plaques in the AD brain [349]. Brain-specific creatine kinase (BB-CK) is significantly inactivated by oxidation in AD patients, which may result in further compromising of the energetic state of neurons and exacerbate the disease process [350]. In addition, recently discovered Cr deposits in the brain of transgenic AD mice, as well as in the hippocampus of AD patients, indicate a direct link between cellular energy levels, mitochondrial function, Cr metabolism, and AD [351]. Neuroprotective effects of Cr have been observed in models of AD with cultured neurons undergoing neurotoxic insults by glutamate excitotoxicity or by exposure to beta-amyloid protein [36]. Hence, it is worth speculating whether Cr supplementation at an early time point of the disease may prevent or delay the course of AD-related neurodegeneration [297]. In fact, a direct connection between AD and MtCK was discovered by showing that MtCK forms a complex with APP family proteins that affects the correct import of MtCK into mitochondria and thus negatively interferes with cellular energetics [352].

Amyotrophic Lateral Sclerosis (ALS)   Amyotrophic lateral sclerosis (ALS) is characterized by progressive loss of motor neurons in the brain and spinal cord [353]. Mitochondrial and energetic defects are implicated in the pathogenesis of motor neuron degeneration in ALS [354]. A marked reduction in cerebral cortex ATP levels was detected in a mouse model of ALS well before symptom onset [355]. Accordingly, reduced CK activity has been reported in transgenic ALS mice [269]. Consequently, it was found that Cr supplementation showed protective properties in mouse models of ALS [356–358]. This neuroprotection could be based on antioxidant effects exerted by Cr, given that Cr has the potential to act as a direct antioxidant against aqueous radical and ROS or RNS species [359], or it could be due to the action of MtCK and Cr in coupling mitochondrial respiration tightly to ATP synthesis, by efficient ADP cycling, and thus suppressing ROS formation in mitochondria [192].

Cr reverted the cholinergic deficit present in some forebrain areas at an intermediate stage of the disease [360]. In a follow-up study, additive neuroprotective effects of oral Cr supplementation together with a cyclooxygenase-2 inhibitor were demonstrated with the same ALS mice [361]. We were able to demonstrate CK immunoreactivity in choline acetyltransferase–expressing neurons in the developing [362] and adult human spinal cord (Andres et al., unpublished), supporting the hypothesis that Cr treatment might be beneficial in ALS or other motor neuron diseases. Despite the promising findings in experimental animal models, first clinical studies have failed to show a relevant benefit of Cr treatment in ALS patients [363, 364]. However, these trials have also posed unanswered questions about the optimal dosage of Cr. It must also be considered that Cr offers potential benefits in terms of facilitating residual muscle contractility in ALS patients, which is a factor that has not been investigated yet [365]. A large placebo-controlled, multi-center trial is currently underway to further investigate the efficacy of Cr supplementation in ALS.

Charcot–Marie–Tooth Disease Charcot–Marie–Tooth disease (CMT) is a common hereditary disorder characterized by slowly progressive sensomotor neuropathy that can lead to lifelong disability in patients. CMT represents a heterogeneous group of genetically distinct disorders with similar clinical presentations and a large number of responsible gene mutations [366]. In a recent study, it was shown that Cr supplementation alters muscle myosin heavy chain (MHC) composition in CMT patients undergoing resistance training and that MHC changes associated with Cr supplementation can improve muscle function [367].

7.5.6
Creatine Supplementation and Acute Neurological Disorders

Cerebral Ischemia Cerebral ischemia rapidly leads to neuronal cell death due to compromised energy metabolism [368], often resulting in disabling neurological sequelae. Neuroprotective effects of Cr supplementation have been reported in an experimental model of ischemia [369, 370]. In a recent study, it was found that Cr treatment of mice suffering from transient focal cerebral ischemia resulted in a reduction in stroke volume in the absence of significant changes in brain Cr, PCr, and ATP levels [371]. The authors presented some evidence that a positive effect of Cr on vasodilatory response in the brain might be responsible for the observed effects.

Traumatic Brain and Spinal Cord Injury Traumatic brain and spinal cord injury is known to initiate a series of cellular and molecular events in the injured tissue, leading to further damage in the surrounding area. This secondary damage is partly due to ischemia and compromised cellular bioenergetics. Cr-mediated neuroprotection has been demonstrated in experimental brain injury [372, 373]. In a prospective randomized study investigating the effects of Cr in children and adolescents suffering from traumatic brain injury, administration of Cr for six months resulted in significantly better clinical recovery in the categories of cognitive, personality/behavior, self-care, and communication aspects [374]. Cr supplementation also has been shown to have protective effects after spinal cord injury [375, 376]. However, exogenous Cr is taken up slowly into CNS tissue; therefore, it seems unlikely that patients suffering from acute brain or spinal cord injury would immediately benefit from Cr supplementation, unless Cr could be brought directly to the sites of injury, e.g., by perifusion of the affected region or intracerebroventricular administration, which leads to a fast and marked increase of Cr levels in the brain [377].

Injury of the Peripheral Nerve   Injury of the peripheral nerve causes denervation of the associated muscle fibers and may be treated by microsurgical nerve repair. Using an experimental paradigm of sciatic nerve transection in the rat, it has been shown that systemic Cr supplementation promotes reinnervation and functional recovery on both surgically repaired and unrepaired nerve injuries [378].

7.5.7
Inborn Errors of Metabolism

Mitochondrial Encephalomyopathies   Mitochondrial encephalomyopathies are a multi-systemic group of disorders that are characterized by a wide range of biochemical and genetic mitochondrial defects and variable modes of inheritance. Among this group of disorders, the "mitochondrial myopathy, encephalopathy, lactic acidosis with stroke-like episodes (MELAS) syndrome" is one of the most frequently occurring, maternally inherited mitochondrial disorders. Clinical trials with Cr supplementation on MELAS patients reported an increase in muscle performance [379, 380]. Interestingly, Cr supplementation led to a reversal of the paracrystalline intramitochondrial inclusions, which were shown to consist mainly of crystallized MtCK [133], as seen in a muscle biopsy from one such patient [190].

Another mitochondrial encephalomyopathy is Leigh syndrome (LS), which results in characteristic focal necrotizing lesions in one or more areas of the central nervous system, including the brainstem, thalamus, basal ganglia, cerebellum, and spinal cord. Clinical symptoms depend on which areas of the central nervous system are involved. The most common underlying cause is a defect in oxidative phosphorylation [381]. In a recent study using oral Cr supplementation, improvement in behavioral and physiological functions in a child with LS [382] was seen. Another study investigating the possible benefit of Cr treatment on an inhomogeneous population of patients with different forms of encephalomyopathies revealed protective effects in all patients, suggesting that targeting the final common pathway of mitochondrial dysfunction favorably influences the course of these diseases [383].

Creatine-deficiency Syndrome with Defects in Creatine Synthesis and Transport
Brain Cr deficiency is involved in the pathogenesis of some severe inheritable neurological phenotypes [384–386]. These so-called Cr-deficiency syndromes form a group of inborn errors in either one of the two enzymes involved in endogenous Cr synthesis, i.e., AGAT or GAMT [387], or, alternatively, in the Cr transporter (CRT), leading to CRT deficiency [388] (Fig. 7.8). Patients with Cr-deficiency syndrome clinically present with graded forms of developmental and speech delay, epileptic seizures, autism, and brain atrophy, leading to varying degrees of mental retardation and suggesting major involvement of cerebral gray matter. Due to the neurotoxic effects of guanidino acetic acetate (GAA) [389], which accumulates in the gray matter, patients with GAMT deficiency additionally show a dystonic-hyperkinetic movement disorder [390]. Cr-deficient patients are characterized by cerebral Cr deficiency, which especially in CRT-deficient patients is completely lacking in the brain, as can be demonstrated by noninvasive ^{1}H-NMR spectroscopy [391]. Measurement of the neurotoxic intermediate metabolite GAA in body fluids can be used to discriminate the extent to which GAMT (high concentration), AGAT (low concentration), and CRT (normal concentration) are affected. CRT-deficient patients present with a significantly elevated Cr:creatinine (Cr:Crn) ratio in the urine. GAMT and AGAT deficiency can be treated by oral Cr supplementation [304], while patients with CRT deficiency do not respond to this type of treatment [392, 393]. Once more, these observations emphasize the importance of the CK–PCr system for the brain and the absolute requirement of Cr for normal brain function.

7.5.8
Psychiatric Disorders

Posttraumatic stress disorder (PTSD) is an anxiety disorder that can develop in persons who have experienced highly traumatic situations. PTSD is linked to structural and neurochemical changes particularly in the limbic system, postulated as a substrate for stress-induced alterations in affective behavior [394]. Decreased Cr levels have been measured in the brains of patients suffering from anxiety disorders [395]. Consequently, Cr supplementation has beneficial effects in treatment-resistant PTSD patients, resulting in relief of symptoms as well as improved sleep and depression parameters [396]. Furthermore, in a patient suffering from PTSD, comorbid depression, and fibromyalgia, Cr treatment led to improvement of symptoms [397].

7.5.9
Neurorestorative Strategies

In order to repair the brain by replacing the neurons lost in acute or chronic pathological processes, transplantation of neuronal precursors or stem cells has attracted great attention. In cell replacement therapy for PD, it has been observed that some transplanted cells show long-term survival and structural and functional integration in the host brain [398, 399]. The grafted tissue is assumed to release dopamine in a regulated fashion and to reverse many of the behavioral deficits seen in animal models of PD [400, 401] as well as in humans [402]. In HD, there is large body of experimental data showing the effectiveness of striatal transplants in experimental models [403, 404], while preliminary studies report that transient recovery can be promoted in patients as well [405, 406]. In stroke, it has been shown that transplanted neuronal stem cells survive, migrate, and differentiate into appropriate neuronal phenotypes [407, 408]. Transplanted motor neurons can survive in spinal cord and reinnervate the denervated target muscle [409]. However, at present, there are some specific obstacles that prevent widespread clinical application of cell replacement techniques. The main problems of neuronal transplantation include the limited availability of donor tissue and the poor survival of transplanted cells, as well as the suboptimal innervation of the targeted structures in host brain. In PD, less than 20% of the implanted cells survive the transplantation procedure [410, 411]. Studies in rats revealed that most implanted embryonic neurons died within one week after transplantation [412], mostly by apoptotic cell death [413]. Strategies for improving cell replacement therapies, therefore, include treatment of the cells with neuroprotective factors.

In this context, we were able to show significant neuroprotective effects of Cr on both dopaminergic mesencephalic and GABA-ergic striatal neuronal precursors [328, 329, 340]. Furthermore, Cr was identified as a potent differentiation factor for striatal precursor neurons, inducing differentiation towards the GABA-ergic phenotype. Interestingly, inhibition of mitogen-activated protein kinase and phosphatidylinositol-3-kinase significantly attenuated the effect on induction of the GABA-ergic phenotype [340]. These data suggest that Cr may play an important role in cell fate decision during development of neuronal cells, a finding that is also supported by a report describing the expression of CK in the developing zebra fish embryo. Expression of CK was shown to be highly dynamic, often being transiently expressed in specific cells for a short time period only, indicating a well-timed, cell type–specific function of CK during brain development [414]. Moreover, in a study describing pattern expression and localization of CK isoforms in fetal rat brain, CK was shown to be present throughout the central and peripheral nervous systems [35, 415, 416]. Research on neuronal precursors and stem cells continues with great hope for the treatment of acute and chronic neurological disorders. In this context, Cr holds a future potential to influence the survival and differentiation of these cells and is hoped to contribute to a better outcome to obtain functional transplants.

7.5.10
Future Prospects of Creatine Supplementation as Adjuvant Therapeutic Strategy

Cr supplementation has been shown to increase intracellular levels of PCr and thereby to stabilize ATP levels, not only in human muscle [281] but also in human brain [417]. Thus, Cr obviously can pass the blood–brain barrier via uptake by endothelial CRT in the brain capillaries [307]. As a result, the potential benefits of Cr supplementation also are likely to be realized in human patients with neurological disorders in which cellular energy metabolism is impaired [294]. Indeed, Cr supplementation has been shown to attenuate neuronal cell loss in various experimental models of neurodegenerative diseases. So far, significant protective effects have been shown in animal models of PD [330], HD [342–344], ALS [356–358], stroke [369, 370], cerebral hypoxia [418], and traumatic injury to the brain [372, 373] and spinal cord [375, 376].

Promising results in first clinical trials with human patients by application of oral Cr supplementation have been reported, particularly in chronic pathological conditions of the CNS such as PD [331, 332], HD [345–347], CMT [367], and inborn errors of metabolism [382, 383, 392]. In a recent study, remarkably beneficial effects were discovered on functional recovery after traumatic brain injury [374]. In addition, positive effects of Cr on mental performance and memory have been reported in healthy volunteers [314]. In conclusion, these results confirm the neuroprotective properties of Cr, particularly on damage due to hypoxia [418] or free ROS or RNS [357]. At present, it seems that there are various mechanisms by which neurons respond to an increase in their Cr levels [36, 419].

So far, however, the general experience with clinical Cr supplementation studies involving human patients with neurological diseases is that, at least for chronic diseases, the beneficial effects seen with Cr are less pronounced or partially disappointing compared with the neuroprotective effects observed in the corresponding animal models. On the other hand, these findings may be a result of Cr dosage, schedule, and time of supplementation, because in these animal studies, the Cr doses were approximately 10 times higher than those used in the clinical trials. In addition, the observation that chronic Cr supplementation failed to protect adult mice from stroke hints that adaptive mechanisms may counteract the beneficial effects of Cr supplementation; thus, further considerations are required if Cr should be used as a long-term nutritional supplement for patients [371]. Nevertheless, Cr remains a promising neuroprotective agent for further studies involving chronic neurological diseases, but its potential for such diseases may only be revealed by large multi-center studies with higher Cr doses over an adequate observation time [331]. This conclusion is supported by the recent futility clinical trial of Cr in early PD patients, where a clear positive effect of Cr on patient mood was discovered [332] that should be assessed in future clinical trials of patients with depression.

7.5.11
Non-energy–related Effects of Creatine

Besides its role in cellular energy metabolism [262] by enhancing cellular energy status [420], Cr is believed to have additional functions in the CNS. For example, there is evidence of the direct antiapoptotic effect of elevated cellular Cr levels. In conjunction with the action of MtCK inside mitochondria, Cr prevents or delays mitochondrial permeability transition pore opening [182, 184], an early event in apoptosis. Moreover, Cr supplementation was demonstrated to have antioxidant properties via a mechanism involving the direct scavenging of reactive oxygen species [421] or, alternatively, reducing the production of mitochondrially generated reactive ROS. The latter is facilitated by the stimulatory effects of Cr on mitochondrial respiration [51], which allows for efficient recycling of ADP inside mitochondria by MtCK, leading to tight coupling of mitochondrial respiration with ATP synthesis and suppression of ROS formation because futile idling of the respiratory chain is avoided [192]. Important with respect to this finding, strong protective effects of Cr against oxidant and UV stress have been directly shown in keratinocytes and on human skin [422]. Furthermore, Cr was reported to normalize mutagenesis of mitochondrial DNA and its functional consequences caused by UV irradiation of skin cells [423], all pointing to effects of Cr on suppression of the generation of ROS and RNS and their congeners that lead to cell damage and inactivation of CK (see Section 7.4).

Further evidence that Cr-mediated neuroprotection can occur independently of changes in the bioenergetic status of brain tissue was reported in another recent study, suggesting an effect of Cr on cerebral vasculature that leads to improved circulation in the brain [371]. A very recent report showed that Cr could be released in an action potential–dependent, excitotic manner (Fig. 7.8) [7], suggesting a novel role for Cr as a neuromodulator [312]. This finding is supported by a study showing attenuation of the acute stress response in chickens by administration of Cr, which has been found to act through GABA-A receptors [424]. Finally, a recent study demonstrated that Cr is able to protect cultured cells from hyperosmotic shock by means of a significant increase in Cr uptake into cells, indicating that Cr can act as a compensatory osmolyte [425]. In contrast, Cr and taurine are released from cortical astrocytes and cortical brain slices after exposure to hypoosmotic perfusion, suggesting a link between brain energy reserve and brain osmoregulation, with Cr as one of the main brain cell osmolytes [426, 427]. An osmotic effect of Cr accompanied by a decrease in myoplasmic ionic strength was postulated to explain some of the beneficial effects of Cr on muscle that are seen before intramuscular PCr levels rise [428]. Thus, the subject of the actions of Cr on biological systems is not yet closed, and future discoveries about this endogenous body substance that has been known for more than 150 years now are likely to follow.

7.5.12
Creatine as a Safe Nutritional Supplement and Functional Food

Because Cr is endogenously synthesized and is a constituent of the regular diet, a potential benefit of Cr administration is assumed to be achieved without any major side effects, even at high dosage. Indeed, preliminary findings on athletes using high-dosage Cr supplementation for improving muscle strength revealed no kidney, liver, or other health problems [429, 430]. For oral dietary supplementation, a loading phase of 5–7 days with four doses of 5 g of Cr per day followed by a maintenance dose of 2–5 g per day for 3–6 months or even longer (up to two years) has shown an excellent safety profile [282, 431]. Even in premature newborns, Cr administration is well tolerated at high dosage [432], which is important for the treatment of inborn metabolic diseases. So far, with the limited research available, the data suggest that chronic Cr administration may be safe. Future studies will be required, however, to further address the potential of Cr supplementation for the treatment of neurological diseases and for improving neurorestorative strategies.

Recent data support the view that humans, born as omnivores, need a certain amount of nutritional Cr, either from fresh fish and meat or from creatine supplementation [294]. Total Cr levels (PCr plus Cr) are significantly lower in serum as well as in muscles of vegetarians [433], and thus Cr is to be highly recommended to vegetarian athletes [434]. Because Cr supplementation has been shown to improve muscle strength and rehabilitation after immobilization and muscle disuse atrophy [435], as well as to enhance mineralization of bone [436] and bone density [284], it should be considered a most valuable supplement within the context of a preventive strategy for seniors to retain muscle strength and bone health, thus extending the time to lead an independent life before entering disability [294]. Because Cr has been shown to improve mental performance [313], memory, learning, and intelligence [314], the beneficial effects of such prevention go beyond skeletomuscular health and positively affect the brain as well.

Acknowledgments

Due to limitations of space, it was not possible to cite many important publications from a number of authors who contributed to development in the field; however, many of them are cited in the reviews referred to herein. We are grateful for intellectual support and numerous discussions with many colleagues and scientists around the world over the last two decades concerning the CK system, as well as for help from and discussions with the members of our laboratories. Special thanks go to Chris Boesch from the NMR Center, University Hospital Insel, Bern, for discussion of NMR-related topics. R.H.A. and H.R.W. would like to thank their collaborator Angelique D. Ducray. This work was financially supported over the years by graduate student stipends from the ETH (to T.W.), by Swiss National Science Foundation grant Nos. 3100AO-102075 (to T.W. and U.S.) and 3100+0-11437/1 (to T.W. and D.N.), by Swiss National Science Foundation grant Nos. 3100-064975.01 and 3100A0-112529 (to H.R.W.), and by the Department of Clinical Research, University of Bern (to R.H.A.). This work was additionally supported by Marie Heim-Vögtlin subsidy grant 3234-069276 (to M.T.-S.), Marie Curie Intraeuropean Fellowship of the European Community (to M.T.-S.), Novartis Stiftung für medizinisch-biologische Forschung (to U.S.), Schweizerische Herzstiftung (to T.W. and U.S.), Wolfermann-Nägeli-Stiftung (to U.S. and T.W.), Schweizer Krebsliga (to U.S. and T.W.), Zentralschweizer Krebsliga (to U.S. and T.W.), and the Swiss Society for Research on Muscle Diseases (to I.A.). Financial support for this work also came from the France–Canada Research Foundation (to R.M.E., as Senior Scientist of the Canadian Institute of Health Sciences).

Abbreviations

AD: Alzheimer's disease; AdK: adenylate kinase; AGAT: L-arginine:glycine amidinotransferase; ALS: amyotrophic lateral sclerosis; AMPK: AMP-activated protein kinase; ANT: adenine nucleotide transporter or mitochondrial ATP/ADP exchange carrier or antiporter; BB-CK: cytosolic brain-type creatine kinase; CMT: Charcot–Marie–Tooth disease; CK: creatine kinase; CL: cardiolipin; CNS: central nervous system; Cr: creatine; Crn: creatinine; CRT: creatine transporter; DPPE: dipalmitoyl PE; DSC: differential scanning calorimetry; GAMT: S-adenosyl-methionine-guanidinoacetate *N*-methyltransferase; GAA: guanidino acetate; GABA: γ-amino butyric acid; GAPDH: glyceraldehyde-3-phosphate dehydrogenase; HD: Huntington's disease; K^+_{ATP}: sarcolemmal ATP-gated K^+-channel; MCL: monolysocardiolipin; IM: inner mitochondrial membrane; MM-CK: cytosolic muscle-type creatine kinase; MS: multiple sclerosis; NDPK-D: mitochondrial nucleoside diphosphate kinase; OM: outer mitochondrial membrane; PFK-1: phosphofructokinase-1; PD: Parkinson's disease; PE: phosphatidylethanolamine; PEP: phosphoenolpyruvate; PCr: phosphocreatine; PN: peroxynitrite; POPE: 1-palmitoyl-2-oleyl PE; RNS: reactive nitrogen species; ROS: reactive oxygen species; sMtCK: sarcomeric mitochondrial creatine kinase; SR: sarcoplasmic reticulum; uMtCK: ubiquitous mitochondrial creatine kinase; VDAC: voltage-dependent anion channel or mitochondrial porin.

References

1 Edwards, M. R. (1996) Metabolite channeling in the origin of life. *J. Theor. Biol. 179*, 313–322.

2 McFarland, E. W., Kushmerick, M. J., Moerland, T. S. (1994) Activity of creatine kinase in a contracting mammalian muscle of uniform fiber type. *Biophys. J. 67*, 1912–1924.

3 Noma, T. (2005) Dynamics of nucleotide metabolism as a supporter of life phenomena. *J. Med. Invest. 52*, 127–136.

4 Wallimann, T., Wyss, M., Brdiczka, D., Nicolay, K., Eppenberger, H. M. (1992) Intracellular compartmentation, structure and function of creatine kinase isoenzymes in tissues with high and fluctuating energy demands: the 'phosphocreatine circuit' for cellular energy homeostasis. *Biochem. J. 281*, 21–40.

5 Wiseman, R. W., Kushmerick, M. J. (1995) Creatine kinase equilibration follows solution thermodynamics in skeletal muscle. 31P NMR studies using creatine analogs. *J. Biol. Chem. 270*, 12428–12438.

6 Saks, V., Dzeja, P., Schlattner, U., Vendelin, M., Terzic, A., Wallimann, T. (2006) Cardiac system bioenergetics: metabolic basis of the Frank-Starling law. *J. Physiol. 571*, 253–273.

7 Schlattner, U., Tokarska-Schlattner, M., Wallimann, T. (2006) Molecular structure and function of mitochondrial creatine kinases, in *Creatine kinase – biochemistry, physiology, structure and function* (Vial, C., Uversky, V. N., Eds.) pp 123–170, Nova Science Publishers, New York, USA.

8 Schlattner, U., Tokarska-Schlattner, M., Wallimann, T. (2006) Mitochondrial creatine kinase in human health and disease. *Biochim. Biophys. Acta 1762*, 164–180.

9 Wallimann, T. (1996) 31P-NMR-measured creatine kinase reaction flux in muscle: a caveat! *J. Muscle Res. Cell Motil. 17*, 177–181.

10 Bandlow, W., Strobel, G., Schricker, R. (1998) Influence of N-terminal sequence variation on the sorting of major adenylate kinase to the mitochondrial intermembrane space in yeast. *Biochem. J. 329*, 359–367.

11 Dzeja, P. P., Terzic, A. (2003) Phosphotransfer networks and cellular energetics. *J. Exp. Biol. 206*, 2039–2047.

12 Dzeja, P. P., Zeleznikar, R. J., Goldberg, N. D. (1996) Suppression of creatine kinase-catalyzed phosphotransfer results in increased phosphoryl transfer by adenylate kinase in intact skeletal muscle. *J. Biol. Chem. 271*, 12847–12851.

13 Pucar, D., Dzeja, P. P., Bast, P., Gumina, R. J., Drahl, C., Lim, L., Juranic, N., Macura, S., Terzic, A. (2004) Mapping hypoxia-induced bioenergetic rearrangements and metabolic signaling by 18O-assisted 31P NMR and 1H NMR spectroscopy. *Mol. Cell. Biochem. 256–257*, 281–289.

14 Ovadi, J., Srere, P. A. (2000) Macromolecular compartmentation and channeling. *Int. Rev. Cytol. 192*, 255–280.

15 Ovadi, J., Saks, V. (2004) On the origin of intracellular compartmentation and organized metabolic systems. *Mol. Cell. Biochem. 256–257*, 5–12.

16 Noble, D. (2006) *The music of life. Biology beyond the genome.*, Oxford, UK.

17 Berridge, M. J., Bootman, M. D., Roderick, H. L. (2003) Calcium signalling: dynamics, homeostasis and remodelling. *Nat. Rev. Mol. Cell Biol. 4*, 517–529.

18 Zaccolo, M., Di Benedetto, G., Lissandron, V., Mancuso, L., Terrin, A., Zamparo, I. (2006) Restricted diffusion of a freely diffusible second messenger: mechanisms underlying compartmentalized cAMP signalling. *Biochem. Soc. Trans. 34*, 495–497.

19 Berridge, M. J. (2006) Calcium microdomains: Organization and function. *Cell Calcium 40*, 405–412.

20 Weiss, J. N., Korge, P. (2001) The cytoplasm: no longer a well-mixed bag. *Circ. Res. 89*, 108–110.

21 Kholodenko, B. N. (2003) Four-dimensional organization of protein kinase signaling cascades: the roles of diffusion, endocytosis and molecular motors. *J. Exp. Biol. 206*, 2073–2082.

22 Boesch, C., Kreis, R. (2001) Dipolar coupling and ordering effects observed in magnetic resonance spectra of skeletal muscle. *NMR Biomed. 14*, 140–148.

23 Joubert, F., Hoerter, J. A., Mazet, J. L. (2001) Discrimination of cardiac subcellular creatine kinase fluxes by NMR spectroscopy: a new method of analysis. *Biophys. J. 81*, 2995–3004.

24 Joubert, F., Mazet, J. L., Mateo, P., Hoerter, J. A. (2002) 31P NMR detection of subcellular creatine kinase fluxes in the perfused rat heart: contractility modifies energy transfer pathways. *J. Biol. Chem. 277*, 18469–18476.

25 Kreis, R., Koster, M., Kamber, M., Hoppeler, H., Boesch, C. (1997) Peak assignment in localized 1H MR spectra of human muscle based on oral creatine supplementation. *Magn. Reson. Med. 37*, 159–163.

26 Wegmann, G., Zanolla, E., Eppenberger, H. M., Wallimann, T. (1992) *In situ* compartmentation of creatine kinase in intact sarcomeric muscle: the actomyosin overlap zone as a molecular sieve. *J. Muscle Res. Cell Motil. 13*, 420–435.

27 Vermathen, P., Boesch, C., Kreis, R. (2003) Mapping fiber orientation in human muscle by proton MR spectroscopic imaging. *Magn. Reson. Med. 49*, 424–432.

28 Ntziachristos, V., Kreis, R., Boesch, C., Quistorff, B. (1997) Dipolar resonance frequency shifts in 1H MR spectra of skeletal muscle: confirmation in rats at 4.7 T *in vivo* and observation of changes postmortem. *Magn. Reson. Med. 38*, 33–39.

29 Schlattner, U., Wallimann, T. (2004) Metabolite channeling: creatine kinase microcompartments, in *Encyclopedia of Biological Chemistry* (Lennarz, W. J.,

Lane, M. D., Eds.) pp 646–651, Academic Press, New York, USA.

30 Hochachka, P. W. (1999) The metabolic implications of intracellular circulation. *Proc. Natl. Acad. Sci. U.S.A.* 96, 12233–12239.

31 Hochachka, P. W. (2003) Intracellular convection, homeostasis and metabolic regulation. *J. Exp. Biol.* 206, 2001–2009.

32 Saks, V., Favier, R., Guzun, R., Schlattner, U., Wallimann, T. (2006) Molecular System Bioenergetics: Regulation of Substrate Supply in Response to Heart Energy Demands. *J. Physiol.* 577, 769–777.

33 Eppenberger, M. E., Eppenberger, H. M., Kaplan, N. O. (1967) Evolution of creatine kinase. *Nature* 214, 239–241.

34 Eppenberger-Eberhardt, M., Riesinger, I., Messerli, M., Schwarb, P., Muller, M., Eppenberger, H. M., Wallimann, T. (1991) Adult rat cardiomyocytes cultured in creatine-deficient medium display large mitochondria with paracrystalline inclusions, enriched for creatine kinase. *J. Cell Biol.* 113, 289–302.

35 Kaldis, P., Hemmer, W., Zanolla, E., Holtzman, D., Wallimann, T. (1996) 'Hot spots' of creatine kinase localization in brain: cerebellum, hippocampus and choroid plexus. *Dev. Neurosci.* 18, 542–554.

36 Brewer, G. J., Wallimann, T. W. (2000) Protective effect of the energy precursor creatine against toxicity of glutamate and beta-amyloid in rat hippocampal neurons. *J. Neurochem.* 74, 1968–1978.

37 Wallimann, T., Wegmann, G., Moser, H., Huber, R., Eppenberger, H. M. (1986) High content of creatine kinase in chicken retina: compartmentalized localization of creatine kinase isoenzymes in photoreceptor cells. *Proc. Natl. Acad. Sci. U.S.A.* 83, 3816–3819.

38 Wegmann, G., Huber, R., Zanolla, E., Eppenberger, H. M., Wallimann, T. (1991) Differential expression and localization of brain-type and mitochondrial creatine kinase isoenzymes during development of the chicken retina: Mi-CK as a marker for differentiation of photoreceptor cells. *Differentiation* 46, 77–87.

39 Shin, J. B., Steijger, F., Beynon, A., Peters, T., Gadzalla, L., McMillen, D., Bystrom, C., Van der Zee, C., Wallimann, T., Gillespie, J. P. (2007) Hair bundles are specialized for ATP delivery via creatine kinase. *Neuron*, 53, 371–386.

40 Ishida, Y., Wyss, M., Hemmer, W., Wallimann, T. (1991) Identification of creatine kinase isoenzymes in the guinea-pig. Presence of mitochondrial creatine kinase in smooth muscle. *FEBS Lett.* 283, 37–43.

41 Guerrero, M. L., Beron, J., Spindler, B., Groscurth, P., Wallimann, T., Verrey, F. (1997) Metabolic support of Na+ pump in apically permeabilized A6 kidney cell epithelia: role of creatine kinase. *Am. J. Physiol.* 272, C697–706.

42 Decking, U. K., Alves, C., Wallimann, T., Wyss, M., Schrader, J. (2001) Functional aspects of creatine kinase isoenzymes in endothelial cells. *Am. J. Physiol.* 281, C320–328.

43 Kaldis, P., Kamp, G., Piendl, T., Wallimann, T. (1997) Functions of creatine kinase isoenzymes in spermatozoa. *Advances in Developmental Biochemistry* 5, 275–312.

44 Schlattner, U., Mockli, N., Speer, O., Werner, S., Wallimann, T. (2002) Creatine kinase and creatine transporter in normal, wounded, and diseased skin. *J. Invest. Dermatol.* 118, 416–423.

45 Meffert, G., Gellerich, F. N., Margreiter, R., Wyss, M. (2005) Elevated creatine kinase activity in primary hepatocellular carcinoma. *BMC Gastroenterol.* 5, 5–9.

46 Wallimann, T., Hemmer, W. (1994) Creatine kinase in non-muscle tissues and cells. *Mol. Cell. Biochem.* 133–134, 193–220.

47 Wyss, M., Smeitink, J., Wevers, R. A., Wallimann, T. (1992) Mitochondrial creatine kinase: a key enzyme of aerobic energy metabolism. *Biochim. Biophys. Acta 1102*, 119–166.

48 Jacobus, W. E., Lehninger, A. L. (1973) Creatine kinase of rat heart mitochondria. Coupling of creatine phosphorylation to electron transport. *J. Biol. Chem.* 248, 4803–4810.

49 Schlattner, U., Forstner, M., Eder, M., Stachowiak, O., Fritz-Wolf, K.,

Wallimann, T. (1998) Functional aspects of the X-ray structure of mitochondrial creatine kinase: a molecular physiology approach. *Mol. Cell. Biochem.* 184, 125–140.

50 Ellington, W. R. (2001) Evolution and physiological roles of phosphagen systems. *Annu. Rev. Physiol.* 63, 289–325.

51 Kay, L., Nicolay, K., Wieringa, B., Saks, V., Wallimann, T. (2000) Direct evidence for the control of mitochondrial respiration by mitochondrial creatine kinase in oxidative muscle cells *in situ*. *J. Biol. Chem.* 275, 6937–6944.

52 Saks, V. A., Kuznetsov, A. V., Vendelin, M., Guerrero, K., Kay, L., Seppet, E. K. (2004) Functional coupling as a basic mechanism of feedback regulation of cardiac energy metabolism. *Mol. Cell. Biochem.* 256–257, 185–199.

53 Ventura-Clapier, R., Kuznetsov, A., Veksler, V., Boehm, E., Anflous, K. (1998) Functional coupling of creatine kinases in muscles: species and tissue specificity. *Mol. Cell. Biochem.* 184, 231–247.

54 Schlattner, U., Gehring, F., Vernoux, N., Tokarska-Schlattner, M., Neumann, D., Marcillat, O., Vial, C., Wallimann, T. (2004) C-terminal lysines determine phospholipid interaction of sarcomeric mitochondrial creatine kinase. *J. Biol. Chem.* 279, 24334–24342.

55 Bessman, S. P., Carpenter, C. L. (1985) The creatine-creatine phosphate energy shuttle. *Annu. Rev. Biochem.* 54, 831–862.

56 Meyer, R. A., Sweeney, H. L., Kushmerick, M. J. (1984) A simple analysis of the "phosphocreatine shuttle". *Am. J. Physiol.* 246, C365–377.

57 Saks, V. A., Rosenshtraukh, L. V., Smirnov, V. N., Chazov, E. I. (1978) Role of creatine phosphokinase in cellular function and metabolism. *Can. J. Physiol. Pharmacol.* 56, 691–706.

58 Saks, V. A., Ventura-Clapier, R., Aliev, M. K. (1996) Metabolic control and metabolic capacity: two aspects of creatine kinase functioning in the cells. *Biochim. Biophys. Acta* 1274, 81–88.

59 Wallimann, T. (1975) Creatine kinase isoenzymes and myofibrillar structure. *PhD thesis Nr 5437. ETH Zurich.*

60 Turner, D. C., Wallimann, T., Eppenberger, H. M. (1973) A protein that binds specifically to the M-line of skeletal muscle is identified as the muscle form of creatine kinase. *Proc. Natl. Acad. Sci. U.S.A.* 70, 702–705.

61 Kraft, T., Hornemann, T., Stolz, M., Nier, V., Wallimann, T. (2000) Coupling of creatine kinase to glycolytic enzymes at the sarcomeric I-band of skeletal muscle: a biochemical study *in situ*. *J. Muscle Res. Cell Motil.* 21, 691–703.

62 Saks, V. A., Ventura-Clapier, R., Huchua, Z. A., Preobrazhensky, A. N., Emelin, I. V. (1984) Creatine kinase in regulation of heart function and metabolism. I. Further evidence for compartmentation of adenine nucleotides in cardiac myofibrillar and sarcolemmal coupled ATPase-creatine kinase systems. *Biochim. Biophys. Acta* 803, 254–264.

63 Rossi, A. M., Eppenberger, H. M., Volpe, P., Cotrufo, R., Wallimann, T. (1990) Muscle-type MM creatine kinase is specifically bound to sarcoplasmic reticulum and can support Ca^{2+} uptake and regulate local ATP/ADP ratios. *J. Biol. Chem.* 265, 5258–5266.

64 Burklen, T. S., Hirschy, A., Wallimann, T. (2006) Brain-type creatine kinase BB-CK interacts with the Golgi Matrix Protein GM130 in early prophase. *Mol. Cell. Biochem.* 297, 53–64.

65 Kaasik, A., Veksler, V., Boehm, E., Novotova, M., Ventura-Clapier, R. (2003) From energy store to energy flux: a study in creatine kinase-deficient fast skeletal muscle. *FASEB J.* 17, 708–710.

66 Wallimann, T., Schnyder, T., Schlegel, J., Wyss, M., Wegmann, G., Rossi, A. M., Hemmer, W., Eppenberger, H. M., Quest, A. F. (1989) Subcellular compartmentation of creatine kinase isoenzymes, regulation of CK and octameric structure of mitochondrial CK: important aspects of the phosphoryl-creatine circuit. *Prog. Clin. Biol. Res.* 315, 159–176.

67 Canonaco, F., Schlattner, U., Pruett, P. S., Wallimann, T., Sauer, U. (2002) Functional expression of phosphagen kinase systems confers resistance to transient stresses in Saccharomyces

cerevisiae by buffering the ATP pool. *J. Biol. Chem. 277*, 31303–31309.

68 Sauer, U., Schlattner, U. (2004) Inverse metabolic engineering with phosphagen kinase systems improves the cellular energy state. *Metab. Eng. 6*, 220–228.

69 Bessman, S. P., Geiger, P. J. (1981) Transport of energy in muscle: the phosphorylcreatine shuttle. *Science 211*, 448–452.

70 Selivanov, V. A., Alekseev, A. E., Hodgson, D. M., Dzeja, P. P., Terzic, A. (2004) Nucleotide-gated KATP channels integrated with creatine and adenylate kinases: amplification, tuning and sensing of energetic signals in the compartmentalized cellular environment. *Mol. Cell. Biochem. 256–257*, 243–256.

71 Tombes, R. M., Shapiro, B. M. (1989) Energy transport and cell polarity: relationship of phosphagen kinase activity to sperm function. *J. Exp. Zool. 251*, 82–90.

72 Vendelin, M., Lemba, M., Saks, V. A. (2004) Analysis of functional coupling: mitochondrial creatine kinase and adenine nucleotide translocase. *Biophys. J. 87*, 696–713.

73 Vendelin, M., Eimre, M., Seppet, E., Peet, N., Andrienko, T., Lemba, M., Engelbrecht, J., Seppet, E. K., Saks, V. A. (2004) Intracellular diffusion of adenosine phosphates is locally restricted in cardiac muscle. *Mol. Cell. Biochem. 256–257*, 229–241.

74 Dzeja, P. P., Bortolon, R., Perez-Terzic, C., Holmuhamedov, E. L., Terzic, A. (2002) Energetic communication between mitochondria and nucleus directed by catalyzed phosphotransfer. *Proc. Natl. Acad. Sci. U.S.A. 99*, 10156–10161.

75 Kaasik, A., Veksler, V., Boehm, E., Novotova, M., Minajeva, A., Ventura-Clapier, R. (2001) Energetic crosstalk between organelles: architectural integration of energy production and utilization. *Circ. Res. 89*, 153–159.

76 Seppet, E. K., Kaambre, T., Sikk, P., Tiivel, T., Vija, H., Tonkonogi, M., Sahlin, K., Kay, L., Appaix, F., Braun, U., Eimre, M., Saks, V. A. (2001) Functional complexes of mitochondria with Ca,MgATPases of myofibrils and sarcoplasmic reticulum in muscle cells. *Biochim. Biophys. Acta 1504*, 379–395.

77 Andrienko, T., Kuznetsov, A. V., Kaambre, T., Usson, Y., Orosco, A., Appaix, F., Tiivel, T., Sikk, P., Vendelin, M., Margreiter, R., Saks, V. A. (2003) Metabolic consequences of functional complexes of mitochondria, myofibrils and sarcoplasmic reticulum in muscle cells. *J. Exp. Biol. 206*, 2059–2072.

78 de Groof, A. J., Fransen, J. A., Errington, R. J., Willems, P. H., Wieringa, B., Koopman, W. J. (2002) The creatine kinase system is essential for optimal refill of the sarcoplasmic reticulum Ca^{2+} store in skeletal muscle. *J. Biol. Chem. 277*, 5275–5284.

79 Spindler, M., Meyer, K., Stromer, H., Leupold, A., Boehm, E., Wagner, H., Neubauer, S. (2004) Creatine kinase-deficient hearts exhibit increased susceptibility to ischemia-reperfusion injury and impaired calcium homeostasis. *Am. J. Physiol. 287*, H1039–1045.

80 Steeghs, K., Benders, A., Oerlemans, F., de Haan, A., Heerschap, A., Ruitenbeek, W., Jost, C., van Deursen, J., Perryman, B., Pette, D., Bruckwilder, M., Koudijs, J., Jap, P., Veerkamp, J., Wieringa, B. (1997) Altered Ca^{2+} responses in muscles with combined mitochondrial and cytosolic creatine kinase deficiencies. *Cell 89*, 93–103.

81 Dzeja, P. P., Terzic, A., Wieringa, B. (2004) Phosphotransfer dynamics in skeletal muscle from creatine kinase gene-deleted mice. *Mol. Cell. Biochem. 256–257*, 13–27.

82 Dorsten, F. A., Wyss, M., Wallimann, T., Nicolay, K. (1997) Activation of sea-urchin sperm motility is accompanied by an increase in the creatine kinase exchange flux. *Biochem. J. 325*, 411–416.

83 Bollard, M. E., Murray, A. J., Clarke, K., Nicholson, J. K., Griffin, J. L. (2003) A study of metabolic compartmentation in the rat heart and cardiac mitochondria using high-resolution magic angle spinning 1H NMR spectroscopy. *FEBS Lett. 553*, 73–78.

84 Smith, S. A., Montain, S. J., Zientara, G. P., Fielding, R. A. (2004) Use of phosphocreatine kinetics to determine the influence of creatine on muscle mitochondrial respiration: an *in vivo* 31P-MRS study of oral creatine ingestion. *J. Appl. Physiol.* 96, 2288–2292.

85 Spindler, M., Niebler, R., Remkes, H., Horn, M., Lanz, T., Neubauer, S. (2002) Mitochondrial creatine kinase is critically necessary for normal myocardial high-energy phosphate metabolism. *Am. J. Physiol.* 283, H680–687.

86 Kemp, G. J., Manners, D. N., Clark, J. F., Bastin, M. E., Radda, G. K. (1998) Theoretical modelling of some spatial and temporal aspects of the mitochondrion/creatine kinase/myofibril system in muscle. *Mol. Cell. Biochem.* 184, 249–289.

87 Saks, V., Kuznetsov, A., Andrienko, T., Usson, Y., Appaix, F., Guerrero, K., Kaambre, T., Sikk, P., Lemba, M., Vendelin, M. (2003) Heterogeneity of ADP diffusion and regulation of respiration in cardiac cells. *Biophys. J.* 84, 3436–3456.

88 Agarkova, I., Perriard, J. C. (2005) The M-band: an elastic web that crosslinks thick filaments in the center of the sarcomere. *Trends Cell Biol.* 15, 477–485.

89 Pask, H. T., Jones, K. L., Luther, P. K., Squire, J. M. (1994) M-band structure, M-bridge interactions and contraction speed in vertebrate cardiac muscles. *J. Muscle Res. Cell Motil.* 15, 633–645.

90 Vinkemeier, U., Obermann, W., Weber, K., Furst, D. O. (1993) The globular head domain of titin extends into the center of the sarcomeric M band. cDNA cloning, epitope mapping and immunoelectron microscopy of two titin-associated proteins. *J. Cell Sci.* 106 (Pt 1), 319–330.

91 Obermann, W. M., Gautel, M., Steiner, F., van der Ven, P. F., Weber, K., Furst, D. O. (1996) The structure of the sarcomeric M band: localization of defined domains of myomesin, M-protein, and the 250-kD carboxy-terminal region of titin by immuno-electron microscopy. *J. Cell Biol.* 134, 1441–1453.

92 Masaki, T., Takaiti, O. (1974) M-protein. *J. Biochem. (Tokyo)* 75, 367–380.

93 Trinick, J., Lowey, S. (1977) M-protein from chicken pectoralis muscle: isolation and characterization. *J. Mol. Biol.* 113, 343–368.

94 Grove, B. K., Kurer, V., Lehner, C., Doetschman, T. C., Perriard, J. C., Eppenberger, H. M. (1984) A new 185,000-dalton skeletal muscle protein detected by monoclonal antibodies. *J. Cell Biol.* 98, 518–524.

95 Wallimann, T., Doetschman, T. C., Eppenberger, H. M. (1983) Novel staining pattern of skeletal muscle M-lines upon incubation with antibodies against MM-creatine kinase. *J. Cell Biol.* 96, 1772–1779.

96 Keller, A., Demeurie, J., Merkulova, T., Geraud, G., Cywiner-Golenzer, C., Lucas, M., Chatelet, F. P. (2000) Fibre-type distribution and subcellular localisation of alpha and beta enolase in mouse striated muscle. *Biol. Cell.* 92, 527–535.

97 Lange, S., Auerbach, D., McLoughlin, P., Perriard, E., Schafer, B. W., Perriard, J. C., Ehler, E. (2002) Subcellular targeting of metabolic enzymes to titin in heart muscle may be mediated by DRAL/FHL-2. *J. Cell Sci.* 115, 4925–4936.

98 Lange, S., Himmel, M., Auerbach, D., Agarkova, I., Hayess, K., Furst, D. O., Perriard, J. C., Ehler, E. (2005) Dimerisation of myomesin: implications for the structure of the sarcomeric M-band. *J. Mol. Biol.* 345, 289–298.

99 Agarkova, I., Ehler, E., Lange, S., Schoenauer, R., Perriard, J. C. (2003) M-band: a safeguard for sarcomere stability? *J. Muscle Res. Cell Motil.* 24, 191–203.

100 Hornemann, T., Stolz, M., Wallimann, T. (2000) Isoenzyme-specific interaction of muscle-type creatine kinase with the sarcomeric M-line is mediated by NH(2)-terminal lysine charge-clamps. *J. Cell Biol.* 149, 1225–1234.

101 Wallimann, T., Moser, H., Eppenberger, H. M. (1983) Isoenzyme-specific localization of M-line bound creatine kinase in myogenic cells. *J. Muscle Res. Cell Motil.* 4, 429–441.

102 Wallimann, T., Pelloni, G., Turner, D. C., Eppenberger, H. M. (1978) Monovalent antibodies against MM-creatine kinase remove the M line from myofibrils. *Proc. Natl. Acad. Sci. U.S.A. 75*, 4296–4300.

103 Hornemann, T., Kempa, S., Himmel, M., Hayess, K., Furst, D. O., Wallimann, T. (2003) Muscle-type creatine kinase interacts with central domains of the M-band proteins myomesin and M-protein. *J. Mol. Biol. 332*, 877–887.

104 Gregor, M., Mejsnar, J., Janovska, A., Zurmanova, J., Benada, O., Mejsnarova, B. (1999) Creatine kinase reaction in skinned rat psoas muscle fibers and their myofibrils. *Physiol. Res. 48*, 27–35.

105 Wallimann, T., Schlosser, T., Eppenberger, H. M. (1984) Function of M-line-bound creatine kinase as intramyofibrillar ATP regenerator at the receiving end of the phosphorylcreatine shuttle in muscle. *J. Biol. Chem. 259*, 5238–5246.

106 Krause, S. M., Jacobus, W. E. (1992) Specific enhancement of the cardiac myofibrillar ATPase by bound creatine kinase. *J. Biol. Chem. 267*, 2480–2486.

107 Ventura-Clapier, R., Mekhfi, H., Vassort, G. (1987) Role of creatine kinase in force development in chemically skinned rat cardiac muscle. *J. Gen. Physiol. 89*, 815–837.

108 Ventura-Clapier, R., Vassort, G. (1985) Role of myofibrillar creatine kinase in the relaxation of rigor tension in skinned cardiac muscle. *Pflugers Arch. 404*, 157–161.

109 Hoerter, J. A., Kuznetsov, A., Ventura-Clapier, R. (1991) Functional development of the creatine kinase system in perinatal rabbit heart. *Circ. Res. 69*, 665–676.

110 van Deursen, J., Heerschap, A., Oerlemans, F., Ruitenbeek, W., Jap, P., ter Laak, H., Wieringa, B. (1993) Skeletal muscles of mice deficient in muscle creatine kinase lack burst activity. *Cell 74*, 621–631.

111 de Haan, A., Koudijs, J. C., Wevers, R. A., Wieringa, B. (1995) The effects of MM-creatine kinase deficiency on sustained force production of mouse fast skeletal muscle. *Exp. Physiol. 80*, 491–494.

112 Ventura-Clapier, R., Kuznetsov, A. V., d'Albis, A., van Deursen, J., Wieringa, B., Veksler, V. I. (1995) Muscle creatine kinase-deficient mice. I. Alterations in myofibrillar function. *J. Biol. Chem. 270*, 19914–19920.

113 Watchko, J. F., Daood, M. J., Wieringa, B., Koretsky, A. P. (2000) Myofibrillar or mitochondrial creatine kinase deficiency alone does not impair mouse diaphragm isotonic function. *J. Appl. Physiol. 88*, 973–980.

114 Veksler, V. I., Kuznetsov, A. V., Anflous, K., Mateo, P., van Deursen, J., Wieringa, B., Ventura-Clapier, R. (1995) Muscle creatine kinase-deficient mice. II. Cardiac and skeletal muscles exhibit tissue-specific adaptation of the mitochondrial function. *J. Biol. Chem. 270*, 19921–19929.

115 LaBella, J. J., Daood, M. J., Koretsky, A. P., Roman, B. B., Sieck, G. C., Wieringa, B., Watchko, J. F. (1998) Absence of myofibrillar creatine kinase and diaphragm isometric function during repetitive activation. *J. Appl. Physiol. 84*, 1166–1173.

116 Boehm, E., Ventura-Clapier, R., Mateo, P., Lechene, P., Veksler, V. (2000) Glycolysis supports calcium uptake by the sarcoplasmic reticulum in skinned ventricular fibres of mice deficient in mitochondrial and cytosolic creatine kinase. *J. Mol. Cell. Cardiol. 32*, 891–902.

117 Steeghs, K., Oerlemans, F., de Haan, A., Heerschap, A., Verdoodt, L., de Bie, M., Ruitenbeek, W., Benders, A., Jost, C., van Deursen, J., Tullson, P., Terjung, R., Jap, P., Jacob, W., Pette, D., Wieringa, B. (1998) Cytoarchitectural and metabolic adaptations in muscles with mitochondrial and cytosolic creatine kinase deficiencies. *Mol. Cell. Biochem. 184*, 183–194.

118 Ventura-Clapier, R., Kaasik, A., Veksler, V. (2004) Structural and functional adaptations of striated muscles to CK deficiency. *Mol. Cell. Biochem. 256–257*, 29–41.

119 Wilding, J. R., Joubert, F., de Araujo, C., Fortin, D., Novotova, M., Veksler, V., Ventura-Clapier, R. (2006) Altered energy transfer from mitochondria to sarcoplasmic reticulum after cytoarchi-

tectural perturbations in mice hearts. *J. Physiol.* 575, 191–200.

120 Bonz, A. W., Kniesch, S., Hofmann, U., Kullmer, S., Bauer, L., Wagner, H., Ertl, G., Spindler, M. (2002) Functional properties and [Ca(2+)](i) metabolism of creatine kinase – KO mice myocardium. *Biochem. Biophys. Res. Commun.* 298, 163–168.

121 Watchko, J. F., Daood, M. J., Sieck, G. C., LaBella, J. J., Ameredes, B. T., Koretsky, A. P., Wieringa, B. (1997) Combined myofibrillar and mitochondrial creatine kinase deficiency impairs mouse diaphragm isotonic function. *J. Appl. Physiol.* 82, 1416–1423.

122 Baskin, R. J., Deamer, D. W. (1970) A membrane-bound creatine phosphokinase in fragmented sarcoplasmic reticulum. *J. Biol. Chem.* 245, 1345–1347.

123 Levitsky, D. O., Levchenko, T. S., Saks, V. A., Sharov, V. G., Smirnov, V. N. (1978) The role of creatine phosphokinase in supplying energy for the calcium pump system of heart sarcoplasmic reticulum. *Membr. Biochem.* 2, 81–96.

124 Wolosker, H., Panizzutti, R., Engelender, S. (1996) Inhibition of creatine kinase by S-nitrosoglutathione. *FEBS Lett.* 392, 274–276.

125 Korge, P., Campbell, K. B. (1994) Local ATP regeneration is important for sarcoplasmic reticulum Ca^{2+} pump function. *Am. J. Physiol.* 267, C357–366.

126 Hornemann, T., Thurnherr, T., Wallimann, T. (2003) Muscle-type creatine kinase binds to a sarcoplasmic reticulum 130–170 kDa polypeptide triplet on overlay blots, in *"Proceedings of the NATO Advanced Workshop on Creatine Kinase and Brain Energy Metabolism" June 14–18, 2001, Tibilisi, Georgia* (Kekelidze, T., Holtzman, D., Eds.) pp 75–83, IOS Press, Amsterdam, Berlin, Oxford, Tokyo, and Washington, DC.

127 Minajeva, A., Ventura-Clapier, R., Veksler, V. (1996) Ca^{2+} uptake by cardiac sarcoplasmic reticulum ATPase *in situ* strongly depends on bound creatine kinase. *Pflugers Arch.* 432, 904–912.

128 Korge, P., Byrd, S. K., Campbell, K. B. (1993) Functional coupling between sarcoplasmic-reticulum-bound creatine kinase and Ca(2+)-ATPase. *Eur. J. Biochem.* 213, 973–980.

129 Duke, A. M., Steele, D. S. (1999) Effects of creatine phosphate on Ca^{2+} regulation by the sarcoplasmic reticulum in mechanically skinned rat skeletal muscle fibres. *J. Physiol.* 517, 447–458.

130 Crozatier, B., Badoual, T., Boehm, E., Ennezat, P. V., Guenoun, T., Su, J., Veksler, V., Hittinger, L., Ventura-Clapier, R. (2002) Role of creatine kinase in cardiac excitation-contraction coupling: studies in creatine kinase-deficient mice. *FASEB J.* 16, 653–660.

131 Kindig, C. A., Howlett, R. A., Stary, C. M., Walsh, B., Hogan, M. C. (2005) Effects of acute creatine kinase inhibition on metabolism and tension development in isolated single myocytes. *J. Appl. Physiol.* 98, 541–549.

132 Field, M. L., Khan, O., Abbaraju, J., Clark, J. F. (2006) Functional compartmentation of glycogen phosphorylase with creatine kinase and Ca^{2+} ATPase in skeletal muscle. *J. Theor. Biol.* 238, 257–268.

133 Stadhouders, A. M., Jap, P. H., Winkler, H. P., Eppenberger, H. M., Wallimann, T. (1994) Mitochondrial creatine kinase: a major constituent of pathological inclusions seen in mitochondrial myopathies. *Proc. Natl. Acad. Sci. U.S.A.* 91, 5089–5093.

134 Tian, R., Halow, J. M., Meyer, M., Dillmann, W. H., Figueredo, V. M., Ingwall, J. S., Camacho, S. A. (1998) Thermodynamic limitation for Ca^{2+} handling contributes to decreased contractile reserve in rat hearts. *Am. J. Physiol.* 275, H2064–2071.

135 Tarnopolsky, M. A., Beal, M. F. (2001) Potential for creatine and other therapies targeting cellular energy dysfunction in neurological disorders. *Ann. Neurol.* 49, 561–574.

136 Momken, I., Lechene, P., Koulmann, N., Fortin, D., Mateo, P., Doan, B. T., Hoerter, J., Bigard, X., Veksler, V., Ventura-Clapier, R. (2005) Impaired voluntary running capacity of creatine kinase-deficient mice. *J. Physiol. 565*, 951–964.

137 Han, J. W., Thieleczek, R., Varsanyi, M., Heilmeyer, L. M., Jr. (1992) Compartmentalized ATP synthesis in skeletal muscle triads. *Biochemistry 31*, 377–384.

138 Xu, K. Y., Zweier, J. L., Becker, L. C. (1995) Functional coupling between glycolysis and sarcoplasmic reticulum Ca2+ transport. *Circ. Res. 77*, 88–97.

139 Sharov, V. G., Saks, V. A., Smirnov, V. N., Chazov, E. I. (1977) An electron microscopic histochemical investigation of the localization of creatine phosphokinase in heart cells. *Biochim. Biophys. Acta 468*, 495–501.

140 Grosse, R., Spitzer, E., Kupriyanov, V. V., Saks, V. A., Repke, K. R. (1980) Coordinate interplay between (Na^+/K^+)-ATPase and creatine phosphokinase optimizes $(Na+/K+)$-antiport across the membrane of vesicles formed from the plasma membrane of cardiac muscle cell. *Biochim. Biophys. Acta 603*, 142–156.

141 Saks, V. A., Lipina, N. V., Sharov, V. G., Smirnov, V. N., Chazov, E., Grosse, R. (1977) The localization of the MM isozyme of creatine phosphokinase on the surface membrane of myocardial cells and its functional coupling to ouabain-inhibited (Na^+/K^+)-ATPase. *Biochim. Biophys. Acta 465*, 550–558.

142 Barrantes, F. J., Mieskes, G., Wallimann, T. (1983) Creatine kinase activity in the Torpedo electrocyte and in the nonreceptor, peripheral v proteins from acetylcholine receptor-rich membranes. *Proc. Natl. Acad. Sci. U.S.A. 80*, 5440–5444.

143 Blum, H., Balschi, J. A., Johnson, R. G., Jr. (1991) Coupled *in vivo* activity of creatine phosphokinase and the membrane-bound (Na^+/K^+)-ATPase in the resting and stimulated electric organ of the electric fish Narcine brasiliensis. *J. Biol. Chem. 266*, 10254–10259.

144 Crawford, R. M., Ranki, H. J., Botting, C. H., Budas, G. R., Jovanovic, A. (2002) Creatine kinase is physically associated with the cardiac ATP-sensitive K+ channel *in vivo*. *FASEB J. 16*, 102–104.

145 Abraham, M. R., Selivanov, V. A., Hodgson, D. M., Pucar, D., Zingman, L. V., Wieringa, B., Dzeja, P. P., Alekseev, A. E., Terzic, A. (2002) Coupling of cell energetics with membrane metabolic sensing. Integrative signaling through creatine kinase phosphotransfer disrupted by M-CK gene knockout. *J. Biol. Chem. 277*, 24427–24434.

146 Dzeja, P. P., Terzic, A. (1998) Phosphotransfer reactions in the regulation of ATP-sensitive K^+ channels. *FASEB J. 12*, 523–529.

147 Zingman, L. V., Alekseev, A. E., Bienengraeber, M., Hodgson, D., Karger, A. B., Dzeja, P. P., Terzic, A. (2001) Signaling in channel/enzyme multimers: ATPase transitions in SUR module gate ATP-sensitive K^+ conductance. *Neuron 31*, 233–245.

148 Inoue, K., Ueno, S., Fukuda, A. (2004) Interaction of neuron-specific K^+Cl^- cotransporter, KCC2, with brain-type creatine kinase. *FEBS Lett. 564*, 131–135.

149 Inoue, K., Yamada, J., Ueno, S., Fukuda, A. (2006) Brain-type creatine kinase activates neuron-specific K+-Cl− cotransporter KCC2. *J. Neurochem. 96*, 598–608.

150 Colanzi, A., Deerinck, T. J., Ellisman, M. H., Malhotra, V. (2000) A specific activation of the mitogen-activated protein kinase kinase 1 (MEK1) is required for Golgi fragmentation during mitosis. *J. Cell Biol. 149*, 331–339.

151 Lowe, M., Rabouille, C., Nakamura, N., Watson, R., Jackman, M., Jamsa, E., Rahman, D., Pappin, D. J., Warren, G. (1998) Cdc2 kinase directly phosphorylates the cis-Golgi matrix protein GM130 and is required for Golgi fragmentation in mitosis. *Cell 94*, 783–793.

152 Sutterlin, C., Lin, C. Y., Feng, Y., Ferris, D. K., Erikson, R. L., Malhotra, V. (2001) Polo-like kinase is required for the fragmentation of pericentriolar Golgi stacks during mitosis. *Proc. Natl. Acad. Sci. U.S.A. 98*, 9128–9132.

153 Tibaldi, E., Arrigoni, G., Brunati, A. M., James, P., Pinna, L. A. (2006) Analysis of a sub-proteome which co-purifies with and is phosphorylated by the Golgi

casein kinase. *Cell. Mol. Life Sci. 63*, 378–389.

154 Suginta, W., Karoulias, N., Aitken, A., Ashley, R. H. (2001) Chloride intracellular channel protein CLIC4 (p64H1) binds directly to brain dynamin I in a complex containing actin, tubulin and 14-3-3 isoforms. *Biochem. J. 359*, 55–64.

155 Schmitt, T. L., Hotz-Wagenblatt, A., Klein, H., Droge, W. (2005) Interdependent regulation of insulin receptor kinase activity by ADP and hydrogen peroxide. *J. Biol. Chem. 280*, 3795–3801.

156 Mahajan, V. B., Pai, K. S., Lau, A., Cunningham, D. D. (2000) Creatine kinase, an ATP-generating enzyme, is required for thrombin receptor signaling to the cytoskeleton. *Proc. Natl. Acad. Sci. U.S.A. 97*, 12062–12067.

157 Wojtas, K., Slepecky, N., von Kalm, L., Sullivan, D. (1997) Flight muscle function in Drosophila requires colocalization of glycolytic enzymes. *Mol. Biol. Cell 8*, 1665–1675.

158 Dillon, P. F., Clark, J. F. (1990) The theory of diazymes and functional coupling of pyruvate kinase and creatine kinase. *J. Theor. Biol. 143*, 275–284.

159 Van Waarde, A., Van den Thillart, G., Erkelens, C., Addink, A., Lugtenburg, J. (1990) Functional coupling of glycolysis and phosphocreatine utilization in anoxic fish muscle. An *in vivo* 31P NMR study. *J. Biol. Chem. 265*, 914–923.

160 Beutner, G., Ruck, A., Riede, B., Brdiczka, D. (1998) Complexes between porin, hexokinase, mitochondrial creatine kinase and adenylate translocator display properties of the permeability transition pore. Implication for regulation of permeability transition by the kinases. *Biochim. Biophys. Acta 1368*, 7–18.

161 Eder, M., Fritz-Wolf, K., Kabsch, W., Wallimann, T., Schlattner, U. (2000) Crystal structure of human ubiquitous mitochondrial creatine kinase. *Proteins 39*, 216–225.

162 Fritz-Wolf, K., Schnyder, T., Wallimann, T., Kabsch, W. (1996) Structure of mitochondrial creatine kinase. *Nature 381*, 341–345.

163 Heidkamper, D., Muller, V., Nelson, D. R., Klingenberg, M. (1996) Probing the role of positive residues in the ADP/ATP carrier from yeast. The effect of six arginine mutations on transport and the four ATP versus ADP exchange modes. *Biochemistry 35*, 16144–16152.

164 Nury, H., Dahout-Gonzalez, C., Trezeguet, V., Lauquin, G. J., Brandolin, G., Pebay-Peyroula, E. (2006) Relations between structure and function of the mitochondrial ADP/ATP carrier. *Annu. Rev. Biochem. 75*, 713–741.

165 Shoshan-Barmatz, V., Israelson, A., Brdiczka, D., Sheu, S. S. (2006) The voltage-dependent anion channel (VDAC): function in intracellular signalling, cell life and cell death. *Curr. Pharm. Des. 12*, 2249–2270.

166 Beyer, K., Klingenberg, M. (1985) ADP/ATP carrier protein from beef heart mitochondria has high amounts of tightly bound cardiolipin, as revealed by 31P nuclear magnetic resonance. *Biochemistry 24*, 3821–3826.

167 Schlattner, U., Dolder, M., Wallimann, T., Tokarska-Schlattner, M. (2001) Mitochondrial creatine kinase and mitochondrial outer membrane porin show a direct interaction that is modulated by calcium. *J. Biol. Chem. 276*, 48027–48030.

168 Khuchua, Z. A., Qin, W., Boero, J., Cheng, J., Payne, R. M., Saks, V. A., Strauss, A. W. (1998) Octamer formation and coupling of cardiac sarcomeric mitochondrial creatine kinase are mediated by charged N-terminal residues. *J. Biol. Chem. 273*, 22990–22996.

169 Brdiczka, D., Beutner, G., Ruck, A., Dolder, M., Wallimann, T. (1998) The molecular structure of mitochondrial contact sites. Their role in regulation of energy metabolism and permeability transition. *Biofactors 8*, 235–242.

170 Brdiczka, D. G., Zorov, D. B., Sheu, S. S. (2006) Mitochondrial contact sites: their role in energy metabolism and apoptosis. *Biochim. Biophys. Acta 1762*, 148–163.

171 Gellerich, F. N., Trumbeckaite, S., Opalka, J. R., Seppet, E., Rasmussen, H. N., Neuhoff, C., Zierz, S. (2000) Function of the mitochondrial outer membrane as a diffusion barrier in

health and diseases. *Biochem. Soc. Trans. 28*, 164–169.

172 Tokarska-Schlattner, M., Wallimann, T., Schlattner, U. (2002) Multiple interference of anthracyclines with mitochondrial creatine kinases: preferential damage of the cardiac isoenzyme and its implications for drug cardiotoxicity. *Mol. Pharmacol. 61*, 516–523.

173 Soboll, S., Brdiczka, D., Jahnke, D., Schmidt, A., Schlattner, U., Wendt, S., Wyss, M., Wallimann, T. (1999) Octamer-dimer transitions of mito-chondrial creatine kinase in heart dis-ease. *J. Mol. Cell. Cardiol. 31*, 857–866.

174 Bernardi, P., Scorrano, L., Colonna, R., Petronilli, V., Di Lisa, F. (1999) Mito-chondria and cell death. Mechanistic aspects and methodological issues. *Eur. J. Biochem. 264*, 687–701.

175 Green, D. R., Kroemer, G. (2004) The pathophysiology of mitochondrial cell death. *Science 305*, 626–629.

176 Crompton, M. (2000) Mitochondrial intermembrane junctional complexes and their role in cell death. *J. Physiol. 529*, 11–21.

177 Marzo, I., Brenner, C., Zamzami, N., Susin, S. A., Beutner, G., Brdiczka, D., Remy, R., Xie, Z. H., Reed, J. C., Kroemer, G. (1998) The permeability transition pore complex: a target for apoptosis regulation by caspases and bcl-2-related proteins. *J. Exp. Med. 187*, 1261–1271.

178 Halestrap, A. P., Doran, E., Gillespie, J. P., O'Toole, A. (2000) Mitochondria and cell death. *Biochem. Soc. Trans. 28*, 170–177.

179 Halestrap, A. P., Clarke, S. J., Javadov, S. A. (2004) Mitochondrial permeability transition pore opening during myocardial reperfusion – a target for cardioprotection. *Cardiovasc. Res. 61*, 372–385.

180 Kokoszka, J. E., Waymire, K. G., Levy, S. E., Sligh, J. E., Cai, J., Jones, D. P., MacGregor, G. R., Wallace, D. C. (2004) The ADP/ATP translocator is not essen-tial for the mitochondrial permeability transition pore. *Nature 427*, 461–465.

181 Miller, K., Sharer, K., Suhan, J., Koretsky, A. P. (1997) Expression of functional mitochondrial creatine kinase in liver of transgenic mice. *Am. J. Physiol. 272*, C1193–1202.

182 Dolder, M., Walzel, B., Speer, O., Schlattner, U., Wallimann, T. (2003) Inhibition of the mitochondrial permeability transition by creatine kinase substrates. Requirement for microcompartmentation. *J. Biol. Chem. 278*, 17760–17766.

183 Dolder, M., Wendt, S., Wallimann, T. (2001) Mitochondrial creatine kinase in contact sites: interaction with porin and adenine nucleotide translocase, role in permeability transition and sensitivity to oxidative damage. *Biol. Signals Recept. 10*, 93–111.

184 O'Gorman, E., Beutner, G., Dolder, M., Koretsky, A. P., Brdiczka, D., Wallimann, T. (1997) The role of creatine kinase in inhibition of mitochondrial permeability transition. *FEBS Lett. 414*, 253–257.

185 Speer, O., Back, N., Buerklen, T., Brdiczka, D., Koretsky, A., Wallimann, T., Eriksson, O. (2005) Octameric mitochondrial creatine kinase induces and stabilizes contact sites between the inner and outer membrane. *Biochem. J. 385*, 445–450.

186 Hatano, E., Tanaka, A., Kanazawa, A., Tsuyuki, S., Tsunekawa, S., Iwata, S., Takahashi, R., Chance, B., Yamaoka, Y. (2004) Inhibition of tumor necrosis factor-induced apoptosis in transgenic mouse liver expressing creatine kinase. *Liver Int. 24*, 384–393.

187 O'Gorman, E., Fuchs, K. H., Tittmann, P., Gross, H., Wallimann, T. (1997) Crystalline mitochondrial inclusion bodies isolated from creatine depleted rat soleus muscle. *J. Cell Sci. 110*, 1403–1411.

188 Lee, C. K., Klopp, R. G., Weindruch, R., Prolla, T. A. (1999) Gene expression profile of aging and its retardation by caloric restriction. *Science 285*, 1390–1393.

189 Heddi, A., Stepien, G., Benke, P. J., Wallace, D. C. (1999) Coordinate induction of energy gene expression in tissues of mitochondrial disease patients. *J. Biol. Chem. 274*, 22968–22976.

190 Tarnopolsky, M. A., Simon, D. K., Roy, B. D., Chorneyko, K., Lowther, S. A., Johns, D. R., Sandhu, J. K., Li, Y., Sikorska, M. (2004) Attenuation of free radical production and paracrystalline inclusions by creatine supplementation in a patient with a novel cytochrome b mutation. *Muscle Nerve 29*, 537–547.

191 Sohal, R. S., Weindruch, R. (1996) Oxidative stress, caloric restriction, and aging. *Science 273*, 59–63.

192 Meyer, L. E., Machado, L. B., Santiago, A. P., da-Silva, W. S., De Felice, F. G., Holub, O., Oliveira, M. F., Galina, A. (2006) Mitochondrial creatine kinase activity prevents reactive oxygen species generation: Antioxidant role of mito-chondrial kinases-dependent ADP re-cycling activity. *J. Biol. Chem. 281*, 37361–37371.

193 Kanemitsu, F., Kawanishi, I., Mizushima, J. (1983) A new creatine kinase found in mitochondrial extracts from malignant liver tissue. *Clin. Chim. Acta 128*, 233–240.

194 Schiemann, S., Schwirzke, M., Brunner, N., Weidle, U. H. (1998) Molecular analysis of two mammary carcinoma cell lines at the transcriptional level as a model system for progression of breast cancer. *Clin. Exp. Metastasis 16*, 129–139.

195 Kornacker, M., Schlattner, U., Wallimann, T., Verneris, M. R., Negrin, R. S., Kornacker, B., Staratschek-Jox, A., Diehl, V., Wolf, J. (2001) Hodgkin disease-derived cell lines expressing ubiquitous mitochondrial creatine kinase show growth inhibition by cyclocreatine treatment independent of apoptosis. *Int. J. Cancer 94*, 513–519.

196 Pratt, R., Vallis, L. M., Lim, C. W., Chisnall, W. N. (1987) Mitochondrial creatine kinase in cancer patients. *Pathology 19*, 162–165.

197 Kanemitsu, F., Kawanishi, I., Mizushima, J., Okigaki, T. (1984) Mitochondrial creatine kinase as a tumor-associated marker. *Clin. Chim. Acta 138*, 175–183.

198 Epand, R. F., Schlattner, U., Wallimann, T., Lacombe, M. L., Epand, R. M. (2007) Novel lipid transfer property of two mitochondrial proteins that bridge the inner and outer membranes. *Biophys. J.*, *92*, 126–137.

199 Epand, R. F., Tokarska-Schlattner, M., Schlattner, U., Wallimann, T., Epand, R. M. (2007) Cardiolipin clusters and membrane domain formation induced by mitochondrial proteins. *J. Mol. Biol.*, *365*, 968–980.

200 Amutha, B., Pain, D. (2003) Nucleoside diphosphate kinase of Saccharomyces cerevisiae, Ynk1p: localization to the mitochondrial intermembrane space. *Biochem. J. 370*, 805–815.

201 Ardail, D., Privat, J. P., Egret-Charlier, M., Levrat, C., Lerme, F., Louisot, P. (1990) Mitochondrial contact sites. Lipid composition and dynamics. *J. Biol. Chem. 265*, 18797–18802.

202 Stachowiak, O., Dolder, M., Wallimann, T. (1996) Membrane-binding and lipid vesicle cross-linking kinetics of the mitochondrial creatine kinase octamer. *Biochemistry 35*, 15522–15528.

203 Vacheron, M. J., Clottes, E., Chautard, C., Vial, C. (1997) Mitochondrial creatine kinase interaction with phospholipid vesicles. *Arch. Biochem. Biophys. 344*, 316–324.

204 Muller, M., Cheneval, D., Carafoli, E. (1986) The mitochondrial creatine phosphokinase is associated with inner membrane cardiolipin. *Adv. Exp. Med. Biol. 194*, 151–156.

205 Muller, M., Moser, R., Cheneval, D., Carafoli, E. (1985) Cardiolipin is the membrane receptor for mitochondrial creatine phosphokinase. *J. Biol. Chem. 260*, 3839–3843.

206 Schlame, M., Augustin, W. (1985) Association of creatine kinase with rat heart mitochondria: high and low affinity binding sites and the involve-ment of phospholipids. *Biomed. Biochim. Acta 44*, 1083–1088.

207 Schlegel, J., Zurbriggen, B., Wegmann, G., Wyss, M., Eppenberger, H. M., Wallimann, T. (1988) Native mitochon-drial creatine kinase forms octameric structures. I. Isolation of two inter-convertible mitochondrial creatine kinase forms, dimeric and octameric mitochondrial creatine kinase: characterization, localization, and structure-function relationships. *J. Biol. Chem. 263*, 16942–16953.

208 Schlattner, U., Wallimann, T. (2000) Octamers of mitochondrial creatine kinase isoenzymes differ in stability and membrane binding. *J. Biol. Chem. 275*, 17314–17320.

209 Rojo, M., Hovius, R., Demel, R., Wallimann, T., Eppenberger, H. M., Nicolay, K. (1991) Interaction of mitochondrial creatine kinase with model membranes. A monolayer study. *FEBS Lett. 281*, 123–129.

210 Muhonen, W. W., Lambeth, D. O. (1995) The compartmentation of nucleoside diphosphate kinase in mitochondria. *Comp. Biochem. Physiol. B Biochem. Mol. Biol. 110*, 211–223.

211 Lascu, L., Giartosio, A., Ransac, S., Erent, M. (2000) Quaternary structure of nucleoside diphosphate kinases. *J. Bioenerg. Biomembr. 32*, 227–236.

212 Milon, L., Meyer, P., Chiadmi, M., Munier, A., Johansson, M., Karlsson, A., Lascu, I., Capeau, J., Janin, J., Lacombe, M. L. (2000) The human nm23-H4 gene product is a mitochondrial nucleoside diphosphate kinase. *J. Biol. Chem. 275*, 14264–14272.

213 Gordon, D. M., Lyver, E. R., Lesuisse, E., Dancis, A., Pain, D. (2006) GTP in the mitochondrial matrix plays a crucial role in organellar iron homoeostasis. *Biochem. J. 400*, 163–168.

214 Lacombe, M. L., Milon, L., Munier, A., Mehus, J. G., Lambeth, D. O. (2000) The human Nm23/nucleoside diphosphate kinases. *J. Bioenerg. Biomembr. 32*, 247–258.

215 Seifert, M., Welter, C., Mehraein, Y., Seitz, G. (2005) Expression of the nm23 homologues nm23-H4, nm23-H6, and nm23-H7 in human gastric and colon cancer. *J. Pathol. 205*, 623–632.

216 Hackenbrock, C. R. (1968) Chemical and physical fixation of isolated mitochondria in low-energy and high-energy states. *Proc. Natl. Acad. Sci. U.S.A. 61*, 598–605.

217 Van Venetie, R., Verkleij, A. J. (1982) Possible role of non-bilayer lipids in the structure of mitochondria. A freeze-fracture electron microscopy study. *Biochim. Biophys. Acta 692*, 397–405.

218 Frey, T. G., Mannella, C. A. (2000) The internal structure of mitochondria. *Trends Biochem. Sci. 25*, 319–324.

219 Vyssokikh, M., Brdiczka, D. (2004) VDAC and peripheral channelling complexes in health and disease. *Mol. Cell. Biochem. 256–257*, 117–126.

220 Vyssokikh, M., Zorova, L., Zorov, D., Heimlich, G., Jurgensmeier, J., Schreiner, D., Brdiczka, D. (2004) The intra-mitochondrial cytochrome *c* distribution varies correlated to the formation of a complex between VDAC and the adenine nucleotide translocase: this affects Bax-dependent cytochrome *c* release. *Biochim. Biophys. Acta 1644*, 27–36.

221 Nakagawa, T., Shimizu, S., Watanabe, T., Yamaguchi, O., Otsu, K., Yamagata, H., Inohara, H., Kubo, T., Tsujimoto, Y. (2005) Cyclophilin D-dependent mitochondrial permeability transition regulates some necrotic but not apoptotic cell death. *Nature 434*, 652–658.

222 Hoffmann, B., Stockl, A., Schlame, M., Beyer, K., Klingenberg, M. (1994) The reconstituted ADP/ATP carrier activity has an absolute requirement for cardiolipin as shown in cysteine mutants. *J. Biol. Chem. 269*, 1940–1944.

223 Rostovtseva, T., Kazemi, N., Bezrukov, S. (2006) The role of cardiolipin in VDAC channel regulation. *Biophys. J. 90*, 333A.

224 Hatch, G. M. (2004) Cell biology of cardiac mitochondrial phospholipids. *Biochem. Cell Biol. 82*, 99–112.

225 Garofalo, T., Giammarioli, A. M., Misasi, R., Tinari, A., Manganelli, V., Gambardella, L., Pavan, A., Malorni, W., Sorice, M. (2005) Lipid microdomains contribute to apoptosis-associated modifications of mitochondria in T cells. *Cell Death Differ. 12*, 1378–1389.

226 Knoll, G., Brdiczka, D. (1983) Changes in freeze-fractured mitochondrial membranes correlated to their energetic state. Dynamic interactions of the boundary membranes. *Biochim. Biophys. Acta 733*, 102–110.

227 Adams, V., Bosch, W., Schlegel, J., Wallimann, T., Brdiczka, D. (1989) Further characterization of contact sites

from mitochondria of different tissues: topology of peripheral kinases. *Biochim. Biophys. Acta 981*, 213–225.

228 Chen, D., Zhang, X. Y., Shi, Y. (2006) Identification and functional characterization of hCLS1, a human cardiolipin synthase localized in mitochondria. *Biochem. J. 398*, 169–176.

229 Schlame, M., Rua, D., Greenberg, M. L. (2000) The biosynthesis and functional role of cardiolipin. *Prog. Lipid Res. 39*, 257–288.

230 Daum, G. (1985) Lipids of mitochondria. *Biochim. Biophys. Acta 822*, 1–42.

231 Garcia Fernandez, M., Troiano, L., Moretti, L., Nasi, M., Pinti, M., Salvioli, S., Dobrucki, J., Cossarizza, A. (2002) Early changes in intramitochondrial cardiolipin distribution during apoptosis. *Cell Growth Differ. 13*, 449–455.

232 Garcia Fernandez, M., Troiano, L., Moretti, L., Pedrazzi, J., Salvioli, S., Castilla-Cortazar, I., Cossarizza, A. (2000) Changes in intramitochondrial cardiolipin distribution in apoptosis-resistant HCW-2 cells, derived from the human promyelocytic leukemia HL-60. *FEBS Lett. 478*, 290–294.

233 Qi, L., Danielson, N. D., Dai, Q., Lee, R. M. (2003) Capillary electrophoresis of cardiolipin with on-line dye interaction and spectrophotometric detection. *Electrophoresis 24*, 1680–1686.

234 Ma, B. J., Taylor, W. A., Dolinsky, V. W., Hatch, G. M. (1999) Acylation of monolysocardiolipin in rat heart. *J. Lipid Res. 40*, 1837–1845.

235 Esposti, M. D., Cristea, I. M., Gaskell, S. J., Nakao, Y., Dive, C. (2003) Proapoptotic Bid binds to monolyso-cardiolipin, a new molecular connection between mitochondrial membranes and cell death. *Cell Death Differ. 10*, 1300–1309.

236 Liu, J., Dai, Q., Chen, J., Durrant, D., Freeman, A., Liu, T., Grossman, D., Lee, R. M. (2003) Phospholipid scramblase 3 controls mitochondrial structure, function, and apoptotic response. *Mol. Cancer Res. 1*, 892–902.

237 Gallet, P. F., Zachowski, A., Julien, R., Fellmann, P., Devaux, P. F., Maftah, A.

(1999) Transbilayer movement and distribution of spin-labelled phospholipids in the inner mitochondrial membrane. *Biochim. Biophys. Acta 1418*, 61–70.

238 Alpy, F., Tomasetto, C. (2005) Give lipids a START: the StAR-related lipid transfer (START) domain in mammals. *J. Cell Sci. 118*, 2791–2801.

239 Arondel, V., Kader, J. C. (1990) Lipid transfer in plants. *Experientia 46*, 579–585.

240 Rueckert, D. G., Schmidt, K. (1990) Lipid transfer proteins. *Chem. Phys. Lipids 56*, 1–20.

241 Esposti, M. D., Erler, J. T., Hickman, J. A., Dive, C. (2001) Bid, a widely expressed proapoptotic protein of the Bcl-2 family, displays lipid transfer activity. *Mol. Cell. Biol. 21*, 7268–7276.

242 Degli Esposti, M. (2002) Sequence and functional similarities between pro-apoptotic Bid and plant lipid transfer proteins. *Biochim. Biophys. Acta 1553*, 331–340.

243 Crimi, M., Astegno, A., Zoccatelli, G., Esposti, M. D. (2006) Pro-apoptotic effect of maize lipid transfer protein on mammalian mitochondria. *Arch. Biochem. Biophys. 445*, 65–71.

244 Denis-Pouxviel, C., Riesinger, I., Buhler, C., Brdiczka, D., Murat, J. C. (1987) Regulation of mitochondrial hexokinase in cultured HT 29 human cancer cells. An ultrastructural and biochemical study. *Biochim. Biophys. Acta 902*, 335–348.

245 Pastorino, J. G., Hoek, J. B. (2003) Hexokinase II: the integration of energy metabolism and control of apoptosis. *Curr. Med. Chem. 10*, 1535–1551.

246 Epand, R. F., Martinou, J. C., Montessuit, S., Epand, R. M. (2002) Membrane perturbations induced by the apoptotic Bax protein. *Biochem. J. 367*, 849–855.

247 Gonzalvez, F., Pariselli, F., Dupaigne, P., Budihardjo, I., Lutter, M., Antonsson, B., Diolez, P., Manon, S., Martinou, J. C., Goubern, M., Wang, X., Bernard, S., Petit, P. X. (2005) tBid interaction with cardiolipin primarily orchestrates mitochondrial dysfunctions and subsequently activates Bax and Bak. *Cell Death Differ. 12*, 614–626.

248 Newmeyer, D. D., Ferguson-Miller, S. (2003) Mitochondria: releasing power for life and unleashing the machineries of death. *Cell 112*, 481–490.

249 Zamzami, N., Kroemer, G. (2003) Apoptosis: mitochondrial membrane permeabilization – the (w)hole story? *Curr. Biol. 13*, R71–73.

250 Lutter, M., Fang, M., Luo, X., Nishijima, M., Xie, X., Wang, X. (2000) Cardiolipin provides specificity for targeting of tBid to mitochondria. *Nat. Cell Biol. 2*, 754–761.

251 Lutter, M., Perkins, G. A., Wang, X. (2001) The pro-apoptotic Bcl-2 family member tBid localizes to mitochondrial contact sites. *BMC Cell Biol. 2*, 22.

252 Epand, R. F., Martinou, J. C., Montessuit, S., Epand, R. M. (2003) Transbilayer lipid diffusion promoted by Bax: implications for apoptosis. *Biochemistry 42*, 14576–14582.

253 Blomgren, K., Hagberg, H. (2006) Free radicals, mitochondria, and hypoxia-ischemia in the developing brain. *Free Radic. Biol. Med. 40*, 388–397.

254 Ferrari, R., Guardigli, G., Mele, D., Percoco, G. F., Ceconi, C., Curello, S. (2004) Oxidative stress during myocardial ischaemia and heart failure. *Curr. Pharm. Des. 10*, 1699–1711.

255 Moro, M. A., Almeida, A., Bolanos, J. P., Lizasoain, I. (2005) Mitochondrial respiratory chain and free radical generation in stroke. *Free Radic. Biol. Med. 39*, 1291–1304.

256 Pacher, P., Schulz, R., Liaudet, L., Szabo, C. (2005) Nitrosative stress and pharmacological modulation of heart failure. *Trends Pharmacol. Sci. 26*, 302–310.

257 Tokarska-Schlattner, M., Zaugg, M., Zuppinger, C., Wallimann, T., Schlattner, U. (2006) New insights into doxorubicin-induced cardiotoxicity: The critical role of cellular energetics. *J. Mol. Cell. Cardiol. 41*, 389–405.

258 Crimi, E., Sica, V., Williams-Ignarro, S., Zhang, H., Slutsky, A. S., Ignarro, L. J., Napoli, C. (2006) The role of oxidative stress in adult critical care. *Free Radic. Biol. Med. 40*, 398–406.

259 Simonian, N. A., Coyle, J. T. (1996) Oxidative stress in neurodegenerative diseases. *Annu. Rev. Pharmacol. Toxicol. 36*, 83–106.

260 Torreilles, F., Salman-Tabcheh, S., Guerin, M., Torreilles, J. (1999) Neuro-degenerative disorders: the role of peroxynitrite. *Brain Res. Brain Res. Rev. 30*, 153–163.

261 Beal, M. F. (2002) Oxidatively modified proteins in aging and disease. *Free Radic. Biol. Med. 32*, 797–803.

262 Wyss, M., Kaddurah-Daouk, R. (2000) Creatine and creatinine metabolism. *Physiol. Rev. 80*, 1107–1213.

263 Banerjee, A., Grosso, M. A., Brown, J. M., Rogers, K. B., Whitman, G. J. (1991) Oxygen metabolite effects on creatine kinase and cardiac energetics after reperfusion. *Am. J. Physiol. 261*, H590–597.

264 Mekhfi, H., Veksler, V., Mateo, P., Maupoil, V., Rochette, L., Ventura-Clapier, R. (1996) Creatine kinase is the main target of reactive oxygen species in cardiac myofibrils. *Circ. Res. 78*, 1016–1027.

265 Barreiro, E., Gea, J., Matar, G., Hussain, S. N. (2005) Expression and carbonyla-tion of creatine kinase in the quadriceps femoris muscles of patients with chronic obstructive pulmonary disease. *Am. J. Respir. Cell Mol. Biol. 33*, 636–642.

266 Mihm, M. J., Coyle, C. M., Schanbacher, B. L., Weinstein, D. M., Bauer, J. A. (2001) Peroxynitrite induced nitration and inactivation of myofibrillar creatine kinase in experimental heart failure. *Cardiovasc. Res. 49*, 798–807.

267 Castegna, A., Aksenov, M., Aksenova, M., Thongboonkerd, V., Klein, J. B., Pierce, W. M., Booze, R., Markesbery, W. R., Butterfield, D. A. (2002) Proteomic identification of oxidatively modified proteins in Alzheimer's disease brain. Part I: creatine kinase BB, glutamine synthase, and ubiquitin carboxy-terminal hydrolase L-1. *Free Radic. Biol. Med. 33*, 562–571.

268 Poon, H. F., Castegna, A., Farr, S. A., Thongboonkerd, V., Lynn, B. C., Banks, W. A., Morley, J. E., Klein, J. B., Butterfield, D. A. (2004) Quantitative proteomics analysis of specific protein

expression and oxidative modification in aged senescence-accelerated-prone 8 mice brain. *Neuroscience 126*, 915–926.

269 Wendt, S., Dedeoglu, A., Speer, O., Wallimann, T., Beal, M. F., Andreassen, O. A. (2002) Reduced creatine kinase activity in transgenic amyotrophic lateral sclerosis mice. *Free Radic. Biol. Med. 32*, 920–926.

270 Callahan, L. A., Supinski, G. S. (2007) Diaphragm and cardiac mitochondrial creatine kinase are impaired in sepsis. *J. Appl. Physiol.*, 102, 44–53.

271 Stachowiak, O., Dolder, M., Wallimann, T., Richter, C. (1998) Mitochondrial creatine kinase is a prime target of peroxynitrite-induced modification and inactivation. *J. Biol. Chem. 273*, 16694–16699.

272 Yen, H. C., Oberley, T. D., Gairola, C. G., Szweda, L. I., St Clair, D. K. (1999) Manganese superoxide dismutase protects mitochondrial complex I against adriamycin-induced cardiomyopathy in transgenic mice. *Arch. Biochem. Biophys. 362*, 59–66.

273 Eaton, P. (2006) Protein thiol oxidation in health and disease: techniques for measuring disulfides and related modifications in complex protein mixtures. *Free Radic. Biol. Med. 40*, 1889–1899.

274 Wendt, S., Schlattner, U., Wallimann, T. (2003) Differential effects of peroxynitrite on human mitochondrial creatine kinase isoenzymes. Inactivation, octamer destabilization, and identification of involved residues. *J. Biol. Chem. 278*, 1125–1130.

275 Koufen, P., Ruck, A., Brdiczka, D., Wendt, S., Wallimann, T., Stark, G. (1999) Free radical-induced inactivation of creatine kinase: influence on the octameric and dimeric states of the mitochondrial enzyme (Mib-CK). *Biochem. J. 344*, 413–417.

276 Reddy, S., Jones, A. D., Cross, C. E., Wong, P. S., Van Der Vliet, A. (2000) Inactivation of creatine kinase by S-glutathionylation of the active-site cysteine residue. *Biochem. J. 347*, 821–827.

277 Klatt, P., Lamas, S. (2000) Regulation of protein function by S-glutathiolation in response to oxidative and nitrosative stress. *Eur. J. Biochem. 267*, 4928–4944.

278 Ingwall, J. S. (2004) Transgenesis and cardiac energetics: new insights into cardiac metabolism. *J. Mol. Cell. Cardiol. 37*, 613–623.

279 Nahrendorf, M., Spindler, M., Hu, K., Bauer, L., Ritter, O., Nordbeck, P., Quaschning, T., Hiller, K. H., Wallis, J., Ertl, G., Bauer, W. R., Neubauer, S. (2005) Creatine kinase knockout mice show left ventricular hypertrophy and dilatation, but unaltered remodeling post-myocardial infarction. *Cardiovasc. Res. 65*, 419–427.

280 Nahrendorf, M., Streif, J. U., Hiller, K. H., Hu, K., Nordbeck, P., Ritter, O., Sosnovik, D., Bauer, L., Neubauer, S., Jakob, P. M., Ertl, G., Spindler, M., Bauer, W. R. (2006) Multimodal functional cardiac MRI in creatine kinase-deficient mice reveals subtle abnormalities in myocardial perfusion and mechanics. *Am. J. Physiol. 290*, H2516–2521.

281 Hespel, P., Eijnde, B. O., Derave, W., Richter, E. A. (2001) Creatine supplementation: exploring the role of the creatine kinase/phosphocreatine system in human muscle. *Can. J. Appl. Physiol. 26 Suppl*, S79–102.

282 Hespel, P., Maughan, R. J., Greenhaff, P. L. (2006) Dietary supplements for football. *J. Sports Sci. 24*, 749–761.

283 Kraemer, W. J., Volek, J. S. (1999) Creatine supplementation. Its role in human performance. *Clin. Sports Med. 18*, 651–666.

284 Chilibeck, P. D., Chrusch, M. J., Chad, K. E., Shawn Davison, K., Burke, D. G. (2005) Creatine monohydrate and resistance training increase bone mineral content and density in older men. *J. Nutr. Health Aging 9*, 352–353.

285 Chetlin, R. D., Gutmann, L., Tarnopolsky, M. A., Ullrich, I. H., Yeater, R. A. (2004) Resistance training exercise and creatine in patients with Charcot-Marie-Tooth disease. *Muscle Nerve 30*, 69–76.

286 Fuld, J. P., Kilduff, L. P., Neder, J. A., Pitsiladis, Y., Lean, M. E., Ward, S. A., Cotton, M. M. (2005) Creatine supplementation during pulmonary rehabilitation in chronic obstructive pulmonary disease. *Thorax 60*, 531–537.

287 Gordon, A., Hultman, E., Kaijser, L., Kristjansson, S., Rolf, C. J., Nyquist, O., Sylven, C. (1995) Creatine supplementation in chronic heart failure increases skeletal muscle creatine phosphate and muscle performance. *Cardiovasc. Res. 30*, 413–418.

288 Louis, M., Lebacq, J., Poortmans, J. R., Belpaire-Dethiou, M. C., Devogelaer, J. P., Van Hecke, P., Goubel, F., Francaux, M. (2003) Beneficial effects of creatine supplementation in dystrophic patients. *Muscle Nerve 27*, 604–610.

289 Tarnopolsky, M., Mahoney, D., Thompson, T., Naylor, H., Doherty, T. J. (2004) Creatine monohydrate supplementation does not increase muscle strength, lean body mass, or muscle phosphocreatine in patients with myotonic dystrophy type 1. *Muscle Nerve 29*, 51–58.

290 Vorgerd, M., Grehl, T., Jager, M., Muller, K., Freitag, G., Patzold, T., Bruns, N., Fabian, K., Tegenthoff, M., Mortier, W., Luttmann, A., Zange, J., Malin, J. P. (2000) Creatine therapy in myophosphorylase deficiency (McArdle disease): a placebo-controlled crossover trial. *Arch. Neurol. 57*, 956–963.

291 Walter, M. C., Lochmuller, H., Reilich, P., Klopstock, T., Huber, R., Hartard, M., Hennig, M., Pongratz, D., Muller-Felber, W. (2000) Creatine monohydrate in muscular dystrophies: A double-blind, placebo-controlled clinical study. *Neurology 54*, 1848–1850.

292 Walter, M. C., Reilich, P., Lochmuller, H., Kohnen, R., Schlotter, B., Hautmann, H., Dunkl, E., Pongratz, D., Muller-Felber, W. (2002) Creatine monohydrate in myotonic dystrophy: a double-blind, placebo-controlled clinical study. *J. Neurol. 249*, 1717–1722.

293 Woo, Y. J., Grand, T. J., Zentko, S., Cohen, J. E., Hsu, V., Atluri, P., Berry, M. F., Taylor, M. D., Moise, M. A., Fisher, O., Kolakowski, S. (2005) Creatine phosphate administration preserves myocardial function in a model of off-pump coronary revascularization. *J. Cardiovasc. Surg. 46*, 297–305.

294 Wallimann, T., Schlattner, U., Guerrero, L., Dolder, M. (1999) The phosphocreatine circuit and creatine supplementation, both come of age, in *Guanidino Compounds in Biology and Medicine* (Mori, A., Ishida, M., Clark, J. F., Eds.) pp 117–129, Blackwell Science Inc., Japan.

295 Wyss, M., Schulze, A. (2002) Health implications of creatine: can oral creatine supplementation protect against neurological and atherosclerotic disease? *Neuroscience 112*, 243–260.

296 Ames, A., 3rd. (2000) CNS energy metabolism as related to function. *Brain Res. Brain Res. Rev. 34*, 42–68.

297 Burklen, T. S., Schlattner, U., Homayouni, R., Gough, K., Rak, M., Szeghalmi, A., Wallimann, T. (2006) The Creatine Kinase/Creatine Connection to Alzheimer's Disease: CK-Inactivation, APP-CK Complexes and Focal Creatine Deposits. *J. Biomed. Biotechnol. 2006*, 35936.

298 Hemmer, W., Wallimann, T. (1993) Functional aspects of creatine kinase in brain. *Dev. Neurosci. 15*, 249–260.

299 Erecinska, M., Silver, I. A. (1989) ATP and brain function. *J. Cereb. Blood Flow Metab. 9*, 2–19.

300 Wyss, M., Wallimann, T. (1994) Creatine metabolism and the consequences of creatine depletion in muscle. *Mol. Cell. Biochem. 133–134*, 51–66.

301 Jost, C. R., Van Der Zee, C. E., In 't Zandt, H. J., Oerlemans, F., Verheij, M., Streijger, F., Fransen, J., Heerschap, A., Cools, A. R., Wieringa, B. (2002) Creatine kinase B-driven energy transfer in the brain is important for habituation and spatial learning behaviour, mossy fibre field size and determination of seizure susceptibility. *Eur. J. Neurosci. 15*, 1692–1706.

302 In 't Zandt, H. J., Renema, W. K., Streijger, F., Jost, C., Klomp, D. W., Oerlemans, F., Van der Zee, C. E., Wieringa, B., Heerschap, A. (2004) Cerebral creatine kinase deficiency influences metabolite levels and morphology in the mouse brain: a quantitative *in vivo* 1H and 31P magnetic resonance study. *J. Neurochem. 90*, 1321–1330.

303 Streijger, F., Oerlemans, F., Ellenbroek, B. A., Jost, C. R., Wieringa, B., Van der Zee, C. E. (2005) Structural and behavioural consequences of double deficiency for creatine kinases BCK and UbCKmit. *Behav. Brain Res. 157*, 219–234.

304 Schulze, A. (2003) Creatine deficiency syndromes. *Mol. Cell. Biochem. 244*, 143–150.

305 Braissant, O., Henry, H., Loup, M., Eilers, B., Bachmann, C. (2001) Endogenous synthesis and transport of creatine in the rat brain: an *in situ* hybridization study. *Brain Res. Mol. Brain Res. 86*, 193–201.

306 Braissant, O., Henry, H., Villard, A. M., Speer, O., Wallimann, T., Bachmann, C. (2005) Creatine synthesis and transport during rat embryogenesis: spatiotemporal expression of AGAT, GAMT and CT1. *BMC Dev. Biol. 5*, 9.

307 Ohtsuki, S., Tachikawa, M., Takanaga, H., Shimizu, H., Watanabe, M., Hosoya, K., Terasaki, T. (2002) The blood-brain barrier creatine transporter is a major pathway for supplying creatine to the brain. *J. Cereb. Blood Flow Metab. 22*, 1327–1335.

308 Nakashima, T., Tomi, M., Katayama, K., Tachikawa, M., Watanabe, M., Terasaki, T., Hosoya, K. (2004) Blood-to-retina transport of creatine via creatine transporter (CRT) at the rat inner blood-retinal barrier. *J. Neurochem. 89*, 1454–1461.

309 Peral, M. J., Garcia-Delgado, M., Calonge, M. L., Duran, J. M., De La Horra, M. C., Wallimann, T., Speer, O., Ilundain, A. (2002) Human, rat and chicken small intestinal Na^+Cl^--creatine transporter: functional, molecular characterization and localization. *J. Physiol. 545*, 133–144.

310 Tachikawa, M., Fukaya, M., Terasaki, T., Ohtsuki, S., Watanabe, M. (2004) Distinct cellular expressions of creatine synthetic enzyme GAMT and creatine kinases uCK-Mi and CK-B suggest a novel neuron-glial relationship for brain energy homeostasis. *Eur. J. Neurosci. 20*, 144–160.

311 Acosta, M. L., Kalloniatis, M., Christie, D. L. (2005) Creatine transporter localization in developing and adult retina: importance of creatine to retinal function. *Am. J. Physiol. 289*, C1015–1023.

312 Almeida, L. S., Salomons, G. S., Hogenboom, F., Jakobs, C., Schoffelmeer, A. N. (2006) Exocytotic release of creatine in rat brain. *Synapse 60*, 118–123.

313 Watanabe, A., Kato, N., Kato, T. (2002) Effects of creatine on mental fatigue and cerebral hemoglobin oxygenation. *Neurosci. Res. 42*, 279–285.

314 Rae, C., Digney, A. L., McEwan, S. R., Bates, T. C. (2003) Oral creatine monohydrate supplementation improves brain performance: a double-blind, placebo-controlled, cross-over trial. *Proc. Biol. Sci. 270*, 2147–2150.

315 McMorris, T., Harris, R. C., Howard, A. N., Langridge, G., Hall, B., Corbett, J., Dicks, M., Hodgson, C. (2007) Creatine supplementation, sleep deprivation, cortisol, melatonin and behavior. *Physiol. Behav. 90*, 21–28.

316 McMorris, T., Harris, R. C., Swain, J., Corbett, J., Collard, K., Dyson, R. J., Dye, L., Hodgson, C., Draper, N. (2006) Effect of creatine supplementation and sleep deprivation, with mild exercise, on cognitive and psychomotor performance, mood state, and plasma concentrations of catecholamines and cortisol. *Psychopharmacology 185*, 93–103.

317 Beal, M. F. (1996) Mitochondria, free radicals, and neurodegeneration. *Curr. Opin. Neurobiol. 6*, 661–666.

318 Browne, S. E., Beal, M. F. (1994) Oxidative damage and mitochondrial dysfunction in neurodegenerative diseases. *Biochem. Soc. Trans. 22*, 1002–1006.

319 Davis, R. E., Miller, S., Herrnstadt, C., Ghosh, S. S., Fahy, E., Shinobu, L. A., Galasko, D., Thal, L. J., Beal, M. F., Howell, N., Parker, W. D., Jr. (1997) Mutations in mitochondrial cytochrome *c* oxidase genes segregate with late-onset Alzheimer disease. *Proc. Natl. Acad. Sci. U.S.A. 94*, 4526–4531.

320 Smith, M. A., Perry, G., Richey, P. L., Sayre, L. M., Anderson, V. E., Beal, M. F., Kowall, N. (1996) Oxidative damage in Alzheimer's. *Nature 382*, 120–121.

321 Marttila, R. J., Rinne, U. K. (1987) Clues from epidemiology of Parkinson's disease. *Adv. Neurol. 45*, 285–288.

322 Lang, A. E., Lozano, A. M. (1998) Parkinson's disease. First of two parts. *N. Engl. J. Med. 339*, 1044–1053.

323 Lang, A. E., Lozano, A. M. (1998) Parkinson's disease. Second of two parts. *N. Engl. J. Med. 339*, 1130–1143.

324 Alam, M., Schmidt, W. J. (2002) Rotenone destroys dopaminergic neurons and induces parkinsonian symptoms in rats. *Behav. Brain Res. 136*, 317–324.

325 Sayre, L. M. (1989) Biochemical mechanism of action of the dopaminergic neurotoxin 1-methyl-4-phenyl-1,2,3,6-tetrahydropyridine (MPTP). *Toxicol. Lett. 48*, 121–149.

326 Schapira, A. H., Cooper, J. M., Dexter, D., Clark, J. B., Jenner, P., Marsden, C. D. (1990) Mitochondrial complex I deficiency in Parkinson's disease. *J. Neurochem. 54*, 823–827.

327 Bonuccelli, U., Del Dotto, P. (2006) New pharmacologic horizons in the treatment of Parkinson disease. *Neurology 67*, S30–38.

328 Andres, R. H., Huber, A. W., Schlattner, U., Perez-Bouza, A., Krebs, S. H., Seiler, R. W., Wallimann, T., Widmer, H. R. (2005) Effects of creatine treatment on the survival of dopaminergic neurons in cultured fetal ventral mesencephalic tissue. *Neuroscience 133*, 701–713.

329 Andres, R. H., Ducray, A. D., Perez-Bouza, A., Schlattner, U., Huber, A. W., Krebs, S. H., Seiler, R. W., Wallimann, T., Widmer, H. R. (2005) Creatine supplementation improves dopaminergic cell survival and protects against MPP+ toxicity in an organotypic tissue culture system. *Cell Transplant. 14*, 537–550.

330 Matthews, R. T., Ferrante, R. J., Klivenyi, P., Yang, L., Klein, A. M., Mueller, G., Kaddurah-Daouk, R., Beal, M. F. (1999) Creatine and cyclocreatine attenuate MPTP neurotoxicity. *Exp. Neurol. 157*, 142–149.

331 Bender, A., Koch, W., Elstner, M., Schombacher, Y., Bender, J., Moeschl, M., Gekeler, F., Muller-Myhsok, B., Gasser, T., Tatsch, K., Klopstock, T.

(2006) Creatine supplementation in Parkinson disease: a placebo-controlled randomized pilot trial. *Neurology 67*, 1262–1264.

332 NINDS-NET-PD-Investigators. (2006) A randomized, double-blind, futility clinical trial of creatine and minocycline in early Parkinson disease. *Neurology 66*, 664–671.

333 Quinn, N., Schrag, A. (1998) Huntington's disease and other choreas. *J. Neurol. 245*, 709–716.

334 Grunewald, T., Beal, M. F. (1999) Bioenergetics in Huntington's disease. *Ann. N. Y. Acad. Sci. 893*, 203–213.

335 Hayden, M. R., Martin, W. R., Stoessl, A. J., Clark, C., Hollenberg, S., Adam, M. J., Ammann, W., Harrop, R., Rogers, J., Ruth, T., et al. (1986) Positron emission tomography in the early diagnosis of Huntington's disease. *Neurology 36*, 888–894.

336 Calabresi, P., Gubellini, P., Picconi, B., Centonze, D., Pisani, A., Bonsi, P., Greengard, P., Hipskind, R. A., Borrelli, E., Bernardi, G. (2001) Inhibition of mitochondrial complex II induces a long-term potentiation of NMDA-mediated synaptic excitation in the striatum requiring endogenous dopamine. *J. Neurosci. 21*, 5110–5120.

337 Gu, M., Gash, M. T., Mann, V. M., Javoy-Agid, F., Cooper, J. M., Schapira, A. H. (1996) Mitochondrial defect in Huntington's disease caudate nucleus. *Ann. Neurol. 39*, 385–389.

338 Tabrizi, S. J., Schapira, A. H. (1999) Secondary abnormalities of mitochondrial DNA associated with neurodegeneration. *Biochem. Soc. Symp. 66*, 99–110.

339 Tabrizi, S. J., Workman, J., Hart, P. E., Mangiarini, L., Mahal, A., Bates, G., Cooper, J. M., Schapira, A. H. (2000) Mitochondrial dysfunction and free radical damage in the Huntington R6/2 transgenic mouse. *Ann. Neurol. 47*, 80–86.

340 Andres, R. H., Ducray, A. D., Huber, A. W., Perez-Bouza, A., Krebs, S. H., Schlattner, U., Seiler, R. W., Wallimann, T., Widmer, H. R. (2005) Effects of creatine treatment on survival and differentiation of GABA-ergic neurons in cultured striatal tissue. *J. Neurochem. 95*, 33–45.

341 Royes, L. F., Fighera, M. R., Furian, A. F., Oliveira, M. S., Myskiw Jde, C., Fiorenza, N. G., Petry, J. C., Coelho, R. C., Mello, C. F. (2006) Effectiveness of creatine monohydrate on seizures and oxidative damage induced by methylmalonate. *Pharmacol. Biochem. Behav. 83*, 136–144.

342 Andreassen, O. A., Dedeoglu, A., Ferrante, R. J., Jenkins, B. G., Ferrante, K. L., Thomas, M., Friedlich, A., Browne, S. E., Schilling, G., Borchelt, D. R., Hersch, S. M., Ross, C. A., Beal, M. F. (2001) Creatine increase survival and delays motor symptoms in a transgenic animal model of Huntington's disease. *Neurobiol. Dis. 8*, 479–491.

343 Ferrante, R. J., Andreassen, O. A., Jenkins, B. G., Dedeoglu, A., Kuemmerle, S., Kubilus, J. K., Kaddurah-Daouk, R., Hersch, S. M., Beal, M. F. (2000) Neuroprotective effects of creatine in a transgenic mouse model of Huntington's disease. *J. Neurosci. 20*, 4389–4397.

344 Shear, D. A., Haik, K. L., Dunbar, G. L. (2000) Creatine reduces 3-nitropropionic-acid-induced cognitive and motor abnormalities in rats. *Neuroreport 11*, 1833–1837.

345 Tabrizi, S. J., Blamire, A. M., Manners, D. N., Rajagopalan, B., Styles, P., Schapira, A. H., Warner, T. T. (2005) High-dose creatine therapy for Huntington disease: a 2-year clinical and MRS study. *Neurology 64*, 1655–1656.

346 Bender, A., Auer, D. P., Merl, T., Reilmann, R., Saemann, P., Yassouridis, A., Bender, J., Weindl, A., Dose, M., Gasser, T., Klopstock, T. (2005) Creatine supplementation lowers brain glutamate levels in Huntington's disease. *J. Neurol. 252*, 36–41.

347 Hersch, S. M., Gevorkian, S., Marder, K., Moskowitz, C., Feigin, A., Cox, M., Como, P., Zimmerman, C., Lin, M., Zhang, L., Ulug, A. M., Beal, M. F., Matson, W., Bogdanov, M., Ebbel, E., Zaleta, A., Kaneko, Y., Jenkins, B., Hevelone, N., Zhang, H., Yu, H., Schoenfeld, D., Ferrante, R., Rosas, H. D. (2006) Creatine in Huntington disease is safe, tolerable, bioavailable in brain and reduces serum 8OH2'dG. *Neurology 66*, 250–252.

348 Kaemmerer, W. F., Rodrigues, C. M., Steer, C. J., Low, W. C. (2001) Creatine-supplemented diet extends Purkinje cell survival in spinocerebellar ataxia type 1 transgenic mice but does not prevent the ataxic phenotype. *Neuroscience 103*, 713–724.

349 Blennow, K., de Leon, M. J., Zetterberg, H. (2006) Alzheimer's disease. *Lancet 368*, 387–403.

350 Aksenov, M., Aksenova, M., Butterfield, D. A., Markesbery, W. R. (2000) Oxidative modification of creatine kinase BB in Alzheimer's disease brain. *J. Neurochem. 74*, 2520–2527.

351 Gallant, M., Rak, M., Szeghalmi, A., Del Bigio, M. R., Westaway, D., Yang, J., Julian, R., Gough, K. M. (2006) Focally elevated creatine detected in amyloid precursor protein (APP) transgenic mice and Alzheimer disease brain tissue. *J. Biol. Chem. 281*, 5–8.

352 Li, X., Burklen, T., Yuan, X., Schlattner, U., Desiderio, D. M., Wallimann, T., Homayouni, R. (2006) Stabilization of ubiquitous mitochondrial creatine kinase preprotein by APP family proteins. *Mol. Cell. Neurosci. 31*, 263–272.

353 Strong, M., Rosenfeld, J. (2003) Amyotrophic lateral sclerosis: a review of current concepts. *Amyotroph. Lateral Scler. Other Motor Neuron Disord. 4*, 136–143.

354 Dupuis, L., Gonzalez de Aguilar, J. L., Oudart, H., de Tapia, M., Barbeito, L., Loeffler, J. P. (2004) Mitochondria in amyotrophic lateral sclerosis: a trigger and a target. *Neurodegener. Dis. 1*, 245–254.

355 Browne, S. E., Yang, L., DiMauro, J. P., Fuller, S. W., Licata, S. C., Beal, M. F. (2006) Bioenergetic abnormalities in discrete cerebral motor pathways presage spinal cord pathology in the G93A SOD1 mouse model of ALS. *Neurobiol. Dis. 22*, 599–610.

356 Dupuis, L., Oudart, H., Rene, F., Gonzalez de Aguilar, J. L., Loeffler, J. P. (2004) Evidence for defective energy

homeostasis in amyotrophic lateral sclerosis: benefit of a high-energy diet in a transgenic mouse model. *Proc. Natl. Acad. Sci. U.S.A. 101*, 11159–11164.

357 Klivenyi, P., Ferrante, R. J., Matthews, R. T., Bogdanov, M. B., Klein, A. M., Andreassen, O. A., Mueller, G., Wermer, M., Kaddurah-Daouk, R., Beal, M. F. (1999) Neuroprotective effects of creatine in a transgenic animal model of amyotrophic lateral sclerosis. *Nat. Med. 5*, 347–350.

358 Zhang, W., Narayanan, M., Friedlander, R. M. (2003) Additive neuroprotective effects of minocycline with creatine in a mouse model of ALS. *Ann. Neurol. 53*, 267–270.

359 Lawler, J. M., Barnes, W. S., Wu, G., Song, W., Demaree, S. (2002) Direct antioxidant properties of creatine. *Biochem. Biophys. Res. Commun. 290*, 47–52.

360 Pena-Altamira, E., Crochemore, C., Virgili, M., Contestabile, A. (2005) Neurochemical correlates of differential neuroprotection by long-term dietary creatine supplementation. *Brain Res. 1058*, 183–188.

361 Klivenyi, P., Kiaei, M., Gardian, G., Calingasan, N. Y., Beal, M. F. (2004) Additive neuroprotective effects of creatine and cyclooxygenase 2 inhibitors in a transgenic mouse model of amyotrophic lateral sclerosis. *J. Neurochem. 88*, 576–582.

362 Andres, R. H., Meiler, F., Huber, A. W., Perez-Bouza, A., Seiler, R. W., Wallimann, T., Widmer, H. R., Schlattner, U. (2002) Human fetal CNS tissue expresses kreatine kinases and the creatine transporter. *FENS 1*, 208.

363 Drory, V. E., Gross, D. (2002) No effect of creatine on respiratory distress in amyotrophic lateral sclerosis. *Amyotroph. Lateral Scler. Other Motor Neuron Disord. 3*, 43–46.

364 Shefner, J. M., Cudkowicz, M. E., Schoenfeld, D., Conrad, T., Taft, J., Chilton, M., Urbinelli, L., Qureshi, M., Zhang, H., Pestronk, A., Caress, J., Donofrio, P., Sorenson, E., Bradley, W., Lomen-Hoerth, C., Pioro, E., Rezania, K., Ross, M., Pascuzzi, R., Heiman-Patterson, T., Tandan, R., Mitsumoto, H., Rothstein, J., Smith-Palmer, T.,

MacDonald, D., Burke, D. (2004) A clinical trial of creatine in ALS. *Neurology 63*, 1656–1661.

365 Ellis, A. C., Rosenfeld, J. (2004) The role of creatine in the management of amyotrophic lateral sclerosis and other neurodegenerative disorders. *CNS Drugs 18*, 967–980.

366 Boerkoel, C. F., Takashima, H., Garcia, C. A., Olney, R. K., Johnson, J., Berry, K., Russo, P., Kennedy, S., Teebi, A. S., Scavina, M., Williams, L. L., Mancias, P., Butler, I. J., Krajewski, K., Shy, M., Lupski, J. R. (2002) Charcot-Marie-Tooth disease and related neuropathies: mutation distribution and genotype-phenotype correlation. *Ann. Neurol. 51*, 190–201.

367 Smith, C. A., Chetlin, R. D., Gutmann, L., Yeater, R. A., Alway, S. E. (2006) Effects of exercise and creatine on myosin heavy chain isoform composition in patients with Charcot-Marie-Tooth disease. *Muscle Nerve 34*, 586–594.

368 Lipton, P., Whittingham, T. S. (1982) Reduced ATP concentration as a basis for synaptic transmission failure during hypoxia in the *in vitro* guinea-pig hippocampus. *J. Physiol. 325*, 51–65.

369 Adcock, K. H., Nedelcu, J., Loenneker, T., Martin, E., Wallimann, T., Wagner, B. P. (2002) Neuroprotection of creatine supplementation in neonatal rats with transient cerebral hypoxia-ischemia. *Dev. Neurosci. 24*, 382–388.

370 Lensman, M., Korzhevskii, D. E., Mourovets, V. O., Kostkin, V. B., Izvarina, N., Perasso, L., Gandolfo, C., Otellin, V. A., Polenov, S. A., Balestrino, M. (2006) Intracerebroventricular administration of creatine protects against damage by global cerebral ischemia in rat. *Brain Res. 1114*, 187–194.

371 Prass, K., Royl, G., Lindauer, U., Freyer, D., Megow, D., Dirnagl, U., Stockler-Ipsiroglu, G., Wallimann, T., Priller, J. (2007) Improved reperfusion and neuroprotection by creatine in a mouse model of stroke. *J. Cereb. Blood Flow Metab. 27*, 432–445.

372 Scheff, S. W., Dhillon, H. S. (2004) Creatine-enhanced diet alters levels of lactate and free fatty acids after experimental brain injury. *Neurochem. Res. 29*, 469–479.

373 Sullivan, P. G., Geiger, J. D., Mattson, M. P., Scheff, S. W. (2000) Dietary supplement creatine protects against traumatic brain injury. *Ann. Neurol. 48*, 723–729.

374 Sakellaris, G., Kotsiou, M., Tamiolaki, M., Kalostos, G., Tsapaki, E., Spanaki, M., Spilioti, M., Charissis, G., Evangeliou, A. (2006) Prevention of complications related to traumatic brain injury in children and adolescents with creatine administration: an open label randomized pilot study. *J. Trauma 61*, 322–329.

375 Hausmann, O. N., Fouad, K., Wallimann, T., Schwab, M. E. (2002) Protective effects of oral creatine supplementation on spinal cord injury in rats. *Spinal Cord 40*, 449–456.

376 Rabchevsky, A. G., Sullivan, P. G., Fugaccia, I., Scheff, S. W. (2003) Creatine diet supplement for spinal cord injury: influences on functional recovery and tissue sparing in rats. *J. Neurotrauma 20*, 659–669.

377 Rebaudo, R., Melani, R., Carita, F., Rosi, L., Picchio, V., Ruggeri, P., Izvarina, N., Balestrino, M. (2000) Increase of cerebral phosphocreatine in normal rats after intracerebroventricular administration of creatine. *Neurochem. Res. 25*, 1493–1495.

378 Ozkan, O., Duman, O., Haspolat, S., Ozgentas, H. E., Dikici, M. B., Gurer, I., Gungor, H. A., Guzide Gokhan, A. (2005) Effect of systemic creatine monohydrate supplementation on denervated muscle during reinnervation: experimental study in the rat. *J. Reconstr. Microsurg. 21*, 573–579.

379 Barisic, N., Bernert, G., Ipsiroglu, O., Stromberger, C., Muller, T., Gruber, S., Prayer, D., Moser, E., Bittner, R. E., Stockler-Ipsiroglu, S. (2002) Effects of oral creatine supplementation in a patient with MELAS phenotype and associated nephropathy. *Neuropediatrics 33*, 157–161.

380 Komura, K., Hobbiebrunken, E., Wilichowski, E. K., Hanefeld, F. A. (2003) Effectiveness of creatine mono-hydrate in mitochondrial encephalomyo-pathies. *Pediatr. Neurol. 28*, 53–58.

381 Dahl, H. H. (1998) Getting to the nucleus of mitochondrial disorders: identification of respiratory chain-enzyme genes causing Leigh syndrome. *Am. J. Hum. Genet. 63*, 1594–1597.

382 Komura, K., Nakano, K., Ishigaki, K., Tarashima, M., Nakayama, T., Sasaki, K., Saito, K., Osawa, M. (2006) Creatine monohydrate therapy in a Leigh syndrome patient with A8344G mutation. *Pediatr. Int. 48*, 409–412.

383 Rodriguez, M. C., Macdonald, J. R., Mahoney, D. J., Parise, G., Beal, M. F., Tarnopolsky, M. A. (2006) Beneficial effects of creatine, CoQ(10), and lipoic acid in mitochondrial disorders. *Muscle Nerve 35*, 235–242.

384 Leuzzi, V., Bianchi, M. C., Tosetti, M., Carducci, C., Cerquiglini, C. A., Cioni, G., Antonozzi, I. (2000) Brain creatine depletion: guanidinoacetate methyl-transferase deficiency (improving with creatine supplementation). *Neurology 55*, 1407–1409.

385 Salomons, G. S., van Dooren, S. J., Verhoeven, N. M., Marsden, D., Schwartz, C., Cecil, K. M., DeGrauw, T. J., Jakobs, C. (2003) X-linked creatine transporter defect: an overview. *J. Inherit. Metab. Dis. 26*, 309–318.

386 van der Knaap, M. S., Verhoeven, N. M., Maaswinkel-Mooij, P., Pouwels, P. J., Onkenhout, W., Peeters, E. A., Stockler-Ipsiroglu, S., Jakobs, C. (2000) Mental retardation and behavioral problems as presenting signs of a creatine synthesis defect. *Ann. Neurol. 47*, 540–543.

387 Sykut-Cegielska, J., Gradowska, W., Mercimek-Mahmutoglu, S., Stockler-Ipsiroglu, S. (2004) Biochemical and clinical characteristics of creatine deficiency syndromes. *Acta Biochim. Pol. 51*, 875–882.

388 deGrauw, T. J., Salomons, G. S., Cecil, K. M., Chuck, G., Newmeyer, A., Schapiro, M. B., Jakobs, C. (2002) Congenital creatine transporter deficiency. *Neuropediatrics 33*, 232–238.

389 Zugno, A. I., Scherer, E. B., Schuck, P. F., Oliveira, D. L., Wofchuk, S., Wannmacher, C. M., Wajner, M., Wyse, A. T. (2006) Intrastriatal administration of guanidinoacetate inhibits Na^+/K^+-ATPase and creatine kinase activities in rat striatum. *Metab. Brain Dis. 21*, 41–50.

390 Sijens, P. E., Verbruggen, K. T., Meiners, L. C., Soorani-Lunsing, R. J., Rake, J. P., Oudkerk, M. (2005) 1H chemical shift imaging of the brain in guanidino methyltransferase deficiency, a creatine deficiency syndrome; guanidinoacetate accumulation in the gray matter. *Eur. Radiol. 15*, 1923–1926.

391 Newmeyer, A., Cecil, K. M., Schapiro, M., Clark, J. F., Degrauw, T. J. (2005) Incidence of brain creatine transporter deficiency in males with developmental delay referred for brain magnetic resonance imaging. *J. Dev. Behav. Pediatr. 26*, 276–282.

392 Mercimek-Mahmutoglu, S., Stoeckler-Ipsiroglu, S., Adami, A., Appleton, R., Araujo, H. C., Duran, M., Ensenauer, R., Fernandez-Alvarez, E., Garcia, P., Grolik, C., Item, C. B., Leuzzi, V., Marquardt, I., Muhl, A., Saelke-Kellermann, R. A., Salomons, G. S., Schulze, A., Surtees, R., van der Knaap, M. S., Vasconcelos, R., Verhoeven, N. M., Vilarinho, L., Wilichowski, E., Jakobs, C. (2006) GAMT deficiency: features, treatment, and outcome in an inborn error of creatine synthesis. *Neurology 67*, 480–484.

393 Stockler, S., Hanefeld, F., Frahm, J. (1996) Creatine replacement therapy in guanidinoacetate methyltransferase deficiency, a novel inborn error of metabolism. *Lancet 348*, 789–790.

394 Shekhar, A., Truitt, W., Rainnie, D., Sajdyk, T. (2005) Role of stress, corticotrophin releasing factor (CRF) and amygdala plasticity in chronic anxiety. *Stress 8*, 209–219.

395 Coplan, J. D., Mathew, S. J., Mao, X., Smith, E. L., Hof, P. R., Coplan, P. M., Rosenblum, L. A., Gorman, J. M., Shungu, D. C. (2006) Decreased choline and creatine concentrations in centrum semiovale in patients with generalized anxiety disorder: relationship to IQ and early trauma. *Psychiatry Res. 147*, 27–39.

396 Amital, D., Vishne, T., Roitman, S., Kotler, M., Levine, J. (2006) Open study of creatine monohydrate in treatment-resistant posttraumatic stress disorder. *J. Clin. Psychiatry 67*, 836–837.

397 Amital, D., Vishne, T., Rubinow, A., Levine, J. (2006) Observed effects of creatine monohydrate in a patient with depression and fibromyalgia. *Am. J. Psychiatry 163*, 1840–1841.

398 Arbuthnott, G., Dunnett, S., MacLeod, N. (1985) Electrophysiological properties of single units in dopamine-rich mesencephalic transplants in rat brain. *Neurosci. Lett. 57*, 205–210.

399 Nikkhah, G., Cunningham, M. G., Cenci, M. A., McKay, R. D., Bjorklund, A. (1995) Dopaminergic microtransplants into the substantia nigra of neonatal rats with bilateral 6-OHDA lesions. I. Evidence for anatomical reconstruction of the nigrostriatal pathway. *J. Neurosci. 15*, 3548–3561.

400 Brundin, P., Bjorklund, A. (1987) Survival, growth and function of dopaminergic neurons grafted to the brain. *Prog. Brain Res. 71*, 293–308.

401 Brundin, P., Nilsson, O. G., Strecker, R. E., Lindvall, O., Astedt, B., Bjorklund, A. (1986) Behavioural effects of human fetal dopamine neurons grafted in a rat model of Parkinson's disease. *Exp. Brain Res. 65*, 235–240.

402 Freed, C. R., Greene, P. E., Breeze, R. E., Tsai, W. Y., DuMouchel, W., Kao, R., Dillon, S., Winfield, H., Culver, S., Trojanowski, J. Q., Eidelberg, D., Fahn, S. (2001) Transplantation of embryonic dopamine neurons for severe Parkinson's disease. *N. Engl. J. Med. 344*, 710–719.

403 Borlongan, C. V., Koutouzis, T. K., Poulos, S. G., Saporta, S., Sanberg, P. R. (1998) Bilateral fetal striatal grafts in the 3-nitropropionic acid-induced hypoactive model of Huntington's disease. *Cell Transplant. 7*, 131–135.

404 Dunnett, S. B., Carter, R. J., Watts, C., Torres, E. M., Mahal, A., Mangiarini, L., Bates, G., Morton, A. J. (1998) Striatal transplantation in a transgenic mouse model of Huntington's disease. *Exp. Neurol. 154*, 31–40.

405 Bachoud-Levi, A. C., Gaura, V., Brugieres, P., Lefaucheur, J. P., Boisse, M. F., Maison, P., Baudic, S., Ribeiro, M. J., Bourdet, C., Remy, P., Cesaro, P., Hantraye, P., Peschanski, M. (2006) Effect of fetal neural transplants in patients with Huntington's disease 6 years after surgery: a long-term follow-up study. *Lancet Neurol. 5*, 303–309.

406 Bachoud-Levi, A. C., Remy, P., Nguyen, J. P., Brugieres, P., Lefaucheur, J. P., Bourdet, C., Baudic, S., Gaura, V., Maison, P., Haddad, B., Boisse, M. F., Grandmougin, T., Jeny, R., Bartolomeo, P., Dalla Barba, G., Degos, J. D., Lisovoski, F., Ergis, A. M., Pailhous, E., Cesaro, P., Hantraye, P., Peschanski, M. (2000) Motor and cognitive improvements in patients with Huntington's disease after neural transplantation. *Lancet 356*, 1975–1979.

407 Bliss, T. M., Kelly, S., Shah, A. K., Foo, W. C., Kohli, P., Stokes, C., Sun, G. H., Ma, M., Masel, J., Kleppner, S. R., Schallert, T., Palmer, T., Steinberg, G. K. (2006) Transplantation of hNT neurons into the ischemic cortex: cell survival and effect on sensorimotor behavior. *J. Neurosci. Res. 83*, 1004–1014.

408 Kelly, S., Bliss, T. M., Shah, A. K., Sun, G. H., Ma, M., Foo, W. C., Masel, J., Yenari, M. A., Weissman, I. L., Uchida, N., Palmer, T., Steinberg, G. K. (2004) Transplanted human fetal neural stem cells survive, migrate, and differentiate in ischemic rat cerebral cortex. *Proc. Natl. Acad. Sci. U.S.A. 101*, 11839–11844.

409 Erb, D. E., Mora, R. J., Bunge, R. P. (1993) Reinnervation of adult rat gastrocnemius muscle by embryonic motoneurons transplanted into the axotomized tibial nerve. *Exp. Neurol. 124*, 372–376.

410 Kordower, J. H., Freeman, T. B., Snow, B. J., Vingerhoets, F. J., Mufson, E. J., Sanberg, P. R., Hauser, R. A., Smith, D. A., Nauert, G. M., Perl, D. P., et al. (1995) Neuropathological evidence of graft survival and striatal reinnervation after the transplantation of fetal mesencephalic tissue in a patient with Parkinson's disease. *N. Engl. J. Med. 332*, 1118–1124.

411 Lindvall, O. (1998) Update on fetal transplantation: the Swedish experience. *Mov. Disord. 13 Suppl 1*, 83–87.

412 Barker, R. A., Dunnett, S. B., Faissner, A., Fawcett, J. W. (1996) The time course of loss of dopaminergic neurons and the gliotic reaction surrounding grafts of embryonic mesencephalon to the striatum. *Exp. Neurol. 141*, 79–93.

413 Zawada, W. M., Zastrow, D. J., Clarkson, E. D., Adams, F. S., Bell, K. P., Freed, C. R. (1998) Growth factors improve immediate survival of embryonic dopamine neurons after transplantation into rats. *Brain Res. 786*, 96–103.

414 Dickmeis, T., Rastegar, S., Aanstad, P., Clark, M., Fischer, N., Plessy, C., Rosa, F., Korzh, V., Strahle, U. (2001) Expression of brain subtype creatine kinase in the zebrafish embryo. *Mech. Dev. 109*, 409–412.

415 Chen, L., Roberts, R., Friedman, D. L. (1995) Expression of brain-type creatine kinase and ubiquitous mitochondrial creatine kinase in the fetal rat brain: evidence for a nuclear energy shuttle. *J. Comp. Neurol. 363*, 389–401.

416 Holtzman, D., Tsuji, M., Wallimann, T., Hemmer, W. (1993) Functional maturation of creatine kinase in rat brain. *Dev. Neurosci. 15*, 261–270.

417 Lyoo, I. K., Kong, S. W., Sung, S. M., Hirashima, F., Parow, A., Hennen, J., Cohen, B. M., Renshaw, P. F. (2003) Multinuclear magnetic resonance spectroscopy of high-energy phosphate metabolites in human brain following oral supplementation of creatine-monohydrate. *Psychiatry Res. 123*, 87–100.

418 Holtzman, D., Togliatti, A., Khait, I., Jensen, F. (1998) Creatine increases survival and suppresses seizures in the hypoxic immature rat. *Pediatr. Res. 44*, 410–414.

419 Brustovetsky, N., Brustovetsky, T., Dubinsky, J. M. (2001) On the mechanisms of neuroprotection by creatine and phosphocreatine. *J. Neurochem. 76*, 425–434.

420 Guerrero-Ontiveros, M. L., Wallimann, T. (1998) Creatine supplementation in health and disease. Effects of chronic creatine ingestion *in vivo*: down-regulation of the expression of creatine transporter isoforms in skeletal muscle. *Mol. Cell. Biochem. 184*, 427–437.

421 Sestili, P., Martinelli, C., Bravi, G., Piccoli, G., Curci, R., Battistelli, M., Falcieri, E., Agostini, D., Gioacchini, A. M., Stocchi, V. (2006) Creatine supplementation affords cytoprotection in oxidatively injured cultured mammalian cells via direct antioxidant activity. *Free Radic. Biol. Med. 40*, 837–849.

422 Lenz, H., Schmidt, M., Welge, V., Schlattner, U., Wallimann, T., Elsasser, H. P., Wittern, K. P., Wenck, H., Stab, F., Blatt, T. (2005) The creatine kinase system in human skin: protective effects of creatine against oxidative and UV damage *in vitro* and *in vivo*. *J. Invest. Dermatol. 124*, 443–452.

423 Berneburg, M., Gremmel, T., Kurten, V., Schroeder, P., Hertel, I., von Mikecz, A., Wild, S., Chen, M., Declercq, L., Matsui, M., Ruzicka, T., Krutmann, J. (2005) Creatine supplementation normalizes mutagenesis of mitochondrial DNA as well as functional consequences. *J. Invest. Dermatol. 125*, 213–220.

424 Koga, Y., Takahashi, H., Oikawa, D., Tachibana, T., Denbow, D. M., Furuse, M. (2005) Brain creatine functions to attenuate acute stress responses through GABAnergic system in chicks. *Neuroscience 132*, 65–71.

425 Alfieri, R. R., Bonelli, M. A., Cavazzoni, A., Brigotti, M., Fumarola, C., Sestili, P., Mozzoni, P., De Palma, G., Mutti, A., Carnicelli, D., Vacondio, F., Silva, C., Borghetti, A. F., Wheeler, K. P., Petronini, P. G. (2006) Creatine as a compatible osmolyte in muscle cells exposed to hypertonic stress. *J. Physiol. 576*, 391–401.

426 Bothwell, J. H., Rae, C., Dixon, R. M., Styles, P., Bhakoo, K. K. (2001) Hypo-osmotic swelling-activated release of organic osmolytes in brain slices: implications for brain oedema *in vivo*. *J. Neurochem. 77*, 1632–1640.

427 Bothwell, J. H., Styles, P., Bhakoo, K. K. (2002) Swelling-activated taurine and creatine effluxes from rat cortical astrocytes are pharmacologically distinct. *J. Membr. Biol. 185*, 157–164.

428 Murphy, R. M., Stephenson, D. G., Lamb, G. D. (2004) Effect of creatine on contractile force and sensitivity in mechanically skinned single fibers from rat skeletal muscle. *Am. J. Physiol. 287*, C1589–1595.

429 Graham, A. S., Hatton, R. C. (1999) Creatine: a review of efficacy and safety. *J. Am. Pharm. Assoc. (Wash.) 39*, 803–810.

430 Kreider, R. B., Melton, C., Rasmussen, C. J., Greenwood, M., Lancaster, S., Cantler, E. C., Milnor, P., Almada, A. L. (2003) Long-term creatine supplementation does not significantly affect clinical markers of health in athletes. *Mol. Cell. Biochem. 244*, 95–104.

431 Mertschenk, B., Gloxhuber, C., Wallimann, T. (2001) Health assessment of creatine as a dietary supplement: Gesundheitliche Bewertung von Kreatin als Nahrungsergaenzungsmittel. *Deutsche Lebensmittel-Rundschau 97*, 250–257.

432 Bohnhorst, B., Geuting, T., Peter, C. S., Dordelmann, M., Wilken, B., Poets, C. F. (2004) Randomized, controlled trial of oral creatine supplementation (not effective) for apnea of prematurity. *Pediatrics 113*, e303–307.

433 Burke, D. G., Chilibeck, P. D., Parise, G., Candow, D. G., Mahoney, D., Tarnopolsky, M. (2003) Effect of creatine and weight training on muscle creatine and performance in vegetarians. *Med. Sci. Sports Exerc. 35*, 1946–1955.

434 Barr, S. I., Rideout, C. A. (2004) Nutritional considerations for vegetarian athletes. *Nutrition 20*, 696–703.

435 Hespel, P., Op't Eijnde, B., Van Leemputte, M., Urso, B., Greenhaff, P. L., Labarque, V., Dymarkowski, S., Van Hecke, P., Richter, E. A. (2001) Oral creatine supplementation facilitates the rehabilitation of disuse atrophy and alters the expression of muscle myogenic factors in humans. *J. Physiol. 536*, 625–633.

436 Gerber, I., ap Gwynn, I., Alini, M., Wallimann, T. (2005) Stimulatory effects of creatine on metabolic activity, differentiation and mineralization of primary osteoblast-like cells in monolayer and micromass cell cultures. *Eur. Cell. Mater. 10*, 8–22.

437 Hornemann, T., Rutishauser, D., Wallimann, T. (2000) Why is creatine kinase a dimer? Evidence for cooperativity between the two subunits. *Biochim. Biophys. Acta 1480*, 365–373.

438 Wallimann, T. (1994) Bioenergetics. Dissecting the role of creatine kinase. *Curr. Biol. 4*, 42–46.

8
Integration of Adenylate Kinase and Glycolytic and Glycogenolytic Circuits in Cellular Energetics

Petras P. Dzeja, Susan Chung, and Andre Terzic

Abstract

Emerging evidence indicates that adenylate kinase and glycolytic/glycogenolytic phosphotransfer enzyme circuits are essential parts of cardiac system bioenergetics, playing a significant role in muscle energetics by delivering high-energy phosphoryls and conveying energy demand signals to ATP-generating pathways and metabolic sensors. Adenylate kinase phosphotransfer promptly responds to metabolic imbalances, facilitating transfer and utilization of both γ- and β-phosphoryls of the ATP molecule and maintaining energy economy. Adenylate kinase–mediated intracellular AMP signaling coupled with AMP-responsive elements such as AMP-sensitive protein kinase (AMPK), ATP-sensitive potassium channels (K_{ATP}), and AMP-sensitive metabolic enzymes, along with adenosine signaling, comprise a key metabolic sensing system regulating vital cellular processes. Localized in close proximity to metabolic sensors, adenylate kinase–catalyzed AMP signal generation and nucleotide exchange regulate the dynamics and frequency of ligand switching in the intimate sensing zone, facilitating the decoding of cellular information. By instigating AMP signaling, adenylate kinase regulates the activity of glycolytic and glycogenolytic enzymes and provides an integrative node for bioenergetic pathways to respond with high fidelity to increased energy demand. Genetic deficiency of the cytosolic AK1 isoform results in defective muscle energetics and AMP signaling, compromising the response to metabolic stress. Spatial extension of the glycolytic pathway indicates that it comprises a network of phosphotransfer circuits and metabolite shuttles, which facilitate high-energy phosphoryl delivery and lactate/pyruvate and P_i shuttling and thus maintain cellular energy and redox balance. The dynamics of glycogen utilization and synthesis, processes that occur close to myofibrillar and mitochondrial compartments, also suggest the existence of a glycogenolytic energetic circuit. In the proposed glycogen energetic network model, mitochondrial metabolic energy, invested into glycogen through locally generated UTP and G-6-P/G-1-P, is released during glycogenolysis at ATP utilization sites in myofibrils and other cellular compartments. Thus, systemic integration of different energetic and metabolic

Molecular System Bioenergetics: Energy for Life. Edited by Valdur Saks
Copyright © 2007 WILEY-VCH Verlag GmbH & Co. KGaA, Weinheim
ISBN: 978-3-527-31787-5

signaling pathways ensures cellular energy homeostasis and an adequate response to a broad range of functional activities and stress challenges.

8.1
Introduction

Phosphotransfer networks play a significant role in cellular life by distributing energy and providing metabolic signals required for the coordination of processes in distinct intracellular compartments [1–10]. In accord with creatine kinase, whose function and energetic significance are described in detail in Chapters 3, 7, and 11, the complementary phosphotransfer enzymes – adenylate kinase and those along the glycolytic/glycogenolytic pathway – play a major role in muscle energy economy and metabolic signaling, coupling mitochondrial energetics with cellular ATPases [5, 7, 11–26]. Accumulating evidence indicates that an intracellular phosphotransfer network, formed by the interaction and complementation between creatine kinase, adenylate kinase, and glycolytic phosphotransfer circuits, is critical for energy supply, spatiotemporal dynamics of cellular ATP, and the response of metabolic sensors to stress signals [5–13, 16–20, 22, 24, 25, 27–36]. The origins of the emerging cardiac system bioenergetics [7, 10, 36] can be traced to Lipmann's "adenylate wire" concept based on a remarkable analogy between the energy-carrying adenine nucleotide system and electrical circuits [37]. The addition of a spatial dimension to the concept of chemical group potential [37] led Peter Mitchell to the chemiosmotic coupling and "proton circuit" theory in mitochondrial energy transduction and to the vectorial metabolism and ligand conduction principle, providing a theoretical basis for a system-wide analysis of metabolic processes [38, 39]. Similarly, the spatial extension of creatine kinase and adenylate kinase bidirectional reactions, as well as the inclusion of the glycolytic pathway, provides the molecular foundation for the intracellular phosphotransfer "circuit" and network concepts [1–3, 5, 7, 9, 24, 40]. The topological organization and spatial directionality of biochemical processes is critical for distinct cellular processes, ranging from DNA replication to morphogenesis, and is one of the major topics in systems biology [41]. Although an initial reductionist approach is essential to characterize the individual components of the system, the molecular properties of separate components need to be integrated to provide a mechanistic understanding of a particular subsystem or overall system behavior. Indeed, the components must be coupled with a network for us to understand how system behavior emerges from their interactions and communication dynamics [7, 10, 36]. This provides the basis for system bioenergetics.

Recent studies provide new evidence of a unique energetic and metabolic signaling role played by the ubiquitous enzyme adenylate kinase, which catalyzes the reaction $2ADP \leftrightarrow ATP + AMP$ [5, 9, 14–19, 22, 25, 33, 42–45]. Adenylate kinase has a critical role in *de novo* adenine nucleotide synthesis and in cell energy economy by regulating nucleotide ratios in different intracellular compartments [5, 15, 18, 46–48]. The energetic role of adenylate kinase gained particular signif-

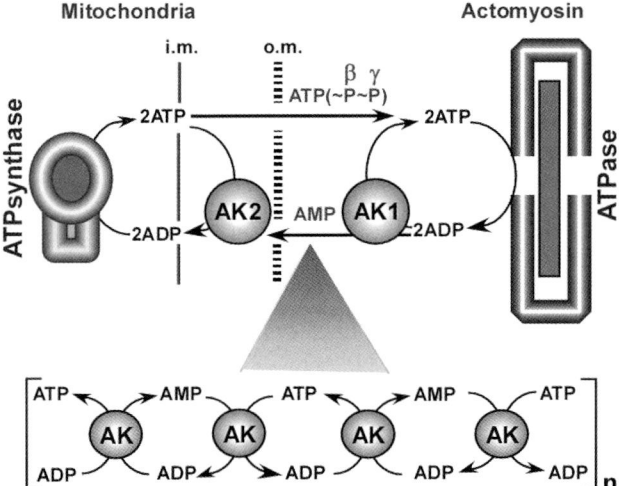

Fig. 8.1 Adenylate kinase shuttle facilitates transfer of ATP β- and γ-phosphoryls from generation to utilization sites. Adenylate kinase (AK), present in mitochondrial and myofibrillar compartments, enables the transfer and makes available the energy of two high-energy phosphoryls, the β- and γ-phosphoryls of a single ATP molecule. In this case, AMP signals feedback to mitochondrial respiration that is amplified by the generation of two molecules of ADP at the mitochondrial intermembrane site. Within the intracellular environment of a cardiomyocyte, the transfer of ATP and AMP between ATP production and ATP consumption sites may involve multiple, sequential, phosphotransfer relays that result in a flux wave propagation along clusters of adenylate kinase molecules (lower panel). Handling of substrates by a "bucket-brigade" or ligand conduction mechanism facilitates metabolic flux without apparent changes in metabolite concentrations.
AK1 and AK2: cytosolic and mitochondrial AK isoforms, respectively; i.m. and o.m.: inner and outer membranes, respectively.

icance after the discovery that this enzyme, through a chain of sequential reactions, facilitates the transfer and utilization of both β- and γ-phosphoryls of the ATP molecule, thereby doubling the energetic potential of ATP and cutting by half the cytosolic diffusional resistance for energy transmission [14]. Based on this distinct property, the concept of an adenylate kinase energy transfer shuttle or ligand conduction network was developed (Fig. 8.1) [5, 7, 14, 17, 25]. This concept is supported by direct biochemical, phosphoryl ^{18}O-labeling, physiological, and gene knockout studies [5, 7, 15–19, 22–25, 46–53] and is broadly used to explain energetic signaling mechanisms in heart or skeletal muscles, in hormone secretion [54, 55], in organ failure [17, 56], in tumor development [57–59], in energy support of the cell nucleus [9, 22, 60], and in sperm and cell motility [61–63]. Moreover, due to a unique property of the catalyzed reaction, adenylate kinase is now recognized as a sensitive reporter of the cellular energy state, translating small changes in the balance between ATP and ADP into relatively large changes in AMP concentration, so that enzymes and metabolic sensors that are affected by AMP can respond with high sensitivity and fidelity to stress signals

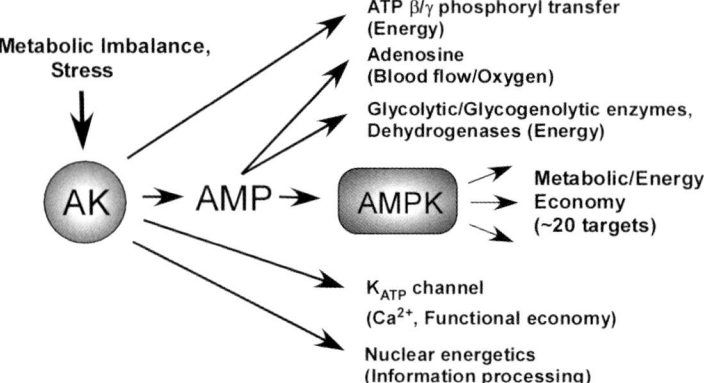

Fig. 8.2 Adenylate kinase–mediated AMP metabolic signaling network. Present throughout different cellular compartments, adenylate kinase (AK) directly senses metabolic imbalance or stress-induced alterations in the ATP:ADP ratio. In response to stress, AK facilitates transfer and utilization of an emergency energy reserve – ATP β-phosphoryls – and, by generating AMP, amplifies ATP/ADP changes into a much larger increase in the AMP:ATP ratio, which triggers and governs metabolic signaling cascades. The result of AK-mediated AMP signaling is activation of adenosine/AMPK/ K_{ATP} signaling that regulates blood flow and oxygen/nutrient supply to the heart. AK-generated AMP by activating metabolic enzymes and AMPK increases ATP production. AK phosphotransfer–governed nucleotide exchange regulates K_{ATP} channels and therefore Ca^{2+} entry and heart functional economy. Communication of energetic signals to the nuclear compartment could facilitate gene transcription and nuclear-cytoplasmic information exchange. AMPK: AMP-activated protein kinase; K_{ATP}: ATP-sensitive potassium channels.

(Fig. 8.2) [5, 7, 19, 23, 28, 43, 45, 53, 64]. Thus, the integrated energetic and metabolic signaling roles place adenylate kinase as a hub within the metabolic regulatory system coordinating components of the cellular bioenergetics network.

Studies of adenylate kinase and creatine kinase gene knockout models have opened new perspectives for the further understanding of how cellular energetic and metabolic signaling networks integrate with genetic, biosynthetic, membrane-electrical, and receptor-mediated signal transduction cellular events [5, 7, 18–20, 23–25, 34, 47, 51–53, 65–69]. Recent evidence indicates that the interaction between adenylate kinase and creatine kinase phosphorelays determines the outcome of metabolic signal communication to the membrane metabolic sensor, the K_{ATP} channel [5, 19, 28, 33, 34, 48, 54], and mediates energetic remodeling in preconditioned [30, 70] and failing hearts [17, 29, 56, 71–73]. Gene knockout studies also have revealed a remarkable plasticity of the cellular phosphotransfer system, where deficiency in an individual enzyme is compensated through the remodeling of the whole energetic network at enzymatic, architectural, and genomic levels [5, 7, 8, 18, 20, 23, 24, 47, 51, 65]. In this regard, one of the major findings resulting from adenylate kinase and creatine kinase knockout studies was the discovery that glycolytic enzymes also have the ability to provide network capacity for transferring and distributing high-energy phosphoryls [5, 7, 20, 24]. This new "spatial" look at glycolytic metabolism, together with the robust phosphotransfer capacity, especially by the GAPDH–PGK couple [74], renders glycolysis an equal partner in the cellular energy distribution network along with the more traditional creatine kinase and adenylate kinase phosphotransfer pathways [7, 24]. The function of the newly discovered adaptor protein DRAL/FHL-2 – which is involved in anchoring adenylate kinase, creatine kinase, and glycolytic enzymes to sites of high energy consumption [75] – further highlights the significance of the topological arrangement of the intracellular phosphotransfer networks in matching cellular energetic needs.

Adenylate kinase–mediated AMP signaling regulates a number of glycolytic and glycogenolytic enzymes conveying information regarding increased energy demand [5, 7, 11, 42, 64]. Regulation of glycogen metabolism is of primary importance in muscle energetics, including that of the heart [6, 11, 21, 26]. Particularly, defective AMP and AMP-activated protein kinase (AMPK) signaling is a primary cause of imbalance in glycogen metabolism and associated disease conditions [11, 76]. The dynamics of glycogen turnover have been suggested to be a primary source of ultrafast local ATP regeneration during the initiation of muscle contraction, and accordingly the "glycogen shunt" hypothesis has been proposed [21]. Here we suggest that the dynamics and topological arrangement of glycogen synthesis and utilization processes could provide an energy transfer network capacity linking mitochondrial energy metabolism with ATP utilization. Thus, glycogenolysis could be an integral part of the complete set of high-energy phosphoryl transfer pathways, along with creatine kinase, adenylate kinase, and glycolysis, providing coordination between cellular sites of ATP consumption and ATP generation.

8.2
The Adenylate Kinase Phosphotransfer System in Cell Energetics and AMP Metabolic Signaling

8.2.1
The Biological Role of Adenylate Kinase

For both its structure and its function, adenylate kinase is one of the most-studied enzymes, serving as a structural model for nucleotide-binding folds, the P-loop, and a lid domain [77–79]. So far, up to seven distinct adenylate kinase isoforms with different intracellular localization have been identified [42, 46, 48, 57, 59–61, 77]. AK1 is the major adenylate kinase isoform, localized in the cytosol and near myofibrils where it serves as a metabolic hub connecting minor adenylate kinase isoforms, namely, AK3 in the mitochondrial matrix, AK2 in the intermembrane space, AK6 in the nucleus, and ecto-AK in the extracellular space [42, 46, 59, 60, 80–82]. The newly discovered AK1 splice variant, AK1β, is targeted to the plasma membrane and has been linked to cell cycle regulation, adenosine signaling, and

K_{ATP} channel gating [48, 57]. Such strategic distribution of adenylate kinase isoforms creates a continuous phosphotransfer network, mediating energy transfer and metabolic signaling between cellular compartments and along the interstitial space [5, 7, 19, 81, 82]. Disruption in the network function, by deletion of the *AK1* gene, lowers the muscle energetic efficiency, compromising relaxation kinetics and increasing vulnerability to stress [18, 23, 47, 51]. In fact, the AK1-deficient phenotype is characterized by a reduced ability to maintain adenine nucleotide pools and by inefficient metabolic signal communication to the nucleus and cytosolic metabolic sensors, including K_{ATP} channels and AMPK [19, 22, 23, 53].

Adenylate kinase's unique energetic function allows the utilization of a second high-energy bond in the ATP molecule, that of the β-phosphoryl group, thereby doubling the energetic potential, a property not shared by other phosphotransfer systems [5, 7, 14, 83]. This function of adenylate kinase is particularly important in tissues with high and fluctuating energy demands and in those under metabolic stress [7, 9, 25]. Moreover, concerted actions of cytosolic and mitochondrial adenylate kinase isoforms facilitate high-energy phosphoryl delivery to cellular ATPases and feedback signal communication to mitochondrial respiration, comprising the "adenylate kinase shuttle," also known as a "ligand conduction phosphorelay" (Fig. 8.1) [5, 14, 17]. These functions render adenylate kinase essential to the integrated cellular phosphotransfer network, sustaining an efficient and vibrant cell energetic economy.

More recently, an important role for adenylate kinase in metabolic signaling has emerged (Fig. 8.2). Because of the nature of the catalyzed reaction, adenylate kinase translates small changes in the balance between ATP and ADP into relatively large changes in the concentration of AMP, enabling enzymes and metabolic sensors affected by AMP to respond with high sensitivity and fidelity to stress signals [5, 19, 28, 33, 43–45]. Adenylate kinase is the major source of cellular AMP and the main AMP signal delivery system [19, 23, 25, 28]. Previously, adenylate kinase–mediated AMP signaling was believed to be restricted to AMP-sensitive enzymes such as phosphofructokinase (PFK-1), glycogen phosphorylase, fructose 1,6-bisphosphate phosphatase, and dehydrogenases [11, 42, 84]. In recent years, adenylate kinase–mediated AMP signaling has gained much broader significance with the discovery of AMP-activated protein kinase (AMPK) – a master metabolic sensor phosphorylating key target proteins that control flux through metabolic pathways [43–45, 53, 71]. Moreover, the discovery of adenylate kinase–mediated signaling to other metabolic sensors, the K_{ATP} channel and the nuclear compartment, along with AK1 knockout studies have further emphasized the critical role of adenylate kinase in processing metabolic signals and maintaining cellular energetic and ionic homeostasis [7, 18, 19, 23, 47, 51, 53]. In particular, adenylate kinase phosphotransfer directly couples with K_{ATP} channels, facilitating the decoding of metabolic signals critical in adjusting cellular excitability-dependent functions in response to demand [19, 28, 33]. Adenylate kinase closely interacts with the parallel creatine kinase phosphotransfer system, conveying positive and negative signals to cellular stress response elements [5, 16, 17, 24]. An imbalance between the activities of phosphotransfer enzymes

compromises signal reception by metabolic sensors and disrupts the maintenance of bioenergetic homeostasis [19, 28, 34, 56, 85, 86].

8.2.2
Adenylate Kinase Isoform-based Metabolic Network

The existence of multiple isoforms of an enzyme usually relates to its different
intracellular distribution with specific targeting to organelles and distinct cellular
compartments and also to its different kinetic properties according to local metabolic requirements [42, 46, 48, 57, 61]. Following the discovery of adenylate kinase more than six decades ago by Kalcar (see [42]), three major isoforms, AK1,
AK2, and AK3, respectively localized in the cytosol, mitochondrial intermembrane space, and matrix, were identified (Fig. 8.3) [42, 46, 48]. They differ in molecular weight, structure, kinetic properties, and nucleotide specificity [77–80].
AK1 and AK2 specifically bind AMP and favor binding to ATP over other nucleotide triphosphates, while AK3 is a GTP:AMP phosphotransferase specific for the
phosphorylation of intramitochondrial AMP that can use only GTP or ITP as a
substrate [42, 87]. Within the AK family there are several conserved regions, including a P-loop, AMP- and MgATP/MgADP-binding domains, and a lid domain
[77–79]. Adenylate kinase isoforms can form dimers and higher-molecular-order
structures [75, 88]. Phosphorylation of adenylate kinase has not been detected,
but acetylation and myristoylation, which could facilitate binding of adenylate
kinase to cell membranes, mitochondria, or the nucleus, have been demonstrated
[48, 57, 89].

   Different polymorphic subforms of AK1 (AK1-1 and AK1-2) and splice variants
of AK2 (AK2A-D), which have distinct electrophoretic mobility and other properties, have been found [90, 91]. AK1-2 occurs only in Caucasian populations and is
common among hemophilia-A patients [90]. Also, the tissue-specific adenylate
kinase isoforms AK4 and AK5, which have mitochondrial matrix and cytosolic localization, respectively, have been cloned [59, 92]. More recently, the existence of
an additional *AK1* gene product, p53-inducible membrane-bound myristoylated
AK1β, has been reported and implicated in p53-dependent cell cycle arrest and
nucleotide exchange in the submembrane space [48, 57]. In this context, the
gene encoding AK1 is downregulated during tumor development, which could
be associated with lower AK1β levels and cell cycle disturbances [93]. Another important step was the identification of the AK6 isoform, localized to the cell nucleus where energy provision and nucleotide channeling into DNA synthesis
play a critical role in processing genetic information [22, 60]. However, there is
still controversy regarding the AK6 isoform [94], suggesting that other adenylate
kinase isoforms can also serve nuclear energetic needs [22, 42, 48]. The high-
molecular-weight isoforms AK6′ and AK7 (Fig. 8.3) may be associated with cell
motility and other processes [95]. Recently, a high level of expression of the AK7
isoform was demonstrated in bronchial epithelium, and it appears to be associated with ciliary function [96]. Thus, multiple adenylate kinase isoforms create a

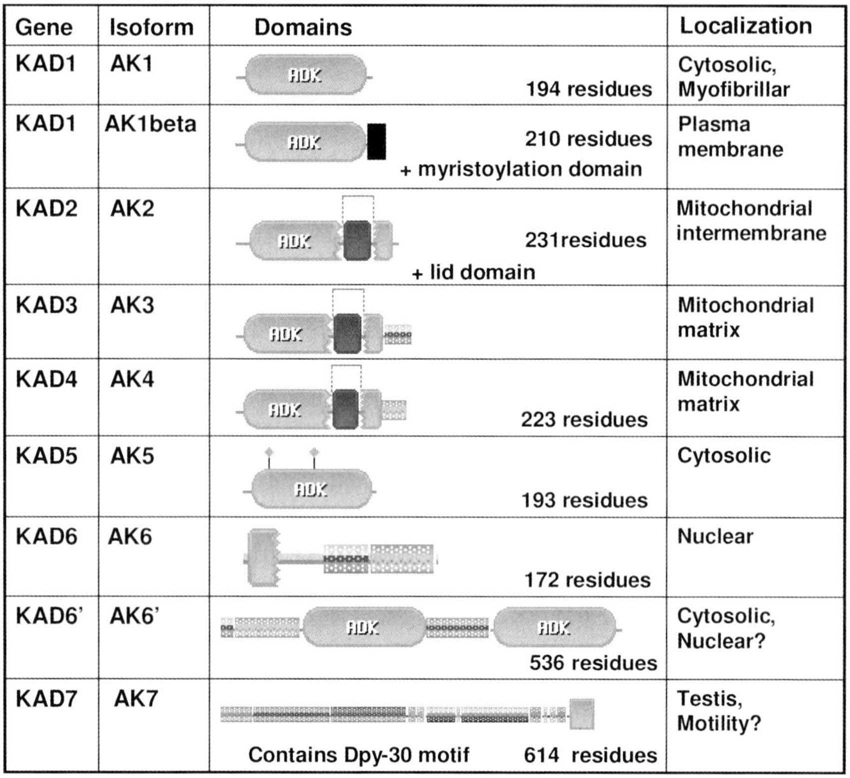

Gene	Isoform	Domains	Localization
KAD1	AK1	194 residues	Cytosolic, Myofibrillar
KAD1	AK1beta	210 residues + myristoylation domain	Plasma membrane
KAD2	AK2	231 residues + lid domain	Mitochondrial intermembrane
KAD3	AK3		Mitochondrial matrix
KAD4	AK4	223 residues	Mitochondrial matrix
KAD5	AK5	193 residues	Cytosolic
KAD6	AK6	172 residues	Nuclear
KAD6'	AK6'	536 residues	Cytosolic, Nuclear?
KAD7	AK7	Contains Dpy-30 motif 614 residues	Testis, Motility?

Fig. 8.3 Adenylate kinase isoforms, domain structure, and intracellular localization. Adenylate kinase isoforms are coded by separate genes, KAD1–KAD7, localized to different chromosomes. The corresponding proteins AK1–AK7 define separate adenylate kinase isoforms with different molecular weights, kinetic properties, and intracellular localization. The AK1 isoform mostly consists of the ADK domain, which is characteristic of the whole protein family. The AK1β splice variant has an additional myristoylation domain that targets the protein to the plasma membrane. The AK2, AK3, and AK4 isoforms have a flexible lid domain that closes over the site of phosphoryl transfer upon ATP binding. A short form of the lid domain also exists in AK1, AK5, and AK6. There is still no consensus agreement regarding AK6 (AK6') isoforms (see text). AK6' isoforms may represent a "polymeric" adenylate kinase with several ADK domains. The AK7 isoform, associated with cell motility, contains a Dpy-30 motif involved in protein dimerization (for more about adenylate kinase isoforms visit http://www.sanger.ac.uk/Software/Pfam/).

phosphotransfer network to serve the needs of different cellular compartments for energetic and metabolic signaling.

Heart muscle harbors about 30–40% of adenylate kinase activity in mitochondria, particularly in the intermembrane space [14, 80]. The mitochondrial AK2 isoform has the highest affinity (lowest K_m) for AMP (≤ 10 µM) among the AMP-metabolizing enzymes and is highly concentrated in the narrow intermembrane space [14, 80, 97]. Virtually all the AMP reaching the mitochondria is converted to ADP and channeled into oxidative phosphorylation, thereby maintaining a low cytosolic AMP concentration [23, 25, 98, 99]. In such a way, adenylate kinase tunes cytosolic AMP signals and guards the cellular adenine nucleotide pool [5, 25, 27, 100]. During intense physical activity or metabolic stress, such as ischemia, the AMP concentration rises, turning on other AMP-metabolizing enzymes such as AMP deaminase and 5′-nucleotidase producing IMP and adenosine [64, 101]. In this regard, a marked elevation of mitochondrial AK2 activity has been demonstrated in hypertrophy in response to increased energy demand and the necessity of maintaining the cellular adenine nucleotide pool [86].

Expression of adenylate kinase isoforms increases in response to muscle exercise, hypoxia, and metabolic stress [102, 103]. Also, muscle exercise performance correlates with adenylate kinase activity, suggesting that this enzyme is an integral part of cellular energetic homeostasis [103]. In humans, deficiency of the AK1 isoform is associated with mental retardation, psychomotor impairment, and congenital anemia [104]. A significant increase in AK1 and other phosphotransfer enzymes in obese/overweight and morbidly obese women could indicate an imbalance in metabolic signaling in the metabolic syndrome [105]. In addition, decreased AK1 activity has been found in a mouse model for muscular dystrophy (*mdx* mice), suggesting a direct relationship between the lack of dystrophin and alteration of AK1 [106].

Although major compartments of adenylate kinase isoform localization are known based on cellular fractionation, intimate intracellular adenylate kinase localization has been studied by immunocytochemistry only in neuronal cells and skeletal muscle [48, 107, 108]. In skeletal muscle myofibrils, adenylate kinase is localized in linear arrays along with creatine kinase and glycolytic enzymes [107]. Similar localization of GFP-tagged AK1 was detected in neonatal cardiomyocytes [75]. Accumulating evidence indicates that sequential arrangement of adenylate kinase molecules provides a bidirectional phosphorelay that links ATP generation with ATP-consuming and ATP-sensing processes (see Figs. 8.1 and 8.2) [5, 7, 9, 14, 17]. Regarding its interacting partners, adenylate kinase was found to co-purify and presumably interact with glycolytic enzymes and associate with myofibrils and cellular and mitochondrial membranes [14, 19, 48, 54, 57, 80, 89, 98, 109]. Adenylate kinase was also found to be engaged in intimate functional-structural interactions with the sarcolemmal K_{ATP} channel, a major metabolic sensor [19, 54]. More recently, a phosphotransfer enzyme anchoring the protein FHL2 was discovered in heart muscle [75]. This protein positions adenylate kinase and other phosphotransfer enzymes close to ATP utilization sites in myofibrils, ensuring that both high-energy phosphate bonds of ATP are efficiently utilized. Mutation of the FHL2 protein is associated with cardiomyopathy [75]. Another adenylate kinase anchoring protein, Oda5p, anchors adenylate kinase in the proximity of the dynein arm, ensuring that both high-energy phosphate bonds of ATP are efficiently utilized at the major site of power production of the microtubule motors involved in diverse cellular movements [62, 110]. A direct interaction between adenylate kinase and several enzymes of the dNTP synthase complex, as well as nucleoside diphosphate kinase, was demonstrated using protein-affinity chromatography and immunoprecipitation [111]. These results identify adenylate kinase as a specific component of the nuclear dNTP synthase complex, where it facilitates synthesis and surveys nucleotide ratios necessary for error-free DNA replication. Although topological positioning of specific proteins is crucial in energetics and signaling processes, the full interactome of adenylate kinase isoforms in cardiac muscle and other tissues is still unknown.

8.2.3
Adenylate Kinase Catalyzed β-phosphoryl Transfer and Energy Economy

Although discovered many years ago, the energetic role of adenylate kinase has remained elusive [5, 42, 77, 80]. The energetic significance of this ubiquitous enzyme has begun to be revealed by the demonstration that adenylate kinase promotes high-energy phosphoryl transfer from mitochondria to ATP consumption sites (Fig. 8.1) and that adenylate kinase phosphotransfer increases in failing hearts in order to provide additional energy resources [14, 17]. Furthermore, adenylate kinase–catalyzed phosphorelay facilitates intracellular energetic communication from mitochondria to myofibrils, the nucleus, and the K_{ATP} channel [5, 7, 17, 19, 22]. Loss of adenylate kinase function is complemented by activation of creatine kinase phosphotransfer [22]. Specific physiological roles of distinct adenylate kinase isoforms could be inferred from their involvement in the regulation of multiple cellular processes [5, 7, 9, 14, 15, 25]. In this regard, activities of adenylate kinase isoforms and intracellular free AMP levels correlate with tissue respiration rates [112, 113]. Expression of adenylate kinase isoforms increases in response to muscle exercise, hypoxia, and metabolic stress [102, 103]. Also, muscle exercise performance correlates with adenylate kinase activity, suggesting that this enzyme is an integral part of cellular energetic homeostasis [103]. Although the significance of energetic and metabolic signaling is growing, little is known regarding the regulation of adenylate kinase phosphotransfer under different metabolic, hormonal, and functional states or about how such regulation affects the ability of the myocardium to generate and respond to stress-related signals.

So far only single AK1 knockout mice have been generated by Wieringa and colleagues [18], where total heart adenylate kinase activity is reduced by ~90% [23, 47]. The remaining AK2 isoform in the mitochondrial intermembrane space partially compensates, so that total adenylate kinase flux is reduced by 60% [23]. In general, AK1-knockout muscles display lower energetic efficiency, slower relaxation kinetics, and a faster drop in contractility upon ischemia [18, 23, 47, 52]. AK1-deficient muscles have an increased vulnerability to metabolic stress associated with reduced high-energy phosphoryl pools upon hypoxia or ischemia–reperfusion, compromised AMP generation, and impaired metabolic signal communication to the membrane metabolic sensor K_{ATP} channel [19, 23, 47]. Absence of AK1 in skeletal muscle also compromises the signaling response of another metabolic sensor, AMPK [53]. Unexpectedly and contrary to common beliefs that adenylate kinase promotes nucleotide degradation, AK1-deficient hearts had less ability to maintain nucleotide pools under metabolic stress [23]. Also, in failing hearts adenylate kinase activity tightly correlates with higher cellular adenine nucleotide content [73], indicating a new function for adenylate kinase to guard the cellular nucleotide pool by rephosphorylating AMP back to ADP and ATP. AK1 deficiency is associated with a wide range of compensatory changes in glycolytic, glycogenolytic, and mitochondrial metabolism and corresponding gene expression to support energy metabolism [23, 47, 114]. Simultaneous double deletion of cytosolic AK1 and M-CK compromises intracellular energetic communication and reduces skeletal muscle ATP turnover under muscle functional load [51]. The remaining glycolytic and glycogenolytic pathways apparently compensate in these double-knockout animals [24]. Thus, adenylate kinase is important for maintaining muscle energy economy, for transducing metabolic signals to K_{ATP} channels and AMPK, for supporting myocardial function after initiation of ischemia, and for safeguarding the cellular nucleotide pool. However, the full significance of AK1 deficiency in functional performance under cardiac functional load, with reduced compensatory energetic pathways, has not yet been established.

In different tissues, adenylate kinase activity depends on the nutritional and hormonal state [115, 116]. In skeletal and cardiac muscle, adenylate kinase phosphotransfer flux increases in direct proportion to the intensity of contractions and to the level of oxygenation [5, 15, 30, 117]. In intact tissues, the highest adenylate kinase phosphotransfer flux is in the kidney, which approximates 98% of ATP turnover, followed by the liver (80%), the heart (15–22%), and contracting (10–17%) and resting (3–5%) skeletal muscles [25]. Adenylate kinase phosphotransfer flux is markedly suppressed by high glucose in insulin-secreting cells, reducing adenylate kinase–mediated AMP signaling to the K_{ATP} channel and likely to AMPK, two regulators of hormone secretion [55, 118]. There is a reciprocal relationship between creatine kinase and adenylate kinase phosphotransfer: reduction of creatine kinase activity promotes high-energy phosphoryl transfer through the adenylate kinase system in creatine kinase–knockout muscles or under hypoxic stress [16, 18, 24, 64].

A mechanistic basis for thermodynamically efficient coupling of cell energetics with cellular functions lies in the unique catalytic property of adenylate kinase: the transfer of both β- and γ-phosphoryls of ATP doubles the energetic potential of ATP as an energy-carrying molecule (Fig. 8.1) [5, 14]. Such coupling is achieved through an intimate relationship of adenylate kinase with mitochondrial respiration and myofibrillar and membrane ATPases, as well as with metabolic sensors such as ATP-sensitive K_{ATP} channels, and secures efficient regulation of energy metabolism, membrane excitability, and cell contraction [5, 7, 19, 22, 25]. Genetic manipulations of adenylate kinases in different cell types impede ATP export from mitochondria; impair nucleotide homeostasis, coordination of cell metabolism, and growth; reduce cellular energetic economy; and increase cell vulnerability to injury [18, 23, 99, 119–121]. Thus, the absence of adenylate kinase dampens intracellular metabolic signaling, disrupting integration and synchronization of ATP-consuming and ATP-producing processes in response to metabolic stress.

In summary, adenylate kinase–facilitated high-energy phosphoryl transfer and coordination between cellular sites of ATP consumption and ATP generation are essential for safeguarding cellular energetic economy. Although significant progress has been made, there are still many unanswered questions concerning adenylate kinase physiology, especially regarding the biological roles of mitochondrial intermembrane AK2, matrix AK3, and other tissue-specific adenylate kinase isoforms.

8.2.4
Adenylate Kinase–instigated AMP Metabolic Signaling and Energy Sensing

AMP signaling is emerging as the most versatile system in the regulation of cellular processes [43–46, 64, 122–124]. The major cellular AMP signal generator is adenylate kinase, which regulates nucleotide ratios in different intracellular compartments and responds rapidly to metabolic imbalances [5, 28, 42, 44, 59, 64]. Because of a unique property of the AK-catalyzed reaction, a small decrease in ATP levels results in a large increase in AMP, making the latter a sensitive indicator, and therefore a suitable signaling molecule, of cellular energetic status [5, 25, 45]. Adenylate kinase–instigated AMP signaling mediated by AMP-responsive components such as AMP-activated protein kinase (AMPK), ATP-sensitive potassium channels (K_{ATP}), and AMP-sensitive metabolic enzymes is a key metabolic sensing system and regulator of vital cellular processes throughout the body and particularly in the heart (Fig. 8.2) [5, 19, 45]. In the intracellular environment, where diffusion is reduced, reactions tend to depend strongly on local rather than global concentrations of metabolites [33]. In this way, adenylate kinase–catalyzed AMP signal generation and nucleotide exchange in the intimate "sensing zone" of metabolic sensors can regulate the dynamics and frequency of ligand switching in order to facilitate decoding of cellular information [9, 10]. Indeed, intracellular measurements using the ^{18}O-assisted ^{31}P-NMR technique indicate that adenylate kinase phosphotransfer displays only a fraction of total capacity and is compartmentalized and not universally at equilibrium, differentially promoting both AMP signal generation and AMP rephosphorylation, thus tuning the magnitude of the AMP signal [5, 15–17, 23–25, 30, 50, 64, 117]. In this way, adenylate kinase through negative and positive feedback governs adenine nucleotide and glycolytic oscillations, providing a dynamic "adenylate-excitable media" component for facilitated intracellular energetic signal communication [5, 9, 122]. Sensing body energy level and corresponding mental and physical strength is important for humans and animals. Such a process could be mediated in part by adenylate kinase, which, through intracellular and extracellular nucleotide-based signaling that affects metabolic sensors, receptors, synaptic transmission, and brain information processing, conveys information about the adenine nucleotide pool status and thus the overall energy balance [9].

In particular, adenylate kinase phosphotransfer directly couples with K_{ATP} channels, facilitating the translation of metabolic signals critical for adjusting cellular excitability-dependent functions in response to demand [5, 19, 28, 33].

An adenylate kinase–induced increase in AMP promotes the generation of adenosine, a powerful metabolic signaling and cardioprotective agent [64, 101]. AMP also triggers AMP-activated protein kinase (AMPK), an essential signaling module in cellular adaptation to stress [43–45, 53, 123]. In the heart, the significance of AMPK is suggested by the findings that AMPK activity is increased in ischemia and preconditioning and that AMPK agonists protect the myocardium against ischemic injury [44, 123, 124]. In this regard, a strong correlation was observed between fasting and higher expression of the adenylate kinase AK3 isoform and UCP3 and increased activity of AMPK and fatty acid oxidation [125], suggesting an interrelated signaling cascade. Moreover, through a series of spatially linked enzymatic reactions, adenylate kinase facilitates propagation of nucleotide signals in the intracellular and extracellular space, thus coordinating the response of metabolic sensors and nucleotide/nucleoside receptor signaling [5, 9, 19, 25, 81, 82]. In this way adenylate kinase appears to provide sustained communication between cytosolic and nuclear processes that coordinate energetic and genetic events [22]. In recent years adenylate kinase–mediated extracellular AMP and ADP signaling has gained significance because of its involvement in HDL endocytosis, adenosine- and AMP-specific receptor signaling, cell differentiation, tumor suppression, and regulation of vascular tone [9, 55, 57, 58, 81, 82, 126, 127]. Finally, adenylate kinase–generated AMP activates glycolysis and glycogenolysis through phosphofructokinase (PFK1) and glycogen phosphorylase [11, 25]. Cell survival during hypoxia largely depends on the energy provided by these pathways [21, 25]. To this end, adenylate kinase phosphotransfer resulting in increased provision of ATP β-phosphoryls for energetic needs and generation of AMP signals provides an integrated mechanism that adjusts energy metabolism and determines cell survival under stress.

Recent evidence indicates that the interaction between adenylate kinase and creatine kinase phosphorelays determines metabolic signal communication to the membrane metabolic sensor, the K_{ATP} channel, and mediates energetic remodeling in preconditioned and failing hearts [19, 28, 29, 33, 34]. Under normal conditions, creatine kinase suppresses adenylate kinase phosphotransfer by scavenging cellular ADP [16], thus maintaining closed K_{ATP} channels with low metabolic signaling through the AK $\rightarrow$ AMP $\rightarrow$ AMPK and AK $\rightarrow$ AMP $\rightarrow$ adenosine axes [19, 34, 43–45, 101]. Hypoxia or metabolic stress diminishes creatine kinase and increases adenylate kinase flux, inducing AMP generation and downstream AMP/adenosine signaling events [5, 44, 64, 117]. Deletion of the AK1 gene shifts this balance towards the creatine kinase system, compromising energetic signaling [19, 23]. Accordingly, AK1 deficiency compromises metabolic signal reception by metabolic sensors, such as K_{ATP} channels and AMPK [19, 53]. Underscoring the significance of phosphotransfer redistribution in metabolic signaling is the observation that altered adenylate kinase and creatine kinase phosphotransfer enzyme activities are associated with hypertrophy, abnormal vascular tone, and hypertension [86, 128, 129]. The role of adenylate kinase in integrating signaling pathways is further indicated by the recent demonstration that AMP-stimulated AMPK regulates vascular response to hypoxia and nitric oxide–dependent vasorelaxation [130, 131]. In this regard, adenylate kinase appears to be the major phosphotransfer system in extraocular muscle, where, together with creatine kinase, it regulates precise eye movements [132]. Thus, the balance between phosphotransfer systems and subsequent signaling events could determine the outcome of the metabolic regulation of myocardial contractility, electrical activity, and vascular tone.

In summary, adenylate kinase phosphotransfer is necessary for facilitating intracellular energetic communication and AMP metabolic signal transduction. Because of signaling by a number of AMP/nucleotide-sensitive cellular and extracellular components, adenylate kinase can sense cellular energetic imbalances caused by physical activity, inadequate oxygenation, or nutrient supply and can generate and transmit feedback to adjust cellular energetics, substrate transport, and vascular blood flow to direct oxygen and nutrient delivery.

8.3
Glycolysis as a Network of Phosphotransfer Circuits and Metabolite Shuttles

The results from the initial characterization of phosphotransfer dynamics in creatine kinase–deficient muscles led us to the unexpected discovery that the "old"

Fig. 8.4 Glycolytic phosphotransfer network: coupling of the glycolytic pathway with mitochondrial energetics. The proposed spatial arrangement of the intracellular phosphotransfer network catalyzed by glycolytic enzymes is designed for efficient high-energy phosphoryl transfer and metabolic signal communication. A spatially extended glycolytic system could serve a similar function to CK and AK in intracellular energy transport by transferring mitochondrially produced high-energy phosphoryls through hexokinase (Hex), phosphofructokinase (PFK), and near-equilibrium glyceraldehyde-3-phosphate dehydrogenase/3-phosphoglycerate kinase (GAPDH/PGK) enzymes, present in the vicinity or bound to the mitochondrial outer membrane, as well as by generating additional ATP molecules as a product of glycolysis. Near-equilibrium reactions catalyzed by the GAPDH/PGK chains also could facilitate transfer of P_i and NADH from remote cellular locales to mitochondria, while chains of reversible lactate dehydrogenase (LDH) reactions could facilitate intracellular lactate/pyruvate and coupled NAD^+/NADH movement, thus maintaining transcellular redox balance. The glycolytic system, which is activated under stress conditions, is envisioned to have a more significant role in phosphoryl transfer by taking advantage of the direct and high-throughput contact sites between mitochondrial membranes, thus bypassing the intracristae space. Binding of glycolytic enzymes to the cytoskeleton and clustering inside myofibrils can provide a scaffold for positioning of the glycolytic phosphotransfer network. Through phosphoglucomutase (PGM)-catalyzed equilibration of G-6-P and G-1-P, the glycolytic system is linked to the glycogen energetic network.
TPI: triosephosphate isomerase; PGM: phosphoglycerate mutase; PGI: phosphoglucoisomerase; ENO: enolase; ANT: adenine nucleotide translocator. Pyruvate and phosphate transporters in the inner membrane as well as porins forming gates in the outer membrane are also represented.

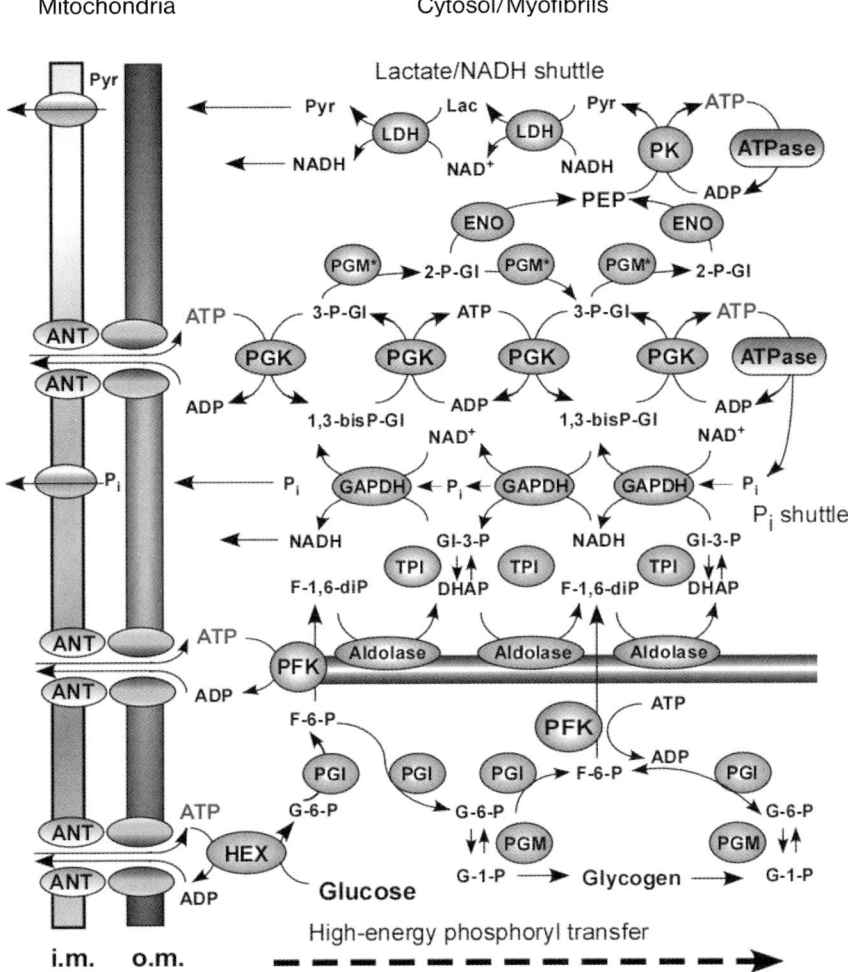

Fig. 8.4 (legend see p. 278)

glycolytic pathway may have a different function: to promote intracellular energy distribution by transferring energy-rich phosphoryls from the ATP, which are used to phosphorylate glucose and fructose-6-phosphate at the mitochondrial site [5]. These high-energy phosphoryls traversing the glycolytic pathway in turn can be used to phosphorylate ADP through the pyruvate kinase (PK)-catalyzed reaction at remote ATP utilization sites (Fig. 8.4). Additional phosphoryls can be transferred through the near-equilibrium GAPDH/PGK couple that catalyzes robust phosphoryl exchange [7, 24]. Such a "spatial" look at glycolytic metabolism is supported by localization of glycolytic enzymes to the I- and M-bands within myofibrils as well as their binding to the mitochondrial outer membrane [11, 40, 107, 133, 134], suggesting a network capability for transferring and distributing high-energy phosphoryls [7, 24]. Indeed, the capacity of glycolytic metabolism as a parallel high-energy phosphoryl transfer pathway is obvious in AK1-, M-CK-, and ScCKmit/M-CK-deficient muscles, where protein and transcript levels of glycolytic enzymes are upregulated and phosphotransfer flux through this pathway is augmented [5, 18, 20, 23, 24, 47, 51, 114, 135]. It was found that glycolysis is a major pathway that supports calcium uptake by the sarcoplasmic reticulum in ventricular fibers of mice deficient in mitochondrial and cytosolic creatine kinase [136]. Moreover, creatine kinase–deficient cardiomyocytes display a higher sensitivity to perturbations in glycolytic metabolism, and inhibition of the glycolytic phosphotransfer pathway impairs metabolic signal communication to sarcolemmal K_{ATP} channels in M-CK knockout, but not wild-type, hearts [34]. Such a boost in glycolytic phosphotransfer capacity in creatine kinase–deficient muscles parallels an increase in mitochondrial abundance, indicating adjustment of the whole energetic system [18, 20, 114].

In this regard, it may not be entirely surprising that glycolytic enzymes have been found clustered in myofibrils and partitioned to cellular membranes and the cytoskeleton [11, 40, 107, 137]. Depending on metabolic needs, there is tight regulation of hexokinase binding to the mitochondrial outer membrane and of pyruvate kinase binding to cellular membranes and other ATP consumption sites [40, 134]. Glycolytic enzymes also have been recognized as important regulators of ATP/ADP-sensitive cellular components such as Ca^{2+}-ATPases and K_{ATP} channels [13, 28, 63, 136, 138], and disruption of the spatial arrangement of glycolytic pathway impairs muscle contractility [139]. In this regard, a network-like distribution of glycolytic enzymes provides a spatiotemporal energy support for sperm motility and the functioning of extended neural cells [63, 140]. Recently, the adaptor protein DRAL/FHL2 that is involved in the anchoring of metabolic enzymes, including phosphofructokinase, to sites of high-energy consumption in the cardiac sarcomere was discovered [75]. Such scaffolding proteins can support spatial organization of the glycolytic phosphotransfer network, thus increasing the efficiency and specificity of high-energy phosphoryl distribution.

The glycolytic network does not function separately. The interaction between glycolytic phosphotransfer and adenylate kinase–creatine kinase systems is critical for energy supply and local ATP regeneration in different cell types [2, 5–12, 20]. Indeed, adenylate kinase can regulate and physically interact with glycolytic enzymes and participate in the formation of larger glycolytic enzyme clusters [5, 11, 84, 109, 117]. The input of ATP from the glycolytic reactions can support and perpetuate phosphotransfer in the creatine kinase system [74, 141]. In muscles with combined AK1 and M-CK deficiency, glycolysis is probably the only remaining major cytosolic phosphotransfer circuit that maintains energetic balance [51]. Simultaneous dissection of all three major phosphotransfer pathways is among the top experimental challenges in future muscle energetics studies.

8.3.1
Glycolytic Phosphotransfer Circuits

In the proposed glycolytic energy network (Fig. 8.4), several phosphotransfer circuits, operating together or semi-independently, can be delineated [133]. The net phosphoryl flux through the glycolytic pathway from hexokinase to pyruvate kinase reactions could be carried out by the linear glycolytic pathway [7]. The network part could include clusters of hexokinase/phosphoglucose isomerase (Hex/ PGI), phosphofructokinase/aldolase/triosephosphate isomerase (PFK/Aldo/TPI), glyceraldehyde-3-phosphate dehydrogenase/phosphoglycerate kinase (GAPDH/ PGK), and phosphoglycerate mutase/enolase/pyruvate kinase (PGM*/ENO/PK), where spatially directed phosphoryl transfer could occur through bidirectional substrate exchange enzymatic chains [7, 9, 142]. However, under high energy demand, glycolytic reactions can become unidirectional, decomposing the network structure and allowing phosphoryl flux through linear chains [7, 133]. The significance of the glycolytic pathway of energy transfer is underscored by the finding that the unidirectional phosphoryl exchange rate through the GAPDH/PGK glycolytic enzyme couple measured by ^{31}P-NMR is high and approaches that of mitochondrial oxidative phosphorylation and creatine kinase [74, 142]. Moreover, ^{32}P metabolic labeling studies also indicate very high phosphoryl turnover of glycolytic intermediates, some of which are even higher than γ-ATP [1, 11]. Because the GAPDH/PGK couple catalyzes a rapidly equilibrating reaction between P_i and γ-ATP, it has been implicated in transferring P_i, as well as NADH and ADP, from myofibrils to mitochondria [7]. In this regard, disequilibrium created at one specific intracellular locale of the glycolytic network would be translated to other cellular compartments [7, 9, 122]. Indeed, "metabolic waves" have been observed to propagate rapidly throughout the entire cell, and oscillations in energy metabolism appear to govern cellular electrical activity, biological information processing, and functional response [9, 143]. High-energy phosphoryls generated by glycolysis can be preferentially delivered and used to support specific cellular functions, such as maintenance of membrane ionic gradients, cell motility, muscle contraction, and nuclear processes [11, 13, 133]. Glycolytic enzymes have also been recognized as important regulators of membrane metabolic sensors, such as K_{ATP} channels, providing energetic signaling between mitochondria and the plasmalemma [13, 28, 138].

To gain further experimental support for the involvement of glycolysis in intracellular phosphotransfer, especially of the near-equilibrium portion of this pathway, a strategy to inhibit glycolysis in creatine kinase–deficient muscles was adapted [24]. We reasoned that if glycolysis operates as an intracellular phosphoryl transfer mechanism, acute chemical inhibition of this pathway would reveal the extent of a possible compensatory shift of phosphoryl transfer function to the adenylate kinase system. Indeed, the increase in the magnitude of adenylate kinase–catalyzed phosphoryl transfer was nearly equivalent to the ATP turnover rate observed in wild-type and creatine kinase–mutant muscles following exposure to an inhibitor of both glycolysis and remaining creatine kinase, i.e., iodo-acetamide [24]. This finding provides direct support for the view that glycolytic metabolism in general and the GAPDH/PGK couple in particular may have a role in directional phosphoryl transfer.

8.3.2
Glycolytic P_i Shuttle

Inorganic P_i and phosphoryl transfer reactions, including protein kinases and phosphatases, play a central role in cell energetics and signal transduction processes [1–10, 24, 43, 144–146]. P_i is an important signaling molecule in the feedback regulation of oxidative phosphorylation and glycolysis/glycogenolysis, and its intracellular free anion level is tightly regulated [7, 145, 147]. Every imbalance in muscle energetics results in P_i elevation, and P_i is considered to be a sensitive error signal that responds to the time-integrated mismatch between ATP usage and ATP production [148]. In the vicinity of ATPases, P_i concentration is a critical determinant of the magnitude of the free energy of ATP hydrolysis, the ultimate driving force of cellular processes, which depends on the ratio $[ATP]/[ADP] \cdot [P_i]$ or the phosphorylation potential [71]. A rise in intracellular P_i corresponds to a precipitous fall in cytosolic phosphorylation potential, free energy of ATP hydrolysis, and heart contractility [149]. Thus, optimal operation of cellular ATPases requires removal of P_i from the intimate reaction environment to avoid kinetic and thermodynamic hindrances [71, 148]. Phosphoryl labeling with oxygen-18 [145] and ^{31}P-NMR analysis [146–150] indicates that cellular P_i metabolism is highly compartmentalized and that its metabolically active pool increases with muscle contraction [145]. Under circumstances of accelerated glycolytic flux, the fraction of total, chemically measurable cellular P_i that is metabolically active becomes reduced [24, 150]. As the GAPDH/PGK couple catalyzes a rapidly equilibrating reaction between P_i and γ-ATP, it has been implicated in relaying P_i, as well as NADH and ADP signals, from myofibrils to mitochondria (Fig. 8.4) [7]. Because P_i is an important signaling molecule in the regulation of oxidative phosphorylation and glycolysis, the observed changes in P_i ^{18}O-labeling kinetics in creatine kinase–deficient muscles indicate alterations in feedback between heart work and respiration [11, 24, 147]. Accelerated GAPDH/PGK and other glycolytic phosphotransfer reactions could be one reason for kinetic entrapment and increased P_i compartmentation/channeling in creatine kinase–deficient hearts [24, 150]. Indeed, an increase in P_i compartmentation, observed in creatine kinase–deficient muscles, and the release of P_i confinement after inhibition of GAPDH strongly support a phosphotransfer function of the near-equilibrium portion of the glycolytic pathway and indicate its augmented role when creatine kinase–catalyzed phosphotransfer is impaired. As is described below, the glycogen metabolic network also could be involved in P_i removal from ATPases, regulation of the cytosolic phosphorylation potential, and P_i intracellular shuttling.

8.3.3
Glycolytic Lactate Shuttle

In general, near-equilibrium reversible reactions could form flux transfer chains along which an incoming flux wave could be transmitted in either direction [7, 151]. Lactate dehydrogenase (LDH), which catalyzes reversible lactate–pyruvate conversions, is among the most active glycolytic enzymes, with total activity far exceeding net glycolytic flux [11, 133]. It has sophisticated tissue-specific isoform composition, with different kinetic and product inhibition properties and the ability to form higher molecular structures, such as dimers, tetramers, and polymeric chains [152, 153]. Kinetic characteristics of the LDH1 (H4) isoform facilitate lactate-to-pyruvate conversion, while LDH5 (M4) is adapted to transform pyruvate to lactate [11]. Intermediate-mixed H-subunit– and M-subunit–containing isoforms could catalyze the lactate–pyruvate equilibrium reaction and translate changes in substrate balance throughout cellular compartments [7, 9]. Thus, in a chain of reactions catalyzed by different LDH isoforms, LDH5 can be at the starting or lactate-producing end, while LDH1 can be positioned at the finishing or lactate-receiving end. As presented in Fig. 8.4, sequentially arranged LDH-catalyzed reactions can form flux transfer chains that facilitate the transfer of pyruvate and NADH from remote cellular sites where pyruvate kinase (PK) resides towards mitochondria. In other non-cardiac muscle cells, lactate can be shuttled into the extracellular space to be removed from the tissue [154].

In this regard, an astrocyte–neuron "lactate shuttle" has been proposed, where specific cellular distribution of LDH5 and LDH1 between astrocytes and neurons suggests that a population of astrocytes is a lactate "source" while neurons may be a lactate "sink" [155]. Because lactate-to-pyruvate conversion is also associated with NAD$^+$ reduction to NADH and *vice versa*, the lactate shuttle could serve as a mobile mediator of the redox state among various compartments both within and between cells [9, 154]. This could be one reason for the high LDH activity in muscle cells. A similar situation apparently exists in the tumor–host metabolic cooperation environment, where distribution of LDH5 and LDH1 isoforms between tumor and tumor-associated fibroblasts and endothelial cells would facilitate lactate extrusion and, consequently, tumor survival and growth [156]. Interestingly, the LDH1:LDH5 ratio correlates with the degree of tumor malignancy [157]. Thus, a dynamic and spatially arranged lactate shuttle with a specific LDH isoform composition could facilitate intracellular metabolite transfer, lactate export, and the cellular redox balance.

8.3.4
Neural Glycolytic Network: Role in Energetics and Information Processing

Although neurons and other cell types have developed mitochondrial networks, brain function specifically depends on glucose metabolism as well [11, 154, 155]. Tight integration of oxidative and glycolytic metabolism is indicated by the high sensitivity of brain activity to both oxygen and glucose deprivation [154]. Hexokinase, which is predominantly bound to brain mitochondria [40, 158], could direct high-energy phosphoryl flow through the glycolytic/glycogenolytic network along dendrites or astrocytic filopodial and lamellipodial extensions, which are too narrow to accommodate mitochondria [140]. Evidence is accumulating that the glycolytic phosphotransfer network, which acts in parallel to creatine kinase and adenylate kinase systems [24], also has important functions in information processing and memory storage [9, 159]. Glucose has been found to improve learning and memory in humans and laboratory animals, but the underlying mechanisms are unknown [9, 159]. Among possible mechanisms by which peripheral glucose might act on memory storage is activation of glycolytic phosphotransfer, which has an important role in energetic signal communication and spatial distribution in cells with large dimensions, such as neurons [9, 122, 140].

Glucose can also affect substrate cycles mediated by membrane-bound glycolytic enzymes [11, 137, 138]. Associated with these cycles are slow oscillations in membrane potential, which could be brought about by the cyclic fluctuations of Ca^{2+} ions and ATP/ADP in the immediate vicinity of the membrane [13, 143]. Memory facilitation and consolidation under glycolytic modifiers also might have been demonstrated in avoidance and discrimination learning trials with honey bees and rats, consistent with the metabolic nature of the slow wave rhythmicity in vertebrate microneurons [160]. Another mechanism by which glucose could affect brain cognitive function is related to the ability of glycolytic phosphotransfer to regulate K_{ATP} channels [13, 28, 138]. In this regard, glucose effect on memory storage appears to be affected by the state of the K_{ATP} channel; pretreatment with minoxidil and glibenclamide, which respectively promote or block ion flow, attenuated or enhanced memory storage [161]. Thus, enhanced glycolytic phosphotransfer could be linked to ATP-sensitive cellular components, including membrane ATPases, ion channels, and nuclear factors regulating neuronal activity and information processing [7, 9, 159].

8.4
Glycogen Energy Transfer Network: Adding a Spatial Dimension to Glycogenolysis

There are still many mysteries surrounding not only glycolysis but also glycogen metabolism. Glycogen's unexplainable dynamics, i.e., its simultaneous synthesis and utilization, contradict the common notion of glycogen as an "energy store" [6, 21, 26, 162, 163]. Also elusive are the coupling of glycogenolysis and glycogen synthesis with mitochondrial oxygen consumption, the specific intracellular localization and compartmentation of glycogen metabolic enzymes, glycogen granule association with mitochondria and their movement along cytoskeletal elements, aerobic lactate production and utilization, and many other aspects of the distinct role of glycogenolysis in cell energetics [11, 21, 76, 140, 162, 164]. Adding to this bewilderment, two separate pathways or compartments for glycolytic and glycogenolytic fluxes have been proposed [165]. Such separation, however, is not required according to the "glycogen shunt" concept, which elegantly connects and explains important aspects of muscle glycogen metabolic dynamics [21].

Here, we provide a different, spatial look at intracellular glycogen metabolism and suggest the novel concept that the topological arrangement and dynamics of glycogen synthesis and utilization provide an energy transfer capacity that couples mitochondrial oxidative phosphorylation with muscular contraction and other energy-dependent cellular functions (Fig. 8.5A). The spatial extension and network representation of energy-investing and ATP-harvesting steps of the glycogen metabolic pathway suggest a unique role in the cardiac energetic system for glycogenolysis acting in partnership with creatine kinase, adenylate kinase, and glycolytic phosphotransfer relays.

Traditionally, glycogen is considered a major energy store in the myocardium, capable of rapidly producing a large amount of ATP to support cardiac function with higher energy demands induced by stress or affected by reduced ATP production in the mitochondria [6, 11]. Approximately 2% of the cell volume of an adult cardiomyocyte is occupied by glycogen, while in fetal heart cells, glycogen can comprise up to 30% of the cell volume [166]. Glycogen provides a significant portion of the glucose utilized by the aerobic working adult heart and is preferentially oxidized compared with exogenous glucose [162, 163]. A simultaneous synthesis and degradation of glycogen is demonstrable in isolated hearts perfused under aerobic conditions with sufficient glucose to cover energetic needs [6, 166]. This argues against "energy storage" or "energy buffer" concepts and points to a more significant role of the dynamic state of glycogen metabolism [6, 164].

Regulation of glycogen metabolism is of primary importance in muscle energetics, including the heart, and defects in this process are associated with a number of diseases [6, 11, 76, 86, 124, 162]. Glycogen content and glycogenolysis also are critical determinants of myocardial survival under stress [6, 163]. Isolated muscle glycogen particles contain glycogenolytic/glycolytic enzymes that form an energetic functional complex capable of quickly replenishing ATP in specific cellular locales [11, 21, 168]. Adenylate kinase–generated AMP is required for activation of glycolytic enzymes and glycogen phosphorylase (GP) through its AMP-binding domain, consequently promoting glycogenolysis [5, 6, 11, 21, 84]. In fact, defective AMP-signaling was found to be the primary cause of an imbalance in glycogen metabolism associated with hypertrophic cardiomyopathy and fatal congenital heart glycogenosis [76, 124, 167, 169]. AMP also activates AMPK, which is bound to glycogen particles through the β1-subunit glycogen-binding domain and can phosphorylate glycogen synthase (GS) and regulate glycogenesis [170]. Mutations of AMPK are associated with arrhythmias and abnormal glycogen storage in the human heart [76, 167, 169].

Thus, it is hardly surprising that glycogen metabolism is required for normal heart development and heart valve formation [166]. This is due in part to the unique role of UDP-glucose, a precursor for glycogen synthesis, which is also required for the production of heparan sulfate, chondroitin sulfate, and hyaluronic acid, important building blocks for cellular structures and glycoproteins [26]. UDP-glucose is also an essential signaling molecule acting inside the cell and

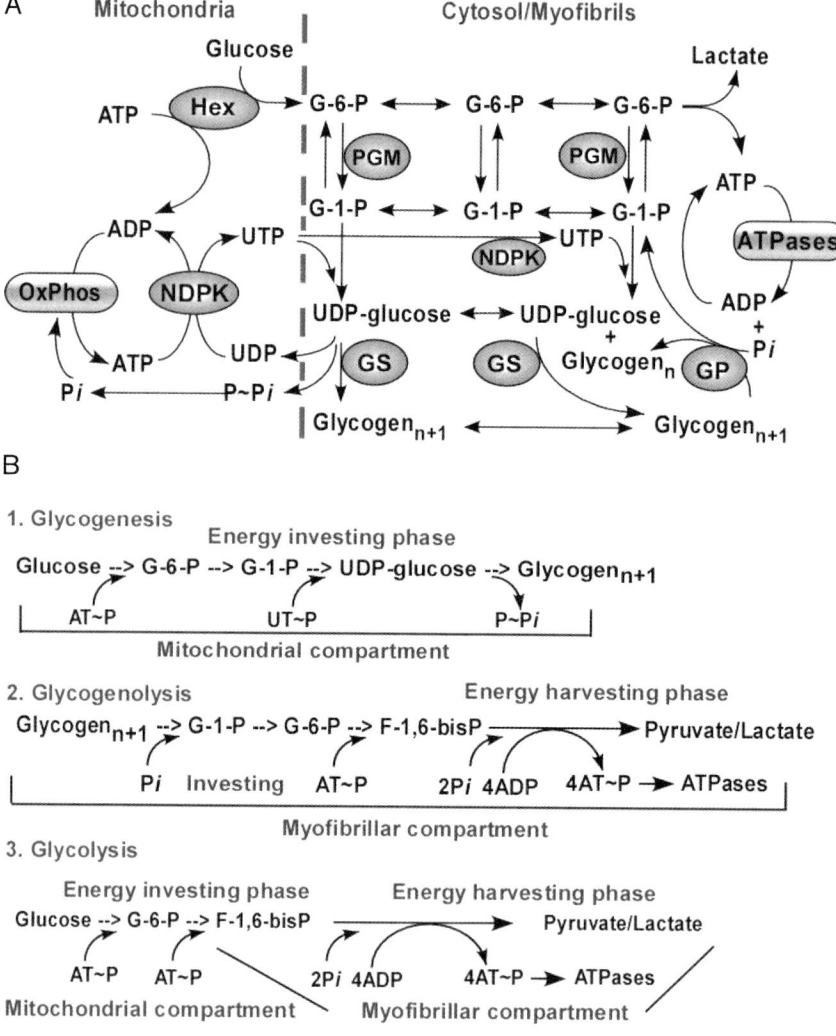

Fig. 8.5 Glycogen energy transfer network. (A) The spatial extension of the glycogen metabolic pathway and the topological positioning of enzymes involved in glycogen synthesis and glycogenolysis suggest a role in intracellular energy transfer. Preferential glycogen synthesis at the mitochondrial site is facilitated by the abundance of G-6-P and UTP, necessary precursors for glycogen synthesis, which are supplied by the hexokinase (Hex) bound to the outer membrane and by nucleoside diphosphate kinase (NDPK) in the intermembrane space. At this step, mitochondrial metabolic energy is invested into G-6-P, G-1-P, UDP-glucose, and, by glycogen synthase (GS), glycogen. These metabolic species as well as UTP can be equilibrated and transferred by phosphoglucomutase (PGM) and NDPK throughout the cell to sites of high ATP consumption to regenerate spent glycogen. Accumulation of glycogen granules around mitochondria and their movement along cytoskeletal elements are characteristic features of many cell types. In such a dynamic system, glycogen utilization at ATP utilization sites by glycogen phosphorylase (GP) occurs to support a sudden increase in energy demand and is coupled with glycogen resynthesis in the immediate vicinity of ATPases and at mitochondrial sites. (B) Compartmental energy balance of glycogenesis and glycogenolysis in comparison to glycolysis (see discussion in the text).

extracellularly through G protein–coupled receptors, inducing the expression of stress proteins required for survival under adverse conditions [26, 171]. Glycogen metabolism is required for cell differentiation and tumor growth [26, 140, 156, 166]. The inability to utilize glycogen stored energy due to phosphorylase deficiency, such as in McArdle's disease, results in exercise intolerance, muscle cramps, pain, and weakness shortly after beginning exercise [11, 26]. Glycogen storage diseases, or glycogenoses, are associated with unbalanced glycogen synthesis and degradation processes resulting from enzymatic deficiencies or aberrant regulatory mechanisms such as Wolff–Parkinson–White syndrome associated with hypertrophic cardiomyopathy and ventricular preexcitation with conduction abnormalities [6, 11, 76, 176, 169]. Wolff–Parkinson–White syndrome represents a systemic metabolic storage disease caused by mutation in the *PRKAG2* gene coding for the energy-sensing kinase AMPK [76, 124]. Thus, the dynamics of glycogen metabolism governed by AMP signaling are essential for cellular life and specialized functions.

Glycogen synthesis starts with the phosphorylation of glucose by hexokinase (Hex), which, by association with or proximity to the outer membrane, has privileged access to mitochondrially generated ATP (Fig. 8.5A). The majority of energy in the ATP molecule is not released but rather invested through phosphotransfer into G-6-P, an important intermediate and metabolic feedback regulator of glycolysis and glycogen synthesis. Subsequently, G-6-P can be converted to G-1-P and transferred/equilibrated throughout the cell by the action of the phosphotransfer enzyme phosphoglucomutase (PGM) [11]. The next step in glycogen synthesis is production of UDP-glucose, an important mobile intermediate and precursor for other molecules [171]. It is produced from G-1-P and UTP in a UDP-glucose pyrophosphorylase–catalyzed reaction. UTP can be provided by NDPK localized in the mitochondrial intermembrane/intracristae space, where it has privileged access to mitochondrial ATP [7, 40]. A reaction byproduct, high-energy pyrophosphate ($P \sim P_i$), can be utilized by mitochondria to generate membrane potential or by other PP_i-dependent reactions [11]. Localization of glycogen synthetic processes close to mitochondria could increase kinetic advantages and efficiency and allow them to be a driving force for both glycolysis and glycogenolysis.

After a specialized initiation step mediated by glycogenin, glycogen synthase catalyzes the addition of glucose residues to the growing glycogen molecule through the formation of α-1,4-glycosidic linkages, the basic polymerizing linkages of the polysaccharide [11, 26]. At this step, mitochondrial metabolic energy is invested into G-6-P, G-1-P, UDP-glucose and, by glycogen synthase (GS), glycogen (Fig. 8.5A). These metabolic species as well as UTP can be equilibrated and transferred by PGM and NDPK throughout the cell to sites of high ATP consumption to regenerate spent glycogen. Accumulation of glycogen granules around mitochondria and their movement along cytoskeletal elements is a characteristic feature of many cell types [164, 172, 173]. Also, a network of branching glycogen aggregates can spread throughout the entire cell cytosol, providing an energetic continuum [174]. Of particular interest is glycogen accumulation in the cell nucleus associated with tumor development [175], as energy support for intense nuclear processes is critical for rapidly proliferating cells [11, 22]. Also, glycogen may serve a structural role as a scaffold for nuclear assembly and sequestration of critical kinases and phosphatases in the nucleus [176]. With respect to the origin of intranuclear glycogen, translocations of cytoplasmic glycogen particles into the nuclear compartment or the synthesis of the polysaccharide within the nucleus itself have been postulated [175, 176]. Because glycogen synthase can translocate or be found in the nucleus [11, 26, 176], transport of UDP-glucose from mitochondria to the intranuclear compartment can be part of the cytosolic-nuclear glycogen energetic shuttle (Fig. 8.5A). To support a sudden increase in energy demand in such a dynamic system, glycogen utilization by GP at ATP utilization sites, including the nucleus, is coupled with glycogen resynthesis in the immediate vicinity of ATPases and at mitochondrial sites. Operation of the proposed glycogen energy transfer network is suggested by labeling studies indicating compartmentation and continuous glycogen utilization and resynthesis even under well-oxygenated conditions and in the presence of other substrates able to cover energy needs [6, 21, 162, 163, 177]. It is additionally suggested by the existence of a tight correlation between aerobic lactate production, which could originate from glycogenolysis [21], and the adenylate kinase phosphotransfer flux [5, 117], suggesting integration of these pathways.

One final discussion involves the energy balance of compartmentalized glycogenesis and glycogenolysis in comparison to glycolysis (Fig. 8.5B). In glycogen synthesis, energy from ATP and UTP is invested to produce UDP-glucose, liberating inorganic pyrophosphate $(P \sim P_i)$, whose energy of hydrolysis is comparable to ATP and subsequently can be utilized in mitochondria and other cellular reactions. Therefore, at this stage ATP and UTP are invested and $P \sim P_i$ is produced, but not much energy is liberated or consumed (disregarding subsequent $P \sim P_i$ hydrolysis). In the mitochondrial intermembrane space, the free energy of ATP hydrolysis could be as high as 63–67 kJ mol^{-1} [7], and only 12–16 kJ mol^{-1} is released during G-6-P synthesis [11]. Even if 1 ATP is utilized per glycosyl unit in glycogen synthesis, this ATP is immediately regenerated in mitochondria without changing the energy balance in the cytosolic compartment. This underscores the advantage of compartmentalized energetics. In glycogenolysis, which can take place in myofibrils in the vicinity of actomyosin ATPases, 1 ATP is invested, 3 P_i are consumed, and 4 ADP are rephosphorylated to ATP, giving a total balance in this compartment of 3P_i consumed and 3 ADP (byproducts of the ATPase reaction) rephosphorylated to ATP. Such simultaneous removal of P_i and ADP can increase the local free energy of ATP hydrolysis and facilitate muscle contraction. In the glycolytic pathway taking place in myofibrils, 1(2) ATP are invested, 2 P_i are consumed, and 4 ADP are rephosphorylated to ATP, giving a total balance in this compartment of 2 P_i consumed and 3(2) ADP rephosphorylated to ATP (in parentheses are the numbers if the cytosolic hexokinase reaction is considered).

Thus, compartmentalized glycogen metabolism may have higher local energetic potential and can act more efficiently in sustaining the free energy of ATP hydrolysis and in promoting muscle contraction. All of the elements above collectively support the concept that the compartmentation and dynamics of glycogen metabolism, and its association with both mitochondrial and myofibrillar ATP synthesis/utilization processes, provide a network capacity for transfer and distribution of cellular high-energy phosphoryl groups to support energy-dependent functions.

8.5
Concluding Remarks: Integration of Phosphotransfer Pathways

Coordination between spatially separated cellular sites of ATP consumption and ATP generation and metabolic signaling is essential for safeguarding cellular energetic economy and adaptation to stress. The evidence discussed here suggests that integrated cellular phosphotransfer circuits catalyzed by adenylate kinase and glycolytic/glycogenolytic phosphotransfer enzymes, acting in concert with the creatine kinase system, contribute to intracellular energetic and metabolic signal communication and promote energetic efficiency (Fig. 8.6). These energy transfer pathways are complementary and interchangeable, and, depending on muscle or tissue type and physiological conditions, energetic communication can proceed through one or another pathway. This commutability is particularly essential for compensation when one phosphotransfer pathway is genetically ablated. Recent studies using the unique ability of the ^{18}O-assisted ^{31}P-NMR technique to monitor adenylate kinase phosphotransfer and AMP signal dynamics in intact hearts, together with measurements of creatine kinase and glycolytic phosphotransfer rates and responses of metabolic sensors, provide a new basis for an integrated, systemic view of the complex cellular energetic system. The data obtained indicate that adenylate kinase–mediated intracellular AMP signaling coupled with AMP-responsive elements such as energy-sensing kinase AMPK, the stress sensor K_{ATP} channel, AMP-sensitive metabolic enzymes, and adenosine signaling is emerging as a key metabolic sensing system and regulator of vital cellular processes. By catalyzing nucleotide exchange and AMP signaling, adenylate kinase regulates the activity of glycolytic and glycogenolytic enzymes and provides an integrative node for both pathways to respond rapidly to increased energy demands. In the cellular environment, adenylate kinase, glycolytic/glycogenolytic, and creatine kinase systems are closely co-localized and interconnected, allowing high-energy phosphoryls to flow from one system to another. Muscles of AK1-knockout mice, with one less phosphotransfer chain, display lower energetic efficiency, slower relaxation kinetics, and a faster drop in contractility upon ischemia associated with compromised AMP generation and impaired metabolic signal communication to the membrane metabolic sensor K_{ATP} channel and the energy-sensing kinase AMPK. Moreover, AK1-deficient skeletal muscles cannot sustain low ADP levels under a functional load despite the presence of the powerful creatine kinase, mitochondrial oxidative phosphorylation, and glycolytic/glycogenolytic ATP-regenerating pathways, whose capacities are

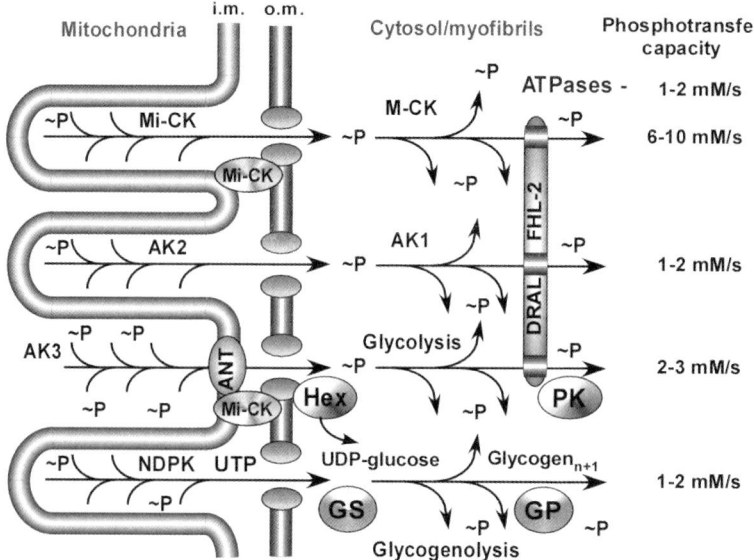

Fig. 8.6 Integration of adenylate kinase with creatine kinase, glycolytic and glycogenolytic circuits in high-energy phosphoryl ($\sim$P) export from mitochondria, and delivery and distribution to cellular ATPases. A model of spatially extended cycles of phosphotransfer reactions catalyzed by adenylate kinase (AK), creatine kinase (CK), and glycolytic and glycogenolytic enzymes providing an energetic continuum between mitochondria and cellular ATPases. Phosphotransfer enzymes can pick up high-energy phosphoryls at the mitochondrial site, while resulting nucleotide diphosphates can be rephosphorylated rapidly in mitochondria without altering the energy balance in the cytosolic compartment. Network infrastructure is maintained by the interaction of phosphotransfer enzymes with cellular constituents mediated by specific protein domains, protein acetylation, myristoylation, and anchor proteins (e.g., DRAL/FHL-2). Mi-CK and M-CK: mitochondrial and cytosolic creatine kinase isoforms, respectively; AK1 and AK2: cytosolic and mitochondrial AK isoforms, respectively; AK3: mitochondrial matrix AK isoform; ANT: adenine nucleotide translocator; Hex: hexokinase; GS: glycogen synthase; GP: glycogen phosphorylase; i.m. and o.m.: inner and outer mitochondrial membranes, respectively.

even heightened, indicating disruption of the energetic network and/or a blunted response to metabolic signals.

Measurements of intracellular phosphotransfer dynamics using ^{18}O-labeling and ^{31}P-NMR techniques indicate that high-energy phosphoryls from mitochondria are distributed to ATP consumption sites through creatine kinase, adenylate kinase, and glycolytic pathways (Fig. 8.6). Here, we further discuss evidence supporting the view that a spatially extended glycolytic network, associated with both mitochondria and ATP consumption sites, has the ability to transfer and distribute high-energy phosphoryls and facilitate intracellular P_i, lactate/pyruvate, and NAD^+/NADH or redox balance. Moreover, a new component in the complete set of high-energy phosphoryl transfer pathways is the glycogen energetic network, suggested by available data regarding the dynamics of glycogen synthesis and utilization and the spatial arrangement and compartmentation of glycogenic and glycogenolytic enzymes in close proximity to mitochondrial and myofibrillar compartments (Fig. 8.6). Such spatial separation and partitioning of energy-investing and ATP-harvesting steps in both glycogen and glycolytic networks provide advantages for intracellular energy flow supporting efficient myofibrillar contractions and other ATP-consuming processes. The range of phosphotransfer capacities of each pathway (not actual net fluxes) presented in Fig. 8.6 were estimated assuming conditions in vivo, i.e., substrate concentrations and their ratios, from ^{31}P- and ^{13}C-NMR, metabolite labeling, and enzymatic studies. Myocardial ATPase and creatine kinase capacities were projected from data presented in [66, 74, 142, 146, 178]. Glycolytic and glycogenolytic capacities, including the phosphotransfer rate in the GAPDH/PGK couple, were estimated from data presented in [74, 142, 162, 163]. Adenylate kinase capacity was estimated from data presented in [17, 23, 24, 30, 100, 101], considering a marked activation of the adenylate kinase phosphotransfer in conditions such as myocardial ischemia [64, 117]. Specifically, the total adenylate kinase enzymatic capacity measured in vitro in murine heart is about 6 mM s^{-1}, while the phosphotransfer flux in vivo, with respect to the ADP utilization rate, can reach 1 mM s^{-1} in hypoxia and is expected to rise in ischemia [23, 47, 64]. In muscles with dissected creatine kinase and glycolytic pathways, adenylate kinase phosphotransfer flux approaches the ATP turnover rate [24]. The adenylate kinase–catalyzed phosphoryl exchange rate measured with ^{31}P-NMR in intact non-muscle cells was reported to be in the range of 1– 1.3 mM s^{-1}, which is about 300 times higher than the ATP utilization rate in low energy-consuming cells [179, 180]. Thus, the phosphotransfer capacities presented in Fig. 8.6 indicate the ability of each pathway to support cellular energetic needs under normal and stress conditions and provide a basis for explaining the remarkable plasticity of energy transfer and distribution networks in the cell.

In summary, systemic integration of complementary energetic and metabolic signaling networks will ensure cellular energy homeostasis and an adequate response to a broad range of functional activities and stress challenges. Moreover, a network and circuit view of the cellular energetic system allows for new perspectives, leading to a better understanding of disease conditions associated with disturbances in adenine nucleotide and glucose metabolism, glycogen storage, and metabolic sensing.

Acknowledgments

This work was supported by the National Institutes of Health (HL64822, HL07111), the Marriott Program for Heart Disease Research, the Marriott Foundation, and the Miami Heart Research Institute. We apologize to our colleagues whose work we were not able to highlight due to space limitations.

References

1 Bessman, S.P., Geiger, P.J. (1981) Transport of energy in muscle: the phosphorylcreatine shuttle. *Science 211*, 448–452.

2 Wallimann, T., Wyss, M., Brdiczka, D., Nicolay, K., Eppenberger, H.M. (1992) Intracellular compartmentation, structure and function of creatine kinase isoenzymes in tissues with high and fluctuating energy demands: the 'phosphocreatine circuit' for cellular energy homeostasis. *Biochem. J. 281*, 21–40.

3 Saks, V.A., Khuchua, Z.A., Vasilyeva, E.V., Belikova, Yu, O., Kuznetsov, A. (1994) Metabolic compartmentation and substrate channeling in muscle cells. Role of coupled creatine kinases *in vivo* regulation of cellular respiration – a synthesis. *Mol. Cell. Biochem. 133/134*, 155–192.

4 Ventura-Clapier, R., Veksler, V., Hoerter, J.A. (1994) Myofibrillar creatine kinase and cardiac contraction. *Mol. Cell. Biochem. 133–134*, 125–144.

5 Dzeja, P.P., Zelenznikar, R.J., Goldberg, N.D. (1998) Adenylate kinase: kinetic behaviour in intact cells indicates it is integral to multiple cellular processes. *Mol. Cell. Biochem. 184*, 169–182.

6 Depre, C., Vanoverschelde, J.L., Taegtmeyer, H. (1999) Glucose for the heart. *Circulation 99*, 578–588.

7 Dzeja, P.P., Terzic, A. (2003) Phosphotransfer networks and cellular energetics. *J. Exp. Biol. 206*, 2039–2047.

8 Ingwall, J.S. (2004) Transgenesis and cardiac energetics: new insights into cardiac metabolism. *J. Mol. Cell. Cardiol. 37*, 613–623.

9 Dzeja, P.P., Terzic, A. (2006) Mitochondrial-nucleus energetic communication: role of phosphotransfer networks in processing cellular information. In *Handbook of Neurochemistry & Molecular Neurobiology: Neural Energy Utilization*, Gibson, G., Dienel, G. (eds.), Kluwer, New York.

10 Saks, V., Dzeja, P., Schlattner, U., Vendelin, M., Terzic, A., Wallimann, T. (2006) Cardiac system bioenergetics:

metabolic basis of Frank-Starling law. *J. Physiol. 571*, 253–273.

11 Ottaway, J.H., Mowbray, J. (1977) The role of compartmentation in the control of glycolysis. *Curr. Top. Cell Regul. 12*, 107–208.

12 Bessman, S.P., Carpenter, C.L. (1985). The creatine-creatine phosphate energy shuttle. *Annu. Rev. Biochem. 54*, 831–862.

13 Weiss, J.N., Lamp, S.T. (1987) Glycolysis preferentially inhibits ATP-sensitive K-channels in isolated guinea-pig cardiac myocytes. *Science 238*, 67–69.

14 Dzeja, P., Kalvenas, A., Toleikis, A., Praskevicius, A. (1985) The effect of adenylate kinase activity on the rate and efficiency of energy transport from mitochondria to hexokinase. *Biochem. Int. 10*, 259–265.

15 Zeleznikar, R.J., Heyman, R.A., Graeff, R.M., Walseth, T.F., Dawis, S.M., Butz, E.A., Goldberg, N.D. (1990) Evidence for compartmentalized adenylate kinase catalysis serving a high energy phosphoryl transfer function in rat skeletal muscle. *J. Biol. Chem. 265*, 300–311.

16 Dzeja, P.P., Zeleznikar, R.J., Goldberg, N.D. (1996) Suppression of creatine kinase-catalyzed phosphotransfer results in increased phosphoryl transfer by adenylate kinase in intact skeletal muscle. *J. Biol. Chem. 271*, 12847–12851.

17 Dzeja, P.P., Vitkevicius, K.T., Redfield, M.M., Burnett, J.C., Terzic, A. (1999) Adenylate kinase-catalyzed phospho-transfer in the myocardium: increased contribution in heart failure. *Circ. Res. 84*, 1137–1143.

18 Janssen, E., Dzeja, P.P., Oerlemans, F., Simonetti, A.W., Heerschap, A., de Haan, A., Rush, P.S., Terjung, R.R., Wieringa, B., Terzic, A. (2000) Adenylate kinase 1 gene deletion disrupts muscle energetic economy despite metabolic rearrangement. *EMBO J. 19*, 6371–6381.

19 Carrasco, A.J., Dzeja, P.P., Alekseev, A.E., Pucar, D., Zingman, L.V., Abraham, M.R., Hodgson, D., Bienengraeber, M., Puceat, M., Janssen,

E., Wieringa, B., Terzic, A. (2001) Adenylate kinase phosphotransfer communicates cellular energetic signals to ATP-sensitive potassium channels. *Proc. Natl. Acad. Sci. U.S.A.* 98, 7623–7628.

20 de Groof, A.J., Oerlemans, F.T., Jost, C.R., Wieringa, B. (2001) Changes in glycolytic network and mitochondrial design in creatine kinase-deficient muscles. *Muscle Nerve 24*, 1188–1196.

21 Shulman, R.G., Rothman, D.L. (2001) The "glycogen shunt" in exercising muscle: A role for glycogen in muscle energetics and fatigue. *Proc. Natl. Acad. Sci. USA.* 98, 457–461.

22 Dzeja, P.P., Bortolon, R., Perez-Terzic, C., Holmuhamedov, E.L., Terzic, A. (2002) Energetic communication between mitochondria and nucleus directed by catalyzed phosphotransfer. *Proc. Natl. Acad. Sci. USA.* 99, 10156–10161.

23 Pucar, D., Bast, P., Gumina, R.J., Lim, L., Drahl, C., Juranic, N., Macura, S., Janssen, E., Wieringa, B., Terzic, A., Dzeja, P.P. (2002) Adenylate kinase AK1 knockout heart: energetics and functional performance under ischemia-reperfusion. *Am. J. Physiol. 283*, H776–H782.

24 Dzeja, P.P., Terzic, A., Wieringa, B. (2004) Phosphotransfer dynamics in skeletal muscle from creatine kinase gene-deleted mice. *Mol. Cell. Biochem. 256–257*, 13–27.

25 Dzeja, P.P., Terzic, A. (2006) Adenylate kinase and creatine kinase phosphotransfer in the regulation of mitochondrial respiration and cellular energetic efficiency. In *Creatine Kinase*, Vial, C. (ed.), In series: *Molecular Anatomy and Physiology of Proteins*, Uversky, V.N. (series ed.), Nova Science, New York, 195–221.

26 Greenberg, C.C., Jurczak, M.J., Danos, A.M., Brady, M.J. (2006) Glycogen branches out: new perspectives on the role of glycogen metabolism in the integration of metabolic pathways. *Am. J. Physiol. 291*, E1–E8.

27 Savabi, F. (1994) Interaction of creatine kinase and adenylate kinase systems in muscle cells. *Mol. Cell. Biochem. 133–134*, 145–152.

28 Dzeja, P.P., Terzic, A. (1998) Phosphotransfer reactions in the regulation of ATP-sensitive K+ channels. *FASEB J. 12*, 523–529.

29 Dzeja, P.P., Redfield, M.M., Burnett, J.C., Terzic, A. (2000) Failing energetics in failing hearts. *Curr. Cardiol. Rep. 2*, 212–217.

30 Pucar, D., Dzeja, P.P., Bast, P., Juranic, N., Macura, S., Terzic, A. (2001) Cellular energetics in the preconditioned state: protective role for phosphotransfer reactions captured by ^{18}O-assisted ^{31}P NMR. *J. Biol. Chem. 276*, 44812–44819.

31 Saks, V.A., Kuznetsov, A.V., Vendelin, M., Guerrero, K., Kay, L., Seppet, E.K. (2004) Functional coupling as a basic mechanism of feedback regulation of cardiac energy metabolism. *Mol. Cell. Biochem. 256–257*, 185–199.

32 Schlattner, U., Wallimann, T. (2004) Metabolite channeling: creatine kinase microcompartments. In *Encyclopedia of Biological Chemistry*, Lennarz, W.J., Lane, M.D. (eds.), Academic Press, New York, USA, 646–651.

33 Selivanov, V.A., Alekseev, A.E., Hodgson, D.M., Dzeja, P.P., Terzic, A. (2004) Nucleotide-gated K_{ATP} channels integrated with creatine and adenylate kinases: Amplification, tuning and sensing of energetics signals in the compartmentalized cellular environment. *Mol. Cell. Biol. 256/257*, 243–256.

34 Abraham, M.R., Selivanov, V.A., Hodgson, D.M., Pucar, D., Zingman, L.V., Wieringa, B., Dzeja, P., Alekseev, A.E., Terzic, A. (2002) Coupling of cell energetics with membrane metabolic sensing. Integrative signaling through creatine kinase phosphotransfer disrupted by M-CK gene knock-out. *J. Biol. Chem. 277*, 24427–24434.

35 Saks, V., Guerrero, K., Vendelin, M., Engelbrecht, J., Seppet, E. (2006) The creatine kinase isoenzymes in organized metabolic networks and regulation of cellular respiration: a new role for Maxwell's demon. In *Creatine Kinase*, Vial, C. (ed.), In series *Molecular Anatomy and Physiology of proteins*,

Uversky, V.N. (series ed.), NovaScience Publisher, New York, 223–267.

36 Weiss, J.N., Yang, L., Qu, Z. (2006) Systems biology approaches to metabolic and cardiovascular disorders: network perspectives of cardiovascular metabolism. *J. Lipid. Res. 47*, 2355–2366.

37 Lipmann, F. (1941) Metabolic generation and utilization of phosphate bond energy. *Adv. Enzymol. 1*, 99–162.

38 Mitchell, P. (1979) Keilin's respiratory chain concept and its chemiosmotic consequences. *Science 206*, 1148–1159.

39 Mitchell, P. (1979) Compartmentation and communication in living systems. Ligand conduction: a general catalytic principle in chemical, osmotic and chemiosmotic reaction systems. *Eur. J. Biochem. 95*, 1–20.

40 Gerbitz, K.D., Gempel, K., Brdiczka, D. (1996) Mitochondria and diabetes. Genetic, biochemical, and clinical implications of the cellular energy circuit. *Diabetes 45*, 113–126.

41 Harold, F.M. (1991) Biochemical topology: from vectorial metabolism to morphogenesis. *Biosci. Rep. 11*, 347–385.

42 Noda, L.H. (1973) Adenylate kinase. In *The Enzymes*, Boyer, P.D. (ed.), 3rd edition, vol. 8, Academic Press, New York, 279–305.

43 Neumann, D., Schlattner, U., Wallimann, T. (2003). A molecular approach to the concerted action of kinases involved in energy homoeostasis. *Biochem. Soc. Trans. 31*, 169–174.

44 Frederich, M., Balschi, J.A. (2002) The relationship between AMP-activated protein kinase activity and AMP concentration in the isolated perfused rat heart. *J. Biol. Chem. 277*, 1928–1932.

45 Hardie, D.G. (2004) The AMP-activated protein kinase pathway – new players upstream and downstream. *J. Cell Sci. 117*, 5479–5487.

46 Tanabe, T., Yamada, M., Noma, T., Kajii, T., Nakazawa, A. (1993) Tissue-specific and developmentally regulated expression of the genes encoding adenylate kinase isozymes. *J. Biochem. 113*, 200–207.

47 Pucar, D., Janssen, E., Dzeja, P.P., Juranic, N., Macura, S., Wieringa, B., Terzic, A. (2000) Compromised energetics in the adenylate kinase *AK1* gene knockout heart under metabolic stress. *J. Biol. Chem. 275*, 41424–41429.

48 Janssen, E., Kuiper, J., Hodgson, D., Zingman, L.V., Alekseev, A.E., Terzic, A., Wieringa, B. (2004) Two structurally distinct and spatially compartmentalized adenylate kinases are expressed from the AK1 gene in mouse brain. *Mol. Cell. Biochem. 256–257*, 59–72.

49 Savabi, F., Geiger, P.J., Bessman, S.P. (1986) Myokinase and contractile function of glycerinated muscle fibers. *Biochem. Med. Metab. Biol. 35*, 227–238.

50 Dzeja, P.P., Pucar, D., Bast, P., Juranic, N., Macura, S., Terzic, A. (2002) Captured cellular adenylate kinase-catalyzed phosphotransfer using ^{18}O-assisted ^{31}P NMR. *MAGMA 14*, 180–181.

51 Janssen, E., Terzic, A., Wieringa, B., Dzeja, P.P. (2003) Impaired intracellular energetic communication in muscles from creatine kinase and adenylate kinase (M-CK/AK1) double knock-out mice. *J. Biol. Chem. 278*, 30441–30449.

52 Hancock, C.R., Janssen, E., Terjung, R.L. (2005) Skeletal muscle contractile performance and ADP accumulation in adenylate kinase-deficient mice. *Am. J. Physiol. 288*, C1287–C1297.

53 Hancock, C.R., Janssen, E., Terjung, R.L. (2006) Contraction-mediated phosphorylation of AMPK is lower in skeletal muscle of adenylate kinase-deficient mice. *J. Appl. Physiol. 100*, 406–413.

54 Elvir-Mairena, J.R., Jovanovic, A., Gomez, L.A., Alekseev, A.E., Terzic, A. (1996) Reversal of the ATP-liganded state of ATP-sensitive K$^+$ channels by adenylate kinase activity. *J. Biol. Chem. 271*, 31903–31908.

55 Olson, L.K., Schroeder, W., Robertson, R.P., Goldberg, N.D., Walseth, T.F. (1996) Suppression of adenylate kinase catalyzed phosphotransfer precedes and is associated with glucose-induced insulin secretion in intact HIT-T15 cells. *J. Biol. Chem. 271*, 16544–16552.

56 Cha, Y.M., Dzeja, P.P., Shen, W.K., Jahangir, A., Hart, C.Y., Terzic, A., Redfield, M.M. (2003) Failing atrial myocardium: energetic deficits accompany structural remodeling and electrical instability. *Am. J. Physiol. 284,* H1313–H1320.

57 Collavin, L., Lazarevic, D., Utrera, R., Marzinotto, S., Monte, M., Schneider, C. (1999) wt p53 dependent expression of a membrane-associated isoform of adenylate kinase. *Oncogene 18,* 5879–5888.

58 Swinnen, J.V., Beckers, A., Brusselmans, K., Organe, S., Segers, J., Timmermans, L., Vanderhoydonc, F., Deboel, L., et al. (2005) Mimicry of a cellular low energy status blocks tumor cell anabolism and suppresses the malignant phenotype. *Cancer Res. 65,* 2441–2448.

59 Noma, T. (2005) Dynamics of nucleotide metabolism as a supporter of life phenomena. *J. Med. Invest. 52,* 127–136.

60 Ren, H., Wang, L., Bennett, M., Liang, Y., Zheng, X., Lu, F., Li, L., Nan, J., et al. (2005) The crystal structure of human adenylate kinase 6: An adenylate kinase localized to the cell nucleus. *Proc. Natl. Acad. Sci. USA. 102,* 303–308.

61 Ginger, M.L., Ngazoa, E.S., Pereira, C.A., Pullen, T.J., Kabiri, M., Becker, K., Gull, K., Steverding, D. (2005) Intracellular positioning of isoforms explains an unusually large adenylate kinase gene family in the parasite Trypanosoma brucei. *J. Biol. Chem. 280,* 11781–11789.

62 Cao, W., Haig-Ladewig, L., Gerton, G.L., Moss, S.B. (2006) Adenylate kinases 1 and 2 are part of the accessory structures in the mouse sperm flagellum. *Biol. Reprod. 75,* 492–500.

63 Ford, W.C. (2006) Glycolysis and sperm motility: does a spoonful of sugar help the flagellum go round? *Hum. Reprod. Update 12,* 269–274.

64 Pucar, D., Dzeja, P.P., Bast, P., Gumina, R.J., Drahl, C., Lim, L., Juranic, N., Macura, S., Terzic, A. (2004) Mapping hypoxia-induced bioenergetic rearrangements and metabolic signaling by ^{18}O-assisted ^{31}P NMR and ^{1}H NMR spectroscopy. *Mol. Cell. Biochem. 256–257,* 281–289.

65 van Deursen, J., Heerschap, A., Oerlemans, F., Ruitenbeek, W., Jap, P., ter Laak, H., Wieringa, B. (1993) Skeletal muscle of mice deficient in muscle creatine kinase lack burst activity. *Cell 74,* 621–631.

66 Saupe, K.W., Spindler, M., Tian, R., Ingwall, J.S. (1998) Impaired cardiac energetics in mice lacking muscle-specific isoenzymes of creatine kinase. *Circ. Res. 82,* 898–907.

67 Crozatier, B., Badoual, T., Boehm, E., Ennezat, P.V., Guenoun, T., Su, J., Veksler, V., Hittinger, L., Ventura-Clapier, R. (2002) Role of creatine kinase in cardiac excitation-contraction coupling: studies in creatine kinase-deficient mice. *FASEB J. 16,* 653–660.

68 Spindler, M., Meyer, K., Stromer, H., Leupold, A., Boehm, E., Wagner, H., Neubauer, S. (2004) Creatine kinase-deficient hearts exhibit increased susceptibility to ischemia-reperfusion injury and impaired calcium homeostasis. *Am. J. Physiol. 287,* H1039–H1045.

69 Momken, I., Lechene, P., Koulmann, N., Fortin, D., Mateo, P., Hoerter, J., Doan, B.T., Bigard, X., Veksler, V., Ventura-Clapier, R.F. (2005) Impaired voluntary running capacity in CK deficient mice. *J. Physiol. 565,* 951–964.

70 Gumina, R.J., Pucar, D., Bast, P., Hodgson, D.M., Kurtz, C.E., Dzeja, P.P., Miki, T., Seino, S., Terzic, A. (2003) Knockout of Kir6.2 negates ischemic preconditioning-induced protection of myocardial energetics. *Am. J. Physiol. 284,* H2106–H2113.

71 Ingwall, J.S. (2002) *ATP and the heart,* Kluwer Academic Publishers, Dordrecht-Boston-London, 1–244.

72 Ventura-Clapier, R., Garnier, A., Veksler, V. (2004) Energy metabolism in heart failure. *J. Physiol. 555,* 1–13.

73 Cha, Y.M., Dzeja, P.P., Redfield, M.M., Shen, W.K., Terzic, A. (2006) Bioenergetic protection of failing atrial and ventricular myocardium by vasopeptidase inhibitor omapatrilat. *Am. J. Physiol. 290,* H1686–H1692.

74 Kingsley-Hickman, P.B., Sako, E.Y., Mohanakrishnan, P., Robitaille, P.M.L., From, A.H.L., Foker, J.E., Ugurbil, K. (1987) ^{31}P NMR studies of ATP synthesis and hydrolysis kinetics in the intact myocardium. *Biochemistry 26,* 7501–7510.

75 Lange, S., Auerbach, D., McLoughlin, P., Perriard, E., Schafer, B.W., Perriard, J.C., Ehler, E. (2002) Subcellular targeting of metabolic enzymes to titin in heart muscle may be mediated by DRAL/FHL-2. *J. Cell Sci. 115*, 4925–4936.

76 Arad, M., Benson, D.W., Perez-Atayde, A.R., McKenna, W.J., Sparks, E.A., Kanter, R.J., McGarry, K., Seidman, J.G., Seidman, C.E. (2002) Constitutively active AMP kinase mutations cause glycogen storage disease mimicking hypertrophic cardiomyopathy. *J. Clin. Invest. 109*, 357–362.

77 Schulz, G.E., Schiltz, E., Tomasselli, A.G., Frank, R., Brune, M., Wittinghofer, A., Schirmer, R.H. (1986) Structural relationships in the adenylate kinase family. *Eur. J. Biochem. 161*, 127–132.

78 Yan, H., Tsai, M.D. (1999) Nucleoside monophosphate kinases: structure, mechanism, and substrate specificity. *Adv. Enzymol. Relat. Areas Mol. Biol. 73*, 103–134.

79 Bae, E., Phillips, G.N. Jr. (2006) Roles of static and dynamic domains in stability and catalysis of adenylate kinase. *Proc. Natl. Acad. Sci. USA. 103*, 2132–2137.

80 Walker, E.J., Dow, J.W. (1982) Location and properties of two isoenzymes of cardiac adenylate kinase. *Biochem. J. 203*, 361–369.

81 Yegutkin, G.G., Henttinen, T., Jalkanen, S. (2001) Extracellular ATP formation on vascular endothelial cells is mediated by ecto-nucleotide kinase activities via phosphotransfer reactions. *FASEB J. 15*, 251–260.

82 Picher, M., Boucher, R.C. (2003) Human airway ecto-adenylate kinase. A mechanism to propagate ATP signaling on airway surfaces. *J. Biol. Chem. 278*, 11256–11264.

83 Schoff, P.K., Cheetham, J., Lardy, H.A. (1989) Adenylate kinase activity in ejaculated bovine sperm flagella. *J. Biol. Chem. 264*, 6086–6091.

84 Peralta, C., Bartrons, R., Riera, L., Manzano, A., Xaus, C., Gelpi, E., Rosello-Catafau, J. (2000) Hepatic preconditioning preserves energy metabolism during sustained ischemia. *Am. J. Physiol. 279*, G163–G171.

85 Hodgson, D.M., Zingman, L.V., Kane, G.C., Perez-Terzic, C., Bienengraeber, M., Ozcan, C., Gumina, R., Pucar, D., O'Coclain, F., Mann, D.L., Alekseev, A.E., Terzic, A. (2003) Cellular remodeling in heart failure disrupts K_{ATP} channel-dependent stress tolerance. *EMBO J. 22*, 1732–1742.

86 Seccia, T.M., Atlante, A., Vulpis, V., Marra, E., Passarella, S., Pirrelli, A. (1998) Mitochondrial energy metabolism in the left ventricular tissue of spontaneously hypertensive rats: abnormalities in both adeninenucleotide and phosphate translocators and enzyme adenylate kinase and creatine phosphokinase activities. *Clin. Exp. Hypertens. 20*, 345–358.

87 Tomasselli, A.G., Schirmer, R.H., Noda, L.H. (1979) Mitochondrial GTP-AMP phosphotransferase. 1. Purification and properties. *Eur. J. Biochem. 93*, 257–262.

88 Wild, K., Grafmuller, R., Wagner, E., Schulz, G.E. (1997) Structure, catalysis and supramolecular assembly of adenylate kinase from maize. *Eur. J. Biochem. 250*, 326–331.

89 Klier, H., Magdolen, V., Schricker, R., Strobel, G., Lottspeich, F., Bandlow, W. (1996) Cytoplasmic and mitochondrial forms of yeast adenylate kinase 2 are N-acetylated. *Biochim. Biophys. Acta. 1280*, 251–256.

90 Luz, C.M., Konig, I., Schirmer, R.H., Frank, R. (1990) Human cytosolic adenylate kinase allelozymes; purification and characterization. *Biochim. Biophys. Acta. 1038*, 80–84.

91 Lee, Y., Kim, J.W., Lee, S.M., Kim, H.J., Lee, K.S., Park, C., Choe, I.S. (1998) Cloning and expression of human adenylate kinase 2 isozymes: differential expression of adenylate kinase 1 and 2 in human muscle tissues. *J. Biochem. 123*, 47–54.

92 Yoneda, T., Sato, M., Maeda, M., Takagi, H. (1998) Identification of a novel adenylate kinase system in the brain: cloning of the fourth adenylate kinase. *Brain. Res. Mol. Brain. Res. 62*, 187–195.

93 Vasseur, S., Malicet, C., Calvo, E.L., Dagorn, J.C., Iovanna, J.L. (2005) Gene expression profiling of tumours derived from rasV12/E1A-transformed mouse embryonic fibroblasts to identify genes required for tumour development. *Mol. Cancer. 4*, 4–11.

94 Santama, N., Ogg, S.C., Malekkou, A., Zographos, S.E., Weis, K., Lamond, A.I. (2005) Characterization of hCINAP, a novel coilin-interacting protein encoded by a transcript from the transcription factor TAFIID32 locus. *J. Biol. Chem. 280*, 36429–36441.

95 Ota, T., Suzuki, Y., Nishikawa, T., Otsuki, T., Sugiyama, T., Irie, R., Wakamatsu, A., Hayashi, K., et al. (2004) Complete sequencing and characterization of 21,243 full-length human cDNAs. *Nat. Genet. 36*, 40–45.

96 Lonergan, K.M., Chari, R., Deleeuw, R.J., Shadeo, A., Chi, B., Tsao, M.S., Jones, S., Marra, M., et al. (2006) Identification of novel lung genes in bronchial epithelium by serial analysis of gene expression. *Am. J. Respir. Cell. Mol. Biol. 35*, 651–661.

97 Dzheia, P.P., Kalvenas, A.A., Toleikis, A.I., Prashkiavichius, A.K. (1983) Functional coupling of creatine phosphokinase and adenylate kinase with adenine nucleotide translocase and its role in regulation of heart mitochondrial respiration. *Biokhimiia 48*, 1471–1478.

98 Gellerich, F.N. (1992) The role of adenylate kinase in dynamic compartmentation of adenine nucleotides in the mitochondrial intermembrane space. *FEBS Lett. 297*, 55–58.

99 Bandlow, W., Strobel, G., Zoglowek, C., Oechsner, U., Magdolen, V. (1988) Yeast adenylate kinase is active simultaneously in mitochondria and cytoplasm and is required for non-fermentative growth. *Eur. J. Biochem. 178*, 451–457.

100 Clark, J.F., Kuznetsov, A.V., Radda, G.K. (1997) ADP-regenerating enzyme systems in mitochondria of guinea pig myometrium and heart. *Am J Physiol. 272*, C399–C404.

101 Nakatsu, K., Drummond, G.I. (1972) Adenylate metabolism and adenosine formation in the heart. *Am. J. Physiol. 223*, 1119–1127.

102 O'Rourke, J.F., Pugh, C.W., Bartlett, S.M., Ratcliffe, P.J. (1996) Identification of hypoxically inducible mRNAs in HeLa cells using differential-display PCR. Role of hypoxia-inducible factor-1. *Eur. J. Biochem. 241*, 403–410.

103 Linossier, M.T., Dormois, D., Perier, C., Frey, J., Geyssant, A., Denis, C. (1997) Enzyme adaptations of human skeletal muscle during bicycle short-sprint training and detraining. *Acta Physiol. Scand. 161*, 439–445.

104 Bianchi, P., Zappa, M., Bredi, E., Vercellati, C., Pelissero, G., Barraco, F., Zanella, A. (1999) A case of complete adenylate kinase deficiency due to a nonsense mutation in AK-1 gene (Arg 107 → Stop, CGA → TGA) associated with chronic haemolytic anaemia. *Br. J. Haematol. 105*, 75–79.

105 Hittel, D.S., Hathout, Y., Hoffman, E.P., Houmard, J.A. (2005) Proteome analysis of skeletal muscle from obese and morbidly obese women. *Diabetes 54*, 1283–1288.

106 Ge, Y., Molloy, M.P., Chamberlain, J.S., Andrews, P.C. (2003) Proteomic analysis of mdx skeletal muscle: Great reduction of adenylate kinase 1 expression and enzymatic activity. *Proteomics 3*, 1895–1903.

107 Wegmann, G., Zanolla, E., Eppenberger, H.M., Wallimann, T. (1992) *In situ* compartmentation of creatine kinase in intact sarcomeric muscle: the acto-myosin overlap zone as a molecular sieve. *J. Muscle Res. Cell Motil. 13*, 420–435.

108 Inouye, S., Yamada, Y., Miura, K., Suzuki, H., Kawata, K., Shinoda, K., Nakazawa, A. (1999) Distribution and developmental changes of adenylate kinase isozymes in the rat brain: localization of ade-nylate kinase 1 in the olfactory bulb. *Biochem. Biophys. Res. Commun. 254*, 618–622.

109 Gerlach, G., Hofer, H.W. (1986) Interaction of immobilized phospho-fructokinase with soluble muscle proteins. *Biochim. Biophys. Acta. 881*, 398–404.

110 Wirschell, M., Pazour, G., Yoda, A., Hirono, M., Kamiya, R., Witman, G.B. (2004) Oda5p, a novel axonemal protein required for assembly of the outer Dynein arm and an associated adenylate kinase. *Mol. Biol. Cell. 15*, 2729–2741.

111 Kim, J., Shen, R., Olcott, M.C., Rajagopal, I., Mathews, C.K. (2005) Adenylate kinase of Escherichia coli, a component of the phage T4 dNTP synthetase complex. *J. Biol. Chem. 280*, 28221–28229.

112 Criss, W.E. (1971) Relationship of ATP:AMP phosphotransferase isozymes to tissue respiration. *Arch. Biochem. Biophys. 144*, 138–142.

113 Bunger, R., Soboll, S. (1986) Cytosolic adenylates and adenosine release in perfused working heart. Comparison of whole tissue with cytosolic non-aqueous fractionation analyses. *Eur. J. Biochem. 159*, 203–213.

114 Janssen, E., de Groof, A., Wijers, M., Fransen, J., Dzeja, P.P., Terzic, A., Wieringa, B. (2003) Adenylate kinase 1 deficiency induces molecular and structural adaptations to support muscle energy metabolism. *J. Biol. Chem. 278*, 12937–12945.

115 Adelman, R.C., Lo, C.H., Weinhouse, S. (1968) Adenylate kinase activity in rat liver. *Adv. Enzyme. Regul. 6*, 425–436.

116 Carvajal, K., El Hafidi, M., Marin-Hernandez, A., Moreno-Sanchez, R. (2005) Structural and functional changes in heart mitochondria from sucrose-fed hypertriglyceridemic rats. *Biochim. Biophys. Acta. 1709*, 231–239.

117 Zelznikar, R.J., Dzeja, P.P., Goldberg, N.D. (1995) Adenylate kinase-catalyzed phosphoryl transfer couples ATP utilization with its generation by glycolysis in intact muscle. *J. Biol. Chem. 270*, 7311–7319.

118 Leclerc, I., Woltersdorf, W.W., da Silva Xavier, G., Rowe, R.L., Cross, S.E., Korbutt, G.S., Rajotte, R.V., Smith, R., Rutter, G.A. (2004) Metformin, but not leptin, regulates AMP-activated protein kinase in pan-creatic islets: impact on glucose-stimulated insulin secretion. *Am. J. Physiol. Endocrinol. Metab. 286*, E1023–E1031.

119 Counago, R., Shamoo, Y. (2005) Gene replacement of adenylate kinase in the gram-positive thermophile Geobacillus stearothermophilus disrupts adenine nucleotide homeostasis and reduces cell viability. *Extremophiles 9*, 135–144.

120 Carrari, F., Coll-Garcia, D., Schauer, N., Lytovchenko, A., Palacios-Rojas, N., Balbo, I., Rosso, M., Fernie, A.R. (2005) Deficiency of a plastidial adenylate kinase in Arabidopsis results in elevated photosynthetic amino acid biosynthesis and enhanced growth. *Plant Physiol. 137*, 70–82.

121 Regierer, B., Fernie, A.R., Springer, F., Perez-Melis, A., Leisse, A., Koehl, K., Willmitzer, L., Geigenberger, P., Kossmann, J. (2002) Starch content and yield increase as a result of altering adenylate pools in transgenic plants. *Nat. Biotechnol. 20*, 1256–1260.

122 Mair, T., Muller, S.C. (1996) Traveling NADH and proton waves during oscillatory glycolysis *in vitro*. *J. Biol. Chem. 271*, 627–630.

123 Dyck, J.R., Lopaschuk, G.D. (2006) AMPK alterations in cardiac physiology and pathology: enemy or ally? *J. Physiol. 574*, 95–112.

124 Young, L.H., Li, J., Baron, S.J., Russell, R.R. (2005) AMP-activated protein kinase: a key stress signaling pathway in the heart. *Trends Cardiovasc. Med. 15*, 110–118.

125 de Lange, P., Ragni, M., Silvestri, E., Moreno, M., Schiavo, L., Lombardi, A., Farina, P., Feola, A., et al. (2004) Combined cDNA array/RT-PCR analysis of gene expression profile in rat gastrocnemius muscle: relation to its adaptive function in energy metabolism during fasting. *FASEB J. 18*, 350–352.

126 Fabre, A.C., Vantourout, P., Champagne, E., Terce, F., Rolland, C., Perret, B., Collet, X., Barbaras, R., Martinez, L.O. (2006) Cell surface adenylate kinase activity regulates the F(1)-ATPase/P2Y (13)-mediated HDL endocytosis pathway on human hepatocytes. *Cell Mol. Life Sci. 63*, 2829–2837.

127 Quillen, E.E., Haslam, G.C., Samra, H.S., Amani-Taleshi, D., Knight, J.A., Wyatt, D.E., Bishop, S.C., Colvert, K.K., et al. (2006) Ectoadenylate kinase and plasma membrane ATP synthase activities of human vascular endothelial cells. *J. Biol. Chem. 281*, 20728–20737.

128 Perez-Terzic, C., Gacy, A.M., Bortolon, R., Dzeja, P.P., Puceat, M., Jaconi, M., Prendergast, F.G., Terzic, A. (2001)

Directed inhibition of nuclear import in cellular hypertrophy. *J. Biol. Chem. 276,* 20566–20571.

129 Brewster, L.M., Mairuhu, G., Bindraban, N.R., Koopmans, R.P., Clark, J.F., van Montfrans, G.A. (2006) Creatine kinase activity is associated with blood pressure. *Circulation* 114, 2034–2039.

130 Morrow, V.A., Foufelle, F., Connell, J.M., Petrie, J.R., Gould, G.W., Salt, I.P. (2003) Direct activation of AMP-activated protein kinase stimulates nitric-oxide synthesis in human aortic endothelial cells. *J. Biol. Chem. 278,* 31629–31639.

131 Rubin, L.J., Magliola, L., Feng, X., Jones, A.W., Hale, C.C. (2005) Metabolic activation of AMP kinase in vascular smooth muscle. *J. Appl. Physiol. 98,* 296–306.

132 Andrade, F.H., Merriam, A.P., Guo, W., Cheng, G., McMullen, C.A., Hayess, K., van der ven, P.F., Porter, J.D. (2003) Paradoxical absence of M lines and downregulation of creatine kinase in mouse extraocular muscle. *J. Appl. Physiol. 95,* 692–699.

133 Masters, C. J., Reid, S., Don, M. (1987) Glycolysis – new concepts in an old pathway. *Mol. Cell. Biochem. 76,* 3–14.

134 Parra, J., Brdiczka, D., Cusso, R., Pette, D. (1997) Enhanced catalytic activity of hexokinase by work-induced mito-chondrial binding in fast-twitch muscle of rat. *FEBS Lett. 403,* 279–282.

135 Ventura-Clapier, R., Kuznetsov, A.V., d'Albis, A., van Deursen, J., Wieringa, B., Veksler, V.I. (1995) Muscle creatine kinase-deficient mice. I. Alterations in myofibrillar function. *J. Biol. Chem. 270,* 19914–19920.

136 Boehm, E., Ventura-Clapier, R., Mateo, P., Lechene, P., Veksler, V. (2000) Glycolysis supports calcium uptake by the sarcoplasmic reticulum in skinned ventricular fibres of mice deficient in mitochondrial and cytosolic creatine kinase. *J. Mol. Cell. Cardiol. 32,* 891–902.

137 Pierce, G.N., Philipson, K.D. (1985) Binding of glycolytic enzymes to cardiac sarcolemmal and sarcoplasmic reticular membranes. *J. Biol. Chem. 260,* 6862–6870.

138 Dhar-Chowdhury, P., Harrell, M.D., Han, S.Y., Jankowska, D., Parachuru, L., Morrissey, A., Srivastava, S., Liu, W., et al. (2005) The glycolytic enzymes, glyceraldehyde-3-phosphate dehydro-genase, triose-phosphate isomerase, and pyruvate kinase are components of the K_{ATP} channel macromolecular complex and regulate its function. *J. Biol. Chem. 280,* 38464–38470.

139 Wojtas, K., Slepecky, N., von Kalm, L., Sullivan, D. (1997) Flight muscle function in Drosophila requires colocalization of glycolytic enzymes. *Mol. Biol. Cell. 8,* 1665–1675.

140 Hertz, L., Peng, L., Dienel, G.A. (2007) Energy metabolism in astrocytes: high rate of oxidative metabolism and spatiotemporal dependence on glycolysis/glycogenolysis. *J. Cereb. Blood. Flow Metab. 27,* 219–249.

141 Kupriyanov, V.V., Seppet, E.K., Emelin, I.V., Saks, V.A. (1980) Phosphocretine production coupled to the glycolytic reactions in the cytosol of cardiac cells. *Biochim. Biophys. Acta. 592,* 197–210.

142 Portman, M.A. (1994) Measurement of unidirectional P(i) → ATP flux in lamb myocardium *in vivo*. *Biochim. Biophys. Acta. 1185,* 221–227.

143 O'Rourke, B., Ramza, B.M., Marban, E. (1994) Oscillations of membrane current and excitability driven by metabolic oscillations in heart cells. *Science 265,* 962–966.

144 Dawis, S.M., Walseth, T.F., Deeg, M.A., Heyman, R.A., Graeff, R.M., Goldberg, N.D. (1989) Adenosine triphosphate utilization and metabolic pool sizes in intact cells measured by transfer of ^{18}O from water. *Biophys. J. 55,* 79–99.

145 Zeleznikar, R.J., Goldberg, N.D. (1991) Kinetics and compartmentation of energy metabolism in intact skeletal muscle determined from ^{18}O labeling of metabolite phosphoryls. *J. Biol. Chem. 266,* 15110–15119.

146 Joubert, F., Mateo, P., Gillet, B., Beloeil, J.C., Mazet, J.L., Hoerter, J.A. (2004) CK flux or direct ATP transfer: versatility of energy transfer pathways evidenced by NMR in the perfused heart. *Mol. Cell. Biochem. 256–257,* 43–58.

147 Bose, S., French, S., Evans, F.J., Joubert, F., Balaban, R.S. (2003) Metabolic network control of oxidative phosphorylation: multiple roles of inorganic phosphate. *J. Biol. Chem. 278*, 39155–39165.

148 Kemp, G.J., Roussel, M., Bendahan, D., Le Fur, Y., Cozzone, P.J. (2001) Interrelations of ATP synthesis and proton handling in ischaemic exercise studied by ^{31}P magnetic resonance spectroscopy. *J. Physiol. 535*, 901–928.

149 Headrick, J.P., Dobson, G.P., Williams, J.P., McKirdy, J.C., Jordan, L., Willis, R.J. (1994) Bioenergetics and control of oxygen consumption in the *in situ* rat heart. *Am. J. Physiol. 267*, H1074–H1084.

150 Bendahan, D., Confort-Gouny, S., Kozak-Reiss, G., Cozzone, P.J. (1990) P_i trapping in glycogenolytic pathway can explain transient P_i disappearance during recovery from muscular exercise. A ^{31}P NMR study in the human. *FEBS Lett. 269*, 402–405.

151 Reich, J.G., Sel'kov, E.E. (1981) *Energy Metabolism of the Cell: A Theoretical Treatise.* Academic Press, London, 95–107.

152 Gottschalk, N., Jaenicke, R. (1987) Chemically crosslinked lactate dehydro-genase: stability and reconstitution after glutaraldehyde fixation. *Biotechnol. Appl. Biochem. 9*, 389–400.

153 Yamamoto, S., Storey, K.B. (1988) Dissociation-association of lactate dehydrogenase isozymes: influences on the formation of tetramers versus dimers of M4-LDH and H4-LDH. *Int. J. Biochem. 20*, 1261–1265.

154 Gladden, L.B. (2004) Lactate metabo-lism: a new paradigm for the third millennium. *J. Physiol. 558*, 5–30.

155 Pellerin, L., Pellegri, G., Bittar, P.G., Charnay, Y., Bouras, C., Martin, J.L., Stella, N., Magistretti, P.J. (1998) Evidence supporting the existence of an activity-dependent astrocyte-neuron lactate shuttle. *Dev. Neurosci. 20*, 291–299.

156 Koukourakis, M.I., Giatromanolaki, A., Harris, A.L., Sivridis, E. (2006) Comparison of metabolic pathways between cancer cells and stromal cells in colorectal carcinomas: a metabolic survival role for tumor-associated stroma. *Cancer Res. 66*, 632–637.

157 Subhash, M.N., Rao, B.S., Shankar, S.K. (1993) Changes in lactate dehydrogenase isoenzyme pattern in patients with tumors of the central nervous system. *Neurochem. Int. 22*, 121–124.

158 Cesar, M.C., Wilson, J.E. (1998) Further studies on the coupling of mitochon-drially bound hexokinase to intramito-chondrially compartmented ATP, generated by oxidative phosphorylation. *Arch. Biochem. Biophys. 350*, 109–117.

159 Hoyer, S., Lannert, H., Latteier, E., Meisel, T. (2004) Relationship between cerebral energy metabolism in parieto-temporal cortex and hippocampus and mental activity during aging in rats. *J. Neural. Transm. 111*, 575–589.

160 Chaplain, R.A. (1979) Metabolic control of neuronal pacemaker activity and the rhythmic organization of central nervous functions. *J. Exp. Biol. 81*, 113–130.

161 Rashidy-Pour, A. (2001) ATP-sensitive potassium channels mediate the effects of a peripheral injection of glucose on memory storage in an inhibitory avoidance task. *Behav. Brain Res. 126*, 43–48.

162 Goodwin, G.W., Ahmad, F., Taegtmeyer, H. (1996) Preferential oxidation of glycogen in isolated working rat heart. *J. Clin. Invest. 97*, 1409–1416.

163 Fraser, H., Lopaschuk, G.D., Clanachan, A.S. (1998) Assessment of glycogen turnover in aerobic, ischemic, and reperfused working rat hearts. *Am. J. Physiol. 275*, H1533–H541.

164 Shearer, J., Graham, T.E. (2004) Novel aspects of skeletal muscle glycogen and its regulation during rest and exercise. *Exerc. Sport. Sci Rev. 32*, 120–126.

165 Hardin, C.D., Kushmerick, M.J. (1994) Simultaneous and separable flux of pathways for glucose and glycogen utilization studied by ^{13}C-NMR. *J. Mol. Cell. Cardiol. 26*, 1197–1210.

166 Pederson, B.A., Chen, H., Schroeder, J.M., Shou, W., DePaoli-Roach, A.A., Roach, P.J. (2004) Abnormal cardiac development in the absence of heart glycogen. *Mol. Cell. Biol. 24*, 7179–7187.

167 Zou, L., Shen, M., Arad, M., He, H., Lofgren, B., Ingwall, J.S., Seidman, C.E., Seidman, J.G., Tian, R. (2005) N488I

mutation of the gamma2-subunit results in bidirectional changes in AMP-activated protein kinase activity. *Circ. Res.* 97, 323–328.

168 Birkel, G., Bauer, H.P., Hofer, H.W. (1986) Phosphofructokinase activity and the binding of enzymes to glycogen particles in the perfused psoas muscle of the rabbit. *Int. J. Biochem.* 18, 79–83.

169 Burwinkel, B., Scott, J.W., Buhrer, C., van Landeghem, F.K., Cox, G.F., Wilson, C.J., Hardie, G.D., Kilimann, M.W. (2005) Fatal congenital heart glycogenosis caused by a recurrent activating R531Q mutation in the gamma 2-subunit of AMP-activated protein kinase (PRKAG2), not by phosphorylase kinase deficiency. *Am. J. Hum. Genet.* 76, 1034–1049.

170 Polekhina, G., Gupta, A., Michell, B.J., van Denderen, B., Murthy, S., Feil, S.C., Jennings, I.G., Camp-bell, D.J. et al. (2003) AMPK beta subunit targets metabolic stress sensing to glycogen. *Curr. Biol.* 13, 867–871.

171 Flores-Diaz, M., Higuita, J.C., Florin, I., Okada, T., Pollesello, P., Bergman, T., Thelestam, M., Mori, K., Alape-Giron, A. (2004) A cellular UDP-glucose deficiency causes overexpression of glucose/oxygen-regulated proteins independent of the endoplasmic reticulum stress elements. *J. Biol. Chem.* 279, 21724–21731.

172 Taylor, C.R., Weibel, E.R., Weber, J.M., Vock, R., Hoppeler, H., Roberts, T.J., Brichon, G. (1996) Design of the oxygen and substrate pathways. I. Model and strategy to test symmorphosis in a network structure. *J. Exp. Biol.* 199, 1643–1649.

173 Weibel, E.R., Hoppeler, H. (2005) Exercise-induced maximal metabolic rate scales with muscle aerobic capacity. *J. Exp. Biol.* 208, 1635–1644.

174 Lo, H.K., Malinin, T.I., Malinin, G.I. (1987) A modified periodic acid-thiocarbohydrazide-silver proteinate staining sequence for enhanced contrast and resolution of glycogen depositions by transmission electron microscopy. *J. Histochem. Cytochem.* 35, 393–399.

175 Granzow, C., Kopun, M., Zimmermann, H.P. (1981) Role of nuclear glycogen synthase and cytoplasmic UDP glucose pyrophosphorylase in the biosynthesis of nuclear glycogen in HD33 Ehrlich-Lettre ascites tumor cells. *J. Cell Biol.* 89, 475–484.

176 Ragano-Caracciolo, M., Berlin, W.K., Miller, M.W., Hanover, J.A. (1998) Nuclear glycogen and glycogen synthase kinase 3. *Biochem. Biophys. Res. Commun.* 249, 422–427.

177 Zhou, L., Yu, X., Cabrera, M.E., Stanley, W.C. (2006) Role of cellular compartmentation in the metabolic response to stress: mechanistic insights from computational Models. *Ann. N. Y. Acad. Sci.* 1080, 120–139.

178 Saupe, K.W., Spindler, M., Hopkins, J.C., Shen, W., Ingwall, J.S. (2000) Kinetic, thermodynamic, and developmental consequences of deleting creatine kinase isoenzymes from the heart. Reaction kinetics of the creatine kinase isoenzymes in the intact heart. *J. Biol. Chem.* 275, 19742–19746.

179 Gupta, R.K. (1979) Saturation transfer ^{31}P NMR studies of the intact. human red blood cell. *Biochim. Biophys. Acta* 586, 189–195.

180 Neeman, M., Rushkin, E., Kaye, A.M., Degani, H. (1987) ^{31}P-NMR studies of phosphate transfer rates in T47D human breast cancer cells. *Biochim. Biophys. Acta.* 930, 179–192.

9
Signaling by AMP-activated Protein Kinase

Dietbert Neumann, Theo Wallimann, Mark H. Rider,
Malgorzata Tokarska-Schlattner, D. Grahame Hardie,
and Uwe Schlattner

Abstract

Intracellular sensors of cellular energy and nutrient status are emerging as key players in the regulation of cell metabolism in health and disease. AMP-activated protein kinase (AMPK) participates in the control of cellular and whole-body energy balance by its exquisite sensitivity to AMP. AMPK is thus able to sense and to react to an increasing AMP:ATP ratio within a complex upstream and downstream signaling network that responds to different energetic and metabolic stresses. The kinase forms heterotrimers with catalytic α- and regulatory β-, and γ-subunits, which exist as different isoforms and splice variants. Global and local cellular ATP:ADP and phosphocreatine:creatine ratios are controlled by the reactions of adenylate kinase and creatine kinase, respectively, which are the first safeguards for keeping cellular ATP:ADP and ATP:AMP ratios high for as long as possible. However, AMPK operates under different metabolic conditions, with different modes of operation and in subsequent time frames. AMP is present at sub-micromolar concentrations in resting cells and increases strongly as a result of the adenylate kinase reaction when ATP is depleted. It is thus an ideal second messenger for reporting cellular energy status to the sensor AMPK. Full activation of AMPK is more complex. In addition to allosteric stimulation by micromolar AMP concentrations, it involves both covalent activation and inhibition via changes in phosphorylation of the Thr172 residue of the α-subunits by different upstream kinases and phosphatases. Such activation depends on different kinds of energy stress as well as on hormones and other signals, including changes in calcium. As a consequence, AMPK aims not only at cellular energy homeostasis but also at whole-body energy balance. Once activated, AMPK induces compensatory measures to maintain cellular ATP:ADP ratios for cell survival. It regulates a large number of downstream targets, shutting down anabolic pathways and stimulating catabolic pathways, thus simultaneously sparing limited energy resources

Molecular System Bioenergetics: Energy for Life. Edited by Valdur Saks
Copyright © 2007 WILEY-VCH Verlag GmbH & Co. KGaA, Weinheim
ISBN: 978-3-527-31787-5

and acquiring extra energy, respectively. AMPK acts at two control levels that are responsible for acute and chronic responses: first by directly affecting the activity of key enzymes, e.g., in glucose and fat metabolism, and second by longer-term transcriptional control of key players of these metabolic pathways. However, new evidence indicates that AMPK also participates in the control of non-metabolic processes such as cell proliferation and cell cycle. Lastly, the kinase complex has been proposed as a potential drug target in treatment of type II diabetes, because its activation both stimulates insulin-independent glucose uptake in the periphery and suppresses glucose production by the liver. More recently, the discovery of AMPK activation by the tumor suppressor LKB1 suggests a strong connection between metabolic signaling and cancer.

9.1
Metabolism and Cell Signaling

9.1.1
A Critical Role for Protein Kinase Signaling

The response of a cell to a changing environment is aimed at maintaining cellular homeostasis for as long as possible in order to avoid cellular damage and cell death. Elaborate mechanisms of allosteric control, including metabolic feedback regulation, have evolved to cope with this challenge. On top of this, another layer of cellular regulation via a network of protein kinases and phosphatases has developed. This network integrates signals from extra- and intracellular messengers and executes rapid responses, as well as longer-term adaptations via modification of gene expression profiles. Thus, the cell can respond to the needs of the surrounding tissue as a result of extracellular stimuli sensed by cell surface receptors, as well as to changes in the intracellular milieu via sensor proteins, in order to mount a coordinate response that controls development, proliferation, growth, adhesion, or metabolic function.

A number of signaling pathways mediated by protein kinases, which regulate both metabolism and cell growth and proliferation, have been characterized in recent years. These include the AMP-activated protein kinase (AMPK), protein kinase B (PKB, Akt), and mammalian target-of-rapamycin (mTOR) pathways (for recent reviews, see [1–5]). These constitute so-called signaling hubs, which respond to multiple signals, such as the availability of ATP, glucose, or amino acids, and cause multiple responses in intermediary metabolism, protein synthesis, cell cycle regulation, and transcriptional control. Because of their importance, mutations in these signaling pathways are involved in multiple pathologies, emphasizing the importance of metabolic dysfunction in a number of unrelated diseases, such as cancer and diabetes. In this chapter, we describe the molecular function of AMPK, the protein kinase most relevant to the regulation of cellular bioenergetics.

9.1.2
At the Interface of Energy Metabolism and Protein Kinase Signaling: AMPK as Receptor for Cellular Energy State

Cellular energy metabolism shows a particularly well-developed capacity to maintain homeostasis of the major cellular energy resource, ATP. Especially in the heart, this has been described in great detail as the "homeostasis paradox" [6]. The cellular metabolic network contains multiple regulatory circuits to account for this. Many of them are addressed in other chapters of this volume, including allosteric regulations and feedback loops, micro-compartmentation, metabolic channeling, and similar mechanisms. Two key players in this immediate energy homeostasis control are creatine kinase (CK) and adenylate kinase (AdK) (see Chapters 3, 7, 8, and 11). At the level of protein kinase signaling, AMPK represents a key sensor of the cellular energy state. It detects changes that are accompanied by a rise in AMP and thus an increase in the AMP:ATP ratio, which in resting cells is extremely low [7]. This kinase was characterized during the last decade as an exquisitely sensitive and versatile metabolic sensor and "energy gauge" [8].

One of the classical concepts of regulation in bioenergetics is the "energy charge" model proposed by Atkinson back in 1968 [9]. It proposed that ratios of adenine nucleotides would oppositely regulate key metabolic enzymes, such that a decrease in ATP would stimulate catabolic ATP-generating pathways, while an increase in ATP would stimulate anabolic ATP-consuming pathways. This concept has inspired a generation of biochemists, although it turned out that direct regulation by adenine nucleotides was true only for a small subset of metabolic enzymes. However, it is now clear that many other metabolic enzymes and other proteins are regulated *indirectly* by energy charge, via the AMPK system [10, 11]. As its name suggests, this kinase is activated by AMP, more specifically by a rise in the AMP:ATP ratio, and in addition by multiple upstream kinases, which have added substantial complexity to this signaling pathway. Recent data implicate AMPK signaling as a key player not only in cellular bioenergetics but also in whole-body energy homeostasis, the metabolic syndrome, and cancer. By phosphorylating target proteins, including key metabolic enzymes, AMPK activation increases the flux through catabolic pathways and decreases anabolic processes, as expected from the energy charge hypothesis of Atkinson, thus preserving ATP levels. In addition, by acting on transcription, AMPK also adjusts the expression level of key enzymes and transporters.

A large number of detailed and recent reviews on AMPK signaling are available, e.g., dealing with AMPK in general [1, 8, 12, 13], with specific aspects of upstream and downstream signaling [14, 15], or with AMPK as a drug target for the metabolic syndrome and cancer [16, 17]. Chapter 17 of this volume is dedicated to the role of AMPK in whole-body energy balance and the metabolic syndrome. We will focus here on selected molecular aspects of AMPK signaling that deserve more detailed consideration. First, we discuss why AMP has been chosen by evolution as an indicator for a challenged cellular energy situation or energy deterioration and how AMP levels in the cell are regulated. After a short description of AMPK structure, we will give an overview of the emerging complexity of mechanisms that activate AMPK as well as of the regulation of its numerous downstream targets.

9.2
Sensing and Signaling of Cellular Energy Stress Situations

9.2.1
Why AMP Represents an Ideal Second Messenger for Reporting Cellular Energy State

ATP is the universal energy currency in all living systems, and the thermodynamic efficiency of biochemical processes that rely on ATP-hydrolysis depends largely on a high local ATP:ADP ratio (at least 100:1) [6]. Therefore, to function economically, healthy cells need to maintain intracellular ratios of ATP:ADP, and concomitantly ATP:AMP, as high as possible. In normal cells, global cellular ATP concentrations ([ATP]) are usually maintained within a narrow range of 3–6 mM, depending on cell type, by the action of CK and AdK preventing a significant decrease in [ATP] under normal cellular workload as depicted in Fig. 9.1 [6]. Both of these kinases display a high affinity for ADP (see below) and have very fast turnover rates. They are therefore able to rephosphorylate and thus very efficiently remove the ADP generated by ATP hydrolysis [6], with CK drawing on the large cellular phosphocreatine (PCr) pool. Tissues and cells with high and fluctuating energy requirements, e.g., skeletal and cardiac muscle, brain, retina and spermatozoa, express high levels of CK [18, 19] and AdK [20, 21]. Assuming that CK and AdK would work at equilibrium in a cell (which may not be true at all, due to compartmentation of these kinases [22]), the AMP:ATP ratio would theoretically vary as the square of the ADP:ATP ratio, as pointed out by Hardie and Hawley [10].

A serious problem in bioenergetics is the fact that although [PCr] and [ATP] can be readily measured in intact cells by chemical methods or even *in vivo* by ^{31}P-NMR, this is more difficult in the case of [ADP] and not possible at all for [AMP] because of their presence in cells at very low concentrations and the limited sensitivity of the methods. Although concentrations can be determined by HPLC or capillary electrophoresis after rapid freezing and acid extraction, these methods ignore cellular compartmentation. In addition, the concentrations of ADP and AMP also may be significantly altered before, during, or after freeze-clamping of the tissues for extraction and analysis. These problems may lead to a deviation of values for [ADP] and [AMP] from the actual *in vivo* situation, e.g., due to mixing of PCr with adenylate pools and rapid equilibration via the CK and AdK reactions, hydrolysis of PCr and ATP, or release of metabolically inactive

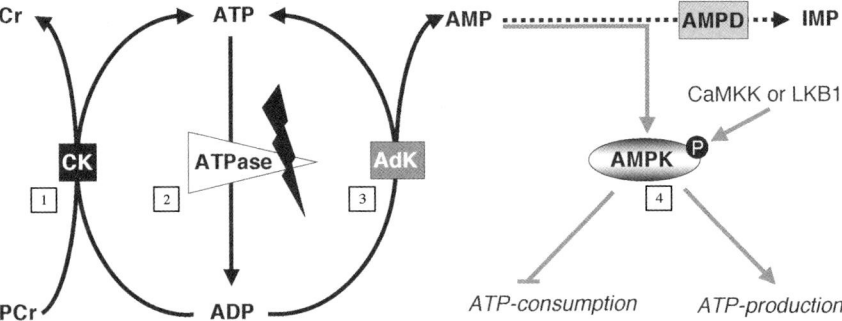

Fig. 9.1 High-energy phosphoryl transfer networks and regulation of cellular energy status. ATP, the universal energy currency, is hydrolyzed by ATPases (2) and the energy is converted, among other purposes, for muscle contraction, ion pumps, and protein synthesis. The cellular ATP:ADP ratio is kept high by the action of two energy-related kinases: creatine kinase (CK) (1) and adenylate kinase (AdK) (3). The ADP generated upon hydrolysis by ATPases is recharged into ATP by the action of CK (1), which draws its energy from a large phosphocreatine (PCr) pool, and by AdK (3), which uses two ADP molecules to regenerate one ATP and to generate one AMP molecule. AMP serves as an indicator for cellular energy stress and stimulates AMPK (4). Phosphorylation by either of the two upstream kinases, i.e., LKB1 or CaMKK, at Thr172 in the α-subunit causes its activation, with AMP inhibiting dephosphorylation. Activated and fully AMP-stimulated AMPK then upregulates catabolic pathways for ATP production and suppresses anabolic pathways that would consume ATP, thus salvaging the cell from irreversible energy deficiency. If AMP concentration is chronically elevated, it becomes deaminated by AMP-deaminase (AMPD) to inosine monophosphate (IMP) and subsequently is dephosphorylated by 5′-nucleotidase to yield inosine. However, inosine can leave the cell, leading to a drop in the total intracellular adenine nucleotide pool. Such a loss of energy-costly building blocks for nucleotide biosynthesis should be avoided if possible.

ADP bound to proteins such as actin [23] (see also [18, 22]). Thus, the "bulk concentrations," if such a parameter exists at all, of ADP and AMP are often calculated from the PCr and ATP concentrations measured directly, assuming complete equilibration of the CK and AdK reactions. Recently, it was possible to measure by *in vivo* ^{31}P-NMR the actual concentration of metabolically active, free ADP (approximately 1.5 mM) in skeletal muscle of an AdK $(^{-/-})$ knockout mouse, after exhaustive stimulation leading to nearly complete PCr depletion [24]. As expected in these animals, the AMP formation during muscle contraction was decreased, resulting in reduced phosphorylation of AMPK [25]. In general, AMP is present in resting cells at extremely low concentrations, probably in the low micromolar range. For example, in normally beating heart muscle, [AMP] has been estimated to be approximately 0.14 μM [26]. Using chemical determination of total metabolite contents by HPLC and assuming that the CK and AdK reactions are in equilibrium, the contents of free AMP and ADP in human skeletal muscle were estimated to be around 7.5 nmol and 57 μmol per kilogram muscle dry weight at rest and 360 nmol and 260 µmol per kilogram after exhaustive exercise, respectively [27]. Under the same conditions, [PCr] decreased from 86 mmol kg^{-1} dry weight to 23 mmol kg^{-1} dry weight and [ATP] remained almost constant, changing from 22 mmol kg^{-1} to 18 mmol kg^{-1} dry weight [27]. Thus, in these *in vivo* experiments with human subjects, the relative increases in [ADP] and [AMP] from rest to exhaustion were 4.5-fold and 48-fold, respectively, while [ATP] dropped by only 20%.

Similar results are consistently observed in various experimental setups, including ^{31}P-NMR methods (see [28] and references therein), always however, with the caveat that concentrations of ADP and AMP have only been estimated [22]. Nevertheless, within these limitations, the data and the resulting relative nucleotide ratios may indicate a trend that is also followed under the much more complex situation *in vivo*. Thus, globally, AMP would appear at significant concentrations in metabolically stressed cells, the AMP:ATP ratio would change much more than the ADP:ATP ratio, and the relative increase in the AMP:ATP ratio from rest to activation or cellular stress would be by orders of magnitude [7]. Therefore, increasing [AMP] or AMP:ATP ratio would be ideally suited to report changes in the cellular energy state. As shown in Fig. 9.2, the changes in PCr, creatine (Cr), P_i, ATP, ADP, and AMP concentrations can also be visualized graphically as a function of the total "energy-rich" phosphorylated compounds, when considering the fast reactions of AdK and CK in equilibrium [29]. Although this may be an oversimplification (see below), the model reflects the decrease in phospho-compounds that occurs during the transition from rest to increasing workload in skeletal muscle (Fig. 9.2A). The resulting relative changes in the PCr:Cr, ADP:ATP, and AMP:ATP ratios are depicted in Fig. 9.2B. They illustrate that the PCr:Cr ratio falls first, then the ADP:ATP ratio begins to rise, and finally the AMP:ATP ratio rises much more sharply, consistent with the above experimental data. However, even initial ATP consumption, which is almost completely replenished from PCr, is immediately translated into a change in [AMP] by the action of AdK, as shown by plotting adenine nucleotide content on a logarithmic scale (Fig. 9.2C).

Again, although the exact figures for free [ADP] and [AMP] *in vivo* are actually unknown and currently immeasurable [26, 29], the above experimental and theoretical approximations allow some insight into their role in energy signaling. Neither PCr and Cr nor ATP would be well suited as energy status indicators *per se* for all physiological situations, in particular those characterized by a depleted [PCr] pool. PCr and Cr are present in muscle and many other excitable cells at concentrations in the millimolar range and change transiently and in relative terms only by a factor of two- to fivefold; ATP then remains stable for a long time after cell activation (Fig. 9.2A). On the other hand, [ADP] would fluctuate sufficiently to serve as a signaling metabolite but may have some limitations due to its binding to cellular proteins (e.g., F-actin [23]) and a multitude of cellular enzymes. The situation is much different for AMP, although it can also bind to some abundant intracellular enzymes such as glycogen phosphorylase and fructose-1,6-bisphosphatase. As outlined above, [AMP] possesses the typical and

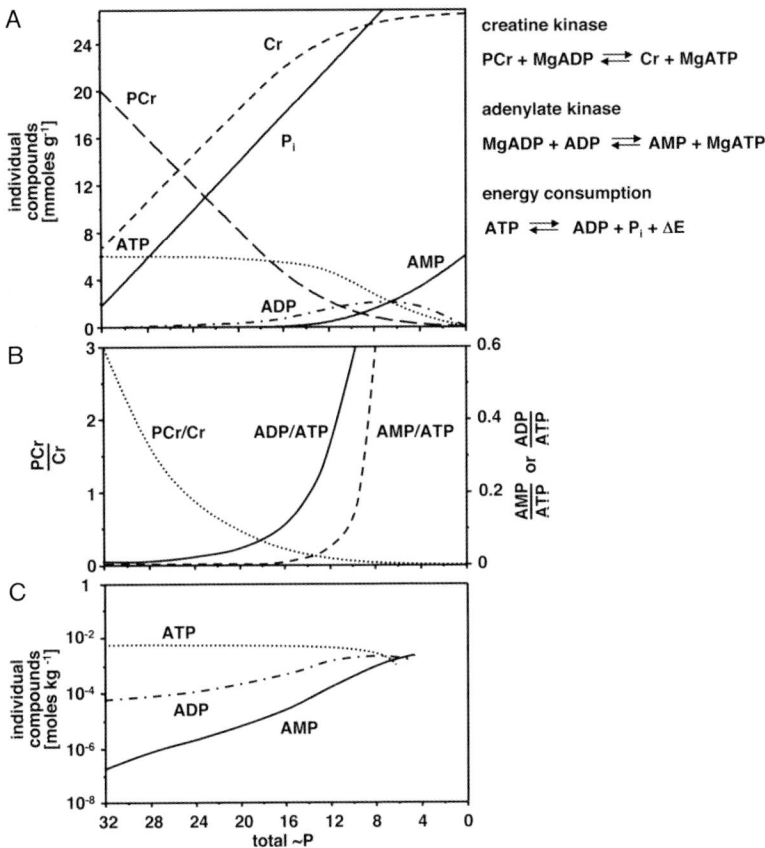

Fig. 9.2 Model calculations for changes in the pool sizes of energy-related metabolites during cell stimulation until exhaustion. (A) Global cellular concentration changes in phosphocreatine [PCr] and adenine nucleotides ([ATP], [ADP] and [AMP]), inorganic phosphate [Pi], and creatine [Cr] calculated from the reactions of CK, AdK, and a generalized ATPase (indicated to the right) at decreasing "high-energy" phosphates, corresponding to a transition from rest to high workload and exhaustion (modified from [29]). At rest, [PCr] and [ATP] are high, while [ADP] is very low and [AMP] is virtually non-measurable. With high-energy phosphate consumption, [ATP] remains constant until more than 80% of the PCr pool is consumed, and only then do [ADP] and [AMP] start to rise dramatically. Note: This simplified model assumes that the CK and AdK reactions work at equilibrium, which may not be true in all

in vivo situations. In addition, it does not account for specific subcellular localizations of CK and AdK isoforms, which lead to local deviations in metabolite concentrations. (B) Changes in global ratios of PCr:Cr, ADP:ATP, and AMP:ATP as calculated from values shown in (A). Note that the adenine nucleotide ratios do not start to rise markedly until the PCr:Cr ratio has decreased by around 80%. (C) Changes in individual adenine nucleotide concentrations taken from (A) but shown on a logarithmic scale. Note: The relative changes in [AMP], which is present under normal conditions at cell rest at low micromolar concentrations only, can be very large. [AMP] increases from the beginning exponentially over several orders of magnitude, as reflected by the approximately linear increase on the logarithmic scale. This will also transmit into large relative changes in the AMP:ATP ratio, which is the signal sensed by AMPK.

ideal characteristics of an energy signaling metabolite and is well suited as a reporter or "second messenger" for sensing cellular energy stress. Of course, there has to be a corresponding AMP sensor such as a protein kinase, which would respond to changes in [AMP] within the low micromolar concentration range and transmit the signal onwards. AMPK perfectly fulfills all these requirements.

9.2.2
A Closer Look at the Role of Creatine Kinase and Adenylate Kinase in AMPK Activation

With their high affinity for ADP and very fast turnover rates, CK and AdK are able to rephosphorylate ADP and thus efficiently remove the ADP generated by ATP hydrolysis [6] (Fig. 9.1). Removal of free ADP in the cell is essential because this nucleotide has been shown to inhibit several ATPases by product inhibition or allosteric regulation (see [18]). CK and AdK are micro-compartmentalized at various subcellular sites of either ATP utilization (ATPases or ATP-gated processes) or ATP generation (mitochondria and glycolysis), where CK and/or AdK form functionally coupled multi-enzyme complexes to constitute an intricate energy-buffering and high-energy phosphoryl transfer system (see also Chapters 3, 7, 8, and 11) [20, 30–33].

Because it is difficult to know the exact concentrations of free ADP and AMP (see above), it may be helpful to analyze the respective apparent K_m values of enzymes participating in the "energy decay cascade" (see Fig. 9.1) to appreciate preferred pathways and possible bottlenecks for accumulation of a substrate or product. As cells increase their basal rate of ATP hydrolysis, e.g., upon activation, the CK and AdK reactions start to run sequentially or in parallel. It is likely that CK, with a K_m for ADP in the range of 35–80 μM [34–36], first draws on the very large PCr energy pool (15–40 mM, depending on cell type). ATP is regenerated by the CK reaction via transphosphorylation of ADP by PCr. However, if PCr is depleted to any significant extent, free ADP (present under resting conditions at an estimated level of approximately 20 μM) will start to rise (Fig. 9.2). Increasing ADP will then be removed efficiently by the action of AdK, which has an apparent K_m for ADP of 30–40 μM [37, 38], converting two molecules of ADP into 1 ATP and 1 AMP, thus resulting in a rise in AMP [7] (Fig. 9.1). In cells and organs that do not express CK, such as liver, AdK takes over the ATP-buffering function in maintaining a high ATP:ADP ratio and keeping free [ADP] low. This may also explain why the apparent K_m values for ADP of CK and AdK are both low and of the same order of magnitude. Similarly, compensatory mechanisms were observed when the CK system was compromised in muscle, e.g., by the rather specific CK inhibitor DNFB [31], or after CK was ablated in knockout mice [39]. Under the latter conditions, a redistribution of phosphotransfer through glycolytic and AdK networks contributes to energy homeostasis in muscles under genetic and metabolic stress (see Chapter 8).

The appearance in cells of significant [AMP], which is normally estimated to be maintained at low micromolar concentrations [29], presumably takes place only if much of the cellular PCr is split, i.e., in conditions of relatively severe metabolic stress, such as exhaustive muscle exercise, osmotic shock, oxidative damage, or chronic Ca^{2+} overload (Fig. 9.2). AMP is then sensed as a second messenger by AMPK [10] (Fig. 9.1), which has very low estimated apparent $A_{0.5}$ (the concentration causing half-maximal activation) for AMP of less than 2 μM [40]. Surprisingly, at around 10 μM AMP, the $\alpha1\beta1\gamma1$ and $\alpha2\beta2\gamma1$ complexes of AMPK are already maximally stimulated [40]. If [AMP] should rise very high, it would be converted into IMP by AMP-deaminase (AMPD), which has a rather high apparent K_m for AMP of approximately 1.5 mM [41], or possibly even higher (7 mM) [42]. Interestingly, when muscle was Cr depleted by chronic administration of β-guanidino propionic acid (GPA; known to cause AMPK activation [43]), the kinetic properties of AMPD were changed into a form with a low apparent K_m in the 30 μM range, presumably by phosphorylation [42]. Both AMP and IMP can be dephosphorylated by 5′-nucleotidases to give adenosine and inosine, respectively, which then can leave the cell via equilibrative nucleoside transporters. To prevent an immediate loss of energy-costly purines, the apparent K_m values for AMP and IMP of 5′-nucleotidase, at around 4 mM [44] or 2–5 mM (depending on the free [ADP] [45]), are also high and thus would allow significant accumulation of either AMP or IMP in the cell. Taken together, AMPK with its extremely high affinity for AMP, will respond first to an energy stress signal long before AdK, AMPD, or 5′-nucleotidase, all with significantly higher apparent K_ms for AMP, would start to operate. The K_m values of the enzymes that participate in AMP metabolism would allow a cell or tissue under prolonged energy stress to persistently elevate [AMP] at a level that chronically stimulates AMPK. This would lead to the well-documented long-term effects on gene expression via transcriptional control (see below) [46].

From what has been said so far, it is obvious that CK and AdK are indirectly connected to AMPK signaling. By maintaining cellular [ATP] under cellular energy stress, CK can delay or prevent a strong increase in free [AMP] via AdK and the resulting activation of AMPK [7]. AdK, although also capable of stabilizing cellular [ATP] to a certain degree, will ultimately generate the [AMP] signal that activates AMPK. However, compared to AMPK, both CK and AdK are working in different metabolic areas with completely different mechanisms of operation and in subsequent time frames. CK and AdK are extremely fast enzymes, responding immediately to changes in high-energy phosphate compounds under normal workload that keep local ATP:ADP ratios high and thus regulate energy buffering and energy flux (see Chapters 3, 7, and 11). AMPK signaling involves its activation by upstream kinases and phosphorylation of downstream targets under cellular stress and probably a number of other stimuli (see below). Although AMP could rise early after cell activation (see Fig. 9.2C), cell signaling by AMPK would be expected to set in at a later point in time compared with the action of CK and AdK. There are also obvious tissue-specific differences with respect to the functions and interplays of these kinases. For example, in heart, where phosphorylated metabolites show an astonishing homeostasis at different workloads (different from what is shown in Fig. 9.2 for a tissue such as skeletal muscle), the CK system provides an exquisite feedback mechanism for energy homeostasis [6], while AMPK may be activated only under more pathological situations such as cardiac ischemia and hypertrophy [47, 48]. By contrast, liver lacks CK (but expresses high levels of AdK), and AMPK plays a prominent regulatory role in this organ [49]. Therefore, the CK/AdK and AMPK systems do not compete with one another but work in a complementary fashion to maintain energy homeostasis.

Some isolated reports have indicated a more direct regulatory crosstalk between the PCr/CK system and AMPK. For example, oral Cr supplementation of rats apparently led to increased glucose uptake, paralleled by increased GLUT4 expression in muscle [50], an event that is also induced after AMPK activation [51]. Using L6 rat skeletal muscle cells, evidence was presented that Cr supplementation led to AMPK phosphorylation [52], which could explain the above *in vivo* finding. In addition, rats that were supplemented with GPA, a strong competitive inhibitor of the creatine transporter [43], showed a marked decrease in total Cr concentration ([PCr] + [Cr]) in muscle that was paralleled by chronic activation of AMPK, indicating some connection between lowered PCr levels or a lowered PCr:Cr ratio and AMPK [53]. Furthermore, evidence was provided for a strong inhibition of AMPK, but not of AMPK kinase, by PCr in muscle that could be alleviated by increasing concentrations of Cr, proposing AMPK as a ratiometric sensor of the PCr:Cr ratio [54]. Conversely, in the same publication, evidence was also provided that AMPK could phosphorylate CK and reduce its activity significantly, that AMPK colocalizes with muscle-type MM-CK at the sarcomeric M-band (see Chapter 7), and that CK and AMPK form a stable complex resistant to high salt [54]. However, a number of laboratories have failed to reproduce some of these experiments. In a detailed investigation, the inhibitory effect of PCr was thrown into doubt by experiments showing that neither PCr nor Cr had a marked direct effect on either AMPK phosphorylation or AMPK activity after activation by LKB1 [55], an upstream AMPK kinase (see below). In addition, PCr and Cr had no effect on LKB1 activity itself. The previously described inhibitory effect of PCr on AMPK may be attributable to the Na^+ salt that was introduced into the AMPK assays along with the PCr preparation, because the disodium salt of PCr had been used [54, 55]. These salt-inhibition experiments have been confirmed with bacterially expressed and highly purified AMPK ([55], Suter and Neumann, unpublished). In addition, PCr as the disodium salt inhibits a number of other protein kinases, suggesting that any effects are nonspecific (Hardie, unpublished). Thus, it now appears that neither PCr nor Cr exerts a direct effect on AMPK or its upstream kinases LKB1 and CaMKK (Suter and Neumann, unpublished). In a human study involving retraining after immobilization with or without Cr supplementation, it was found that retraining activated AMPK to the same extent in both groups; thus, AMPK appears not to be involved in the beneficial effects of oral Cr supplementation during rehabilitation [56].

As to the proposed stable complex between AMPK and CK [54], in a cell *in vivo* this would lead to permanent complex formation of AMPK, a protein of rather low abundance, with the very abundant CK. This would not make much physiological sense for the manifold proposed functions of AMPK [57]. In fact, no direct interaction between recombinant CK and AMPK isoforms could be detected by surface plasmon resonance spectroscopy (Schlattner et al., unpublished). Concerning the phosphorylation of CK by AMPK, confirmation of the stoichiometry of phosphorylation and identification of the AMPK phosphosite on CK, as well as a demonstration that phosphorylation also occurs *in vivo*, e.g., by using anti-CK phosphosite-specific antibodies, are still lacking, and this issue remains a matter of uncertainty. Therefore, considering the present knowledge, no more than an indirect connection between CK, AdK, and AMPK can be considered an established fact (see Figs. 9.1 and 9.2).

9.2.3
Some Considerations on the Sensing and Signaling Properties of AMPK

Although it is well known that AMPK binds AMP at concentrations in the low micromolar range and is activated by elevated AMP:ATP ratios [12, 58], only recently exact data for allosteric activation of $\alpha 1\beta 1\gamma 1$ and $\alpha 2\beta 2\gamma 1$ AMPK complexes have become available [40]. Using purified recombinant AMPK and a novel HPLC-based AMPK activity *in vitro* assay, it was possible to determine half-maximal and maximal stimulation of AMPK at concentrations as low as approximately 2 μM and 10 μM AMP, respectively [40]. In addition, it was found that AMPK was stimulated by a factor of 1000-fold by the combined action of upstream kinase and AMP [40]. Obviously, AMP and AMPK are extremely well suited to function together as a cellular energy indicator and sensor pair at the physiologically relevant AMP concentrations. AMPK can thus be portrayed as a "ratiometric energy sensor" monitoring the AMP:ATP ratio, which is extremely low at rest and rises exponentially under cellular stress conditions (Fig. 9.2B,C).

Such a paradigm can be very effective for rapid and stringent regulation of cellular processes, as also exemplified in Ca^{2+} signaling, where relatively small changes in the very low resting $[Ca^{2+}]$ during cell activation, in combination with local Ca^{2+} sparks and Ca^{2+} waves, produce strong dynamic and versatile signaling effects [59]. Thus, *in vivo* functioning of AMP/AMPK signaling may be analogous to that observed in Ca^{2+} signaling and Ca^{2+} sensing by Ca^{2+}-responsive proteins. The free cytosolic $[Ca^{2+}]$ corresponds to about 0.05 μM at rest, but increases up to 50 μM after cell activation. Ca^{2+} sensors are able to bind Ca^{2+} specifically at sub-micromolar concentrations, even in the presence of the millimolar concentrations of free Mg^{2+} typically found in eukaryotic cells [60]. AMPK is also able to sense sub-micromolar concentrations of AMP in the presence of millimolar concentrations of ATP [61]. As in Ca^{2+} signaling, a small absolute change in [AMP] can exert a large effect on cell metabolism, transmitted via AMPK. As observed in Ca^{2+} signaling [59] or cyclic AMP signaling [62], AMP/AMPK signaling is also likely to be organized in microdomains in specific subcellular AMP/AMPK compartments, such that metabolic "AMP sparks" and "AMP waves" that are either contained locally or propagated through the cell, respectively, may lead to selective temporal and spatial activation of AMPK and phosphorylation of its downstream targets within a cell.

9.3
Mammalian AMPK Is a Member of an Ancient, Conserved Protein Kinase Family

9.3.1
AMPK in the Eukaryotic Kinome

AMPK is a serine/threonine protein kinase that is not particularly closely related to, but still sometimes is confused with, the cyclic AMP-dependent protein kinase (PKA). In fact, the AMPK system was one of the first protein kinase cascades described back in the 1970s. However, it was not until 1987 that Hardie and Carling showed that AMPK was a multi-substrate kinase that was activated by AMP [63, 64]. With the increasing number of genomes that have been sequenced, two important facts have become apparent. Firstly, AMPK is part of a structurally related family of serine/threonine protein kinases (the AMPK-related kinases [ARKs]) comprising around 14 members [65]. The function of most of these members is not completely clear, although some appear to have a role in cell polarity, cell proliferation, and cell communication (see below). Secondly, AMPK homologues occur in all eukaryotic genomes that have been sequenced to date, from protozoa and yeast to plants and human [66]. AMPK is therefore phylogenetically one of the most ancient eukaryotic protein kinases. It appears to have evolved at a very early stage during the development of eukaryotes and may fulfill basic functions specific to the highly compartmentalized eukaryotic cell.

9.3.2
Structure of AMPK

The basic functional unit of AMPK, from yeast [67] to mammals, seems to be a heterotrimeric complex consisting of a catalytic α-subunit and regulatory β- and γ-subunits, as schematically depicted in Fig. 9.3 (for reviews, see [64, 68]). Further complexity is added by the existence of multiple subunit genes encoding each subunit, giving rise to different isoforms and splice variants. In mammals, there are two α-, two β-, and three γ-subunit genes, with the $\gamma2$ and $\gamma3$ genes also giving rise to short and long splice variants.

The α-subunit bears a classical protein kinase catalytic domain, whose X-ray structure has been solved [69], and an inhibitory activation loop involved in its regulation [70, 71]. Different phosphorylation sites that depend on upstream kinases have been mapped, including the "activation loop" or "T-loop" Thr172, which is the key site for AMPK activation by different upstream kinases (see below). The β-subunit seems to interact with both the α- and the γ-subunits [72, 73] and also bears a glycogen-binding domain [72, 74] an N-terminal myristoyl moiety, and some specific phosphorylation sites. While the function of the latter modifications is not clear, the X-ray structure of the glycogen-binding domain has been solved [75], and functions in subcellular targeting to glycogen granules and regulation of glycogen metabolism have been proposed. The γ-subunits have attracted most of the recent interest. These subunits show a large variability, with three isoforms encoded by distinct genes ($\gamma1$, $\gamma2$, and $\gamma3$) and with $\gamma2$ and $\gamma3$ exist-

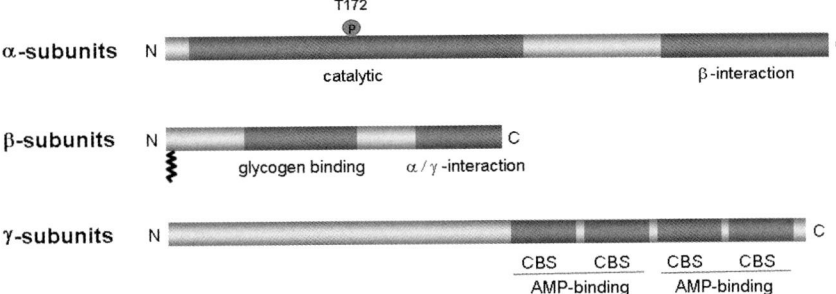

Fig. 9.3 Domain structure of mammalian AMPK. Schematic representation of the domain structure of the three AMPK subunits with selected secondary modifications (activating phosphorylation of the α-subunit T172 residue and N-terminal myristoylation of the β-subunit). Isoforms exist for each subunit, with α1/α2 and β1/β2 showing the most homology, while the γ-subunit isoforms (γ1–3) and their splice variants (γ2/γ3 long and short) differ at their N-terminus (here, a long splice variant is shown).

ing as splice variants, having in their long forms large N-terminal extensions. Most importantly, each γ-subunit contains four so-called cystathionine β-synthetase (CBS) motifs that are organized in tandem pairs to form two AMP-binding sites, the so-called Bateman domains [76]. Several mutations have been mapped in these domains that lead to glycogen storage disorders and a related hereditary heart disease (see below) [77–80].

Although there are crystal structures for the kinase domain of the yeast homologue Snf1 [69], the human α2 subunit (PDB: 2H6D), and a small glycogen-binding domain of rat β1 [75] and human β2 (PDB: 2F15), the molecular structure of the heterotrimeric complex remains unknown. Based on indirect evidence, a current model proposes that the C-terminal domain of the β-subunit binds the C-terminal region of the α-subunit and an unknown region of the γ-subunit [72, 73]. This topology is consistent with a recent X-ray structure of the AMPK orthologue of the yeast *S. pombe* [81]. However, the yeast complex is possibly not activated by AMP, and has been highly truncated for successful crystallization, thus limiting functional and mechanistic insight. Only a structure of full-length heterotrimeric complex at molecular resolution will be able to settle these issues.

9.4
Regulation of AMPK

One important way to transmit extra- and intracellular signals is by reversible phosphorylation of proteins that participate in a cell signaling pathway. Phosphorylation and dephosphorylation of proteins is achieved by protein kinases and phosphatases, respectively. Among the >500 protein kinases encoded by the human genome [82], AMPK plays an important role in the regulation of mam-

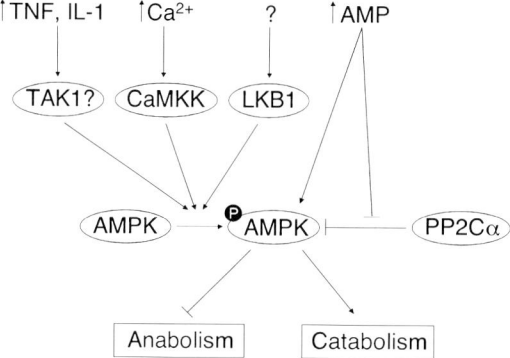

Fig. 9.4 Regulation of AMP-activated protein kinase (AMPK). LKB1 and Ca^{2+}/calmodulin-dependent protein kinase kinase (CaMKK) are capable of phosphorylating and activating AMPK in intact cells or *in vivo*, but the ability of transforming growth factor β–activated protein kinase (TAK1) to activate AMPK so far has been demonstrated only *in vitro*. Protein phosphatase 2Cα (PP2Cα) can dephosphorylate and deactivate AMPK. AMP allosterically activates AMPK and inhibits its dephosphorylation by PP2Cα. Ca^{2+} or tumor necrosis factor (TNF) and interleukin-1 (IL-1) are upstream signals that activate CaMKK and TAK1, respectively, but regulation of LKB1, if any, is not known. After activation, AMPK inhibits anabolic pathways to preserve cellular energy and activates catabolism to replenish ATP.

malian cellular and whole-body energy homeostasis [1, 83]. The requirement of biological systems to maintain high cellular ATP concentrations at all times, even under conditions of high metabolic workload, is reflected by dynamic and stringent control of AMPK activity. Regulation of AMPK activity is an elaborate process involving at least two upstream kinases and at least one protein phosphatase, as well as allosteric mechanisms [1, 8]. The complex regulatory network and metabolic signaling that influence the activity of AMPK is depicted schematically in Fig. 9.4. In general, following activation, AMPK inhibits anabolic pathways, whereas catabolic pathways are activated, although recent studies show that the kinase also regulates processes other than energy metabolism (see Section 9.5).

9.4.1
Allosteric Regulation of AMPK

The molecular mechanism of AMPK activation involves binding of AMP to the regulatory γ-subunit. The γ-subunits contain the two tandem Bateman domains [76], each of which binds one molecule of AMP or ATP (in a mutually exclusive manner) when expressed independently in bacteria [79]. When the tandem Bateman domains are expressed together, they bind two molecules of AMP or ATP with a Hill coefficient of 2, indicating strongly cooperative binding [79]. In the γ2 isoform, which is expressed at particularly high levels in cardiac muscle, several dominant-acting mutations are associated with heart diseases of varying degrees of severity, involving cardiac hypertrophy, contractile dysfunction, and arrhythmias. These mutations have been found to impair both the binding of AMP to the isolated Bateman domains and the activation of the heterotrimeric complex by AMP [79, 84], proving that the Bateman domains represent the regulatory nucleotide-binding sites of the complex. The disease-causing mutations can occur in either of the tandem domains, suggesting that AMP has to bind to both domains for activation to occur. At least nine different point mutations have now been reported. Five of them (R302Q, H383R, R384T, R531G, and R531Q) convert positively charged side chains of conserved basic residues in the first, second, and fourth CBS motifs (which are probably involved in binding the negatively charged phosphate groups of AMP and ATP) to amino acids with uncharged side chains [79, 84].

Although defects in AMP activation represent a loss-of-function effect, the mutations also cause reduced binding of the inhibitory nucleotide, ATP [79, 84], and this appears to increase the basal activity of the mutant complex [79, 84–86]. This gain-of-function effect probably explains why these mutations are dominant in nature. The primary cause of the heart diseases associated with these mutations appears to be excessive storage of glycogen in the cardiac myocytes [84, 85], possibly because the increased basal activity of the AMPK complex leads to increased basal glucose uptake. In cardiac myocytes, this excessive glycogen storage is harmful and leads to improper development and function of the heart muscle. Interestingly, a similar mutation (R200Q) affecting a conserved basic residue in the first CBS motif of the $\gamma3$ isoform (which is expressed at the highest levels in skeletal muscle) leads to abnormal deposition of glycogen in skeletal muscle in pigs [77]. The high glycogen content affects meat quality, but muscle function in the pigs does not appear to be adversely affected.

Recent studies in Hardie's laboratory have suggested a novel mechanism to explain allosteric activation by AMP [87]. The second CBS motif in the γ-subunits of all species contains a "pseudosubstrate" sequence that resembles the sequence at target sites for AMPK, except that it has a non-phosphorylatable hydrophobic residue in place of serine or threonine. Evidence was provided that in the absence of AMP, this sequence interacts with and occludes the substrate-binding groove on the kinase domain of the α-subunit [87]. However, binding of AMP prevents this interaction, thus relieving inhibition of the kinase domain. Intriguingly, basic residues in the second CBS motif (H383 and R384 in human $\gamma2$) proposed to be involved in the interaction of the pseudosubstrate with the substrate-binding groove are identical with those thought to be involved with binding the α phosphate group of AMP. If this model is correct, inhibition of the kinase domain by the pseudosubstrate sequence and binding of AMP would be mutually exclusive, providing an elegant explanation as to how AMP causes allosteric activation [87]. Although the details remain unclear, inhibition of the kinase domain by the pseudosubstrate sequence could act in concert with inhibition by the proposed autoinhibitory domain on the α-subunit itself, because the latter is believed to bind on the opposite face of the kinase domain to the substrate-binding groove [71].

Besides allosterically stimulating AMPK activity, binding of AMP also protects the complex against dephosphorylation of Thr172 by protein phosphatases, especially PP2Cα [40, 58]. It was originally reported that AMP also promoted phosphorylation of AMPK by an upstream kinase [88], later identified as LKB1 [89, 95]. However, this has not been observed in highly purified, reconstituted systems using AMPK and LKB1 complexes expressed in bacteria [40, 90]. This raises the possibility that the effects of AMP to promote phosphorylation observed previously, with preparations of upstream and downstream kinases purified from mammalian cells [88, 89], may have been due to contamination of one of the preparations with PP2C. Protein phosphatase 2Cα has been detected in preparations of AMPK and LKB1 purified from rat liver, and the effect of AMP was shown to imply protection against α-subunit Thr172 dephosphorylation [90]. In any case the net result, i.e., that AMP stimulates phosphorylation of Thr172 and hence activation of AMPK in the intact cell, remains the same irrespective of the exact mechanism.

Two other pairs of metabolites affected by cellular energy or redox status, namely, PCr/Cr [54] and NADH/NAD$^+$ [91], also have been claimed to be allosteric regulators of AMPK. However, these findings have not been substantiated by more recent studies [40, 55]. Thus, AMP and ATP, exerting their effects both by allosteric regulation and by modulating dephosphorylation, remain the primary regulators of AMPK activity.

9.4.2
Regulation of AMPK by Upstream Kinases

AMPK complexes are active only after phosphorylation at the T-loop Thr172 of the catalytic α-subunit by upstream kinases [92, 93]. There are at least two protein kinases capable of phosphorylating this residue *in vivo*, namely, LKB1 [89, 94, 95] and Ca^{2+}/calmodulin-dependent kinase kinase, especially the β isoform (CaMKKβ) [96–98]. LKB1 is a serine/threonine kinase with a tumor suppressor function. Mutations in the *LKB1* gene (also known as *STK11*) cause an inherited cancer susceptibility termed Peutz-Jeghers cancer syndrome (PJS) and are also observed in some sporadic cancers. In lower eukaryotes such as *Drosophila melanogaster* and *Caenorhabditis elegans*, genetic studies of LKB1 orthologs identified them as playing important roles in cell polarity (reviewed in [2, 99]). Thus, the involvement of LKB1 in the regulation of energy metabolism by activation of AMPK was unexpected and suggested a new connection between AMPK and cancer [17, 100]. Besides phosphorylating AMPK, LKB1 may exert its effects by phosphorylating and activating at least 12 other protein kinases, all members of the AMPK-related kinase family [65, 101]. Interestingly, some of these appear to have roles in the establishment of cell polarity. However, the available evidence suggests that the tumor suppressor function of LKB1 is related to its ability to activate AMPK, at least in part via the ability of the latter to inhibit mTOR signaling linked to cell growth and proliferation [2, 102, 103].

The active form of LKB is a heterotrimeric complex with an armadillo repeat-containing protein, called mouse protein 25 (MO25), and a catalytically inactive protein kinase, a pseudokinase called STRAD, both of which exist as two different isoforms, α and β [104]. In cell-free assays the LKB1–MO25α–STRADα complex is more potent in activating AMPK compared with complexes containing the β isoforms [89]. STRAD contains a pseudokinase domain related to those in conventional serine/threonine protein kinases, but it appears to be catalytically inactive. Interestingly, this domain still binds ATP, although this does not appear to be necessary for its ability to activate LKB1 [105].

While evidence is compelling that LKB1 is indeed the primary upstream kinase for AMPK in many tissues including muscle [106] and liver [107], it remains unclear whether LKB1 is itself regulated. Interestingly, it does not require phosphorylation of its own T-loop residue to be active [105], although binding of LKB1 to STRAD–MO25 is essential [99]. LKB1 is phosphorylated at multiple sites by several kinases [2, 99], but mutation of these sites does not alter the activity of the kinase, and it has therefore been proposed that the LKB1 complex is constitutively active [2]. Because it acts upstream of at least 14 downstream kinases that appear to have distinct functions, it may actually make sense for this to be the case, with any regulatory molecules such as AMP binding to the downstream kinases instead. The idea that LKB1 is constitutively active is also consistent with findings that active complexes were obtained after co-expression of LKB1, MO25α, and STRADα in *E. coli* ([40] and Neumann et al. (2007), Mol. Biotechnol., in press). Because bacteria do not perform most of the post-translational protein modifications seen in higher eukaryotes (such as phosphorylation), this suggests that prior covalent modifications are not required for LKB1 activity. When expressed on its own in mammalian cells, LKB1 is exclusively nuclear, but co-expression with MO25 and STRAD causes it to relocalize from the nucleus to the cytoplasm [104]. The nuclear form that occurs in the absence of MO25 and STRAD appears to be completely inactive, but the question of whether this form has any physiological relevance remains unclear.

CaMKKβ and, to a lesser extent, also CaMKKα are capable of phosphorylating and activating AMPK, thereby linking cytoplasmic Ca^{2+} levels to AMPK activity [96–98]. These CaMKKs were originally identified as the upstream kinases in the Ca^{2+}/calmodulin-dependent protein kinase-I/IV (CaMKI/CaMKIV) signaling pathways [108]. In response to various stimuli, intracellular Ca^{2+} levels increase from around 100 nM to 1–2 μM [108]. One of the key binding proteins that convey signals elicited by elevated Ca^{2+} levels is calmodulin (CaM). Ca^{2+} binding to CaM induces a local conformational change, causing a global structural change that allows the Ca^{2+}–CaM complex to bind and activate several enzymes, including CaMKKs [109]. CaMKKs in turn phosphorylate and activate CaMKI and CaMKIV, which are themselves Ca^{2+}–CaM dependent [109–111].

The tissue distribution of CaMKKs, with the highest levels in brain, suggests that they may play important roles in the nervous system. Indeed, mutant mice lacking CaMKKβ were impaired in some hippocampal-related types of long-term memory formation [112]. A role for AMPK in regions of the hypothalamus that are involved in the regulation of food intake has been described recently [113]. Whether AMPK has crucial functions in other regions of the brain has not been investigated, but given the ubiquitous expression of AMPK, coupled with the high energy demands and the importance of Ca^{2+}-dependent signaling in excitable cells [114], one would expect the AMPK pathway to be of particular importance in the nervous system. Indeed, several brain-specific potential downstream-targets of AMPK have been recently identified based on a new proteomic screening approach (MudSeeK) (R. Turk et al. (2007), J. Proteome Res., in press). Although CaMKKα is largely restricted to neuronal tissue, CaMKKβ is also expressed in cells of the endothelial/hematopoietic lineage. It has recently been reported that AMPK is activated by thrombin in endothelial cells [115] and by stimulation of the antigen receptor in T cells [116], and in both cases this appeared to involve the Ca^{2+}–CaMKK–AMPK pathway.

Very recently, mammalian transforming growth factor β–activated kinase (TAK1), in complex with its accessory protein TAB1, was identified as a third possible upstream kinase capable of activating AMPK, using expression of a human kinase library in a yeast strain in which the three endogenous upstream kinases had been knocked out [117]. This suggests that TAK1 can activate the yeast AMPK ortholog but does not prove that this is physiologically relevant. A role for TAK1 upstream of AMPK in the heart was also recently suggested [118], although in that case it was proposed that TAK1 was in fact upstream of LKB1, which is not consistent with the data obtained from the yeast system [117]. The role of TAK1 in the AMPK system therefore remains uncertain.

AMPK is also a direct target of Akt/PKB [119, 120] and PKA [121], which appear to phosphorylate the same sites on the catalytic α-subunit, i.e., Ser485 in α1 or Ser491, the equivalent site in α2. Phosphorylation of this site, which also may be catalyzed by AMPK itself in an autophosphorylation reaction, appears to reduce the accessibility of Thr172 to upstream kinases [119, 121]. CaMKK is also phosphorylated and inhibited by PKA [109], while LKB1 is phosphorylated by PKA at its C-terminal Ser428 residue. In the latter case this does not appear to affect its activity or its localization [122]. Nevertheless, these observations suggest that possible crosstalk between the AMPK pathway and other signaling pathways warrants further investigation.

9.4.3
Inactivation of AMPK by Protein Phosphatases

Although much progress has been made in identifying the upstream kinases that activate AMPK by phosphorylating Thr172, much less is known about the protein phosphatases that reverse this process. Protein phosphatase 2Cα (PP2Cα) and protein phosphatase 2A (PP2A) are both capable of dephosphorylating AMPK at Thr172 *in vitro*, whereas protein phosphatase 1 (PP1) is much less effective. However, PP2A does not seem likely to dephosphorylate AMPK *in vivo*, as has been shown by the use of inhibitors of this protein phosphatase family in cultured cells

[58, 123, 124]. Unfortunately, specific inhibitors of PP2Cα have not been described and supportive genetic evidence is lacking, thus leaving open the possibility that protein phosphatases other than or apart from PP2Cα might be responsible for the inactivation of AMPK *in vivo*. However, AMP clearly attenuates the dephosphorylation of AMPK by PP2Cα, which can be attributed to binding of AMP to the regulatory γ-subunit of AMPK rather than direct inhibition of PP2Cα by AMP [40, 58, 90]. Moreover, a recent study showing an inverse relationship between PP2C expression and the phosphorylation state of AMPK at Thr172 supports the conclusion that PP2C plays a major role in AMPK deactivation *in vivo* [125].

9.5
Signaling Downstream of AMPK

9.5.1
Role of AMPK

The role of AMPK as a sensor of cellular energy status was proposed in part on the basis that its first described metabolic targets were in energy-consuming pathways and were inactivated upon phosphorylation, whereas enzymes of catabolic processes were activated by phosphorylation [1, 12, 64, 126]. AMPK is activated in response to hypoxia or ischemia in several tissues and in response to an increase in energy demand during exercise in skeletal muscle. The phosphorylation-induced inactivation of acetyl-CoA carboxylase (ACC), 3-hydroxy-3-methylglutaryl-CoA reductase (HMGR), and glycogen synthase (GS) by AMPK inhibits *de novo* fatty acid, cholesterol, and glycogen synthesis, respectively. AMPK activation also inhibits protein synthesis, a major consumer of ATP, both via the phosphorylation of eukaryotic elongation factor-2 (eEF2) kinase (eEF2K), leading in turn to the phosphorylation and inactivation of eEF2, and via a reduction in mTOR signaling. In contrast, the AMPK-induced increase in glucose uptake in skeletal muscle and 6-phosphofructo-2-kinase activation in cardiac muscle stimulate glycolysis, thereby helping to maintain intracellular ATP levels. ACC phosphorylation (particularly phosphorylation of the ACC2 isoform) is also important in this respect, because its inactivation by AMPK is one factor that determines cytosolic malonyl-CoA concentrations [47]. A decrease in malonyl-CoA stimulates fatty acid oxidation by relieving inhibition of carnitine palmitoyl-CoA acyl transferase-1 (CPT1), via which fatty acids enter mitochondria.

It has recently become apparent that AMPK is also involved in whole-body energy homeostasis, stimulating energy expenditure by promoting fatty acid and glucose oxidation in the periphery, while at the same time inhibiting energy intake via effects on appetite in the hypothalamus [12]. The role of AMPK probably needs to be broadened even further following the discovery of upstream activating AMPK kinases other than LKB1, i.e., CaMKKs [96–98] and perhaps also TAK1 [117]. AMPK activation via these pathways would not necessarily occur only during energy stress. While the primary targets for AMPK have been involved mainly in energy homeostasis, AMPK now seems to control non-metabolic processes such as cell growth, progression through the cell cycle, and organization of the cytoskeleton.

9.5.2
The AMPK Recognition Motif

A list of some AMPK targets and their phosphorylation site sequences is given in Table 9.1. The core substrate recognition motif for AMPK is $\phi\,(X, \beta)\,X\,X\,S/T\,X\,X\,X\,\phi$, where ϕ is a hydrophobic residue (M, V, L, I, or F), β is a basic residue (R, K, or H), and the parentheses indicate that the order of residues at the P-4 and P-3 positions is not critical [127]. For ACC1 and some other substrates, an amphipathic helix stretching from P-16 to P-5 is also a positive determinant [128]. It can be seen in Table 9.1 that AMPK is predominantly a serine-directed protein kinase and that in some proposed targets the consensus is not strictly adhered to. For example, the site in eEF2K does not contain the basic residue at the −3 position, and in eEF2K, mTOR, and the Ser1176 site of NOS III, the +4 position is not occupied by a hydrophobic amino acid. However, the P+4 site is less critical and NOS III contains a glutamine residue that can substitute for a hydrophobic residue at this position [128]. An interesting possibility is that the various AMPK isoforms might differ slightly in substrate recognition. Without casting doubt on the results published in the literature, some caution needs to be exercised before a target protein can be considered a *bona fide* AMPK substrate (see below). For example, some AMPK targets have been reported on the basis of overexpression of recombinant proteins in eukaryotic cells incubated with pharmacological activators of AMPK, such as AICA riboside, or subjected to treatments that deplete intracellular ATP. These methods are not necessarily completely specific for AMPK, and in any case this approach does not prove that the protein is *directly* phosphorylated by AMPK. Also in some cases, the sites for AMPK phosphorylation have not been identified formally by direct sequencing or mass spectrometry. Instead, this has been performed by searching target protein sequences for the AMPK consensus, then mutating putative phosphorylation sites to negatively charged (to mimic phosphorylation) or non-phosphorylatable residues, and looking at effects on phosphorylation or function. This method is prone to error because mutations can lead to conformational effects at sites remote from the mutation itself.

9.5.3
A Note of Caution: The Krebs–Beavo Criteria

In a classic review of reversible phosphorylation of enzymes, Krebs and Beavo proposed four criteria for establishing that changes in enzyme activity as a result of phosphorylation/dephosphorylation would be of physiological relevance [129]. These criteria can be extended to any changes in protein function resulting from reversible phosphorylation, but sadly they have been met for only a few AMPK targets and, indeed, for targets of most other protein kinases. The four criteria are as follows:

1. The enzyme (protein) should be phosphorylated stoichiometrically "at a significant rate" in cell-free assays by an appropriate protein kinase and dephosphorylated by a protein phosphatase.

2. The functional properties of the enzyme (protein) should undergo meaningful changes that correlate with the extent of phosphorylation.

3. The enzyme (protein) should be shown to be phosphorylated and dephosphorylated *in vivo* or in an intact cell system with accompanying functional changes. This criterion is easier to satisfy than it was at the time when Krebs and Beavo wrote their review. In those days, to establish that a protein was phosphorylated *in vivo* in response to activation of a kinase, labeling of the ATP pool of isolated cells or organs with millicurie amounts of ^{32}P-labeled inorganic phosphate had to be used. Nowadays, changes in the phosphorylation states of proteins at specific sites can easily be assessed without radioactive labeling by using phosphorylation site–specific antibodies.

4. There should be a correlation between cellular levels of protein kinase effectors and the extent of phosphorylation of the enzyme. When Krebs and Beavo proposed their criteria for establishing physiologically relevant phosphorylation/dephosphorylation of enzymes, this was based mainly on the study of protein phosphorylation in response to changes in cyclic AMP. To verify that AMPK is the physiologically relevant kinase for a particular target, changes in cellular AMPK activity should be correlated with the extent of phosphorylation of the substrate protein at the relevant site. This could be done by incubating cells with a selective AMPK inhibitor, such as "compound C" [130], by transfecting cells with vectors overexpressing constitutively active and/or dominant-negative forms of AMPK [131], by using genetic approaches [132, 133], by knockdown of the AMPK subunits by transfecting small interfering RNAs [134], or by using mouse knockout models [132, 133]. The catalytic subunit knockouts ($\alpha1^{-/-}$ and $\alpha2^{-/-}$) have proved useful in this respect. Unfortunately, a double knockout is embryonic lethal, although mouse embryo fibroblasts in which both catalytic subunit isoforms have been deleted are now available. In addition, tissue-specific, conditional double knockouts are also now being generated.

All of the approaches mentioned under criterion 4 have some drawbacks (e.g., lack of isoform specificity with the dominant-negative approach, incomplete AMPK inhibition using the siRNA approach, or compensatory changes in the expression of other AMPK catalytic subunit isoforms or indeed of other related kinases when using knockdowns or knockouts of AMPK [see below]). The overexpression of a dominant-negative construct may not target endogenous AMPK in the right intracellular compartment, or it may titrate out signaling molecules

Table 9.1 Phosphorylation sites in AMPK substrate proteins. The phosphorylated serine/threonine residue is indicated in red, hydrophobic residues that form the AMPK consensus are in blue and basic residues of the consensus are in green.

Target	Phosphorylation site sequence	Comments A: Activation I: Inactivation or inhibition	Refs.
Metabolic targets			
Ser79 of rat acetyl-CoA carboxylase (ACC1)	HMRSSMSGLHLVK	I. Inhibitory for fatty acid synthesis.	[164]
Ser221 of human acetyl-CoA carboxylase (ACC2)	TMRPSMSGLHLVK	I/A. Inhibits ACC2 but stimulates fatty acid oxidation by lowering [malonyl-CoA]. Site identified by homology with the site in rat ACC1.	[165]
Ser871 of rat 3-hydroxy-3-methylglutaryl-CoA reductase (HMGR)	HMVHNRSKINLQD	I. Inhibitory for cholesterol biosynthesis.	[166]
Ser565 of rat hormone-sensitive lipase (HSL)	SMRRSVSEAALAQ	I. Ser563 upstream of Ser565 is phosphorylated by PKA.	[144]
Ser7 of rabbit muscle glycogen synthase (GS)	PLSRTLSVSSLPG	I. This is the so-called 'site 2' that is also phosphorylated by PKA *in vitro*.	[167]
Ser466 of bovine heart 6-phosphofructo-2-kinase (PFK2FB2)	VRMRRNSFTPLSS	A. Site can also be phosphorylated by PKA, PKCs, and insulin-stimulated protein kinases such as PKB.	[147]
Ser461 of human inducible 6-phosphofructo-2-kinase (PFK2FB3)	PLMRRNSVTPLAS	A. Stimulatory for glycolysis in activated monocytes subjected to hypoxia.	[168]
Transcription factors			
Ser171 of rat transducer of regulated CREB activity (TORC2)	ALNRTSSDSALHT	I. This site is phosphorylated by the AMPK-related Salt-Inducible Kinase-2 (SIK2). Phosphorylation sequesters TORC in the cytoplasm where it cannot co-activate CREB.	[136, 137]

Ser568 of rat carbohydrate-response element-binding protein (ChREBP)	LLRPPESPDAVPE	I. Decreases DNA-binding activity to the promoter of the L-type pyruvate kinase gene.	[169]
Ser304 of human hepatic nuclear factor a (HNF4α)	KIKRLRSQVQVSL	I. Decreases formation of homodimers and DNA binding, and promotes degradation.	[170]
Ser89 of human transcriptional co-activator p300	ELLRSGSSPNLNM	I. Blocks its ability to interact with nuclear receptors such as PPARγ.	[171]

Proteins involved in the control of translation/cell growth

Thr2446 of the human mammalian target of rapamycin (mTOR)	KRSRTRTDSYSAG	I. Ser2448 downstream of this site can be phosphorylated in response to insulin by PKB or p70S6K.	[159]
Thr1227 of rat tuberous sclerosis complex 2 (TSC2)	TLPRSNTVASFSS	A. This, or the next site, may stabilize the TSC2–TSC1 complex.	[103]
Ser1345 of rat tuberous sclerosis complex 2 (TSC2)	PLSKSSSPELQT	A. This, or the previous site, may stabilize the TSC2–TSC1 complex.	[103]
Ser398 of human eukaryotic elongation factor 2 kinase (eEF2K)	SLPSPSSATPHS	A. Inhibits protein synthesis elongation by increasing eEF2 Thr56 phosphorylation.	[172]

Signaling proteins

Thr494 of human endothelial NO synthase (NOS III)	GITRKKTFKEVAN	A. Increases activity at low [Ca^{2+}/calmodulin].	[148]
Ser1176 of human endothelial NO synthase (NOS III)	SRIRTQSFSLQER	A. This site can be phosphorylated by PKB.	[148]
Ser789 of rat insulin receptor substrate-1 (IRS-1)	HLRLSSSSGRLRY	A. This site is also phosphorylated by Qin-induced kinase (QIK, also known as SIK2).	[173]
Ser621 of human/rat Raf-1	KINRSASEPSLHR	I?	[174]

necessary for the activation of other kinase cascades that converge on AMPK targets. Clearly, where possible, several approaches should be tested in controlled experiments where AMPK activity is monitored in cells, e.g., by measuring phosphorylation of ACC.

9.5.4
The AMPK-related Protein Kinases

Thirteen purified AMPK-related protein kinases (ARKs) in the human kinome, produced by bacterial overexpression, can phosphorylate the AMARA peptide often used to assay AMPK [65], indicating that they have similar specificity and that there may be some overlap between targets for AMPK and ARKs. All the ARKs except maternal embryonic leucine-zipper kinase (MELK) are phosphorylated in their T-loop and activated by LKB1. However, it is not yet known whether CaMKKs can activate any of the ARKs. The AMPK-related kinase Qin-induced kinase (QIK), also known as salt-inducible kinase-2 (SIK2), phosphorylates the same residue in human IRS1 (Ser794) [135] as that phosphorylated by AMPK in rat IRS1 (see Table 9.1) and which increases insulin-stimulated PI 3-kinase activity. Also, SIK2 phosphorylates the Ser171 site in TORC2 [136] that is phosphorylated in response to AMPK activation [137]. The substrate specificities and roles of many of the ARKs remain poorly defined at present [65], although the microtubule affinity-regulating kinases (MARKs) are known to phosphorylate microtubule-associated proteins and are involved in regulating cell polarity [138]. Interestingly, NUAK2 (also known as SNARK, SNF1/AMPK-related kinase), has been reported to be activated, like AMPK, by cellular stresses such as glucose deprivation, ATP depletion, and hyperosmolarity [139]. In rat skeletal muscle, contraction, phenformin treatment, and AICA riboside treatment all increased the activity of the $\alpha2$ isoform of AMPK, but without increasing the activities of QSK, QIK, MARK2/3, or MARK4 [140].

9.5.5
Multi-site/Hierarchical Phosphorylation and Convergence of Signaling Pathways on AMPK Targets

Virtually all known AMPK targets are multi-site phosphorylated proteins containing either more than one site for AMPK (e.g., ACC, TSC2, and NOS III) or sites for other protein kinases, including PKA (e.g., ACC, glycogen synthase, HSL, heart PFK-2, and eEF2K) and PKB (e.g., heart PFK-2, mTOR, TSC2, and NOS III). Hierarchical phosphorylation of adipocyte HSL by AMPK at Ser565 has an antilipolytic effect [141–143] by decreasing subsequent phosphorylation and activation of HSL by PKA in response to β-agonists [144]. The major regulatory site for PKA was originally thought to be Ser563, adjacent to the AMPK site, although it now appears that Ser559 and/or Ser660 are more important for controlling HSL activity [145]. In muscle cells, but not adipocytes, AMPK activation appears to inhibit adrenaline-induced HSL activation via a decrease in Ser660 rather than a decrease in Ser563 phosphorylation [146]. AMPK phosphorylation sites are often adjacent to sites phosphorylated by other protein kinases. In other cases, the AMPK site itself is also phosphorylated by a protein kinase from another signaling pathway (see Table 9.1). For example the Ser466 site in heart PFK-2 is phosphorylated not only by AMPK but also by PKA, PKC, and insulin-stimulated protein kinases such as PKB [147]. There are cases where the AMPK and insulin signaling (PKB) pathways converge on the same target at the same site and with the same effects, such as heart PFK-2 [147] and NOS III [148, 149]. Both AMPK activation and insulin also stimulate glucose uptake in muscle via the recruitment of GLUT4 transporters to the plasma membrane. The Rab GTPase-activating protein Akt substrate (AS160) regulates GLUT4 translocation and is phosphorylated at six PKB sites in response to insulin [150]. Moreover, AICA riboside treatment and contraction increased the phosphorylation of AS160 in skeletal muscle [151, 152]; however, the phosphorylation sites for AMPK in AS160 have yet to be identified.

In contrast to the stimulation of glucose uptake, heart PFK-2, and NOS III activation, the insulin and AMPK pathways have opposite effects on protein synthesis. Insulin stimulates protein synthesis by activating the mTOR pathway, which controls translation initiation, and by inactivating eEF2K, allowing elongation to proceed by decreasing eEF2 phosphorylation levels. AMPK activation decreases protein synthesis via eEF2K activation [153] and also by inhibiting mTOR signaling [154], but in some cases the latter effect was evident only after the mTOR pathway had first been stimulated by growth factors/insulin or amino acids [155, 156]. PKB and AMPK have been reported to phosphorylate distinct sites on both TSC2 (Ser939/1086/1088/Thr1422 for PKB [157] and Thr1227/Ser1345 for AMPK [103]) and mTOR (Ser2448 [158] for PKB and Thr2446 [159] for AMPK) with opposing effects. Therefore, the possibility of control via hierarchical phosphorylation by PKB and AMPK of these target proteins obviously needs further investigation.

In conclusion, while the number of papers published on AMPK is still on the rise, the number of *bona fide* target proteins is lagging behind. There are likely to be many more targets to be identified in view of the fact that AMPK is an ancient eukaryotic kinase and that its role extends beyond metabolism and control of cellular and whole-body energy homeostasis.

9.6
Conclusions and Perspectives

Recent years have seen a rapid development in our understanding of the function of AMPK. From a simple "energy sensor" concept, the AMPK signaling pathway has developed into a complex machinery for controlling metabolic stress responses, which integrates signals from within the cell as well as from the cellular environment and the whole organism. In addition to its task in sensing cellular energy status via the high AMP affinity of its allosteric activation mechanism, additional activating pathways that involve a growing number of upstream kinases are now being defined. Two of the major drugs widely used for treating type II diabetes, metformin and thiazolidinediones [130, 160], activate AMPK, and their therapeutic effects, particularly in the case of metformin, appear to be largely mediated by AMPK signaling [107]. Furthermore, the identification of a known tumor suppressor, i.e., LKB1 [89, 94, 95], as a critical upstream kinase for AMPK has suggested a role for impaired AMPK in the development of cancer. In fact, activated AMPK negatively regulates proliferation and the cell cycle, mediated by mTOR and p53, respectively [2, 103, 161, 162]. Consequently, AMPK has been proposed as a drug target in the treatment of cancer as well as diabetes [16, 163]. The design of activators and inhibitors of AMPK will be greatly facilitated by a better understanding of the molecular structure of the AMPK complex, with only the X-ray structures of certain individual domains or the truncated heterotrimer of AMPK having been solved to date [69, 75, 81].

Acknowledgments

The authors were supported by the EXGENESIS Integrated Project (LSHM-CT-2004-005272) funded by the European Commission. Work by the authors has also been financially supported in the past by ETH graduate student stipends; by the Swiss National Science Foundation (grant no. 3100AO-102075 to T.W. and U.S.; grant no. 3100+0-11437/1 to T.W. and D.N.; Marie Heim-Vögtlin grant no. 3234-069276 to M.T.-S.); by an EU Marie Curie Fellowship (M.T.-S.); by Novartis Stiftung für medizinisch-biologische Forschung (U.S.); by Schweizerische Herzstiftung, Wolfermann-Nägeli-Stiftung, Schweizer Krebsliga, and Zentralschweizer Krebsliga (U.S. and T.W.); by the French Agence National de Recherche ("chaire d'excellence" awarded to U.S.); by the Novartis Foundation No. 05A07 and Helmut Horten Foundation (D.N. and T.W.); by the Belgian Fund for Medical Scientific Research (M.H.R.); by the Interuniversity Poles of Attraction Belgian Science Policy P5/05 (M.H.R.); and by the French Community of Belgium "Actions de Recherche Concertées" (M.H.R.). We acknowledge the work of many other researchers whose work could not be referenced due to space limitations.

References

1 Hardie, D. G., Hawley, S. A., Scott, J. W. (2006) AMP-activated protein kinase–development of the energy sensor concept. *J. Physiol. 574*, 7–15.

2 Alessi, D., Sakamoto, K., Bayascas, J. (2006) LKB1-Dependent Signaling Pathways. *Annu. Rev. Biochem. 75*, 137–163.

3 Sarbassov, D. D., Ali, S. M., Sabatini, D. M. (2005) Growing roles for the mTOR pathway. *Curr. Opin. Cell Biol. 17*, 596–603.

4 Inoki, K., Guan, K. L. (2006) Complexity of the TOR signaling network. *Trends Cell Biol. 16*, 206–212.

5 Wullschleger, S., Loewith, R., Hall, M. N. (2006) TOR signaling in growth and metabolism. *Cell 124*, 471–484.

6 Saks, V., Dzeja, P., Schlattner, U., Vendelin, M., Terzic, A., Wallimann, T. (2006) Cardiac system bioenergetics: metabolic basis of the Frank-Starling law. *J. Physiol. 571*, 253–273.

7 Neumann, D., Schlattner, U., Wallimann, T. (2003) A molecular approach to the concerted action of kinases involved in energy homoeostasis. *Biochem. Soc. Trans. 31*, 169–174.

8 Hardie, D. G., Sakamoto, K. (2006) AMPK: a key sensor of fuel and energy status in skeletal muscle. *Physiology 21*, 48–60.

9 Atkinson, D. E. (1968) The energy charge of the adenylate pool as a regulatory parameter. Interaction with feedback modifiers. *Biochemistry 7*, 4030–4034.

10 Hardie, D. G., Hawley, S. A. (2001) AMP-activated protein kinase: the energy charge hypothesis revisited. *Bioessays 23*, 1112–1119.

11 Hardie, D. G. (2000) Metabolic control: a new solution to an old problem. *Curr. Biol. 10*, R757–759.

12 Kahn, B. B., Alquier, T., Carling, D., Hardie, D. G. (2005) AMP-activated protein kinase: ancient energy gauge provides clues to modern understanding of metabolism. *Cell Metab. 1*, 15–25.

13 Fryer, L. G., Carling, D. (2005) AMP-activated protein kinase and the metabolic syndrome. *Biochem. Soc. Trans. 33*, 362–366.

14 Witters, L. A., Kemp, B. E., Means, A. R. (2006) Chutes and Ladders: the search for protein kinases that act on AMPK. *Trends Biochem. Sci. 31*, 13–16.

15 Hallows, K. R. (2005) Emerging role of AMP-activated protein kinase in coupling membrane transport to cellular metabolism. *Curr. Opin. Nephrol. Hypertens. 14*, 464–471.

16 Hardie, D. G. (2007) AMP-Activated Protein Kinase as a Drug Target. *Annu. Rev. Pharmacol. Toxicol. 47*, 185–210.

17 Luo, Z., Saha, A. K., Xiang, X., Ruderman, N. B. (2005) AMPK, the metabolic syndrome and cancer. *Trends Pharmacol. Sci. 26*, 69–76.

18 Wallimann, T., Wyss, M., Brdiczka, D., Nicolay, K., Eppenberger, H. M. (1992) Intracellular compartmentation, structure and function of creatine kinase isoenzymes in tissues with high and fluctuating energy demands: the 'phosphocreatine circuit' for cellular energy homeostasis. *Biochem. J. 281*, 21–40.

19 Wallimann, T., Hemmer, W. (1994) Creatine kinase in non-muscle tissues and cells. *Mol. Cell. Biochem. 133–134*, 193–220.

20 Noma, T. (2005) Dynamics of nucleotide metabolism as a supporter of life phenomena. *J. Med. Invest. 52*, 127–136.

21 Kinukawa, M., Nomura, M., Vacquier, V. D. (2007) A sea urchin sperm flagellar adenylate kinase with triplicated catalytic domains. *J. Biol. Chem. 282*, 2947–2955.

22 Wallimann, T. (1996) 31P-NMR-measured creatine kinase reaction flux in muscle: a caveat! *J. Muscle Res. Cell Motil. 17*, 177–181.

23 Martonosi, A., Gouvea, M. A., Gergely, J. (1960) Studies on actin. III. G-F transformation of actin and muscular contraction (experiments *in vivo*). *J. Biol. Chem. 235*, 1707–1710.

24 Hancock, C. R., Brault, J. J., Wiseman, R. W., Terjung, R. L., Meyer, R. A. (2005) 31P-NMR observation of free ADP during fatiguing, repetitive contractions of murine skeletal muscle lacking AK1. *Am. J. Physiol. 288*, C1298–1304.

25 Hancock, C. R., Janssen, E., Terjung, R. L. (2006) Contraction-mediated phosphorylation of AMPK is lower in skeletal muscle of adenylate kinase-deficient mice. *J. Appl. Physiol. 100*, 406–413.

26 Ingwall, J. S. (2002) Is creatine kinase a target for AMP-activated protein kinase in the heart? *J. Mol. Cell. Cardiol. 34*, 1111–1120.

27 Hellsten, Y., Richter, E. A., Kiens, B., Bangsbo, J. (1999) AMP deamination and purine exchange in human skeletal muscle during and after intense exercise. *J. Physiol. 520*, 909–920.

28 Wadley, G. D., Lee-Young, R. S., Canny, B. J., Wasuntarawat, C., Chen, Z. P., Hargreaves, M., Kemp, B. E., McConell, G. K. (2006) Effect of exercise intensity and hypoxia on skeletal muscle AMPK signaling and substrate metabolism in humans. *Am. J. Physiol. 290*, E694–702.

29 McGilvery, R. W., Murray, T. W. (1974) Calculated equilibria of phosphocreatine and adenosine phosphates during utilization of high energy phosphate by muscle. *J. Biol. Chem. 249*, 5845–5850.

30 Wallimann, T., Schnyder, T., Schlegel, J., Wyss, M., Wegmann, G., Rossi, A. M., Hemmer, W., Eppenberger, H. M., Quest, A. F. (1989) Subcellular compartmentation of creatine kinase isoenzymes, regulation of CK and octameric structure of mitochondrial CK: important aspects of the phosphoryl-creatine circuit. *Prog. Clin. Biol. Res. 315*, 159–176.

31 Dzeja, P. P., Zeleznikar, R. J., Goldberg, N. D. (1996) Suppression of creatine kinase-catalyzed phosphotransfer results in increased phosphoryl transfer by adenylate kinase in intact skeletal muscle. *J. Biol. Chem. 271*, 12847–12851.

32 Schlattner, U., Wallimann, T. (2004) Metabolite channeling: creatine kinase microcompartments, in *Encyclopedia of Biological Chemistry* (Lennarz, W. J., Lane, M. D., Eds.) pp 646–651, Academic Press, New York, USA.

33 Schlattner, U., Tokarska-Schlattner, M., Wallimann, T. (2006) Mitochondrial creatine kinase in human health and disease. *Biochim. Biophys. Acta 1762*, 164–180.

34 Matthews, P. M., Bland, J. L., Gadian, D. G., Radda, G. K. (1982) A 31P-NMR saturation transfer study of the regulation of creatine kinase in the rat heart. *Biochim. Biophys. Acta 721*, 312–320.

35 Wyss, M., Smeitink, J., Wevers, R. A., Wallimann, T. (1992) Mitochondrial creatine kinase: a key enzyme of aerobic energy metabolism. *Biochim. Biophys. Acta 1102*, 119–166.

36 Winnard, P., Cashon, R. E., Sidell, B. D., Vayda, M. E. (2003) Isolation, characterization and nucleotide sequence of the muscle isoforms of creatine kinase from the Antarctic teleost Chaenocephalus aceratus. *Comp. Biochem. Physiol. B Biochem. Mol. Biol. 134*, 651–667.

37 Sheng, X. R., Li, X., Pan, X. M. (1999) An iso-random Bi Bi mechanism for adenylate kinase. *J. Biol. Chem. 274*, 22238–22242.

38 Li, X., Han, Y., Pan, X. M. (2001) Cysteine-25 of adenylate kinase reacts with dithiothreitol to form an adduct upon aging of the enzyme. *FEBS Lett. 507*, 169–173.

39 Dzeja, P. P., Terzic, A., Wieringa, B. (2004) Phosphotransfer dynamics in skeletal muscle from creatine kinase gene-deleted mice. *Mol. Cell. Biochem. 256–257*, 13–27.

40 Suter, M., Riek, U., Tuerk, R., Schlattner, U., Wallimann, T., Neumann, D. (2006) Dissecting the role of 5′-AMP for allosteric stimulation, activation and deactivation of AMP-activated protein kinase. *J. Biol. Chem. 281*, 32207–32216.

41 Rush, J. W., Tullson, P. C., Terjung, R. L. (1998) Molecular and kinetic alterations of muscle AMP deaminase during chronic creatine depletion. *Am. J. Physiol. 274*, C465–471.

42 Janero, D. R., Yarwood, C. (1995) Oxidative modulation and inactivation of rabbit cardiac adenylate deaminase. *Biochem. J. 306*, 421–427.

43 Zong, H., Ren, J. M., Young, L. H., Pypaert, M., Mu, J., Birnbaum, M. J., Shulman, G. I. (2002) AMP kinase is required for mitochondrial biogenesis in skeletal muscle in response to chronic energy deprivation. *Proc. Natl. Acad. Sci. USA 99*, 15983–15987.

44 Newby, A. C. (1988) The pigeon heart 5′-nucleotidase responsible for ischaemia-induced adenosine formation. *Biochem. J. 253*, 123–130.

45 Yamazaki, Y., Truong, V. L., Lowenstein, J. M. (1991) 5′-Nucleotidase I from rabbit heart. *Biochemistry 30*, 1503–1509.

46 Leff, T. (2003) AMP-activated protein kinase regulates gene expression by direct phosphorylation of nuclear proteins. *Biochem. Soc. Trans. 31*, 224–227.

47 Saks, V., Favier, R., Guzun, R., Schlattner, U., Wallimann, T. (2006) Molecular system bioenergetics: regulation of substrate supply in response to heart energy demands. *J. Physiol. 577*, 769–777.

48 Dyck, J. R., Lopaschuk, G. D. (2006) AMPK alterations in cardiac physiology and pathology: enemy or ally? *J. Physiol. 574*, 95–112.

49 Viollet, B., Foretz, M., Guigas, B., Horman, S., Dentin, R., Bertrand, L., Hue, L., Andreelli, F. (2006) Activation of AMP-activated protein kinase in the liver: a new strategy for the management of metabolic hepatic disorders. *J. Physiol. 574*, 41–53.

50 Ju, J. S., Smith, J. L., Oppelt, P. J., Fisher, J. S. (2005) Creatine feeding increases GLUT4 expression in rat skeletal muscle. *Am. J. Physiol. 288*, E347–352.

51 Jorgensen, S. B., Treebak, J. T., Viollet, B., Schjerling, P., Vaulont, S., Wojtaszewski, J. F., Richter, E. A. (2007) Role of {alpha}2-AMPK in basal, training- and AICAR-induced GLUT4, hexokinase II and mitochondrial protein expression in mouse muscle. *Am. J. Physiol. 292*, E331–339.

52 Ceddia, R. B., Sweeney, G. (2004) Creatine supplementation increases glucose oxidation and AMPK phosphorylation and reduces lactate production in L6 rat skeletal muscle cells. *J. Physiol. 555*, 409–421.

53 Bergeron, R., Ren, J. M., Cadman, K. S., Moore, I. K., Perret, P., Pypaert, M., Young, L. H., Semenkovich, C. F., Shulman, G. I. (2001) Chronic activation of AMP kinase results in NRF-1 activation and mitochondrial biogenesis. *Am. J. Physiol. 281*, E1340–1346.

54 Ponticos, M., Lu, Q. L., Morgan, J. E., Hardie, D. G., Partridge, T. A., Carling, D. (1998) Dual regulation of the AMP-activated protein kinase provides a novel mechanism for the control of creatine kinase in skeletal muscle. *EMBO J. 17*, 1688–1699.

55 Taylor, E. B., Ellingson, W. J., Lamb, J. D., Chesser, D. G., Compton, C. L., Winder, W. W. (2006) Evidence against regulation of AMP-activated protein kinase and LKB1/STRAD/MO25 activity by creatine phosphate. *Am. J. Physiol. 290*, E661–669.

56 Eijnde, B. O., Derave, W., Wojtaszewski, J. F., Richter, E. A., Hespel, P. (2005) AMP kinase expression and activity in human skeletal muscle: effects of immobilization, retraining, and creatine supplementation. *J. Appl. Physiol. 98*, 1228–1233.

57 Hardie, D. G., Scott, J. W., Pan, D. A., Hudson, E. R. (2003) Management of cellular energy by the AMP-activated protein kinase system. *FEBS Lett. 546*, 113–120.

58 Davies, S. P., Helps, N. R., Cohen, P. T., Hardie, D. G. (1995) 5′-AMP inhibits dephosphorylation, as well as promoting phosphorylation, of the AMP-activated protein kinase. Studies using bacterially expressed human protein phosphatase-2C alpha and native bovine protein phosphatase-2AC. *FEBS Lett. 377*, 421–425.

59 Berridge, M. J. (2006) Calcium microdomains: Organization and function. *Cell Calcium 40*, 405–412.

60 Berridge, M. J., Bootman, M. D., Roderick, H. L. (2003) Calcium signalling: dynamics, homeostasis and remodelling. *Nat. Rev. Mol. Cell Biol. 4*, 517–529.

61 Hardie, D. G., Carling, D. (1997) The AMP-activated protein kinase – fuel gauge of the mammalian cell? *Eur. J. Biochem. 246*, 259–273.

62 Zaccolo, M., Di Benedetto, G., Lissandron, V., Mancuso, L., Terrin, A., Zamparo, I. (2006) Restricted diffusion of a freely diffusible second messenger: mechanisms underlying compartmentalized cAMP signalling. *Biochem. Soc. Trans. 34*, 495–497.

63 Carling, D., Clarke, P. R., Zammit, V. A., Hardie, D. G. (1989) Purification and characterization of the AMP-activated protein kinase. Copurification of acetyl-CoA carboxylase kinase and 3-hydroxy-3-methylglutaryl-CoA reductase kinase activities. *Eur. J. Biochem. 186*, 129–136.

64 Hardie, D. G., Carling, D., Carlson, M. (1998) The AMP-activated/SNF1 protein kinase subfamily: metabolic sensors of the eukaryotic cell? *Annu. Rev. Biochem. 67*, 821–855.

65 Lizcano, J. M., Goransson, O., Toth, R., Deak, M., Morrice, N. A., Boudeau, J.,

Hawley, S. A., Udd, L., Makela, T. P., Hardie, D. G., Alessi, D. R. (2004) LKB1 is a master kinase that activates 13 kinases of the AMPK subfamily, including MARK/PAR-1. *EMBO J. 23*, 833–843.

66 Hardie, D. G. (2003) Minireview: the AMP-activated protein kinase cascade: the key sensor of cellular energy status. *Endocrinology 144*, 5179–5183.

67 Elbing, K., Rubenstein, E. M., McCartney, R. R., Schmidt, M. C. (2006) Subunits of the Snf1 kinase hetero-trimer show interdependence for association and activity. *J. Biol. Chem. 281*, 26170–26180.

68 Carling, D. (2004) The AMP-activated protein kinase cascade – a unifying system for energy control. *Trends Biochem. Sci. 29*, 18–24.

69 Rudolph, M. J., Amodeo, G. A., Bai, Y., Tong, L. (2005) Crystal structure of the protein kinase domain of yeast AMP-activated protein kinase Snf1. *Biochem. Biophys. Res. Commun. 337*, 1224–1228.

70 Crute, B. E., Seefeld, K., Gamble, J., Kemp, B. E., Witters, L. A. (1998) Functional domains of the alpha1 catalytic subunit of the AMP-activated protein kinase. *J. Biol. Chem. 273*, 35347–35354.

71 Pang, T., Xiong, B., Li, J. Y., Qiu, B. Y., Jin, G. Z., Shen, J. K., Li, J. (2007) Conserved alpha helix acts as autoinhibitory sequence in AMP-activated protein kinase alpha subunits. *J. Biol. Chem. 282*, 495–506.

72 Hudson, E. R., Pan, D. A., James, J., Lucocq, J. M., Hawley, S. A., Green, K. A., Baba, O., Terashima, T., Hardie, D. G. (2003) A Novel Domain in AMP-Activated Protein Kinase Causes Glycogen Storage Bodies Similar to Those Seen in Hereditary Cardiac Arrhythmias. *Curr. Biol. 13*, 861–866.

73 Iseli, T. J., Walter, M., van Denderen, B. J., Katsis, F., Witters, L. A., Kemp, B. E., Michell, B. J., Stapleton, D. (2005) AMP-activated Protein Kinase beta Subunit Tethers alpha and gamma Subunits via Its C-terminal Sequence (186–270). *J. Biol. Chem. 280*, 13395–13400.

74 Polekhina, G., Gupta, A., Michell, B. J., van Denderen, B., Murthy, S., Feil, S. C., Jennings, I. G., Campbell, D. J., Witters, L. A., Parker, M. W., Kemp, B. E., Stapleton, D. (2003) AMPK beta Subunit Targets Metabolic Stress Sensing to Glycogen. *Curr. Biol. 13*, 867–871.

75 Polekhina, G., Gupta, A., van Denderen, B. J., Feil, S. C., Kemp, B. E., Stapleton, D., Parker, M. W. (2005) Structural basis for glycogen recognition by AMP-activated protein kinase. *Structure 13*, 1453–1462.

76 Kemp, B. E. (2004) Bateman domains and adenosine derivatives form a binding contract. *J. Clin. Invest. 113*, 182–184.

77 Milan, D., Jeon, J. T., Looft, C., Amarger, V., Robic, A., Thelander, M., Rogel-Gaillard, C., Paul, S., Iannuccelli, N., Rask, L., Ronne, H., Lundstrom, K., Reinsch, N., Gellin, J., Kalm, E., Roy, P. L., Chardon, P., Andersson, L. (2000) A mutation in PRKAG3 associated with excess glycogen content in pig skeletal muscle. *Science 288*, 1248–1251.

78 Gollob, M. H., Seger, J. J., Gollob, T. N., Tapscott, T., Gonzales, O., Bachinski, L., Roberts, R. (2001) Novel PRKAG2 mutation responsible for the genetic syndrome of ventricular preexcitation and conduction system disease with childhood onset and absence of cardiac hypertrophy. *Circulation 104*, 3030–3033.

79 Scott, J. W., Hawley, S. A., Green, K. A., Anis, M., Stewart, G., Scullion, G. A., Norman, D. G., Hardie, D. G. (2004) CBS domains form energy-sensing modules whose binding of adenosine ligands is disrupted by disease mutations. *J. Clin. Invest. 113*, 274–284.

80 Blair, E., Redwood, C., Ashrafian, H., Oliveira, M., Broxholme, J., Kerr, B., Salmon, A., Ostman-Smith, I., Watkins, H. (2001) Mutations in the gamma(2) subunit of AMP-activated protein kinase cause familial hypertrophic cardiomyopathy: evidence for the central role of energy compromise in disease pathogenesis. *Hum. Mol. Genet. 10*, 1215–1220.

81 Townley, R., Shapiro, L. (2007) Crystal structures of the adenylate sensor from fission yeast AMP-activated protein kinase. *Science 315*, 1726–1729.

82 Manning, G., Whyte, D., Martinez, R., Hunter, T., Sudarsanam, S. (2002) The protein kinase complement of the human genome. *Science 298*, 1912–1934.

83 Carling, D. (2005) AMP-activated protein kinase: balancing the scales. *Biochimie 87*, 87–91.

84 Burwinkel, B., Scott, J. W., Buhrer, C., van Landeghem, F. K., Cox, G. F., Wilson, C. J., Grahame Hardie, D., Kilimann, M. W. (2005) Fatal congenital heart glycogenosis caused by a recurrent activating R531Q mutation in the gamma 2-subunit of AMP-activated protein kinase (PRKAG2), not by phosphorylase kinase deficiency. *Am. J. Hum. Genet. 76*, 1034–1049.

85 Arad, M., Benson, D. W., Perez-Atayde, A. R., McKenna, W. J., Sparks, E. A., Kanter, R. J., McGarry, K., Seidman, J. G., Seidman, C. E. (2002) Constitutively active AMP kinase mutations cause glycogen storage disease mimicking hypertrophic cardiomyopathy. *J. Clin. Invest. 109*, 357–362.

86 Hamilton, S. R., Stapleton, D., O'Donnell, J. B., Jr., Kung, J. T., Dalal, S. R., Kemp, B. E., Witters, L. A. (2001) An activating mutation in the gamma1 subunit of the AMP-activated protein kinase. *FEBS Lett. 500*, 163–168.

87 Scott, J. W., Ross, F. A., Liu, J. K. D., Hardie, D. G. (2007) Regulation of AMP-activated protein kinase by a pseudosubstrate sequence on the gamma subunit. *EMBO J. 26*, 806–815.

88 Hawley, S. A., Selbert, M. A., Goldstein, E. G., Edelman, A. M., Carling, D., Hardie, D. G. (1995) 5′-AMP activates the AMP-activated protein kinase cascade, and Ca^{2+}/calmodulin activates the calmodulin-dependent protein kinase I cascade, via three independent mechanisms. *J. Biol. Chem. 270*, 27186–27191.

89 Hawley, S. A., Boudeau, J., Reid, J. L., Mustard, K. J., Udd, L., Makela, T. P., Alessi, D. R., Hardie, D. G. (2003) Complexes between the LKB1 tumor suppressor, STRAD alpha/beta and MO25 alpha/beta are upstream kinases

in the AMP-activated protein kinase cascade. *J. Biol. 2*, 28.

90 Sanders, M. J., Grondin, P. O., Hegarty, B. D., Snowden, M. A., Carling, D. (2006) Investigating the mechanism for AMP activation of the AMP-activated protein kinase cascade. *Biochem. J. 403*, 139–148.

91 Rafaeloff-Phail, R., Ding, L., Conner, L., Yeh, W. K., McClure, D., Guo, H., Emerson, K., Brooks, H. (2004) Biochemical regulation of mammalian AMP-activated protein kinase activity by NAD and NADH. *J. Biol. Chem. 279*, 52934–52939.

92 Hawley, S. A., Davison, M., Woods, A., Davies, S. P., Beri, R. K., Carling, D., Hardie, D. G. (1996) Characterization of the AMP-activated protein kinase kinase from rat liver and identification of threonine 172 as the major site at which it phosphorylates AMP-activated protein kinase. *J. Biol. Chem. 271*, 27879–27887.

93 Stein, S. C., Woods, A., Jones, N. A., Davison, M. D., Carling, D. (2000) The regulation of AMP-activated protein kinase by phosphorylation. *Biochem. J. 345*, 437–443.

94 Shaw, R. J., Kosmatka, M., Bardeesy, N., Hurley, R. L., Witters, L. A., DePinho, R. A., Cantley, L. C. (2004) The tumor suppressor LKB1 kinase directly activates AMP-activated kinase and regulates apoptosis in response to energy stress. *Proc. Natl. Acad. Sci. U.S.A. 101*, 3329–3335.

95 Woods, A., Johnstone, S. R., Dickerson, K., Leiper, F. C., Fryer, L. G., Neumann, D., Schlattner, U., Wallimann, T., Carlson, M., Carling, D. (2003) LKB1 is the upstream kinase in the AMP-activated protein kinase cascade. *Curr. Biol. 13*, 2004–2008.

96 Hawley, S. A., Pan, D. A., Mustard, K. J., Ross, L., Bain, J., Edelman, A. M., Frenguelli, B. G., Hardie, D. G. (2005) Calmodulin-dependent protein kinase kinase-beta is an alternative upstream kinase for AMP-activated protein kinase. *Cell Metab. 2*, 9–19.

97 Woods, A., Dickerson, K., Heath, R., Hong, S. P., Momcilovic, M., Johnstone, S. R., Carlson, M., Carling, D. (2005) Ca^{2+}/calmodulin-dependent protein

kinase kinase-beta acts upstream of AMP-activated protein kinase in mammalian cells. *Cell Metab. 2*, 21–33.

98 Hurley, R. L., Anderson, K. A., Franzone, J. M., Kemp, B. E., Means, A. R., Witters, L. A. (2005) The Ca^{2+}/calmodulin-dependent protein kinase kinases are AMP-activated protein kinase kinases. *J. Biol. Chem. 280*, 29060–29066.

99 Boudeau, J., Sapkota, G., Alessi, D. R. (2003) LKB1, a protein kinase regulating cell proliferation and polarity. *FEBS Lett. 546*, 159–165.

100 Motoshima, H., Goldstein, B. J., Igata, M., Araki, E. (2006) AMPK and cell proliferation – AMPK as a therapeutic target for atherosclerosis and cancer. *J. Physiol. 574*, 63–71.

101 Jaleel, M., McBride, A., Lizcano, J. M., Deak, M., Toth, R., Morrice, N. A., Alessi, D. R. (2005) Identification of the sucrose non-fermenting related kinase SNRK, as a novel LKB1 substrate. *FEBS Lett. 579*, 1417–1423.

102 Shaw, R. J., Bardeesy, N., Manning, B. D., Lopez, L., Kosmatka, M., DePinho, R. A., Cantley, L. C. (2004) The LKB1 tumor suppressor negatively regulates mTOR signaling. *Cancer Cell 6*, 91–99.

103 Inoki, K., Zhu, T., Guan, K. L. (2003) TSC2 mediates cellular energy response to control cell growth and survival. *Cell 115*, 577–590.

104 Boudeau, J., Baas, A. F., Deak, M., Morrice, N. A., Kieloch, A., Schutkowski, M., Prescott, A. R., Clevers, H. C., Alessi, D. R. (2003) MO25alpha/beta interact with STRADalpha/beta enhancing their ability to bind, activate and localize LKB1 in the cytoplasm. *EMBO J. 22*, 5102–5114.

105 Boudeau, J., Scott, J. W., Resta, N., Deak, M., Kieloch, A., Komander, D., Hardie, D. G., Prescott, A. R., van Aalten, D. M., Alessi, D. R. (2004) Analysis of the LKB1-STRAD-MO25 complex. *J. Cell Sci. 117*, 6365–6375.

106 Sakamoto, K., McCarthy, A., Smith, D., Green, K. A., Grahame Hardie, D., Ashworth, A., Alessi, D. R. (2005) Deficiency of LKB1 in skeletal muscle prevents AMPK activation and glucose uptake during contraction. *EMBO J. 24*, 1810–1820.

107 Shaw, R. J., Lamia, K. A., Vasquez, D., Koo, S. H., Bardeesy, N., Depinho, R. A., Montminy, M., Cantley, L. C. (2005) The kinase LKB1 mediates glucose homeostasis in liver and therapeutic effects of metformin. *Science 310*, 1642–1646.

108 Hook, S. S., Means, A. R. (2001) Ca(2+)/CaM-dependent kinases: from activation to function. *Annu. Rev. Pharmacol. Toxicol. 41*, 471–505.

109 Soderling, T. R. (1999) The Ca-calmodulin-dependent protein kinase cascade. *Trends Biochem. Sci. 24*, 232–236.

110 Soderling, T. R., Chang, B., Brickey, D. (2001) Cellular signaling through multifunctional Ca^{2+}/calmodulin-dependent protein kinase II. *J. Biol. Chem. 276*, 3719–3722.

111 Soderling, T. R., Stull, J. T. (2001) Structure and regulation of calcium/calmodulin-dependent protein kinases. *Chem. Rev. 101*, 2341–2352.

112 Peters, M., Mizuno, K., Ris, L., Angelo, M., Godaux, E., Giese, K. P. (2003) Loss of Ca2+/calmodulin kinase kinase beta affects the formation of some, but not all, types of hippocampus-dependent long-term memory. *J. Neurosci. 23*, 9752–9760.

113 Andersson, U., Filipsson, K., Abbott, C. R., Woods, A., Smith, K., Bloom, S. R., Carling, D., Small, C. J. (2004) AMP-activated protein kinase plays a role in the control of food intake. *J. Biol. Chem. 279*, 12005–12008.

114 Anderson, K. A., Noeldner, P. K., Reece, K., Wadzinski, B. E., Means, A. R. (2004) Regulation and function of the calcium/calmodulin-dependent protein kinase IV/protein serine/threonine phosphatase 2A signaling complex. *J. Biol. Chem. 279*, 31708–31716.

115 Stahmann, N., Woods, A., Carling, D., Heller, R. (2006) Thrombin activates AMP-activated protein kinase in endothelial cells via a pathway involving Ca^{2+}/calmodulin-dependent protein kinase kinase beta. *Mol. Cell. Biol. 26*, 5933–5945.

116 Tamas, P., Hawley, S. A., Clarke, R. G., Mustard, K. J., Green, K., Hardie, D. G.,

Cantrell, D. A. (2006) Regulation of the energy sensor AMP-activated protein kinase by antigen receptor and Ca²⁺ in T lymphocytes. *J. Exp. Med. 203*, 1665–1670.

117 Momcilovic, M., Hong, S. P., Carlson, M. (2006) Mammalian TAK1 activates Snf1 protein kinase in yeast and phosphorylates AMP-activated protein kinase *in vitro. J. Biol. Chem. 281*, 25336–25343.

118 Xie, M., Zhang, D., Dyck, J. R., Li, Y., Zhang, H., Morishima, M., Mann, D. L., Taffet, G. E., Baldini, A., Khoury, D. S., Schneider, M. D. (2006) A pivotal role for endogenous TGF-beta-activated kinase-1 in the LKB1/AMP-activated protein kinase energy-sensor pathway. *Proc. Natl. Acad. Sci. U.S.A. 103*, 17378–17383.

119 Horman, S., Vertommen, D., Heath, R., Neumann, D., Mouton, V., Woods, A., Schlattner, U., Wallimann, T., Carling, D., Hue, L., Rider, M. H. (2006) Insulin antagonizes ischemia-induced Thr172 phosphorylation of AMP-activated protein kinase alpha-subunits in heart via hierarchical phosphorylation of Ser485/491. *J. Biol. Chem. 281*, 5335–5340.

120 Kovacic, S., Soltys, C. L., Barr, A. J., Shiojima, I., Walsh, K., Dyck, J. R. (2003) Akt activity negatively regulates phosphorylation of AMP-activated protein kinase in the heart. *J. Biol. Chem. 278*, 39422–39427.

121 Hurley, R. L., Barre, L. K., Wood, S. D., Anderson, K. A., Kemp, B. E., Means, A. R., Witters, L. A. (2006) Regulation of AMP-activated protein kinase by multi-site phosphorylation in response to agents that elevate cellular cAMP. *J. Biol. Chem. 281*, 36662–36672.

122 Collins, S. P., Reoma, J. L., Gamm, D. M., Uhler, M. D. (2000) LKB1, a novel serine/threonine protein kinase and potential tumour suppressor, is phosphorylated by cAMP-dependent protein kinase (PKA) and prenylated *in vivo. Biochem. J. 345*, 673–680.

123 Marley, A. E., Sullivan, J. E., Carling, D., Abbott, W. M., Smith, G. J., Taylor, I. W., Carey, F., Beri, R. K. (1996) Biochemical characterization and deletion analysis of recombinant human protein phosphatase 2C alpha. *Biochem. J. 320*, 801–806.

124 Moore, F., Weekes, J., Hardie, D. G. (1991) Evidence that AMP triggers phosphorylation as well as direct allosteric activation of rat liver AMP-activated protein kinase. A sensitive mechanism to protect the cell against ATP depletion. *Eur. J. Biochem. 199*, 691–697.

125 Wang, M. Y., Unger, R. H. (2005) Role of PP2C in cardiac lipid accumulation in obese rodents and its prevention by troglitazone. *Am. J. Physiol. 288*, E216–221.

126 Hue, L., Beauloye, C., Bertrand, L., Horman, S., Krause, U., Marsin, A.-S., Meisse, D., Vertommen, D., Rider, M. H. (2003) New targets of AMP-activated protein kinase. *Biochem. Soc. Trans. 31*, 213–215.

127 Dale, S., Wilson, W. A., Edelman, A. M., Hardie, D. G. (1995) Similar substrate recognition motifs for mammalian AMP-activated protein kinase, higher plant HMG-CoA reductase kinase-A, yeast SNF1, and mammalian calmodulin-dependent protein kinase I. *FEBS Lett. 361*, 191–195.

128 Scott, J. W., Norman, D. G., Hawley, S. A., Kontogiannis, L., Hardie, D. G. (2002) Protein kinase substrate recognition studied using the recombinant catalytic domain of AMP-activated protein kinase and a model substrate. *J. Mol. Biol. 317*, 309–323.

129 Krebs, E. G., Beavo, J. A. (1979) Phosphorylation-dephosphorylation of enzymes. *Annu. Rev. Biochem. 48*, 923–959.

130 Zhou, G., Myers, R., Li, Y., Chen, Y., Shen, X., Fenyk-Melody, J., Wu, M., Ventre, J., Doebber, T., Fujii, N., Musi, N., Hirshman, M. F., Goodyear, L. J., Moller, D. E. (2001) Role of AMP-activated protein kinase in mechanism of metformin action. *J. Clin. Invest. 108*, 1167–1174.

131 Woods, A., Azzout-Marniche, D., Foretz, M., Stein, S. C., Lemarchand, P., Ferre, P., Foufelle, F., Carling, D. (2000) Characterization of the role of AMP-activated protein kinase in the

regulation of glucose-activated gene expression using constitutively active and dominant negative forms of the kinase. *Mol. Cell. Biol. 20*, 6704–6711.

132 Viollet, B., Andreelli, F., Jorgensen, S. B., Perrin, C., Geloen, A., Flamez, D., Mu, J., Lenzner, C., Baud, O., Bennoun, M., Gomas, E., Nicolas, G., Wojtaszewski, J. F., Kahn, A., Carling, D., Schuit, F. C., Birnbaum, M. J., Richter, E. A., Burcelin, R., Vaulont, S. (2003) The AMP-activated protein kinase alpha2 catalytic subunit controls whole-body insulin sensitivity. *J. Clin. Invest. 111*, 91–98.

133 Jorgensen, S. B., Viollet, B., Andreelli, F., Frosig, C., Birk, J. B., Schjerling, P., Vaulont, S., Richter, E. A., Wojtaszewski, J. F. (2004) Knockout of the alpha2 but not alpha1 5′-AMP-activated protein kinase isoform abolishes 5-aminoimidazole-4-carboxamide-1-beta-4-ribofuranosidebut not contraction-induced glucose uptake in skeletal muscle. *J. Biol. Chem. 279*, 1070–1079.

134 Cidad, P., Almeida, A., Bolanos, J. P. (2004) Inhibition of mitochondrial respiration by nitric oxide rapidly stimulates cytoprotective GLUT3-mediated glucose uptake through 5′-AMP-activated protein kinase. *Biochem. J. 384*, 629–636.

135 Horike, N., Takemori, H., Katoh, Y., Doi, J., Min, L., Asano, T., Sun, X. J., Yamamoto, H., Kasayama, S., Muraoka, M., Nonaka, Y., Okamoto, M. (2003) Adipose-specific expression, phosphorylation of Ser794 in insulin receptor substrate-1, and activation in diabetic animals of salt-inducible kinase-2. *J. Biol. Chem. 278*, 18440–18447.

136 Screaton, R. A., Conkright, M. D., Katoh, Y., Best, J. L., Canettieri, G., Jeffries, S., Guzman, E., Niessen, S., Yates, J. R., 3rd, Takemori, H., Okamoto, M., Montminy, M. (2004) The CREB coactivator TORC2 functions as a calcium- and cAMP-sensitive coincidence detector. *Cell 119*, 61–74.

137 Koo, S. H., Flechner, L., Qi, L., Zhang, X., Screaton, R. A., Jeffries, S., Hedrick, S., Xu, W., Boussouar, F., Brindle, P., Takemori, H., Montminy, M. (2005) The CREB coactivator TORC2 is a key regulator of fasting glucose metabolism. *Nature 437*, 1109–1111.

138 Baas, A. F., Smit, L., Clevers, H. (2004) LKB1 tumor suppressor protein: PARtaker in cell polarity. *Trends Cell Biol. 14*, 312–319.

139 Lefebvre, D. L., Rosen, C. F. (2005) Regulation of SNARK activity in response to cellular stresses. *Biochim. Biophys. Acta 1724*, 71–85.

140 Sakamoto, K., Goransson, O., Hardie, D. G., Alessi, D. R. (2004) Activity of LKB1 and AMPK-related kinases in skeletal muscle: effects of contraction, phenformin, and AICAR. *Am. J. Physiol. 287*, E310–317.

141 Daval, M., Diot-Dupuy, F., Bazin, R., Hainault, I., Viollet, B., Vaulont, S., Hajduch, E., Ferre, P., Foufelle, F. (2005) Anti-lipolytic action of AMP-activated protein kinase in rodent adipocytes. *J. Biol. Chem. 280*, 25250–25257.

142 Corton, J. M., Gillespie, J. G., Hawley, S. A., Hardie, D. G. (1995) 5-aminoimidazole-4-carboxamide ribonucleoside. A specific method for activating AMP-activated protein kinase in intact cells? *Eur. J. Biochem. 229*, 558–565.

143 Sullivan, J. E., Brocklehurst, K. J., Marley, A. E., Carey, F., Carling, D., Beri, R. K. (1994) Inhibition of lipolysis and lipogenesis in isolated rat adipocytes with AICAR, a cell-permeable activator of AMP-activated protein kinase. *FEBS Lett. 353*, 33–36.

144 Garton, A. J., Campbell, D. G., Carling, D., Hardie, D. G., Colbran, R. J., Yeaman, S. J. (1989) Phosphorylation of bovine hormone-sensitive lipase by the AMP-activated protein kinase. A possible antilipolytic mechanism. *Eur. J. Biochem. 179*, 249–254.

145 Anthonsen, M. W., Ronnstrand, L., Wernstedt, C., Degerman, E., Holm, C. (1998) Identification of novel phosphorylation sites in hormone-sensitive lipase that are phosphorylated in response to isoproterenol and govern activation properties *in vitro*. *J. Biol. Chem. 273*, 215–221.

146 Watt, M. J., Holmes, A. G., Pinnamaneni, S. K., Garnham, A. P.,

Steinberg, G. R., Kemp, B. E., Febbraio, M. A. (2006) Regulation of HSL serine phosphorylation in skeletal muscle and adipose tissue. *Am. J. Physiol. 290,* E500–508.

147 Rider, M. H., Bertrand, L., Vertommen, D., Michels, P. A., Rousseau, G. G., Hue, L. (2004) 6-phosphofructo-2-kinase/fructose-2,6-bisphosphatase: head-to-head with a bifunctional enzyme that controls glycolysis. *Biochem. J. 381,* 561–579.

148 Chen, Z. P., Mitchelhill, K. I., Michell, B. J., Stapleton, D., Rodriguez-Crespo, I., Witters, L. A., Power, D. A., Ortiz de Montellano, P. R., Kemp, B. E. (1999) AMP-activated protein kinase phosphorylation of endothelial NO synthase. *FEBS Lett. 443,* 285–289.

149 Dimmeler, S., Fleming, I., Fisslthaler, B., Hermann, C., Busse, R., Zeiher, A. M. (1999) Activation of nitric oxide synthase in endothelial cells by Akt-dependent phosphorylation. *Nature 399,* 601–605.

150 Sano, H., Kane, S., Sano, E., Miinea, C. P., Asara, J. M., Lane, W. S., Garner, C. W., Lienhard, G. E. (2003) Insulin-stimulated phosphorylation of a Rab GTPase-activating protein regulates GLUT4 translocation. *J. Biol. Chem. 278,* 14599–14602.

151 Treebak, J. T., Glund, S., Deshmukh, A., Klein, D. K., Long, Y. C., Jensen, T. E., Jorgensen, S. B., Viollet, B., Andersson, L., Neumann, D., Wallimann, T., Richter, E. A., Chibalin, A. V., Zierath, J. R., Wojtaszewski, J. F. (2006) AMPK-mediated AS160 phosphorylation in skeletal muscle is dependent on AMPK catalytic and regulatory subunits. *Diabetes 55,* 2051–2058.

152 Kramer, H. F., Witczak, C. A., Fujii, N., Jessen, N., Taylor, E. B., Arnolds, D. E., Sakamoto, K., Hirshman, M. F., Goodyear, L. J. (2006) Distinct signals regulate AS160 phosphorylation in response to insulin, AICAR, and contraction in mouse skeletal muscle. *Diabetes 55,* 2067–2076.

153 Horman, S., Beauloye, C., Vertommen, D., Vanoverschelde, J. L., Hue, L., Rider, M. H. (2003) Myocardial ischemia and increased heart work modulate the phosphorylation state of eukaryotic elongation factor-2. *J. Biol. Chem 278,* 41970–41976.

154 Bolster, D. R., Crozier, S. J., Kimball, S. R., Jefferson, L. S. (2002) AMP-activated protein kinase suppresses protein synthesis in rat skeletal muscle through down-regulated mammalian target of rapamycin (mTOR) signaling. *J. Biol. Chem. 277,* 23977–23980.

155 Horman, S., Browne, G. J., Krause, U., Patel, J. V., Vertommen, D., Bertrand, L., Lavoinne, A., Hue, L., Proud, C. G., Rider, M. H. (2002) Activation of AMP-activated protein kinase leads to the phosphorylation of elongation factor 2 and an inhibition of protein synthesis. *Curr. Biol. 12,* 1419–1423.

156 Krause, U., Bertrand, L., Hue, L. (2002) Control of p70 ribosomal protein S6 kinase and acetylCoA carboxylase activity by AMP-activated protein kinase and protein phosphatases in isolated rat hepatocytes. *Eur. J. Biochem. 269,* 3751–3759.

157 Inoki, K., Li, Y., Zhu, T., Wu, J., Guan, K. L. (2002) TSC2 is phosphorylated and inhibited by Akt and suppresses mTOR signalling. *Nat. Cell Biol. 4,* 648–657.

158 Nave, B. T., Ouwens, M., Withers, D. J., Alessi, D. R., Shepherd, P. R. (1999) Mammalian target of rapamycin is a direct target for protein kinase B: identification of a convergence point for opposing effects of insulin and amino-acid deficiency on protein translation. *Biochem. J. 344,* 427–431.

159 Cheng, S. W., Fryer, L. G., Carling, D., Shepherd, P. R. (2004) Thr2446 is a novel mammalian target of rapamycin (mTOR) phosphorylation site regulated by nutrient status. *J. Biol. Chem. 279,* 15719–15722.

160 Fryer, L. G., Parbu-Patel, A., Carling, D. (2002) The Anti-diabetic drugs rosiglitazone and metformin stimulate AMP-activated protein kinase through distinct signaling pathways. *J. Biol. Chem. 277,* 25226–25232.

161 Jones, R. G., Plas, D. R., Kubek, S., Buzzai, M., Mu, J., Xu, Y., Birnbaum, M. J., Thompson, C. B. (2005) AMP-activated protein kinase induces a p53-

dependent metabolic checkpoint. *Mol. Cell 18*, 283–293.

162 Kimura, N., Tokunaga, C., Dalal, S., Richardson, C., Yoshino, K., Hara, K., Kemp, B. E., Witters, L. A., Mimura, O., Yonezawa, K. (2003) A possible linkage between AMP-activated protein kinase (AMPK) and mammalian target of rapamycin (mTOR) signalling pathway. *Genes Cells 8*, 65–79.

163 Motoshima, H., Goldstein, B. J., Igata, M., Araki, E. (2006) AMPK and cell proliferation – AMPK as a therapeutic target for atherosclerosis and cancer. *J. Physiol. 574*, 63–71.

164 Munday, M. R., Campbell, D. G., Carling, D., Hardie, D. G. (1988) Identification by amino acid sequencing of three major regulatory phosphorylation sites on rat acetyl-CoA carboxylase. *Eur. J. Biochem. 175*, 331–338.

165 Winder, W. W., Wilson, H. A., Hardie, D. G., Rasmussen, B. B., Hutber, C. A., Call, G. B., Clayton, R. D., Conley, L. M., Yoon, S., Zhou, B. (1997) Phosphorylation of rat muscle acetyl-CoA carboxylase by AMP-activated protein kinase and protein kinase A. *J. Appl. Physiol. 82*, 219–225.

166 Ching, Y. P., Davies, S. P., Hardie, D. G. (1996) Analysis of the specificity of the AMP-activated protein kinase by site-directed mutagenesis of bacterially expressed 3-hydroxy 3-methylglutaryl-CoA reductase, using a single primer variant of the unique-site-elimination method. *Eur. J. Biochem. 237*, 800–808.

167 Carling, D., Hardie, D. G. (1989) The substrate and sequence specificity of the AMP-activated protein kinase. Phosphorylation of glycogen synthase and phosphorylase kinase. *Biochim. Biophys. Acta 1012*, 81–86.

168 Marsin, A.-S., Bouzin, C., Bertrand, L., Hue, L. (2002) The stimulation of glycolysis by hypoxia in activated monocytes is mediated by AMP-activated protein kinase and inducible 6-phosphofructo-2-kinase. *J. Biol. Chem. 277*, 30778–30783.

169 Kawaguchi, T., Osatomi, K., Yamashita, H., Kabashima, T., Uyeda, K. (2002) Mechanism for fatty acid "sparing" effect on glucose-induced transcription: regulation of carbohydrate-responsive element-binding protein by AMP-activated protein kinase. *J. Biol. Chem. 277*, 3829–3835.

170 Hong, Y. H., Varanasi, U. S., Yang, W., Leff, T. (2003) AMP-activated protein kinase regulates HNF4alpha transcriptional activity by inhibiting dimer formation and decreasing protein stability. *J. Biol. Chem. 278*, 27495–27501.

171 Yang, W., Hong, Y. H., Shen, X. Q., Frankowski, C., Camp, H. S., Leff, T. (2001) Regulation of transcription by AMP-activated protein kinase: phosphorylation of p300 blocks its interaction with nuclear receptors. *J. Biol. Chem. 276*, 38341–38344.

172 Browne, G. J., Finn, S. G., Proud, C. G. (2004) Stimulation of the AMP-activated protein kinase leads to activation of eukaryotic elongation factor 2 kinase and to its phosphorylation at a novel site, serine 398. *J. Biol. Chem. 279*, 12220–12231.

173 Jakobsen, S. N., Hardie, D. G., Morrice, N., Tornqvist, H. E. (2001) 5′-AMP-activated protein kinase phosphorylates IRS-1 on Ser-789 in mouse C2C12 myotubes in response to 5-aminoimidazole-4-carboxamide riboside. *J. Biol. Chem. 276*, 46912–46916.

174 Sprenkle, A. B., Davies, S. P., Carling, D., Hardie, D. G., Sturgill, T. W. (1997) Identification of Raf-1 Ser621 kinase activity from NIH 3T3 cells as AMP-activated protein kinase. *FEBS Lett. 403*, 254–258.

10

Developmental and Functional Consequences of Disturbed Energetic Communication in Brain of Creatine Kinase–deficient Mice: Understanding CK's Role in the Fuelling of Behavior and Learning

Femke Streijger, René in 't Zandt, Klaas Jan Renema, Frank Oerlemans,
Arend Heerschap, Jan Kuiper, Helma Pluk, Caroline Jost, Ineke van der Zee,
and Bé Wieringa

Abstract

Maintenance of a tight coupling between energy production and consumption of energy-metabolites in different cellular locales, i.e., energy homeostasis regulation, is crucial for proper functioning and viability control of every cell. Balance in global and local concentrations of ATP, the key energy metabolite of life, is hereto essential. A better understanding of how cellular energetics – and in particular, the aspect of compartmentalization – is regulated is therefore relevant for health and disease. Reversible exchange of high-energy phosphoryl groups between ATP and phosphocreatine (PCr) in the creatine kinase (CK) circuit is a mechanism of the physiological regulation of storage and adequate compartmentalization of cellular energy. The CK–PCr system is particularly important in contractile cells and in cell types with high cell shape plasticity, such as muscle or brain cells. Energy requirements in brain are special because both neurons and glial cells (astrocytes and microglia) are highly dynamic, with elaborate peripheral processes that form distinct locales in the cell. During development as well as in adulthood, brain cells have to invest a high percentage of their ATP not only for "normal" maintenance of housekeeping functions but also for fuelling of molecular transport, intracellular communication, actomyosin-based dynamics during synaptogenesis, and continuous dendritic spine or glial endfeet remodeling. In addition, neurons and glial cells have to invest high and sudden amounts of ATP for fuelling of Ca^{2+}- and Na^+/K^+-ATPases involved in electrogenic or ion homeostasis activity and for neurotransmitter cycling or neuroendocrine signaling. In the brain of mammals, the various creatine kinase enzymes that are coupled to these energy needs are expressed differently but generally consist of the dimeric brain-type creatine kinase (B-CK), distributed mainly throughout the cytosol, and the octameric/dimeric ubiquitous creatine kinase (UbCKmit) isoform, located in the mitochondria. B-CK is found throughout the brain but generally occurs at higher levels in glial cells than in neuronal cell types. UbCKmit

Molecular System Bioenergetics: Energy for Life. Edited by Valdur Saks
Copyright © 2007 WILEY-VCH Verlag GmbH & Co. KGaA, Weinheim
ISBN: 978-3-527-31787-5

expression is also widespread and is correlated to mitochondrial activity, mainly in neurons. Here we give an overview of studies into the biological significance of the CK system in the CNS. By using gene knockout in embryonic stem cells, we have generated mouse strains that carry null mutations for either B-CK (B-CK$^{-/-}$ mice) or UbCKmit (UbCKmit$^{-/-}$ mice) or for both enzymes (CK$^{--/--}$ mice). For phenotyping of brain functions in these mice, histo-anatomical analyses and biophysical approaches with MR spectroscopy or MR imaging were used. Neurobehavioral approaches for monitoring of brain functions also were employed. For this purpose, a battery of different behavioral and learning paradigms was developed and used to study the relationship between phosphocreatine–creatine energetics and task performance. Tests used included habituation behavior and spatial learning tests, which specifically address hippocampal-based brain functions, as well as the acoustic startle reflex challenge, which probes the auditory brainstem circuit. In addition, the response to seizure induction and other integral physiological features, including body temperature control, were monitored. Mice with complete CK deficiency (CK$^{--/--}$) were studied most intensely. CK$^{--/--}$ mice appeared viable; however, their growth and development was affected, resulting in an overall smaller body posture. Furthermore, the complete lack of CK in the brains of these animals severely impaired spatial learning and was coupled to abnormal startle responses and hearing problems. In contrast, the visual and motor functions, exploration behavior, and anxiety-related responses were not changed in CK$^{--/--}$ mice, suggesting no global deficit in sensorimotor function or motivation. Various effects on behavior were consistent with underlying anatomical changes in the hippocampus. Others can best be explained by loss of functional integrity that may have occurred in development. Parallel to the anatomical and behavioral abnormalities, CK$^{--/--}$ mice also had problems with thermoregulation, presumably due to an altered hypothalamic thermostat. Furthermore, their response to chemical seizure induction appeared suppressed. When combined, our findings suggest that the lack of creatine kinase isoforms renders the synaptic circuitry in brain less efficient for coping with (sudden) changes in sensory or cognitive activity or for handling of intrinsic and extrinsic signals that affect whole-body physiology. Metabolic changes and cellular changes due to (signaling to) altered growth and pruning in development or turnover of synaptic spines during adulthood may have an important role, but further research is needed to couple CK function at the molecular and cellular levels to events at the system level.

10.1
Use of Reverse Genetics to Study CK Function in Mouse Models

During the past two to three decades, the use of genetically altered mice has opened up a broad range of entirely new possibilities for studying the physiological and structural role of proteins in cell growth, signaling, and metabolism. Although different approaches for reverse genetics are possible (with the design of

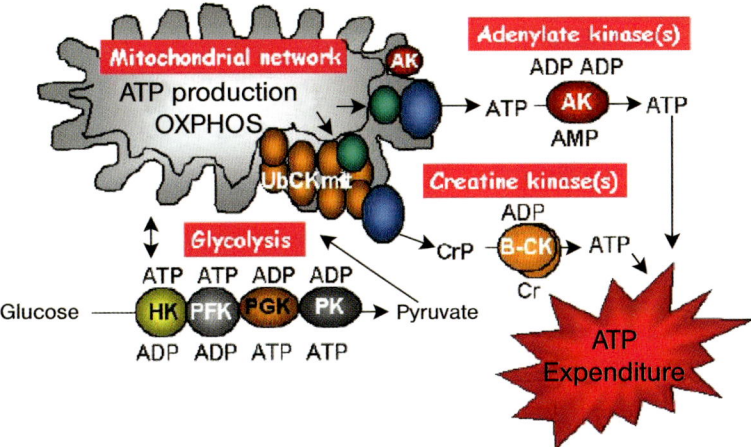

Fig. 10.1 Mammalian cells have different systems for high-energy phosphoryl ($\sim$P) transfer between the sites of production and actual use of ATP. (i) The mitochondrial network appears as a reticular network in most cells and not only is important for the mere production of energy by OXPHOS but also serves as an elaborate transport and distribution system for cellular energy. (ii) The glycolytic pathway, a multiple-enzyme circuit, is evolutionarily the oldest system in which investment of two ATP molecules at the level of hexokinase (HK) and phosphofructokinase (PFK) is needed to harvest four ATP molecules, two ATPs each at the level of phosphoglycerokinase (PGK) and pyruvate kinase (PK), later in the pathway. Distinct locations of enzymes help here in the generation and distribution of ATP. Finally, two main accessory systems for direct exchange of high-energy phosphoryl groups are important. One system (iii) involves the family of adenylate kinases found in the mitochondrial inner membrane or matrix space, the cytosol or the nucleus. AKs (which are related to the larger family of nucleoside diphosphate kinases [NDPKs]) catalyze the reaction ATP + AMP = 2ADP and can harvest $\sim$P energy in the β-P position of ADP, contributing to [ATP] homeostasis in different locales of the cell. The other system (iv) involves the creatine kinases (CK), of which UbCKmit in the inner membrane space of mitochondria and B-CK in the cytosol are shown. CK isoenzymes (EC 2.7.3.2) are cytosolic and mitochondrial enzymes that catalyze the reaction $MgADP^- + PCr^{2-} + H^+ \Leftrightarrow Cr + MgATP^{2-}$ and contribute to pool forming and distribution of ATP (for details, see other chapters in this book or see [6, 7, 29] for review).

transgenic models that overexpress protein, express mutant protein, or express no protein at all), our group has focused mainly on the generation of models with complete or partial enzyme deficiency. To address the role of high-energy phosphoryl transfer enzymes (Fig. 10.1) in mammals, we therefore used homologous DNA recombination in *in vitro*–cultured mouse embryonic stem (ES) cells in order to precisely engineer a pre-designed mutation into each of the four distinct genes for creatine kinases, i.e., brain-type creatine kinase (B-CK), muscle-type creatine kinase (M-CK), ubiquitous mitochondrial creatine kinase (UbCKmit), and sarcomeric mitochondrial kinase (ScCKmit) [1–8]. Complete or partial knockout of gene expression was thereby achieved by replacing an essential gene region (usually a coding segment, an essential promoter, or RNA-processing signals, but sometimes also intron tracts) with a selectable DNA cassette in one of the two parental alleles. The identity of the mutation had always been studied when still at the ES cell stage in cell culture. Only if the mutation was genuine and no other undesired alterations had been introduced were the ES cells used for reintroduction *in vivo* by injection into blastocysts *ex vivo*, followed by intrauterine implantation into foster mothers and growth of the embryos to term. By breeding of chimeric animals that gave germ line transmission of the mutation and by subsequent crossing of offspring with knockout alleles, we established heterozygous, homozygous, and compound homozygous knockout models. Although a high level of sophistication is possible, including built-in features to render expression of the transgene conditionally or specifically in cell types or tissues (see [9] for details), our approaches thus far have been limited to the design of simple hypomorphs or complete knockouts for the individual M-CK, ScCKmit, B-CK, and UbCKmit genes [2, 4, 8, 10–15]. Until now no other CK models had been made. Therefore, our mice can still be considered unique tools (*in vivo* veritas!) for the study of the physiological role of the CK–PCr circuit and its coupling to whole-body energy metabolism. By following similar strategies, our group also has ablated another phosphotransfer enzyme, the adenylate kinase type 1 (AK1) enzyme [16]. Details on the generation of phosphotransfer-mutant models with a muscle phenotype, i.e., mice with single or compound CK or CK and adenylate kinase (AK1) double-deficient models, can be found in previous publications from our group [4, 14, 17]. Mice with knockout of brain isoforms of CK, which are especially relevant for this chapter, have been described in greater detail elsewhere [15, 18, 19]. For more specific technical background on gene knockout technology, we refer the reader to the many excellent reviews and commentaries on this topic (see [9] and [20] and references therein). For additional reading on transgenic and knockout models for the neural basis of behavior or physical capacity at the "systems" level, we refer the reader to [21–26].

10.2
Expression Distribution of Brain-type CK mRNA and Protein Isoforms

By using zymogram assays and CK–enzymatic activity essays in extracts of whole brain [15, 18, 19], we previously demonstrated the specific presence of dimeric cytosolic brain-type creatine kinase (B-CK) and octameric and dimeric mitochondrial UbCKmit protein but the complete absence of muscle-type creatine kinase isoforms (MCK or ScCK) in the wild-type brain. The enzyme profiles confirmed the reliability of the assay and helped us to verify that UbCKmit$^{-/-}$ and B-CK$^{-/-}$ single knockout brains lack only UbCKmit or B-CK, respectively, whereas the CK$^{--/--}$ double-knockout mouse brain homogenates completely lack both B-CK and UbCKmit. Importantly, we also found that (1) UbCKmit activity determined by zymogram analysis was essentially similar in wild-type and B-CK$^{-/-}$ brain ex-

Arbitrary units/mg protein

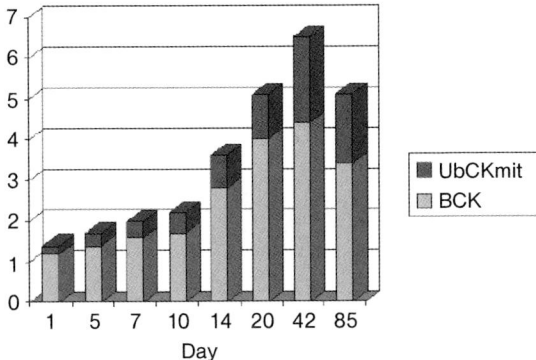

Fig. 10.2 Increase in overall CK activity during postnatal development. Total CK activity was determined in pooled brain extracts taken from 1-, 7-, 10-, 14-, 20-, 42-, and 85-day-old wild-type (strains C57BL/6 and 129/Sv combined) and B-CK$^{-/-}$ mutant mice (plotted as arbitrary units per milligram protein in total brain extract), and fractional contribution of B-CK and UbCKmit activity was calculated (dark = fractional UbCmit activity; light = fractional B-CK activity).

tracts and (2) B-CK zymogram activity was similar in UbCKmit$^{-/-}$ and wild-type brain extracts (for details, see [13, 15]).

Based on these observations and a comparison of total (remaining) creatine kinase (CK) enzyme activity between brains from wild-type and knockout mice, we were also able to estimate the relative expression levels of B-CK and UbCKmit in brains of mice at different time points after birth. Our data (Fig. 10.2) demonstrate a postnatal increase in total CK enzyme activity, with a peak between 3 and 6 weeks of age and an estimated contribution of B-CK and UbCKmit of around 67% and 33%, respectively, to total CK activity. Of note, the peak of highest CK activity coincides with the time period of highest developmental activity of the mouse brain, characterized by a dramatic increase in cell number of both neurons and astrocytes, with highly dynamic pathfinding, shaping, and pruning activity [27, 28].

The existence of a CK shuttle, with CK isoforms located at both the mitochondrial and cytosolic endpoints of a cellular energy relay pathway, is now a commonly accepted concept [6, 29]. Unfortunately, this model is rather uncritically presented as one metabolic entity in many reviews on cell energy metabolism, and the question of whether both endpoints indeed coexist in CK-expressing cells is hardly ever asked. It could thus very well be that capacities of the CK reactions in the mitochondrial inner membrane space and at locations in the cytosol are severely unbalanced in various cell types. Literature data on the cell type dependence and regional distribution of CK isoenzymes throughout the mammalian brain were until recently not very helpful in clarifying this point, because the

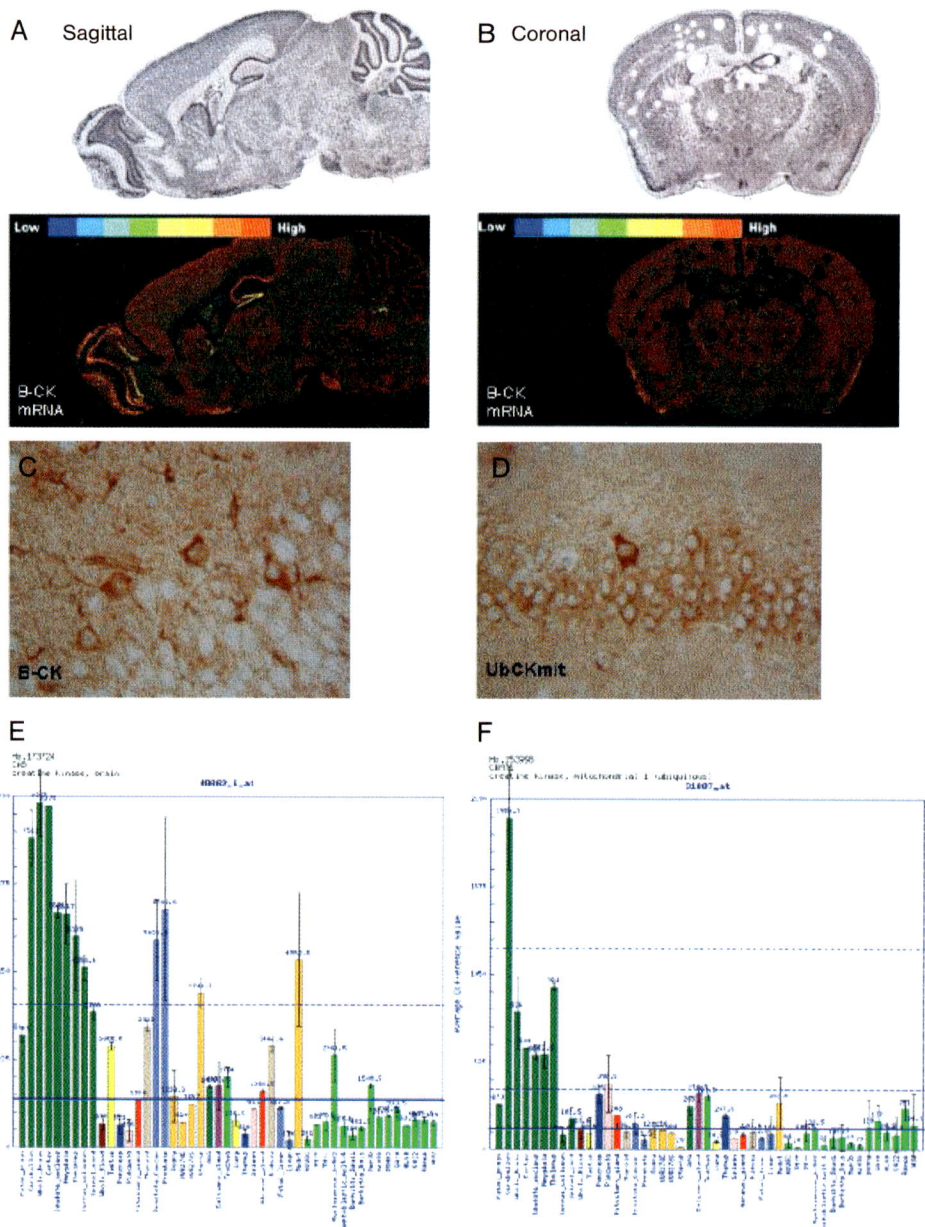

Fig. 10.3 (legend see p. 345)

data available were incomplete and could not be easily compared because different brain regions, cells of different species, or cells with different degrees of oncogenic transformation were studied. Developmental aspects were not even considered in these studies.

For B-CK and UbCKmit in mouse, this situation has recently been improved by the publication of a detailed immunohistochemical and *in situ* hybridization (ISH) survey by Tachikawa et al. [30] and by the availability of ISH data on CK mRNA distribution in the Allen Brain Atlas project (see [31] and http://www .brain-map.org/welcome.do;jsessionid=49C421996BB75A83AB95875C93A5779A) and in the Gene Expression Nervous System Atlas (GENSAT) at http://www .ncbi.nlm.nih.gov/projects/gensat/. Also, the availability of microarray-based gene expression data [32, 33] has proved helpful. Compilation of ISH data (examples of which are shown in Fig. 10.3A,B) confirms that mRNA for the B-CK subunit is expressed across almost all brain regions, including the cerebral cortex, olfactory bulb, striatum, hippocampus, hypothalamus, midbrain, and cerebellum. ISH for UbCKmit revealed a clearly distinct pattern (signals in the Allen Brain Atlas are weak and therefore are more difficult to interpret [not shown]; see Ref. [30] and GENSAT data). Transcriptome data obtained from regional brain samples (at http://expression.gnf.org/cgi-bin/index.cgi#Q) (Fig. 10.3E,F) confirm that B-CK and UbCKmit mRNAs are certainly not always expressed at constant ratios in different brain areas.

Study of the cell type–dependent distribution of B-CK and UbCKmit by immunocytochemistry is of course complementary to ISH but has been long problematic due to difficulties with CK subunit–isoform specificity and reliability of available antibodies. We and others have contributed to the development of new antibodies [30, 34–36]. Immunostaining with our anti–B-CK antibody 20H3B revealed that B-CK was present throughout the wild-type brain, including in the hippocampus, with strong staining of neurons and astrocytes and small blood vessels aligned with astrocytic endfeet (Fig. 10.3C) [10, 29]. A small subpopula-

---

Fig. 10.3 Creatine kinase distribution in the brain. ISH pictures of B-CK mRNA expression distribution in sagittal (A) and coronal (B) sections of the mouse brain showed moderate (yellow) to high (red) levels in areas such as the cerebral cortex, olfactory bulb, striatum, hippocampus, hypothalamus, midbrain, and cerebellum. Pictures are taken from the Allen Brain Atlas project ([31] published on the Web at http://www.brain-map.org/welcome.do;jsessionid= 49C421996BB75A83AB95875C93A5779A). (C) Immunohistochemical staining with monospecific antibodies for B-CK (20H3B [36]) demonstrated CA1 neuronal staining as well as strong labeling of astrocytes and their endfeet alignment of small blood vessels. (D) Staining with anti-UbCKmit (peptide antiserum [34]) revealed clear staining of CA1 pyramidal neurons in the hippocampus. Bottom panels show transcriptome data obtained by microarray quantification assays with B-CK (E) and UbCKmit (F) probes. Average difference values (relative expression levels) are given, showing large variation in ratios of B-CK and UbCKmit mRNA levels for different brain areas, non-brain tissues, and single cell lines. Data are taken from http:// expression.gnf.org/cgi-bin/index.cgi#Q.

tion of neurons in the deeper cortical layers, purkinje cells and glomerular synapses in the granular layer in the cerebellum, and neurons in several other nuclear groups show a very high level of expression in their soma. Glia (astrocytes) localized around blood vessels, in the cortex and throughout the midbrain, also express high levels of B-CK in both the soma and protrusions. No reaction whatsoever was found on control sections of brains of B-CK$^{-/-}$ mice, indicating the specificity of the antibody. By using an antiserum that was raised against an UbCKmit-specific peptide (AASERRRLYPPSA, residues 1–13), strong staining of many different neuronal nuclei was demonstrated (Fig. 10.3D). Positive staining was found mainly in the soma of certain neurons in layer V in the cortex and in several nuclear groups in the brain stem and midbrain. In the cerebellum, UbCKmit is located in neurons in the lateral cerebellar nucleus and in glomerular synapses in the cerebellum granular layer, but it is hardly detectable in the hippocampal area. Immunohistochemical comparison of wild-type and knockout mice revealed that the distribution profile of UbCKmit was not affected by the absence of the cytosolic counterpart B-CK. In a limited number of cross-sections through the forebrain, midbrain, and cerebellum of a B-CK$^{-/-}$ mouse, we found essentially no differences with UbCKmit localization in wild-type mice. The converse is also true, i.e., B-CK expression appears virtually unchanged in UbCKmit$^{-/-}$ brain, indicating that there is no compensational overexpression of the remaining CK isoform in brain with a single CK deficiency. Because the presence of the cytosolic isotype M-CK had been reported in a limited number of cell types in the brain, including human hippocampal neurons [37] and purkinje cells of the chicken cerebellum [38], we tried to immunolocalize M-CK in wild-type brain sections. With an antibody raised against an M-CK–specific peptide (QKIEEIFKKAGHP, residues 258–270 [34]), we found only occasional a-specific labeling along the plasma membrane of a very limited number of neurons in the cortex (data not shown), in accordance with the absence of M-CK activity in the brain extracts on zymograms.

These data on M-CK and our findings on B-CK and UbCKmit localization were confirmed in a recent study by Tachikawa et al. [30], who compared mRNA distribution patterns in mouse brain, determined with ISH, directly with cellular and ultrastructural staining patterns. This study, for which newly developed monospecific antibodies against B-CK and UbCKmit were used, is certainly the most elaborate survey currently available. The authors demonstrated that B-CK is highly expressed in the olfactory bulb, thalamus, pontine nuclei, superior olive, cerebellar molecular layer, and various other brain regions as well. Interestingly, in different brain regions such as the cerebral cortex, hippocampus, and cerebellum, B-CK was found selectively expressed in astrocytes among glial populations and in inhibitory but not excitatory neurons. UbCKmit expression was largely gray matter– and mitochondria-specific. In cortex, hippocampus, and brainstem, UbCKmit was confined to neuronal cell populations and located in mitochondria in dendrites, axons, and cell bodies of neurons.

Taken together, the currently available ISH data [30, 31] strongly suggest that B-CK and UbCKmit mRNA levels vary profoundly across cell types. New immunohistochemical staining data (data in Fig. 10.3C,D [15, 30, 36]) confirm this picture and demonstrate that enzymatic activities of B-CK at the cytosolic and UbCKmit at the mitochondrial endpoints of the CK shuttle vary and even are not coupled at all in certain cell types. For a better understanding of the validity of the CK shuttle concept or the existence of any form of intercellular cooperation of Cr–PCr metabolism, we need to study regional and cell type–dependent variation in B-CK versus UbCKmit expression levels in more detail. Therefore, we first need better classification of neuronal and glial cell types, of which several hundred subtypes may exist in various brain structures.

10.3
CK$^{--/--}$ Mice Have Lower Body Weights

During our studies of the whole-body physiological effects of combined B-CK and UbCKmit deficiency, we noted that the absence of these CK isoforms affects brain as well as cells in tissues such as kidney, vasculature (smooth muscle cells), and circulation (macrophages; data not shown). The absence of other CK isoforms, i.e., M-CK and ScCKmit, from skeletal muscle had no such dramatic anatomic, physiological, and metabolic effects [4, 14]. Weight measurements at various time points during growth and young adulthood indicated that both male and female CK$^{--/--}$ double-knockout mice gained weight more slowly than their male and female wild-type littermates, as shown by Boltzman sigmoidal curve fitting (Fig. 10.4; see also [19]). At 3 months of age, the average body weight of CK$^{--/--}$ double-knockout mice (19–22 g, for females and males respectively) was 13–20% lower than that of wild-type mice (22–28 g). Food intake was also lower in CK$^{--/--}$ mice (4.4 g versus 5.3 g of food per day) but, if corrected for body weight, was not overtly abnormal and was comparable to that of young

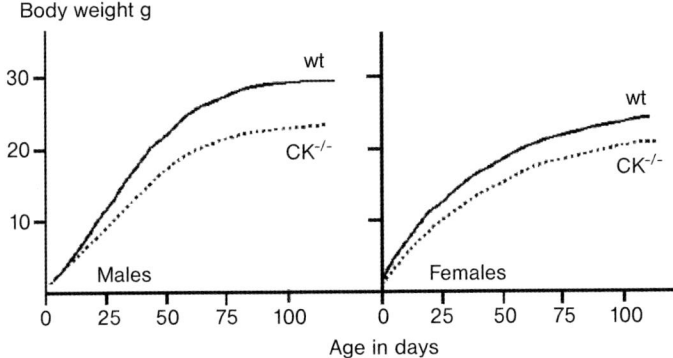

Fig. 10.4 Average body weights of male and female wild-type (wt; closed lines) and CK$^{--/--}$ double-knockout (dashed lines) mice during growth and adulthood (0–4 months). Note that CK mutants gain weight more slowly than their male and female wild-type littermates.

mice with comparable body weight. We therefore conclude that combined B-CK and UbCKmit deficiency has a selective and major impact on growth, development, and whole-body physiology. Here, we will continue with a review of CNS-related findings only.

10.4
^{31}P Magnetic Resonance Spectroscopy

To investigate the metabolic consequences of a lack of CK activity in the CNS, ^{31}P magnetic resonance (MR) spectroscopy was performed on a voxel area ($7 \times 7 \times 5$ mm^3) containing the entire brain, except the cerebellum, of a number of wild-type and our B-CK-, UbCKmit-, and double-CK mutant mice. Typical representative spectra obtained from these genotype groups are shown in Fig. 10.5, with resonances for PCr (0 ppm), ATP (γ, α, β: -2.5, -7.5, and -16.1 ppm, respectively),

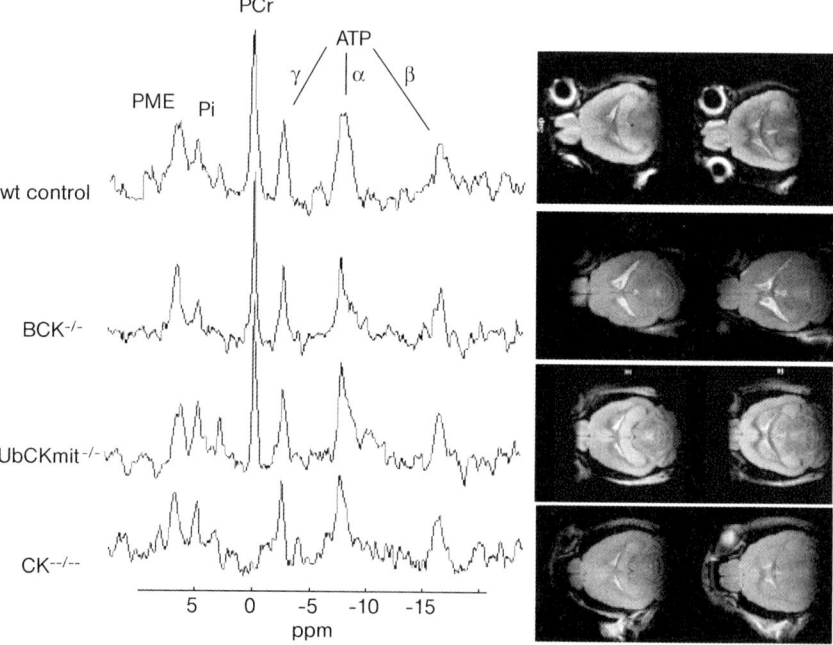

Fig. 10.5 Left: SIS-localized 31P-MR spectra from brains from CK-deficient and control mice. The PCr resonance peak at 0 ppm is lacking in the spectrum of CK$^{--/--}$ double-knockout mice, but no other differences are apparent. Other resonances visible in the profiles shown are from the three phosphate groups of ATP (-2.5 ppm, -7.5 ppm, -16.1 ppm), inorganic phosphate (P$_i$; 5 ppm), and phospomonoesters (PME; 6.7 ppm). Right: T$_2$-weighted spin-echo images (horizontal plane) of brain from wild type (top), B-CK$^{-/-}$ mutant (second), UbCKmit$^{-/-}$ (third), and CK$^{--/--}$ mice (bottom), showing the eyes and various brain structures including ventricles (compare to pictures in *The Mouse Brain in Stereotaxic Coordinates* by K.B. Franklin and G. Paxinos, 1997, Academic Press). (Details in [41]).

inorganic phosphate (P_i, +5 ppm), and phosphomonoester (PME, +6.7 ppm) indicated. Clearly, in the spectra for $CK^{--/--}$ mice the PCr peak disappeared entirely, indicating that phosphotransfer capacity was diminished and steady-state PCr levels were reduced to below the level of detection in brain. Subsequently, localized ^{1}HMR spectroscopy was used to study concomitant metabolic changes in functionally different regions of the brain, including the thalamus, hippocampus, and cerebellum (not shown). Most conspicuous findings were the decrease in the level of total creatine by 20–30% and a similar increase in the level of the neuronal/axonal marker metabolite N-acetyl-aspartic acid across all brain regions in the double mutant mice compared to wild type (for details, see [39, 40]). From these spectral analyses we drew several conclusions. Firstly, it was clear that the lack of PCr in brain was obviously not compensated by import of PCr from the circulation, a hypothetical mechanism that could occur after release from other tissues that still have a normal complement of CK isoforms, such as muscle, which expresses M-CK and ScCKmit. Furthermore, it is noteworthy that ATP and P_i levels and the (intracellular) pH calculated from the P_i-αATP shift in resonance position did not reveal any significant differences between wild-type and $CK^{--/--}$ mice. Hence, we may conclude that gross metabolic changes in glucose–lactate metabolism did not occur in our model. Finally, further study is necessary to explain the increase in NAA level, as this may indicate a developmental shift in cell type distribution or equally well point to adaptation in metabolic pathways involved in lipid synthesis or a decline in energetic state (see [39] for review on NAA function) [40]. This phenomenon may also be related to the partial, but quantitatively similar, decrease in the total creatine level.

10.5
Altered Brain Morphology: Involvement of the Intra-infra-pyramidal Mossy Fiber Field in $CK^{--/--}$ Mice

Indication of developmental adaptation in brain architecture was also obtained in other studies. In accordance with smaller body posture, we observed an overall decrease of ~6% in wet brain weight for $CK^{--/--}$ double-knockout brains compared with wild-type brains. Magnetic resonance imaging (MRI) via acquisition of high-resolution T_2-weighted images from brains of individual animals (Fig. 10.5, rightmost pictures) was used to provide us with morphological–anatomical details regarding this weight reduction so that we could determine whether the small weight reduction of brains of $CK^{--/--}$ mice was correlated to a global reduction in brain size or to more regional changes in brain structure. Unfortunately, this technique appeared not sensitive enough due to problems with individual background heterogeneity. Previous MRI analysis by our group had already revealed that some individual variation in gross brain anatomy, including variation in ventricle size, occurred in mice with single B-CK or UbCKmit deficiency [41]. From these analyses and subsequent comparison with morphometric data on mouse brain anatomy of different mouse strains, we learned that the morphological variation in our mutant cohorts was most easily explained by Mendelian variation in the genetic background of individual animals, as all CK mutants in our "collection" were deliberately kept on a mixed background with contribution from C57BL and 129/Ola parental strains (see [42] for information on background-related features). We therefore continued to use MRI only for a search of consistent anatomical abnormalities in CK single- and double-mutant brains, outside the range covered by background variation. Unfortunately, no clues were found with this technique, probably due to lack of capacity to resolve substructures in sufficient detail in the small mouse brain. Subsequent global histological survey of thick vibratome or thin paraffin sections using light microscopy in combination with hematoxylin and eosin staining also revealed no size changes or obvious anatomical defects in the gross anatomy and layering or composition of the different neuronal nuclei/clusters in the brains of CK-deficient mice. Only when we adopted the use of more sophisticated histological staining procedures for in-depth analysis of various distinct substructures in the histochemical atlas of the brain did we consistently observe subtle but specific changes in the mossy fiber projection fields of the hippocampus and in structures of the hypothalamus of B-CK$^{-/-}$ and CK$^{--/--}$ animals [15, 19]. The mossy fibers located inside the hippocampus are axons and axon collaterals of the dentate granule cells projecting to the hilus and the CA3 pyramidal cell layer. The terminal field of the hippocampal mossy fibers located superficially to the pyramidal cell layer is called the suprapyramidal mossy fiber area (SP-MF). In addition, a small extent terminates within or below the pyramidal cell layer; this is called the intra-infra-pyramidal mossy fiber (IIP-MF) area. In Fig. 10.6 it is shown how the mossy fibers, which terminate either within or below the pyramidal cell layer, are grouped together. Interestingly, Timm staining of the zinc-containing mossy fibers in CK$^{--/--}$ double-knockout brains revealed no differences in the absolute values of the hilus, SP-MF, or IIP-MF area between both genotype groups. However, the size of the regio inferior (CA3 area plus the hilus) was approximately ~15% smaller in the CK$^{--/--}$ double-knockout brains than in wild-type brains. As a result, the relative size of the SP-MF area of the CK$^{--/--}$ double-knockout brains was 15–20% larger than that of wild-type brains. The relative size of the IIP-MF area seemed larger as well (about 15%) in CK$^{--/--}$ mice, but this observation did not reach significance. Besides a reduced regio inferior size, the total hippocampal area within the same brain sections appeared smaller and was reduced in size by approximately 7% in the CK$^{--/--}$ double-knockout brains. It should be mentioned here that earlier work in our group on single mutants had demonstrated that SP-MF/IIP-MF abnormalities also occur in B-CK$^{-/-}$ mice. Visual comparison of Timm's sulfide silver–stained hippocampal sections at the midseptotemporal level in these animals revealed a larger IIP-MF area than in wild-type control animals (Fig. 10.6), whereas the SP-MF, the hilus, and the total area of the regio inferior was not overtly different. Computer-assisted image analysis of the absolute sizes of these different projection fields confirmed that the IIP-MF area in B-CK$^{-/-}$ brain was ~50% larger than in wild-type brain. When taken combined, these results suggest that mutation type–dependent changes occur in the mossy fiber area of the hippocampus in B-CK$^{-/-}$ and CK$^{--/--}$ animals. Subtle anatom-

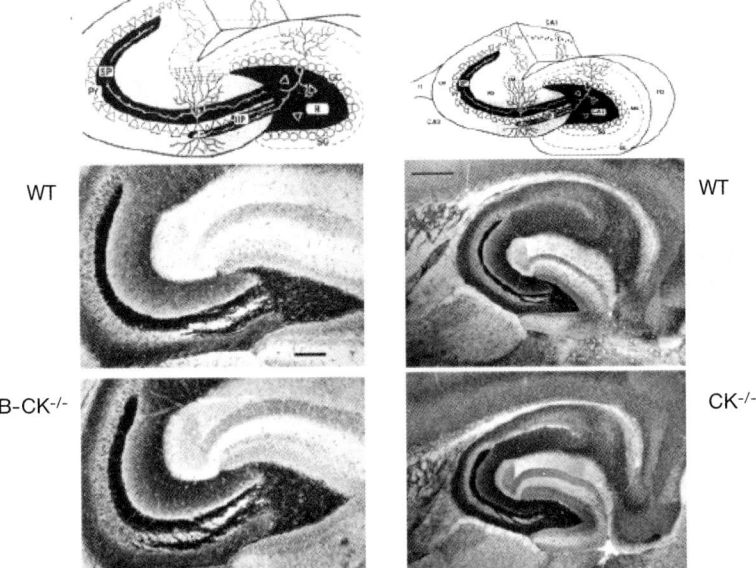

Fig. 10.6 Hippocampal mossy fiber projection fields in brains of wild-type, B-CK$^{-/-}$, and CK$^{--/--}$ mice. Top (left/right): Schematic drawings of the hippocampus containing the CA3 pyramidal (PY) cell layer area and the dentate gyrus with the supragranular layer and the mossy fiber (dentate gyrus granule cell [GC] axon) terminal projection fields (in black). Shown are the intra-infra-pyramidal (IIP) mossy fiber area, the supra-pyramidal (SP) mossy fiber zone, and the hilus (H or CA4). Other hippocampal areas indicated are the fascia dentate (FD), the fimbria (FI), the stratum lacunosum-moleculare (LM), the dentate gyrus middle and outer molecular layers (ML and OL), the stratum oriens (OR), the stratum radiatum (RD), and the supragranular layer (SG). Courtesy of W.E. Crusio and H.P. Lipp. Left panels: Representative examples of horizontal Timm-stained hippocampal sections at the midseptotemporal level show that B-CK$^{-/-}$ mice have a larger IIP-MF (mossy fiber) projection field than do wild-type animals, whereas the SP-MF, the hilus, and the total area of the regio inferior (CA3 area plus the hilus) are very similar in both genotype groups. Bar: 220 μm. Right panels: Similarly stained sections from CK$^{--/--}$ double-knockout and wild-type mice show a smaller regio inferior (CA3 area plus the hilus) but a relatively larger IIP-MF and SP-MF area in CK$^{--/--}$ mice. Bar: 440 μm.

ical variations in the hippocampus mossy fiber fields have been shown to co-vary with behavioral variations in a number of tasks [43]. Therefore, the performance of CK$^{--/--}$ mice in learning and cognition and habituation tasks was assessed in much detail (see below).

Other regions that strongly immunostained for B-CK and UbCK were also analyzed in detail. Among these were the arcuate nucleus (Arc), the medial preoptic area (POA), and the paraventricular nucleus (PVN) of the hypothalamus, the brain regions known to be implicated in the regulation of neuroendocrine processes and in the central control of body temperature, in which neuropeptide Y (NPY) expression is though to have a central role [44]. Because we had indications suggesting that our animals had a defect at the hypothalamic endpoint in the brain–fat axis for thermoregulation (data not shown; see below), we decided to focus on the distribution of NPY-rich regions of the hypothalamus. In control animals, NPY immunoreactivity was predominantly located in the PVN. In CK$^{--/--}$ mice, although the intensity of labeling for NPY in the PVN was not overtly different from controls, we observed that the total area of NPY expression observed in CK$^{--/--}$ mice was approximately 10% greater ($212 \pm 8.9 \times 10^3\ \mu m^2$ versus $191 \pm 5.2 \times 10^3\ \mu m^2$) than in wild types.

The findings regarding the increased size of the SP-MF and IIP-MF areas in the hippocampus and the NPY-positive PVN area of the hypothalamus of CK$^{--/--}$ mice are somewhat counterintuitive and are not in accordance with the global decrease in hippocampal size or the overall decrease in brain weight of these mice. Thus, our results suggest that changes in brain anatomy are not indiscriminate and that specific changes in the organization of hippocampal substructures and hypothalamic nuclei probably occur as an active response to specific changes in CK profile in specific cell types. How, when, and why these changes occur is a subject for further study.

10.6
CK$^{--/--}$ Mice Show Impaired Spatial Learning in Wet and Dry Maze Tests

The Morris water maze test is a well-established paradigm for assessment of CNS task performance and is used to examine spatial learning, i.e., the ability of individual mice to orientate in a novel environment using surrounding distal cues (spatial map [22, 24, 25]). In the spatial learning test of the Morris water maze, mice have to locate a submerged hidden platform while swimming in a circular pool with turbid water (Fig. 10.7). Performance of mutant and wild-type mice was compared in groups of eight animals each. Training sessions included several acquisition trials per day during three consecutive days (for details, see [18, 19]), and the latency to find the submerged platform was determined. For the wild-type group the escape latencies decreased over the training sessions, with an average escape latency of 100 ± 6 s in the first two trials and 14 ± 3 s during the last trial, indicating spatial learning. The CK$^{--/--}$ double-knockout mice also demonstrated a decrease in average latency over time; however, the escape latencies remained significantly higher compared with the wild-type group. The average escape latencies in the last trial for the CK$^{--/--}$ mice (65 ± 8 s) were 4–5 times higher than those observed in wild-type mice, indicating impaired spatial learning. The same learning trend was also seen in another experiment with 6 instead of 12 trials per day (Fig. 10.7, right panel).

The probe test, a 2-min swimming and searching trial with the platform removed from the pool (not shown), serves to assess the spatial learning ability of the animals in a complementary manner. The wild-type animals spent $53 \pm 4.5\%$ of their time searching in the quadrant where the platform was initially situated, indicating a preferable search in the training quadrant. The CK$^{--/--}$ mice spent

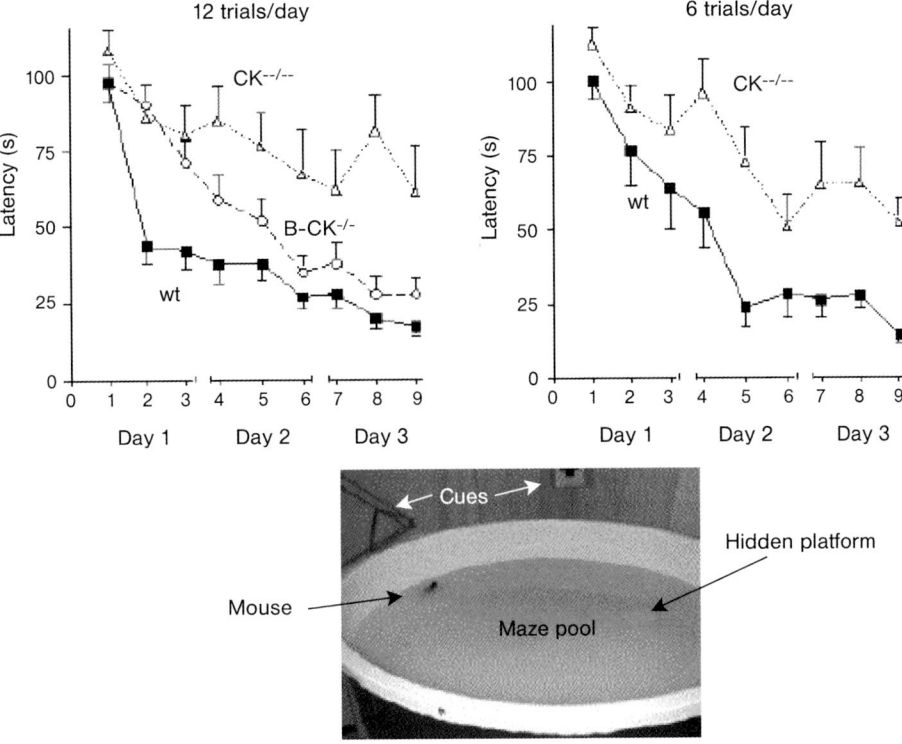

Fig. 10.7 Spatial learning in the Morris water maze is impaired in adult B-CK$^{-/-}$ and CK$^{--/--}$ mice. Latencies (given in seconds) involved in finding the submerged platform in the pool at various time points in the training period are given. Note that wild-type mice exhibit fast learning and that the learning curve is much slower in B-CK$^{-/-}$ mice. CK$^{--/--}$ mice show hardly any learning ability. Left: Results from a test with 12 trials per day (3 blocks, 4 trials each). Right: Results from a test with 6 trials per day (3 blocks, 2 trials each). (For details, see [19]).

only 22.9 ± 3.9% of their trial time in this quadrant, which was significantly lower compared with the wild-type group and equal to the 25% chance level. Taken together, the higher escape latencies in the acquisition phase and the non-specific searching performance in the probe trial demonstrate impaired spatial learning in the wet maze for CK double knockouts. For comparison, the performance of B-CK$^{-/-}$ mice is also given in Fig. 10.7. Clearly, these mutants performed better than the CK$^{--/--}$ mice, but learning acquisition was significantly slower than in wild types (for further details, compare data in [15, 18, 19]).

In order to confirm the impaired spatial learning ability of the CK$^{--/--}$ double-knockout mice and to account for effects from several swimming-related (stress) factors [45], a dry version of the analysis of spatial orientation learning was performed on a circular hole board (a dry maze surrounded by distal guiding cues [46, 24]). When latencies to find the escape hole and the number of errors made before finding the correct hole were measured, wild-type mice showed a learning curve with an initial latency of 19 ± 5 s for trial day 1 and an average latency of 9 ± 2 s on trial day 9 (data not shown; see [19]). Also, wild-type mice became more accurate over time in the localization of the exit hole, demonstrated by their decreasing number of visits to closed holes, which were registered as errors. The $CK^{--/--}$ knockout animals revealed no clear decline in latency. Moreover, they demonstrated significantly increased escape latencies during the nine session days when compared with the wild-type animals [19]. The $CK^{--/--}$ mice were less accurate over time in finding the escape hole, i.e., they made significantly more errors than the wild-type mice. During the last session, $CK^{--/--}$ double-knockout mice still visited about seven other holes before they reached the escape hole, which is two times more errors than made by wild-type mice (3 ± 1 errors).

To confirm that the mice showed spatial learning in locating the escape hole, a probe trial was again conducted (involving a board search with all holes closed). The number of holes visited in the quadrant where the escape hole had been located and in the three other quadrants was recorded. The wild-type animals visited significantly more holes in the relevant quadrant ($38.4 \pm 4\%$) than in the other areas of the board. However, the $CK^{--/--}$ mice appeared not to search specifically and made only $29.4 \pm 3\%$ of the total number of hole visits in the right quadrant, which was not significantly different from the 25% chance level. These results revealed that the $CK^{--/--}$ mice did not learn the position of the escape hole, indicating impaired spatial learning and corroborating the findings of the water maze test. It is important to note here, as mentioned above, that single B-CK- and UbCKmit-mutant mice also showed impaired learning ability [18, 19]. However, the tests used monitored only quantitative effects and were not stringent enough to reveal anything about possible mutation-dependent differences in "sub-aspects" of learning.

Aging is known to cause a general decay in the homeostatic energy reserve in cells and is associated with a variety of learning and memory deficits [47, 48]. Indeed, for both wild-type mice and B-$CK^{-/-}$ mice, we found that aging exacerbated spatial learning impairment (data not shown; [15]), bringing B-$CK^{-/-}$ mice into a situation where they hardly showed a learning curve. Inter-comparison of mice showed that the performance of $CK^{--/--}$ mice was by far the most abnormal, showing hardly any learning curve at even a young age. Therefore, we were unable to study the additive effects of aging in $CK^{--/--}$ mice but feel that aging effects are almost surely superimposes upon effects of energy stress in our model.

10.7
Cued Performance and Motor Coordination Are Normal in $CK^{--/--}$ Mice

To analyze whether the observed learning impairment could be due to a performance defect involving visual or motor dysfunction, the wild-type and $CK^{--/--}$ double-knockout mice conducted a water maze visual platform swimming task and the so-called rotarod task. During the visual task, the platform in the water maze was raised above the surface of the water. The latencies for the mice in the wild-type group and the CK$^{--/--}$ double-knockout group were not overtly different (data not shown; [19]). The rotarod test is routinely used to study motor coordination and balance. The total time the animals could keep their balance and stay walking on top of the rotating rod was 333 ± 22 s ($n = 9$) for CK$^{--/--}$ double-knockout animals, which was not different from the wild-type group (331 ± 24 s, $n = 9$). These findings indicate that the reduced spatial learning ability of the CK$^{--/--}$ double-knockout mice is not caused by visual or motor deficiencies.

10.8
CK$^{--/--}$ Mice Show Normal Open-field Exploration and Habituation

Another aspect of behavioral testing is the recording of activity and habituation in open-field areas. Surprisingly, wild-type mice and CK$^{--/--}$ double-knockout mice demonstrated similar active exploration and locomotion activity. The time spent walking, wall leaning, rearing, and sitting during the 30-min test periods was similar for both groups (for details, see [19]). Only grooming was decreased for CK$^{--/--}$ double-knockout mice compared with wild-type mice (77.7 ± 24.5 s versus 176.4 ± 42 s; $n = 9$), but this difference did not reach statistical significance. During the test periods, both groups displayed similar walking patterns, covered similar walking distances, and showed similar home base formation behavior, an indication of habituation (data not shown). The overall results demonstrate that complete CK deficiency does not influence explorative behavior. This result is striking, as earlier studies by our group on single B-CK mutant mice did reveal abnormal open-field activity and lack of habituation, with altered walking patterns and thigmotactic behavior. Again, we must conclude that specific changes in behavior are coupled to specific changes in CK profile and therefore correlate to mutation type.

To verify whether habituation and anxiety are normal in CK$^{--/--}$ mice, we also tested them in the light–dark box (not shown). This test is based on the tendency of mice to prefer and explore a dimly lit, concealed area and avoid a brightly lit, open area. Wild-type control mice ($n = 8$) preferred the dark compartment to the light compartment and spent 78% of the test period (10 min) in the dark compartment and only 22% in the light compartment. After the first transition between the two chambers (latency L-D), wild-type mice repeatedly entered the light and dark chambers (indicated as number of crossings) and showed rearing behavior in the light chamber. The CK$^{--/--}$ mice ($n = 7$) demonstrated a similar preference for the dark chamber (75% of the total time), and no abnormalities in their transitions from light to dark or in their behavior in the light chamber were observed. CK$^{--/--}$ and wild-type controls were also compared by testing in the T-maze continuous alternation task [49], which is based on the phenomenon of spontaneous alternation by mice. Animals in both genotype groups performed equally well and demonstrated an alternation rate of ~60%. The average time to complete the 15 trials was 19 ± 3 min for the wild-type animals and 17 ± 3 min (data not shown) for the CK$^{--/--}$ double-knockout mice, which was not significantly different. Combined, the results suggested that there was no difference between wild-type and CK$^{--/--}$ double-knockout mice in their habituation, exploratory behavior, or anxiety-related responses, which is an unanticipated outcome given the profound architectural changes in the hippocampal area of the brain.

10.9
CK$^{--/--}$ Mice Show Abnormal Thermogenesis

Whereas the hippocampus is considered the principal steering center for learning, cognition, and behavior, brain structures such as the hypothalamus are in control of sympathetic regulation of body functions. Therefore, during our phenotyping studies of CK$^{--/--}$ mice, we studied various parameters of whole-body physiology and noted abnormal body temperature regulation. The average body temperature profiles for both wild-type and CK$^{--/--}$ mice showed a circadian periodicity with equal rhythmicity (data not shown). However, whereas wild-type mice, housed at room temperature, showed average body temperatures alternating between 36.7 ± 0.1 °C and 37.3 ± 0.1 °C during the day and night period, respectively, the body temperature of CK$^{--/--}$ mice varied between 35.9 ± 0.2 °C and 37.3 ± 0.1 °C. Consistently, the body temperature of CK$^{--/--}$ animals was 0.5–1.0 °C lower than that of wild types during the early morning to early afternoon period but normalized thereafter. This pattern of abnormally low morning body temperature was confirmed by repeated temperature measurements over a period of 12 consecutive days at different ages, which always showed an average body temperature of 36.8 ± 0.1 °C for wild-type and 35.8 ± 0.1 °C for CK$^{--/--}$ mice during the morning interval. Subsequent studies of the effects of cold challenge revealed that CK$^{--/--}$ mice indeed have a reduced capacity for body temperature maintenance (data not shown; Streijger et al., submitted). However, neither studies into the possible involvement of the abnormal uncoupling of brown fat mitochondria, or a general transcription–regulation abnormality in the machinery for cellular energetics as recently described for a Huntington mouse model [50], nor the monitoring of locomotor activity and food intake have revealed clues to help us explain the thermogenic problems in CK$^{--/--}$ animals. Currently, our best explanation is that the defect is at the brain end of the brain–body axis and may be caused by abnormal synaptic connectivity or neuropeptide Y (NPY) neurosecretory activity in the hypothalamus. As outlined above, the NPY-positive area is clearly abnormal in CK$^{--/--}$ animals.

10.10
Altered Acoustic Startle Reflex Response and Hearing Problems in CK$^{--/--}$ Mice

A complete lack of CK could have an impact on sensory functions of the CNS and peripheral neural systems. To investigate whether animals with a deficit in their brain creatine metabolism show an impairment in their immediate reaction to sudden stimuli, separate cohorts of wild-type ($n = 17$) and CK$^{--/--}$ double-knockout mice ($n = 13$) were tested in the acoustic startle reflex setup [18, 19]. Special attention was focused on the assessment of prepulse inhibition of the acoustic startle response, a process of sensorimotor activity integration in which the hippocampus plays an important role [51] but also involves other brain circuits. Startle response amplitudes as a response to 120-dB startle pulses differed vastly different between the two groups (average maximal startle amplitude of ∼500 arbitrary units for wild-type and ∼130 arbitrary units for CK$^{--/--}$ animals). The wild-type mice showed increasing startle amplitudes following increasing sound levels. Relatively low amplitudes were observed with 70–105-dB stimulus levels, but this was followed by a rapid increase in the startle response amplitude following stimuli ranging from 110 dB to 150 dB. Our recordings revealed that the CK$^{--/--}$ double-knockout mice had significantly reduced startle amplitudes, in the range of 70–150 dB, compared with the wild-type group and showed startle responses that were almost 3.5 times lower than that of the wild-type mice upon increasing sound levels. Prepulse inhibition is defined as the percentage of reduction in the 120-dB startle stimulus response when a weak prepulse (+2 dB to +16 dB above background) precedes the 120-dB stimulus pulse(s). The prepulse itself is too low to elicit a startle response. Briefly, the wild-type mice revealed a clear-cut prepulse inhibition, which increased with increasing prepulse intensities. The strongest inhibition (56 ± 9%) was shown following a 16-dB-above-background prepulse. CK$^{--/--}$ double-knockout mice also showed an increasing prepulse inhibition of the startle response, with a prepulse inhibition of 40 ± 10 % for the strongest prepulse of +16 dB. Thus, our data revealed that both genotype groups demonstrated prepulse inhibition. Differences between the groups did not reach significance, however (see [18, 19] for more details).

More interestingly, subsequent hearing studies revealed that the sensitivity of the auditory system to pure tones was reduced substantially in CK$^{--/--}$ mice. Thresholds were elevated by 20–30 dB, which indicated that the sound intensity required to reach a detectable signal in the brainstem was 10- to 30-fold greater in CK$^{--/--}$ mice. Recently, in collaboration with the group of P. Gillespie [52], we found that this threshold elevation can be best explained by impaired mechano-transduction in hair bundle cells in the vestibular hearing system in the absence of CK. Exactly how CK-mediated ATP delivery feeds into the process of mechanotransduction at the cellular level is a subject for future study.

10.11
Pentylenetetrazole-induced Seizures Occur Later in B-CK$^{--/--}$ Mice

To obtain an independent measure of the efficiency of the synaptic communication network in CK mutant mice and the defective CK–PCr energy system, we exposed the animals to a situation of acute high-energy demand, using chemically (pentylenetetrazole [PTZ]) induced seizures. PTZ treatment blocks GABA inhibitory function probably by reducing the coupling between GABA and the benzodiazepine recognition sites of the $GABA_A$ receptor [53] and therefore results in increased and prolonged excitation, which in turn may lead to a full-blown seizure. Previous studies showed during seizure activity an increase in local cerebral glucose and cellular energy (ATP) utilization, a higher CK rate constant, and a high ATP turnover [54–56]. This large amount of quickly provided energy is required for the fuelling of seizure progression. At different time points after PTZ administration, metabolic activity in brain regions such as the thalamus, the retrosplenial cortex, and the dentate gyrus peaks highly (as shown with functional MRI [fMRI]) [57], but probably the entire brain network gets involved at some time point during the seizure period. Response to PTZ injections can therefore serve as an index of global neuronal network excitability. In this PTZ-induced process, both neuronal and glial cell types play a role, as changes in glial-neuronal interactions and metabolism have been reported [58].

An analysis of multiple videotaped seizures in our study revealed that wild-type controls and $CK^{--/--}$ animals showed similar active exploration and locomotion in the 5-min period prior to the first PTZ injection. The exploration mainly consisted of walking, rearing, and, to a lesser extent, sitting. When wild-type and $CK^{--/--}$ mice received five or six PTZ injections, respectively, the animals began to display subtle jerks consisting of muscle spasms mainly of the face and the skin on their backs. As the experiment proceeded and more PTZ was administered, myoclonic jerking of the neck and forelimb was observed. Prior to seizure onset, both wild-type and $CK^{--/--}$ mice displayed short, repetitive periods (~2–5 s) of vigorous jerking of the complete body with increasing frequency. In each wild type ($n = 12$ out of 12), a seizure was characterized by a period of vigorous clonic movements in all limbs and the trunk, as righting reflexes could be temporarily lost. Severe clonic seizures could also develop into a brief episode of tonic limb extension. Typically, 8–10 injections of PTZ evoked a generalized seizure episode of ~26 s, which was followed by complete physical inactivity. The number of $CK^{--/--}$ mice experiencing a generalized convulsive episode was significantly smaller that the number of wild types with full-blown seizure (3 out of 10 $CK^{--/--}$ mice versus 12 out of 12 controls).

To be able to correlate the difference in seizure progression with characteristic features in electroencephalographic (EEG) activity, cortical electrodes were implanted two weeks prior to simultaneous seizure and EEG recording. Prior to testing, the baseline cortical activity showed low-amplitude desynchronized EEG activity with an initial baseline amplitude of 148 ± 10 μV for wild-type and 123 ± 13 μV for $CK^{--/--}$ mice. About 60 min after the first PTZ injection, abnormal brain activity and muscular responses were observed in both groups (wild types: 60 ± 2 min; $CK^{--/--}$: 64 ± 2 min). These abnormalities consisted of spontaneous single high-voltage spikes accompanied by a jerk. The shape and firing duration of a single spike was similar between wild types and $CK^{--/--}$ mice (respectively 191 ± 4 ms and 193 ± 4 ms), whereas the interspike interval obtained from $CK^{--/--}$ mice (25 ± 5 s) was slightly increased compared with wild-type mice (15 ± 2 s). Prior to the seizure onset, the average duration between two successive single spikes became shorter, indicating increasing frequency as the seizure progressed. Conspicuously, the amount of single spikes was significantly reduced in CK$^{--/--}$ mice (74.0 ± 1.1% versus 86.4 ± 1.3% of the total spikes classified as single spikes in CK$^{--/--}$ and wild types, respectively), and the spike pattern was altered (CK$^{--/--}$ mice displayed a twofold increase in multi-spike complexes, with minimally three successive high-amplitude spikes). Seizure onset in all wild-type mice was characterized by high-amplitude firing activity in EEG signals, i.e., the ictal episode, starting with a short period (~3–4 s) of irregular high-amplitude spikes (Fig. 10.8; 12 out of 12 animals).

A wildtype (n = 12 out of 12)

B CK--/-- (n = 3 out of 10)

C CK--/-- (n = 7 out of 10)

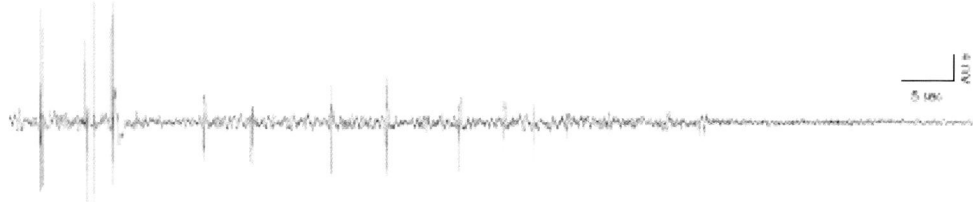

Fig. 10.8 Typical electroencephalographic (EEG) tracings of generalized absence seizures in wild-type and CK$^{--/--}$ mice. Shown are spike discharge patterns. The black arrows depict the onset of a full seizure. Only 3 out of 10 CK$^{--/--}$ mice showed EEG activity with similarity to that of wild-type animals. Seven out of 10 CK$^{--/--}$ mice demonstrated transition to EEG inactivity characterizing postictal depression. Vertical black bar represents 500 μV; horizontal bar represents 5 s.

After seizure initiation, but still relatively early, an organized rhythmic activity with discharge signals with frequencies in the range of 4–6 Hz. was seen. The seizure episode ended with the onset of an EEG suppression or isoelectric postictal period, during which time the mice remained completely motionless. Although 3 out of 10 CK$^{--/--}$ mice showed an initial ictal episode, none of them demonstrated the characteristic 4–6-Hz rhythmic high-amplitude spikes, and the entire EEG trace revealed irregular firing. Interestingly, the 7 out of 10 CK$^{--/--}$ mice that did not display a behavioral seizure all demonstrated a sudden EEG depression ~90 min after the initial PTZ injection, mimicking a postictal depression. Despite these differences, no differences between wild-type and CK$^{--/--}$ mice were observed for the latency and duration of the high-amplitude burst activity during the seizure intervals. In a previous study, we had already reported that differences also existed in seizure vulnerability between B-CK single knockouts and wild-type controls, indicating a slower chemically induced seizure development in B-CK animals. Because B-CK single and CK$^{--/--}$ double mutants were not directly compared in the more elaborate test with simultaneous EEG recording, we do not know whether these mice would react differently. What we can conclude, however, is that the partial or complete absence of CK has a dramatic impact on overall neural network excitability. Again, future study is needed to reveal whether this is a direct effect of cell-intrinsic metabolic changes or an indirect consequence of altered cellular wiring and developmental adaptation. More in-depth studies are also needed to reveal whether neuronal cell loss or changes in phosphotransfer activity in the CK circuit in neurons or astrocytes, or both, are dominant in the blocking of seizure development in animals without the CK–PCr circuit.

10.12
Conclusions and Future Outlook

Energy metabolism in mammalian brain correlates with neuronal activity and is characterized by a high rate of glucose import, active lactate shuttling between astrocytes and neurons, high levels of enzymes for glucose combustion, a high rate of oxidative phosphorylation (OXPHOS) in mitochondria, and very high ATP consumption in most cell types of the brain [59–63]. As in all tissues, ATP in brain is needed for common housekeeping functions, but in addition it is needed for electrogenic activity, neurotransmitter cycling, axoplasmic cellular transport, and continuous lipid turnover and actomyosin-based dynamics (see [64, 65] for review) involved in dendrite development, ongoing synaptic spine remodeling, and cell contact (endfeet) reshaping. Because of the special cytoarchitecture of neurons and glial cells, their intriguing structural-functional cooperation, and their highly complex spatiotemporal regulation of energy metabolism in these two cell types, brain tissue is particularly difficult to study. Neurons are highly specialized cells, each with its own individual role in the complex cellular society of the brain. Astrocytes are particularly well suited for interacting with neurons. Old ideas about "active roles" for neurons and "passive bystander" functions for astrocytes have long disappeared, and according to modern concepts neuron–astrocyte cooperation and reciprocal crosstalk are essential for almost all aspects of brain function (for more details, we refer the reader to the extensive literature on this point). Also, given that aspects of energy and creatine metabolism are shared and perhaps can be considered a "continuum" in these two cell types and that neurons and astrocytes occur as hundreds of different cell types, each with its own distinct structural and functional properties in different regions of the brain, it is evident that it is a most challenging task to study the role of separate enzymes in this enormous network.

Here we have described the first steps on a long road towards the delineation of the biological role that creatine kinases play in brain energetics. From the overview of data given here, it is apparent that CK isoenzymes may have an accessory role if we consider global brain physiology in a system-level context. Complete genetic ablation of B-CK and UbCKmit, the only two brain-expressed CK isoforms, is not directly lethal for the whole animal. Brain tissue is highly adaptive and has extraordinarily high plasticity; therefore, it is still possible that – as in neurodegeneration – CK deficiency is directly or indirectly coupled to the disruption of molecular pathways, synapses, neuronal or glial subpopulations, and local circuits in specific brain regions and/or higher-order networks. Our current phenotyping abilities are not powerful enough to discriminate between these possibilities. One future approach that will help to deconvolute the physiological role of CK at the system level will be to compare behavioral phenotypes between our CK mice and mice with knockout mutations in partner proteins for CKs. To our knowledge, currently only one such model is available: the hypomorphic KCC2-deficient mouse [66], which has lost one allele of the potassium chloride co-transporter KCC2. KCC2 is a membrane protein that interacts with B-CK and has a role in hyperpolarizing fast inhibitory neurotransmission [67]. Interestingly, KCC2 hypomorphs are smaller and exhibit altered habituation behavior and impaired learning ability but no altered motor coordination, like our CK$^{--/--}$ mice. However, PTZ sensitivity is increased and not impaired as in our CK mutants. Perhaps parallel testing should be done to reveal more similarities and differences, but ultimately this may help to interpret the relative importance of components in the CK circuit itself and in the structurally or functionally coupled systems.

For now we cannot discriminate very well between functions, but some findings point to the direction of cell loss. CK-related late-prenatal and postnatal developmental impairment of brain growth is evident if we consider the overall smaller brain size. Also, the altered seizure susceptibility, involving a general depolarization of large brain areas, may point to a general loss-of-function or cell connectivity effect. We cannot, however, correlate this to specific loss of neuronal or glial subtypes or local cellular wiring. Both CK isoforms are abundantly expressed in neuron and/or glial cell populations in *in vitro* cultures, but ISH and immunohistochemical assays have revealed extensive regional cell type differences. B-CK is certainly prominently expressed in glial subtypes and is most abundant in inhibitory neurons. UbCKmit expression appears to be confined to neurons only. Both glia and neurons could therefore be equally involved in $CK^{--/--}$ abnormalities. Moreover, we could not discriminate between the roles of these cell types by comparing the phenotypic appearance of B-CK and UbCKmit single-mutant mice, simply because the tests used are too crude. This, and the emerging knowledge that tight cooperation of neurons and glial cells is essential for almost all aspects of brain growth as well as electrogenic and endocrine activity, is also not very helpful in pinpointing the cellular origin of defects. Most direct indications of subselective and discriminate effects came from behavior and task typing and from regional (immuno)histochemical studies. Behavior and learning tasks are affected in $CK^{--/--}$ mice, but cued performance and motor coordination is intact. This may imply that some brain regions are functionally more affected than others, but whether this could be due to neuronal cell loss or cell function impairment remains controversial. Evidence for regional effects was also obtained from analysis of the mossy fiber area of the hippocampus and the NPY region of the hypothalamus. Here, regions with subpopulations of cells increased while the surrounding areas shrank in volume. Most easily, these findings can be explained by the absence of pruning or by abnormal neuron–astrocyte cooperation during subregional brain development, possibly in the postnatal period. Less likely is that the extended SP-MF/IIP-MF structures are caused by improved dendrite development in the absence of CK. Interestingly, even in the alteration of the global process of seizure induction observed in our mice, the regional effects of changes in the NPY region of the hypothalamus may play a role, as NPY has a significant effect on seizure latency [68].

Further study is thus necessary. On the one hand, we must delineate whether the patterns of CK expression – or perhaps better, CK activity – correlate with regional effects on brain development and the role of neurons and glial cells therein. Cell biological comparison of viability parameters, ion homeostasis, and/or actomyosin dynamics in primary neurons or astrocytes obtained from these regions may ultimately help us to reveal the critical role of the CK system at the cellular level. On the other hand, we need brain-imaging techniques such as fMRI to delineate the most vulnerable nodal regions in the cellular brain network, regions that lose functional integrity in the absence of CK. Here a general decline in the homeostatic ATP reserve or more specific metabolic loss-of-function effects may be involved, with similarity to the effects of aging or to neurodegenerative disease such as Alzheimer's, Parkinson's and Huntington's diseases, type 2 diabetes, or mitochondrial disorders. Thus, research on CK function in brain ultimately may not only contribute to the topic of CK–Cr significance itself but also help us to better understand how [ATP] functions within the normal physiological range or is associated with cell death or pathophysiological dysfunction in disease.

References

1 Mariman, E.C.M., Schepens, J.T.G., Wieringa, B. (1989) *Nucl. Acid. Res.* 17, 6385. Complete nucleotide sequence of the human Creatine kinase B gene.

2 Van Deursen, J., Wieringa, B. (1992) *Nucl. Acids Res.* 20, 3815–3820. Targeting of the creatine kinase M gene in embryonic stem cells using isogenic and nonisogenic vectors.

3 Van Deursen, J., Schepens, J., Peters, W., Meijer, D., Grosveld, G., Hendriks, W., Wieringa, B. (1992) *Genomics* 12, 340–349. Genetic variability of the murine creatine kinase B gene locus and related pseudogenes in different inbred strains of mice.

4 Van Deursen, J., Heerschap, A., Oerlemans, F., Ruitenbeek, W., Jap, P., Laak, H. ter, Wieringa, B. (1993) Skeletal Muscles of Mice Deficient in Muscle Creatine Kinase Lack Burst Activity. *Cell* 74, 621–631.cfv

5 Qin, W., Khuchua, Z., Cheng, J., Boero, J., Payne, R.M., Strauss, A.W. (1998) Molecular characterization of the creatine kinases and some historical perspectives. *Mol. Cell. Biochem.* 184, 153–167.

6 Wyss, M., Kadurah-Daouk, R. (2000) Creatine and creatinine metabolism. *Physiol. Rev.* 80, 1107–1213.

7 Wallimann, T., Wyss, M., Brdiczka, D., Nicolay, K., Eppenberger, H.M. (1992). Intracellular compartmentation, structure and function of creatine kinase isoenzymes in tissues with high and fluctuating energy demands: the "phosphocreatine circuit" for cellular energy homeostasis. *Biochem. J.* 28, 21–40.

8 Dzeja, P.P., Terzic, A., Wieringa, B. (2004) Phosphotransfer dynamics in skeletal muscle from Creatine kinase gene-deleted mice. *Mol. Cell. Biochem.* 256, 13–27.

9 Glaser, S., Anastassiadis, K., Steward, A.F. (2005) Current Issues in mouse genome engineering. *Nature Genetics* 37, 1187–1193.

10 Van Deursen, J., Lovell-Badge, R., Oerlemans, F., Schepens, J., Wieringa, B. (1991) *Nucl. Acids Res.* 19, 2637–2643. Modulation of gene activity by consecutive gene targeting of one creatine kinase M allele in mouse embryonic stem cells.

11 Van Deursen, J., Ruitenbeek, W., Heerschap, A., Jap, P., Laak, H. ter, Wieringa, B (1994). Creatine kinase (CK) in skeletal muscle energy metabolism: A study of mouse mutants with graded reduction in muscle CK expression. *Proc. Natl. Acad Sci USA* 91, 9091–9095.

12 Steeghs, K., Peters, W., Brückwilder, M., Croes, H., van Alewijk, D., Wieringa, B. (1995) Mouse ubiquitous mitochondrial Creatine kinase gene organization and consequences of inactivation in mouse embryonic stem cells. *DNA and Cell Biol.* 14, 539–553.

13 Steeghs, K., Oerlemans, F., Wieringa, B. (1995) Mice deficient in ubiquitous mitochondrial creatine kinase are viable and fertile. *Biochim. Biophys. Acta.* 1230, 130–138.

14 Steeghs, K., Benders, A., Oerlemans, F., de Haan, A., Heerschap, A., Ruitenbeek, W., Jost, C., van Deursen, J., Perryman, B., Pette, D., Brückwilder, M., Koudijs, J., Jap, P., Veerkamp, J., Wieringa. B. (1997) Altered Ca^{2+} responses in muscles with combined mitochondrial and cytosolic creatine kinase deficiencies. *Cell* 89, 93–103.

15 Jost, C., van der Zee, C.E.E.M., in 't Zandt, H.J.A., Oerlemans, F., Verheij, M., Streijger, F., Fransen, J., van Deursen, J., Heerschap, A., Cools, A., Wieringa, B. (2002) Creatine kinase B-driven energy transfer in the brain is important for habituation and spatial learning behaviour, mossy fibre field size and determination of seizure susceptibility. *Eur. J. Neurosci.* 15, 1692–1706.

16 Janssen, E., Dzeja, P.P., Oerlemans, F., Simonetti, A.W., Heerschap, A., de Haan, A., Rush, P.S., Terjung, R.R., Wieringa, B. and Terzic, A. (2000) Adenylate kinase 1 gene deletion disrupts muscle energetic economy despite metabolic rearrangement. *EMBO J.* 19, 6371–6381.

17 Janssen, E., Wieringa, B., Terzic A., Dzeja P.P. (2003) Impaired intracellular energetic communication in muscles of mice with combined Creatine kinase and

adenylate kinase (M-CK/AK1) deficiency. *J. Biol. Chem.* 278(33): 30441–30449.

18 Streijger, F., Jost, C.R., Oerlemans, F., Ellenbroek, B.A., Cools, A.R., Wieringa, B., Van der Zee, I. (2004) Mice lacking the UbCKmit isoform of creatine kinase reveal slower spatial learning acquisition, diminished exploration and habituation, and reduced acoustic startle reflex responses. *Mol. Cell. Biochem.* 256, 305–318.

19 Streijger, F., Oerlemans, F., Ellenbroek, B.A., Jost, C.R., Wieringa, B., Van der Zee, C.E.E.M. (2005) Structural and behavioural consequences of double deficiency for Creatine kinases B-CK and UbCKmit. *Behavioural Brain Res.* 157, 219–234.

20 Moreadith, R.W., Butwell, Radford, N. (1997) Gene targeting in embryonic stem cells: The new physiology and metabolism. *J. Mol. Med.* 75, 208–216.

21 Britton, S.L., Koch, L.G. (2001) Animal genetic models for complex traits of physical capacity. Exerc. *Sport Sci. Rev.* 29, 7–14.

22 Lipp, H.-P., Wolfer, D.P. (1998) Genetically modified mice and cognition. *Curr. Op. Neurobiol.* 8, 272–280.

23 Crawley, J.N., Belknap, J.K., Collins, A. et al. (1997) Behavioral phenotypes of inbred mouse strains: Implications and recommendations for molecular studies. *Psychopharmacol.* 132, 107–124.

24 Crawley, J.N., Paylor, R. (1997) A proposed test battery and constellation of specific behavioral paradigms to investigate the behavioral phenotypes of transgenic and knockout mice. *Horm. Behav.* 31, 197–211.

25 Gerlai, R. (1996) Gene-targeting studies of mammalian behavior: is it the mutation of the background phenotype? *Trends Neurosci.* 19, 177–181.

26 Jucker, M., Ingram, D.K. (1997) Murine models of brain aging and age-related neurodegenerative diseases. *Behav. Brain Res.* 85, 1–25.

27 Clancy, B., Darlington, R.B., Finlay, B.L. (2001) Translating developmental time across mammalian species. *Neurosci.* 105, 7–17.

28 Bhide, P.G., Day, M., Sapp, E., Schwarz, C., Sheth, A., Kim, J., Young, A.B., Penney, J., Golden, J., Aronin, N., DiFiglia, M. (1996) Expression of normal and mutant huntingtin in the developing brain. *J. Neurosci.* 16, 5523–5535.

29 Dzeja, P.P., Terzic, A. (2003) Phosphotransfer networks and cellular energetics. *J Exp Biol.* 206, 2039–47.

30 Tachikawa, M., Fukaya, M., Terasaki, T., Ohtsuki, S., Watanabe, M. (2004) Distinct cellular expressions of creatine synthetic enzyme GAMT and creatine kinases uCK-Mi and CK-B suggest a novel neuron-glial relationship for brain energy homeostasis. *Eur. J. Neurosci.* 20, 144–160.

31 Lein, E.S. et al. (2007) Genome-wide atlas of gene expression in the adult mouse brain. *Nature* 445, 168–176.

32 Su, A.I. et al. (2002) Large-scale analysis of the human and mouse transcriptomes. *Proc. Natl. Acad. Sci USA* 99, 4465–4470.

33 Barett, T., Suzek, T.O., Troup, D.B., Wilhite, S.E., Ngau, W.-C., Ledoux, P., Rudnev, D., Lash, A.E., Tujibuchi, W., Edgar, R. (2005) NCBI-GEO: Mining millions of expression profiles – databases and tools. *Nucl. Acid. Res.* 33, D562–D566.

34 Friedman, D.L., Perryman, M.B. (1991) Compartmentation of multiple forms of creatine kinase in the distal nephron of the rat kidney. *J. Biol. Chem.* 266, 22404–22410.

35 De Kok, Y.J.M., Geurds, M.P.A., Sistermans, E.A., Usmany, M., Vlak, J.M., Wieringa, B. (1995) Production of native creatine kinase B in insect cells using a baculovirus expression vector. *Molec. Cell. Biochem.* 143, 59–65.

36 Sistermans, E.A., de Kok, Y.J.M., Peters, W., Ginsel, L.A., Jap, P.H.K., Wieringa, B. (1995) Tissue- and cell-specific distribution of creatine kinase B: A new and highly specific monoclonal antibody for use in immunohistochemistry. *Cell Tissue Res* 280, 435–446.

37 Hamburg, R.J., Friedman, D.L., Olson, E.N., Ma, T.S., Cortez, M.D., Goodman, C., Puleo, P.R., Perryman, M.B. (1990). Muscle creatine kinase isoenzyme expression in adult human brain. *J. Biol. Chem.* 265, 6403–6409.

38 Hemmer, W., Zanolla, E., Furter-Graves, E.M., Eppenberger, H.M., Wallimann, T.

(1994). Creatine kinase isoenzymes in chicken cerebellum: Specific localization of brain-type creatine kinase in Bergmann glial cells and muscle-type creatine kinase in Purkinje neurons. *Eur. J. Neurosci.* 6, 538–549.

39 Baslow, M.H. (2000) Functions of N-acetyl-L-aspartate and N-Acetyl-L-aspartylglutamate in the vertebrate brain: Role in glial cell-specific signaling. *J. Neurochem.* 75, 453–459.

40 Pan, J.W., Takahakshi, K. (2004) Interdependence of N-Acetyl aspartate and high-energy phosphates in healthy human brains. *Ann. Neurol.* 57, 92–97.

41 In 't Zandt, H., J., A., Renema, W., K.J., Streijger, F., Jost, C., Klomp, D.W.J., Oerlemans, F., van der Zee, C.E.E.M., Wieringa, B., Heerschap, A. (2004) Cerebral Creatine kinase deficiency influences metabolite levels and morphology in the mouse brain: A quantitative *in vivo* 1H and 31P magnetic resonance study. *J. Neurochem.* 90, 1321–1330.

42 Brayton, C., Justice, M., Montgomery, C.A. (2001). Animal models: Evaluating mutant mice: Anatomic pathology. *Vet. Pathol.* 38, 1–19.

43 Henze, D.A., Urban, N.N., Barrionuevo, G. (2000) The mutifarious hippocampal mossy fiber pathway: A review. *Neurosci.* 98, 407–427.

44 Paul, M.J., Freeman, D.A., Park, J.H., Dark, J. (2005) Neuropeptide Y induces torpor-like hypothermia in Siberian hamsters. *Brain Res.* 1055, 83–92.

45 Hölscher, C. (1999). Stress impairs performance in spatial water maze learning tasks. Behav. *Brain Res.* 32, 56–62.

46 Gerlai, R. (2001) Behavioral tests of hippocampal function: simple paradigms, complex problems (review). *Behav. Brain Res.* 25, 269–277.

47 Jucker, M., Ingram, D.K. (1997) Murine models of brain aging and age-related neurodegenerative diseases. *Behav. Brain Res.* 85, 1–26.

48 Murphy, G.G., Rahnama, N.P., Silva, A.J. (2006) Investigation of age-related cognitive decline using mice as a model system: behavioral correlates. *Am. J. Geriatr. Psychiatry* 14, 1004–1011.

49 Gerlai, R. (1998). A new continuous alternation task in T-maze detects hippocampal dysfunction in mice. A strain comparison and lesion study. *Behav. Brain Res.* 95, 91–101.

50 Weydt, P., Pineda, V., Torrence, A., Libby, R.T. et al. (2006) Thermoregulation and metabolic defects in Huntington's disease transgenic mice implicate PGC-1α in Huntington's disease neurodegeneration. *Cell Metabolism* 4, 349–362.

51 Bast, T., Feldon, J. (2003) Hippocampal modulation of sensorimotor processes. *Prog. Neurobiol.* 70, 319–345.

52 Shin, J.-B., Streijger, F., Beynon, A., Peters, T., Gadzala, L., McMillen, D., Bystrom, C., Van der Zee, C.E.E.M., Wallimann, T., Gillespie, P.G. (2007) Hair bundles are specialized for ATP delivery via creatine kinase. *Neuron* 53, 371–386.

53 Walsh, L.A., Li, M., Zhao, T.J., Chiu, T.H., Rosenberg, H.C. (1999) Acute pentylenetetrazol inhection reduces rat GABAA receptor mRNA levels and GABA stimulation of benzodiazepine binding with no effects on benzodiazepine binding site density. *J. Pharmacol. Exp. Ther.* 289, 1626–1623.

54 Holtzman, D., Mulkern, R., Meyers, R., Cook, C., Khait, I., Jensen, F., Tsuji, M., Laussen, P. (1998) *In vivo* phosphocreatine and ATP in piglet cerebral gray and white matter during seizures. *Brain Res.* 783, 19–27.

55 Holtzman, D., Togliatti, A., Khait, I., Jensen, F. (1998). Creatine increases survival and suppresses seizures in the hypoxic immature rat. *Pediatr. Res.* 44, 410–414.

56 Bonan, C.D., Amaral, O.B., Rockenbach, I.C., Walz, R., Battastini, A.M., Izquierdo, I., Sarkis, J.J. (2000) Altered ATP hydrolysis induced by pentylenetetrazol kindling in rat brain synaptosomes. *Neurochem. Res.* 25, 775–779.

57 Brevard, M.E., Kulkarni, P., King, J.A., Ferris, C.F. (2006). Imaging the neural substrates involved in the genesis of pentylenetetrazol-induced seizures. *Epilepsia* 47, 745–754.

58 Eloqayli, H., Dahl, C.B., Gotestam, K.G., Unsgard, G., Sonnewald, U. (2004). Changes of glial-neuronal interaction and metabolism after a subconvulsive dose of pentylenetetrazole. *Neurochem. Int.* 45, 739–745.

59 Attwell, D., Laughlin, S.B. (2001) An energy budget for signaling in the grey matter of the brain. *J.Cereb. Blood Flow Metab.* 21, 1133–1145.

60 Chih, C.-P., Roberts, E.L. (2001). Energy substrates for neurons during neural activity: A critical review of the astrocyte-neuron lactate shuttle hypothesis. *Trends Neurosci.* 24, 573–578. Review.

61 Hoyer, S., Lannert, H., Latteier, E., Meisel, Th. (2004). Relationship between cerebral energy metabolism in parieto-temporal cortex and hippocampus and mental activity during aging in rats. *J. Neural. Transm.* 111, 575–589.

62 Magistretti, P.J. (1999) *Brain energy metabolism. In Fundamental neuroscience* (Zigmond, M.J., Bloom, F.E., Landis, S.C., Roberts, J.L., Squire, I. eds.) p. 389. San Diego, Academic.

63 Mehrabian, Z., Liu, L.I., Fiskum, G., Rapoport, S.I., Chandrasekaran, K. (2005)

Regulation of mitochondrial gene expression by energy demand in neural cells. *J. Neurochem.* 93, 850–860.

64 Bernstein, B.W., Bamburg, J.R. (2003). Actin-ATP hydrolysis is a major energy drain for neurons. *J. Neurosci.* 23, 1–6.

65 Hertz, L., Peng, L., Dienel, G.A. (2007). Energy metabolism in astrocytes: high rate of oxidative metabolism and spatiotemporal dependence on glycolysis/ glycogenolysis. *J. Cerebr. Blood Flow Metab.* 27, 219–249.

66 Tornberg, J., Volkar, V., Savilahti, H., Rauvala, H., Airaksinen, M.S. (2005) Behavioural phenotypes of hypomorphic KCC2-deficient mice. *Eur. J. Neurosci.* 21, 1327–1337.

67 Inoue, K., Yamada, J., Ueno, S., Fukuda, A. (2006) Brain-type creatine kinase activates neuron-specific K^+-Cl^- co-transporter KCC2. *J. Neurochem.* 96, 598–608.

68 Wolbye, D.P. (1998). Antiepileptic effects of NPY on pentylenetetrazole seizures. *Regu. Pept.* 75–76, 279–282.

11

System Analysis of Cardiac Energetics–Excitation–Contraction Coupling: Integration of Mitochondrial Respiration, Phosphotransfer Pathways, Metabolic Pacing, and Substrate Supply in the Heart

Valdur Saks, Petras P. Dzeja, Rita Guzun, Mayis K. Aliev, Marko Vendelin, André Terzic, and Theo Wallimann

Abstract

Because of high and changing energy demands and perfect structural, metabolic and functional organization of the cells (cardiomyocytes), heart muscle is a classical object of intense bioenergetic studies, and yet many unanswered questions remain. Recent progress in studies of cellular microcompartmentation, organized energetic units, and phosphotransfer networks, culminating in the concept of cardiac system bioenergetics, has significantly increased our understanding of how mitochondrial respiration and energy fluxes throughout cellular compartments are regulated, how energy is sensed, and how precise mechano-energetic coupling is maintained over a broad rage of functional activity. System analysis indicates that integration of cellular energetic systems with ion currents during action potential, Ca^{2+} signaling, and changes in sarcomere length during muscle extension constitute the mechanisms of the basic Frank–Starling law of the heart. The changes in energy demand are transmitted to mitochondria by the networks of metabolic signaling via phosphotransfer pathways to regulate ATP production and thus maintain the energy homeostasis of the cell. Because of the metabolic sensors at the sarcolemma, coupled phosphotransfer reactions provide a high-fidelity regulation of ion fluxes and excitation–contraction coupling. Here, we describe in quantitative terms the basic intracellular mechanisms of integration of energetics with calcium and magnesium signaling systems as a basis of metabolic pacing, synchronizing cellular electrical and mechanical activities with energy supply and substrate oxidation. Such analysis of cell energy metabolism as a whole functional unit at systemic level provides new insights about function of the most marvelous nature's created engine – the heart.

11.1
Introduction

Under normoxia, cardiac energy metabolism relies on aerobic oxidation of fatty acids and carbohydrate substrates in mitochondria, with the majority of ATP con-

Molecular System Bioenergetics: Energy for Life. Edited by Valdur Saks

sumed by contractile machinery and ion pumps [1–3]. ATP is used during contractions with the production of ADP and inorganic phosphate in the actomyosin ATPase reaction [4–8]. In brain cells, aerobically produced ATP is used mostly to support ion gradients by Na^+/K^+-ATPase and for processing of neurotransmitters [9]. In cardiac cells, the specific contractile function is controlled and regulated by processes of excitation–contraction coupling [1, 10–12]. An active role in the regulation of cytoplasmic calcium can also be attributed to the mitochondrial calcium cycle [13–18]. These events have been found to be localized in the form of calcium sparks as visualized by confocal microscopy [18–20]. Normal muscle function depends on the fine interplay, or interaction, of energy metabolism and calcium metabolism inside the cells. Analysis of the integrated energy metabolism described in many chapters of this book, by applying the methods of molecular system bioenergetics, shows that highly organized phosphotransfer pathways play a central role in intracellular communication and signaling. This chapter describes the mechanisms of metabolic feedback regulation of respiration and substrate supply, as well as the mechanisms of energy sensing by which the energy state of the cell controls ion fluxes, including those of calcium.

11.2
Cardiac Energetics: The Frank-Starling Law and Its Metabolic Aspects

The heart maintains normal blood circulation under a wide range of workloads, a function governed by the Frank–Starling law [1, 21–25], originally described by Otto Frank [22] and Ernest Starling [23, 24]. This law states that cardiac performance increases with an increase in end-diastolic ventricular volume. In this way, the heart responds to increases in venous filling pressure [22–26]. In their original experiments, Starling and coworkers used heart–lung preparations and measured the rate of oxygen consumption, which was taken as a measure of "the total energy set free in the heart during its activity," following the definition of biological oxidation by Lavoisier and Laplace (see the Introduction to this volume) [24, 26]. Some of the results of these historical experiments are reproduced in Fig. 11.1. Evans, Starling, and their coworkers found that cardiac work (see Fig. 11.1A) increased linearly with an increase in left ventricular end-diastolic volume followed by an increase in respiration rates (Fig. 11.1B). They concluded that "any increase in the work demanded of the heart is met by corresponding increase in the oxygen consumption and in the amount of chemical changes taking place" [24]. This is the metabolic underpinning of the Frank–Starling law, and elucidation of its molecular and cellular mechanism is still a central question in cardiac cellular bioenergetics [27, 28]. By increasing the rate of left ventricular filling, Williamson et al. showed that the rate of respiration in isolated working hearts can be changed by more than an order of magnitude, about 15–20 times, from unloaded VO_2 of around 8–12 μmol min^{-1} g^{-1} dry weight to a maximal value of 170 μmol min^{-1} g^{-1} dry weight (see Fig. 11.2A) [29]. In these experiments, Williamson et al. also measured the redox state of the $NADH/NAD^+$ system in heart cells [29]. As illustrated in Fig. 11.2C, an increase in the workload

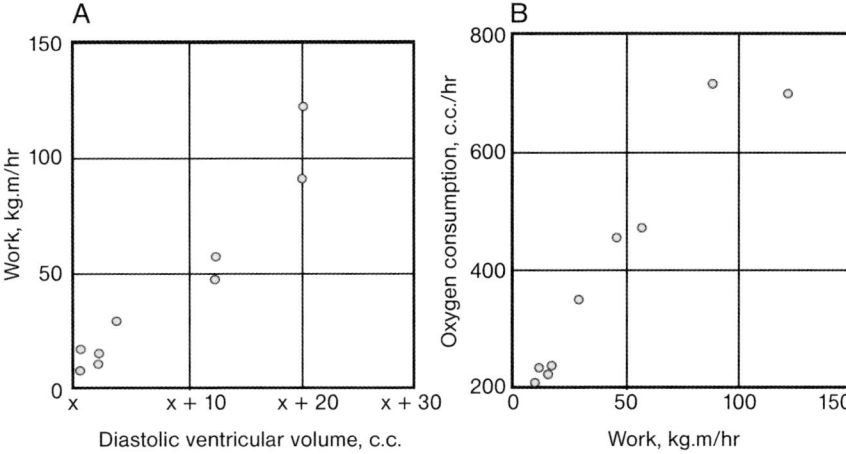

Fig. 11.1 Classical experiments showing the metabolic aspect of the Frank–Starling law of the heart, published by Starling and Visscher in 1926. Linear dependence of work (A) and oxygen consumption (B) upon left ventricular filling. (Reproduced from [24] with permission of the *Journal of Physiology*).

was accompanied by profound oxidation of NADH in these experiments. Another principal characteristic of heart energetics is that the linear increase in VO_2 with workload is observed in the absence of measurable changes in the ATP and phosphocreatine (PCr) cellular content (Fig. 11.2B) [30, 31]. This remarkable metabolic stability also has been called "metabolic homeostasis" [31]. In the face of large changes in muscle work and respiration, this observed metabolic stability or homeostasis underlying the Frank–Starling law is referred to as the "stability paradox" [32].

At present, there are sufficient experimental data, described in different chapters of this book, to make detailed conclusions about the nature of the mechanisms regulating cardiac cell respiration, energy fluxes, and substrate supply and thus to explain in molecular terms the metabolic basis of the Frank–Starling law of the heart [27, 28]. In mitochondria, the respiration rate is regulated by the availability of ADP for adenine nucleotide translocase (ANT) [33]. This is the classical respiratory control mechanism discovered by Lardy and Wellmann [34] and by Chance [35]. Jacobus et al. have found that even in the presence of ATP in high (physiological) concentrations, mitochondrial respiration follows the changes in ADP but not in the ratio of ADP to ATP [36]. In cells *in vivo* the main problem is the way in which signals are transmitted in the organized intracellular medium from MgATPases, which produce ADP and P_i, to mitochondria, which consume these substrates, to match ATP production to its demand. This signaling is controlled by multiple processes. It will be shown in this and related chapters that the integrated mechanisms of the system-level regulation of cardiac muscle energetics, respiration rate, and substrate supply under normal physiological conditions, governed by the Frank–Starling law, are described by the complex

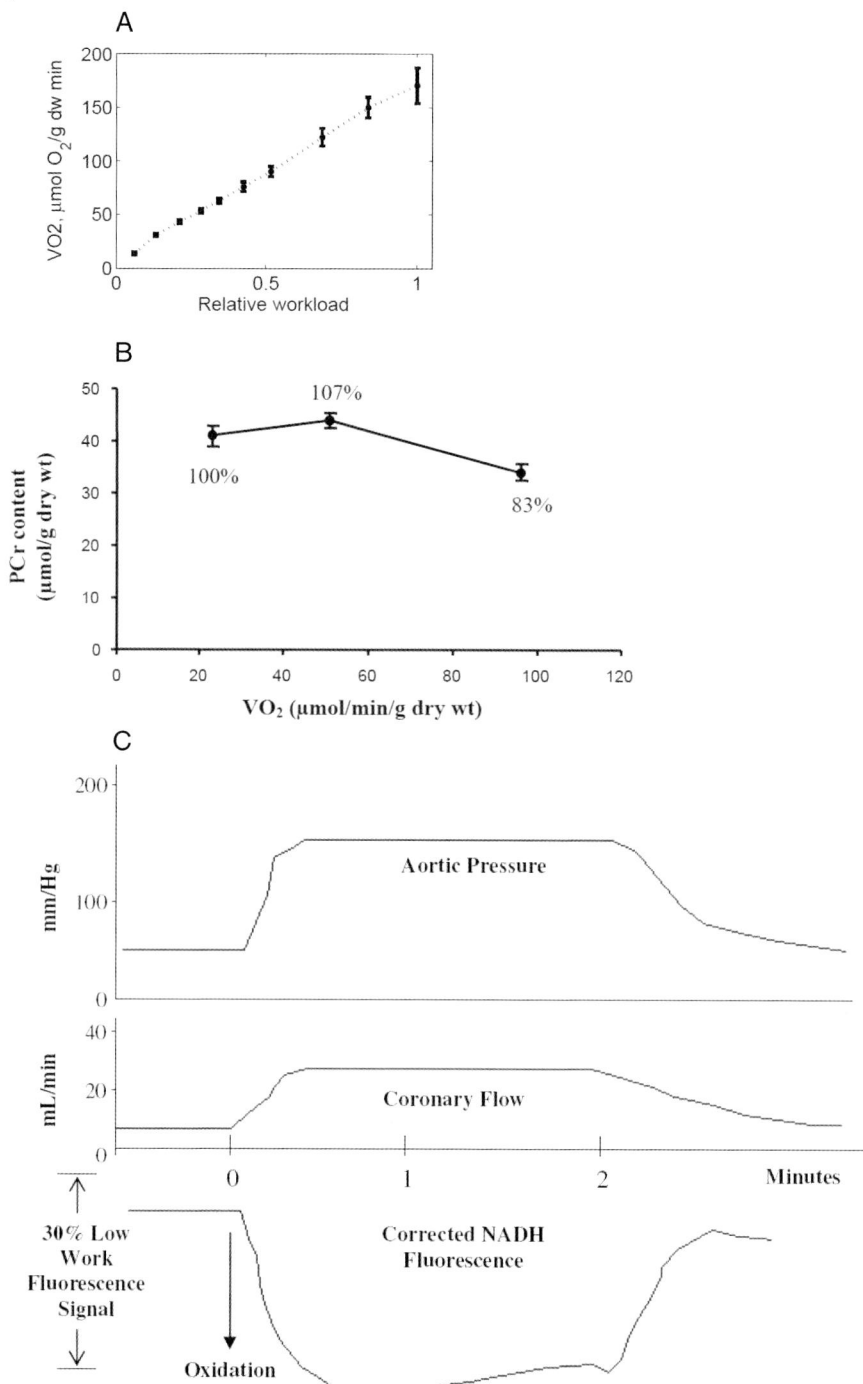

Fig. 11.2 (legend see p. 371)

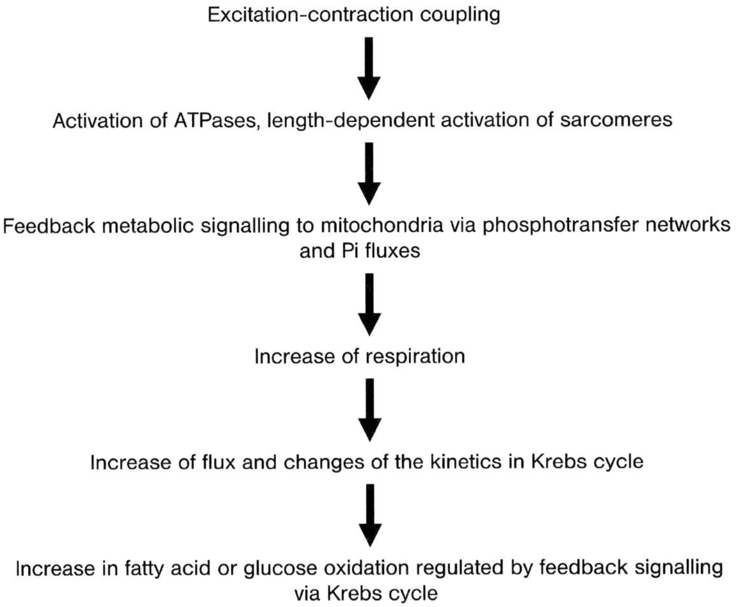

Excitation-contraction coupling

↓

Activation of ATPases, length-dependent activation of sarcomeres

↓

Feedback metabolic signalling to mitochondria via phosphotransfer networks
and Pi fluxes

↓

Increase of respiration

↓

Increase of flux and changes of the kinetics in Krebs cycle

↓

Increase in fatty acid or glucose oxidation regulated by feedback signalling
via Krebs cycle

Fig. 11.3 Schematic representation of sequences of processes of
workload-dependent regulation of cardiac energetics and mitochondrial
respiration *in vivo*.

sequence of processes shown in Fig. 11.3. These sequences can be modified by
changes in contraction frequency or hormonal stimulation, which are described
in Chapter 13. All steps described in Fig. 11.3 will be briefly described below.

11.3
Excitation–Contraction Coupling and Calcium Metabolism

11.3.1
Excitation–Contraction Coupling

The cardiac contraction is initiated and regulated by the process of excitation–
contraction coupling. In 1883 Sidney Ringer discovered that contraction of iso-

←———

Fig. 11.2 (A) Dependence of the rate of
oxygen consumption by isolated perfused
working rat heart on the work regulated
according to the Frank–Starling law. Data are
from the work of Williamson et al. [29]. The
maximal workload corresponded to 0.6 kg m

(g d.w. min^{-1}) at the ventricular filling rate
of 56 mL min^{-1} [29]. (B) The remarkable
metabolic stability of the heart. (Reproduced
from [28] with permission. (C) The workload
dependence of oxidation of NADH in the
heart. Redrawn from [29]).

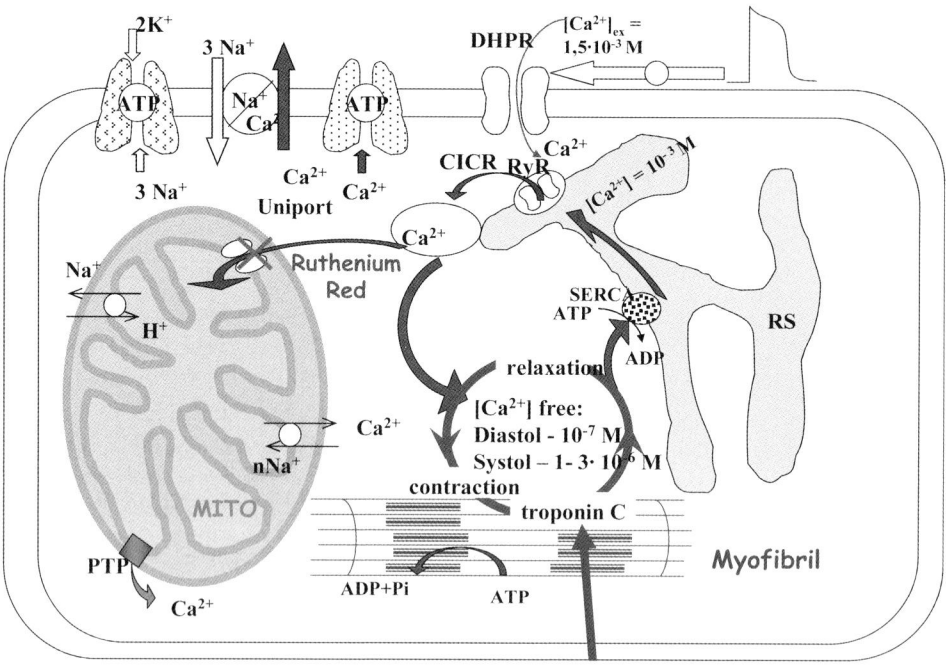

Fig. 11.4 Schematic representation of excitation–contraction coupling in cardiac cells. The arrow points to the importance of length-dependent activation of sarcomeres as a main regulatory mechanism under conditions of the Frank–Starling law. DHPR: dihydropyridine receptor; CICR: calcium-induced calcium release; SERCA: sarcoplasmic–endoplasmic reticulum calcium-dependent ATPase; RS: sarcoplasmic reticulum; Mito: mitochondrion; PTP: permeability transition pore.

lated perfused heart requires the presence of calcium in the perfusate [37, 38]. More than 100 years after these studies, an explanation for this phenomenon can now be given at almost atomic resolution (for reviews, see [10–12]). A simplified and general version of these events is given in Fig. 11.4. Calcium enters the cardiac cells via slow L-type calcium channels (also called dihydropyridine receptors [DHPRs]) in sarcolemma during the plateau phase of action potential. The amount of calcium entering the cell is dependent on the duration of this plateau phase, but it is not sufficient to directly activate contraction [12]. This small calcium current, however, activates a massive release of calcium from its large pools in the sarcoplasmic reticulum via ryanodine-sensitive calcium release channels [10–12]. In skeletal muscle, calcium release from the sarcoplasmic reticulum is initiated by a sarcolemmal voltage sensor protein that is a homologue of the DHPR; therefore, contraction of skeletal muscle does not depend upon the extracellular calcium concentration [10, 39]. Calcium liberated from local intracellular stores by calcium-induced calcium release (CICR) mechanisms during excitation–contraction coupling activates the contraction cycle by binding to troponin C in the troponin–tropomyosin complex of thin filaments [4–8]. Subsequent uptake of calcium by the Ca/MgATPase of the sarcoplasmatic reticulum terminates this calcium cycle.

The force developed by the myofibrils in the contractile cycle in heart cells is determined mostly by two parameters: duration and rate of calcium entry and release by a CICR mechanism (these depend on the duration of the action potential) and the length of sarcomeres, because of their length-dependent activation (arrow in Fig. 11.4). Under normal physiological conditions, when cardiac contraction is regulated by the Frank–Starling law, the latter mechanism is the central one (see below).

Direct measurements of the cyclic changes in calcium concentration (calcium transients) in the cytoplasm of cardiomyocytes became possible after application of fluorescent or bioluminescent calcium probes [40, 41]. However, the use of confocal microscopy and imaging techniques revealed the unique, discrete nature of calcium transients as consisting of localized changes within calcium release units (CRUs), structurally organized sites of Ca^{2+} microdomains (Ca^{2+} sparks) that form a discrete, stochastic system of intracellular calcium signaling in cardiac cells [19]. In these studies the theory of intracellular compartmentation of metabolites including calcium (see Chapter 3) found its brilliant confirmation.

11.3.2
The Mitochondrial Calcium Cycle

The mitochondrial calcium cycle, whose elucidation began with the pioneering works of Lehninger, Hunter, and Carafoli (for a review, see Refs. [13–18]), includes a calcium entry-and-export system that

1. regulates calcium concentration in the mitochondrial matrix, where calcium is an important activator of the Krebs cycle dehydrogenases (such as pyruvate dehydrogenase [PDH], isocitrate dehydrogenase, and α-ketoglutarate dehydrogenase [42]), thus increasing the capacity of oxidative phosphorylation [42–45];
2. allows active regulation of compartmentalized cytoplasmic calcium fluctuations and related signaling [14, 46]; and
3. protects mitochondria from calcium overload [15–17, 47].

Rizzuto and Pozzan have discovered a direct communication between the SR and mitochondria and calcium transfer from the SR into the mitochondrial intermembrane space [15, 16], evidencing the important role of mitochondria in the control of the cellular calcium cycle.

It is assumed by some authors that calcium signals can match the demand for ATP in the cell to its production in mitochondria and thus be entirely responsible for the control of respiration [18, 31, 48–52]. However, quantitative estimates of calcium effects in mitochondria are in conflict with the magnitude of changes in the respiratory rate *in vivo* [27, 28]. Both experimental studies of calcium effects on mitochondrial respiration *in vitro* [50] and mathematical modeling of mitochondrial metabolism [53] have shown that changes in calcium concentration can at most double mitochondrial respiration. The problem with this hypothesis is that there is already a high respiration rate at zero calcium [27, 28, 50, 53]. This degree of activation of mitochondrial respiration is far too small to explain the energy flux changes of more than an order of magnitude that are observed in muscle cells *in vivo* (see also below). In cells with small fluctuations in energy fluxes, direct regulation of mitochondrial activity by calcium may be sufficient [49], but for excitable cells with high and rapidly fluctuating energy fluxes – such as heart, skeletal muscle, brain, and other cells – this is not the case. The calcium hypothesis of respiration regulation always predicts an increase in NADH production in response to the elevation of workload as an immediate mechanism of activation of the respiratory chain [31, 54]. However, in actual experiments the opposite effects are seen: an increase in the workload of the heart, including that by the Frank–Starling mechanism, always results in a very clear and significant decrease in the ratio of NADH:NAD$^+$ in the mitochondrial matrix, as is shown in Fig. 11.2C [3, 29, 55–58]. This may be explained only by a metabolic feedback regulatory mechanism (see also Chapter 13). Interestingly, the nature of this mechanism of respiration regulation was elucidated mostly in cardiac physiological studies concerning the cellular basis of the Frank–Starling law and was found not to be directly related to the calcium cycle [27, 28]. The changes in the NADH/NAD$^+$ redox state are more complicated when contraction frequency is increased: in this case initial oxidation of NADH is followed by its reduction due to mitochondrial calcium accumulation [58].

11.4
Length-dependent Activation of Contractile System

It was discovered by Hibberd and Jewell and then confirmed by many others that the cellular mechanism behind the Frank–Starling law is the force–length relationship, a length-dependent activation of myofilaments that results from increased sensitivity of the thin filaments to calcium at greater sarcomere length [5, 21, 25, 59–65]. Intracellular calcium probes have revealed that practically no changes in the intracellular calcium transients accompany the stretching of sarcomeres sufficient to activate myofilaments [64, 65]. Only a slow increase in calcium transients after a length change was observed in the experiments of Allen and Kurihara [60]. However, it has clearly been shown in Suga's laboratory that the Frank–Starling mechanism does not affect intracellular calcium recirculation between the sarcoplasm and the sarcoplasmic reticulum [66]. The number of strongly bound cross-bridges and the force generated after calcium binding depend upon the length of sarcomeres [5–8, 21, 22, 63]. The mechanism of this length-dependent activation is complex. It includes changes in myofilament lattice spacing with possible involvement of titin [67], resulting in a decrease in the distance between actin and myosin filaments and an increased probability of cross-bridge formation with an increase in sarcomere length [67, 68], positive co-operativity of cross-bridge binding to actin [63], and an increase in the affinity of the troponin complex for calcium, induced by strong binding of cross-bridges [5, 61–63]. Because of these mechanisms, sarcomere stretch at submaximal calcium concentrations in the cytoplasm results in an increase in the number of active cross-bridges, and thus increased release of the products of ATP hydrolysis, first P_i during the power stroke and then ADP [4–8, 61–63]. By this mechanism of length-dependent activation of sarcomeres, the steady-state rates of ATP consumption can change by an order of magnitude without any changes in the calcium cycle. This is also confirmed by mathematical modeling [69] (see also Chapter 13). Under these conditions, the rate of ATP production is increased by a factor of 15–20 to match the rates of ATP consumption in cardiomyocytes, without the need of alteration of calcium signals. Theoretically, the feedback mechanism of this regulation should respond linearly to the ADP and P_i liberation within the contraction cycle. At the same time, MgADP, as a close structural analogue of MgATP, is an efficient, competitive inhibitor of ATPases, e.g., with a K_i close to 200 μM for filament sliding [70]. In addition, MgADP induces the actomyosin–ADP complex, which cooperatively promotes strong-binding cross-bridges such as the rigor complexes and increases myofilament Ca^{2+} sensitivity, thus masking the length-dependent activation [71]. The reversible ion pumps of the sarcoplasmic reticulum and sarcolemma are also very sensitive to inhibition by MgADP [72]. Therefore, accumulation of MgADP in the vicinity of any ATPase slows down the contraction cycle [70, 73] and impairs the Frank–Starling mechanism [63, 67, 71, 74].

Intracellular energy and metabolic signaling phosphotransfer networks within a defined cellular structural organization have the potential both (1) to protect cells from an excess of cytosolic free calcium and ADP and (2) to regulate respiratory ATP production in close correspondence to ATP consumption [27].

11.5
Integrated Phosphotransfer and Signaling Networks in Regulation of Cellular Energy Homeostasis

The CK-PCr circuit represents an efficient regulator of energy flux and uses metabolite channeling as a fine-tuning device for local ATP levels (see Chapters 3 and 7). The significance of such a regulated channeling circuit operating at high total PCr and Cr pools lies in its high sensitivity towards ADP that prevents, especially in excitable cells, the accumulation of ADP and, consequently, AMP through the adenylate kinase reaction [75, 76], unless severe stress, such as hypoxia or ischemia, is imposed. In the latter case, AMP-activated protein kinase (AMPK) and other AMP-sensitive components would be activated by free AMP, initiating signaling cascades that would turn on compensatory mechanisms for increasing energy supply and reducing energy consumption [77, 78].

11.5.1
Evidence for the Role of MtCK in Respiration Regulation in Permeabilized Cells
in Situ

Studies of the regulation of mitochondrial respiration in permeabilized muscle cells and fibers have provided important information on the structure–function relationship and the role of creatine kinases in this process, as well as on calcium effects on the respiration [79–87].

First, Fig. 11.5A shows further that calcium in its physiological concentrations induces a decrease in apparent K_m for exogenous ADP in permeabilized cardio-

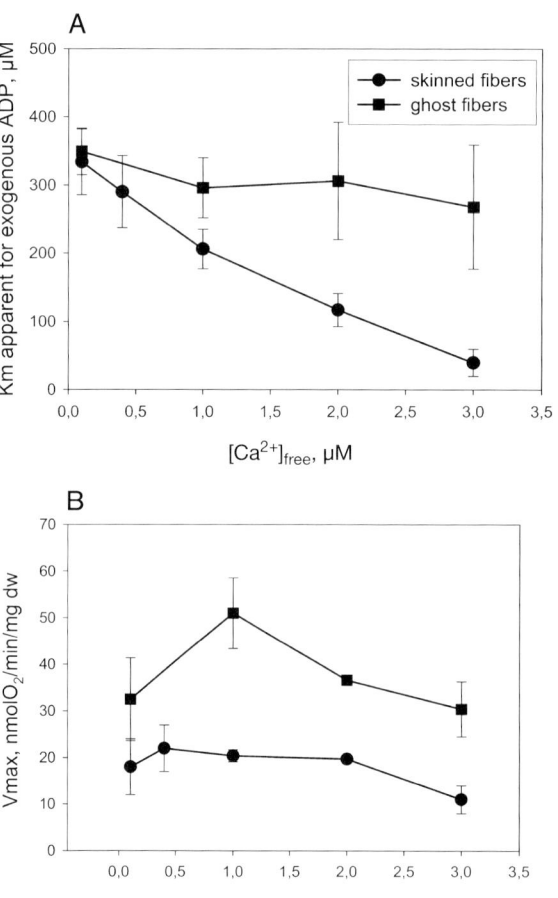

Fig. 11.5 Effects of the elevation of free calcium concentration in the medium on apparent K_m for exogenous ADP (A) and maximal rate of respiration (B) in skinned cardiac fibers and their ghost fibers (after extraction of myosin). (Redrawn from [86]).

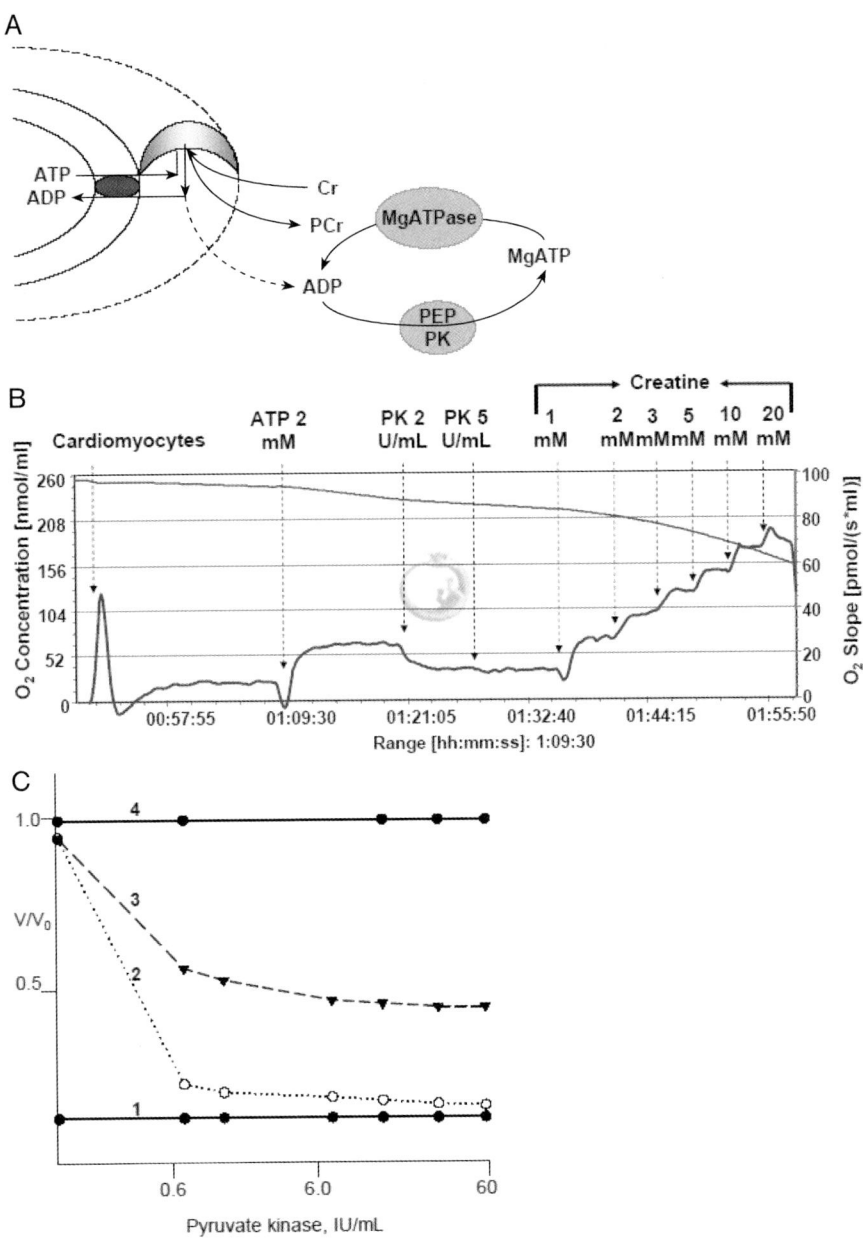

Fig. 11.6 (legend see p. 378)

myocytes due to supercontraction-related structural changes [86], because in ghost cardiomyocytes obtained after removal of myosin, no changes were observed in this parameter. However, in both cases, an increase in calcium concentration induced only a transitory increase in the maximal respiration (Fig. 11.5B). This quantitatively confirms the earlier results of Territo and Cortassa discussed above [50, 53] and conforms to the point of view that calcium is not a major direct regulator of respiration in cardiac cells but may regulate respiration via activation of Ca-dependent ATPases [88].

The permeabilized cell technique also allows us to demonstrate the role of MtCK in the regulation of mitochondrial respiration in the cells *in situ* and reveals some of the important aspects of functional coupling between MtCK and ANT discussed in Chapter 3 (Fig. 11.6). In these experiments, endogenous ADP production was induced by addition of MgATP (2 mM) to activate mitochondrial respiration in permeabilized cardiomyocytes. Then the powerful competing pyruvate kinase–phosphoenolpyruvate system was added to trap the ADP (Fig. 11.6A), this decreasing the respiration to 25% of the initial value (in these experiments, the calcium concentration was zero and endogenous ADP production was activated mostly by MgATPases) (Fig. 11.6B). The addition of creatine, however, rapidly activated respiration up to almost maximal levels, despite the presence of a powerful ADP-trapping system (Fig. 11.6B). This means that the ADP produced locally in the MtCK reaction is not at all available to the PEP–PK system. Earlier,

Fig. 11.6 Kinetics of respiration regulation in permeabilized cardiomyocytes recorded using a two-channel, high-resolution respirometer (Oroboros oxygraph-2k, Oroboros, Innsbruck, Austria) in respirometry medium Mitomed [87].
(A) Competition between the pyruvate kinase (PK)–phosphoenolpyruvate (PEP) system and mitochondria for endogenous ADP in permeabilized cardiomyocytes, by which the strong control of respiration by the mitochondrial creatine kinase reaction can be seen.
(B) Oxygraph recordings. Respiration was activated by endogenous ADP after addition of 2 mM MgATP. The protein concentration was 0.0621 mg mL^{-1}. In the presence of 3 mM PEP, addition of PK in increasing amounts effectively removes this ADP; at 20 IU mL^{-1} of PK, about 25% of the initial ADP-dependent respiration is observed. In the presence of this powerful ADP-consuming PEP–PK system, activation of the MtCK reaction by stepwise addition of creatine rapidly increases the respiration up to maximal values: the acceptor control ratio is about 7 and is close to that seen with saturating concentrations of ADP [87].
(C) Comparison of the effects of the PEP–PK system of ADP trapping on the respiration of isolated heart mitochondria (1–3) and permeabilized cardiomyocytes (4). (1) State 2 respiration of isolated mitochondria; (2) respiration of isolated mitochondria activated by MgATP and the hexokinase–glucose system of regeneration of ADP in medium; (3) respiration of isolated heart mitochondria controlled by the MtCK reaction in the presence of creatine (20 mM); (4) respiration of permeabilized cardiomyocytes in the presence of MgATP (2 mM) and creatine (20 mM). In cardiomyocytes, ADP generated in the MtCK reaction is not accessible for the PEP–PK system, while in isolated heart mitochondria, about 50% of ADP produced by MtCK is used up by the PK–PEP system. This shows the very limited permeability of the outer mitochondrial membrane for ADP in permeabilized cells *in situ*. Curves 1–3 are redrawn from [89]. Curve 4: unpublished results.

Gellerich et al. used the same system in experiments with isolated heart mitochondria to find that in this case the PEP–PK system inhibited more than 50% of creatine-stimulated respiration (curve 3 in Fig. 11.6C) [89]. The same degree of inhibition was seen in permeabilized cardiomyocytes after their treatment with trypsin (results not shown), in contrast to permeabilized intact cardiomyocytes in which the PK–PEP system was not at all able to reduce MtCK-stimulated respiration (curve 4 in Fig. 11.6C) [87]. These data favor the hypothesis that in cardiac cells *in vivo*, mitochondrial outer membrane permeability is limited and controlled by some cytoskeletal proteins [81–84] and clearly show the restrictions of diffusion of adenine nucleotides within the highly organized structure of ICEUs (see Chapter 3). In any case, the limited permeability of the outer mitochondrial membrane for ADP strengthens the functional coupling between MtCK and ANT and increases the role of MtCK in communication between mitochondria and the cytoplasm [81, 82–87].

Thus, the results of studies on respiration regulation in permeabilized cardiac cells strongly support the metabolic feedback regulation of respiration via phosphotransfer networks.

11.5.2
In Vivo Kinetic Evidence

Further insights into and key support for the current understanding of metabolic signaling networks in their full complexity have come with the application of new methodologies in investigations of the *in vivo* kinetics of energy transfer [75–78, 90–93]. High-energy phosphoryl fluxes through creatine kinase, adenylate kinase, and glycolytic phosphotransfer, captured with ^{18}O-assisted ^{31}P-NMR, tightly correlate with the performance of the myocardium under various conditions of stress load (Fig. 11.7) [77, 91], implicating phosphotransfer reactions as indispensable routes that direct the flow of high-energy phosphoryls between cellular ATPases and the ATP production machinery in mitochondria. This new methodology allows quantitative evaluation of cellular ATP turnover and the distribution of energy fluxes between different phosphotransfer pathways *in vivo*. In the experiments with Langendorff-perfused hearts shown in Fig. 11.7, the creatine kinase flux increased linearly with cardiac performance (evaluated as the rate–pressure product) and reached a value close to 300 nmoles min^{-1} mg^{-1} of protein with the heart performing work at about 30,000 mmHg min^{-1}, corresponding to a rate of oxygen consumption of about 40 µmoles of O_2 per minute per gram dry weight [94]. Taking into account the CK flux in the equilibrium state (in arrested heart) (Fig. 11.7), protein content of tissue of about 150 mg g^{-1} wet weight, and a wet weight:dry weight ratio of 5, these data show a PCr:O_2 ratio close to 5. This supports the notion that the CK pathway carries the major part of the energy flux out of mitochondria to the ATPases under normal physiological conditions in heart cells, as described in Chapters 3 and 7 and in agreement with the mathematical modeling of the compartmentalized energy transfer [27–95]. About 10–15% of cellular high-energy phosphoryls can be carried by the adenylate kinase

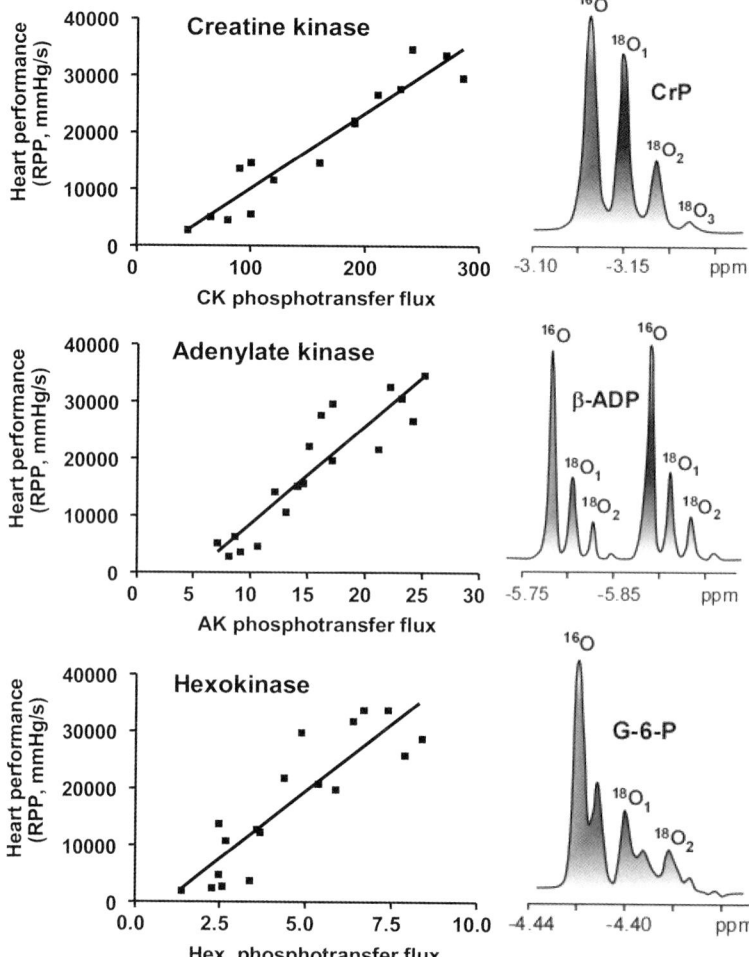

Fig. 11.7 Phosphotransfer networks mediate coupling between heart functional activity and ATP generation. Processing of high-energy phosphoryls through creatine kinase (CK), adenylate kinase (AK), and glycolytic (hexokinase) systems correlates linearly with heart functional activity (left panels). Phosphotransfer flux measurements obtained from ^{18}O-induced shift in ^{31}P-NMR spectra of phosphocreatine (CrP), β-ADP, and glucose-6-phosphate (G-6-P) (right panels) [77, 91]. Appearance of $^{18}O_1$, $^{18}O_2$, and $^{18}O_3$ peaks in corresponding phosphoryl-containing metabolites indicates phosphotransfer flux through the CK, AK, and hexokinase steps in the glycolytic system [77, 93]. Control hearts and contractile function modulated by ischemia–reperfusion are included in calculations [91]. RPP: rate–pressure product. Phosphotransfer flux is expressed as nmol min^{-1} mg^{-1} of protein of corresponding metabolite.

(AK)–glycolytic systems, whose contribution increases with muscle contraction and in failing hearts [27, 75–77, 90–93]. Such a distribution of contributions between CK, AK, and glycolytic systems is based on the assumption of parallel phosphotransfer pathways. However, in the cellular environment, these pathways are closely co-localized and interconnected, allowing high-energy phosphoryls to flow from one system to another [90–98]. The linear relationship between the creatine kinase flux and the workload (and thus the respiration rate) is also consistent with its measurements by the ^{31}P-NMR saturation transfer method [96, 99].

The high values of the CK fluxes and their linear relationship with workload confirm the central role of the CK system in the energo-mechanical coupling underlying the heart's Frank–Starling law.

11.5.3
Mathematical Modeling of Metabolic Feedback Regulation

Because of the high complexity of the processes involved, mathematical modeling is being increasingly and very effectively used to analyze the mechanisms of regulation of respiration and cellular energy fluxes [81, 82, 85, 94, 95, 100–108]. For this purpose, different models with distinct levels of detail are used, depending on the aim and point of view of the authors [82, 102, 107]. There is an increasing consensus among different groups that mathematical modeling supports the theory of metabolic feedback signaling from ATPases to mitochondria [95, 100, 102, 108], in agreement with experimental observations [109]. In the literature, different mathematical models of energy metabolism of muscle cells are available. The mathematical model of compartmentalized energy transfer was initially developed by Aliev and Saks [82, 95] and was modified further by Vendelin et al. [100]. These groups of models with similar structure may be called Vendelin–Aliev–Saks [102] or VAS models. Interesting models have been developed by Beard [102], by Cortassa et al. [101], and by Matsuoka and Noma (Kyoto model) [103], and then there are Korzeniewski's models of respiratory chain and "parallel activation" [104–106]. In the latter models, however, compartmentalized energy transfer is not yet analyzed in detail [106], as is done in VAS models [82].

Our VAS models (reviewed in [82]) quantitatively describe the results of *in vivo* studies of respiration regulation in ICEUs (see Chapter 3) under physiological conditions of the Frank–Starling law. The latest version of one of these models is available at http://cens.ioc.ee/~markov/etransfer/current_model.pdf). These models are based on the concepts of ICEUs described in Chapter 3 and include

- the kinetics of ATP hydrolysis by actomyosin ATPase during the contraction cycle,
- the diffusional exchange of metabolites between myofibril and mitochondrial compartments,
- VDAC-restricted diffusion of ATP and ADP across the mitochondrial outer membrane,
- the mitochondrial synthesis of ATP by ATP synthase,

- ΔpH- and ΔΨ-controlled P_i and ADP transport into the mitochondrial matrix, and
- PCr production in the coupled mitochondrial CK reaction and its utilization in the cytoplasmic CK reaction.

These events are considered in a system consisting of a myofibril with a radius of 1 μm, a mitochondrion, and a thin layer of cytoplasm interposed between them [81, 82, 94, 95]. The computations of diffusion and chemical events were first performed for every 0.1-μm segment of chosen diffusion path at each 0.01-ms time step [95]. This allowed the simulation of space-dependent changes throughout the entire cardiac cycle. The mitochondrial section of the model was first based on a simple kinetic scheme of mitochondrial ATP synthase with parameters allowing the description of experimental ADP and P_i dependences of oxidative phosphorylation in isolated mitochondria [95]; in the latest version, Korzeniewski's kinetic scheme [104] for the respiratory chain was adapted [100]. Mitochondrial oxidative phosphorylation is activated by ADP and P_i produced by ATP hydrolysis in the myofibrillar compartment. The kinetics of ATP hydrolysis by myosin in contracting muscle was predicted from dP/dt changes in isovolumic rat heart: a linear increase in the ATP hydrolysis rate up to 30 ms, followed by its linear decrease to zero at 60 ms into the contraction–relaxation cycle. The total duration of this cycle was taken to be 180 ms [95]. In mitochondria, the ATP/ADP translocase (ANT) and the P_i carrier (PiC) regulate the matrix concentrations of ATP, ADP, and P_i available for the ATP synthase. These carriers establish constant positive ADP and P_i gradients between the matrix and the mitochondrial intermembrane space. The model of cellular events considers CK compartmentation: the molecules of the cytoplasmic isoenzyme of CK (MM-CK), 69% of total activity, are taken to be freely distributed in the myofibrillar and cytoplasmic spaces. A remaining part of cellular CK, the mitochondrial isoenzyme of CK (MtCK), is localized in the mitochondrial compartment. The resulting close proximity of MtCK and ANT allows direct channeling of adenine nucleotides between their adjacent active centers; this channeling is the actual base for shifting the MtCK reaction toward the synthesis of PCr from translocase-supplied ATP even at high levels of ATP in the myoplasm of *in vivo* heart cells. In the modeling, we accounted for the limitation of the outer mitochondrial membrane for ADP in cardiac cells *in vivo*.

The results of the modeling show cyclic changes in the concentration of ADP in the core of myofibrils in ICEUs (Fig. 11.8A) in a microcompartment containing myofibrillar-bound MM-CK, where ADP is first produced by actomyosin MgATPase during the contraction cycle of cross-bridges and then rephosphorylated by CK due to the non-equilibrium state of the CK reaction [95, 100]. Interestingly, these calculated cyclic changes in PCr, ATP, and Cr, which are in the range of 5–10% of their cellular contents, are in good agreement with the multiple observations of the cyclic changes of these compounds in the contraction cycle published in the literature [110, 111]. These changes in Cr, PCr, and total ATP are, however, close to the experimental errors of their detection, thus giving an overall impression of metabolic stability [27, 28, 100]. Changes in ADP and P_i concentrations are relatively much more significant because of very low initial values (Fig. 11.8A). Without CK, the changes in local ADP concentrations in these microcompartments are much more dramatic [82, 94, 95].

Within the whole contraction cycle, these coupled reactions are in the steady state, in which the rates of ADP and ATP cycling, and thus the respiration in mitochondria coupled to PCr production, are increased with elevation of the workload. Increasing cyclic changes in the local ADP production in myofibrils immediately displace the myofibrillar MM-CK reaction in the direction of local ATP regeneration (Fig. 11.8B). The amplitude of displacement of CK from equilibrium is proportionally increased with workload (Fig. 11.8B,C) [100]. In this regard, CK, adenylate kinase, and other phosphotransfer isoenzymes in different intracellular compartments are "pushed" or "pulled" from the equilibrium in opposite directions, depending on the activity of an associated process that drives steady-state, high-energy phosphoryl flux [27, 83, 92, 112].

If myoplasmic CK is structurally organized and bound to the cytoskeleton, these cyclic changes may be channeled to mitochondria by a mechanism of so-called vectorial ligand conduction – spatially directed group translocation down-gradients of group potential – in accordance with the theories of metabolic networks [76, 82, 83, 90, 91, 113]. The stimulatory effect of these CK ligands (Cr and/or ADP) on mitochondrial respiration is amplified by the functional coupling between MtCK and ANT [28, 81, 82]. Therefore, MtCK always catalyzes the unidirectional PCr and ADP production from mitochondrial ATP and cytoplasmic creatine (Fig. 11.8C) [95, 100]. Local ADP produced in this reaction controls the rate of respiration. Figure 11.8D shows that the model quantitatively describes the observations of Williamson et al. [29] on the dependence of the respiration rate upon the workload.

Under conditions of metabolic stability, an increase in the rate of ATP synthesis and of the coupled respiration (oxidative phosphorylation) is possible only if the supply of ADP and P_i to mitochondria is increased, making both metabolites good candidates for metabolic feedback signals controlling respiration [27, 28, 114]. ADP is supplied by the metabolic signaling networks described above, to overcome the possible restriction of its diffusion and to avoid the inhibition of MgATPases. Additionally, the CK reaction, if running in the direction of ATP regeneration, compensates for pH changes due to MgATPase hydrolysis [98]. Application of metabolic control analysis to the mathematical model developed showed that in parallel to the cyclic changes in ADP, at least at low workloads, P_i flux to mitochondria plays an important regulatory role (Fig. 11.8E) [108]. This conclusion has been confirmed experimentally [109] and, recently, by models [102]. Thus, the feedback metabolic signal has a complex nature; several of its components, such as P_i, ADP, and Cr, may act in parallel, and their relative contribution changes with workload [27, 28, 76, 82, 92, 108]. Instead of the parallel activation of all ATP-producing and -utilizing systems proposed by Korzeniewski [106], there seems to be a self-regulatory metabolic feedback signaling via phosphotransfer pathways and P_i fluxes. In any case, as a response to sarcomere stretch

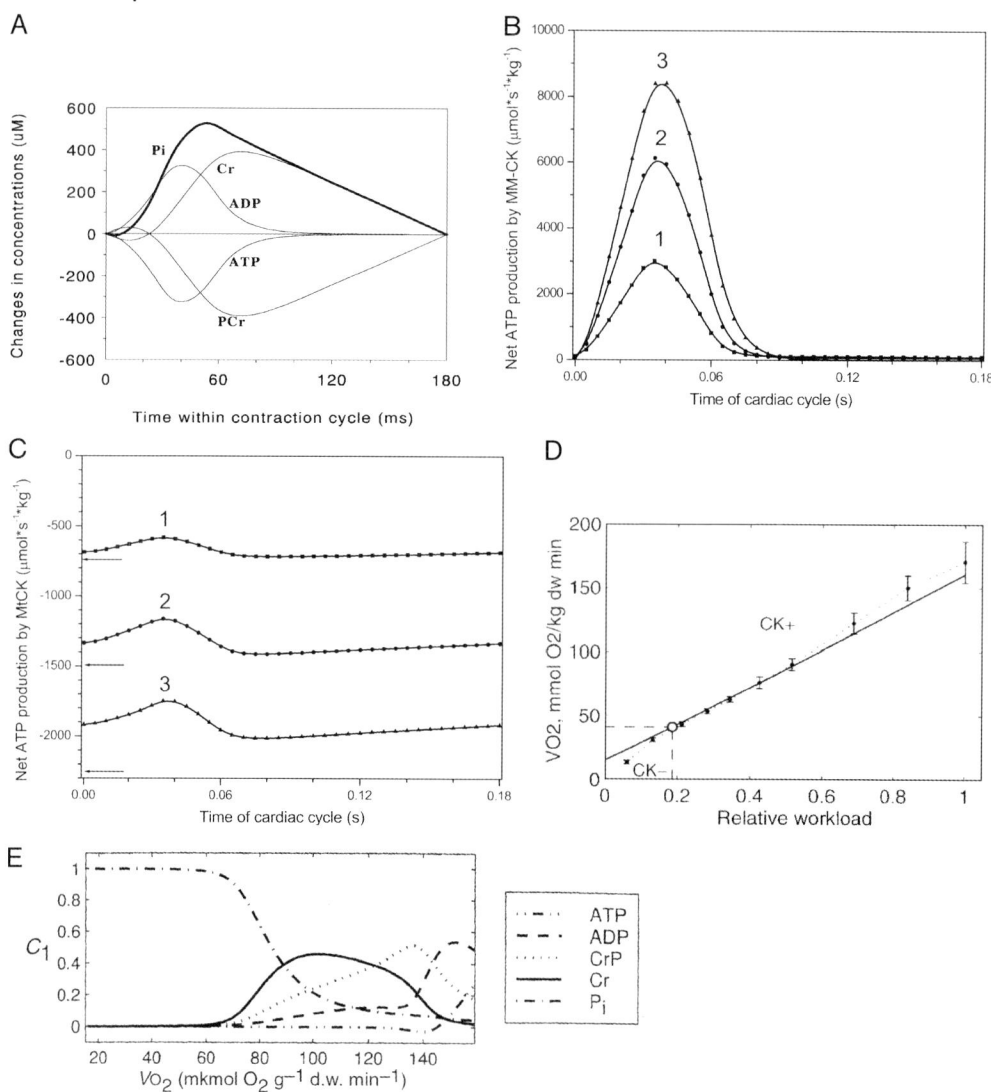

Fig. 11.8 (legend see p. 385)

and cross-bridge cycling, this complex signal leads to increases in respiration rate, coupled mitochondrial PCr production, and energy flux via the CK–PCr system, while also maintaining metabolic stability (homeostasis) at elevated workload.

Thus, there is no need to search for some mysterious agent, such as the "parallel activator" of ATPases and the respiratory chain proposed by Korzeniewski [106]: the metabolic feedback mechanisms for regulating respiration via phosphotransfer pathways and P_i fluxes are there, and they carry out the task of matching ATP production to its demand very effectively.

11.6
"Metabolic Pacing": Synchronization of Electrical and Mechanical Activities With Energy Supply

The existence of two interrelated systems regulating mitochondrial respiration and energy fluxes in the cells is increasingly being recognized (Fig. 11.9). The first system is composed of structurally organized enzymatic modules and networks of the CK–AK–glycolytic systems communicating flux changes from cellular ATPases to mitochondrial oxidative phosphorylation in non-equilibrium steady state [27, 28, 75, 92, 116], and the secondary amplifying system is based on cellular and mitochondrial calcium cycles, which adjust the capacity of substrate oxidation and energy-transducing processes to meet increasing cellular en-

---

Fig. 11.8 Mathematical modeling of the regulation of mitochondrial respiration in cardiac cells *in vivo* under physiological conditions controlled by the Frank–Starling mechanism.
(A) Phasic changes in metabolite concentrations in the core of myofibrils of ICEUs over the cardiac contraction cycle. Reproduced from [82] with permission.
(B) Workload dependence of the dynamics of net ATP production by MM-CK in myofibrils. The reaction rates for workloads of 750 (curve 1), 1500 (curve 2), and 2250 (curve 3) μmol ATP s^{-1} kg^{-1} are shown. Net rate of the reaction means displacement from the equilibrium position of the creatine kinase reaction (observed in the diastolic phase) as a result of local ADP production in the contraction cycle.
(C) Workload dependence of the dynamics of net PCr production by MtCK. Workload values are also indicated by the arrows on the respective curves. Results were obtained by complete model of Dos Santos et al. [94] used in simulations for [82]. Mitochondrial $\Delta\Psi = -160$ mV; V_{max} of ATP synthase $= 4269.3$ μmol ATP s^{-1} kg^{-1}. PCr export by mitochondria is 90.9%, 89.1%, and 85.8% of total mitochondrial energy export for workloads of 750, 1500, and 2250 μmol ATP s^{-1} kg^{-1}, respectively. Due to the functional coupling between ANT and MtCK, the reaction always runs out of equilibrium in the direction of PCr synthesis [100], the steady-state values of the rates of this coupled reaction are increased by increasing the workload (cyclic changes in MgATP and MM-CK reactions in myofibrils), and there is metabolic signaling of these changes to mitochondria by the creatine kinase–phosphocreatine energy transfer system. Reproduced from [27] with permission.
(D) Computed (solid line) and experimental (points with standard deviations, from [29]) oxygen consumption rates of working cardiac muscle. Relative workload is the fraction of maximal workload applied; in computations this is the rate of ATP hydrolysis by actomyosin MgATPase. CK+ shows the calculations for the active creatine kinase system; CK− shows the calculations for the cells with inactivated creatine kinases (results shown within the square). The model includes the diffusion restriction of ADP through the mitochondrial outer membrane. Reproduced from [27, 100] with permission.
(E) Metabolic control analysis of the mathematical model of compartmentalized energy transfer. Flux control coefficients for different metabolites are shown. Reproduced from [108] with permission.

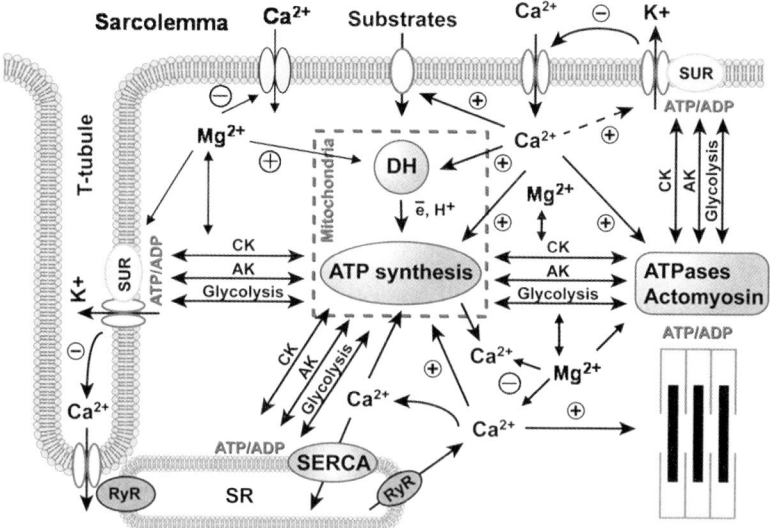

Fig. 11.9 Cardiac excitation–contraction energy coupling: synchronization of electrical and metabolic pacing. Electrical pacing–induced action potential and membrane depolarization cause Ca^{2+} influx through L-type Ca^{2+} channels in the sarcolemma and T-tubules, triggering Ca^{2+} release from intracellular stores in the sarcoplasmic reticulum (SR) through ryanodine receptor channels (RyR) and cardiomyocyte contraction. Interplay between CK, AK, and glycolytic phosphotransfer relays; energetic modules (mitochondria and ATPases); metabolic sensors (K_{ATP} channel); and Ca^{2+} and Mg^{2+} transients generates metabolic pacing signals in synchrony with the electrical and functional activity to ensure cellular energetic homeostasis. Intracellular and mitochondrial Mg^{2+} activates metabolic enzymes and tunes Ca^{2+} signaling. After contraction, intracellular Ca^{2+} is sequestrated by the SR Ca^{2+}-ATPase (SERCA) and by mitochondria and is removed from the cell by the Na^+/Ca^{2+} exchanger and a Ca^{2+} pump in the sarcolemma (see details in the text). (Modified from [27] with permission).

ergy demands [31, 42, 53]. Moreover, the coupled CK, AK, and glycolytic reactions by communicating signals to the K_{ATP} channel, a sarcolemmal metabolic sensor, provide fine-tuned regulation of excitation–contraction and thus of the calcium cycle within a cell [75, 115, 117–119] (see Chapter 3). Such integration of energetic and ion signaling systems provides the basis for "metabolic pacing," the synchronizing of electrical and mechanical activities with energy supply processes, which is fundamental for optimal heart function and for sustaining an adequate force–frequency relationship during stress [27, 91–93]. The significance of such a relationship is indicated by a number of studies demonstrating that inhibition of CK or abnormal regulation of K_{ATP} channels disturbs the heart force–frequency relationship and inotropic responses to stress challenges [117–125]. In this regard, one of the earliest signs of heart failure is the disturbed force–frequency relationship [121, 126] indicating developing defects in the cardiac energetic signaling system and mechano-energetic coupling [27, 73, 90, 116, 127].

In recent years, accumulating evidence indicates that phosphotransfer-governed nucleotide exchange and metabolic sensing are essential for processing cellular information and sustaining energy, ionic, and hormonal homeostasis [76, 128–130]. By synchronizing electrical, ionic, and metabolic oscillations, phosphotransfer reactions ensure the pacing of cellular energy metabolism to adjust ATP production strictly according to demands, thus facilitating muscle energy economy [27, 92, 112]. As presented in Fig. 11.9, during the cardiac cycle, an electrical impulse–induced action potential and membrane depolarization cause Ca^{2+} influx through L-type Ca^{2+} channels in the sarcolemma and T-tubules, triggering Ca^{2+} release from intracellular stores in the sarcoplasmic reticulum (SR) through ryanodine receptor channels (RyR) and cardiomyocyte contraction [10, 115]. ATP consumed during myofibrillar contraction and Ca^{2+} pumping is rapidly replenished by creatine kinase (CK), adenylate kinase (AK), and glycolytic phosphotransfer relays, which maintain optimal local ATP:ADP ratios and free energy of ATP hydrolysis [28, 92, 98, 116]. Simultaneously, these systems generate and translate feedback signals to stimulate ATP production in mitochondria and to convey information to energy demand sensors, such as K_{ATP} channels and AMPK [27, 28, 75, 76, 93, 117]. The amount of Ca^{2+} entering a cell is controlled by the action potential duration regulated by K_{ATP} channels located in the sarcolemma and T-tubules [117–120]. Conversely, the activity of this metabolic sensor is regulated by phosphotransfer reactions and Ca^{2+} feedback [119, 130]. Ca^{2+} entering the cell or released from intracellular stores activates substrate transport (recruitment of glucose transporters), dehydrogenase (DH) activity (such as pyruvate dehydrogenase [PDH], an entry point into the mitochondrial Krebs cycle), and glycogenolysis to meet incoming energy needs [31, 44, 45, 53]. Another player in metabolic pacing is intracellular Mg^{2+}, the most abundant divalent cation in the cell, which activates a large number of metabolic enzymes and regulates Ca^{2+} entry, release from intracellular stores, and interaction with binding proteins (Fig. 11.9) [131–136]. Intracellular free Mg^{2+} levels are regulated by adenylate kinase, and other phosphotransfer enzymes govern equilibrium of adenine nucleotide species, the Na^+/Mg^{2+} exchanger, TRPM7, and the Mrs2 family of cation channels [131–135]. Ca^{2+} entering the cell and ATP consumption –induced metabolic oscillations can trigger parallel Mg^{2+} oscillations, which could affect dehydrogenases and a number of other critical Mg^{2+}-sensitive cellular components [132]. Previous and more recent studies indicate that Mg^{2+} indeed could serve as a feedback signal from the adenine nucleotide pool to ATP-generating processes, simultaneously tuning Ca^{2+} signals and reducing Ca^{2+} accumulation in mitochondria [134–136]. Both Ca^{2+} and Mg^{2+} signals can activate the PDH complex and promote substrate oxidation in heart mitochondria [125]. Such simultaneous recruitment of different steps in the cellular energetic system and ligand conduction–type handling of substrates in sequential phosphotransfer systems can increase metabolic flux many times without apparent changes in substrate concentrations [75, 76, 82, 112, 113].

The first system – metabolic feedback signaling – quantitatively explains the metabolic aspect of the classical Frank–Starling law regarding regulation of cardiac function and respiration under conditions of metabolic stability and unchanged calcium transients. This system is based on compartmentalized energy transfer, a deficit in which explains the rapid fall of contractile force in the first minutes of total ischemia (see Chapter 3). The second system explains adrenergic modulation of cardiac cell function and energetics under stress [14, 53, 137]. By regulating the sarcolemmal metabolic sensor K_{ATP} channels, the CK–AK–glycolytic network affects the excitation–contraction process and thus the calcium cycle of the cell (Fig. 11.9). Both systems may be activated simultaneously, as is observed in the case of positive inotropy induced by β-adrenergic agents, when Frank–Starling curves are shifted upward [1, 24]. These two more or less independent systems regulating cellular energetics have been described separately by mathematical models [100, 137]. Integration of all this information into quantitative models of whole-cell functioning and metabolism is an interesting and challenging perspective of molecular system bioenergetics.

Thus, both phosphotransfer-mediated metabolic feedback and Ca^{2+}/Mg^{2+} signaling are important components in regulated signal transmission and coordinated dehydrogenase activation, producing an optimal mitochondrial response to increased cellular energy demands.

11.7
Metabolic Channeling Is Needed for Protection of the Cell from Functional Failure, Deleterious Effects of Calcium Overload, and Overproduction of Free Radicals

The physiological mechanism of respiration regulation described above has the important advantage of ensuring effective control of free energy conversion across the whole physiological range of workloads, without requiring a severe increase in cytoplasmic calcium and ADP concentrations. It thus avoids any danger of mitochondrial calcium overload that would open the mitochondrial permeability transition pore and thus lead to cell death [47, 52]. The functioning of the coupled MtCK–ANT system in mitochondria prevents reactive oxygen species (ROS, oxygen free radicals) formation in the mitochondrial respiratory chain (see Fig. 3.9 in Chapter 3) and helps to avoid many problems related to ROS production, such as PTP opening, necrosis, apoptosis, and rapid aging [138, 139]. In this way, the CK–PCr network may significantly contribute to the positive effects of physical exercise on human health: exercise-induced increases in fluxes via this pathway increase ADP–ATP turnover in coupled MtCK–ANT reactions in mitochondria and keep ROS production low.

Because of the central importance of this efficient and protective metabolic feedback signaling system, its alteration in pathological states of the cell contributes significantly to the development of cardiac failure, resulting from a decrease in total creatine content [140, 141], from alterations in CK expression [122, 123, 142], or from changes in the coupling of CK in mitochondria and with MgATPases [143] (for a review, see [144–146]. Ingwall et al. have shown that the product of CK activity and total creatine content, CK × [Cr], decreases by a factor of 3–6 in cardiomyopathy and in failing human hearts [140]. Figure 11.10 shows some

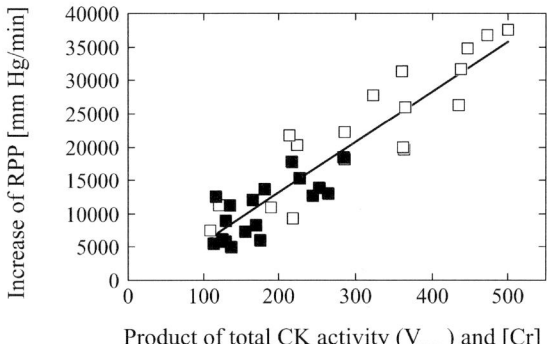

Fig. 11.10 Relationship between total creatine content, creatine kinase activity, and cardiac function in perfused rat hearts. RPP: rate–pressure product. (Reproduced from [146] with permission). Open squares: normal hearts; closed squares: cardiomyopathic hearts.

interesting data summarized by Wyss and Kaddurah-Daouk [146] on the relationship between creatine content, CK activity, and contractile performance. These results are consistent with earlier data showing that replacement of creatine by its much less productive analogue guanidino propionic acid (GPA), given to rats in diet, decreases by more than half the maximal work capacity and, correspondingly, the rate of respiration [147]. Also, significant washout of creatine during long periods of perfusion of frog hearts has been shown to lead to a hypodynamic state with a two- to threefold decrease in contractile force, which can be completely reversed by uptake of creatine [148, 149]. It was shown in Neubauer's laboratory that hearts with undetectable levels of creatine and phosphocreatine due to knock-out of a key enzyme of creatine biosynthesis [123] lost the ability to respond to inotropic stimulation even at low workload, and showed markedly impaired recovery of heart function during ischemia–reperfusion (see also [122, 141, 142]). Momken et al. found recently that double M-CK and MtCK knockout mice showed only 10% of work capacity in voluntary exercise in running wheels, as compared with that of normal wild-type mice (see also Chapter 3) [150].

Thus, effective cardiac work and fine metabolic regulation of respiration and energy fluxes require the organized and interconnected energy transfer and metabolic signaling systems. Direct transfer of ATP and ADP between mitochondria and different cellular compartments is not able to fulfill this important task efficiently.

11.8
Molecular System Analysis of Integrated Mechanisms of Regulation of Fatty Acid and Glucose Oxidation

Molecular systems analysis as a method is also useful for elucidation of the mechanisms of regulation of substrate supply for the heart [151]. In muscle cells, contractile function and cellular energetics are fuelled by oxidation of carbohy-

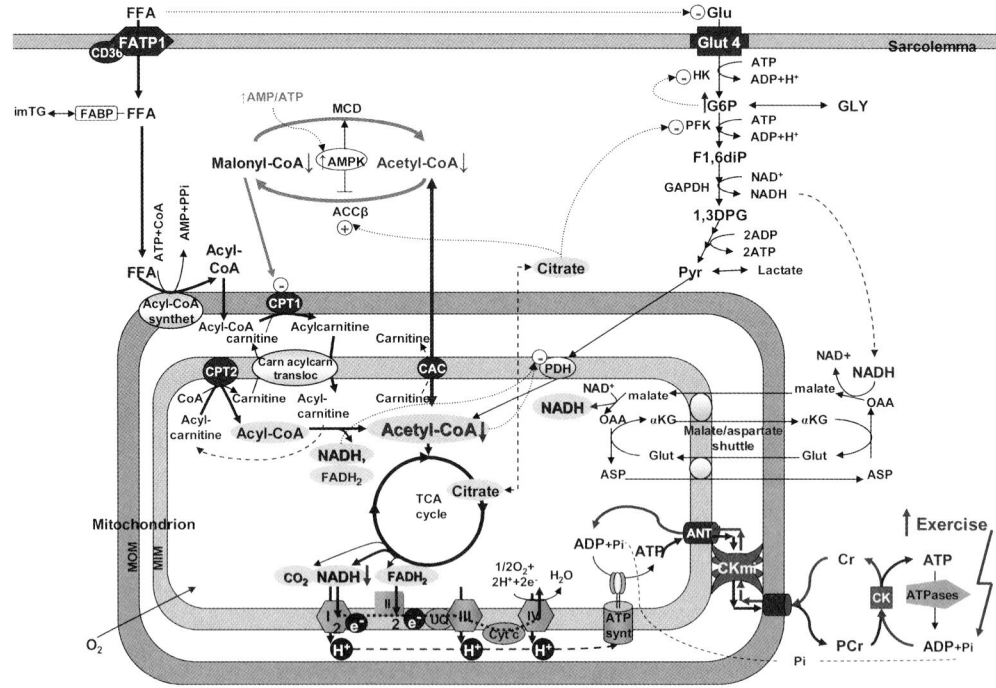

Fig. 11.11 The scheme of substrate supply for mitochondrial respiration and the mechanisms of feedback regulation of fatty acid and glucose oxidation during workload elevation in oxidative muscle cells: central role of TCA cycle intermediates. FFAs are taken up by a family of plasma membrane proteins (fatty acid transporter protein [FATP1], fatty acid translocase [CD36], and fatty acid binding protein [FABP]) and are esterified to acyl-CoA via fatty acyl-CoA synthetase. The resulting acyl-CoA is then transported through the inner membrane of the mitochondrion, via the exchange of CoA for carnitine by carnitine palmitoyltransferase I (CPT I). Acylcarnitine is then transported by carnitine–acylcarnitine translocase into the mitochondrial matrix, where a reversal exchange takes place through the action of carnitine palmitoyltransferase II (CPT II). Once inside, the mitochondrion acyl-CoA is a substrate for the beta-oxidative pathway, resulting in acetyl-CoA production. Each round of beta-oxidation produces one mole of NADH, one mole of $FADH_2$, and one mole of acetyl-CoA. Acetyl-CoA enters the TCA cycle, where it is further oxidized to CO_2 with the concomitant generation of three moles of NADH, one mole of $FADH_2$, and one mole of ATP. Acetyl-CoA, which is formed in the mitochondrial matrix, can be transferred into the cytoplasm with the participation of carnitine, carnitine acetyltransferases, and carnitine acetyltranslocase (carnitine acetylcarnitine carrier complex, CAC). Glucose (GLU) is taken up by glucose transporter-4 (GLUT4) and enters the Embden–Meyerhof pathway, which converts glucose via a series of reactions into two molecules of pyruvate (PYR). As a result of these reactions, a small amount of ATP and NADH is produced. G6P: glucose 6-phosphate; HK: hexokinase; PFK: phosphofructokinase; GLY: glycogen; F1,6diP: fructose-1,6-bisphosphate; GAPDH: glyceraldehyde phosphate dehydrogenase; 1,3DPG: 1,3 diphosphoglycerate. The redox potential of NADH is transferred into the mitochondrial matrix via the malate–aspartate shuttle. OAA: oxaloacetate; Glut: glutamate; αKG: alpha-ketoglutarate; ASP: aspartate. Malate generated in the cytosol enters the matrix in exchange for αKG and can be used to produce matrix NADH. Matrix

drate substrates and fatty acids [1–3, 27, 28, 55, 151–154]. The choice of sub-strates depends upon their availability [1–3, 55, 127, 152–155], and the rates of their utilization are very precisely regulated by multiple interactions between the intracellular compartmentalized and integrated bioenergetic systems of glycoly-sis, fatty acid oxidation, and the Krebs cycle in the mitochondrial matrix, linked directly to the activity of the respiratory chain and the phosphorylation process catalyzed by the ATP synthase complex [1–3, 127, 151–156]. The rates of all these processes are geared to the workload, mostly by the mechanism of feedback met-abolic regulation [3, 55, 151, 156].

In 1931 Clark et al. showed in isolated frog heart that the oxidation of carbohy-drates explains not more than 40% of oxygen uptake [157]. In 1954, Bing and coworkers demonstrated, by using coronary sinus catheterization, the absolute re-

Fig. 11.11 (cont.)
OAA is returned to the cytosol by conversion to ASP and exchange with Glut. Most of the metabolic energy derived from glucose can come from the entry of pyruvate into the citric acid cycle and oxidative phos-phorylation via acetyl-CoA production. NADH and FADH$_2$ are oxidized in the respiratory chain (complexes I–IV). These pathways occur under aerobic conditions. Under anaerobic conditions, pyruvate can be converted to lactate. *Feedback regulation.* Glucose–fatty acid cycle (Randle hypothesis): if glucose and FFAs are both present, FFAs inhibit the transport of glucose across the plasma membrane, acyl-CoA oxidation increases the mitochondrial ratios of acetyl-CoA:CoA and of NADH:NAD$^+$, which inhibit the pyruvate dehydrogenase (PDH) complex, and increased citrate (produced in the TCA cycle) can inhibit phosphofructokinase (PFK). These changes would slow down oxidation of glucose and pyruvate and increase glucose-6-phosphate, which would inhibit hexokinase (HK) and decrease glucose transport. The mitochondrial creatine kinase (MtCK) catalyzes the direct transphosphorylation of intramitochondrially produced ATP and cytosolic creatine (Cr) into ADP and phospho-creatine (PCr). ADP enters the matrix space to stimulate oxidative phosphoryla-tion, while PCr is transferred via the cytosolic Cr–PCr shuttle to functional coupling of CK to ATPases (actomyosin ATPase and ion

pumps), resulting in the release of high free energy of ATP hydrolysis. If the workload increases, ATP production and respiration are increased due to feedback signaling via the creatine kinase (CK) system, leading to a decrease in mitochondrial acetyl-CoA content, which is transferred into the cyto-plasm with the participation of the carnitine acetyl carrier (CAC). Acetyl-CoA carboxylase (ACC) is responsible for converting acetyl-CoA to malonyl-CoA, a potent inhibitor of CPT I, with the aim of avoiding overloading the mitochondria with fatty acid oxidation intermediates when the workload is decreased. Inactivation of ACC occurs via phosphorylation catalyzed by AMP-activated protein kinase (AMPK). Phosphorylation and inactivation of ACC lead to a decrease in the concentration of malonyl-CoA. A fall in malonyl-CoA levels disinhibits CPT I, resulting in increased fatty acid oxidation. Malonyl-CoA is also converted back into acetyl-CoA in the malonyl-CoA decarboxylase (MCD) reaction. An increase in the workload increases the rate of acetyl-CoA consumption, and this automatically decreases the malonyl-CoA content. Regulation of ACC and MCD occurs under stress conditions when the AMP:ATP ratios are increased but is unlikely to occur under normal workload conditions of the heart. Thus, AMPK may be envisaged as a modulator, under situations of cellular stress, rather than as a master on/off switch of fatty acid oxidation. (Reproduced with permission from [151]).

quirement of the human heart for free fatty acids (FFA) as fuel [152, 158]. The mechanisms of regulation of substrate uptake have been intensively studied by Opie et al., Neely, Morgan, Williamson, Randle, and many others [1–3, 27, 28, 127, 152–156, 159–165]. In general, the results of all these studies show that in heart, in the presence of both carbohydrate substrates and FFA, about 60–90% of the oxygen consumed is used for oxidation of FFAs and that the rates of both oxygen consumption and fatty acid oxidation increase linearly with the elevation of the workload [1–3, 139, 158, 159]. In cardiac muscle, all these important changes occur at unchanged global levels of ATP, phosphocreatine (PCr), ADP, and AMP [30–32, 159, 160].

The network of reactions of substrate supply for mitochondrial respiration in muscle cells and their multiple interactions and feedback mechanisms of regulation are illustrated in Fig. 11.11. This network emerged from 70 years of research on muscle energy metabolism. The choice of substrates for oxidation depends on their availability, and if glucose and FFAs are both present, FFAs strongly inhibit the transport of glucose across the plasma membrane in both heart and skeletal muscle [3, 153–155]. At relatively low workloads, mitochondrial acetyl-CoA and NADH produced by beta-oxidation tend to inhibit the pyruvate dehydrogenase complex in the mitochondrial inner membrane. Citrate, whose production is increased in the Krebs cycle after transport across the inner mitochondrial membrane into the cytoplasm, inhibits PFK [2, 3, 55, 164]. All these regulatory mechanisms explain the preference for FFAs for respiration in oxidative muscle cells. An important limitation of aerobic glycolysis is the necessity to maintain the rather low NADH:NAD$^+$ ratio in the cytoplasm needed for the high steady-state flux through the GAPDH reaction step that is achieved by transfer of reducing equivalents into the mitochondrial matrix by the malate–aspartate shuttle [55, 165]. However, this shuttle becomes the rate-limiting step at even medium workloads [165]. The fatty acid pathway is free of this kind of limitation. During contraction, the regulation of all reactions of substrate supply starts with an increase in the workload (see Fig. 11.3), which under normal physiological conditions is usually governed by the Frank–Starling law [27, 28]. The sequence of regulatory signals from cellular ATPases, such as the actomyosin ATPase for muscle contraction, to the mitochondrial matrix, which controls fatty acid oxidation during workload changes, is shown in Figs. 11.3 and 11.11. A workload-dependent increase in the rate of the actomyosin ATPase reaction results in the release of increasing amounts of ADP and P_i; this metabolic signal is transmitted to the mitochondrial adenine nucleotide translocase (ANT) by phosphotransfer networks under conditions of apparent metabolic stability as described above [27]. P_i enters the mitochondria by a special phosphate carrier [27, 33, 109]. These signals activate the ATP synthase and the utilization of the proton transmembrane electrochemical gradient ($\Delta\mu_{H+}$) for ATP synthesis in mitochondria [1–3, 27, 33, 56–58, 165]. This process always leads to an increase in electron transfer to oxygen via the respiratory chain and NADH and FADH$_2$ oxidation. Independent of the substrate used, an increased workload always decreases the NADH:NAD$^+$ ratio in the mitochondrial matrix (see Fig. 11.2) [27, 28, 55–58]. This decrease in the

NADH:NAD$^+$ ratio, on the other hand, increases the rates of the dehydrogenase reactions in the Krebs cycle and gears the latter to the rate of electron transfer via the respiratory chain [3, 55]. It is the Krebs cycle – which is on the crossroads between the metabolic pathways of glucose and fatty acid oxidation – and its intermediates that play a very important role in feedback metabolic regulation of upstream pathways of substrate oxidation. There is a mitochondrial matrix pool of important intermediates of this cycle, which plays a role in feedback regulation of substrate oxidation [55]. One of the results of the activation of the reactions of the Krebs cycle is the strong decrease in the level of acetyl-CoA in the mitochondrial matrix at high fluxes through this cycle when the rate-controlling steps are shifted towards the acyl-carnitine transporter and beta-oxidation of fatty acids [159]. Acetyl-CoA, which is formed in the mitochondrial matrix space, is transferred into the cytoplasm with the participation of carnitine, carnitine acetyltransferases, and carnitine acetyltranslocase (carnitine acetyl carnitine carrier, CAC in Fig. 11.11). These reactions seem to be in rapid equilibrium [161–163, 166–171]. In the cytoplasm, acetyl-CoA is used to produce malonyl-CoA through the acetyl-CoA carboxylase (ACC) reaction, which is in the malonyl-CoA decarboxylase (MCD) reaction converted back into acetyl-CoA [161–163, 168–172]. Malonyl-CoA is an effective inhibitor of carnitine palmitoyltransferase I (CPT I) and thus of the transfer of acyl groups into mitochondria for beta-oxidation (for reviews, see [2, 172, 173]). ACC is inhibited upon its phosphorylation by AMP-activated protein kinase (AMPK) [2, 173–177]. This reaction sequence served as the basis for a hypothesis that was advocated in numerous recent works, according to which malonyl-CoA is a key regulator of fatty acid oxidation [153, 154, 161, 172–177]. However, as an alternative, it seems much more reasonable to consider malonyl-CoA as a regular negative (inhibitory) metabolic feedback signal, as described elsewhere [1, 27]. The key regulator of respiration rate and energy fluxes, then, is the actual energy demand, the workload, and this also concerns fatty acid oxidation (see Fig. 11.3). Indeed, analysis of the changes in the intermediates of fatty acid oxidation showed that the workload-dependent increase in palmitate uptake and oxidation is accompanied by a significant decrease in acetyl-coenzyme A content [159]. This reflects the kinetics of the reactions in the Krebs cycle in response to an increase in the rate of respiration and NADH oxidation relative to acyl-carnitine transport into mitochondria and beta-oxidation [3]. A decrease in the content of acetyl-CoA leaves ACC in the cytosol with much less substrate, thus resulting in a significant decrease in malonyl-CoA in the presence of active MCD. ACC in the cytoplasm has a rather low affinity for acetyl-CoA [163, 169], and a decrease in this substrate due to changes in the kinetics of reactions in the Krebs cycle, as well as in acyl-carnitine transport and beta-oxidation, naturally results in decreased malonyl-CoA production [161, 171]. Therefore, malonyl-CoA decreases rapidly because of its decarboxylation by MCD back into acetyl-CoA, which is consumed in the Krebs cycle in the mitochondrial matrix [161, 171]. This decrease in malonyl-CoA then releases CPT I automatically from any inhibition, and FFA oxidation proceeds at a rate necessary for the energy supply of contraction and ion pumps. An additional interesting possibility, worthy of detailed studies, is the probable compartmentation of AMPK at the sites of AMP production close to the acyl-CoA synthetase. This may be an additional local mechanism of keeping the level of malonyl-CoA low and releasing CPT I from inhibition during fatty acid oxidation.

Thus, the decrease in malonyl-CoA is not the reason for but rather the consequence of the increases in workload and fatty acid oxidation. If at some point the workload is decreased again and less energy is needed, the reaction sequence described is reversed (see Fig. 11.11), and malonyl-CoA may increase to fulfill its role as a negative feedback regulator of FFA oxidation, as was correctly pointed out by Opie [1], by inhibiting CPT I to safeguard mitochondria from overload by acyl-CoA, which has detergent properties, and also may degrade to FFA and, via the uncoupling protein (UCP), may uncouple oxidative phosphorylation [3, 177, 178].

The integrated network shown in Fig. 11.11 explains all of the main experimental observations on the regulation of respiration and substrate utilization in the heart. Malonyl-CoA clearly seems to be at the end of this sequence of events, and the AMPK signaling pathway apparently modifies only this last step of regulation under cellular stress conditions under which the [AMP]:[ATP] ratio increases significantly.

For oxidative skeletal muscle cells, the integrated regulatory mechanisms described in Fig. 11.11 for heart and the sequence of the reactions given above are probably valid as well, especially for certain muscle types under certain conditions (endurance exercise training, etc.) [179, 180]. However, it should be emphasized that the mechanisms of regulation of skeletal muscle energy metabolism generally are different from those for the heart and are very different between distinct muscle fiber types, as described in Chapter 14 (see also [179, 180]).

11.9
Concluding Remarks and Future Directions

In this chapter we have demonstrated the advantages of molecular systems analysis to address the problems of regulation of energetics of the cell as a whole integrated system. This is opposite to the reductionist approach, which tends to use only the results of studies of the isolated system, such as mitochondria, or to focus on small segments of the whole system. Thus, the effects of calcium on the Krebs cycle dehydrogenases and isolated mitochondria led to overestimation of the role of this divalent cation in respiration regulation in the cell, in particular when the experimental data on the whole heart were not accounted for [54]. The conclusion of the key role of malonyl-CoA in the regulation of fatty acid oxidation was made in studies of only one fragment of the system (see gray area in Fig. 11.11). While justified in separate experiments, these conclusions are no longer valid for explaining regulatory mechanisms in the integrated system *in vivo*, when interactions between the parts of the system lead to completely new and unexpected mechanisms, the elucidation of which needs a molecular systems approach. As we have shown in this book, molecular system bioenergetics, a new approach to studying the integrated cellular systems of energy metabolism, is very helpful in the study and identification of cellular mechanisms of such complex phenomena as the acute contractile failure of the heart in the first minutes of total ischemia (see Chapter 3), regulation of respiration under the conditions of the Frank–Starling law, and the mechanisms of substrate supply to the heart.

These achievements and conclusions are in concord with and are supported by recent work published by Weiss et al. [181]. These authors presented a holistic view of cardiovascular metabolism, considering it from the perspective of a physical network in which various metabolic modules are spatially distributed throughout the interior of the cell to optimize ATP delivery to specific ATPases [181]. In addition to the mitochondrial module (which is represented in our works by ICEUs; see Chapter 3), the authors also considered a module consisting of glycolytic enzyme complexes that provide energy channeling to molecular complexes in the sarcolemma and sarcoplasmic reticulum and modules of calcium cycling (which Wang et al. called calcium release units [19]). These modules were further analyzed from the abstract perspective of fundamental concepts in network theory [182, 183] and the dynamic perspective of interactions between modules [181]. The authors emphasized that understanding the nature of these interactions within hierarchical modular structures is a major challenge of research into cardiac metabolism to gain a deeper understanding of possible mechanisms of cardioprotection [181].

Thus, in this new field of molecular system bioenergetics, structural, functional, genomic, and computational analysis of enzymatic clusters and networks is yet another future challenge in the area of cellular energetics. One of the most intriguing questions concerns the nature of local restrictions on the diffusion of adenine nucleotides in a highly structured cytosol despite the relatively high rates of their diffusion in the intracellular bulk water phase. These diffusion restrictions result in compartmentation of adenine nucleotides and kinetic and thermodynamic inefficiency of energy-dependent processes, which may explain the necessity of energy transfer and metabolic signaling networks. Future developments to verify predictions made by theoretical considerations and by increasingly complex mathematical models should include new experimental methods allowing detection of small changes in the transduced metabolic signals (Cr, ADP, AMP, P_i) in critical cellular compartments and new means of independent manipulation of energetic signals. These advances would provide a broader and more molecular understanding of the regulation of cellular energetics and metabolic signaling, a rapidly growing area in cellular systems biology.

Finally, to understand the complex network of cellular regulatory systems, it seems essential to include all layers of regulation in a systems biology approach, including, e.g., classical metabolic regulation together with more recent advances in cellular signaling cascades. It is clear, however, that further quantitative analyses of all these metabolic interactions in the network of substrate supply and utilization, by both experiments and mathematical modeling, are challenging and urgent tasks for molecular system bioenergetics in the future. The importance of this approach is evident from clinical studies by Neubauer's and Ingwall's groups [184, 185] (see also Chapter 7).

Acknowledgments

This work was supported by INSERM, France; by grants from the Estonian Science Foundation (Nos. 7117 and 6142 to V.S.), the Swiss National Science Foundation (No. 3100AO-102075), the Swiss Heart Foundation (to both T.W. and U.S.), the National Institutes of Health (HL64822, HL07111), the Marriott Program for Heart Disease Research, the Marriott Foundation, and the Miami Heart Research Institute (A.T. and P.D); by grants from Conseil Scientifique de l'Association Nationale de Traitement à Domicile Innovation et Recherche, Conseil Scientifique of AGIRàDom, Comité départemental de Lutte contre les Maladies Respiratoires de l'Isère, Délégation à la Recherche Clinique du Centre Hospitalier Universitaire de Grenoble, Centre d'Investigation Clinique, Inserm, CHU Grenoble, and Programme Interdisciplinaire Complexité du Vivant et Action STIC-Inserm (R.G.); and by grant 06-04-48620 from the Russian Foundation for Basic Research (M.A.). Due to space limitations and restrictions on cited references, we apologize to all those colleagues and researchers in the field whose work is not directly cited here, although they significantly contributed to this synopsis through their work and discussions.

References

1 Opie, L. H. (1998). In *The Heart. Physiology, from cell to circulation*, pp. 43–63, Lippincott-Raven Publishers, Philadelphia.

2 Stanley, W. C., Recchia, F. A., Lopaschuk, G. D. (2005) Myocardial substrate metabolism in the normal and failing heart. *Physiol. Rev. 85*, 1093–1129.

3 Neely, J. R., Morgan, H. E. (1974) Relationship between carbohydrate and lipid metabolism and the energy balance of heart muscle. *Annu. Rev. Physiol. 63*, 413–459.

4 Goldman, Y. (1987) Kinetics of the actomyosin ATPase in muscle fibers. *Annu. Rev. Physiol. 49*, 637–654.

5 Gordon, A. M., Regnier, M., Homsher, E. (2001) Skeletal and cardiac muscle contractile activation: tropomyosin "rocksandrolls". *News Physiol. Sc. i 16*, 49–55.

6 Rayment, I., Holden, H. M., Whittaker, M., Yohn, C. B., Lorenz, M., Holmes, K. C., Milligan, R. A. (1993) Structure of the actin-myosin complexandits implications for muscle contraction. *Science 261*, 58–65.

7 Cooke, R. (1997) Actomyosin interaction in striated muscle. *Physiol. Rev. 77*, 671–697.

8 Gordon, A. M., Homsher, E., Regnier, M. (2000) Regulation of contraction in striated muscle. *Physiol. Rev. 80*, 853–924.

9 Ames, A. (2000) CNS energy metabolism as related to function. *Brain Research Reviews 34*, 42–68.

10 Bers, D. M. (2002) Cardiac excitation-contraction coupling. *Nature 415*, 198–205.

11 Endoh, M. (2006) Signal transduction and Ca2+ signaling in intact myocardium. *J. Pharmacol. Sci. 100*, 525–37.

12 Bers, D. (2001) *Excitation – contraction coupling and cardiac contraction*. Kluwer academic Publishers, Dordrecht.

13 Carafoli, E. (2003) Historical review: mitochondria and calcium: ups and downs of an unusual relationship. *Trends Biochem. Sci. 28*, 175–181.

14 Berridge, M. J., Bootman, M. D., Roderick, H. L. (2003) Calcium signalling: dynamics, homeostasis and

remodelling. *Nat. Rev. Mol. Cell. Biol. 4*, 517–529.

15 Rizzuto, R., Bernardi, P., Pozzan, T. (2000) Mitochondria as all-round players of the calcium game. *J. Physiol. 529 Pt 1*, 37–47.

16 Rizzuto, R., Pozzan, T. (2006) Microdomains of intracellular Ca^{2+}: molecular determinants and functional consequences. *Physiol. Rev. 86*, 369–408.

17 Jacobson, J., Duchen, M. R. (2004) Interplay between mitochondria and cellular calcium signalling. *Mol. Cell. Biochem. 256–257*, 209–218.

18 Bianchi, K., Rimessi, A., Prandini, A., Szabadkai, G., Rizzuto, R. (2004) Calcium and mitochondria: mechanisms and functions of a troubled relationship. *Biochim. Biophys. Acta. 1742*, 119–131.

19 Wang, S. Q., Wei, C., Zhao, G., Brochet, D. X., Shen, J., Song, L. S., Wang, W., Yang, D., Cheng, H. (2004) Imaging microdomain Ca^{2+} in muscle cells. *Circ. Res. 94*, 1011–1022.

20 Niggli, E. (1999) Ca^{2+} Sparks in Cardiac Muscle: Is There Life Without Them? *News Physiol. Sci. 14*, 129–134.

21 Fuchs, F., Smith, S. H. (2001) Calcium, cross-bridges, and the Frank-Starling relationship. *News Physiol. Sci. 16*, 5–10.

22 Frank, O. (1885) Zur Dynamik des Herzmuskels. *Zeitschrift für Biologie 32*, 370–447.

23 Patterson, S. W., Piper, H., Starling, E. H. (1914) The regulation of the heartbeat. *J. Physiol. 48*, 357–379.

24 Starling, E. H. And Visscher, M. B. (1926) The regulation of the energy output of the heart. *J. Physiol. 62*, 243–261.

25 Katz, A. M. (2002) Ernest Henry Starling, his predecessors,and the "Law of the Heart". *Circulation 106*, 2986–2992.

26 Evans, C. L., Matsuoka, Y. (1915) The effects of various mechanical conditions on gaseous metabolism and efficiency of the mammalian heart. *J. Physiol. 49*, 379–405.

27 Saks, V., Dzeja, P., Schlattner, U., Vendelin, M., Terzic, A., Wallimann, T. (2006b) Cardiac system bioenergetics: metabolic basis of Frank-Starling law. *J. Physiol. 571 (2)*, 253–273.

28 Saks, V. A., Kuznetsov, A. V., Vendelin, M., Guerrero, K., Kay, L., Seppet, E. K. (2004) Functional coupling as a basic mechanism of feedback regulation of cardiac energy metabolism. *Mol. Cell. Biochem. 256–257*, 185–199.

29 Williamson, J. R., Ford, C., Illingworth, J., Safer, B. (1976) Coordination of citric acid cycle activity with electron transport flux. *Circ. Res. 38*, 39–51.

30 Neely, J. R., Denton, R. M., England, P. J., Randle, P. J. (1972) The effects of increased heart work on the tricarboxylate cycle and its interactions with glycolysis in the perfused rat heart. *Biochem. J. 128*, 147–159.

31 Balaban, R. S. (2002) Cardiac energy metabolism homeostasis: role of cytosolic calcium. *J. Mol. Cell. Cardiol. 34*, 1259–1271.

32 Hochachka, P. W. (2003) Intracellular convection, homeostasis and metabolic regulation. *J. Exp. Biol. 206*, 2001–2009.

33 Nicholls, D., Ferguson, S. J. (2002) *Bioenergetics* Academic Press, London-New York.

34 Lardy, H. A., Wellman, H. (1952) Oxidative phosphorylations: role of inorganic phosphate and acceptor systems in control metabolic rates. *J. Biol. Chem. 195*, 215–224.

35 Chance, B., Williams, G. R. (1956) Regulatory chain and oxidative phosphorylation. *Adv. Enzymol. 17*, 65–134.

36 Jacobus, W. E., Moreadith, R. W., Vandegaer, K. M. (1982) Mitochondrial respiratory control. Evidence against the regulation of respiration by extramitochondrial phosphorylation potentials or by [ATP]/[ADP] ratios. *J. Biol. Chem. 257*, 2397–2402.

37 Ringer, S. (1883) A further contribution regarding the influence of the different constituents of the blood on the contraction of the heart. *J. Physiol. 4*, 29–42.

38 Miller, D. J. (2004) Sydney Ringer; physiological saline, calcium and the contraction of the heart. *J. Physiol. 555*, 585–587.

39 Dulhunty, A. F. (2006) Excitation-contraction coupling from the 1950s

into the new millennium. *Clin. Exp. Pharmacol. Physiol. 33*, 763–772.

40 Blinks, J. R., Wier, W. G., Hess, P., Prendergast, F. G. (1982) Measurement of Ca^{2+} concentrations in living cells. *Prog. Biophys. Mol Biol. 40*, 1–114.

41 Minta, A., Kao, J. P., Tsien, R. Y. (1989) Fluorescent indicators for cytosolic calcium based on rhodamine and fluorescein chromophores. *J. Biol. Chem. 264*, 8171–8178.

42 Hansford, R. (1985) Relation between mitochondrial calcium transport and control of energy metabolism. *Rev. Physiol. Biochem. Pharmacol. 102*, 1–72.

43 Denton, R. M., Richards, D. A., Chin, J. G. (1978) Calcium ions and the regulation of NAD^+-linked isocitrate dehydrogenase from the mitochondria of rat heart and other tissues. *Biochem. J. 176*, 899–906.

44 McCormack, J. G., Halestrap, A. P., Denton, R. M. (1990) Role of calcium ions in regulation of mammalian intramitochondrial metabolism. *Physiol. Rev. 70*, 391–425.

45 Hansford, R. G., Zorov, D. (1998) Role of mitochondrial calcium transport in the control of substrate oxidation. *Mol. Cell. Biochem. 184*, 359–369.

46 Mackenzie, L., Roderick, H. L., Berridge, M. J., Conway, S. J., Bootman, M. D. (2004) The spatial pattern of atrial cardiomyocyte calcium signalling modulates contraction. *J. Cell. Sci. 117*, 6327–6337.

47 Bernardi, P. (1999) Mitochondrial transport of cations: channels, exchangers, and permeability transition. *Physiol. Rev. 79*, 1127–1155.

48 Gunter, T. E., Yule, D. I., Gunter, K. K., Eliseev, R. A., Salter, J. D. (2004) Calcium and mitochondria. *FEBS Lett. 567*, 96–102.

49 Jouaville, L. S., Pinton, P., Bastianutto, C., Rutter, G. A., Rizzuto, R. (1999) Regulation of mitochondrial ATP synthesis by calcium: evidence for a long-term metabolic priming. *Proc. Natl. Acad. Sci. USA 96*, 13807–13812.

50 Territo, P. R., French, S. A., Dunleavy, M. C., Evans, F. J., Balaban, R. S. (2001) Calcium activation of heart mitochondrial oxidative phosphorylation: rapid

kinetics of mVO2, NADH, and light scattering. *J. Biol. Chem. 276*, 2586–2599.

51 Brini, M. (2003) Ca^{2+} signalling in mitochondria: mechanism and role in physiology and pathology. *Cell. Calcium. 34*, 399–405.

52 Brookes, P. S., Yoon, Y., Robotham, J. L., Anders, M. W., Sheu, S. S. (2004) Calcium, ATP, and ROS: a mitochondrial love-hate triangle. *Am. J. Physiol. Cell. Physiol. 287*, C817–833.

53 Cortassa, S., Aon, M. A., Marban, E., Winslow, R. L., O'Rourke, B. (2003) An integrated model of cardiac mitochondrial energy metabolism and calcium dynamics. *Biophys. J. 84*, 2734–2755.

54 Schaub, M. C., Hefti, M. A., Zaugg, M. (2006) Integration of calcium with the signaling network in cardiac myocytes. *J. Mol. Cell. Cardiol. 41*, 183–214.

55 Williamson, J. R. (1979) Mitochondrial function in the heart. *Annu. Rev. Physiol. 41*, 485–506.

56 Hassinen, I. E., Hiltunen, K. (1975) Respiratory control in isolated perfused rat heart. Role of the equilibrium relations between the mitochondrial electron carriers and the adenylate system. *Biochim. Biophys. Acta. 408*, 319–330.

57 Hassinen, I. E. (1986) Mitochondrial respiratory control in the myocardium. *Biochim. Biophys. Acta. 853*, 135–151.

58 Brandes, R., Bers, D. M. (1999) Analysis of the mechanisms of mitochondrial NADH regulation in cardiac trabeculae. *Biophysical J. 77*, 1666–1682.

59 Hibberd, M. G., Jewell, B. R. (1982) Calcium and length dependent – force production in rat ventricular muscle. *J. Physiol. 329*, 527–540.

60 Allen, D. D. G., Kurihara, S. (1982) The effects of muscle length on intracellular calcium transients in mammalian cardiac muscle. *J. Physiol. 327*, 79–94.

61 Landesberg, A. (1996) End-systolic pressure-volume relationship and intracellular control of contraction. *Am. J. Physiol. 270*, H338–349.

62 Landesberg, A., Sideman, S. (1999) Regulation of energy consumption in cardiac muscle: analysis of isometric contractions. *Am. J. Physiol. 276*, H998–H1011.

63 Robinson, J. M., Wang, Y., Kerrick, W. G., Kawai, R., Cheung, H. C. (2002) Activation of striated muscle: nearest-neighbor regulatory-unit and cross-bridge influence on myofilament kinetics. *J. Mol. Biol. 322*, 1065–1088.

64 Kentish, J. C., Wrzosek, A. (1998) Changes in force and cytosolic Ca2+ concentration after length changes in isolated rat ventricular trabeculae. *J. Physiol. 506 Pt 2*, 431–444.

65 Shimizu, J., Todaka, K., Burkhoff, D. (2002) Load dependence of ventricular performance explained by model of calcium-myofilament interactions. *Am. J. Physiol. Heart Circ. Physiol. 282*, H1081–1091.

66 Mizuno, J., Araki, J., Mohri, S., Minami, H., Doi, Y., Fujinaka, W., Miyaji, K., Kiyooka, T., Oshima, Y., Iribe, G., Hirakawa, M., Suga, H. (2001) Frank-Starling mechanism retains recirculation fraction of myocardial Ca^{2+} in the beating heart. *Jpn. J. Physiol. 51*, 733–743.

67 Fukuda, N., Sasaki, D., Ishiwata, S., Kurihara, S. (2001) Length dependence of tension generation in rat skinned cardiac muscle: role of titin in the Frank-Starling mechanism of the heart. *Circulation 104*, 1639–1645.

68 Millman, B. M. (1998) The filament lattice of striated muscle. *Physiol. Rev. 78*, 359–391.

69 Vendelin, M., Bovendeerd, P. H., Engelbrecht, J., Arts, T. (2002) Optimizing ventricular fibers: uniform strain or stress, but not ATP consumption, leads to high efficiency. *Am. J. Physiol. Heart. Circ. Physiol. 283*, H1072–1081.

70 Yamashita, H., Sata, M., Sugiura, S., Momomura, S., Serizawa, T., Iizuka, M. (1994) ADP inhibits the sliding velocity of fluorescent actin filaments on cardiac and skeletal myosins. *Circ. Res. 74*, 1027–1033.

71 Fukuda, N., Kajiwara, H., Ishiwata, S., Kurihara, S. (2000) Effects of MgADP on length dependence of tension generation in skinned rat cardiac muscle. *Circ. Res. 86*, E1–6.

72 de Meis, L., Inesi, G. (1982) ATP synthesis by sarcoplasmic reticulum ATPase following Ca^{2+}, pH, temperature, and water activity jumps. *J. Biol. Chem. 257*, 1289–1294.

73 Sata, M., Sugiura, S., Yamashita, H., Momomura, S., Serizawa, T. (1996) Coupling between myosin ATPase cycle and creatinine kinase cycle facilitates cardiac actomyosin sliding *in vitro*. A clue to mechanical dysfunction during myocardial ischemia. *Circulation 93*, 310–317.

74 Fukuda, N., Fujita, H., Fujita, T., Ishiwata, S. (1998) Regulatory roles of MgADP and calcium in tension development of skinned cardiac muscle. *J. Muscle Res. Cell. Motil. 19*, 909–921.

75 Dzeja, P. P., Terzic, A. (1998) Phosphotransfer reactions in the regulation of ATP-sensitive K+ channels. *Faseb J. 12*, 523–529.

76 Dzeja, P. P., Terzic, A. (2005) Mitochondrial-nucleus energetic communication: role of phospho-transfer networks in processing cellular information. In *Handbook of Neurochemistry and Molecular Neurobiology: Neural Energy Utilization*, Gibson G., Dienel G. Kluwer (eds), New York.

77 Pucar, D., Dzeja, P. P., Bast, P., Gumina, R. J., Drahl, C., Lim, L., Juranic, N., Macura, S., Terzic, A. (2004) Mapping hypoxia-induced bioenergetic rearrangements and metabolic signaling by ^{18}O-assisted ^{31}P NMR and ^{1}H NMR spectroscopy. *Mol. Cell. Biochem. 256–257*, 281–289.

78 Kahn, B. B., Alquier, T., Carling, D., Hardie, D. G. (2005) AMP-activated protein kinase: ancient energy gauge provides clues to modern understanding of metabolism. *Cell. Metab. 1*, 15–25.

79 Andrienko, T., Kuznetsov, A. V., Kaambre, T., Usson, Y., Orosco, A., Appaix, F., Tiivel, T., Sikk, P., Vendelin, M., Margreiter, R., Saks, V. A. (2003) Metabolic consequences of functional complexes of mitochondria, myofibrils and sarcoplasmic reticulum in muscle cells. *J. Exp. Biology 206*, 2059–2072.

80 Appaix, F., Kuznetsov, A. V., Usson, Y., Andrienko, T., Olivares, J., Kaambre, T., Sikk, P., Margreiter, R., Saks, V. (2003) Possible role of cytoskeleton in

intracellular arrangement and regulation of mitochondria. *Exp. Physiology 88*, 175–190.

81 Saks, V., Kuznetsov, A., Andrienko, T., Usson, Y., Appaix, F., Guerrero, K., Kaambre, T., Sikk, P., Lemba, M., Vendelin, M. (2003) Heterogeneity of ADP diffusion and regulation of respiration in cardiac cells. *Biophys. J. 84*, 3436–3456.

82 Saks, V. A., Vendelin, M., Aliev, M. K., Kekelidze, T., Engelbrecht, J. (2007) Mechanisms and modeling of energy transfer between and among intracellular compartments. In: *Handbook of Neurochemistry and Molecular Neurobiology, 3rd edition*, volume 5, *Brain Energetics*, Dienel G., Gibson G. (eds), Springer Science, New York-Boston, pp. 815–860.

83 Saks, V. A., Khuchua, Z. A., Vasilyeva, E. V., Belikova, Y. O., Kuznetsov, A. (1994) Metabolic compartmentation and substrate channelling in muscle cells. Role of coupled creatine kinases *in vivo* regulation of cellular respiration. A Synthesis. *J. Molec. Cell. Biochem. 133/134*, 155–192.

84 Saks, V. A., Kuznetsov, A. V., Khuchua, Z. A., Vasilyeva, E. V., Belikova, J. O., Kesvatera, T., Tiivel, T. (1995) Control of cellular respiration *in vivo* by mitochondrial outer membrane and by creatine kinase. A new speculative hypothesis: possible involvement of mitochondrial-cytoskeleton interactions. *J. Mol. Cell. Cardiol. 27*, 625–645.

85 Saks, V. A., Aliev, M., Guzun, R., Beraud, N., Monge, C., Anmann, T., Kuznetsov, A. V., Seppet, E. (2006) Biophysics of the organized metabolic networks in muscle and brain cells. In *Recent Research Developments in Biophysics*, 5, 269–318, Transworld Research Network 37/661, Kerala, India.

86 Anmann, T., Eimre, M., Kuznetsov, A. V., Andrienko, T., Kaambre, T., Sikk, P., Seppet, E., Tiivel, T., Vendelin, M., Seppet, E., Saks, V. A. (2005) Calcium-induced contraction of sarcomeres changes regulation of mitochondrial respiration in permeabilized cardiac cells. *FEBS Journal 272*, 3145–3161.

87 Anmann, T., Guzun, R., Beraud, N., Pelloux, S., Kuznetsov, A. V., Kogerman, L., Kaambre, T., Sikk, P., Paju, K., Peet, N., Seppet, E., Ojeda, C., Tourneur, Y., Saks, V. A. (2006) Different kinetics of the regulation of respiration in permeabilized cardiomyocytes and in HL-1 cardiac cells. Importance of cell structure/organization for respiration regulation. *Biochim. Biophys. Acta*, 1757, 1597–1606.

88 Khuchua, Z., Belikova, Y., Kuznetsov, A. V., Gellerich, F. N., Schild, L., Neumann, H. W., Kunz, W. S. (1994) Caffeine and Ca^{2+} stimulate mitochondrial oxidative phosphorylation in saponin-skinned human skeletal muscle fibers due to activation of actomyosin ATPase. *Biochim. Biophys. Acta. 1188*, 373–379.

89 Gellerich, F., Saks, V. A. (1982) Control of heart mitochondrial oxygen consumption by creatine kinase: the importance of enzyme localization. *Biochem. Biophys. Res. Commun. 105*, 1473–1481.

90 Dzeja, P. P., Vitkevicius, K. T., Redfield, M. M., Burnett, J. C., Terzic, A. (1999) Adenylate kinase-catalyzed phosphotransfer in the myocardium: increased contribution in heart failure. *Circ. Res. 84*, 1137–1143.

91 Pucar, D., Dzeja, P. P., Bast, P., Juranic, N., Macura, S., Terzic, A. (2001) Cellular energetics in the preconditioned state: protective role for phosphotransfer reactions captured by ^{18}O-assisted ^{31}P NMR. *J. Biol. Chem. 276*, 44812–44819.

92 Dzeja, P. P., Terzic, A. (2003) Phosphotransfer networks and cellular energetics. *J. Exp. Biol. 206*, 2039–2047.

93 Dzeja, P. P., Terzic, A., Wieringa, B. (2004) Phosphotransfer dynamics in skeletal muscle from creatine kinase gene-deleted mice. *Mol. Cell. Biochem. 256–257*, 13–27.

94 Dos Santos, P., Aliev, M. K., Diolez, P., Duclos, F., Besse, P., Bonoron-Adele, S., Sikk, P., Canioni, P., Saks, V. A. (2000) Metabolic control of contractile performance in isolated perfused rat heart. Analysis of experimental data by reaction: diffusion mathematical model. *J. Mol. Cell. Cardiol. 32*, 1703–1734.

95 Aliev, M. K., Saks, V. A. (1997) Compartmentalized energy transfer in cardiomyocytes: use of mathematical modeling for analysis of *in vivo* regulation of respiration. *Biophys. J. 73*, 428–445.

96 Kupriyanov, V. V., Ya Steinschneider, A., Ruuge, E. K., Kapel'ko, V. I., Yu Zueva, M., Lakomkin, V. L., Smirnov, V. N., Saks, V. A. (1984) Regulation of energy flux through the creatine kinase reaction *in vitro* and in perfused rat heart. ^{31}P-NMR studies. *Biochim. Biophys. Acta. 805*, 319–331.

97 Bessman, S. P., Carpenter, C. L. (1985) The creatine-creatine phosphate energy shuttle. *Annu. Rev. Biochem. 54*, 831–862.

98 Wallimann, T., Wyss, M., Brdiczka, D., Nicolay, K., Eppenberger, H. M. (1992) Intracellular compartmentation, structure and function of creatine kinase isoenzymes in tissues with high and fluctuating energy demands: the 'phosphocreatine circuit' for cellular energy homeostasis. *Biochem. J. 281 Pt 1*, 21–40.

99 Bittl, J. A., Ingwall, J. S. (1985). Reaction rates of creatine kinase and ATP synthesis in the isolated rat heart. A ^{31}P NMR magnetization transfer study. *J. Biol. Chem. 260*, 3512–3517.

100 Vendelin, M., Kongas, O., Saks, V. (2000) Regulation of mitochondrial respiration in heart cells analyzed by reaction-diffusion model of energy transfer. *Am. J. Physiol. Cell. Physiol. 278*, C747–764.

101 Cortassa, S., Aon, M. A., O'Rourke, B., Jacques, R., Tseng, H. J., Marban, E., Winslow, R. L. (2006) A computational model integrating electrophysiology, contraction, and mitochondrial bio-energetics in the ventricular myocyte. *Biophys. J. 91*, 1564–89.

102 Beard, D. A. (2006) Modeling of oxygen transport and cellular energetics explains observations on *in vivo* cardiac energy metabolism. *PLoS Comput. Biol. 2*, 1093–1106.

103 Matsuoka, S., Sarai, N., Jo, H., Noma, A. (2004) Simulation of ATP metabolism in cardiac excitation-contraction coupling. *Prog. Biophys. Mol. Biol. 85*, 279–99.

104 Korzeniewski, B. (1998) Regulation of ATP supply during muscle contraction: theoretical studies. *Biochem. J. 330*, 1189–1195.

105 Korzeniewski, B., Zoladz, J. A. (2001) A model of oxidative phosphorylation in mammalian skeletal muscle. *Biophys. Chem. 92*, 17–34.

106 Korzeniewski, B. (2006) Oxygen consumption and metabolite concentrations during transitions between different work intensities in heart. *Am. J. Physiol. 291*, H799–H804.

107 Jafri, M. S., Dudycha, S. J., O'Rourke, B. (2001) Cardiac energy metabolism: models of cellular respiration. *Annu. Rev. Biomed. Eng. 3*, 57–81.

108 Saks, V. A., Kongas, O., Vendelin, M., Kay, L. (2000) Role of the creatine/phosphocreatine system in the regulation of mitochondrial respiration. *Acta. Physiol. Scand. 168*, 635–641.

109 Bose, S., French, S., Evans, F. J., Joubert, F., Balaban, R. S. (2003) Metabolic network control of oxidative phosphorylation: multiple roles of inorganic phosphate. *J. Biol. Chem. 278*, 39155–39165.

110 Spindler, M., Illing, B., Horn, M., de Groot, M., Ertl, G., Neubauer, S. (2001) Temporal fluctuations of myocardia high-energy phosphate metabolite with the cardiac cycle. *Basic. Res. Cardiol. 96*, 553–556.

111 Honda, H., Tanaka, K., Akita, N., Haneda, T. (2002) Cyclical changes in high-energy phosphates during the cardiac cycle by pacing-Gated ^{31}P nuclear magnetic resonance. *Circ. J. 66*, 80–6.

112 Dzeja, P. P., Zeleznikar, R. J., Goldberg, N. D. (1998) Adenylate kinase: kinetic behavior in intact cells indicates it is integral to multiple cellular processes. *Mol. Cell. Biochem. 184*, 169–182.

113 Mitchell, P. (1979) The ninth Sir Hans Krebs lecture. Compartmentation and communication in living systems. Ligand conduction: a general catalytic principle in chemical, osmotic and chemiosmotic reaction systems. *Eur. J. Biochem. 95*, 1–20.

114 Garlid, K. D. (2001). Physiology of mitochondria. *In cell physiology*

sourcebook: a molecular approach, 3[e]
edition, Sperelakis N. (ed.), pp. 139–151,
Academic Press, San Diego-New York-
London.

115 Foell, J. D., Balijepalli, R. C., Delisle,
B. P., Yunker, A. M., Robia, S. L.,
Walker, J. W., McEnery, M. W., January,
C. T., Kamp, T. J. (2004) Molecular
heterogeneity of calcium channel beta-
subunits in canine and human heart:
evidence for differential subcellular
localization. *Physiol. Genomics 17*, 183–
200.

116 Ingwall, J. S. (2004) Transgenesis and
cardiac energetics: new insights into
cardiac metabolism. *J. Mol. Cell. Cardiol.
37*, 613–623.

117 Kane, G. C., Behfar, A., Yamada, S.,
Perez-Terzic, C., O'Cochlain, F., Reyes,
S., Dzeja, P. P., Miki, T., Seino, S.,
Terzic, A. (2004) ATP-sensitive K$^+$
channel knockout compromises the
metabolic benefit of exercise training,
resulting in cardiac deficits. *Diabetes 53*,
S169–S175.

118 Alekseev, A. E., Hodgson, D. M., Karger,
A. B., Park, S., Zingman, L. V., Terzic,
A. (2005) ATP-sensitive K$^+$ channel/
enzyme multimer: metabolic gating in
the heart. *J Mol Cell Cardiol 38*, 895–905.

119 Selivanov, V. A., Alekseev, A. E.,
Hodgson, D. M., Dzeja, P. P., Terzic, A.
(2004) Nucleotide-gated K(ATP)
channels integrated with creatine and
adenylate kinases: amplification, tuning
and sensing of energetic signals in the
compartmentalized cellular environment. *Mol. Cell. Biochem. 256–257*,
243–256.

120 Szigligeti, P., Pankucsi, C., Banyasz, T.,
Varro, A., Nanasi, P. P. (1996) Action
potential duration and force-frequency
relationship in isolated rabbit, guinea
pig and rat cardiac muscle. *J Comp
Physiol*, 166, 150–5.

121 Trost, S., LeWinter, M. (2001) Diabetic
Cardiomyopathy. *Curr Treat Options
Cardiovasc Med*. 6, 481–492.

122 Spindler, M., Meyer, K., Stromer, H.,
Leupold, A., Boehm, E., Wagner, H.,
Neubauer, S. (2004) Creatine kinase-
deficient hearts exhibit increased
susceptibility to ischemia-reperfusion
injury and impaired calcium homeo-

stasis. *Am J Physiol Heart Circ Physiol*.
287, H1039–45.

123 ten Hove, M., Lygate, C. A., Fischer, A.,
Schneider, J. E., Sang, A. E., Hulbert, K.,
Sebag-Montefiore, L., Watkins, H.,
Clarke, K., Isbrandt, D., Wallis, J.,
Neubauer, S. (2005) Reduced inotropic
reserve and increased susceptibility to
cardiac ischemia/reperfusion injury in
phosphocreatine-deficient guanidino-
acetate-N-methyltransferase-knockout
mice. *Circulation*. 111, 2477–85.

124 Zingman, L. V., Hodgson, D. M., Bast,
P. H., Kane, G. C., Perez-Terzic, C.,
Gumina, R. J., Pucar, D., Bienengraeber,
M., Dzeja, P. P., Miki, T., Seino, S.,
Alekseev, A. E., Terzic, A. (2002) Kir6.2
is required for adaptation to stress. *Proc
Natl Acad Sci U S A*. 99, 3278–83.

125 Depre, C., Vanoverschelde, J. L.,
Taegtmeyer H. (1999) Glucose for the
heart. *Circulation*. 99, 578–88.

126 Yamanaka, T., Onishi, K., Tanabe, M.,
Dohi, K., Funabiki-Yamanaka, K.,
Fujimoto, N., Kurita, T., Tanigawa, T.,
Kitamura, T., Ito, M., Nobori, T.,
Nakano, T. (2006) Force- and relaxation-
frequency relations in patients with
diastolic heart failure. *Am Heart J*. 152,
966.e1–7.

127 Taegtmeyer, H., Wilson, C. R., Razeghi,
P., Sharma, S. (2005) Metabolic
energetics and genetics in the heart.
Ann. NY Acad. Sci. 1047, 208–18.

128 Neumann, D., Schlattner, U.,
Wallimann, T. (2003) A molecular
approach to the concerted action of
kinases involved in energy
homoeostasis. *Biochem. Soc. Trans. 31*,
169–174.

129 Sasaki, N., Sato, T., Marban, E.,
O'Rourke, B. (2001) ATP consumption
by uncoupled mitochondria activates
sarcolemmal K(ATP) channels in cardiac
myocytes. *Am. J. Physiol. Heart Circ.
Physiol. 280*, H1882–1888.

130 Weiss, J. N., Venkatesh, N. (1993)
Metabolic regulation of cardiac ATP-
sensitive K$^+$ channels *Cardiovasc. Drugs
Ther. 7 Suppl 3*, 499–505.

131 Gwanyanya, A., Sipido, K. R., Vereecke,
J., Mubagwa, K. (2006) ATP and PIP2
dependence of the magnesium-
inhibited, TRPM7-like cation channel in

cardiac myocytes. *Am J Physiol Cell Physiol.* 291, C627–35.

132 Fatholahi, M., LaNoue, K., Romani, A., Scarpa, A. (2000) Relationship between total and free cellular Mg(2+) during metabolic stimulation of rat cardiac myocytes and perfused hearts. *Arch Biochem Biophys.* 374, 395–401.

133 Igamberdiev, A. U., Kleczkowski, L. A. (2003) Membrane potential, adenylate levels and Mg^{2+} are interconnected via adenylate kinase equilibrium in plant cells. *Biochim Biophys Acta.* 1607, 11–9.

134 Blair, J. M. (1970) Magnesium, potassium, and the adenylate kinase equilibrium. Magnesium as a feedback signal from the adenine nucleotide pool. *Eur J Biochem.* 13, 384–90.

135 Leyssens, A., Nowicky, A. V., Patterson, L., Crompton, M., Duchen, M. R. (1996) The relationship between mitochondrial state, ATP hydrolysis, $[Mg^{2+}]i$ and $[Ca^{2+}]i$ studied in isolated rat cardiomyocytes. *J Physiol.* 496, 111–28.

136 Mooren, F. C., Turi, S., Gunzel, D., Schlue, W. R., Domschke, W., Singh, J., Lerch, M. M. (2001) Calcium-magnesium interactions in pancreatic acinar cells. *FASEB J.* 15, 659–72.

137 Saucerman, J. J., Brunton, L. L., Michailova, A. P., McCulloch, A. D. (2003) Modeling beta-adrenergic control of cardiac myocyte contractility in silico. *J. Biol. Chem.* 278, 47997–48003.

138 Meyer, L. E., Machado, L. B., Santiago, A. P. S. A., da-Silva, S., De Felice, F. G., Holub, O., Oliviera, M., Galina, A. (2006) Mitochondrial creatine kinase activity prevents reactive oxygen species generation: antioxidant role of mitochondrial kinases-dependent ADP re-cycling activity. *J. Biol. Chem.* 281, 29916–29928.

139 Jezek, P., Hlavata, L., (2005) Mitochondria in homeostasis of reactive oxygen species in cell, tissues and organism. *Intl. J. Biochem.* 37, 2478–2503.

140 Nascimben, L., Ingwall, J. S., Pauletto, P., Friedrich, J., Gwathmey, J. K., Saks, V., Pessina, A. C., Allen, P. D. (1996) Creatine kinase system in failing and nonfailing human myocardium. *Circulation* 94, 1894–1901.

141 ten Hove, M., Chan, S., Lygate, C., Monfared, M., Boehm, E., Hulbert, K., Watkins, H., Clarke, K., Neubauer, S. (2005) Mechanisms of creatine depletion in chronically failing rat heart. *J. Mol. Cell. Cardiol.* 38, 309–313.

142 Nahrendorf, M., Spindler, M., Hu, K., Bauer, L., Ritter, O., Nordbeck, P., Quaschning, T., Hiller, K. H., Wallis, J., Ertl, G., Bauer, W. R., Neubauer, S. (2005) Creatine kinase knockout mice show left ventricular hypertrophy and dilatation, but unaltered remodeling post-myocardial infarction. *Cardiovasc. Res.* 65, 419–427.

143 Zoll, J., Ponsot, E., Doutreleau, S., Mettauer, B., Piquard, F., Mazzucotelli, J. P., Diemunsch, P., Geny, B. (2005) Acute myocardial ischaemia induces specific alterations of ventricular mitochondrial function in experimental pigs. *Acta. Physiol. Scand.* 185, 25–32.

144 Ventura-Clapier, R., Garnier, A., Veksler, V. (2004) Energy metabolism in heart failure. *J. Physiol.* 555, 1–13.

145 Schlattner, U., Tokarska-Schlattner, M., Wallimann, T. (2005) Mitochondrial creatine kinase in human health and disease. *Biochim. Biophys. Acta* 1762, 164–80.

146 Wyss, M., Kaddurah-Daouk, R. (2000) Creatine and creatinine metabolism. *Physiol. Rev. 80*, 1107–1213.

147 Kapelko, V. I., Kupriyanov, V. V., Novikova, N. A., Lakomkin, V. L., Steinschneider, A., Severina, M., Veksler, V. I., Saks, V. A. (1988) The cardiac contractile failure induced by chronic creatine and phosphocreatine deficiency. *J. Mol. Cell. Cardiol.* 20, 465–479.

148 Saks, V. A., Rosenshtraukh, L. V., Undrovinas, A. I., Smirnov, V. N., Chazov, E. I. (1976) Studies of energy transport in heart cells. Intracellular creatine content as a regulatory factor of frog heart energetics and force of contraction. *Biochem. Med.* 16, 21–36.

149 Ventura-Clapier, R., Vassort, G. (1980) The hypodynamic state of the frog heart. Further evidence for a phosphocreatine-creatine pathway. *J. Physiol. (Paris)* 76, 583–589.

150 Momken, I., Lechene, P., Koulmann, N., Fortin, D., Mateo, P., Hoerter, J., Doan,

B. T., Bigard, X., Veksler, V., Ventura-Clapier, R. F. (2005) Impaired voluntary running capacity in CK deficient mice. *J. Physiol. 565*, 951–964.

151 Saks, V. A., Favier, R., Guzun, R., Schlattner, U., Wallimann, T. (2006) Molecular system bioenergetics: regulation of substrate supply in response to heart energy demands. *J. Physiol. 577*, 769–777.

152 Bing, R. J. (1965) Cardiac metabolism. *Physiol. Rev. 45*, 171–213.

153 Rasmussen, B. B., Wolfe, R. R. (1999) Regulation of fatty acid oxidation in skeletal muscle. *Ann. Rev. Nutr. 19*, 463–484.

154 Kiens, B. (2006) Skeletal muscle lipid metabolism in exercise and insulin resistance. *Physiol. Rev. 86*, 205–243.

155 Roden, M. (2004) How free fatty acids inhibit glucose utilization in human skeletal muscle. *News Physiol. 19*, 92–96.

156 Randle, P. J. (1998) Regulatory interactions between lipids, carbohydrates: the glucose fatty acid cycle after 35 years. *Diabetes Metab. Rev. 14*, 263–83.

157 Clark, A. J., Graddie, R., Stewart, C. P. (1931) The metabolism of the isolated heart of the frog. *J. Physiol. 72*, 443.

158 Bing, R. J., Siegel, A., Ungar, I., Gilbert, M. (1954) Metabolism of the human heart. II Metabolism of fats, proteins and ketones. *Am. J. Med. 16*, 504.

159 Oram, J. F., Bennetch, S. L., Neely, J. R. (1973) Regulation of fatty acid utilization in isolated perfused hearts. *J. Biol. Chem. 248*, 5299–5309.

160 Beauloye, C., Marsin, A. S., Bertrand, L., Vanoverschelde, J. L., Rider, M. H., Hue, L. (2002) The stimulation of heart glycolysis by increased workload does not require AMP activated protein kinase but a wortmannin–sensitive mechanism. *FEBS Lett. 531*, 324–328.

161 Reszko, A. E., Kasumov, T., David, F., Thomas, K. R., Jobbins, K. A., Cheng, J. F., Lopaschuk, G., Dyck, J. R. B., Diaz, M., Rosiers, C. D., Stanley, W. C., Brunengraber, H. (2004) Regulation of malonyl–CoA concentration and turnover in the normal heart. *J. Biol. Chem. 279*, 34298–34301.

162 King, K. L., Okere, I. C., Sharma, N., Dyck, J. R. B., Reszko, A. E., McElfresh, T. A., Kerner, J., Chandler, M. P., Lopaschuk, G., Stanley, W. C. (2005) Regulation of manonyl–CoA content and fatty acid oxidation during increased cardiac power. *Am. J. Physiol. 289*, H1033–1037.

163 Poirier, M., Vincent, G., Reszko, A. E., Bouchard, B., Kelleher, J. K., Brunengraber, H., Des Rosiers, C. (2002) Probing the link between citrate and malonyl-CoA in perfused rate heart. *Am. J. Physiol. 283*, 1379–1386.

164 Passoneau, J. V., Lowry, O. H. (1963) Phosphofructokinase and the control of the citric acid cycle. *Biochem. Biophys. Res. Comm. 13*, 372–379.

165 Kobayashi, K., Neely, J. R. (1979) Control of maximum rates of glycolysis in rat cardiac muscle. *Circ. Res. 4*, 166–175.

166 Marquis, N. R., Fritz, I. B. (1965) The distribution of carnitine, acetylcarnitine and carnitine acetyltransferase in rat tissues. *J. Biol. Chem. 240*, 2193–2196.

167 Fritz, I. B., Marquis, N. R. (1965) *Proc. Natl. Acad. Sci. US. 54*, 1226–1233.

168 Lysiak, W., Lilly, K., DiLisa, F., Toth, P., Bieber, L. L. (1988) Quantitation of the effect of L-carnitine on the levels of acid-soluble short-chain acyl-CoA and CoASH in rat heart and liver mitochondria. *J. Biol. Chem. 263*, 1151–1156.

169 Saddik, M., Gamble, J., Witters, L. A., Lopaschuk, G. D. (1993) Acetyl–CoA carboxylase regulation of fatty acid oxidation in heart. *J. Biol. Chem. 268*, 25836–25845.

170 DiLisa, F., Barbato, R., Menabo, R., Siliprandi, N. (1995) Carnitine and carnitine esters in mitochondrial metabolism and function. In *The carnitine system. Developments in Cardiovascular medicine*, pp 21–38, DeJong J. W., Ferrari R. (eds.), Kluwer Academic Publishers, Dordrecht/Boston/London.

171 Schonekess, B. O., Lopaschuk, G. D. (1995) The effects of carnitine on myocardial carbohydrate metabolism. In *The carnitine system. Developments in Cardiovascular medicine*, pp 39–52,

DeJong J. W., Ferrari R. (eds.), Kluwer Academic Publishers, Dordrecht/Boston/London.

172 Cuthbert, K. D., Dyck, J. R. (2005) Malonyl–CoA decarboxylase is a major regulator of myocardial fatty acid oxidation. *Curr. Hyertens. Rep. 7*, 407–411.

173 Hardie, D. G., Sakamoto, K. (2006) AMPK: a key sensor of fuel and energy status in skeletal muscle. *Physiology 21*, 48–60.

174 Winder, W. W., Hardie, D. G. (1996) Inactivation of acetyl-CoA carboxylase and activation of AMP-activated protein kinase in muscle during exercise. *Am. J. Physiol. 270*, E299–304.

175 Winder, W. W., Wilson, H. A., Hardie, D. G., Rasmussen, B. B., Hutber, C. A., Call, G. B., Clayton, R. D., Conley, L. M., Yoon, S., Zhou, B. (1997) Phosphorylation of rat muscle acetyl-CoA carboxylase by AMP-activated protein kinase and protein kinase A. *J. Appl. Physiol. 82*, 219–25.

176 Dyck, J. R. B., Lopaschuk, G. D. (2002) Malonyl CoA control fatty acid oxidation in the ischemic heart. *J. Mol. Cell. Cardiol. 34*, 1099–1109.

177 Dyck, J., Lopaschuk, G. D. (2006) AMPK alterations in cardiac physiology and pathology: enemy or ally? *J. Physiol. 574*, 95–112.

178 Korge, P., Honda, H. M., Weiss, J. N. (2003) Effects of fatty acids in isolated mitochondria: implications for ischemic injury and cardioprotection. *Am. J. Physiol. 285*, H259–H269.

179 Kushmerick, M. J., Meyer, R. A., Brown, T. R. (1992) Regulation of oxygen consumption in fast- and slow-twitch muscle. *Am. J. Physiol. 263*, C598–606.

180 Jorgensen, S. B., Richter, E. A., Wojtaszewski, J. F. P. (2006) Role of AMPK in skeletal muscle metabolic regulation and adaptation in relation to exercise. *J. Physiol. 574*, 17–31.

181 Weiss, J. N., Yang, L., Qu, Z. (2006) Network perspectives of cardiovascular metabolism. *J. Lipid Research 47*, 2355–2366.

182 Ravasz, E., Somera, A. L., Mongru, D. A., Oltvai, Z. N., Barabasi, A. L. (2002) Hierarchical organization of modularity in metabolic networks. *Science 297*, 1551–1555.

183 Barabasi, A. L., Oltvai, Z. N. (2004) Network biology: understanding the cell's functional organization. *Nature Rev. Genet. 24*, 101–113.

184 Ingwall, J., Weiss, R. (2004) Is the failing heart energy starved? *Circ. Res. 95*, 135–145.

185 Neubauer, S. (2007) The failing heart – an engine out of fuel. *New Engl. J. Med. 356*, 1140–1151.

12
Principles of Mathematical Modeling and *in Silico* Studies of Integrated Cellular Energetics

Marko Vendelin, Valdur Saks, and Jüri Engelbrecht

Abstract

Mathematical modeling of biological processes means describing mathematically the physiological phenomena and functional behavior of living cells, organs, tissues, neuronal networks, and many other biological systems. Proper modeling of energy metabolism includes quantitative description of energy exchange between different structural elements (compartments), chemical reactions, and different time scales. Knowledge of enzyme kinetics, biochemistry, thermodynamics, structure of the medium, computational methods, etc., should be interwoven into a whole. Modeling is a tool not only for describing biological processes but also for performing *in silico* experiments. Structural and functional hierarchies are important, and in this context the concept of internal variables used in continuum mechanics has been extremely useful. In this chapter, our aim is to introduce these methods to newcomers to the discipline and to non-experienced but interested readers in the field of modeling of cellular bioenergetics. We also demonstrate how thermodynamic constraints can be applied to learn the details of interaction between proteins. Two systems will be considered on the opposite ends of energy transfer networks in cardiac muscle cells. First, mitochondrial creatine kinase (MtCK) and adenine nucleotide translocase (ANT) coupling will be analyzed and the free energy profile will be proposed. This coupled system is the last step in mitochondrial conversion of the metabolic energy of substrate into that of ATP and phosphocreatine in these cells. Second, the properties of actomyosin interaction will be studied on the basis of macroscopical measurements of force development and energy consumption. This is an example that allows us to link mechanics and biochemistry. General approaches and problems of the mathematical modeling of cellular energy metabolism are discussed.

12.1
Introduction

There are several ways of understanding phenomena in the physical world, including theories, measurements, and modeling. The latter includes mostly math-

Molecular System Bioenergetics: Energy for Life. Edited by Valdur Saks
Copyright © 2007 WILEY-VCH Verlag GmbH & Co. KGaA, Weinheim
ISBN: 978-3-527-31787-5

ematical interpretation, i.e., casting a real-world system, process, or phenomenon into a mathematical representation. This enhances our abilities to understand and predict, and possibly also control, the behavior of the system being modeled.

Mathematical modeling of biological processes means describing mathematically the physiological phenomena and structural behavior of living tissues, organs, cells, neuronal networks, and many other biological systems. It is not easy and in most cases is much more difficult than, say, everyday physical processes, for the following reasons.

- Biological systems need energy exchange with the surrounding environment and represent the systems far from the thermodynamic equilibrium.
- Biological systems involve many chemical reactions and transfer mechanisms where the constituents should be analyzed on the molecular level.
- The processes operate over different time scales, are spatially extended, and include many hierarchies.
- Adaptivity should be taken into account.
- In mathematical terms, the biological systems can often be described by different types of mathematical equations, which causes difficulties in solving them.
- In physical terms, one should account for nonlinearities (additivity is lost), dissipation, activity/excitability, spatiotemporal coupling, etc.

Contemporary understanding of the behavior of biological systems takes a large number of heterogeneous components into account, and the main aim is to make a whole from these parts. On that basis, an academic field called systems biology integrates different levels of systems and their functions in order to understand the behavior of a whole system [1].

Despite the complexity characterized above, mathematical modeling of biological systems as an important part of systems biology gains more and more importance and has progressed during the last decade, supported by contemporary computing facilities and computational methods [2–6]. Here we enlarge our earlier analysis [6] and focus on the central problems of the modeling of energy transfer in heart.

12.2
Mathematical Modeling

12.2.1
Basic Principles

Mathematical modeling as a tool for understanding reality should be based on clear rules on the one hand, but on the other hand should be flexible enough to grasp possible changes. In terms of systems biology, the basic principles could be formulated as follows:

- In addition to the physical/chemical/geometrical, etc., properties of biosystems being modeled, the space and time scales are important, reflecting the structure

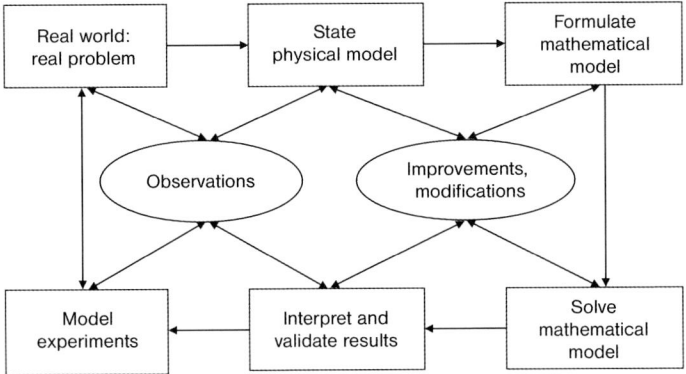

Fig. 12.1 A possible flowchart of modeling reflecting the validation and iterative improvement of mathematical models. (Reproduced from [6] with permission).

of the system and its behavior. The existence of scales leads to certain hierarchies of processes, substructures, and variables ordered by scales.

- Whatever the mathematical apparatus is, a model cannot reflect the behavior of the system over all the time scales and conditions but usually only its basic (backbone) or specific (under special conditions) features. Therefore, a model should always be validated by new experimental data and, if necessary, improved on the basis of validation.

- In systems biology, other fields of knowledge such as enzyme kinetics, biochemistry, thermodynamics, knowledge of the structure of intracellular medium, computational methods, etc., should be interwoven into a whole on the basis of biological functions and physiology.

A possible flowchart of modeling is shown in Fig. 12.1, also reflecting the validation and iterative improvement of mathematical models. In the final stage, modeling describes the systems or processes in the most economical ways using mathematical language.

12.2.2
Hierarchies

The complexity of biological systems is reflected by the existence of scales and, consequently, hierarchies. One should distinguish two possible types of hierarchies: (1) a structural hierarchy, involving strong dependence on length scales, and (2) a functional hierarchy, meaning that at various levels (of scale), various dynamical processes are of importance, all of which influence macrobehavior.

Structural hierarchies actually reflect the enormously rich architecture of biological tissues. The fundamental structural hierarchy is atom → molecule

$\rightarrow$ cell $\rightarrow$ tissue $\rightarrow$ organ $\rightarrow$ human [7], genes $\rightarrow$ mRNA $\rightarrow$ proteins $\rightarrow$ cells $\rightarrow$ tissue $\rightarrow$ organ $\rightarrow$ body [8], or sarcomeres $\rightarrow$ myofibrils $\rightarrow$ fibers $\rightarrow$ myocardium $\rightarrow$ heart. Clearly, the biomechanics should take structural hierarchy into account when the stresses and strains in tissues are calculated. In some sense, this can be compared with the mechanics of microstructured material [9].

Functional hierarchies reflect the complexity of functioning biosystems. In principle, one can distinguish between two types of functioning: a sequence in series or a sequence with parallel connections. The first type of functioning includes the usual behavior of the heart: oxygen consumption $\rightarrow$ energy transfer $\rightarrow$ Ca^{2+} signals $\rightarrow$ cross-bridge motion $\rightarrow$ contraction. The second type includes the cellular model of ischemia: electrical activity $\rightarrow$ creatine kinase system/sodium-potassium pump/intracellular calcium concentration $\rightarrow$ arrhythmic contracture [5].

Whatever the hierarchical model or its subsystem, the number of variables in biosystems is high because of the complexity of biological systems. Again, knowledge from other fields of science is useful. In the analysis of functional hierarchies, the comparatively novel concept of continuum mechanics – the concept of internal variables – can be used. The concept of internal variables has its origin in thermodynamics, the first ideas being in the description of reacting chemical systems. Contemporary understanding [10, 11] rests upon the assumption that the thermodynamic state is determined by two types of variables: observable and internal. In brief, it means distinguishing between macroscopic and microscopic behavior. Macroscopic behavior is governed by the observable variables like stress or potential. Microscopic behavior is hidden from the external observer and is governed by internal variables. In physiology, such variables are called hidden, auxiliary, or phenomenological.

The special features of biological systems, especially for functional hierarchy with sequences in series, require the concept of internal variables to be generalized. This leads to the concept of hierarchical internal variables [12, 13]. The idea is as follows: any observable variable depends directly on the first-level internal variable, which depends on the second-level internal variable, which depends on the third-level internal variable, etc. For interested readers, the theory of internal variables is described in the Appendix.

12.3
Modeling of Energy Metabolism

12.3.1
Enzyme Kinetics

Inside the cell, all metabolic processes are catalyzed by enzymes. Therefore, the principal basis for mathematical modeling of metabolism is enzyme kinetics [14–20]. The first kinetic equation for enzyme reaction was written by Victor Henri more than 100 years ago [21]. His basic assumption was that in the enzymatic reaction the first step is the association–dissociation equilibrium of an enzyme–substrate complex formation, followed by a monomolecular reaction of product formation from this complex:

$$E + S \overset{K_s}{\leftrightarrow} ES \overset{k_{cat}}{\longrightarrow} E + P$$

where K_s is the dissociation constant for ES, and k_{cat} is the rate constant according the mass action law. This gives the reaction rate equation in the following form, known as the Henri equation [15]:

$$v = \frac{V_m \dfrac{[S]}{K_s}}{1 + \dfrac{[S]}{K_s}} \tag{1}$$

Eleven years later, Leonor Michaelis and Maud Menten modified this equation to give its final famous form, known to all students of biochemistry and enzymology [22]:

$$v = \frac{V_m[S]}{K_m + [S]} \tag{2}$$

$$K_m = K_s; \quad V_m = k_{cat} \cdot [E]_t \tag{3}$$

K_m is the same as K_s, so written to pay respect to Michaelis and Menten for their contribution to enzyme kinetics. $[E]_t$ is the total concentration of enzyme. Its popularity is based on the simple fact that in double-reciprocal plots it gives a linear function of $1/v$ of $1/[S]$, which allows one to easily find two parameters of the reaction, V_m and K_m. Equation (2) is known as the Michaelis–Menten equation, but for the historical reasons discussed above, it is more correct and justified to call it the Henri–Michaelis–Menten equation [15].

While popular, this equation is not yet sufficient to be applied to metabolic studies. Indeed, this equation is valid only for an enzyme with one substrate and describes only the initial rate of this reaction [16, 17]. If the whole reaction mechanism of an enzyme with one substrate and one product is accounted for, the kinetic scheme of the reversible reaction is written as:

$$E + S \overset{K_s}{\leftrightarrow} ES \underset{k_{cat}^{-1}}{\overset{k_{cat}^{1}}{\rightleftarrows}} EP \overset{K_p}{\leftrightarrow} E + P$$

In this case, the net reaction rate is:

$$v_{net} = v^1 - v^{-1} = \frac{V_m^1 \dfrac{[S]}{K_s} - V_m^{-1} \dfrac{[P]}{K_p}}{1 + \dfrac{[S]}{K_s} + \dfrac{[P]}{K_p}} \tag{4}$$

where K_p is the dissociation constant for *EP*. In the absence of reaction product *P*, this equation automatically reduces to the Henri–Michaelis–Menten equation for the initial reaction rate. The same is true for the reverse reaction, when *S* is absent. All reactions proceed spontaneously in the direction of equilibrium, when

$$\frac{[P]_\infty}{[S]_\infty} = K_{eq} \tag{5}$$

$$K_{eq} = \frac{V_m^1}{V_m^{-1}} \frac{K_p}{K_m} = \exp\left(-\frac{\Delta G^0}{RT}\right) \tag{6}$$

This is the famous Haldane relationship [23], telling us that the kinetic constants are not independent of each other but are related to the standard free energy change of the reaction, on the basis of the van't Hoff equation (see Chapter 3).

The reactions described above were analyzed according to the proposition by Henri that the enzyme–substrate binding reactions are in equilibrium [15, 21]. However, it rapidly became clear that this is not always the case. In 1913 Max Bodenstein [24] proposed that non-equilibrium reactions can be analyzed when considered to be in steady state for some period of time, when only the rate constants and mass action law, but not the equilibrium reactions, are applicable. The steady-state (Bodenstein's) principle means that for some time interval the concentrations of all the reaction intermediates can be considered to be constant [14–17, 24–26]. This principle was applied for the first time to enzymatic reactions by Briggs and Haldane in 1925 [25]. In this case, the completely reversible enzymatic reaction with one substrate can be rewritten as:

$$E + S \overset{k_1}{\underset{k_{-1}}{\leftrightarrow}} ES \overset{k_{cat}^1}{\underset{k_{cat}^{-1}}{\rightleftarrows}} EP \overset{k_2}{\underset{k_{-2}}{\leftrightarrow}} E + P$$

Equations (2) and (4) in their general form are still applicable, but the meaning of the constants changes:

$$K_s = \frac{k_{-1} + k_{cat}^1}{k_1}; \quad K_p = \frac{k_{cat}^{-1} + k_2}{k_{-2}} \tag{7}$$

In the steady state, the reaction rate is described in molecular terms of rate constants, but not dissociation constants.

The explanation of the rate constants for chemical reactions is given by Eyring's transition state theory (TST) [26, 27]:

$$k = \frac{k_B T}{h} \exp\left(-\frac{\Delta G^\ddagger}{RT}\right) \tag{8}$$

In this expression, k_B and h are the Boltzmann and Planck constants, respectively, and $\Delta G^\ddagger$ is the free energy change needed for the formation of the transition state. The TST was developed from quantum mechanical calculation of the surfaces of potential energy for chemical reaction [27]. Pauling used it to explain the high efficiency of enzyme catalysis, supposing that,

> the only reasonable picture of catalytic activity of enzymes is that which involves an active region of the surface of the enzyme which is closely complementary in structure not to the substrate molecule itself, but rather to the substrate in a strained configuration, corresponding to the "activated complex": the substrate molecule is attracted to the enzyme and caused by the forces of attraction to assume the strained state which favors the chemical reaction – that is, the activation energy of the reaction is decreased by the enzyme to such an extent as to cause the reaction to proceed at an appreciably greater rate than it would in the absence of the enzyme [28].

The reaction always follows the path of minimal potential energy of the system during transition state formation, and this is true for the reactions in both directions. This later gave rise to the principle of thermodynamic reversibility, which says that the transition state is the same for both direct and reverse reactions, and in equilibrium the forward and reverse reactions pass the transition state with equal frequency [29]. Microscopic reversibility enables us to make statements about the forward reaction even if we can experimentally observe only the back reaction (and vice versa).

While all these theories form the firm basis of the kinetics of enzymatic catalysis, their application for cell metabolism is limited by the simple fact that about 60% of the metabolic reactions have not one but two or more substrates and products and most substrates are multi-substrate reactions [15–17]. In the 1960s Cleland systematically analyzed this complex situation and gave a simple and clear classification of these reactions [18–20]. It is most interesting that the Henri–Michaelis–Menten equations are still valid in these cases for the initial reaction rates, with the exception that the kinetics constants V_m and K_m are the apparent ones and are functions of the concentration of the second substrate [15–20]. In the final precise forms, the equations for the completely reversible reaction rates with substrates and products present are rather complicated, especially for steady-state conditions. One rather simple case is the creatine kinase reaction, whose role in cellular energy metabolism is described in many chapters of this book. The creatine kinase reaction

$$\text{MgATP} + \text{Cr} \overset{\text{CK}}{\rightleftharpoons} \text{MgADP} + \text{PCr}$$
$$\quad\text{(A)}\qquad\text{(B)}\quad\text{(E)}\quad\text{(C)}\qquad\text{(D)}$$

is of the Bi-Bi quasi-equilibrium random type, according to Cleland's classification [30, 31] and is described in Fig. 12.2. The constants K_{ia}, K_a, K_{ib}, K_b, K_{ic}, K_c, K_{id}, and K_d are the dissociation constants of the corresponding complexes. For the equilibrium state of binding and dissociation of substrates,

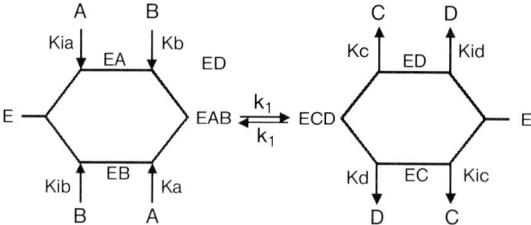

Fig. 12.2 Kinetic scheme for the creatine kinase reaction. For explanation, see the text.

$$K_{ia}K_b = K_{ib}K_a \tag{9}$$

$$K_{ic}K_d = K_{id}K_c \tag{10}$$

The rate of the reaction in the presence of substrate and products is given by the following equation (formation of dead-end complexes is not accounted for):

$$v = \frac{V_m^1 \dfrac{[A][B]}{K_{ia}K_b} - V^{-1} \dfrac{[C][D]}{K_{ic}K_d}}{1 + \dfrac{[A]}{K_{ia}} + \dfrac{[B]}{K_{ib}} + \dfrac{[A][B]}{K_{ia}K_b} + \dfrac{[C]}{K_{ic}} + \dfrac{[D]}{K_{id}} + \dfrac{[C][D]}{K_{ic}K_d}} \tag{11}$$

In a similar way, the rate equations can be written for all enzymatic reactions for which the reaction mechanism is known. They also can be written for the trans-membrane transporters, such as the members of the mitochondrial carrier family including the adenine nucleotide translocator (ANT), ion channels, etc.

12.3.2
Modeling of Intracellular Events

Mathematical modeling of cellular events began with the pioneering work of Alan Hodgkin and Andrew Huxley on quantitative descriptions of ion currents in squid giant axon and conduction of nerve impulses [32]. These models were developed for and applied to the heart by Denis Noble [33, 34]. Ion currents through cardiac cell sarcolemma are described in Chapter 13 of this volume. Modeling of cellular energy metabolism was initiated by Garfinkel, Hess, Higgins, and Selkov [35–37], and metabolic control analysis was created by Heinrich, Rapoport, Burns, and Kacser (reviewed in [14]). At present, numerous models for most cellular processes, including energy metabolism, can be found in the literature [38–53]. Mathematical models of energy metabolism usually contain several to several hundred rate equations for different reactions and processes involved in the free energy transition [48, 49, 52]. Computational methods described in the literature are used for solutions to these systems of equations with the aim of finding the quantitative relationships between input and output parameters. The art of creation of meaningful mathematical models always requires (1) knowledge of the real mechanisms of reactions and processes involved and (2) the use of reliable and firm experimental data as a basis for model building, determining the values of model parameters, and verifying of model correctness by comparison of model predictions with the new sets of experimental data. Without model verification by a firm set of experimental data, the model loses its value and may be easily converted into a type of computer game studying not the real world but a virtual one.

12.3.3
Thermodynamically Consistent Models

We have seen that modeling of metabolic systems is based on the quantitative description of enzymatic reactions. It is very useful to simultaneously analyze two aspects of the enzymatic reaction – its formal kinetics and the free energy profile. The simplest scheme for reversible reaction with one substrate S and one product P is shown in Fig. 12.3, when the substrate and product binding and dissociation are considered to be in equilibrium. Its rate is given in Eq. (4). Figure 12.3 also shows the free energy profile for this reaction – the transition state for the chemical reaction in the active center of the enzyme and changes in free energy levels during substrate and product binding and dissociation. According to the principle of microscopic reversibility, the back reaction will follow exactly the same path as the forward reaction [27]. The free energy profile shown in Fig. 12.3 is a useful basis for building a correct model of enzymatic reactions that respects both kinetic and thermodynamic laws, the so-called thermokinetic approach to building thermodynamically consistent models. For that, an interesting and useful way of

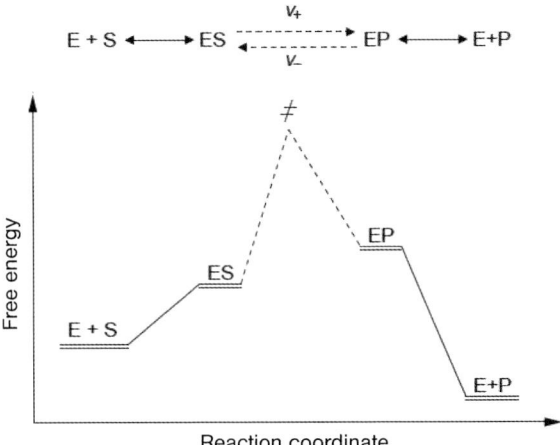

Fig. 12.3 Simple enzyme reaction (top) with the corresponding free energy profile (bottom). All reactions, except transition from ES to EP, are considered to be in equilibrium. The overall rate of the reaction is $v_+ - v_-$.

writing the rate equation is to use the parameters of free energy profile and the free energies of transition. In the simplest case, according to Eq. (8), the rates of monomolecular transition between two states, noted as A and B, are governed by the following equation,

$$v_+ = [A] \cdot \alpha \cdot e^{-(G^\ddagger - G_A)/RT}$$
$$v_- = [B] \cdot \alpha \cdot e^{-(G^\ddagger - G_B)/RT} \tag{12}$$

where v_+ and v_- are the rates of transitions between states A and B in forward and backward directions, respectively; $[A]$ and $[B]$ are normalized concentrations; α is a factor that depends on the nature of the transition and also includes $k_B T/h$; $G^\ddagger$ is free energy in the transition state; G_A and G_B are the free energies of states A and B, respectively; R is the gas constant; and T is the absolute temperature. For the enzymatic reaction shown in Fig. 12.3, the transition state theory may be used for finding the maximal reaction rates. Assuming the fast binding of substrate S and product P to the enzyme E, similar expressions can be used for these steps of reaction. Indeed, using the expression given by van't Hoff for $K_{eq} = \exp(-\Delta G^0/RT)$, we get the following equation:

$$[ES] = [E] \cdot [S] \exp(-(G_{ES} - G_E - G_S)/RT)$$
$$[EP] = [E] \cdot [P] \exp(-(G_{EP} - G_E - G_P)/RT) \tag{13}$$

where G_x is the free energy of x and $[X]$ denotes the concentration of X (metabolite or enzyme state). As one can see, we used the free energies instead of the dissociation constants in these equations. Using such formalism and taking into account that the total enzyme concentration is known ($[E_{tot}] = [E] + [ES] + [EP]$), we can find the concentration of free enzyme by using Eq. (13):

$$Den = \frac{[E_{tot}]}{[E]} = 1 + [S] \exp(-(G_{ES} - G_E - G_S)/RT)$$
$$+ [P] \exp(-(G_{EP} - G_E - G_P)/RT) \tag{14}$$

On the basis of the transition state theory, the reaction rates are

$$v_+ = \alpha \cdot [ES] \cdot \exp(-(G^\ddagger - G_{ES})/RT)$$
$$v_- = \alpha \cdot [EP] \cdot \exp(-(G^\ddagger - G_{EP})/RT)$$

Taking into account the expressions for $[ES]$, $[EP]$, and $[E]$ derived earlier, we get

$$v_+ = \alpha \frac{[E_{tot}]}{Den} [S] \exp(-(G_{ES} - G_E - G_S)/RT) \exp(-(G^\ddagger - G_{ES})/RT)$$

$$v_- = \alpha \cdot \frac{[E_{tot}]}{Den} [P] \exp(-(G_{EP} - G_E - G_P)/RT) \exp(-(G^\ddagger - G_{EP})/RT)$$

which can be simplified to:

$$v_+ = \alpha \cdot \frac{[E_{tot}]}{Den} [S] \exp(-(G^\ddagger - G_E - G_S)/RT)$$

$$v_- = \alpha \cdot \frac{[E_{tot}]}{Den} [P] \exp(-(G^\ddagger - G_E - G_P)/RT) \tag{15}$$

Is the form of the equations derived for the rate expressions any better than one with the binding constants? No, both formulations are equal. While it seems that there are more constants to be specified for a formulation that uses the free energy profile of the reaction, this first impression is incorrect. Namely, the absolute values of free energies are not important, but the free energy difference is. By assuming $G_E + G_S = 0$, it is easy to show that G_{ES}, G_{EP}, $G^\ddagger$, and $G_P - G_S$ determine the system completely. The same number of parameters is needed to describe the reaction rate using binding constants: dissociation constants for S and P, the rate constant in one direction, and the equilibrium constant of the reaction. The advantages of using the free energy profile to describe enzyme reactions become apparent when we analyze the enzymes that can reach the same state through different pathways. For example, creatine kinase binds MgATP and creatine (Cr) in random order, leading to the complex with MgATP and Cr (Fig. 12.2). When the enzyme kinetics using dissociation constants are described, as usually is done, one has to ensure the thermodynamic consistency by taking into account that one of the dissociation constants can be derived on the basis of the others and is not an independent parameter. Indeed, to ensure that the proportion of the enzyme in the bound state CK.ATP.Cr in equilibrium would not depend on the path, the product of the dissociation constants corresponding to the binding of ATP and Cr through the upper pathway in Fig. 12.2 should be the same as for the lower one (see above). While it is easy to detect such situations and to take the relationship between dissociation constants for simple enzyme reaction schemes, the situation may easily become nontrivial for heavily branching kinetic schemes. As soon as several enzymes are analyzed with the direct interaction between them, the application of the free energy profile becomes significantly easier and allows the models to remain thermodynamically consistent. In the following section, we will describe how we used the free energy profile formalism to analyze interactions between mitochondrial creatine kinase and adenine nucleotide translocase.

12.4
Interaction Between Enzymes

In muscle and brain cells, phosphocreatine and adenylate kinase shuttles provide a link between ATP-producing and ATP-consuming sites [54–58], as is described in detail in many chapters of this book. As a part of the phosphocreatine shuttle, the functional coupling between mitochondrial creatine kinase (MtCK) and adenine nucleotide translocase (ANT) has been identified by stimulating oxidative phosphorylation with creatine (Cr) [54] and has been examined further with kinetic and structural studies [54, 59–65]. This functional coupling between MtCK and ANT is described in detail in Chapter 3. Several possible approaches to the mathematical modeling of this coupling have been used and described in the literature, depending on the purpose of the modeling.

12.4.1
Phenomenological Modeling

When we are not interested in detailed mechanisms of functional coupling but the aim is the formal description of the behavior of the whole cellular system of energy transfer, a rather simple and easy method of phenomenological modeling can be used [6]. This means that to describe the functional coupling of an enzyme with other systems (transporters or other enzymes), the kinetic equation corresponding to the mechanism of the reaction catalyzed by this enzyme in the isolated state is also used to describe its behavior in the coupled state. In the latter case, however, the parameters of equations are found by fitting it with the experimental data obtained in the coupled state [6]. The more experimental data obtained in different conditions, the better. Thus, instead of the real kinetic con-

---

Fig. 12.4 Scheme of interaction between mitochondrial creatine kinase (MtCK) and adenine nucleotide translocase (ANT). The interaction between the proteins is considered the sum of two interaction modes: ATP and ADP are transferred through solution (A) or directly channeled between the proteins (B).
(A) In the first mode of MtCK–ANT interaction, after transport from the matrix to the intermembrane space by ANT (link between ANTx.ATP and ANTi.ATP in the scheme), ATP is released to solution. Next, the released ATP participates in the MtCK reaction according to the random Bi-Bi mechanism with fast equilibrium between all states of MtCK. ADP produced by the MtCK reaction is released into solution by MtCK and picked up by ANT for transport to the mitochondrial matrix (link between ANTi.ADP and ANTx.ATP in the scheme).
(B) In direct transfer mode, ATP is transferred from ANT to MtCK without leaving the two-protein complex to solution. Because MtCK has only one binding site for ATP and ADP, such transfer is possible only if this site is free, i.e., MtCK is either free (CK) or has only Cr or PCr bound (CK.Cr and CK.PCr). In the scheme, we grouped the states of MtCK according to whether ATP or ADP is bound to enzyme or not (white boxes in the scheme with three states of MtCK in each group). During direct transfer of ATP from ANT to MtCK, MtCK is transferred from states CK, CK.Cr, and CK.PCr to states CK.MgATP, CK.Cr.MgATP, and CK.PCr.MgATP, respectively. In the scheme, this transfer is shown as a link between ANTi.ATP and two corresponding groups of MtCK states. Next, after the MtCK reaction (link between states CK.Cr.MgATP and CK.PCr.MgADP in the scheme), ADP is transferred directly to ANT. Note that MtCK operates with Mg-bound ATP and ADP and that ANT requires Mg-free ATP and ADP forms. Thus, during direct transfer between MtCK and ANT, Mg is either bound or released, as is shown in the scheme. (The figure is reproduced from [66] with permission).

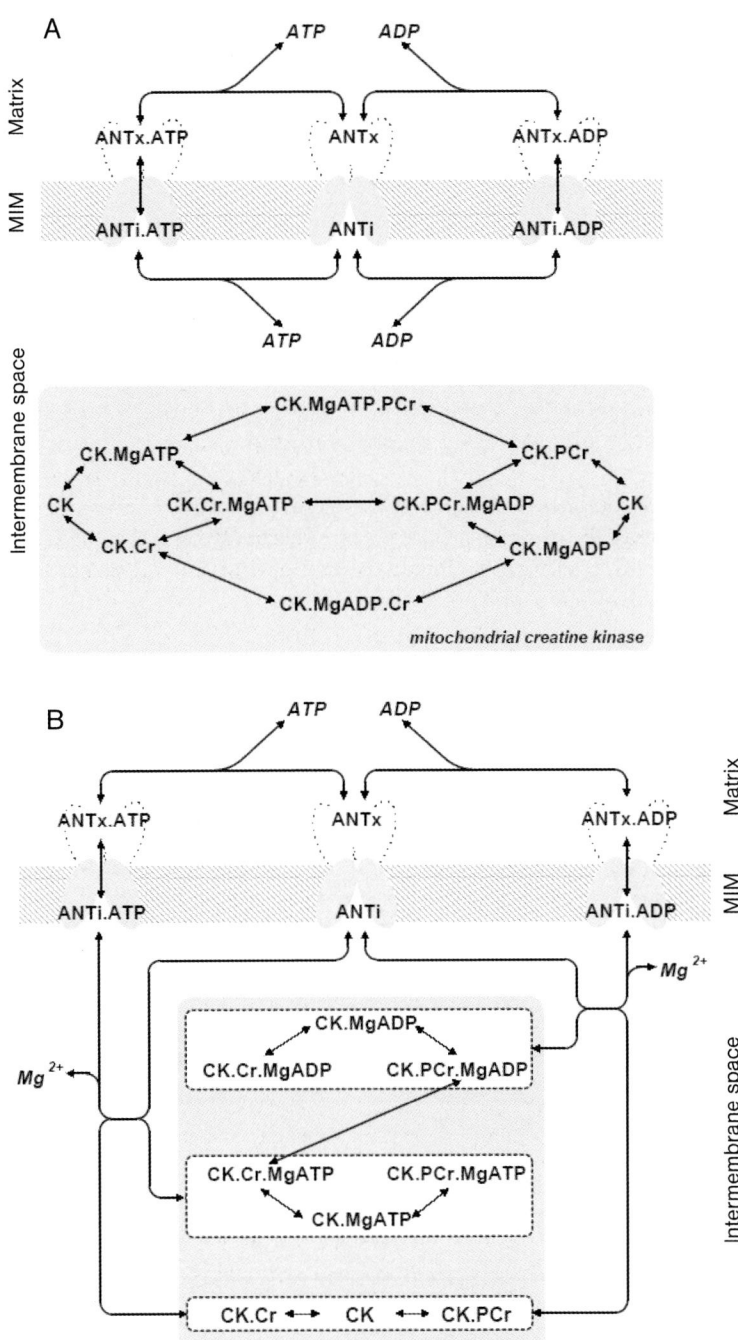

Fig. 12.4 (legend see p. 418)

stants of the enzyme, the phenomenological constants are used to describe the changes induced by the coupled reactions [52, 53]. We used this approach to describe the behavior of the compartmentalized energy transfer system and the regulation of mitochondrial respiration by metabolic signaling within a whole energetic unit, the ICEU [51–53]; the results of its use are shown in Chapter 11. By fitting the creatine kinase rate equation with multiple experimental data on its behavior in the coupled state with ANT, the phenomenological dissociation constants for adenine nucleotides were found and used in modeling of the whole system [53]. Another method of phenomenological modeling is to introduce into the equation of the individual enzyme the values of changed local concentrations of substrates that are assumed to depend on the neighboring systems. This also conforms to the model of existence of the microcompartments within the complexes [51]. In this way, the direct transfer mechanism between MtCK and ANT was first modeled by assuming high local ATP concentrations in the vicinity of the active center of MtCK created by ANT [51]. These phenomenological approaches simplify the calculations but at the same time are potential sources of errors, if the conditions of the reaction are too far from those used for finding the values of the phenomenological parameters. Therefore, the corresponding checks are required.

12.4.2
Modeling the Mechanism of Functional Coupling Between MtCK and ANT

More challenging is to explain the mechanism of effective interaction between MtCK and ANT in heart mitochondria. Here, the thermokinetic approach developed by us may be particularly interesting and important. Two principal mechanisms of this coupling have been proposed: (1) dynamic compartmentation of ATP and ADP [62] and (2) the direct transfer of ATP and ADP between the proteins (see Fig. 12.4) [55, 56, 59, 60]. According to the first mechanism, functional coupling between MtCK and ANT can be explained by differences between the concentrations of ATP and ADP in the intermembrane space and those in the surrounding solution due to some limitation of their diffusion across the outer mitochondrial membrane [62]. According to the second mechanism of coupling, ATP and ADP are directly transferred between MtCK and ANT without leaving the complex of proteins [55, 56, 59, 60]. To test the dynamic compartmentation mechanism, we composed the model based on it and tried to reproduce the measurements [65]. In this model, ANT and MtCK kinetics was considered to be the same as without any coupling between proteins, with one notable exception: we added a compartment shared by MtCK and ANT that was isolated from the rest of the intermembrane space by two diffusion restrictions [65]. These diffusion restrictions were set for ADP and ATP [65] and were varied together with the maximal relative activity of ANT to determine whether any combination of model parameters would be able to describe the kinetic data on functional coupling between MtCK and ANT described in detail in Chapter 3. The results of the simulations are shown in Fig. 12.5. As is clear from the figure, the combination

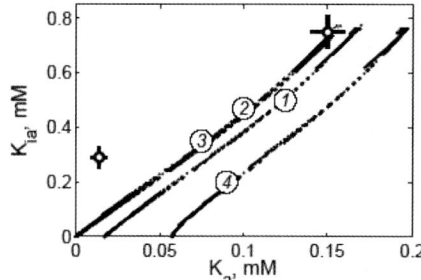

Fig. 12.5 Calculated apparent dissociation constants of ATP from both the ternary and binary complex with MtCK (K_a and K_{ia}, respectively) in the presence of oxidative phosphorylation. Coupling between MtCK and oxidative phosphorylation was modeled according to the dynamic compartmentation hypothesis. Here, apparent dissociation constants K_a and K_{ia} (represented by small dots in the figure) were computed in the case of different combinations of values of ATPase activity in the solution and exchange constants characterizing diffusion of ATP and ADP between the vicinity of MtCK–ANT and surrounding solution. In the figure, the measured values are shown by open circles in the upper right corner (no oxidative phosphorylation) and in the lower left corner (with oxidative phosphorylation). When using the default kinetic constants of ANT transport, all combinations of computed K_a and K_{ia} are aligned along the line with index 1 (indexes are shown inside circles in the figure). By increasing the maximal activity of ANT by 10 or 100 times, this line can be shifted to the left (lines with indexes 2 and 3, respectively). When the apparent dissociation constant of ATP is increased instead, the line shifts to the right (line with index 4). Note that regardless of the used values of ANT kinetic constants, all computed combinations of K_a and K_{ia} were considerably adrift from the measured values of these constants in the presence of oxidative phosphorylation. Experimental data are described in Chapter 3. (The figure is reproduced from [65] with permission).

of dissociation constants measured in the presence of oxidative phosphorylation, i.e., when the functional coupling between MtCK and ANT was active, cannot be reproduced by a dynamic compartmentation mechanism. Thus, this mechanism alone is not yet sufficient to explain the functional coupling between MtCK and ANT.

The second proposed mechanism – the direct channeling of ATP and ADP between MtCK and ANT – has been analyzed mathematically by Aliev and Saks [63, 64] by using a probability approach. In these works, the model was produced on the basis of coupled kinetic schemes and by determining the analytical form of probability of each reaction. The resulting system took into account the transfer of ATP and ADP between MtCK and ANT, leading to a rather complex formulation, even when the reverse reaction was not accounted for [64].

To gain better insight into the mechanism of this functional coupling and to describe the changes induced by oxidative phosphorylation in the kinetics of the MtCK reaction, we composed the thermokinetic model of the MtCK–ANT interaction by using a free energy profile that corresponded to the proposed kinetic scheme of the interaction (Fig. 12.4). As a first approximation, we assumed that the free energies of the coupled MtCK–ANT system in each of the states were equal to the sum of the free energies of isolated MtCK and ANT in the corresponding state. For example, the free energy G describing the state of the coupled system with ATP bound to ANT that is oriented towards the intermembrane space and PCr bound to MtCK (state of the complex $Ni.T - CK.PCr$) is equal to

$$G_{Ni.T-CK.PCr} = G_{Ni.T} + G_{CK.PCr} \tag{16}$$

Here, ANT oriented towards the intermembrane space is denoted by Ni, ATP by T, and MtCK by CK. In addition, the free energy of transition for reactions that involve the states of ANT only is independent of the current state of MtCK. Thus, the binding of ATP by ANT in matrix is independent of whether MtCK has any of the substrates bound or not. The same principle holds for reactions involving MtCK only: the ANT state does not influence these reactions. To simulate the influence of oxidative phosphorylation on the MtCK reaction, we had to introduce a direct transfer of ATP and ADP between ANT and MtCK. This introduced additional parameters into the model: free energies of several new states of the MtCK–ANT complex and free energies of transitions for several reactions. By fitting the model solution to the measurements, these free energies were found.

In the case of MtCK–ANT interaction, the free energy profile found by us is shown in Fig. 12.6. From a comparison of coupled (boxes with solid border) and uncoupled (boxes with dashed border) free energy profiles shown in Fig. 12.6, the following changes were introduced into the free energy profile to fit the experimental data. First, the free energy of ANT with bound ATP directed towards the intermembrane space (first column in Fig. 12.6) is considerably increased for the coupled system. Thus, while transferring ATP to MtCK, ANT is doing so in an energetically advantageous mode (note the drop in MtCK–ANT free energy when moving from the states in the first column to the states in the middle column of Fig. 12.6). Second, the free energy of complex with ADP attached to ANT on the side of the intermembrane space is somewhat smaller than in the uncoupled case. Because of both changes in the free energy profile, the synthesis of PCr from ATP transferred directly from ANT to MtCK is energetically advantageous. According to our analysis, the change in the free energy of ATP binding to ANT is required to reproduce all of the analyzed experiments, and the change in the free energy of ADP binding to ANT just improves the quality of the fit [65]. There is an important technical point that we would like to stress: while the kinetic scheme looks complicated, the reaction rates can be found automatically. For that, one has just to write a set of programs that take into account the free energies of the states between which the reaction occurs and the free energy of transition for that particular reaction. As a result, a thermodynamically consistent set of equations can be found. An additional advantage is that it is easy to modify the schemes and compare model solutions obtained for different possible interaction modes. Finally, this approach is scalable, i.e., it can be used for very complex interactions between different sets of enzymes and proteins.

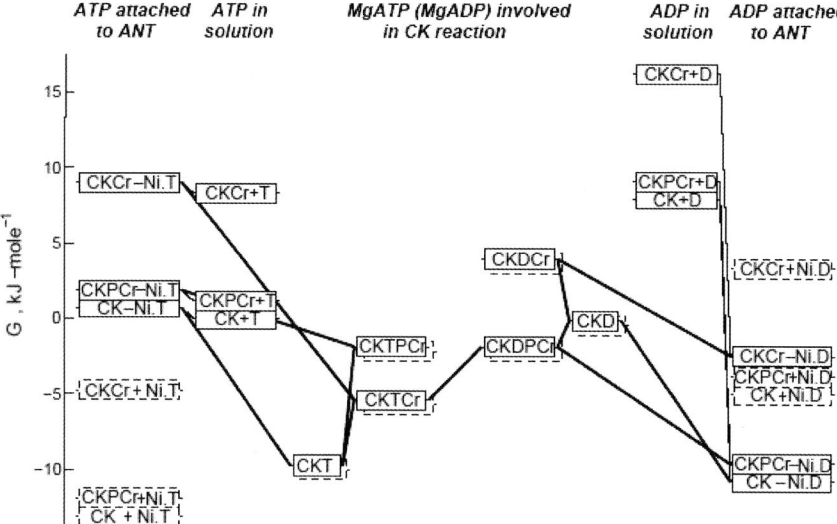

Fig. 12.6 Partial free energy profile of the MtCK reaction coupled to ANT. The free energies of the coupled system (boxes with solid border) are compared with the free energies of uncoupled MtCK and ANT (boxes with dashed border). In the first column, three states of two-protein complexes are shown, with ATP (T in the scheme) attached to ANT directed to the intermembrane space (Ni). The total free energy of the ANT–MtCK complex depends on the state of coupled MtCK, as is shown in the first column. ATP, attached to ANT, is either released to solution (second column) or transferred from ANT to MtCK (third column). After the MtCK reaction, ADP (D in the scheme) is transferred to ANT (last column) and then either released to solution (fourth column) or transported to the mitochondrial matrix (not shown). In the scheme, all reactions that are in the pathway leading to synthesis of PCr after the transfer of ATP from ANT to MtCK are shown by thick lines. Note that there are several simplifications made to keep the profile as simple as possible. First, the free energy changes indicated in the profile are induced by differences in the free energies of the complex states as well as by changes in solution due to binding and release of the substrates and magnesium. However, in the scheme release and binding of ATP and ADP are indicated only. Second, possible binding of ADP and ATP by ANT during the MtCK reaction is not indicated in this profile. (The figure is reproduced from [66] with permission).

12.5
Linking Mechanics and Free Energy Profile

12.5.1
Terrell Hill Formalism

Another important area of application for the thermodynamic and thermokinetic approaches to the analysis of protein–protein interactions is the modeling of the muscular contraction cycle. When the mechanical force is developed by interac-

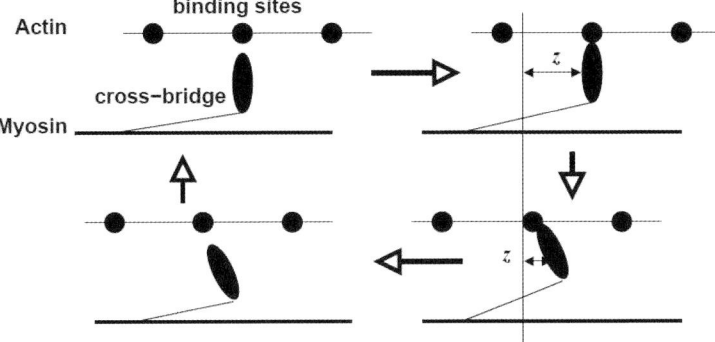

Fig. 12.7 Scheme of actomyosin interaction. The main sequence of events is indicated with arrows. First, the cross-bridge leaves the "weak binding" state (top left), attaches to actin, and forms a "strong binding" state (top right). Next, the cross-bridges in the attached state shift actin and make mechanical work, leading to a tilted configuration of the cross-bridge (bottom right). Then, the cross-bridge detaches from the binding site. The mechanical force produced by the cross-bridge is assumed to depend on the distance z between the equilibrium position of the cross-bridge and the closest binding site.

tion between molecules (such as actomyosin interaction), the free energy profile should reflect the mechanical work that can be performed by the molecules. Here, we will give a short description of Hill formalism applied to actomyosin interaction. Full description is given by Terrell L. Hill [66, 67].

The interaction of actin and myosin during a power stroke is illustrated in Fig. 12.7. The cross-bridge can be in so-called weak (left side of figure) or strong (right side of figure) binding states. The mechanical force produced by the cross-bridge is assumed to come from the elastic deformation. Thus, there is a configuration at which no force is produced even if the cross-bridge is attached to actin. The position of the binding site that corresponds to such a configuration will be called here an equilibrium position, and the distance between the actual position of the binding site and the equilibrium between is denoted by z (Fig. 12.7). Assuming that z fully determines a mechanical force F_i of the cross-bridge in state i, the possible mechanical work W_i has to be taken into account when finding the free energy for a cross-bridge attached at z_0:

$$G_i(z_0) = G_i(0) + W_i \tag{17}$$

With the work equal to

$$W_i = \int_0^{z_0} F_i(z)\, dz \tag{18}$$

we get the following relations between the free energy and the mechanical force:

$$G_i(z_0) = G_i(0) + \int_0^{z_0} F_i(z)\,dz \qquad (19)$$

$$F_i(z) = \frac{\partial G_i}{\partial z} \qquad (20)$$

As is clear from these equations, as soon as the dependency of mechanical force F_i on z and the free energy at $z = 0$ are given, the free energy dependency on z is fully determined. These relationships between the free energy of the state and the mechanical force of the state have to be taken into account when models of acto-myosin interaction are composed. As a result, a thermodynamically sound model can be produced to analyze the development of the force and the kinetics of the reactions. This approach is not the only one possible, and alternative ways to take into account the link between the developed force and the free energy profile are discussed in Refs. [68–70]. The link between mechanics and free energy profile is important not only for motor proteins: ATP synthase is another enzyme that uses the mechanical energy of torsion to synthesize ATP. As a result, the free energy profile of the synthesis depends on elastic energy stored in the protein [71, 72].

12.5.2
Actomyosin Interaction

As an example, we will give an overview of our model of the mechanoenergetics of cardiac muscle. This model was based on T. L. Hill formalism and was com-posed to analyze the relation between energy (ATP) consumption and developed mechanical force by actomyosin [73]. Experimentally, it has been shown that oxygen consumption of the ventricle is linearly related to a specific area in the pressure–volume diagram [74]. Similar results were obtained at the tissue level by Hisano and Cooper [75], who related oxygen consumption to the stress-strain area (SSA) (Fig. 12.8). The experimental data relating energy consumption and

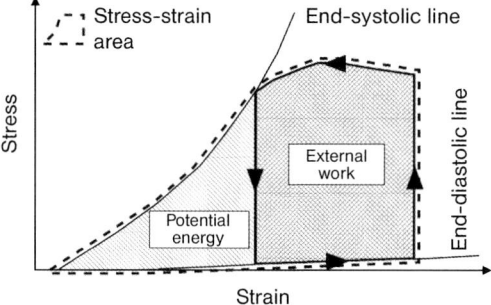

Fig. 12.8 Schematic representation of the stress–strain area (SSA). The SSA is a specific area in the stress–strain (SS) diagram surrounded by the end-systolic SS line, the end-diastolic SS line, and the systolic seg-ment of the SS trajectory during ejection. According to the definition, the SSA is the sum of external work and so-called "potential energy."

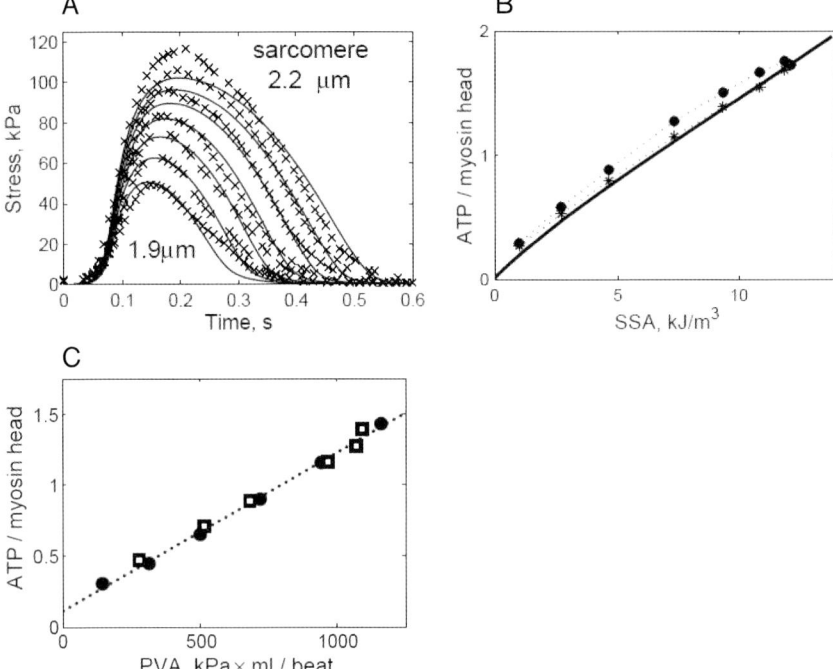

Fig. 12.9 Simulation results obtained by a Huxley-type model of cardiac muscle (A, B) and a finite element model of the left ventricle (C). (A) Isometric stress is shown during a twitch at sarcomere lengths of 1.9 µM, 1.95 µM, 2.0 µM, 2.05 µM, 2.1 µM, 2.15 µM, and 2.2 µM. The simulation results are compared with measurements (crosses) by Janssen and Hunter [79]. (B) Relation between the total amount of consumed ATP molecules per myosin head during a cardiac cycle and SSA for isometric (bold line) and two types of shortening contractions (thin dashed lines with dots and stars). (C) Computed ATP consumption by the left ventricle is related to the pressure–volume area for isovolumetric (solid dots) and ejecting (open squares) contractions. (Adapted from [78] with permission).

mechanical work are also reviewed in Chapter 11. Regardless of the experimental data, the computations performed using earlier Huxley-type models have predicted a nonlinear relationship between oxygen (or ATP) consumption and the SSA. In isometric contraction, the calculated ATP consumption to SSA ratio was found to decrease as a function of sarcomere length [76, 77]. In isotonic contraction, the computed SSA to ATP consumption ratio decreased with an increase in the applied afterload [77].

To reproduce the linear oxygen consumption–SSA relationship, we used a new approach. Instead of computing ATP consumption using prescribed cross-bridge cycling rates, as in [76, 77], we treated the linear relationship between ATP consumption and the SSA as a fundamental property of the muscle and determined the cross-bridge cycling rate constants and the activation parameters with which this macroscopical property of the muscle could be reproduced. When this approach was applied, the following properties of the cardiac muscle were reproduced [73]: (1) the relationship between ATP consumption and the SSA is linear, with contractile efficiency close to the measured one; (2) the computed isometric active stress during a beat replicates the measured stress in isosarcometric contractions at different sarcomere length values [79]; (3) the contraction duration is smaller in the isotonic case compared with the isometric case, in agreement with isotonic contraction experiment results [80]; and (4) the end-systolic point in the stress–strain diagram in isotonic contraction lies close to the end-systolic line computed for the isometric case [75]. Some of the simulation results are shown in Fig. 12.9. It is important to note that the set of parameters found by optimization may be not unique, as discussed in [73]. For example, one can find cross-bridge rates with different shapes and still obtain good results. In our work, we tried to find the rates with as simple shapes as possible to fit the desired data.

We used the found description of muscle contraction to simulate contraction of the left ventricle [78]. The computed relationship between ATP consumption and the pressure–volume area was found to be linear (Fig. 12.9C), in correspondence with the relationship found by Suga [74]. This relationship describes the length-dependent activation of cardiac sarcomeres as a basis of the Frank–Starling mechanism and metabolic feedback regulation of respiration described in Chapter 11.

12.6
Concluding Remarks

The examples of thermokinetic analysis of protein–protein functional interactions given above concern only two important steps of free energy conversion and transfer in muscle cells: functional coupling between MtCK and actin-myosin interactions in the muscular contraction cycle. Similar analyses of the numerous other coupled systems are needed in the future for their inclusion into more complete models of energy metabolism [48]. These coupled systems are described in many chapters of this book. Obviously, we are only beginning to understand the complexity of cellular life [81] and thus the complexity of the task of modeling of the whole cellular metabolism [82–84]. It is of fundamental importance to account for the reaction kinetics in real intracellular environments with macromolecular crowding [85] and metabolic channeling [86]. Regarding the modeling of cellular energy metabolism, one of the most important tasks is to understand the nature and to describe quantitatively the heterogeneity and local restriction of the diffusion of adenine nucleotides in microdomains of the cell [53, 87]. For this, simulation of both stochastic and diffusive effects in small subcellular volumes will be necessary [88]. At the same time, it will be necessary to avoid creating incomprehensively complex models; to accomplish this, organizing the models into supply-and-demand blocks linked by metabolic products and cofactor cycles will be very useful [89, 90].

Appendix: Theory of Internal Variables

Based on continuum mechanics, there is a clear difference between observable and internal variables. Observable variables are governed by balance laws and, in terms of mechanics, have kinetic energy, i.e., possess inertia, while internal variables do not possess inertia and are governed by kinetic equations. The formalism advocated in [11] includes the Helmholtz free energy ψ (or Lagrangian) from which the governing equations for observable variables are derived and the dissipation potential, leading to constitutive equations for internal variables. Briefly, following [11] and [12], the free energy is

$$\psi = \psi(\chi, \alpha, \nabla\alpha), \tag{21}$$

where χ and α are the observable and internal variables, respectively. For the sake of simplicity, only one observable and one internal variable are taken into account. In principle, the number of internal variables is usually higher. The dependent variable, e.g., stress σ, is then calculated by

$$\sigma = \frac{\partial\psi}{\partial\chi} = \sigma(\chi, \alpha). \tag{22}$$

The dissipative potential is

$$\mathscr{D} = \mathscr{D}(\sigma, \chi, \alpha, \dot{\alpha}, \nabla\alpha), \tag{23}$$

where the dot denotes derivative with respect to time. The governing equation for α is then an evolution law:

$$\frac{\delta\psi}{\delta\alpha} + \frac{\partial\mathscr{D}}{\partial\dot{\alpha}} = 0, \tag{24}$$

or finally:

$$\dot{\alpha} = f(\chi, \alpha) + g(\chi, \alpha)\dot{\chi} \tag{25}$$

Here $\delta/\delta\alpha$ is the Euler Lagrange derivative [11] and $f(..)$, $g(..)$ are certain functions.

Hierarchical internal variables can be formalized in the following way [12]. The hierarchy of internal variables is $\alpha, \beta, \gamma, \ldots$. Then,

1. the constitutive equation for a dependent variable σ depends on the observable variable χ and the *first-level* internal variable α

$$\sigma = \sigma(\chi, \alpha) \tag{26}$$

2. the evolution law for α is

$$\dot{\alpha} = f_1(\chi, \alpha, \beta) \tag{27}$$

where β is the next, *second-level* internal variable influencing σ only through dynamics of α;

3. the evolution law for β is

$$\dot{\beta} = f_2(\chi, \alpha, \beta, \gamma) \tag{28}$$

where γ is the *third-level* internal variable;

4. the evolution law for γ is

$$\dot{\gamma} = f_3(\chi, \alpha, \beta, \gamma, \ldots) \tag{29}$$

etc.

The concept of internal variables is used in [10] for description of nerve pulse dynamics where the action potential is dependent on ion currents treated as internal variables. The concept of hierarchical internal variables is used in [10, 12, 91] for modeling cardiac contraction using the Huxley model. There the first-level internal variables are n_A and n_B, which are relative amounts of cross-bridges producing force; the second-level internal variable is A, the amount of all activated cross-bridges; and the third-level internal variable is the calcium signal $[Ca^{2+}]$. All internal variables are governed by their evolution laws (kinetic equations), and their solution starts from the third-level internal variable up to the first-level internal variable and then to calculating active stress in myocardium.

References

1 Kitano, H. (2001) *Foundations of Systems Biology.* MIT Press, Boston.

2 Gavaghan, D., A. Garny, P. K. Maini, P. Kohl. (2006) Mathematical models in biology. *Phil Trans. R. Soc. London* A 364:1099–1106.

3 Humphrey, J. D. (2000) Continuum biomechanics of soft biological tissues. *Proc. R. Soc. London* A 459:3–46.

4 Hunter, P. J. (2004) The IUPS physiome project: a framework for computational physiology. *Progr. Biophys. Mol. Biol* 85:551–569.

5 Kohl, P., D. Noble, R. L. Winslow, P. T. Hunter. (2000) Computational modelling of biological systems: tools and visions. *Phil Trans. R. Soc. London* A 352:578–610.

6 Saks V.A., Vendelin M., Aliev M.K., Kekelidze T., Engelbrecht J. (2006) Mechanisms and modeling of energy transfer between and among intracellular compartments. In: *Handbook of Neurochemistry and Molecular Neurobiology, 3 edition,* volume 5, *Brain Energetics,* Dienel G., Gibson G. (Eds.) Springer Science, New York-Boston, pp. 815–860.

7 Kolston, P. J. (2000) Finite-element modelling: a new tool for the biologist. *Proc. R. Soc. London* A 358:611–631.

8 Fernandez, J. W., H. Schmid, P. J. Hunter. (2006) A framework for soft tissue and musculo-sceletal modelling: clinical uses and future challenges.

In Mechanics of Biological Tissue, Proc. IUTAM Symp., G.A. Holzapfel, K.W. Ogden (Eds.), pp. 339–354, Springer, Berlin.

9 Engelbrecht, J., A. Berezovski, F. Pastrone, M. Braun. (2005) Waves in microstructured materials and dispersion. *Phil. Mag.* 85:33–35, 4127–4141.

10 Maugin, G. A., J. Engelbrecht. (1994) A thermodynamical viewpoint on nerve pulse dynamics. *J. Non. Equilib. Thermodyn.* 19:9–23.

11 Maugin, G. A., W. Muschik. (1994) Thermodynamics with internal variables. Part I: general concepts, part ii: Applications. *J. Non-Equilib. Thermodyn.* 19:217–249, 250–289.

12 Engelbrecht, J., M. Vendelin. 2000. Microstructure described by hierarchical internal variables. *Rend. Sem. Mat.* 58:83–91.

13 Engelbrecht, J., M. Vendelin, G. A. Maugin. (2000) Hierarchical internal variables reflecting microstructural properties: application to cardiac muscle contraction. *J. Non-Equilib. Thermodyn.* 25:119–130.

14 Heinrich R., S. Schuster (1996) *The regulation of cellular systems*. Chapman & Hall, New York.

15 Segel I.H. (1975) *Enzyme kinetics. Behaviour and analysis of rapid equilibrium and steady state enzyme systems.* Wiley Interscience publication. John Wiley & Sons, New York–London.

16 Cornish-Bowden A. (2004) *Fundamentals of enzyme kinetics*. Portland Press, London.

17 Cornish-Bowden A., Jamin M., Saks V. (2005) *Cinétique Enzymatique*. EDP Sciences, Les Ulis.

18 Cleland W.W. (1963) The kinetics of enzyme-catalyzed reactions with two or more substrates or products. III. Prediction of initial velocity and inhibition patterns by inspection. *Biochim Biophys Acta*. 67, 188–196.

19 Cleland W.W. (1963) The kinetics of enzyme-catalyzed reactions with two or more substrates or products. II. Inhibition: nomenclature and theory. *Biochim Biophys Acta* 67, 173–187.

20 Cleland W.W. (1963) The kinetics of enzyme-catalyzed reactions with two or more substrates or products. I. Nomenclature and rate equations. *Biochim Biophys Acta*. 67, 104–137.

21 Henri V. (1902) Théorie générale de l'action de quelques diastases. *Comp. Rend. Acad. Sci.* 135, 916–919.

22 Michaelis L., Menten M.L. (1913) Die Kinetik der Invertinwirkung, *Biochem. Z.* 49, 333–369.

23 Haldane J.B.S. (1930) *Enzymes*. Longmans, Green and Co., London.

24 Bodenstein M. (1913) Eine Theorie der photochemischen Reactionsgeschwindigkeiten. *Z. Phys. Chem.* 85, 329–397.

25 Briggs G.E., Haldane J.B.S. (1925) A note on the kinetics of enzyme action. *Biochemical Journal* 19, 338–339.

26 Gutfreund H. (1995) *Kinetics for life sciences. Receptors, transmitters and catalysts*. Cambridge University Press, Cambridge, UK.

27 Eyring H. (1935) The activated complex in chemical reactions. *J. Chem. Phys.* 3, 107–115.

28 Pauling L. (1946) Molecular architecture and biological reactions. Chemical and Engeneering News 24, 1375–1377.

29 Keizer, J. 1987. *Statistical Thermodynamics of Nonequilibrium Processes*. Springer-Verlag, New York.

30 Kenyon, G. L., Reed, G. H. (1983) Creatine kinase: structure-activity relationships. *Adv. Enzymol.* 54, 367–426.

31 McLeish, M. J., Kenyon, G. L. (2005) Relating structure to mechanism in creatine kinase. *Crit. Rev. Biochem. Mol. Biol.* 40, 1–20.

32 Hodgkin A.L., Huxley A.F. (1952) A quantitative description of membrane current and its application to conduction and excitation in nerve. *J. Physiol.* 117, 500–544.

33 Noble D. (1962) A modification of the Hodgkin – Huxley equations applicable to Purkinje fiber action and pace-maker potentials. *J. Physiol.* 160, 317–352.

34 Noble D. (2006) *The music of life. Biology beyond the genome*. Oxford University Press, UK.

35 Garfinkel D., Frenkel R.A., Garfinkel L. (1968) Simulation of the detailed regulation of glycolysis in a heart supernatant preparation. *Comput. Biomed. Res.* 2, 68–91.

36 Selkov E.E. (1968) Self – oscillations in glycolysis. *Eur. J. Biochem.* 4, 79–86.

37 Chance B., Garfinkel D., Higgins J., Hess B. (1960) Metabolic control mechanisms. V A solution for the equations representing interaction between glycolysis and respiration in ascites tumor cells. *J. Biol. Chem.* 235, 2426–2439.

38 Bohnensack R. (1981) Control of energy transformation in mitochondria. Analysis by a quantitative model. *Biochim. Biophys. Acta* 634, 203–218.

39 Korzeniewski, B., Zoladz, J. A. (2001) A model of oxidative phosphorylation in mammalian skeletal muscle. *Biophys. Chem. 92*, 17–34.

40 Korzeniewski, B. (2006) Oxygen consumption and metabolite concentrations during transitions between different work intensities in heart. *Am. J. Physiol. 291*, H799–H804.

41 Jafri M.S., Dudycha S.J., Brian O'Rourke (2001) Cardiac energy metabolism: models of cellular respiration. *Annu. Rev. Biomed. Eng.* 3, 57–81.

42 Magnus G., Keizer J. (1997) Minimal model of β – cell mitochondrial Ca^{2+} handling. *Am. J. Physiol.* 273, C717–C733.

43 Puglisi J.L., Wang F., Bers D.M. (2004) Modeling the isolated cardiac myocyte. *Progr. Biophys. Mol. Biol.* 85, 163–178.

44 Coutu P., Metzger J.M. (2005) Genetic manipulation of calcium-handling proteins. II Mathematical modeling studies. *Am. J. Physiol.* 288, H613–H631.

45 Saucerman J.J., Brunton L.L., Michailova A.P., McCulloch A.D. (2003) Modeling β – adrenergetic control of cardiac myocyte contractility in silico. *J. Biol. Chem.* 278, 47997–48003.

46 Schaff J., Fink C.C., Slepchenko B., Carson J.H., Loew L. (1997) A general computational framework for modeling cellular structure and function. *Biophys. J.* 73, 1135–1146.

47 Kushmerick M. (1998) Energy balance in muscle activity: simulations of ATPse coupled to oxidative phosphorylation and to creatine kinase. *Comp. Biochem. Physiol.* 120, 109–123.

48 Beard D. (2006) Modeling of oxygen transport and cellular energetics explains observations on *in vivo* cardiac energy metabolism. *PLoS Computational Biology.* 2, 1093–1106.

49 Cortassa S., Aon M.A., O'Rourke B., Jacques R., Tseng H.J., Marban E., Winslow R. (2006) A computational model integrating electrophysiology, contraction and mitochondrial bioenergetics in the ventricle myocyte. *Biophys. J.* 91, 1564–1589.

50 Cramplin E.J., Schnell S., McSharry P.E. (2004) Mathematical and computational techniques to deduce complex biochemical reaction mechanisms. *Progr. Biophys. Mol. Biol.* 86, 77–112.

51 Aliev MK, Saks VA. (1997) Compartmentalized energy transfer in cardiomyocytes: use of mathematical modeling for analysis of *in vivo* regulation of respiration. *Biophys. J.* 73, 428–445.

52 Vendelin M., Kongas O., Saks V. (2000). Regulation of mitochondrial respiration in heart cells analyzed by reaction-diffusion model of energy transfer. *Am. J. Physiol.* 278, C747–764.

53 Saks V., Kuznetsov A., Andrienko T., Usson Y., Appaix F., Guerrero K., Kaambre T., Sikk P., Lemba M., Vendelin M. (2003) Heterogeneity of ADP diffusion and regulation of respiration in cardiac cells. *Biophys. J.,* 84, 3436–3456.

54 Bessman, S. P., A. Fonyo (1966) The possible role of the mitochondrial bound creatine kinase in regulation of mitochondrial respiration. *Biochem. Biophys. Res. Commun.* 22:597–602.

55 Jacobus, W. E., A. L. Lehninger. (1973) Creatine kinase of rat heart mitochondria. coupling of creatine phosphorylation to electron transport. *J. Biol. Chem.* 248:4803–4810.

56 Saks, V. A., G. B. Chernousova, D. E. Gukovsky, V. N. Smirnov, E. I. Chazov (1975) Studies of energy transport in heart cells. Mitochondrial isoenzyme of creatine phosphokinase: kinetic properties and regulatory action of mg2+ ions. *Eur. J. Biochem.* 57:273–290.

57 Wallimann, T., M. Wyss, D. Brdiczka, K. Nicolay, H. M. Eppenberger (1992) Intracellular compartmentation, structure and function of creatine kinase isoenzymes in tissues with high and fluctuating energy demands: the 'phosphocreatine circuit' for cellular energy homeostasis. *Biochem. J.* 281: 21–40.

58 Bessman, S. P., P. J. Geiger (1981) Transport of energy in muscle: the phosphorylcreatine shuttle. *Science* 211:448–452.

59 Jacobus, W. E., V. A. Saks (1982) Creatine kinase of heart mitochondria: changes in its kinetic properties induced by coupling to oxidative phosphorylation. *Arch. Biochem. Biophys.* 219:167–178.

60 Barbour, R. L., J. Ribaudo, S. H. Chan (1984) Effect of creatine kinase activity on mitochondrial ADP/ATP transport. Evidence for a functional interaction. *J. Biol. Chem.* 259:8246–8251.

61 Gellerich, F., V. A. Saks (1982) Control of heart mitochondrial oxygen consumption by creatine kinase: the importance of enzyme localization. *Biochem. Biophys. Res. Commun.* 105:1473–1481.

62 Gellerich, F. N., M. Schlame, R. Bohnensack, W. Kunz (1987) Dynamic compartmentation of adenine nucleotides in the mitochondrial intermembrane space of rat-heart mitochondria. *Biochim. Biophys. Acta.* 890:117–126.

63 Aliev, M. K., V. A. Saks (1993) Quantitative analysis of the 'phosphocreatine shuttle': I. a probability approach to the description of phosphocreatine production in the coupled creatine kinase-atp/adp translocase-oxidative phosphorylation reactions I. *Biochim. Biophys. Acta.* 1143:291–300.

64 Aliev, M. K., V. A. Saks (1994) Mathematical modeling of intracellular transport processes and the creatine kinase systems: a probability approach. *Mol. Cell. Biochem.* 133–134:333–346.

65 Vendelin, M., M. Lemba, V. Saks (2004) Analysis of functional coupling: Mitochondrial creatine kinase and adenine nucleotide translocase. *Biophys. J.* 87:696–713.

66 Hill, T. L. 1974. Theoretical formalism for the sliding filament model of contraction of striated muscle. Part I. *Prog. Biophys. Mol. Biol.* 28:267–340.

67 Hill, T. L., E. Eisenberg, Y. D. Chen, R. J. Podolsky. 1975. Some self-consistent two-state sliding filament models of muscle contraction. *Biophys. J.* 15:335–372.

68 Baker, LaConte, Brust-Mascher, Thomas. 1999. Mechanochemical coupling in spin-labeled, active, isometric muscle. *Biophys. J.* 77:2657–2664.

69 Baker, J. E., D. D. Thomas. 2000. Thermodynamics and kinetics of a molecular motor ensemble. *Biophys. J.* 79:1731–1736.

70 Highsmith, S. 2000. Muscle cross-bridge chemistry and force. *Biophys. J.* 79:1686–1687.

71 Sun, S., D. Chandler, A. R. Dinner, G. Oster. 2003. Elastic energy storage in beta-sheets with application to F_1-ATPase. *Eur. Biophys. J.* 32:676–683.

72 Sun, S. X., H. Wang, G. Oster. 2004. Asymmetry in the F_1-ATPase and its implications for the rotational cycle. *Biophys. J.* 86:1373–1384.

73 Vendelin, M., P. H. M. Bovendeerd, T. Arts, J. Engelbrecht, D. H. van Campen. 2000. Cardiac mechanoenergetics replicated by cross-bridge model. *Ann. Biomed. Eng.* 28:629–640.

74 Suga, H. 1990. Ventricular energetics. *Physiol. Rev.* 70:247–277.

75 Hisano, R., G. Cooper. 1987. Correlation of force-length area with oxygen consumption in ferret papillary muscle. *Circ. Res.* 61:318–328.

76 Taylor, T. W., Y. Goto, K. Hata, T. Takasago, A. Saeki, T. Nishioka, H. Suga. 1993a. Comparison of the cardiac force-time integral with energetics using a cardiac muscle model. *J. Biomech.* 26:1217–1225.

77 Taylor, T. W., Y. Goto, H. Suga. 1993b. Variable cross-bridge cycling-atp coupling accounts for cardiac mechanoenergetics. *Am. J. Physiol.* 264:H994–1004.

78 Vendelin, M., P. H. Bovendeerd, J. Engelbrecht, T. Arts. 2002. Optimizing ventricular fibers: uniform strain or stress, but not ATP consumption, leads to high efficiency. *Am. J. Physiol.* 283:H1072–H1081.

79 Janssen, P. M., W. C. Hunter. 1995. Force, not sarcomere length, correlates with prolongation of isosarcometric contraction. *Am. J. Physiol.* 269:H676–H685.

80 Brutsaert, D. L., N. M. de Clerck, M. A. Goethals, P. R. Housmans. 1978. Relaxation of ventricular cardiac muscle. *J. Physiol.* 283:469–480.

81 Prigogine I., Strengers I. (1986) *La nouvelle alliance.* Collection Folio. Les Editions Gallimard, Paris.

82 Weiss J.N., Yang L., Qu Z. (2006) Systems biology approaches to metabolic and cardiovascular disorders: network perspectives of cardiovascular metabolism. *J. Lipid. Res.* 47, 2355–2366.

83 Saks, V., Dzeja, P., Schlattner, U., Vendelin, M., Terzic, A., Wallimann, T. (2006b) Cardiac system bioenergetics: metabolic basis of Frank-Starling law. *J. Physiol. 571 (2)*, 253–273.

84 Saks V., Favier R., Guzun R., Schlattner U., Wallimann T. (2006) Molecular System Bioenergetics: regulation of substrate supply in response to heart energy demands. *J. Physiol.* 577, 769–777.

85 Schell S., Turner T.E. (2004) Reaction kinetics in intracellular environments with macromolecular crowding: stimulation and rate laws. *Progr. Biophys. Mol. Biol.* 85, 235–260.

86 Degenring D., Rohl M., Uhrmacher A.M. (2004) Discret-event, multi-level simulation of metabolite channeling. *BioSystems* 75, 29–41.

87 Vendelin, M., Eimre, M., Seppet, E., Peet, N., Andrienko, T., Lemba, M., Engelbrecht, J., Seppet, E. K., Saks, V. A. (2004) Intracellular diffusion of adenosine phosphates is locally restricted in cardiac muscle *Mol. Cell. Biochem. 256/257*, 229–241.

88 Bhalla U.S. (2004) Signaling in small subcellular volumes. I. Stochastic and diffusion effects on individual pathways. *Biophys. J.* 87, 733–744.

89 Hofmeyr J.H., Cornish-Bowden, A. (2000) Regulating the cellular economy of supply and demand. *FEBS Letters* 476, 47–51.

90 Kholodenko B.N., Schuster S., Rohwer J.M., Cascante M., Westerhoff H. (1995) Composite control of cell function: metabolic pathways behaving as single control units. *FEBS Let.* 368, 1–4.

91 Engelbrecht, J., M. Vendelin. 2006. Mathematical modelling of cardiac mechanoenergetics. *In Mechanics of Biological Tissue, Proc. IUTAM Symp.*, editors G. Holzapfel, K. Ogden, 369–378, Springer-Verlag, Berlin.

13

Modeling Energetics of Ion Transport, Membrane Sensing and Systems Biology of the Heart

Satoshi Matsuoka, Hikari Jo, Masanori Kuzumoto, Ayako Takeuchi, Ryuta Saito, and Akinori Noma

Abstract

The heart is a continuously working pump, and the cardiac myocyte depends on efficient ATP production by mitochondrial oxidative phosphorylation. Most of the ATP produced is used for contraction and for primary active transport of ions, such as Na^+/K^+-ATPase and sarcoplasmic reticulum Ca^{2+}-ATPase. Moreover, the activities of some of channels are maintained by ATP, (e.g., the ATP-sensitive K^+ channel and the L-type Ca^{2+} channel). Feedback control by ADP and P_i, which are released by ATPases during excitation–contraction coupling, is the key mechanism controlling mitochondrial function. Additionally, Ca^{2+} regulates the mitochondrial NADH level. To quantitatively understand the ATP balance during cardiac excitation–contraction coupling, we have developed a computer model of the cardiac myocyte (Kyoto model) that implements membrane excitation and changes in intracellular ions, contraction, and the above ATP-related systems. The model reproduced experimental data of cardiac muscle energetics and demonstrated the key role of feedback control by ADP and P_i when the sarcomere length was changed under the condition of isometric contraction. It was also revealed that feedback control and Ca^{2+} activation of dehydrogenases synchronously regulate mitochondrial NADH levels and produce complicated changes in NADH when the heart rate frequency is altered.

13.1
Introduction

The heart is a continuously working pump, and its pumping function is maintained by converting chemical energy stored in substrates, such as fatty acids and glucose, to mechanical energy and work. ATP is a common intracellular reservoir of chemical energy. In heart, ATP is produced mainly by mitochondrial oxidative phosphorylation and is consumed by several ATPases, such as Na^+/K^+-

ATPase, sarcoplasmic reticulum Ca^{2+}-ATPase (SERCA), and myosin ATPase. ATP consumption of the heart changes dramatically depending on the workload of the heart from moment to moment. Therefore, cardiomyocytes need to synthesize ATP to meet the ATP demand. It has been proposed that the heart produces large amounts of ATP. The estimated amount of ATP produced in a day by human heart is more than its own weight ($\sim$300 g): $\sim$35 kg [1], $\sim$6 kg [2], and 3.5$\sim$5 kg [3]. Thus, the heart produces and uses ATP at an extremely high rate and the ATP balance is maintained by the extraordinarily dynamic processes.

Computer simulation of cardiac energetics based on the experimental data is a powerful tool to quantitatively understand the ATP balance of heart. Several computer models of cardiac excitation–contraction coupling have been developed [4–8]. However, only a few computer models deal with the cardiac excitation–contraction–ATP metabolism coupling [9–11].

In this chapter, we will introduce our way of modeling cardiac energetics using a comprehensive model of cardiac excitation–contraction coupling (Kyoto model) [7, 10, 12]. The ventricular cell model used contains sarcolemmal channels and transporters, the sarcoplasmic reticulum (Ca^{2+} pump and Ca^{2+} release channel), the cross-bridge cycle, and mitochondrial oxidative phosphorylation. Details of the model were described in our previous studies [7, 10, 12] and the source code of the Kyoto model is available from http://www.sim-bio.org/. The model used for this study is a modified version of Matsuoka et al. [10]. In the following sections, we will describe only ATP-related systems.

13.2
Modeling ATP-related Systems

In the cardiac myocyte, most of the ATP is used by myosin ATPase for contraction and by sarcolemmal Na^+/K^+-ATPase and sarcoplasmic reticulum Ca^{2+}-ATPase (SERCA) for transporting ions. According to experiments by Schramm et al. [13], using ventricular trabeculae isolated from guinea pig heart, contributions of myosin ATPase, sarcolemmal Na^+/K^+-ATPase, and SERCA to myocardial energy expenditure are 76%, 9%, and 15%, respectively. The sarcolemmal Ca^{2+} pump (PMCA) also expresses in the heart, but its ATP consumption is probably small, because the relative contribution of Ca^{2+} removal by PMCA compared with that by Na^+/Ca^{2+} exchange is only 6–25% [14, 15]. The primary means of ATP production in the heart is oxidative phosphorylation within the mitochondria ($>$98%), and only a small amount ($<$2%) of ATP comes from glycolysis [16]. The heart maintains a constant ATP level to preserve the viability of cardiomyocytes during the normal variations in cardiac workload [17]. Additionally, the activity of some channels, such as the L-type Ca^{2+} channel and the ATP-sensitive K^+ channel, is maintained by cytoplasmic ATP. In our computer simulation model, these major ATP-related processes are incorporated as shown in Fig. 13.1. Creatine kinase, phosphocreatine, and adenylate kinase act as phosphotransfer systems connecting mitochondria to ATPases [18]. The details of the phospho-

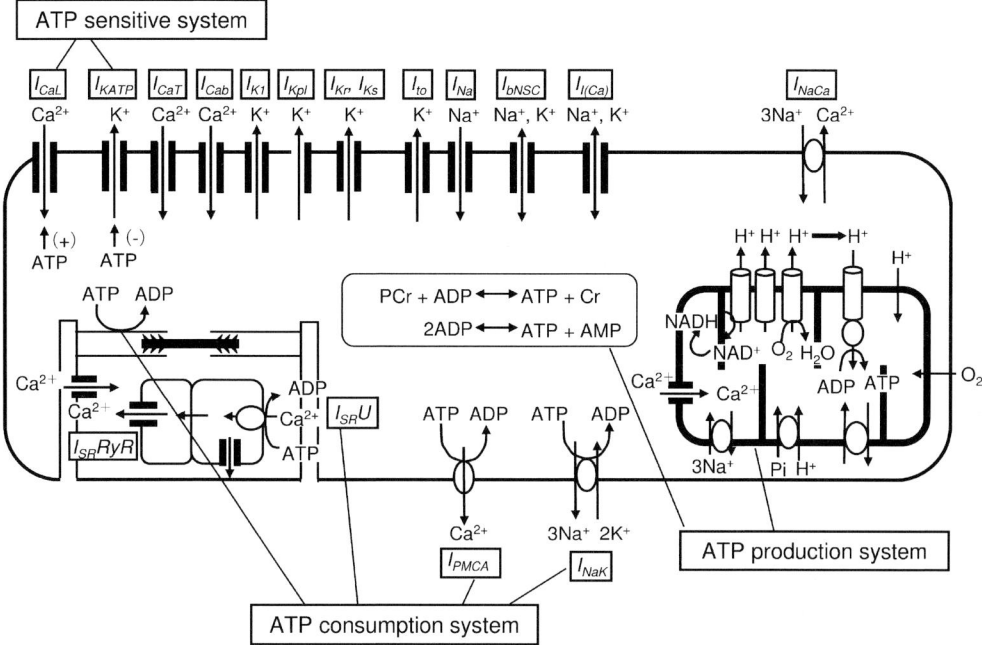

Fig. 13.1 Scheme of the comprehensive cardiac cell model (Kyoto model). ATP-sensitive, consumption, and production systems in the Kyoto model are indicated. I_{CaL}: L-type Ca^{2+} current (pA); I_{KATP}: ATP-sensitive K^+ current (pA); I_{CaT}: T-type Ca^{2+} current (pA); I_{Cab}: background Ca^{2+} current (pA); I_{K1}: inward rectifier K^+ current (pA); I_{Kpl}: voltage-dependent K^+ current (plateau current) (pA); I_{Kr}: delayed rectifier K^+ current, rapid component (pA); I_{Ks}: delayed rectifier K^+ current, slow component (pA); I_{to}: transient outward current (pA); I_{Na}: Na^+ current (pA); I_{bNSC}: background non-selective cation current (pA); $I_{l(Ca)}$: Ca^{2+}-activated background cation current (pA); I_{NaCa}: Na^+/Ca^{2+} exchange current (pA); I_{NaK}: Na^+/K^+ pump current (pA); I_{PMCA}: PMCA current (pA); $I_{SR}RyR$: Ca^{2+} release through ryanodine (RyR) channel in the SR (pA); $I_{SR}U$: Ca^{2+} uptake into the SR (pA).

transfer systems are not considered, and these reactions are expressed as simple chemical reactions in the present computer model [10]. However, these systems are not rate limiting in the majority of simulation protocols in this study, where relatively low workload is induced. It is certainly valuable to study how the phosphotransfer systems work in our computer model by incorporating the Vendelin model [19, 20]. The ATP consumption by other kinases is not considered because its amount is small [3].

13.2.1
Na$^+$/K$^+$-ATPase

Sarcolemmal Na$^+$/K$^+$-ATPase, also known as the sodium pump or the Na$^+$/K$^+$ pump, is the active transport system and is present in all animal cells [21, 22].

The Na^+/K^+-ATPase as well as SERCA and PMCA belong to the P-type ATPase family, which forms a phosphoprotein intermediate and undergoes conformational changes during the course of ATP hydrolysis. The Na^+/K^+-ATPase pumps out intracellular Na^+ in exchange for extracellular K^+ by using the energy of ATP hydrolysis, and it is primarily responsible for creating the gradients of Na^+ and K^+ across the sarcolemmal membrane. Therefore, the Na^+/K^+-ATPase contributes substantially to the maintenance of the membrane potential, the cell volume, and the activity of secondary active transport systems. The Na^+/K^+-ATPase is electrogenic and it generates outward membrane current during the Na^+-K^+ exchange. It is well recognized that 1 ATP is hydrolyzed for each exchange cycle of $3Na^+$ for $2K^+$. Thus, the stoichiometry is $3Na^+:2K^+:1ATP$.

The Na^+/K^+-ATPase has two conformational states: the E_1 and E_2 states. In the E_1 state, the cation-binding sites face the intracellular side of the sarcolemmal membrane, and Na^+ and ATP bind with high affinities. In the E_2 conformation, the cation-biding sites face the extracellular side of the membrane, and the affinity of the enzyme for Na^+ decreases and K^+ increases. Several models of the Na^+/K^+-ATPase with detailed kinetics have been proposed [23–25] based on the Post–Albers model. We have simplified the Post–Albers model of Na^+/K^+-ATPase (see Table 13.1).

ATP has two effects on the Na^+/K^+-ATPase: phosphorylation and stimulation of K^+ de-occlusion with high ($K_d(ATP) = 0.1{\sim}0.2$ μM) and low ($K_d(ATP) = 150{\sim}450$ μM) apparent affinities [21, 22]. Our Na^+/K^+-ATPase model, as well as the SERCA and PMCA models, includes only the lower ATP dependence. Figure 13.2A demonstrates the cytoplasmic ATP concentration ($[ATP]_i$) dependence of the Na^+/K^+ pump model. The ATP consumption rate ($ATPuse_{NaK}$, mM ms^{-1}) is calculated based on the stoichiometry as follows.

$$ATPuse_{NaK} = \frac{I_{NaK}}{F \cdot V_i} \tag{1}$$

where I_{NaK} is the membrane current generated by the Na^+/K^+-ATPase (pA), F is Faraday's constant (96.4867 C mmol^{-1}), T is absolute temperature (310 K), and V_i is cytosol volume ($100 \times 20 \times 8 \times 0.5 = 8000$ μm^3).

13.2.2
SERCA and PMCA

The sarcoplasmic reticulum Ca^{2+}-ATPase (SERCA) uses the chemical energy produced by ATP hydrolysis to transport Ca^{2+} across the sarcoplasmic reticulum (SR) membrane from cytosol to the SR against the concentration gradient of Ca^{2+}. Details about the Ca^{2+}-transporting mechanism and its molecular physiology are available in recent reviews [26, 27]. In the E_1 state, the Ca^{2+}-binding sites are of high affinity and face the cytoplasm, whereas in the E_2 state, the Ca^{2+}-binding sites are of low affinity and face the lumenal side. Cytoplasmic $2Ca^{2+}$ and 1ATP bind to the E_1 conformation. ATP is hydrolyzed to form ADP and the

Table 13.1 Equations for the ATP-related processes.

Na$^+$/K$^+$ pump

$$E_2\,Na \xrightleftharpoons{K_d Na_o} E_2 \xrightleftharpoons{K_d K_o} E_2\,K \qquad (1-\gamma)$$

$$k_1 \Big\uparrow\Big\downarrow k_2 \qquad\qquad k_3 \Big\uparrow\Big\downarrow k_4 \quad \beta\Big\uparrow\Big\downarrow \alpha$$

$$E_1\,Na \xrightleftharpoons{K_d Na_i} E_1 \xrightleftharpoons{K_d K_i} E_1\,K \qquad \gamma$$

$K_d Na_o = 69.8$, $K_d K_o = 0.258$, $K_d Na_i = 4.05$, $K_d K_i = 32.88 \;(mM)$

$k_1 = 0.37 \cdot (1/(1 + 0.094/[ATP]_i))$, $k_2 = 0.04$, $k_3 = 0.01$, $k_4 = 0.165 \;(ms^{-1})$

$p(E_1 Na) = 1/(1 + (KdN_{ai}/[Na]_i)^{1.06} \cdot (1 + ([K]_i/KdK_i)^{1.12}))$: probability of $E_1 Na$

$p(E_1 K) = 1/(1 + (KdK_i/[K]_i)^{1.12} \cdot (1 + ([Na]_i/KdN_{ai})^{1.06}))$: probability of $E_1 K$

$p(E_2 Na) = 1/(1 + (KdN_{ao}/[Na]_{eff})^{1.06} \cdot (1 + ([K]_o/KdK_o)^{1.12}))$: probability of $E_2 Na$

$p(E_2 K) = 1/(1 + (KdK_o/[K]_o)^{1.12} \cdot (1 + ([Na]_{eff}/KdN_{ao})^{1.06}))$: probability of $E_2 K$

$[Na]_{eff} = [Na]_o \cdot \exp(-0.82 \cdot F \cdot V_m/R/T)$: effective concentration of $[Na^+]_o$ (mM)

F: Faraday's constant ($96.4867 \;C \;mmol^{-1}$), Vm: membrane potential (mV), R: gas constant ($8.3143 \;C \;mV \;K^{-1} \;mmol^{-1}$), T: absolute temperature (310 K)

In the reduced two-state model

$I_{NaK} = 12.5 \cdot C_m \cdot (k_1 \cdot p(E_1 Na) \cdot \gamma - k_2 \cdot p(E_2 Na) \cdot (1-\gamma))$ (pA)

$\alpha = k_2 \cdot p(E_2 Na) + k_4 \cdot p(E_2 K)$, $\beta = k_1 \cdot p(E_1 Na) + k_3 \cdot p(E_1 K)$ (ms^{-1})

$dy/dt = \alpha \cdot (1-\gamma) - \beta \cdot \gamma$

SERCA

$$E_2\,Ca \xrightleftharpoons{K_d Ca_{SR}} E_2 \qquad 1-\gamma$$

$$k_1 \Big\uparrow\Big\downarrow k_2 \qquad k_3 \Big\uparrow\Big\downarrow k_4 \quad \beta\Big\uparrow\Big\downarrow \alpha$$

$$E_1\,Ca \xrightleftharpoons{K_d Ca_i} E_1 \qquad \gamma$$

$K_d Ca_{SR} = 3$, $K_d Ca_i = 0.00065 \;(mM)$

$k_1 = \dfrac{1}{1 + \dfrac{0.1}{[ATP]_i}}$, $K_2 = 0.01$, $k_3 = 0.01$, $k_4 = 1 \;(ms^{-1})$

$p(E_2 Ca) = 1/(1 + (K_d Ca_{SR}/[Ca^{2+}]_{up})^2)$: probability of $E_2 Ca$

$p(E_2) = 1 - p(E_1 Ca)$: probability of E_2

$p(E_1 Ca) = 1/(1 + (K_d Ca_i/[Ca^{2+}]_i)^2)$: probability of $E_1 Ca$

$p(E_1) = 1 - p(E_2 Ca)$: probability of E_1

In the reduced two-state model

$I_{SR}U = 3192.75 \cdot (k_1 \cdot p(E_1 Ca) \cdot \gamma - k_2 \cdot p(E_2 Ca) \cdot (1-\gamma))$ (pA)

$dy/dt = \alpha \cdot (1-\gamma) - \beta \cdot \gamma$

$\alpha = k_2 \cdot p(E_2 Ca) + k_4 \cdot p(E_2)$, $\beta = k_1 \cdot p(E_1 Ca) + k_3 \cdot p(E_1)$ (ms^{-1})

PMCA

$$E_2\,Ca \xrightleftharpoons{K_d Ca_o} E_2 \qquad 1-\gamma$$

$$k_1 \Big\uparrow\Big\downarrow k_2 \qquad k_3 \Big\uparrow\Big\downarrow k_4 \quad \beta\Big\uparrow\Big\downarrow \alpha$$

$$E_1\,Ca \xrightleftharpoons{K_d Ca_i} E_1 \qquad \gamma$$

Table 13.1 (continued)

$K_d Ca_o = 2$, $K_d Cai = ((180.0 - 6.4)/(1 + CaCaM/0.00005) + 6.4)/100\,000$ (mM)
$A_{PMCA} = 10.56 \cdot CaCaM/(CaCaM + 0.00005) + 1.2$
CaCaM: Ca^{2+} bound calmodulin

$$k_1 = \frac{1}{1 + \dfrac{0.1}{[ATP]_i}}, \; k_2 = 0.001, \; k_3 = 0.001, \; k_4 = 1.0 \; (ms^{-1})$$

$p(E_1 Ca) = 1/(1 + K_d Ca_i/[Ca^{2+}]_i)$: probability of $E_1 Ca$
$p(E_1) = 1 - p(E_1 Ca)$: probability of E_1
$p(E_2 Ca) = 1/(1 + K_d Ca_o/[Ca^{2+}]_o)$: probability of $E_2 Ca$
$p(E_2) = 1 - p(E_2 Ca)$: probability of E_2

In the reduced two-state model
$I_{PMCA} = A_{PMCA} \cdot (k_1 \cdot p(E_1 Ca) \cdot y - k_2 \cdot p(E_2 Ca) \cdot (1 - y))$ (pA)
$dy/dt = \alpha \cdot (1 - y) - \beta \cdot y$
$\alpha = k_2 \cdot p(E_2 Ca) + k_4 \cdot p(E_2)$, $\beta = k_1 \cdot p(E_1 Ca) + k_3 \cdot p(E_1)$ (ms^{-1})

Contraction

$\alpha_1 = 39 \; (mM \cdot m\,s^{-1})$, $\beta_1 = 0.03 \; (m\,s^{-1})$

$$\alpha_2 = 0.0039 \times \left(0.54 \times \frac{KdPI_i}{KdPI_i + PI_i} + 0.46\right) \; (m\,s^{-1})$$

$$\beta_2 = 0.0039 \cdot \frac{1}{1 + \left(\dfrac{KdATP_i}{ATP_i}\right)^3} \; (m\,s^{-1})$$

$\alpha_3 = 0.06 \; (m\,s^{-1})$, $\beta_3 = 1248 \; (mM \cdot m\,s^{-1})$

$$\alpha_4 = 0.12 \cdot \frac{1}{1 + \left(\dfrac{KdATP_i}{ATP_i}\right)^3} \; (m\,s^{-1}), \; \alpha_5 = 0.027 \cdot \frac{1}{1 + \left(\dfrac{KdATP_i}{ATP_i}\right)^3} \; (m\,s^{-1})$$

$KdPI_i = 1.83 \; mM$, $KdATP_i = 0.1 \; mM$
$dX/dt = -B \cdot (h - h_c)$, $B = 1.2 \; m\,s^{-1}$, $h_c = 0.005 \; \mu m$
$effectiveTCa = e^{-20 \cdot (hSML - 1.17)^2}$
$Q_1 = \alpha_1 \cdot Ca_i \cdot P(T) - \beta \cdot P(TCa)$
$Q_2 = \alpha_2 \cdot P(TCa) \cdot effectiveTCa - \beta_2 \cdot P(TCa^*)$
$Q_3 = \alpha_3 \cdot P(TCa^*) - \beta_3 \cdot Ca_i \cdot P(T^*)$
$Q_4 = \alpha_4 \cdot P(T^*) + \alpha_5 \cdot (dX/dt)^2 \cdot P(T^*)$
$Q_5 = \alpha_5 \cdot (dX/dt)^2 \cdot P(TCa^*)$

Table 13.1 (continued)

$dP(TCa) = Q_1 - Q_2$
$dP(TCa^*) = Q_2 - Q_3 - Q_5$
$dP(T^*) = Q_3 - Q_4$
$P(T) = 1 - P(TCa) - P(TCa^*) - P(T^*)$
$ForceCB$: cross − bridge force (mN/mm^2)
$ForceCB = 1800000 \cdot TroponinC \cdot ([TCa^*] + [T^*]) \cdot (hSML - X)$
$TroponinC = 0.07\ mM$

$ForceEcomp$: elastic component of force (mN/mm^2)
$ForceEcomp = 140000 \cdot (0.97 - hSML)^5 + 200 \cdot (0.97 - hSML)$

Mitochondrial Ca^{2+} uniporter

$$V_{Cauni} = \frac{Cm_{mito} \cdot V\max_{Cauni} \cdot VD_{Cauni} \cdot (MWC - Ca_{mito} \cdot e^{-VD_{Cauni}})}{(1 - e^{-VD_{Cauni}})}$$

$V\max_{Cauni} = 1.44 \cdot 10^{-7}\ (mM\ ms^{-1}\ pF^{-1})$, $Cm_{mito} = 1.812\ (pF)$

$K_{trans} = 0.006\ mM$, $K_{act} = 3.8 \cdot 10^{-4}\ mM$

$$MWC = \frac{\dfrac{Ca_i}{K_{trans}} \cdot \left(1 + \dfrac{Ca_i}{K_{trans}}\right)^3}{\left(1 + \dfrac{Ca_i}{K_{trans}}\right)^4 + \dfrac{50.0}{\left(1 + \dfrac{Ca_i}{K_{act}}\right)^{2.8}}}$$

$$VD_{Cauni} = -\frac{2F}{1000 \cdot R \cdot T} \cdot (\Delta\Psi + 91),$$

$\Delta\Psi$: membrane potential of inner membrane (mV)

Mitochondrial Na/Ca exchange

$$V_{mitoNCX} = Cm_{mito} \cdot V\max_{mitoNCX} \cdot \frac{e^{-(F/2R \cdot T) \cdot (\Delta\Psi + 91)}}{\left(1 + \dfrac{K_d Na}{Na_i}\right)^3 \cdot \left(1 + \dfrac{K_d Ca}{Ca_{mito}}\right)}$$

$V\max_{mitoNCX} = 9.68 \cdot 10^{-7}\ mM\ ms^{-1}$
$K_d Na = 9.4\ mM$, $K_d Ca = 3.75 \cdot 10^{-4}\ mM$

Mitochondrial Ca

$$\frac{dCa_{mito}total}{dt} = \frac{(V_{Cauni} - V_{mitoNCX})}{Rmc},\ Rmc = 0.23$$

Ca_m was calculated by solving an equation below

$$Ca_m total = Ca_m + \frac{B_{max}}{1 + \dfrac{K_d Ca}{Ca_m}}$$

$B_{max} = 0.05\ mM$, $K_d Ca = 0.001\ mM$

Mitochondrial substrate dehydrogenation

$$v_{DH} = 5 \times 0.0004679 \frac{1}{\left(1 + \dfrac{100}{NAD/NADH}\right)^{0.8}} \cdot \left(0.2 + \frac{0.6}{1 + \left(\dfrac{K_d Ca}{Ca_{mito}}\right)^3}\right)$$

$K_d Ca = 0.0002\ mM$

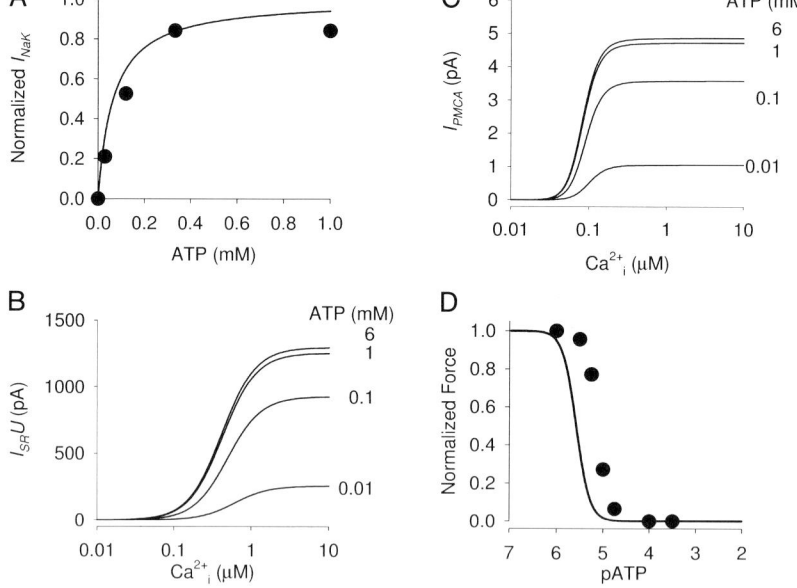

Fig. 13.2 ATP dependences. (A) [ATP]$_i$ dependence of Na$^+$/K$^+$-ATPase. Circles indicate data from Collins et al. [72]. (B) [Ca^{2+}]$_i$ dependence of SERCA at various levels of [ATP]$_i$. (C) [Ca^{2+}]$_i$ dependence of PMCA at various levels of [ATP]$_i$. (D) [ATP]$_i$ dependence of contraction (developed tension). Ca^{2+}$_i$ = 1 nM and P$_i$ = 0 mM. Data (circles) are from Mekhfi and Ventura-Clapier [73].

SERCA is phosphorylated. The conformational change from E$_1$ to E$_2$ causes the Ca^{2+}-binding site to face the lumen, followed by the release of Ca^{2+}. A proton is countertransported in exchange for Ca^{2+} during the reaction cycle. Therefore, the stoichiometry is 2Ca^{2+}:2H$^+$:1ATP.

Three genes encode SERCA1–3. SERCA2a is the cardiac/slow-twitch muscle isoform. In the heart, phospholamban inhibits Ca^{2+} transport of SERCA by decreasing affinity for Ca^{2+}, and this inhibition is attenuated by the phosphorylation of phospholamban by protein kinase A (PKA). The K_d(Ca) value in the submicromolar range was reported without PKA phosphorylation of phospholamban to be ∼0.7 μM [28] and ∼0.5 μM [29]. SERCA has high and low affinities for ATP. It is assumed that the high affinity reflects the binding of ATP to the catalytic site of the enzyme and that the lower affinity reflects the binding of ATP to a regulatory site that leads to an increase in the rate of transition between E$_2$ and E$_1$. In cardiac muscle preparations, the high and low K_d(ATP) is 4 μM and 192 μM, respectively [30].

SERCA has been modeled by a simple equation [4, 5, 8, 31] or a states model [7, 10, 32, 33]. In the Kyoto model, we used a four-state model, which was further reduced to the two-state mode by assuming an instantaneous Ca^{2+} binding/

unbinding (Table 13.1). Figure 13.2B demonstrates the dependence of the Ca^{2+} uptake rate ($I_{SR}U$, pA) on different $[ATP]_i$. With decreasing $[ATP]_i$, the Ca^{2+} uptake rate is attenuated. The ATP consumption rate ($ATPuse_{SERCA}$, mM ms^{-1}) is calculated based on the stoichiometry as follows:

$$ATPuse_{SERCA} = \frac{I_{SR}U}{2 \cdot F \cdot V_i} \tag{2}$$

The sarcolemmal Ca^{2+} pump (PMCA) is also a member of the P-type ATPase. PMCA1 and PMCA4 express in the heart [34]. The important feature of PMCA is its stimulation by calmodulin (CaM). CaM decreases the K_d(Ca) from 1800 nM to 64 nM and increases the V_{max} 8.8 times [35]. The K_d(CaM) is about 40~50 nM in the cloned PMCA1 and PMCA4 [36]. The apparent K_d(ATP) for Ca^{2+} pumping by cardiac PMCA is about 100 μM, ranging from 30 μM [37] to 120 μM [35]. It is reported that ATP interacts with the PMCA at two sites [38], one site having high affinity (K_d(ATP) = 1~2.5 μM) and the other having lower affinity (K_d(ATP) = ~100 μM). It has been demonstrated that the PMCA operates as an electrogenic Ca^{2+}/H^+ exchange [39] and that the net stoichiometry is $1Ca^{2+}:1H^+:1ATP$. The ATP consumption rate ($ATPuse_{PMCA}$, mM ms^{-1}) is calculated based on the stoichiometry as follows:

$$ATPuse_{PMCA} = \frac{I_{PMCA}}{F \cdot V_i} \tag{3}$$

where I_{PMCA} is the membrane current generated by PMCA (pA).

The PMCA and the sarcolemmal membrane Na^+/Ca^{2+} exchange are the Ca^{2+} extrusion mechanisms in the heart as well as in other type of cells. The PMCA has a high affinity for cytoplasmic Ca^{2+}, but Ca^{2+} transport capacity is low [38]. In the cardiac myocyte, the relative contribution of Ca^{2+} removal by PMCA compared with that by Na^+/Ca^{2+} exchange is small, although it depends on the species: 13~25% (rat), 6% (rabbit), 14% (ferret), and 8% (guinea pig) [14, 15].

The detailed kinetics of PMCA has been modeled previously [40–42]. We modified the four-state SERCA model to fit the experimental data. The dissociation constant for cytoplasmic Ca^{2+} and the amplitude factor (A_{PMCA}) were formulated to fit the data by Dixon and Haynes [35] (Table 13.1). Figure 13.2C demonstrates the dependence of PMCA on cytoplasmic Ca^{2+}. With decreasing $[ATP]_i$, Ca^{2+} extrusion by PMCA is attenuated.

13.2.3
Contraction

Several models of cardiac contraction have been proposed [43]. We have adopted the model proposed by Negroni and Lascano [44]. Their model is composed of inextensible thick and thin filaments in parallel with an elastic element. Attached cross-bridges act as force generators whose force is linearly related to the elongation of their elastic structure. The Ca^{2+} kinetics is described by a four-state system of sites on the thin filament associated with troponin C. This model reconstructs well many of the mechanical properties of cardiac muscle, while the pCa-tension relation is relatively shallow (Hill coefficient = ~1). We modified this model to incorporate ATP and inorganic phosphate (P_i) dependences [10]. Because ATP binding to a myosin head detaches the cross-bridge formation between myosin and actin, we assumed that all transition steps from cross-bridge-formed states ([T^*] and [TCa^*]) to cross-bridge-released states ([T] and [TCa]) are ATP-dependent. The ATP consumption rate (ATPuse$_{contraction}$, mM ms^{-1}) was calculated from the transition rate and its amplitude factor was determined by the cardiac oxygen consumption rate:

$$ATPuse_{contraction} = 14 \cdot TroponinC \cdot (\beta_2 \cdot P(TCa^*) + Q4 + Q5)$$

$$Q_4 = \{\alpha_4 \cdot + \alpha_5 \cdot (dX/dt)^2\} \cdot P(T^*) \tag{4}$$

$$Q_5 = \alpha_5 \cdot (dX/dt)^2 \cdot P(TCa^*)$$

where $P(TCa^*)$ is a fraction of the contraction unit with troponin C bound to Ca^{2+} and attached cross-bridge, $P(T^*)$ is a fraction of the contraction unit with attached cross-bridge and free troponin C, X is a length composed of half of the thick filament and the free portion of the thin filament, and α_4, α_5, and β_2 are rate constants. Figure 13.2D shows the ATP dependence of force used in the model. A set of equations for the calculation of contraction can be found in Table 13.1.

13.2.4
ATP-sensitive K$^+$ Channel and L-type Ca^{2+} Channel

The ATP-sensitive K$^+$ (K_{ATP}) channel was first found in cardiac myocyte by Noma [45]. It is now known that this channel presents in many tissues, including pancreatic β-cells, skeletal and smooth muscle cells, and neurons. Intracellular ATP closes this channel, and a decrease in ATP concentration is associated with the opening of the channel. Thus, the K_{ATP} channel functions as a sensor of bioenergetic state. The opening of the K_{ATP} channel results in the shortening of the action potential duration and cellular K$^+$ loss. The shortening of the action potential duration may attenuate Ca^{2+} influx through the L-type Ca^{2+} channel and cellular Ca^{2+} overload. A K_d(ATP) value of ~0.1 mM to inhibit the K_{ATP} channel has been reported in the inside-out patches, and the Hill coefficient varies between 1 and 5 [46].

Intracellular ATP is required to maintain the activity of the L-type Ca^{2+} channel [47]. The K_d(ATP) is ~1.4 mM. The mechanisms underlying ATP-induced activation have not yet been clarified, but cooperation between ATP and "cytoplasmic factor(s)" not yet been identified appears to be required to maintain the basal activity of the L-type Ca^{2+} channel [48].

The ATP-dependent factors (F_{ATP}) of these channels are expressed as a Hill equation, and the open probability of the channels is multiplied by the factor.

$$F_{ATP\_KATP} = \cfrac{1}{1 + \left(\cfrac{[ATP]_i}{0.1}\right)^2}$$

$$F_{ATP\_ICaL} = \cfrac{1}{1 + \left(\cfrac{1.4}{[ATP]_i}\right)^3}$$

(5)

13.2.5
Mitochondrial Oxidative Phosphorylation

In the heart, ATP is produced mainly by oxidative phosphorylation in mitochondria. The volume of cardiac mitochondria differs in various species, ranging between 22 and 37% [49], but smaller animals tend to have a larger mitochondrial fraction.

Korzeniewski's group has extensively elaborated on modeling oxidative phosphorylation in mammalian skeletal muscle. We have adopted their model [50] and incorporated it into the Kyoto model with minor modifications [10]. The volume of mitochondria was increased from 6.7 to 23%, and all the rate constants of reactions were increased by 5 times to maintain steady-state concentrations of cytoplasmic ATP and to fit the NADH value measured in the isolated guinea pig ventricular cells [51]. In this cardiac cell model, NADH production (citric acid cycle) is expressed by a simple numerical expression, and β-oxidation and shuttle systems are not included for simplicity. A H^+/ATP stoichiometry of 2.5 is assumed based on the Korzeniewski and Zoladz original model [50], although its value is still controversial [2]. The details of the model are described elsewhere [10, 50].

Mitochondrial Ca^{2+} plays an important role in regulating the activity of Ca^{2+}-sensitive mitochondrial dehydrogenases, i.e., pyruvate, isocitrate, and 2-oxoglutarate dehydrogenases [52]. Ca^{2+} enters mitochondria through the Ca^{2+} uniporter and is excluded through mitochondrial Na^+/Ca^{2+} exchange. We have adopted the models of these Ca^{2+} carriers developed by Fall and Keizer [53] and assumed an empirical Ca^{2+} buffer to fit the change in mitochondrial Ca^{2+} upon changing the beating frequency [51]. The parameters were adjusted so that Ca^{2+} in mitochondria at a 2.5-Hz beating rate was less than 1.0 μM (~0.8 μM) based on measurements by Griffiths et al. [54]. The activation of substrate dehydrogenation by Ca^{2+} was incorporated by assuming a K_d(Ca) of 0.2 μM [52]. Equations and parameters of the mitochondria model related to Ca^{2+} are listed in Table 13.1.

13.3
ATP Balance in the Kyoto Model

Figure 13.3 demonstrates membrane potential (Fig. 13.3A), cytoplasmic Ca^{2+} concentration (Fig. 13.3B), and force of contraction (Fig. 13.3C) during isometric

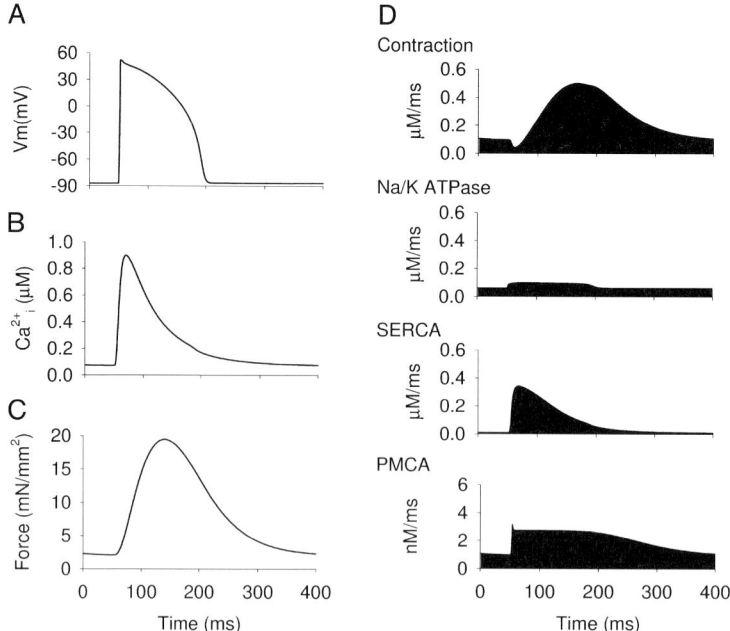

Fig. 13.3 ATP usage during isometric contraction. Membrane voltage
(A), cytoplasmic Ca^{2+} concentration (B), and force of cross-bridge
(C) during isometric contraction of the Kyoto model are plotted
(2.5 Hz electrical stimulation, hSML = 1.17 μm). (D) ATP usages by
contraction, Na$^+$/K$^+$-ATPase, SERCA, and PMCA.

contraction of the Kyoto model. The half-sarcomere length (hSML) is 1.17 μm,
which induces maximum developed tension in this model. These time courses
are in good agreement with experimental data of papillary muscle, where these
parameters were simultaneously measured [55]. Figure 13.3D demonstrates ATP
uses by contraction, Na$^+$/K$^+$-ATPase, SERCA, and PMCA. ATP use by contraction accounts for a major fraction of the total ATP use. ATP use by Na$^+$/K$^+$-
ATPase increases by 70% during the action potential, because of the activation of
Na$^+$/K$^+$-ATPase by membrane depolarization. ATP use by SERCA is parallel to
cytoplasmic Ca^{2+} concentration. PMCA uses only a minute amount of ATP because Ca^{2+} efflux through PMCA is only ~10% of that through Na$^+$/Ca^{2+} exchange in this model.

ATP consumption by Na$^+$/K$^+$-ATPase is largely determined by the amplitudes
of Na$^+$ influx through ion channels and the secondary Na$^+$-dependent transporters. In this model, the amplitude of the fast Na$^+$ current was accurately determined to give the maximum rate of rise of the action potential determined in
experiments. The Ca^{2+} flux through the L-type Ca^{2+} channel is largely balanced
with Ca^{2+} efflux through the Na$^+$/Ca^{2+} exchange (Na$^+$ influx). The amplitude
of the L-type Ca^{2+} current was adjusted according to action potential clamp
experiments [56], and that of the Na$^+$/Ca^{2+} exchange was also determined by

experimental data [57]. The other Na$^+$ flux is through the slow component of the delayed rectifier K$^+$ current (I_{Ks}) and the background non-selective cation current. These amplitudes are also determined by experimental data. The amplitude of SERCA as well as of the ryanodine channel of the SR is determined so as to reproduce a physiological Ca^{2+} transient ranging from 0.1 to 1 μM during the excitation–contraction coupling in the presence of the Ca^{2+} buffering system (50 μM calmodulin and 70 μM troponin C).

With a hSML of 1.17 μm, which produces maximum developed tension in this model, %ATP use by contraction, Na$^+$/K$^+$-ATPase, SERCA, and PMCA is 60%, 18.5%, 20.8%, and 0.5%, respectively. These values are close to the experimental estimation by Schramm et al. [13]: contraction, 76%; Na$^+$/K$^+$-ATPase, 9%; and SERCA, 15%. The maximum O$_2$ consumption of this model cell is 32.5 μM per beat (hSML = 1.17 μm). This value is in agreement with experimental data using guinea pig heart (∼30 μM per beat) [58] or guinea pig ventricular trabeculae (∼31 μM per beat) [13].

The total ATP use depends on the workload of the myocyte. In Fig. 13.4, the workload was changed by adding various external loads or by changing the half-

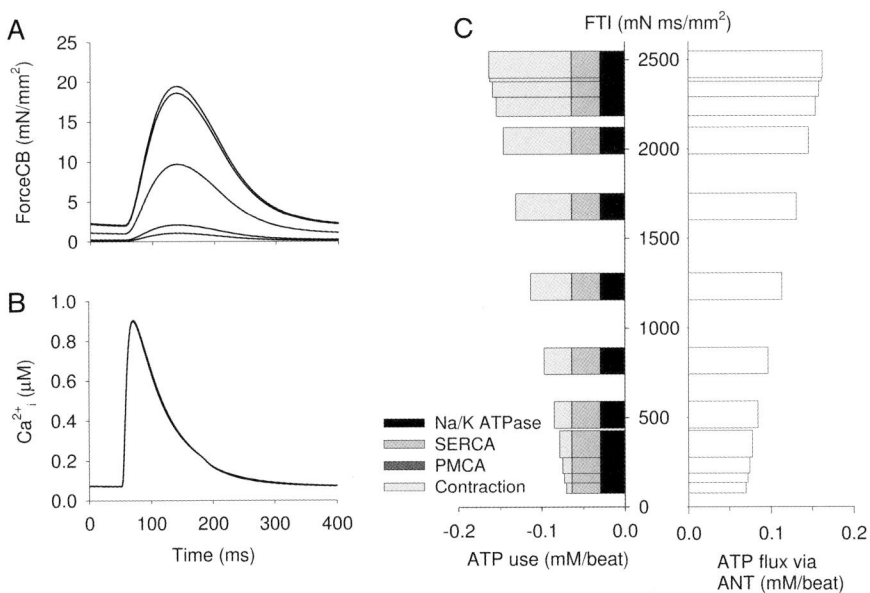

Fig. 13.4 Workload dependence. Force of contraction (A) and cytoplasmic Ca^{2+} (B) at various workloads are plotted. The workload was altered by changing the half-sarcomere length (hSML). Traces in (A) correspond to hSML = 1.17 μm, 1.12 μm, 0.97 μm, 0.82 μm, and 0.77 μm from top to bottom, respectively. All traces overlap in (B). (C) ATP usage and ATP supply from mitochondria. ATP usages by Na$^+$/K$^+$-ATPase, SERCA, PMCA, and contraction (from right to left) are plotted against the force–time integral (FTI) as an index of workload on the left side. FTI was calculated as the area under the twitch force but above the diastolic force. On the right side, ATP flux through the mitochondrial adenine nucleotide translocator (ANT) was plotted against FTI.

sarcomere length (hSML). Increasing the hSML augmented the force of the cross-bridge (Fig. 13.4A) due to the increase in the overlap between thick and thin filaments (see [44] for details). However, the changes in hSML under the isometric contraction did not cause a remarkable change in the cytoplasmic Ca^{2+} transient (Fig. 13.4B), consistent with experimental findings [59–61]. In the left part of Fig. 13.4C, ATP use by Na^+/K^+-ATPase, SERCA, PMCA, and contraction were plotted against the force–time integral (FTI) as an index of the workload. ATP use by contraction increases while that by Na^+/K^+-ATPase, SERCA, and PMCA remains unchanged. The augmentation of ATP use by contraction is effectively compensated by the increase in ATP flux through the mitochondrial adenine nucleotide translocator (ANT) (Fig. 13.4C, right side), which exchanges matrix ATP for cytoplasmic ADP, and by the increase in ATP synthesis by F_1F_0ATPase, as discussed in the following section. This automatic compensation maintains a nearly constant cytoplasmic ATP level and normal excitation–contraction coupling over a wide range of workloads.

13.4
Feedback Control and Ca^{2+}-dependent Regulation of Mitochondria Function

The cardiac myocyte automatically adjusts the rate of ATP production to match the ATP utilization under various physiological conditions. However, the mechanisms underlying this adjustment have not been fully clarified [17, 18]. Classically, Chance and Williams [62] asserted that cytoplasmic ADP and P_i take the key role in regulating the mitochondrial ATP production (feedback control theory). Namely, an increase in workload results in a rise in cytoplasmic ADP ($[ADP]_c$) and P_i ($[P_i]_c$) concentrations, and then the rise in these substrate concentrations within mitochondria ($[ADP]_m$ and $[P_i]_m$) augments ATP synthesis by F_1F_0-ATPase. However, recent ^{31}P NMR studies have demonstrated that ATP as well as creatine phosphate, P_i, and ADP remain constant when the cardiac workload is modified [17]. Ca^{2+} has drawn attention as being one of the key molecules coupling mitochondrial ATP production with muscle workload. When heart rate increases, the cytoplasmic Ca^{2+} transient accompanying the membrane excitation increases in both frequency and amplitude, and thereby mitochondrial Ca^{2+} concentration ($[Ca^{2+}]_m$) increases via the Ca^{2+} uniporter on the mitochondrial inner membrane. The increase in $[Ca^{2+}]_m$ then activates the Ca^{2+}-sensitive mitochondrial dehydrogenases, such as pyruvate, isocitrate, and 2-oxoglutarate dehydrogenases, as demonstrated by biochemical studies [52]. In fact, the heart rate-dependent increases in mitochondrial Ca^{2+} and NADH were experimentally demonstrated by Brandes and Bers [63] and Jo et al. [51].

In Fig. 13.5, we studied experiments by Jo et al. [51] with this model cell. The simulation was performed with (black lines) and without (gray lines) the Ca^{2+} activation of substrate dehydrogenation. The model cell was first electrically stimulated at 0.1 Hz (9000-ms interval). The percentage of NADH (Fig. 13.5A) at 0.1 Hz was 22%, which is within the experimental data [51]. An increase in stim-

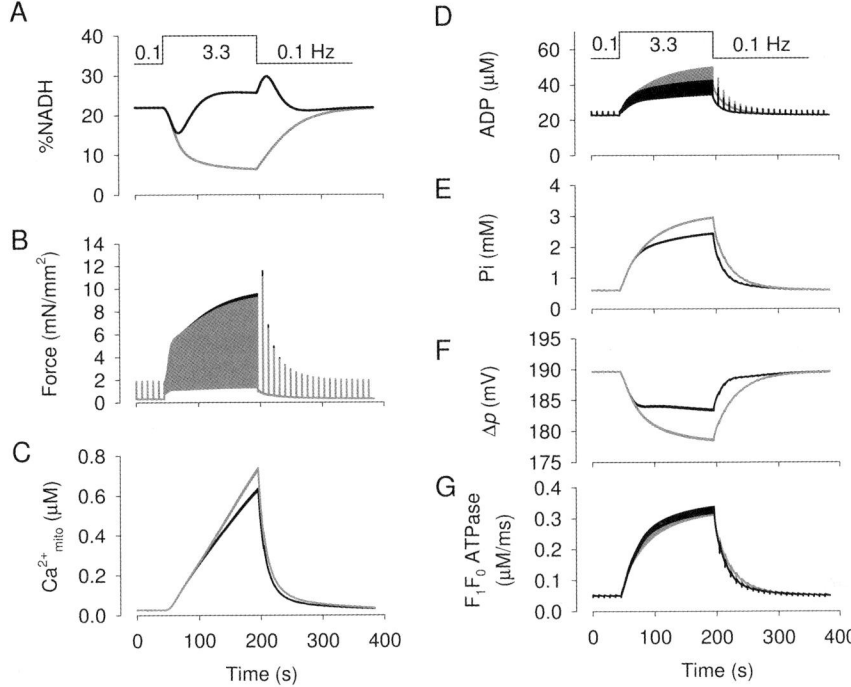

Fig. 13.5 Heart rate dependence. Percentage of NADH (A), force of contraction (B), mitochondrial Ca²⁺ (C), cytoplasmic ADP (D), cytoplasmic P_i (E), protonmotive force of mitochondria (F), and rate of F_1F_0-ATPase (G) are plotted. Black and gray traces are simulation results with and without Ca²⁺ activation of substrate dehydrogenation, respectively. The electrical stimulation rate was changed as indicated above the traces.

ulation frequency to 3.3 Hz (300-ms interval) gradually augmented the cytoplasmic Ca²⁺ transient, resulting in an increase in force of contraction (Fig. 13.5B; positive staircase phenomenon) followed by an increase in mitochondrial Ca²⁺ (Fig. 13.5C). The mitochondrial NADH exerted a biphasic change: initial decrease and late increase. Feedback control by $[ADP]_c$ and $[P_i]_c$ caused the initial NADH decrease. Namely, the sudden increase in heart rate frequency augmented the mean ATP consumption by approximately three times, followed by an increase in $[ADP]_c$ (Fig. 13.5D) and $[P_i]_c$ (Fig. 13.5E). $[ADP]_m$ and $[P_i]_m$ increased via the adenine nucleotide translocators (ANT) and the phosphate carrier. As a result, ATP synthesis was facilitated (Fig. 13.5G) and the rate of NADH consumption by respiratory chain was increased. However, the delayed increase in mitochondrial Ca²⁺ gradually activated the NADH production, overcoming the increased NADH consumption. Therefore, NADH increased after the initial drop. The feedback control without Ca²⁺ activation induced a monotonic decrease in NADH and more increases in $[ADP]_c$ and $[P_i]_c$ (Fig. 13.5A,D,E, gray lines). The increases in $[ADP]_c$ and $[P_i]_c$ augmented the rate of F_1F_0-ATPase (Fig. 13.5G), but depolarized the inner membrane potential and decreased the protonmotive force (Δp, Fig. 13.5F) and free energy changes of synthesizing ATP from ADP and P_i (ΔG_p) (not shown, see Jo et al. [51]). No remarkable difference was found in the rate of F_1F_0-ATPase (Fig. 13.5G) with or without Ca^{2+} activation. But, the decreases in both Δp and ΔG_p were attenuated by 43 and 34%, respectively, by the implementation of Ca^{2+} activation. If we assume that ΔG_p is an index of feedback control [51], the Ca^{2+} activation provides $\sim$34% of the driving force of the ATP synthesis and attenuates the contribution of the feedback control. This value is close to our previous estimation ($\sim$20%) [51]. Saks et al. [64] also proposed a minor contribution of Ca^{2+} activation. Upon returning to 0.1 Hz, NADH shows an overshoot before declining to the control level. This overshoot is also consistent with experimental data [51, 63]. The sudden decrease in ATP consumption by contraction and delayed deactivation of NADH production causes this NADH overshoot. $[ATP]_i$ does not change during high-rate beating under either simulation condition because of the increase in ATP synthesis and the action of creatine phosphate. The slight decrease in the force of contraction when the Ca^{2+}-dependent activation is omitted is due to the attenuation of contraction by P_i. The above simulations demonstrate that the feedback control and the activation of the mitochondrial dehydrogenases by Ca^{2+} are the key mechanisms explaining the frequency-dependent change in mitochondrial NADH.

The feedback control depends on the rate of ATP consumption by contraction. In other words, it depends on the workload. In Fig. 13.6, the frequency-dependent change in mitochondrial NADH was studied at the higher workload by increasing the half-sarcomere length (hSML). As the workload increases, mitochondrial NADH at 0.1 Hz decreases. The decrease is more remarkable during high-rate beating, resulting in deeper initial dip and lower NADH at 3.3 Hz. The late level of NADH at 3.3 Hz is similar to that at 0.1 Hz when the force–time integral (FTI) is at a maximum (hSML = 1.17 µm). Larger overshoot was observed upon returning to 0.1 Hz because of the larger decrease in the ATP consumption rate by contraction. In the previous experiments, a variety of patterns in NADH change were reported when the cardiac workload was changed [65–69]. The ap-

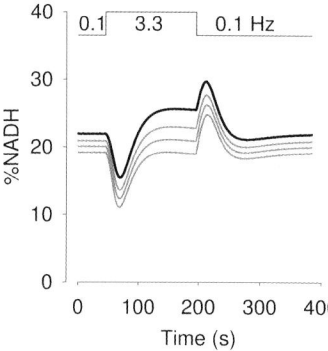

Fig. 13.6 Workload dependence of heart rate–dependent NADH change. Workload was altered by changing the hSML. hSML was 0.97 µm, 1.02 µm, 1.06 µm, and 1.17 µm from top to bottom, respectively. The electrical stimulation rate was changed as indicated above the traces.

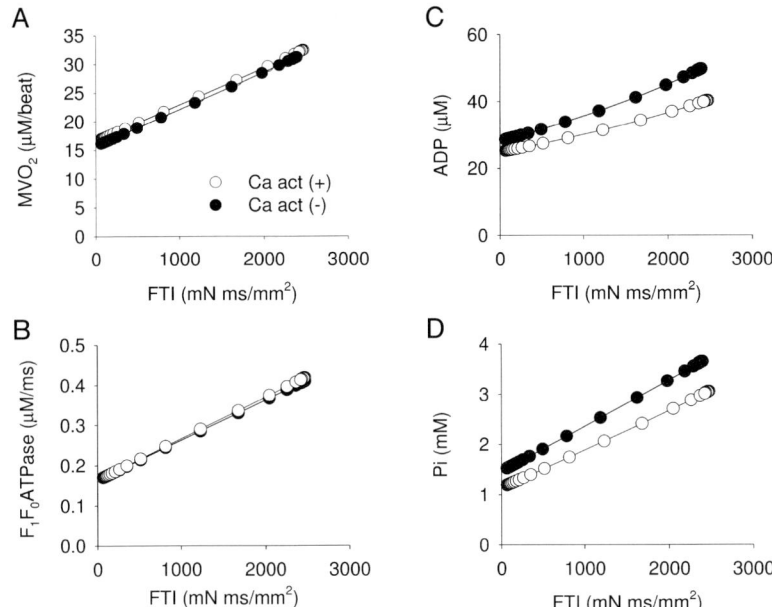

Fig. 13.7 The workload dependence of MVO_2 (A), F_1F_0-ATPase activity (B), cytoplasmic ADP (C), and cytoplasmic P_i (D) is plotted against the FTI. Workload was altered by changing the hSML, and FTI was used as an index of workload. Open and filled circles are simulation results with and without Ca^{2+} activation of substrate dehydrogenation, respectively.

parently inconsistent results are probably due, at least in part, to the different workload and mitochondrial Ca^{2+} levels.

In Fig. 13.7, we further studied the feedback control and the Ca^{2+} activation of substrate dehydrogenation in a wide range of workloads under the condition of isometric contraction. The model cell has a linear relationship between the FTI and the myocardial oxygen consumption (MVO_2) (Fig. 13.7A), which is consistent with experimental findings by Hisano and Cooper [70]. It is notable that intracellular and mitochondrial Ca^{2+} does not change remarkably when the workload is changed (see Fig. 13.4B); therefore, the degree of Ca^{2+} activation of substrate dehydrogenation is nearly the same at all FTI levels. The turnover rate of F_1F_0-ATPase (Fig. 13.7B) increases as the workload or FTI increases. This augmentation of ATP synthesis is due to the increase in $[ADP]_c$ (Fig. 13.7C) and $[P_i]_c$ (Fig. 13.7D), which rise two- to threefold (feedback control) [62]. These results are consistent with a recent study by Beard [71] indicating that the cytoplasmic inorganic phosphate level is a key regulator of the rate of mitochondrial respiration.

This model demonstrated that the relationships between FTI and MVO_2 (Fig. 13.7A) and between FTI and F_1F_0-ATPase (Fig. 13.7B) do not remarkably change with (open circles) or without (closed circles) Ca^{2+} activation. However, the implementation of the Ca^{2+} activation of substrate dehydrogenation remarkably attenuated the increases in $[ADP]_c$ and $[P_i]_c$ (Fig. 13.7C,D, open circles). The Ca^{2+} activation of substrate dehydrogenation may be important for preventing excessive rises in $[ADP]_c$ and $[P_i]_c$ at high workloads of the heart; otherwise, these metabolites attenuate the activities of ATPase, such as myosin ATPase.

In this study, we used the comprehensive computer model of cardiac excitation–contraction coupling (Kyoto model) to analyze the complex interactions among the ATP-related proteins within a single cardiomyocyte. The model is constructed based on a considerable amount of experimental data of our own and from others. It is increasingly being realized that whole systems of cardiac energetics cannot be elucidated by intuition alone because of the complex networks among ions, substrates, and proteins. Computer modeling is a powerful strategy for integrating experimental data and for quantitative understanding of the networks within the cardiomyocyte. This comprehensive mathematical modeling sheds light on complex biological phenomena and provides a new working hypothesis. Experimental validation of the hypothesis will in turn refine the model. Thus, the systems biology approach in the tight collaboration between computer simulation and wet experiment enables us to gain profound insights into the mechanisms underlying the biological phenomena.

Acknowledgments

This study was supported by a Grant-in-Aid for Scientific Research and the Leading Project for Biosimulation at the Ministry of Education, Culture, Sports, Science, and Technology, Japan.

References

1 Taegtmeyer, H. (1994) Energy metabolism of the heart: from basic concepts to clinical applications. *Curr. Probl. Cardiol. 19*, 59–113.

2 Ingwall, J.S. (2002) *ATP and the Heart*, Kluwer Academic Publishers, Netherlands..

3 Opie, L.H., Lopaschuk, G.D. (2004) Fuels: aerobic and anaerobic metabolism. In *Heart Physiology: From cell to circulation*. 4th edn. Opie L.H. (eds). Lippincott Williams & Wilkins, Philadelphia.

4 Luo, C.H., Rudy, Y. (1994) A dynamic model of the cardiac ventricular action potential. I. Simulations of ionic currents and concentration changes. *Circ. Res. 74*, 1071–96.

5 Noble, D., Varghese, A., Kohl, P., Noble, P. (1998) Improved guinea-pig ventricular cell model incorporating a diadic space, IKr and IKs, and length- and tension-dependent processes. *Can. J. Cardiol. 14*, 123–34.

6 Winslow, R.L., Rice, J., Jafri, S., Marban, E., O'Rourke, B. (1999) Mechanisms of altered excitation-contraction coupling in canine tachycardia-induced heart failure, II: model studies. *Circ. Res. 84*, 571–86.

7 Matsuoka, S., Sarai, N., Kuratomi, S., Ono, K., Noma, A. (2003) Role of individual ionic current systems in ventricular cells hypothesized by a model study. *Jpn. J. Physiol. 53*, 105–23.

8 Shannon, T.R., Wang, F., Puglisi, J., Weber, C., Bers, D.M. (2004) A mathematical treatment of integrated Ca dynamics within the ventricular myocyte. *Biophys. J. 87*, 3351–71.

9 Ch'en, F.F., Vaughan-Jones, R.D., Clarke, K., Noble, D. (1998) Modelling myocardial ischaemia and reperfusion. *Prog. Biophys. Mol. Biol.* 69, 515–38.

10 Matsuoka, S., Sarai, N., Jo, H., Noma, A. (2004) Simulation of ATP metabolism in cardiac excitation-contraction coupling. *Prog. Biophys. Mol. Biol.* 85, 279–99.

11 Cortassa, S., Aon, M.A., O'Rourke, B., Jacques, R., Tseng, H.J., Marban, E., Winslow, R.L. (2006) A computational model integrating electrophysiology, contraction, and mitochondrial bioenergetics in the ventricular myocyte. *Biophys. J.* 91, 1564–89.

12 Takeuchi, A., Tatsumi, S., Sarai, N., Terashima, K., Matsuoka, S., Noma, A. (2006) Ionic mechanisms of cardiac cell swelling induced by blocking Na^+/K^+ pump as revealed by experiments and simulation. *J. Gen. Physiol.* 128, 495–507.

13 Schramm, M., Klieber, H.G., Daut, J. (1994) The energy expenditure of actomyosin-ATPase, Ca^{2+}-ATPase and Na^+,K^+-ATPase in guinea-pig cardiac ventricular muscle. *J. Physiol.* 481, 647–62.

14 Bers, D.M. (2003) Na/Ca exchange and the sarcolemmal Ca-pump. In Excitation-contraction coupling and cardiac contractile force, Bers, D.M., 2nd edn. Kluwer Academic Publishers, Netherlands.

15 Mackiewicz, U., Lewartowski, B. (2006) Temperature dependent contribution of Ca^{2+} transporters to relaxation in cardiac myocytes: important role of sarcolemmal Ca^{2+}-ATPase. *J. Physiol. Pharmacol.* 57, 3–15.

16 Stanley, W.C., Chandler, M.P. (2002) Energy metabolism in the normal and failing heart: potential for therapeutic interventions. *Heart Fail. Rev.* 7, 115–30.

17 Balaban, R.S. (2002) Cardiac energy metabolism homeostasis: role of cytosolic calcium. *J. Mol. Cell. Cardiol.* 34, 1259–71.

18 Saks, V.A., Kuznetsov, A.V., Vendelin, M., Guerrero, K., Kay, L., Seppet, E.K. (2004) Functional coupling as a basic mechanism of feedback regulation of cardiac energy metabolism. *Mol. Cell. Biochem.* 256–257, 185–99.

19 Vendelin, M., Eimre, M., Seppet, E., Peet, N., Andrienko, T., Lemba, M., Engelbrecht, J., Seppet, E.K., Saks, V.A. (2004) Intracellular diffusion of adenosine phosphates is locally restricted in cardiac muscle. *Mol. Cell. Biochem.* 256–257, 229–41.

20 Vendelin, M., Lemba, M., Saks, V.A. (2004) Analysis of functional coupling: mitochondrial creatine kinase and adenine nucleotide translocase. *Biophys. J.* 87, 696–713.

21 Kaplan, J.H. (2002) Biochemistry of Na,K-ATPase. *Annu. Rev. Biochem.* 71, 511–35.

22 Scheiner-Bobis, G. (2002) The sodium pump. Its molecular properties and mechanics of ion transport. *Eur. J. Biochem.* 269, 2424–33.

23 Apell, H.J. (1989) Electrogenic properties of the Na,K pump. *J. Membr. Biol.* 110, 103–14.

24 Heyse, S., Wuddel, I., Apell, H.J., Sturmer, W. (1994) Partial reactions of the Na,K-ATPase: determination of rate constants. *J. Gen. Physiol.* 104, 197–240.

25 Smith, N.P., Crampin, E.J. (2004) Development of models of active ion transport for whole-cell modelling: cardiac sodium-potassium pump as a case study. *Prog. Biophys. Mol. Biol.* 85, 387–405.

26 MacLennan, D.H., Rice, W.J., Green, N.M. (1997) The mechanism of Ca^{2+} transport by sarco(endo)plasmic reticulum Ca^{2+}-ATPases. *J. Biol. Chem.* 272, 28815–8.

27 Wuytack, F., Raeymaekers, L., Missiaen, L. (2002) Molecular physiology of the SERCA and SPCA pumps. *Cell. Calcium.* 32, 279–305.

28 Odermatt, A., Kurzydlowski, K., MacLennan, D.H. (1996) The vmax of the Ca^{2+}-ATPase of cardiac sarcoplasmic reticulum (SERCA2a) is not altered by Ca^{2+}/calmodulin-dependent phosphorylation or by interaction with phospholamban. *J. Biol. Chem.* 271, 14206–13.

29 Sasaki, T., Inui, M., Kimura, Y., Kuzuya, T., Tada, M. (1992) Molecular mechanism of regulation of Ca^{2+} pump ATPase by phospholamban in cardiac sarcoplasmic reticulum. Effects of synthetic

phospholamban peptides on Ca^{2+} pump ATPase. *J. Biol. Chem. 267*, 1674–9.

30 Engelender, S., De Meis, L. (1996) Pharmacological differentiation between intracellular calcium pump isoforms. *Mol. Pharmacol. 50*, 1243–52.

31 ten Tusscher, K.H., Noble, D., Noble, P.J., Panfilov, A.V. (2004) A model for human ventricular tissue. *Am. J. Physiol. Heart Circ. Physiol. 286*, H1573–89.

32 Yano, K., Petersen, O.H., Tepikin, A.V. (2004) Dual sensitivity of sarcoplasmic/endoplasmic Ca^{2+}-ATPase to cytosolic and endoplasmic reticulum Ca^{2+} as a mechanism of modulating cytosolic Ca^{2+} oscillations. *Biochem. J. 383*, 353–60.

33 Higgins, E.R., Cannell, M.B., Sneyd, J. (2006) A buffering SERCA pump in models of calcium dynamics. *Biophys. J. 91*, 151–63.

34 Strehler, E.E., Zacharias, D.A. (2001) Role of alternative splicing in generating isoform diversity among plasma membrane calcium pumps. *Physiol. Rev. 81*, 21–50.

35 Dixon, D.A., Haynes, D.H. (1989) Kinetic characterization of the Ca^{2+} pumping ATPase of cardia sarcolemma in four states of activation. *J. Biol. Chem. 264*, 13612–22.

36 Guerini, D. (1998) The significance of the isoforms of plasma membrane calcium ATPase. *Cell Tissue Res. 292*, 191–7.

37 Caroni, P., Carafoli, E. (1981) The Ca^{2+} pumping ATPase of heart sarcolemma. Characterization, calmodulin dependence, and partial purification. *J. Biol. Chem. 256*, 3263–70.

38 Carafoli, E. (1991) The calcium pumping ATPase of the plasma membrane. *Annu. Rev. Physiol. 53*, 531–47.

39 Salvador, J.M., Inesi, G., Rigaud, J.L., Mata, A.M. (1998) Ca^{2+} transport by reconstituted synaptosomal ATPase is associated with H^+ countertransport and net charge displacement. *J. Biol. Chem. 273*, 18230–4.

40 Penheiter, A.R., Bajzer, Z., Filoteo, A.G., Thorogate, R., Torok, K., Caride, A.J. (2003) A model for the activation of plasma membrane calcium pump isoform 4b by calmodulin. *Biochemistry. 42*, 12115–24.

41 Caride, A.J., Penheiter, A.R., Filoteo, A.G., Bajzer, Z., Enyedi, A., Penniston, J.T. (2001) The plasma membrane calcium pump displays memory of past calcium spikes. Differences between isoforms 2b and 4b. *J. Biol. Chem. 276*, 39797–804.

42 Graupner, M., Erler, F., Meyer-Hermann, M. (2005) A theory of plasma membrane calcium pump stimulation and activity. *J. Biological Physics. 31*, 183–206.

43 Gibbs, C.L. (2003) Cardiac energetics: sense and nonsense. *Clin. Exp. Pharmacol. Physiol. 30*, 598–603.

44 Negroni, J.A., Lascano, E.C. (1996) A cardiac muscle model relating sarcomere dynamics to calcium kinetics. *J. Mol. Cell. Cardiol. 28*, 915–29.

45 Noma, A. (1983) ATP-regulated K^+ channels in cardiac muscle. *Nature. 305*, 147–8.

46 Carmeliet, E. (1999) Cardiac ionic currents and acute ischemia: from channels to arrhythmias. *Physiol. Rev. 79*, 917–1017.

47 Noma, A., Shibasaki, T. (1985) Membrane current through adenosine-triphosphate-regulated potassium channels in guinea-pig ventricular cells. *J. Physiol. 363*, 463–80.

48 Yamaoka, K., Kameyama, M. (2003) Regulation of L-type Ca^{2+} channels in the heart: overview of recent advances. *Mol. Cell Biochem. 253*, 3–13.

49 Barth, E., Stammler, G., Speiser, B., Schaper, J. (1992) Ultrastructural quantitation of mitochondria and myofilaments in cardiac muscle from 10 different animal species including man. *J. Mol. Cell. Cardiol. 24*, 669–81.

50 Korzeniewski, B., Zoladz, J.A. (2001) A model of oxidative phosphorylation in mammalian skeletal muscle. *Biophys. Chem. 92*, 17–34.

51 Jo, H., Noma, A., Matsuoka, S. (2006) Calcium-mediated coupling between mitochondrial substrate dehydrogenation and cardiac workload in single guinea-pig ventricular myocytes. *J. Mol. Cell. Cardiol. 40*, 394–404.

52 McCormack, J.G., Halestrap, A.P., Denton, R.M. (1990) Role of calcium ions in regulation of mammalian intramitochondrial metabolism. *Physiol. Rev. 70*, 391–425.

53 Fall, C.P., Keizer, J.E. (2001) Mitochondrial modulation of intracellular Ca^{2+} signaling. *J. Theor. Biol. 210*, 151–65.

54 Griffiths, E.J., Stern, M.D., Silverman, H.S. (1997) Measurement of mitochondrial calcium in single living cardiomyocytes by selective removal of cytosolic indo 1. *Am. J. Physiol. 273*, C37–44.

55 Kurihara, S. (1994) Regulation of cardiac muscle contraction by intracellular Ca^{2+}. *Jpn. J. Physiol. 44*, 591–611.

56 Linz, K.W., Meyer, R. (1998) Control of L-type calcium current during the action potential of guinea-pig ventricular myocytes. *J. Physiol. 513*, 425–42.

57 Lin, X., Jo, H., Sakakibara, Y., Tambara, K., Kim, B., Komeda, M., Matsuoka, S. (2006) Beta-adrenergic stimulation does not activate Na$^+$/Ca^{2+} exchange current in guinea pig, mouse, and rat ventricular myocytes. *Am. J. Physiol. Cell Physiol. 290*, C601–8.

58 Bardenheuer, H., Schrader, J. (1983) Relationship between myocardial oxygen consumption, coronary flow, and adenosine release in an improved isolated working heart preparation of guinea pigs. *Circ. Res. 52*, 263–71.

59 Allen, D.G., Kurihara, S. (1982) The effects of muscle length on intracellular calcium transients in mammalian cardiac muscle. *J. Physiol. 327*, 79–94.

60 Backx, P.H., Ter Keurs, H.E. (1993) Fluorescent properties of rat cardiac trabeculae microinjected with fura-2 salt. *Am. J. Physiol. 264*, H1098–110.

61 Shimizu, J., Todaka, K., Burkhoff, D. (2002) Load dependence of ventricular performance explained by model of calcium-myofilament interactions. *Am. J. Physiol. Heart Circ. Physiol. 282*, H1081–91.

62 Chance, B., Williams, G.R. (1956) The respiratory chain and oxidative phosphorylation. *Adv. Enzymol. Relat. Subj. Biochem. 17*, 65–134.

63 Brandes, R., Bers, D.M. (2002) Simultaneous measurements of mitochondrial NADH and Ca^{2+} during increased work in intact rat heart trabeculae. *Biophys. J. 83*, 587–604.

64 Saks, V., Dzeja, P., Schlattner, U., Vendelin, M., Terzic, A., Wallimann, T. (2006) Cardiac system bioenergetics: metabolic basis of the Frank-Starling law. *J. Physiol. 571*, 253–73.

65 Heineman, F.W., Balaban, R.S. (1993) Effects of afterload and heart rate on NAD(P)H redox state in the isolated rabbit heart. *Am. J. Physiol. 264*, H433–440.

66 Wan, B., Doumen, C., Duszynski, J., Salama, G., Vary, T.C., LaNoue, K.F. (1993) Effects of cardiac work on electrical potential gradient across mitochondrial membrane in perfused rat hearts. *Am. J. Physiol. 265*, H453–460.

67 Ashruf, J.F., Coremans, J.M., Bruining, H.A., Ince, C. (1995) Increase of cardiac work is associated with decrease of mitochondrial NADH. *Am. J. Physiol. 269*, H856–862.

68 Griffiths, E.J., Lin, H., Suleiman, M.S. (1998) NADH fluorescence in isolated guinea-pig and rat cardiomyocytes exposed to low or high stimulation rates and effect of metabolic inhibition with cyanide. *Biochem. Pharmacol. 56*, 173–179.

69 White, R.L., Wittenberg, B.A. (2000) Mitochondrial NAD(P)H, ADP, oxidative phosphorylation, and contraction in isolated heart cells. *Am. J. Physiol. Heart Circ. Physiol. 279*, H1849–1857.

70 Hisano, R., Cooper, G. 4th. (1987) Correlation of force-length area with oxygen consumption in ferret papillary muscle. *Circ. Res. 61*, 318–28.

71 Beard, D.A. (2006) Modeling of oxygen transport and cellular energetics explains observations on *in vivo* cardiac energy metabolism. *PLoS Comput. Biol. 2*, 1093–1106.

72 Collins, A., Somlyo, A.V., Hilgemann, D.W. (1992) The giant cardiac membrane patch method: stimulation of outward Na$^+$-Ca^{2+} exchange current by MgATP. *J. Physiol. 454*, 27–57.

73 Mekhfi, H., Ventura-Clapier, R. (1988) Dependence upon high-energy phosphates of the effects of inorganic phosphate on contractile properties in chemically skinned rat cardiac fibres. *Pflugers. Arch. 411*, 378–385.

Part III
Applied Molecular System Bioenergetics

Molecular System Bioenergetics: Energy for Life. Edited by Valdur Saks
Copyright © 2007 WILEY-VCH Verlag GmbH & Co. KGaA, Weinheim
ISBN: 978-3-527-31787-5

14
Mitochondrial Adaptation to Exercise and Training:
A Physiological Approach

Kent Sahlin

Abstract

Mitochondrial function has traditionally been an issue of biochemistry and bio-physics, but during recent years, the study of mitochondria has become a tool of exercise physiology. The transition from rest to maximal exercise is associated with a large increase in energy demand (up to 400 times) and oxygen utilization (up to 100 times) in the working muscle and is a major challenge for bio-energetics and system control. The increased rate of oxidative phosphorylation can in part be explained by feedback signaling by increases in ADP and creatine, which, due to parallel activation of glycolysis and the creatine kinase system, links anaerobic metabolism to aerobic metabolism. The peripheral capacity to utilize oxygen exceeds the capacity to transport oxygen to the working muscle. Despite this overcapacity, there is a further increase in muscle oxidative power (i.e., mito-chondrial volume density) of about 40% after 1–2 months of training. The training-induced increase in muscle oxidative power has a minor influence on whole-body maximal oxygen uptake but affects the metabolic control of oxidative phosphorylation during exercise. The reduced formation of lactic acid and in-creased reliance on fat oxidation during exercise after training are signs of an in-creased metabolic fitness, which increases performance in endurance sports and is of benefit to one's health. Mitochondrial efficiency, defined as the coupling between ATP formation and oxygen utilization, is influenced by the type of fuel oxidized and the rate of respiration and may also decrease by activation of un-coupling proteins and by elevated concentrations of fatty acids. It is hypothesized that changes in mitochondrial efficiency have a role in the control of oxidative phosphorylation during transition from rest to exercise. Despite its major physiological importance, there is considerable controversy about mitochondrial efficiency *in vivo*, and further studies are therefore required.

Molecular System Bioenergetics: Energy for Life. Edited by Valdur Saks
Copyright © 2007 WILEY-VCH Verlag GmbH & Co. KGaA, Weinheim
ISBN: 978-3-527-31787-5

14.1
Introduction

During physical exercise, energy turnover in skeletal muscle may increase by 400 times compared to that at rest, and muscle oxygen consumption may increase up to 100 times [1]. Although during certain conditions ATP is produced transiently by anaerobic processes, it is clear that the energy expenditure ultimately is covered by oxidative phosphorylation. The control of oxygen transport (respiration and circulation) has been a major field of research in exercise physiology since the first decade of the 1900s. However, a number of issues related to the kinetics of oxygen utilization are related to the function and control of mitochondrial respiration. Furthermore, mitochondria have a key role in the control of fuel utilization and the interaction between aerobic and anaerobic energy production. Knowledge of mitochondrial function is therefore crucial to understanding phenomena within exercise physiology and the etiology of metabolic disorders.

Research on mitochondria has traditionally been a matter for biochemists, but during recent years the study of mitochondria has become a tool of exercise physiology as well. It is now recognized that understanding of mitochondrial function during exercise and training is central to our understanding of bioenergetics during whole-body exercise.

Determination of mitochondrial function in skeletal muscle has been hampered by methodological restraints; however, during the last 10–15 years the techniques have been improved, and it is now possible to isolate mitochondria from small muscle samples obtained by needle biopsy from human muscle and to measure both ATP production and respiration with different substrates. Furthermore, the development of the permeabilized fiber (skinned fiber) technique has provided a tool for investigating mitochondrial function *in situ*, where the interaction with cellular components is maintained.

A general theme of this book is to describe the control of mitochondrial energetics, and several chapters in this book discuss this issue from different perspectives. The present report will give a brief review of recent research related to the adaptation of mitochondrial function in skeletal muscle to acute exercise and to endurance training. The review has a major focus on studies in human muscle with the purpose of providing a physiological perspective on the role of mitochondria in muscle energetics.

14.2
Control of Oxidative Phosphorylation

Mitochondrial ATP production in skeletal muscle is controlled by the energy demand. Increases in the products of ATP hydrolysis, i.e., ADP and P_i, activate the rate of oxidative phosphorylation (OXPHOS) by feedback control. Increased ADP was identified as an important stimulator of mitochondrial respiration in the early 1950s by classical studies on isolated mitochondria [3] and has a role in practically all presented models of respiratory control. The control of respiration in isolated mitochondria follows Michaelis–Menten kinetics, and the apparent K_m of ADP has been determined to be in the range of 10–30 μM. The total concentration of ADP in skeletal muscle is about 1 mM, but the major part (>95%) is considered to be bound to proteins (or otherwise unavailable). Free ADP is the metabolically active form, which takes part in enzymatic reactions and in the control of metabolic pathways. The current practice is to calculate the concentration of free ADP from the creatine kinase (CK) reaction, which is considered to be close to equilibrium.

$$PCr + ADP_{free} + H^+ \leftrightarrow Cr + ATP$$

$$[ADP_{free}] = ([Cr][ATP])/([PCr][H^+]) \times 1/K_{eq}$$

where PCr denotes phosphocreatine, Cr denotes creatine, and K_{eq} denotes the equilibrium constant. Calculation of ADP_{free} from the CK equilibrium involves some critical assumptions whose validity has been questioned [36]. Considering the critical importance of ADP in muscle energetics and metabolic control, it is important to establish alternative methods to estimate free ADP in muscle tissue during different metabolic conditions. The concentration of ADP_{free} in human skeletal muscle has been calculated from the CK equilibrium to be about 20 μM at rest and 138 μM at fatigue [39]. The concentration of ADP_{free} at rest is thus similar to the apparent K_m value and would elicit about 50% of maximum respiration. The low rate of O_2 consumption in resting muscle *in vivo* (about 2% of V_{max}) thus is not compatible with the low K_m value observed in isolated mitochondria.

Studies on permeabilized fibers have shown that their ADP sensitivity is considerably lower than that observed in isolated mitochondria [41]. The low ADP sensitivity of fiber respiration is more compatible with the low rate of oxygen utilization at rest. The high ADP sensitivity in isolated mitochondria has been attributed to damage of the outer mitochondrial membrane during the preparation of isolated mitochondria [41]. The apparent K_m was determined to about 300 μM in fibers from rat heart and 350 μM in fibers from rat soleus muscle, whereas the ADP sensitivity was much higher in glycolytic fibers (14 μM in rat white gastrocnemius [20]). Another important finding in this study was that an increase in creatine, which is a characteristic feature of increased energy demand, increases the sensitivity of mitochondrial respiration to ADP [20]. The respiratory control exerted by creatine was present in oxidative but not glycolytic fibers. Human vastus lateralis muscle has a mixture of fiber types, and its apparent K_m has been determined to be 120 μM [51]. The ADP sensitivity is dependent on the fiber-type composition in human muscle being higher (lower apparent K_m) in glycolytic fibers than in oxidative fibers [44].

PCr has an important role in buffering the ATP store and the acid load during high-intensity exercise. However, studies on permeabilized fibers from human skeletal muscle demonstrate that both PCr and Cr have important roles in the

VO$_2$ (% of maximal)

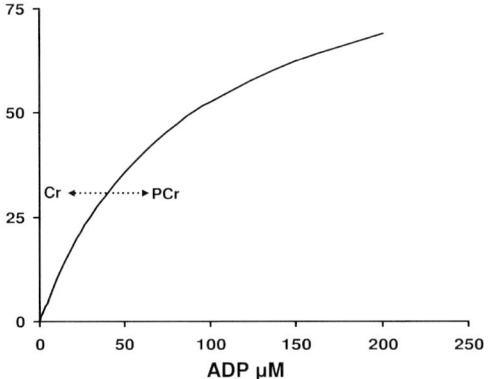

Fig. 14.1 The dependence of skinned fiber respiration on the concentration of ADP and the phosphocreatine–creatine system. The curve was constructed from data obtained from skinned fibers prepared from human muscle [52] assuming Michaelis–Menten kinetics. An increase in creatine (Cr) increases ADP sensitivity and shifts the curve to the left. An increase in phosphocreatine (PCr) reduces ADP sensitivity and shifts the curve to the right.

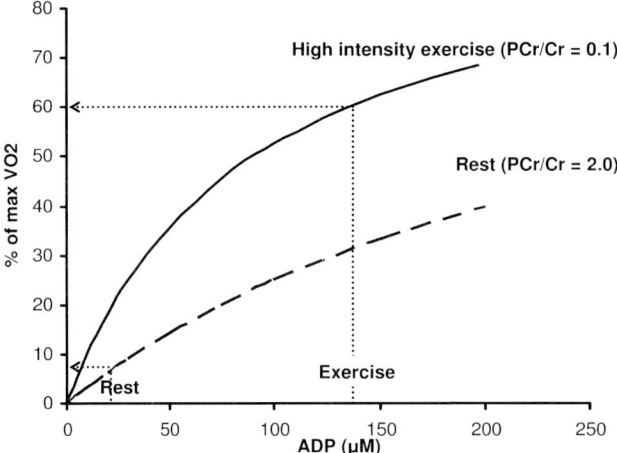

Fig. 14.2 Model of respiratory control during transition from rest to strenuous exercise. The curves were constructed from the dependence of fiber respiration on ADP and PCr/Cr (Fig. 14.1), from estimated free ADP (calculated from the creatine kinase equilibrium), and from measured concentrations of PCr and Cr at rest and during strenuous exercise [39]. The changes in free ADP and PCr/Cr explain the 10-fold increase in muscle VO$_2$. Substrate availability and feed-forward control by Ca^{2+} affect the redox pressure (i.e., NADH:NAD$^+$ ratio) on the electron transport chain and provide an additional mode of control not included in the model.

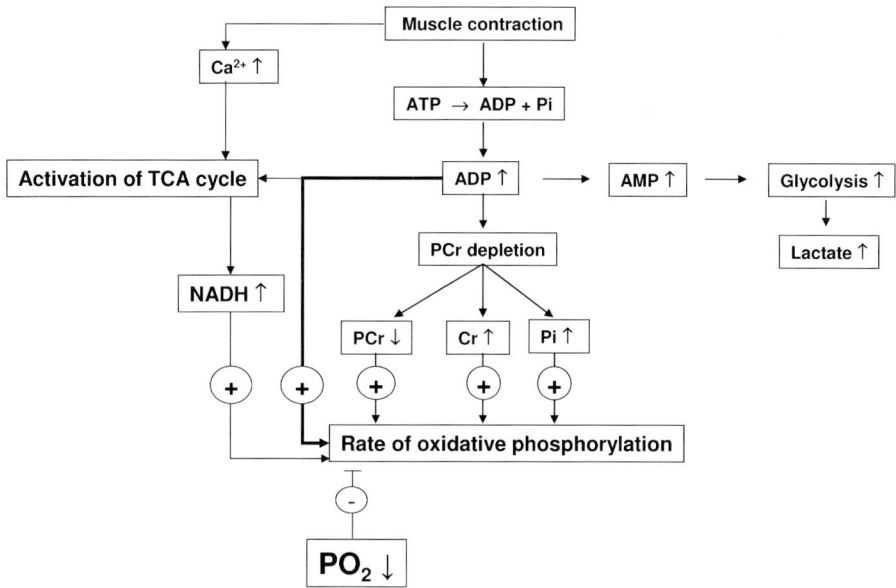

Fig. 14.3 Schematic view of the metabolic control of oxidative phosphorylation. The increase in ADP is a master controller of both aerobic and anaerobic metabolism. Modulation of oxygen tension can affect oxidative phosphorylation during both maximal and moderate submaximal exercise. However, the rate of oxidative phosphorylation covers the energy demand during submaximal exercise due to adjustments of the other controllers (cf. Fig. 14.4).

control of OXPHOS [52]. In addition to the stimulating role of Cr in mitochondrial respiration, as described above, PCr has an opposite effect, being an inhibitor of respiration at submaximal ADP concentration (Fig. 14.1). The decrease in PCr that occurs during exercise reflects increases in ADP_{free} according to the creatine kinase reaction equilibrium. The decrease in PCr and the corresponding increase in Cr amplify the stimulation of respiration by ADP and minimize the requirements of large changes in ADP concentration (Fig. 14.2). The effect of Cr/PCr on mitochondrial respiration is mediated by mitochondrial creatine kinase (CKmit), which is located in the mitochondrial intermembrane space and is, as discussed above, fiber-type specific. Control of OXPHOS by the PCr/Cr system appears to be an additional control system developed in oxidative fibers.

The recognition of Cr and PCr as essential regulators of mitochondrial ATP production in human skeletal muscle demonstrates that there is a tight interaction between aerobic and anaerobic metabolism. What could be the physiological advantage with this type of control? Firstly, modulation of oxidative phosphorylation by the PCr–Cr system would reduce the amplitude of ADP fluctuation during exercise. ADP is a metabolite, which reflects the energy state of the cell and is in equilibrium with ATP and AMP by means of the adenylate kinase reaction.

$$ADP + ADP \leftrightarrow AMP + ATP$$

Increases in ADP and AMP reflect a condition of energetic stress and affect the activity or the equilibrium state of a number of other enzymes involved in energy metabolism, including glycogen phosphorylase, phosphofructokinase, creatine kinase AMP kinase, and AMP deaminase. Increases in ADP and AMP are key intermediates in metabolic control and important triggers of glycolytic rate and lactic acid formation. The involvement of PCr/Cr in the control of OXPHOS reduces the amplitude of exercise-induced increases in ADP and thus reduces the glycolytic flux and formation of lactic acid (Fig. 14.3). This is of considerable physiological importance because of the central importance of lactic acid in fatigue and hence exercise performance.

14.2.1
Influence of Oxygen Availability on the Control of Oxidative Phosphorylation

Anaerobic processes produce ATP without using oxygen and at a rate that can exceed that of OXPHOS [40]. Glycolysis and PCr utilization are therefore important for muscle energetics during high-intensity exercise and during hypoxic/anoxic conditions. However, it is well known that increases in blood lactate also occur during submaximal exercise. The increase in blood lactate is closely related to an increased muscle lactate level and to increased lactate efflux from muscle and thus to an increased production of lactate [18]. The exercise intensity at which blood lactate shows a clear increase has been termed anaerobic threshold or lactate threshold. Because of its physiological importance, the mechanism for lactate formation during submaximal exercise has been discussed for a long time. There has been considerable controversy about the role of oxygen availability. The term anaerobic threshold has led to some confusion, because there is convincing evidence that oxygen tension within the lactate-producing working muscle is maintained at a level where oxidative phosphorylation (OXPHOS) can proceed [11, 35]. However, because glycolysis is a process by which ATP is produced without oxygen utilization, the term is semantically correct.

The minimum oxygen requirement (apparent K_m) for maintenance of OXPHOS has been determined to be 0.1–0.7 μM [42, 54] in isolated mitochondria and is higher during high rates of respiration than during low rates. The oxygen tension in intact working human muscle has been estimated from measurements of myoglobin saturation with magnetic resonance spectroscopy [35]. During one-legged exercise, PO_2 was 3.1 torr (4.3 μM), which is much higher than the apparent K_m for O_2. Despite the relatively high oxygen concentration, there was a sevenfold increase in lactate efflux from muscle when the exercise intensity increased from 50% to 77% VO_{2max} [35]. These data indicate that OXPHOS is not compromised during submaximal exercise, although a high rate of lactate formation occurs. Therefore, it appears highly unlikely that the increased lactate production during submaximal exercise is a consequence of the oxygen limitation of OXPHOS.

The oxygen dependence of metabolism has been investigated in humans by having subjects exercise at the same absolute workload but with different fractions of inspired oxygen (FIO$_2$). Firstly, it has been known for a long time that reducing FIO$_2$ during submaximal exercise (hypoxic exercise) results in increased levels of lactate in blood [7] and muscle [21] despite maintenance of the same oxygen uptake. Measurements *in vivo* with the magnetic resonance technique demonstrated that PCr is inversely related to the availability of oxygen during steady-state submaximal exercise [13]. PCr decreased when oxygen tension was reduced but reverted back to the original level when normoxic conditions were reestablished. Similar findings have been observed in human muscle during submaximal hypoxic exercise [17]. Subjects cycled at the same submaximal workload during hypoxic conditions (FIO$_2$ = 11%) and normoxic conditions. VO$_2$ was similar during the two conditions but there was an increase in muscle NADH, ADP, and P$_i$ (the latter two indicated by lower PCr) during hypoxic conditions (Fig. 14.4).

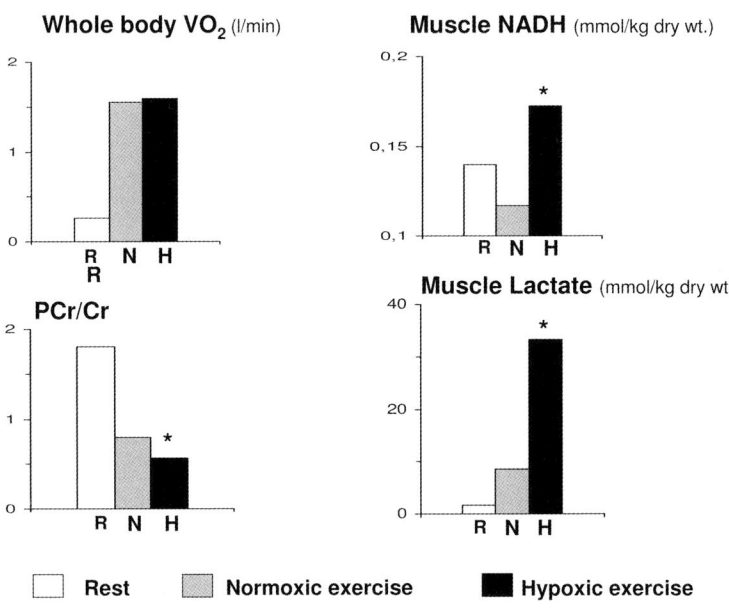

Fig. 14.4 Effect of oxygen availability on metabolism. Subjects cycled during normoxic (N: 21% O$_2$) and hypoxic (H: 11% O$_2$) conditions at the same absolute workload (50% of normoxic VO$_2$). Muscle biopsies were taken before (rest; R) and after 5 min of exercise (N, H) and analyzed for metabolites. Asterisk (*) denotes $P < 0.05$ vs. N. The figure is based on results from [17] and demonstrates that metabolism can be dependent on oxygen availability despite maintained rates of oxidative phosphorylation.

Although the results support the view that OXPHOS was not limited by PO$_2$ during hypoxia, it is clear that the decrease in oxygen availability has a major influence on metabolism.

There is an abundance of research supporting the view that oxygen availability may affect metabolism during submaximal exercise although OXPHOS is unaffected. Wilson and colleagues [8, 53] have presented evidence that oxygen availability may affect metabolism at an oxygen pressure as low as 30 torr (~40 μM), and they have emphasized that there is a difference between oxygen dependency of metabolism and oxygen limitation of OXPHOS, which occurs at PO$_2$ < 1 μM. According to the near-equilibrium hypothesis, OXPHOS is controlled by the redox state (NADH/NAD$^+$), the phosphorylation potential (ATP/(ADP × P$_i$)), and PO$_2$ [19, 53]. The tissue maintains OXPHOS during conditions of decreased PO$_2$ by compensatory increases in NADH, ADP, and P$_i$. The metabolic changes (increased NADH, ADP, and P$_i$) observed in human muscle during hypoxic exercise (Fig. 14.4) are consistent with this theoretical model [17]. Additional control is exerted by a decreased PCr:Cr ratio. Furthermore, during incremental exercise the increased energy demand is matched by an increased stimulation of OXPHOS by increased levels of ADP, P$_i$, and NADH (Fig. 14.5) and a decreased PCr:Cr ratio.

The current view, accepted by most but not all groups, is that metabolism is oxygen dependent during submaximal exercise but oxygen limited only when the intensity exceeds VO_{2max}. ADP is an important stimulator of OXPHOS (see

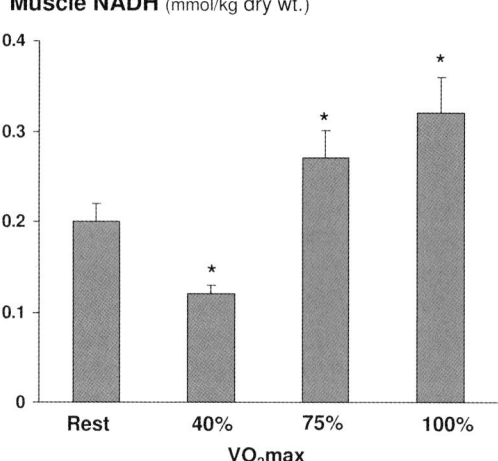

Fig. 14.5 Muscle NADH during incremental exercise. Muscle NADH was measured in muscle biopsies with luminometric technique. Muscle biopsies were taken at rest, after 10 min exercise at 40% and 75% VO_{2max}, and at exhaustion, i.e., an average of 4.8 min exercise at 100% VO_{2max}. Asterisk (*) denotes $P < 0.05$ vs. rest. (The figure is based on results presented in [37]).

above), and an increase in [ADP] is necessary during conditions of increased metabolic rate and/or decreased oxygen availability. Increases in ADP are intimately related to increases in AMP (see above) and P_i (combination of ATP hydrolysis and creatine kinase reaction), which all are important allosteric regulators of glycogenolysis and glycolysis. Lactate formation during submaximal exercise is, according to this view, a consequence of increased stimulation of OXPHOS [4]. During these conditions lactate formation has a negligible importance for covering the ATP balance but is instead a consequence of metabolic control [4].

14.3
Mitochondrial Uncoupling and Mitochondrial Efficiency

The efficiency of mitochondrial energy transfer (produced ATP per oxygen consumed or P:O ratio) is a matter of controversy. The theoretical maximal P:O ratio has been estimated to be 2.6 with a carbohydrate-derived substrate (i.e., pyruvate) and about 10% lower with a fat-derived substrate (palmitate) due to the formation of complex II substrate (FADH) in β-oxidation [2]. The mitochondrial efficiency may be reduced by back-leakage of protons, electron slip in the electron transport chain (reduced proton pumping per flux of electrons), and/or altered stoichiometry of ATP synthase (Fig. 14.6). These processes increase non-coupled respiration (NCR) and reduce mitochondrial efficiency. NCR has been estimated in perfused rat skeletal muscle using CN^- to block the electron transport chain and oligomycin to block OXPHOS. With this technique NCR was estimated to be about 40% of the basal metabolic rate and to have a P:O ratio as low as 1.1 [2]. However mitochondrial efficiency has also been estimated by noninvasive methods by which ATP consumption was measured with 31P MRS and oxygen utilization with near-

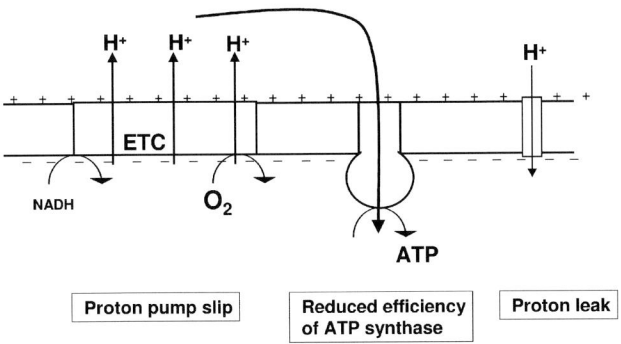

Fig. 14.6 Schematic view of factors influencing the efficiency of oxidative phosphorylation (P:O ratio). The P:O ratio is reduced if there is a proton leak, a slip in the pumping of protons (reduced proton efflux/electron flux), or reduced efficiency of ATP synthase (reduced ATP/proton influx).

infrared spectroscopy [26]. With this technique the P:O ratio was determined at rest to be about 2.2 in both mouse [26] and human muscle [25], whereas in aged mouse muscle the P:O ratio was 50% lower [27]. Obviously, there is a large discrepancy between estimates of mitochondrial efficiency with different techniques, and further studies are required to determine the P:O ratio and NCR *in vivo*.

An approximate measure of NCR is the oxygen consumption of isolated mitochondria and permeabilized fibers during non-phosphorylating conditions. State 4 respiration is defined as the mitochondrial respiration in the presence of substrate but in the absence of ADP. Under certain conditions (e.g., maintained electrochemical proton gradient, negligible ATPase activity) state 4 respiration reflects proton leak in isolated mitochondria. If an increased proton leakage during state 4 is maintained, this results in a reduced P:O ratio during state 3. However, the extent of proton leakage is influenced by the electrochemical proton gradient, which is lower during state 3 than during state 4. Proton leak is therefore expected to be lower during state 3 than during state 4. An increased proton leak during state 4 is therefore not necessarily expressed during state 3. Furthermore, P:O ratio is also influenced by electron slip and by the efficiency of the ATP synthase complex (Fig. 14.6), as well as by oxygen utilization by processes other than the electron transport chain. Changes in these processes are not included by estimating NCR from state 4 respiration in isolated mitochondria. The link between increased state 4 respiration and decreased P:O ratio is therefore indirect and not compulsory, and a dissociation between altered state 4 and P:O ratio has been observed during acidosis [47] and after extreme endurance exercise [9].

The respiration rate *in vivo* depends on the energy demand and corresponds to a respiration rate between state 4 and state 3. We have determined the P:O ratio in isolated mitochondria during different rates of respiration achieved by using different rates of ADP infusion [29]. The results demonstrate that the P:O ratio is not a fixed value but is dependent on the rate of respiration [29]. The P:O ratio was relatively stable between 100% and 50% of state 3 but decreased by 23% when the respiration rate decreased to 20% of that during state 3 (Fig. 14.7). The P:O ratio *in vivo* therefore is lower than that determined during state 3 in isolated mitochondria.

It has been known for a long time that elevated concentrations of fatty acid (FA) reduce the coupling between oxygen consumption and ATP formation in isolated mitochondria. The uncoupling potency is highest for FA with 12–18 C atoms and for unsaturated FA [55]. Studies in perfused intact hearts from guinea pigs show that circulating FA may uncouple mitochondria in intact cardiac muscle as well and thus may be of physiological importance [34]. We have investigated the effect of endurance training on the sensitivity of NCR to FA in mitochondria isolated from muscle biopsies [45] obtained before and after six weeks of endurance training. FA-induced uncoupling was observed at a concentration of free FA of 21 nM NCR, and the extent of uncoupling was twofold higher after training than before training. The training-induced increase in FA-induced un-

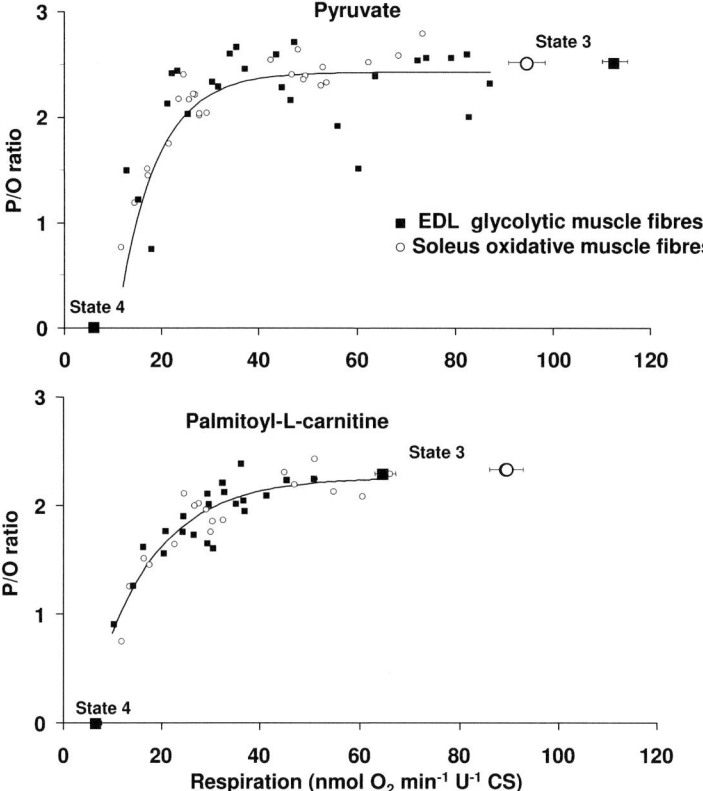

Fig. 14.7 Influence of respiration rate on mitochondrial efficiency. The respiration rate was varied in isolated mitochondria from rat extensor digitorum longus (EDL) (■) and soleus muscle (○) by maintaining different concentrations of ADP (different rates of ADP infusion). The P:O ratio was calculated at steady state as the ratio between the rate of ADP addition and the rate of oxygen utilization. (The figure is modified from [29] and is published with permission from Blackwell Publishing Ltd.).

coupling may be of physiological importance. Whole-body oxygen consumption is increased after food intake (diet-induced thermogenesis), and the increase is higher in endurance-trained subjects, being proportional to their VO_{2max} [22]. An intriguing possibility is that this relates to the increased mitochondrial uncoupling in the presence of FA and that endurance training may improve the mechanisms to maintain stability in body weight.

The extent of NCR may be an important factor in muscle energetics and as such important for body weight regulation. However, proton leak may also be important in the control of OXPHOS. Transition from rest to maximal exercise is a challenge to muscle energetics in that the rate of OXPHOS may increase up to 100 times. There are considerable problems in regulating a metabolic pathway

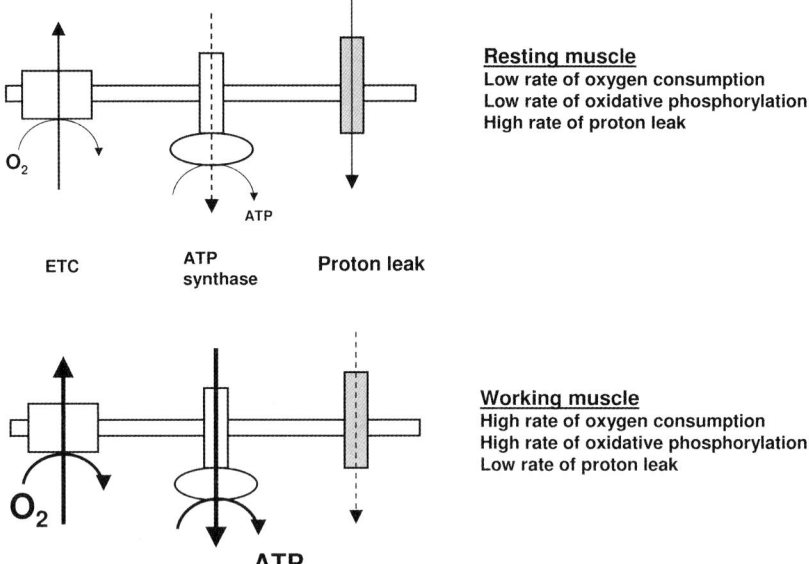

Fig. 14.8 Proton pumping and proton leak as a substrate cycle to control oxidative phosphorylation. Proton leak is high at low rates of respiration and functions as a brake on oxidative phosphorylation. The electrochemical gradient of protons is reduced when the respiration rate increases and proton leak therefore decreases (reduced brake) (Fig. 14.7). The simultaneous increase in respiration (gas) and decrease in proton leak (brake) are important parts in the control of oxidative phosphorylation.

within this range without large perturbations of cellular homeostasis. Alterations in substrate/product concentration and allosteric regulation are important mechanisms for regulating the flux in a metabolic pathway. However, additional control mechanisms such as covalent modification of proteins and substrate cycles are required to enable large variations in flux during rest–work transitions [31]. The proton pumping related to the electron transport chain and the proton leak may be regarded as a substrate cycle and serves as a control mechanism of OXPHOS (Fig. 14.8). In resting muscle, where the energy demand is low, mitochondrial respiration is close to state 4 respiration, with a high rate of proton leak (brake) and a low rate of OXPHOS. However, during exercise, when the energy demand increases, proton leak is reduced because of reduced membrane potential, simultaneous with an increased stimulation of respiration by feedback control (ADP) and substrate availability (NADH). The concomitant decreased brake and increased gas may provide the necessary requirements for attaining a 100-fold increase in OXPHOS. NCR thus may be a parameter under physiological control that can be both upregulated (overfeeding) and downregulated (exercise).

14.4
Quantitative and Qualitative Adaptations of Mitochondria to Training

It has been known for a long time that mitochondrial density rapidly adapts to endurance training. In humans, 6–8 weeks of training may result in a 30–40% increase in mitochondrial enzyme activities [14] and respiration [48]. Recent findings demonstrate that in addition to increased mitochondrial density there are changes in mitochondrial quality [48]. The adaptation of mitochondria to training alters the metabolic response during exercise and improves endurance.

14.4.1
Effect of Training on Mitochondrial Quantity

Mitochondrial density has a high plasticity in muscle tissue, and it is well known that there are large increases in oxidative enzymes [15] and mitochondrial volume [16] with training. A six-week endurance training program in previously untrained subjects increased muscle oxidative potential by about 40% when measured in both isolated mitochondria [48] and permeabilized fibers [51]. The adaptation in muscle oxidative potential was more pronounced than that of whole-body VO_{2max}. However, the gain in mitochondrial mass is short-lived and reverts back to the pre-training level with a half-time of two weeks or less when training is interrupted [14].

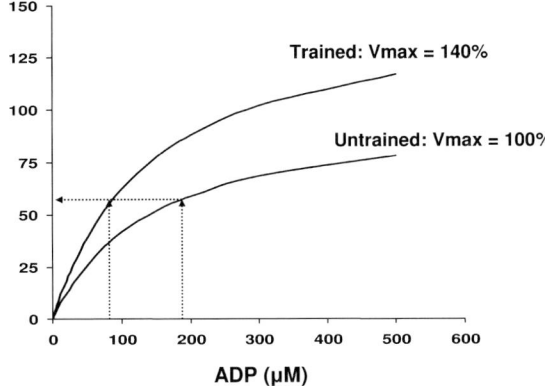

Fig. 14.9 The effect of increased mitochondrial density on the control of oxidative phosphorylation. The figure shows the effect of a 40% increase in mitochondrial density [51] on the required ADP necessary to maintain a certain absolute VO_2 (60% of the pre-training muscle V_{max}). The increased mitochondrial density increases the ADP sensitivity of muscle respiration, and a certain rate of muscle respiration can be achieved by less stimulation after training (more mitochondria = less respiration per mitochondrion).

In cross-sectional studies there is a correlation between peak values of muscle respiration and whole-body respiration [46, 57]. However, there is convincing evidence that the maximal mitochondrial oxidative power (V_{max}) in human skeletal muscle is in excess of what is required during exercise with large muscle groups (e.g., two-legged exercise) [1]. Despite this apparent overcapacity, there is a further increase in mitochondrial density during endurance training. The increase in volume density of mitochondria after training increases the sensitivity of respiratory control to ADP in the whole working muscle (Fig. 14.9) [5, 15]. Although the increased mitochondrial density has little influence on whole-body maximal oxygen utilization (VO_{2max}), it plays a major role in reducing the metabolic perturbations and increasing endurance during submaximal exercise (see above).

14.4.2
Effect of Training on Mitochondrial Quality

The phenomenon of a training-induced rise in V_{max} is well understood and extensively confirmed by morphological studies and observations at the level of single mitochondrial enzymes. However, much less is known about how training influences the qualitative aspects of mitochondrial function, such as the efficiency of oxidative energy transfer, the control of mitochondrial respiration to effectors (e.g., ADP, fatty acids, creatine [Cr], phosphocreatine [PCr]), and mitochondrial fuel utilization. In a longitudinal study sedentary subjects were investigated before and after six weeks of endurance training. In addition to an increased maximal respiration due to increased mitochondrial mass, there was a change in respiratory control in that ADP sensitivity decreased and the effect of creatine was enhanced [51]. Similar findings have been observed in elite cross-country skiers after four months of training [30] and, even more markedly, in rat muscle [56]. Furthermore, a similar change in metabolic control was observed in endurance-trained athletes after hypoxic training [32] and in rats bred for high aerobic running capacity for 15 generations [50]. These studies demonstrate that endurance training results in qualitative adaptation of mitochondria in skeletal muscle towards a more oxidative type of control (decreased ADP sensitivity and increased stimulation by creatine). These qualitative changes in mitochondrial function may work in combination with increased mitochondrial mass to provide a tighter coupling between ATP demand and mitochondrial ATP supply.

14.5
Effect of Acute Exercise on Mitochondrial Function

Acute exercise is associated with physical and chemical perturbations of the intracellular environment. Some of the exercise-induced changes serve as regulating signals for the adjustment of the rate of ATP synthesis to the increased energy demand. However, disturbances in the intracellular homeostasis during acute exercise, such as acidosis and increased generation of reactive oxygen species, not only may affect the respiration rate of mitochondria but also may elicit lasting alterations in the intrinsic properties of mitochondrial function.

14.5.1
Effect of Strenuous Exercise on Maximal Respiration

Early studies on rat muscle showed that respiration in isolated mitochondria was not affected by two hours' running to exhaustion [43], whereas other studies showed that maximal respiration in homogenates of thoroughbred horse muscle was severely reduced after gallops [12]. Our own studies in humans demonstrate that maximal respiration of skinned muscle fibers in the presence of creatine increases after both 1–2 h exhaustive exercise [44] and high-intensity exercise [49]. The mechanism for the upregulation of creatine-stimulated respiration is unknown. Measurements in isolated mitochondria from humans show that maximal ADP-stimulated respiration (state 3) was not changed after 1–2 h exhaustive exercise [10, 24], high-intensity exercise [33, 49], or intermittent static contractions [38]. As is evident from the unchanged or increased maximal respiration rate, maximal oxidative potential in human muscle is well protected from the potentially destructive influence of changes in the physical and chemical milieu during exercise.

14.5.2
Effect of Acute Exercise on Non-coupled Respiration (State 4) and Mitochondrial Efficiency

After prolonged cycling at moderate intensity, non-coupled respiration (NCR) measured in human skinned muscle fibers increased by 18% [44]. Measurements in isolated mitochondria have given different results. In one study state 4 respiration was unchanged at exhaustion but increased by 23% after 30–60 min of recovery [23]. In another study state 4 respiration was higher both immediately post-exercise (+14%) and after 3 h of recovery (+19%), but the differences did not reach statistical significance [10]. In a recent study we observed increased state 4 respiration with palmitoyl carnitine (+21%) but not with pyruvate after extreme-endurance exercise [9].

The increased state 4 respiration observed in some but not all studies after endurance exercise indicates that proton leak is augmented. If maintained during state 3, this might reduce mitochondrial coupling. Experiments in rats have shown that 2 h of exhaustive running had no effect on P:O ratio and state 4 respiration in isolated mitochondria [43]. In a recent study of extreme-endurance exercise, P:O ratio decreased by 9% (FA-derived substrate) after 24 h of exercise, which was consistent with the increased state 4 respiration (with FA) [9]. However, after 28 h of recovery the P:O ratio remained reduced despite decreased state 4 respiration. This demonstrates that the relation between state 4 respiration and P:O ratio is complex and that changes in proton leak during state 4 are not necessarily expressed during active phosphorylating conditions.

Studies of human muscle after high-intensity exercise demonstrate that NCR measured in skinned fibers [49] and in isolated mitochondria (state 4) [33, 49] is unchanged. Mitochondrial efficiency was also unchanged immediately after high-intensity exercise [33, 49], but a decrease was observed after 3 h of recovery [49].

14.5.3
Work Efficiency and Mitochondrial Efficiency

Work efficiency (WE) during cycling is defined as the ratio between performed work and exercise-induced energy expenditure as measured by indirect calorimetry. Work efficiency is highly dependent on biomechanical factors and varies between 0% (isometric contraction) and 30% (cycling). Recently, we showed that WE measured during submaximal exercise was not significantly different in trained ($28.0 \pm 0.5\%$) and untrained ($27.7 \pm 0.8\%$) subjects [28]. Mitochondrial efficiency was measured during both maximal (state 3) and during submaximal ADP-stimulated respiration, but there was no relation between WE measured *in vivo* and individual mitochondrial efficiency (P:O ratio) measured *in vitro* [28]. However, an intriguing finding was that WE was inversely related to the protein content of uncoupling protein UCP3. It was speculated that the inverse correlation between WE and UCP3 is caused by activation of UCP3 by factors present *in vivo* but not *in vitro*, e.g., superoxide and fatty acids [6].

14.6
Integrated View of the Role of Mitochondria in Exercise Physiology

Endurance training is known to increase whole-body VO_{2max} and is also associated with a number of metabolic adaptations during submaximal exercise (reduced anaerobic energy production, shift of fuel utilization from carbohydrates to fat). There is convincing evidence that the increase in VO_{2max} is caused mainly by cardiorespiratory adaptations resulting in increased delivery of blood and oxygen to the working muscles. Although an increased oxidative potential of the muscle may increase oxygen extraction and thus contribute to the training-induced increase in VO_{2max}, it is clear that there is an overcapacity in muscle oxidative potential during exercise with large muscle groups (i.e., running or cycling). Despite the limited importance of mitochondrial mass for VO_{2max}, there is a large and rapid adaptation to the degree of physical activity, and the increase in muscle oxidative potential is more pronounced than that of whole-body VO_{2max}. The adaptation of muscle oxidative function to endurance training includes both an increased volume density of mitochondria and changes in the control of respiration. Both of these changes have an impact on metabolic control and may to a large extent explain the metabolic changes during submaximal exercise. The reduced concentrations of ADP and AMP brought about by mitochondrial adaptations to endurance exercise affect a number of key enzymes in metabolic control and result in a reduction in PCr degradation, glycolytic flux, lactic acid formation, and carbohydrate utilization. These changes have a major impact on performance and enhance endurance during submaximal exercise.

Acknowledgments

The author's research is supported financially by grants from the Swedish Research Council (Project 13020), the Swedish National Centre for Research in Sport (CIF), Statens Sundhedsvidenskabelige Forskningsråd (Denmark), Kulturministeriets Udvalg for Idraetsforskning (KIF, Denmark), and the Swedish School of Sport and Health Sciences (GIH).

References

1 Andersen, P., Saltin, B. (1985) Maximal perfusion of skeletal muscle in man. *J Physiol. 366*, 233–249.

2 Brand, M.D., Chien, L.F., Ainscow, E.K., Rolfe, D.F., Porter, R.K. (1994) The causes and functions of mitochondrial proton leak. *Biochim Biophys Acta. 1187*, 132–139.

3 Chance, B., Williams, G.R. (1955) Respiratory enzymes in oxidative phosphorylation. I. Kinetics of oygen utilisation. *J Biol Chem. 217*, 383–393.

4 Connett, R.J., Sahlin, K. (1996) Control of glycolysis and glycogen metabolism. In *Handbook of Physiology. Section 12. Exercise: regulation and integration of multiple systems.* ed. Rowell, L.B.S., J.T., pp. 870–911. Oxford University Press, New York.

5 Dudley, G.A., Tullson, P.C., Terjung, R.L. (1987) Influence of mitochondrial content on the sensitivity of respiratory control. *J Biol Chem. 262*, 9109–9114.

6 Echtay, K.S., Roussel, D., St-Pierre, J., Jekabsons, M.B., Cadenas, S., Stuart, J.A., Harper, J.A., Roebuck, S.J., *et al.* (2002) Superoxide activates mitochondrial uncoupling proteins. *Nature. 415*, 96–99.

7 Edwards, H.T. (1936) Lactic acid in rest and work at high altitude. *Am. J. Physiol. 116*, 367.

8 Erecinska, M., Wilson, D.F., Nishiki, K. (1978) Homeostatic regulation of cellular energy metabolism: experimental characterization *in vivo* and fit to a model. *Am J Physiol. 234*, C82–89.

9 Fernstrom, M., Bakkman, L., Tonkonogi, M., Shabalina, I.G., Rozhdestvenskaya, Z., Mattsson, M., Enqvist, J., Ekblom, B., et al. (2007) Reduced efficiency but increased fat oxidation in mitochondria from human skeletal muscle after 24 hours ultra-endurance exercise. *J. Appl. Physiol. 102*, 1844–1849.

10 Fernstrom, M., Tonkonogi, M., Sahlin, K. (2004) Effects of acute and chronic endurance exercise on mitochondrial uncoupling in human skeletal muscle. *J Physiol. 554*, 755–763.

11 Gayeski, T.E., Connett, R.J., Honig, C.R. (1987) Minimum intracellular PO2 for maximum cytochrome turnover in red muscle *in situ*. *Am J Physiol. 252*, H906–915.

12 Gollnick, P.D., Bertocci, L.A., Kelso, T.B., Witt, E.H., Hodgson, D.R. (1990) The effect of high-intensity exercise on the respiratory capacity of skeletal muscle. *Pflugers Arch. 415*, 407–413.

13 Haseler, L.J., Richardson, R.S., Videen, J.S., Hogan, M.C. (1998) Phosphocreatine hydrolysis during submaximal exercise: the effect of FIO$_2$. *J Appl Physiol. 85*, 1457–1463.

14 Henriksson, J., Reitman, J.S. (1977) Time course of changes in human skeletal muscle succinate dehydrogenase and cytochrome oxidase activities and maximal oxygen uptake with physical

activity and inactivity. *Acta Physiol Scand.* *99*, 91–97.

15 Holloszy, J.O. (1967) Biochemical adaptations in muscle. Effects of exercise on mitochondrial oxygen uptake and respiratory enzyme activity in skeletal muscle. *J Biol Chem.* *242*, 2278–2282.

16 Hoppeler, H., Fluck, M. (2003) Plasticity of skeletal muscle mitochondria: structure and function. *Med Sci Sports Exerc.* *35*, 95–104.

17 Katz, A., Sahlin, K. (1987) Effect of decreased oxygen availability on NADH and lactate contents in human skeletal muscle during exercise. *Acta Physiol Scand.* *131*, 119–127.

18 Katz, A., Sahlin, K. (1988) Regulation of lactic acid production during exercise. *J Appl Physiol.* *65*, 509–518.

19 Kushmerick, M.J. (1983) Energetics of muscle contraction, *Handbook of physiology*, Peachey, L.D., Adrian, R.H., Geiger, S.R. (eds.) Vol. 10. Am Physiol Soc, Bethesda.

20 Kuznetsov, A.V., Tiivel, T., Sikk, P., Kaambre, T., Kay, L., Daneshrad, Z., Rossi, A., Kadaja, L., *et al.* (1996) Striking differences between the kinetics of regulation of respiration by ADP in slow-twitch and fast-twitch muscles *in vivo*. *Eur J Biochem.* *241*, 909–915.

21 Linnarsson, D., Karlsson, J., Fagraeus, L., Saltin, B. (1974) Muscle metabolites and oxygen deficit with exercise in hypoxia and hyperoxia. *J Appl Physiol.* *36*, 399–402.

22 Lopez, P., Ledoux, M., Garrel, D.R. (2000) Increased thermogenic response to food and fat oxidation in female athletes: relationship with VO(2 max). *Am J Physiol Endocrinol Metab.* *279*, E601–607.

23 Madsen, K., Ertbjerg, P., Djurhuus, M.S., Pedersen, P.K. (1996) Calcium content and respiratory control index of skeletal muscle mitochondria during exercise and recovery. *Am J Physiol.* *271*, E1044–1050.

24 Madsen, K., Ertbjerg, P., Pedersen, P.K. (1996) Calcium content and respiratory control index of isolated skeletal muscle mitochondria: effects of different isolation media. *Anal Biochem.* *237*, 37–41.

25 Marcinek, D.J. (2004) Mitochondrial dysfunction measured *in vivo*. *Acta Physiol Scand.* *182*, 343–352.

26 Marcinek, D.J., Schenkman, K.A., Ciesielski, W.A., Conley, K.E. (2004) Mitochondrial coupling *in vivo* in mouse skeletal muscle. *Am J Physiol Cell Physiol.* *286*, C457–463.

27 Marcinek, D.J., Schenkman, K.A., Ciesielski, W.A., Lee, D., Conley, K.E. (2005) Reduced mitochondrial coupling *in vivo* alters cellular energetics in aged mouse skeletal muscle. *J Physiol.* *569*, 467–473.

28 Mogensen, M., Bagger, M., Pedersen, P.K., Fernstrom, M., Sahlin, K. (2006) Cycling efficiency in humans is related to low UCP3 content and to type I fibres but not to mitochondrial efficiency. *J Physiol.* *571*, 669–681.

29 Mogensen, M., Sahlin, K. (2005) Mitochondrial efficiency in rat skeletal muscle: influence of respiration rate, substrate and muscle type. *Acta Physiol Scand.* *185*, 229–236.

30 Nemirovskaia, T.L., Shenkman, B.S., Nekrasov, A.N., Kuznetsov, A.V., Saks, V.A. (1993) [Effect of training on the structural-metabolic indicators in athletes' skeletal muscles]. *Biokhimiia.* *58*, 471–479.

31 Newsholme, E., Start, C. (1974). Regulation in metabolism (eds.). Wiley, London.

32 Ponsot, E., Dufour, S.P., Zoll, J., Doutrelau, S., N'Guessan, B., Geny, B., Hoppeler, H., Lampert, E., *et al.* (2006) Exercise training in normobaric hypoxia in endurance runners. II. Improvement of mitochondrial properties in skeletal muscle. *J Appl Physiol.* *100*, 1249–1257.

33 Rasmussen, U.F., Krustrup, P., Bangsbo, J., Rasmussen, H.N. (2001) The effect of high-intensity exhaustive exercise studied in isolated mitochondria from human skeletal muscle. *Pflugers Arch.* *443*, 180–187.

34 Ray, J., Noll, F., Daut, J., Hanley, P.J. (2002) Long-chain fatty acids increase basal metabolism and depolarize mitochondria in cardiac muscle cells. *Am J Physiol Heart Circ Physiol.* *282*, H1495–1501.

35 Richardson, R.S., Noyszewski, E.A., Leigh, J.S., Wagner, P.D. (1998) Lactate efflux from exercising human skeletal muscle: role of intracellular PO2. *J Appl Physiol. 85*, 627–634.

36 Sahlin, K. (1991) Control of energetic processes in contracting human skeletal muscle. *Biochem Soc Trans. 19*, 353–358.

37 Sahlin, K., Katz, A., Henriksson, J. (1987) Redox state and lactate accumulation in human skeletal muscle during dynamic exercise. *Biochem J. 245*, 551–556.

38 Sahlin, K., Nielsen, J.S., Mogensen, M., Tonkonogi, M. (2006) Repeated static contractions increase mitochondrial vulnerability toward oxidative stress in human skeletal muscle. *J Appl Physiol. 101*, 833–839.

39 Sahlin, K., Soderlund, K., Tonkonogi, M., Hirakoba, K. (1997) Phosphocreatine content in single fibers of human muscle after sustained submaximal exercise. *Am J Physiol. 273*, C172–178.

40 Sahlin, K., Tonkonogi, M., Soderlund, K. (1998) Energy supply and muscle fatigue in humans. *Acta Physiol Scand. 162*, 261–266.

41 Saks, V.A., Kuznetsov, A.V., Khuchua, Z.A., Vasilyeva, E.V., Belikova, J.O., Kesvatera, T., Tiivel, T. (1995) Control of cellular respiration *in vivo* by mitochondrial outer membrane and by creatine kinase. A new speculative hypothesis: possible involvement of mitochondrial-cytoskeleton interactions. *J Mol Cell Cardiol. 27*, 625–645.

42 Sugano, T., Oshino, N., Chance, B. (1974) Mitochondrial functions under hypoxic conditions. The steady states of cytochrome c reduction and of energy metabolism. *Biochim Biophys Acta. 347*, 340–358.

43 Terjung, R.L., Baldwin, K.M., Mole, P.A., Klinkerfuss, G.H., Holloszy, J.O. (1972) Effect of running to exhaustion on skeletal muscle mitochondria: a biochemical study. *Am J Physiol. 223*, 549–554.

44 Tonkonogi, M., Harris, B., Sahlin, K. (1998) Mitochondrial oxidative function in human saponin-skinned muscle fibres: effects of prolonged exercise. *J Physiol. 510*, 279–286.

45 Tonkonogi, M., Krook, A., Walsh, B., Sahlin, K. (2000) Endurance training

increases stimulation of uncoupling of skeletal muscle mitochondria in humans by non-esterified fatty acids: an uncoupling-protein-mediated effect? *Biochem J. 351 Pt 3*, 805–810.

46 Tonkonogi, M., Sahlin, K. (1997) Rate of oxidative phosphorylation in isolated mitochondria from human skeletal muscle: effect of training status. *Acta Physiol Scand. 161*, 345–353.

47 Tonkonogi, M., Sahlin, K. (1999) Actively phosphorylating mitochondria are more resistant to lactic acidosis than inactive mitochondria. *Am J Physiol. 277*, C288–293.

48 Tonkonogi, M., Walsh, B., Svensson, M., Sahlin, K. (2000) Mitochondrial function and antioxidative defence in human muscle: effects of endurance training and oxidative stress. *J Physiol. 528 Pt 2*, 379–388.

49 Tonkonogi, M., Walsh, B., Tiivel, T., Saks, V., Sahlin, K. (1999) Mitochondrial function in human skeletal muscle is not impaired by high intensity exercise. *Pflugers Arch. 437*, 562–568.

50 Walsh, B., Hooks, R.B., Hornyak, J.E., Koch, L.G., Britton, S.L., Hogan, M.C. (2006) Enhanced mitochondrial sensitivity to creatine in rats bred for high aerobic capacity. *J Appl Physiol. 100*, 1765–1769.

51 Walsh, B., Tonkonogi, M., Sahlin, K. (2001) Effect of endurance training on oxidative and antioxidative function in human permeabilized muscle fibres. *Pflugers Arch. 442*, 420–425.

52 Walsh, B., Tonkonogi, M., Soderlund, K., Hultman, E., Saks, V., Sahlin, K. (2001) The role of phosphorylcreatine and creatine in the regulation of mitochondrial respiration in human skeletal muscle. *J Physiol. 537*, 971–978.

53 Wilson, D.F. (1994) Factors affecting the rate and energetics of mitochondrial oxidative phosphorylation. *Med Sci Sports Exerc. 26*, 37–43.

54 Wilson, D.F., Rumsey, W.L., Green, T.J., Vanderkooi, J.M. (1988) The oxygen dependence of mitochondrial oxidative phosphorylation measured by a new optical method for measuring oxygen concentration. *J Biol Chem. 263*, 2712–2718.

55 Wojtczak, L., Schonfeld, P. (1993) Effect of fatty acids on energy coupling processes in mitochondria. *Biochim Biophys Acta. 1183*, 41–57.

56 Zoll, J., Koulmann, N., Bahi, L., Ventura-Clapier, R., Bigard, A.X. (2003) Quantitative and qualitative adaptation of skeletal muscle mitochondria to increased physical activity. *J Cell Physiol. 194*, 186–193.

57 Zoll, J., Sanchez, H., N'Guessan, B., Ribera, F., Lampert, E., Bigard, X., Serrurier, B., Fortin, D., *et al.* (2002) Physical activity changes the regulation of mitochondrial respiration in human skeletal muscle. *J Physiol. 543*, 191–200.

15
Mitochondrial Medicine: The Central Role of Cellular Energetic Depression and Mitochondria in Cell Pathophysiology

Enn Seppet, Zemfira Gizatullina, Sonata Trumbeckaite,
Stephan Zierz, Frank Striggow, and Frank Norbert Gellerich

Abstract

Dysfunction of mitochondria affects a cell's life in many ways. It causes cellular energetic suppression, as insufficient mitochondrial synthesis of ATP directly impairs the functions of all ATP-dependent enzymes and proteins. Associated intracellular Ca^{2+} and ROS overload activates mitochondria to initiate and promote cell death via apoptosis or necrosis, depending on cellular levels of ATP. The clinical manifestation of mitochondrial dysfunction may be very different because of the large variety of underlying mechanisms. Acquired or inherited mutations of the mitochondrial genome result in a number of syndromes known as mitochondrial diseases. In addition, there is a second large group of diseases in which mitochondrial impairments are initiated and promoted by extramitochondrial pathological processes. This group includes neurodegenerative diseases and syndromes resulting from typical pathological processes such as hypoxia/ischemia, inflammation, intoxications, cancerogenesis, and abnormal protein aggregations. In these conditions, mitochondria may be the source and targets of different intracellular signaling cascades. At mitochondrial outer membrane permeabilization (MOMP), apoptosis-inducing factor (AIF) and cytochrome c are liberated, inducing the apoptotic cell death. If oxidative phosphorylation is impaired, cytosolic phosphorylation potentials decrease and the ability for cell work is reduced. As a result the $[Ca^{2+}]_{cyt}$ increases. The impaired energy metabolism of the affected cell is called *cellular energetic depression* (CED). At pronounced CED, the cell dies from necrosis. Both the apoptotic and necrotic pathways, and all others that include mitochondrial impairments, initiate *mitochondrial cell death*. Regarding tumor progression, it is now widely accepted that mitochondria are responsible for the avoidance of cell death by apoptosis, a typical feature of cancer. To understand the pathogenetic mechanisms of mitochondrial dysfunction and to cope with it via adequate medical cure, it is important to effectively diagnose the altered functions of mitochondria. We provide data showing that permeabilized cell techniques in combination with high-resolution respirometry and multi-substrate

Molecular System Bioenergetics: Energy for Life. Edited by Valdur Saks
Copyright © 2007 WILEY-VCH Verlag GmbH & Co. KGaA, Weinheim
ISBN: 978-3-527-31787-5

inhibitor titrations provide a novel and effective set of methods for the diagnosis of mitochondrial impairments.

15.1
Introduction

Otto Warburg was the first to recognize the importance of impaired mitochondrial function in cellular pathophysiology by proposing that development of cancer is causally related to altered energetic metabolism by suppression of oxidative phosphorylation and activation of glycolysis [1]. In 1962 Rolf Luft demonstrated that a clinically expressed hypermetabolic state resulted from alterations in mitochondrial structure and function [2]. Later, different encephalomyopathic syndromes (Table 15.1) were shown to be determined by defects in mitochondrial metabolism due to mutations in mitochondrial DNA (mtDNA) and nuclear DNA (nDNA) [3–5]. Based on these studies, the concept of mitochondrial diseases [6] was introduced, which considers these diseases to be manifested in many different tissues but primarily caused by defects in oxidative phosphorylation (OXPHOS) due to mutations of mitochondrial proteins (Table 15.1). Recently, mutations in genes, responsible for the expression of cardiolipin (Barth syndrome [7]) and coenzyme Q [8] metabolism, also have been discovered.

Table 15.1 Genetic classification of mitochondrial diseases.

Mutations of mtDNA	Mutated nuclear genes controlling the stability of mtDNA	Mutated nuclear genes encoding respiratory chain proteins	Mutated nuclear genes encoding proteins of non-respiratory chain metabolism	Mutated nuclear genes indirectly involved in the respiratory chain
Large-scale deletions: KSS, Pearson's syndrome, PEO	ANT1, Twinkle, POLGI1, TP, TH2, DGUOK, deoxynucleotide carrier	Complex I, NDUFS1, NDUFS2, NDUSFS4, NDUFS8, NDUFV1	Coenzyme Q deficiency, cardiolipin deficiency	FRDA1 (Friedreich's ataxia), X-linked deafness ataxia, sideroblastic anemia
Point mutations: MELAS, MERRF, NARP, LHON		Complex II, SDHA, SDHB, SDHC, SDHD		Hereditary spastic paraplegia, X-linked deafness-dystonia syndrome
		Complex III, UQCRB, subunit VII		Autosomal dominant optic atrophy

The concept of mitochondrial medicine was successful in promoting a massive search for new DNA mutations underlying the pathogenesis of different diseases. An impressive number of different mutations have been discovered (see http://www.mitomap.org and recent reviews: [9, 10]). However, it has become increasingly clear that the genetic approach suffers from serious limitations.

Firstly, the genotype–phenotype relationships registered so far are too poor to explain the enormous phenotypic variety of clinical manifestations of mitochon-

---

Fig. 15.1 The central role of mitochondria in mitochondrial diseases, neurodegenerative diseases, inflammation, ischemia, intoxication, cancer, and aging. The mitochondrial pathophysiology of these clinical entities arises from very different disease-specific impairments. Most importantly, mitochondria produce ATP for cell work. The phosphocreatine (PCr) shuttle adjusts high phosphorylation potentials for cell work and effectively transports PCr (the ATP equivalent) to the ATP-utilizing enzymes and Cr (the ADP equivalent) back into the intermembrane space (IMS). Note that the mitochondrial and extramitochondrial creatine kinases (CK) work in opposite directions and are functionally coupled to mi-CK and MM(BB)-CK, respectively. In *ischemia* the mitochondrial inner membrane (IM) becomes permeable due to opening of the permeability transition (PT) pore, and the outer membrane (OM) becomes leaky, which results in the release of the apoptosis-inducing factor (AIF) and cytochrome *c*, thus inducing apoptosis and dysfunction of the OXPHOS. From the intermembrane space (IMS), ADP is transported via the ADN translocator into the matrix, where F_0F_1-ATPase rephosphorylates it to ATP driven by the proton gradient that is regenerated by the action of the respiratory chain. The respiratory chain oxidizes the reducing equivalents (e.g., NADH) formed by dehydrogenases, such as pyruvate dehydrogenase, which reduce NAD^+. A high redox pressure ($NADH/NAD^+\uparrow$) is required for the reduction of oxygen, forming water via the cytochrome *c* oxidase (COX). The energy of this reaction is used for pumping the protons into the IMS. The K_m of COX for O_2 is very low (0.3 μM). With increasing O_2 levels, the non-enzymatical reduction of O_2 by NADH increases, resulting in formation of reactive oxygen species (ROS). Normally the ROS can be eliminated by mitochondrial Mn-dependent superoxide dismutase (MnSOD), a component of the antioxidant defense system. When the ROS formation exceeds the defense capacity, a dangerous ROS attack on all biomolecules occurs (oxidative stress), acutely reducing the activity of respiratory chain enzymes but chronically impairing the mitochondrial and nuclear DNA. In *inflammation* increased NO production acutely reinforces the oxidative stress in mitochondria via the formation of peroxynitrite ($ONOO^-$) from O_2 and NO. Chronic ROS is assumed to cause mutations responsible for *aging* and *cancerogenesis*. Hereditary mutations in the mitochondrial genome cause mitochondrial cytopathies (mt-cytopathies), e.g., due to impairment of respiratory chain complexes. In *neurodegenerative diseases* hereditary mutations in non-mitochondrial genes cause the formation of cytotoxic proteins that give rise to mitochondrial dysfunction. Ca^{2+} ions can be accumulated by mitochondria, e.g., via the uniporter. Excessive Ca^{2+} in the matrix can induce the opening of the PT pore. In the case of reversible PT, the mitochondria release a fraction of Ca^{2+} that serves as a signaling messenger, but in conditions of irreversible PT, the mitochondria deteriorate and die, leading to serious pathophysiological consequences. Finally, the *intoxication* of mitochondria by medicaments or by specific toxins can cause acute impairments and in some cases symptomatic neurodegenerative diseases. Impaired mitochondrial function results in cellular energetic depression (CED) (Fig. 15.2).

drial dysfunction [11]. Moreover, mitochondrial disturbances have been found to be linked to an increasing number of disorders, but in most of them, either the mitochondrial dysfunction is not primarily causal for the corresponding disease or its primary causal role has not been disclosed to date [12–25].

Secondly, for about 50% of adult patients with mitochondrial disorders, the underlying genetic defect is still unknown. Among children, this percentage is even higher, between 80% and 90% [9]. The incidences of such functional disorders of mitochondria probably will increase as the methods of their appropriate detection improve further.

Thirdly, gene-based strategies neglect the fact that mitochondria are cell organelles with complex physiological properties (impermeable membranes except for water, NO, CO, and CO_2; special osmotic behavior; concentration gradients for substrates and regulators across the inner and outer membranes [26]; fission–fusion equilibrium [27]) that underlie a large network of relations between mitochondria and cells. Many of the mitochondrial functions are still awaiting discovery, as mitochondria consist of more than 900 proteins, 160 for which the function is unknown [28]. However, it is clear that mitochondria play a universal role in responding to and participating in different pathological processes (Fig. 15.1).

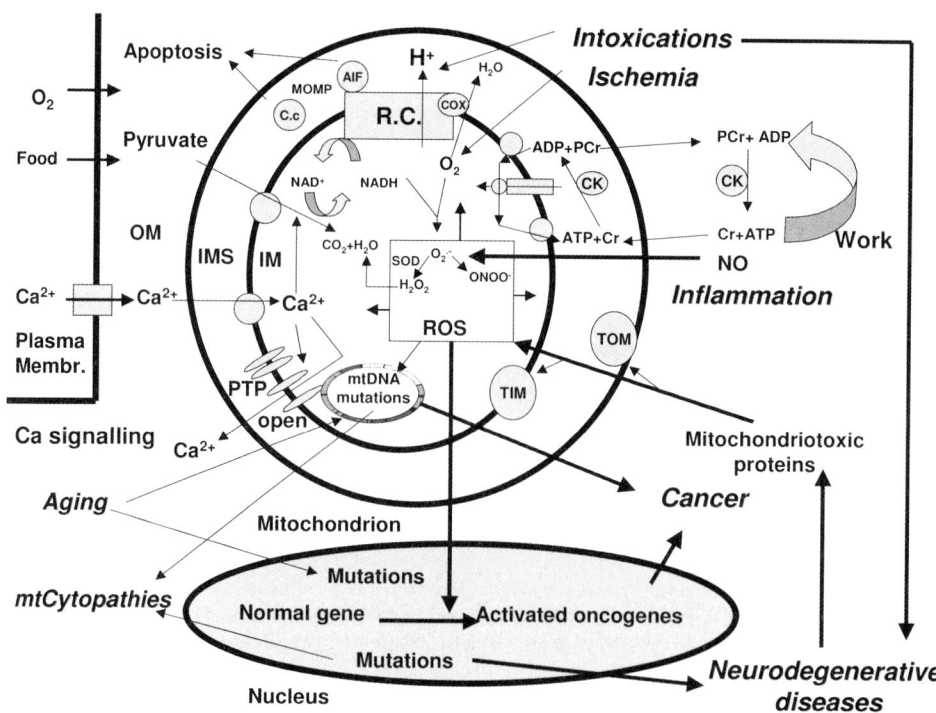

Fig. 15.1 (legend see p. 481)

15.2
The Concept and Molecular Mechanisms of Cellular Energetic Depression and Mitochondrial Cell Death

Neurodegenerative, infectious, traumatic, ischemic, and metabolic disorders are associated with progression of different cell death programs. There are at least three different forms of programmed cell death (PCD): type I, also known as nuclear or apoptotic cell death; type 2, an autophagic cell death; and type III, also known as cytoplasmic cell death [29]. Others even distinguish between six subtypes of cell death (see [29] and references therein): apoptosis, autophagy, paraptosis, calcium-mediated cell death, poly(ADP-ribose)polymerase (PARP)-dependent cell death, and oncosis. Despite the complexity of the pathways of these different forms of cell death, they all have one common property in that mitochondria play a crucial role in each of them, with the possible exception of extrinsic apoptotic and autophagic cell death [29, 30]. Moreover, it is probably not wise to assume that mitochondria act as an all-or-nothing master switch between the different death pathways without regarding their bioenergetic and other complex tasks in cell metabolism (Fig. 15.1) [31]. One should also consider that mitochondrial bioenergetic functions may be acutely impaired in the course of different pathological processes such as inflammation, hypoxia/ischemia, intoxications, metabolic blockade, and oxidative stress (Fig. 15.1 and Table 15.2). As a result of all these conditions, cytosolic phosphorylation potential declines, limiting the cell's ability for work. There is a development of intracellular Ca^{2+} overload due to reduced efficiency of the sarco/endoplasmic reticulum (S/EPR) and sarcolemmal Ca^{2+} pumps (Figs. 15.2 and 15.3). Under this mismatch be-

Table 15.2 Mechanisms that impair mitochondrial function.

Oxidative stress by ROS and peroxynitrite

Decreased cytosolic adenine nucleotide concentrations

Leaks in the mitochondrial outer membrane, loss of cytochrome *c*

Leaks in the mitochondrial inner membrane, opening of PTP

Decreased capacity of oxidative phosphorylation due to diminished activities of respiratory chain complexes and other enzymes resulting from mutations and inhibitions

Ca^{2+} overload

Decreased resistance against Ca^{2+}-induced stress

Decreased substrate supply (O_2, fatty acids, pyruvate, etc.)

Impaired ADP transport into mitochondria

Impaired protein transport into mitochondria

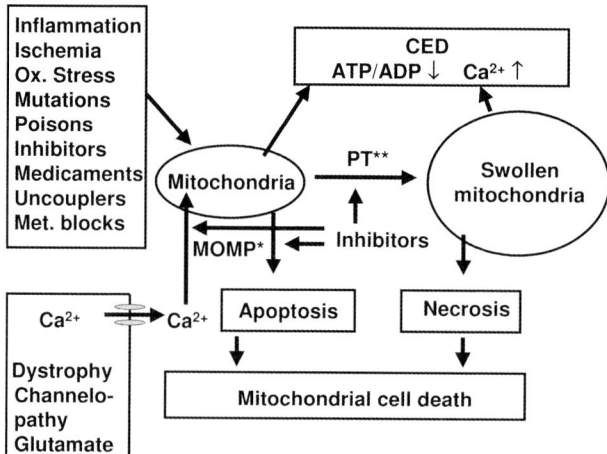

Fig. 15.2 Mitochondrial cell death. Several mechanisms cause impairments of mitochondrial function. At mitochondrial outer membrane permeabilization (MOMP), AIF and cytochrome *c* are liberated, inducing apoptotic cell death. If OXPHOS is impaired, cytosolic phosphorylation potentials decrease and the ability for cell work is reduced. As a result the cytosolic Ca^{2+} concentration increases. The impaired energy metabolism of the affected cell is called cellular energetic depression (CED). At pronounced CED the cell dies from necrosis. Both pathways and all others that include mitochondrial impairments initiate *mitochondrial cell death*. The molecular mechanisms of MOMP and PT are potential targets for preventing mitochondrial cell death.

tween ATP production and utilization (for which we propose the term cellular energetic depression [CED]), mitochondria accumulate significant amounts of Ca^{2+}, which induces the permeability transition (PT) of the mitochondrial inner membrane [32]. This process includes mitochondrial swelling, collapse of the membrane potential ($\Delta\Psi$), and splitting of ATP by mitochondrial F_0F_1-ATPase instead of forming it. If mitochondria shift into PT, the cell must die by necrosis. The CED pathway also can start after Ca^{2+} overload caused by degenerative changes in the cell membrane, or channelopathies [33]. Whereas all of these disturbances lead to the cell's death, the functional status of mitochondria plays a key role in determining which type of mitochondrial cell death will prevail. If the pathogenic signals induce MOMP [29, 30] and CED is relatively mild (i.e., OXPHOS is not yet markedly compromised), cytochrome *c* and AIF are released, initiating apoptotic cell death [29] (Figs. 15.2 and 15.3). However, along with progression of CED, the cellular ATP level may become less than required for supporting apoptosis, and the cell will enter the necrotic pathway of death. We propose that both apoptotic and necrotic cell death, as well as all other pathways that include mitochondrial impairments, should be called mitochondrial cell death (MCD). MCD contributes to degeneration or atrophy of the affected tissue and

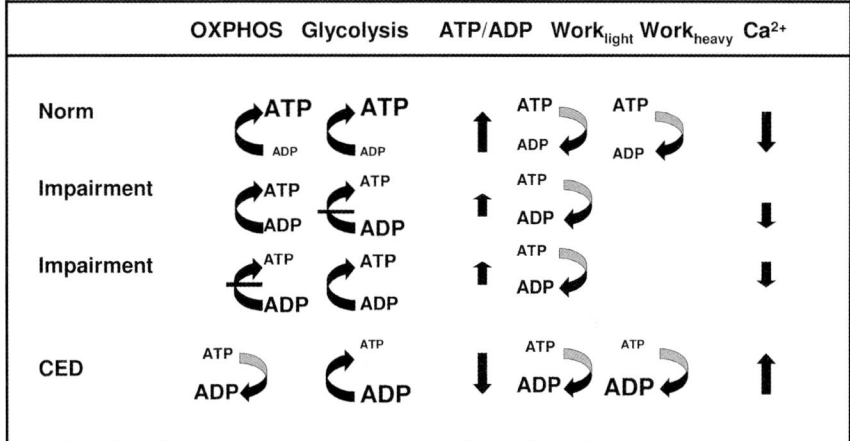

Fig. 15.3 Cellular energetic depression (CED) as a result of an imbalance between ATP-producing and ATP-utilizing processes. Under normal conditions ATP formation by OXPHOS and glycolysis exceeds the ATP-utilizing capacity even at heavy work. The phosphorylation potential remains sufficiently high because $[ATP]_{cyt}$ is high and $[ADP]_{cyt}$ is low, as indicated by the letters' size. Under these conditions $[Ca^{2+}]_{cyt}$ is low. When the mitochondrial or glycolytic functions become impaired, the ATP-regenerating capacity decreases, but it allows performing of minimal cellular work without significant accumulation of cytosolic Ca^{2+}. At increased cell work, however, CED occurs due to decreased phosphorylation potentials and increasing cytosolic Ca^{2+}. After PT pore opening, mitochondria split glycolytic-formed ATP. When the workload reaches extreme levels, even normal cells can develop CED.

can occur slowly, as in the case of neurodegenerative diseases, or rapidly, as in rhabdomyolysis. Because of the central importance of MCD, the mitochondria should be regarded as an important target for pharmacological treatments. Indeed, several potent inhibitors of PT exist, such as cyclosporin A (CsA) [29] and sanglifehrin A [34], that make it possible to attenuate or cure the atrophic processes. It has been shown that CsA can increase the lifespan and decrease the symptoms of degeneration in collagen VI–deficient mice [35]. Also, MOMP is realized via a cascade of molecular reactions [29, 30] and therefore should be inhibitable. To be effective in pharmacological intervention, it is necessary to discriminate between cell death pathways where mitochondria participate and those that occur without mitochondrial involvement [29]. To our knowledge, the term MCD was first proposed by Kroemer et al., but they used it only for the mitochondrial apoptotic pathway including MOMP [28, 36], a phenomenon newly discovered at that time. We now propose to use the term MCD for all modes of mitochondrial cell death irrespective of the primary reason, including acute and chronic events of PT, MOMP, and any other mitochondrial impairment.

15.2.1
Interaction of Mitochondria and Sarco/Endoplasmic Reticulum in Regulation of Cytosolic Ca^{2+}

It has been known for a long time that Ca^{2+} overload impairs mitochondrial function [37]. Later, it was found that mitochondrial function can be regulated by [Ca^{2+}]$_{cyt}$ [38] and that mitochondria are actively involved in controlling cellular Ca^{2+} signaling [39] (Fig. 15.4). We investigated the problem of whether or not mitochondria accumulate and release Ca^{2+} during Ca^{2+} signaling in a cell-free system, consisting of isolated sarcoplasmatic reticulum vesicles, an ATP-regenerating system (creatine kinase, 10 mM PCr, and 5 mM ATP), and heart mitochondria in agarose gel. We found that after addition of mitochondria to the system, the velocity of Ca^{2+} waves traveling along the gel increased by 48% [40] but decreased after addition of antimycin A to values registered without mitochondria. These data clearly show that mitochondrial control over Ca^{2+} signaling is exerted by active accumulation and release of Ca^{2+} as well as by the adjustment of sufficiently high phosphorylation potentials [40, 41]. As shown in Fig. 15.4, Ca^{2+} enters the matrix through the uniporter located in the inner membrane [42]. This process is driven by $\Delta\Psi$ and is sensitive to ruthenium red [42]. Because mitochondrial enzymes such as pyruvate dehydrogenase, isocitrate dehydrogenase, and 2-oxoglutarate dehydrogenase can be stimulated by Ca^{2+}, thus forcing

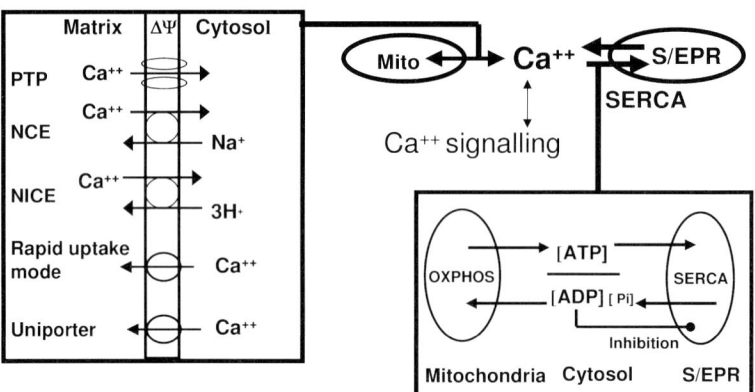

Fig. 15.4 Involvement of mitochondria and the sarco/endoplasmic reticulum (S/EPR) in Ca^{2+} signaling. Mitochondria and the S/EPR are involved mainly in cellular Ca^{2+} signaling. Ca^{2+} accumulation by S/EPR is realized by SERCA, which requires ATP at sufficiently high phosphorylation potentials. Because increasing ADP would inhibit SERCA, the PCr shuttle transports it into the mitochondria (see Fig. 15.1). Mitochondria accumulate Ca^{2+} via the uniporter and by the rapid uptake mode. Ca^{2+} releasing rates via the Na$^+$-independent pathway for Ca^{2+} efflux (NICE) or the Na$^+$-dependent pathway (NCE) for Ca^{2+} efflux are low in comparison to the fast reversible and irreversible Ca^{2+} efflux via the PT pore [32].

the processes of $\Delta\Psi$ generation [38], mitochondrial energy metabolism and Ca^{2+} signaling are obviously linked to each other. It is known that the $[Ca^{2+}]_{cyt}$ normally varies between 100 nM and 1 µM, whereas $[Ca^{2+}]_{mit}$ accumulation starts significantly only if $[Ca^{2+}]_{cyt}$ exceeds 400 nM [42]. Based on these data, the effectiveness of mitochondrial Ca^{2+} uptake appears to be rather low and therefore unimportant in the regulation of $[Ca^{2+}]_{cyt}$ levels. However, Ca^{2+} in the cytosol is strongly compartmentalized, and spatiotemporal changes to values of its concentration higher than 40 µM can be achieved even in normal cellular circumstances [43]. This situation confers to mitochondria an important task in taking up Ca^{2+} that locally accumulates at very high concentrations (e.g., in the close proximity of mitochondria to Ca^{2+} channels of the cell membrane or of the S/EPR) in order to ensure a proper metabolic response to this signal and to redistribute Ca^{2+} over the cell by dynamic fusion–fission conversion [42].

However, in the case of decreased phosphorylation potentials and CED, reduced availability of mitochondrial ATP suppresses the activity of the S/EPR Ca^{2+} pump (SERCA), which in turn results in increased $[Ca^{2+}]_{cyt}$ and prolonged duration of Ca^{2+}-mediated signals. Indeed, in fibroblast lines from patients with Leigh disease, reduced SERCA function was detectable as the Ca^{2+} influx into the Ca^{2+}-depleted endoplasmic reticulum was significantly reduced [45].

15.2.2
Role of Altered Energy Compartmentation in Cellular Energetic Depression

CED is related not only to injuries of mitochondrial membranes but also to alterations in the intracellular integration of mitochondria with the cytoskeleton and ATP-consuming enzymes such as SERCA and contractile proteins (ATPases). Our previous studies addressing the processes of energy metabolism in its whole cellular context have revealed that in cardiomyocytes, mitochondria, ATPases, creatine kinases (CK), adenylate kinases (AK), and adenine nucleotides are compartmentalized into functional complexes termed *intracellular energy units* (ICEUs), which represent the basic modules of cellular energy metabolism. The structure of ICEUs formed by cytoskeletal proteins ensures effective energy transfer and feedback regulation between mitochondria and ATPases via specific isoenzymes of CK, AK, and direct ATP/ADP channeling (see Fig. 15.1) [46–50]. There are several studies showing that intracellular energy transfer via the CK phosphotransfer network becomes compromised in the case of cardiac diseases, e.g., in heart failure [51–53]. This finding was evident from a reduced intracellular PCr:ATP ratio and increased ADP concentrations detected by ^{31}P-NMR in the failing heart [51]. Because the tissue PCr:ATP ratio decreases in parallel to clinical progression of heart failure, a change in this parameter has been suggested to predict the mortality of patients with dilated cardiomyopathy [51]. The mechanisms underlying decreased PCr synthesis include reduced myocardial creatine and creatine transporter levels [52], downregulated expression of mitochondrial CK (mi-CK) [52, 53], and decoupling between the mi-CK and adenine nucleotide translocase

(ANT) due to oxidation of mi-CK by reactive oxygen species (ROS) and/or nitrosylation of its SH groups [54]. On the other hand, reduced expression of extra-mitochondrial CK (cytosolic and MM-CK or BB-CK bound to different structures) and their inhibition by AMP kinase [53, 55, 56] decrease the effectiveness of ADP rephosphorylation in the closest vicinity of ATP-consuming enzymes [52]. Impaired interaction between mi-CK and ANT might favor opening of the mitochondrial PT pore, thus leading to MCD. This hypothesis is based on studies showing that mi-CK forms tight complexes with ANT. The functional coupling between both enzymes not only ensures effective production of PCr but also protects the mitochondria from PT pore opening [57]. This protective mechanism may be based on the local accumulation of ADP in the cleft between the interacting ANT and mi-CK, thus promoting the locking of ANT in the m-state. This condition favors less PT pore opening compared with an ANT in the c-state, which predominates in the absence of coupling between mi-CK and ANT [58]. This assumption is further substantiated by the observation that ADP is able to prevent the Ca^{2+}-induced opening of the PT pore [59].

Whenever intracellular CK-mediated energy transfer fails, the role of the adenylate kinase (AK) system in shuttling ADP into the mitochondria increases [60, 61]. This happens because the mitochondrial AK2 isoform is functionally coupled to ANT and accumulation of ADP near ATP-utilizing enzymes favors using its β-phosphoryls by AK for ATP formation [60]. If the CK phosphotransfer fades, the direct ADP diffusion to ANT comes increasingly into play. For example, it is activated when control of mitochondrial respiration via mi-CK weakens, as in dystrophin-deficient mice heart [62] or in muscles deficient in cytoplasmic and mitochondrial CK [63], probably due to increased permeability of the mitochondrial outer membrane (MOM) for adenine nucleotides [62].

In the case of neurodegenerative diseases (e.g., Parkinson's disease, Alzheimer's disease, Huntington's disease, and amyotrophic lateral sclerosis [ALS]), mitochondria may be impaired due to toxic and disease-specific proteins. Mutated huntingtin binds to the MOM, suppresses the mitochondrial capacity of Ca^{2+} accumulation [64], and hinders the axonal mobility of mitochondria [65]. In this regard, the function of mitochondrial kinases such as hexokinase (HK) and mi-CK may become altered. In normal brain cells, the function of HK depends on its interaction with the MOM and the cytoskeletal network [68, 69]. On the other hand, mi-CK localizing in the intermembrane space is functionally coupled to ANT [70]. Both of these mitochondrial kinases are expected to participate in the formation of the PT pore [71, 72]. Both of them also suppress reactive oxygen species (ROS) production in mitochondria, due to their stimulation of OXPHOS [73, 74]. In the case of Alzheimer's disease, CK activity is decreased, caused by structural alterations in the brain cell's cytoskeleton due to hyperphosphorylated tau protein [67]. Likewise, a G93A transgenic mice model of familial ALS exhibited decreased CK activity in the spinal cord [75]. These data point to the possibility that neurodegenerative diseases may be associated with impaired functional coupling of mitochondrial kinases with OXPHOS, which could contribute to CED and ROS production in these diseases [76, 77].

15.3
Involvement of Mitochondria in Aging and Disease

15.3.1
Mitochondria and Aging

It is well established that large-scale deletions and point mutations in protein and tRNA genes of mtDNA accumulate with aging. Recently, an increasing number of mutations in control regions of the mtDNA have been identified, i.e., T414G in cultured fibroblasts, C150T in white blood cells, and A189G and T408 in muscle [78]. Heteroplasmy levels of these mutations correlate with a decline in mitochondrial function as reflected by decreased cytochrome c oxidase (COX) activity [79]. However, these hotspot mutations have not been observed in the brain [80]. New evidence for the mitochondrial contribution to aging came from the mtDNA mutator mice generated by Larsson et al. in which the 3'- to 5'-exonuclease activity of the mtDNA polymerase-γ (POLG) was diminished [81]. Because the POLG proofreading activity was eliminated while the polymerase activity was preserved in these animals, the mtDNA mutations accumulated due to a summation of uncorrected errors during replication. At each investigated stage of life, the POLG$^{-/-}$ mice had many more point mutations in mtDNA than did wild-type mice [82]. These increased point mutations caused a drastic reduction in respiration rates and ATP production [83]. Furthermore, the lifespan of the POLG$^{-/-}$ mice was shortened to about 50 weeks from more than 100 weeks in normal mice. POLG mutations cause parkinsonism, ophthalmoplegia, and myopathy in humans [84]. The importance of mitochondrial ROS production in aging is demonstrated by the observation that exogenous enhancements of mitochondrial antioxidant defense resources can increase longevity. For example, overexpression of mitochondrial MnSOD and methionine sulfoxide reductase prolonged the lifespan in short-living but not in long-living strains of *Drosophila* [85]. Recently, a mitochondrial catalase could be shown to increase the lifespan of a mouse strain by 20% [86]. These observations demonstrate that aging may be associated with excessive accumulation of ROS. Most likely, ROS stimulates aging via promotion of mutations in mtDNA. In support of this idea, it was demonstrated that increased mitochondrial ROS production accelerated the mtDNA mutation rate in mice with deficient nDNA-encoded heart/muscle and brain isoforms of the ANT gene. Because the mitochondria in these tissues could not import ADP, the main user of reducing equivalents was missing. Therefore, increased amounts of NADH and oxygen were used for ROS production, causing higher rates of mtDNA mutations compared with normal animals [87].

The involvement of mitochondrial ROS production in aging is congruent with the life-extending impact of caloric restriction. In rodent studies, oxidation products of mtDNA bases were increased with age, whereas caloric restriction inhibited this process [88, 89]. Furthermore, age-induced alterations in genes encoding for enzymes of both mitochondrial energy metabolism and antioxidant defense in muscle and brain were largely prevented by caloric restriction [90]. All of these data are consistent with the hypothesis that oxidative damage to mitochondria and mtDNA is of great importance in aging.

15.3.2
Mitochondria in Neurodegenerative Diseases

It is widely accepted that neurodegenerative diseases such as Alzheimer's disease, Parkinson's disease, Huntington's disease, ALS, and prion diseases have common cellular and molecular mechanisms that are based on abnormal protein aggregation and inclusion body formation [91]. Protein aggregates usually consist of fibers of misfolded proteins with a β-sheet conformation, called amyloids. Although the altered proteins are believed to exert a neurotoxic effect, it is not clear in which form (soluble monomers, oligomers, or larger aggregates) they are most dangerous [92]. Mostly, as in the case of synuclein, tau, and huntingtin, the mutated proteins or their chemically modified forms exhibit cytotoxic properties and aggregate more rapidly than the wild-type proteins [91]. However, as discussed below, the soluble cytotoxic proteins rather than the amyloids can interfere with numerous cellular proteins and structures [93], especially with mitochondrial membranes (for a review, see [84]). Indeed, evidence has been accumulated indicating that despite a large clinical and pathophysiological heterogeneity, the neurodegenerative diseases have a second common factor: the involvement of impaired mitochondria in their pathophysiology [84].

15.3.2.1 **Huntington's Disease**
Huntington's disease (HD) is a progressive neurodegenerative disorder [94] caused by a CAG repeat expansion in the coding region of the IT15 gene, resulting in an elongated polyglutamine stretch in the huntingtin (htt) protein [95, 96]. The length of the CAG repeat is inversely correlated with age at disease onset [97]. The symptoms of HD are motor abnormalities including chorea and psychiatric disturbances with gradual dementia, and autopsies have revealed atrophic changes in the striatum [94].

Observations that HD patients lose body weight despite normal or above-average food intake [99–101] support the assumption that energy metabolism must be impaired in HD. Alterations in energy metabolism have been found in the brain of HD patients [102–107]. Accordingly, NMR studies have demonstrated elevated lactate in the cerebral cortex of HD patients, suggesting defects in OXPHOS [102, 103]. Decreased activities of the respiratory chain complexes were found postmortem in HD striatum [104, 105] and basal ganglia [103]. In addition, abnormal mitochondrial morphology was detected in cortical biopsies obtained from patients suffering from HD [106, 107]. The animal models of HD also confirm the mitochondrial dysfunction. In striatal cells from mutant htt knock-in mice embryos, mitochondrial respiration and ATP production were significantly decreased [108]. Toxicological investigations showed that ingestion of 3-nitro-propionic acid, an inhibitor of the mitochondrial enzyme succinate dehydrogenase, results in selective striatal lesions accompanied by chorea and dystonia

[109, 110]. Because htt is expressed in many cerebral and non-cerebral cell types [96], HD is a disease not only of the striatum but of other tissues as well. It is known that skeletal muscle atrophy in association with impaired energy metabolism is common in HD patients [111–113]. In skeletal muscle reduced phosphorylation potentials and reduced recovery rates of PCr [112, 113] are valid markers of the occurrence of CED. A detailed gene expression study revealed that CED is probably related to the altered metabolic profile [111].

We recently found in skeletal muscle of 14-week-old transgenic R6/2 mice [114] that the respirometric properties of isolated HD mitochondria are not different from that in wild-type mitochondria. After addition of 20 μM Ca^{2+}, however, a significant impairment of mitochondrial function was detectable in transgenic R6/2 mice. Complex I–dependent respiration rates were sensitively inhibited by increasing $[Ca^{2+}]$. In parallel, we found a decreased threshold against Ca^{2+}-induced PT as revealed by mitochondrial swelling measurements in comparison to wild-type mitochondria [114]. Therefore, we conclude that HD is likely associated with a decreased tolerance of mitochondria against conditions leading to a cytosolic Ca^{2+} overload.

15.3.2.2 **Parkinson's Disease**

A possible role for mitochondrial dysfunction in Parkinson's disease (PD) was suggested by the discovery that 1-methyl-4 phenyl-1,2,3,6-tetrahydropyridine (MPTP) causes parkinsonian syndromes by acting through inhibition of complex I of the respiratory chain [115]. This effect was observed in the substantia nigra [116, 117] and platelets of PD patients [118]. In skeletal muscle biopsies of PD patients, the activities of the different complexes of the respiratory chain were found to be reduced. This finding was accompanied by an increased number of point mutations in mtDNA [119]. The functional impact of these alterations was further confirmed by demonstrating increased flux control coefficients of complexes I and IV in HD patients [119].

The etiology of the distinct mitochondrial dysfunction in PD is still unclear. This dysfunction may be caused by MPTP [120], rotenone [121], paraquat [122], endogenous ROS [123], or isoquinolines [124]. So far several mutations have been shown to be related to PD, but these mutations cover only about 10% of all patients. Mutations or polymorphisms in mtDNA [125, 126] and in at least nine nuclear genes were identified as causes of PD, or at least as risk factors for PD. Correspondingly, the mutated proteins include α-synuclein, the ubiquitin E3 ligase parkin, the antioxidant protein DJ-1, the tensin homologue (PTEN)-induced kinase 1 (PINK1), the leucine-rich repeat kinase (LRRK2), and the serine protease HTRA2, which are directly or indirectly connected to mitochondrial function [93].

α-Synuclein is a major component of the Lewy bodies. Specific mutations in α-synuclein are associated with autosomal dominant familial PD. The primary effect of PD-related mutations of α-synuclein is an increased formation of oligomeric and fibrillar aggregates that promotes abnormal protein accumulation or degradation with oxidative stress and mitochondrial dysfunction. In transgenic mice, overexpression of α-synuclein impairs mitochondrial function, increases oxidative stress and enhances the MPTP-induced pathology of the substantia nigra [127]. Moreover, overexpression of the A53T mutant α-synuclein gene causes direct damage to mitochondria [92, 128]. In contrast, α-synuclein knockout mice were resistant against MPTP and mitochondrial toxins, e.g., malonate and 3-nitropropionic acid [84, 129].

Mutations in parkin are associated with autosomal recessive juvenile PD. Parkin-deficient *Drosophila* [130] and mice [131] strains also exhibit an impairment of mitochondrial function as well as increased oxidative stress. Leucocytes from patients with parkin mutations showed decreased complex I activities [132]. Parkin can associate with the MOM and prevent mitochondrial swelling and cytochrome *c* release. These positive effects can be abolished by specific parkin mutations [133].

Mutations in DJ-1 are also linked to autosomal recessive juvenile PD. The function of DJ-1 seems to be the protection of cells against oxidative stress. DJ-1 can act as a redox sensor of oxidative stress that causes its translocation into mitochondria. The C106 mutation of DJ-1 prevents this translocation and induces mitochondrial dysfunction [134]. DJ-1 knockout results in a normal mice phenotype. Nevertheless, these animals exhibit an increased loss of dopaminergic neurons in response to MPTP [135].

PINK1 is a kinase that is localized in mitochondria, where it is somehow involved in neuroprotection. Overexpression of wild-type PINK1 prevents apoptosis under basal and staurosporine-induced conditions by hindering cytochrome *c* release. On the other hand, mutated PINK1 antagonizes this effect [136]. PINK1 deficiency causes increased sensitivity to the specific complex I inhibitor rotenone in *Drosophila* [137].

A number of studies have shown that degeneration of dopaminergic neurons in PD is associated with microglial-mediated inflammation and neurotoxicity (reviewed in [138] and also below). Activation of inflammation is suggested by the finding that PD patients as well as animal models of PD that were treated with lipopolysaccharide (LPS), MPTP, rotenone, or 6-hydroxydopamine exhibit elevated antibody levels against proteins modified by dopamine oxidation products, increased concentrations of cytokines (IL-1, IL-6, IL-10, and TNF-α), and augmented ROS production [138]. All these changes are associated with impaired function of complex I of the respiratory chain in dopaminergic neurons. It has been suggested that modifications of biomolecules by ROS and dopamine quinones trigger microglia activation, which in turn further promotes neurotoxicity [138] (see also below).

15.3.2.3 Alzheimer's Disease

Alzheimer's disease (AD) is the most common neurodegenerative disease, with age being the most important risk factor [139]. About 5–10% of all AD cases are familial, with early onset in an autosomal-dominant manner. AD is characterized by two major histopathological hallmarks: extracellular plaques of fibrillar Aβ peptides and intracellular neurofibrillary tangles (NFTs) composed of hyperphosphorylated tau protein [140]. Point mutations in three protein-encoding genes –

the Aβ precursor protein (APP), presenilin 1 (PSEN1), and presenilin 2 (PSEN2) – are known to associate with AD onset and/or progress (for recent reviews, see [25, 84]).

^{31}P-NMR spectroscopy analysis has revealed substantial bioenergetic defects in AD brains compared to age-matched controls [141–143]. Enzymatic measurements demonstrated decreased activities of COX in platelets from Alzheimer patients and in postmortem brains [144]. With the cybrid technique it was found that a COX defect is carried by the mtDNA of Alzheimer patients [145], although AD-specific mutations have not yet been identified in mtDNA [149]. Furthermore, disturbed Ca^{2+} signaling was detected in cybrids with mtDNA [146] as well as in fibroblasts [147] from AD patients. Oxidative damage occurs early in the AD brain, even before the onset of significant plaque formation [148], and thus is probably the cause of mitochondrial dysfunction.

Recently, APP was detected in the mitochondrial OM [150]. Similarly, the widely expressed γ-secretase complex, consisting of the four integral membrane proteins PSEN, nicastrin, Aph-1, and Pen-2, was identified in the mitochondrial IM [151]. This enzyme complex is responsible for the proteolytic processing of a variety of membrane proteins. Whether the γ-secretase also can process APP in the MOM is uncertain because of the different localization and additional requirements for Aβ production by β-secretase. Within APP, sequences for targeting S/EPR and mitochondria have been detected. Overexpression of APP causes mitochondrial dysfunction by disturbances of the protein import machinery [150]. Aβ can bind to one of the mitochondrial matrix enzymes, the Aβ-binding alcohol dehydrogenase (ABAD) [152], and it has been shown that the blockade of the interaction between Aβ and ABAD favors apoptosis and ROS formation. Aβ also binds to other mitochondrial enzymes such as cytochrome c oxidase [153, 154] and α-ketoglutarate dehydrogenase [155], thereby decreasing their activities in parallel with increased ROS production [156].

In patients with frontotemporal dementia (FTH), another neurodegenerative disease, tau-containing tangles (but not amyloid plaques) are also detectable [157]. In these patients mutations in the tau gene were identified [158]. However, mutations have not yet been detected in the tau protein gene of AD patients. Consequently, intensive attempts are being made to search for tau mutations in AD as well. Mice overexpressing the P301L mutant of the human tau protein were developed as a model of taupathy [159]. By using proteomics, a reduction of complex V of the mitochondrial respiratory chain was detected in brains of P301L mice and FTH patients [160]. Functional analysis demonstrated a mitochondrial dysfunction in P301L tau mice together with reduced complex I activity and, with age, impaired mitochondrial respiration and ATP synthesis. Moreover, mitochondrial dysfunction was associated with higher levels of ROS in aged transgenic mice [160].

15.3.2.4 Hypoxia

Tissue hypoxia resulting from oxygen supply–demand mismatch can develop in exercising skeletal muscle, especially in a high-altitude, hypoxic environment

[161], and in conditions of tissue hypoperfusion, as in ischemic myocardium or in the core part of solid tumors [162]. In any of these circumstances, the mitochondria represent both sensors and targets of hypoxia. It has been proposed that the electron transport chain reacts with hypoxia as an O_2 sensor by releasing ROS [163]. This process is activated as a result of inhibition of electron flow along the respiratory chain, which reduces the cytochromes and increases the lifetime of the ubisemiquinone radical in complex III [163]. The ROS also activate the hypoxia-inducible transcription factor 1α (HIF-1α) [163–170], probably via its stabilization mediated by p38 mitogen-activated protein kinase (MAPK) [168]. HIF-1α is a potent inducer of gene transcription (all genes encoding glycolytic enzymes, glucose transporters, vascular endothelial growth factor [VEGF], erythropoietin, and insulin-like growth factor [IGF-2]) [169, 170], which enables the cells to survive during the hypoxic period. In parallel, HIF-1α downregulates mitochondrial O_2 consumption by inhibiting pyruvate oxidation through phosphorylation of the pyruvate dehydrogenase (PDH) E1α subunit [171, 172] and suppression of the succinate dehydrogenase [173] (Fig. 15.5). This shift allows a new

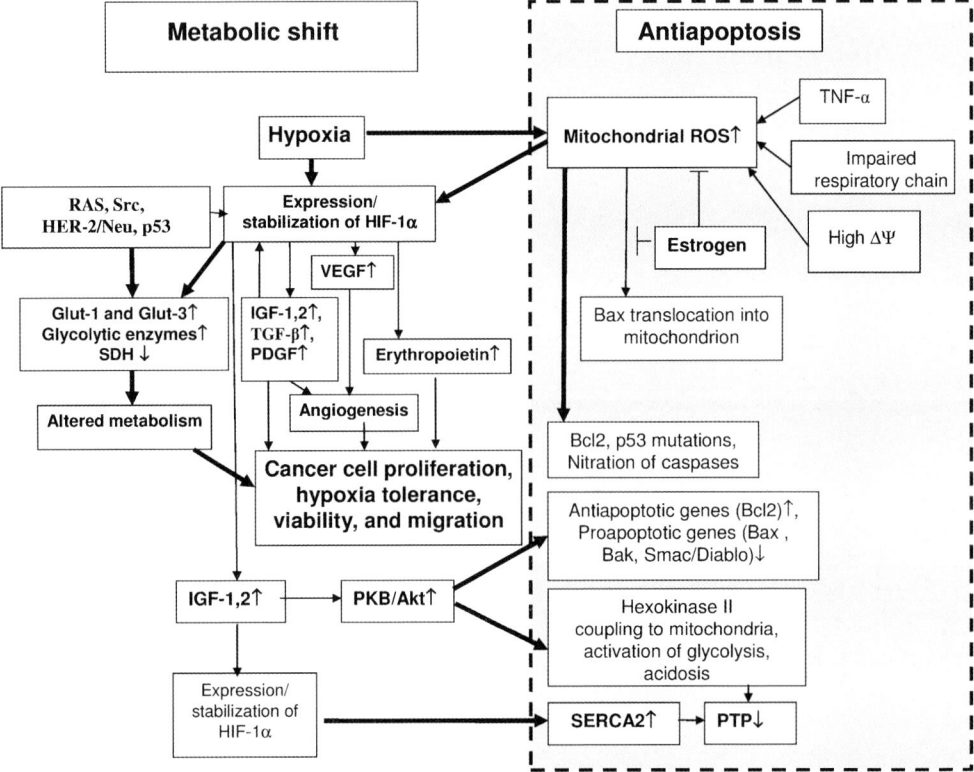

Fig. 15.5 Mechanisms of mitochondrial involvement in hypoxia and cancer development.

balance to be reached between the cellular energy and redox states, in which ATP is produced predominantly via glycolysis owing to increased activities of glycolytic enzymes, whereas the NAD^+ necessary for glycolytic ATP production is regenerated through the lactate dehydrogenase (LDH) reaction and mitochondrial ROS production is reduced. Neumann et al. [174] have shown that constitutive stabilization of HIF-1α in murine thymocytes is also associated with overexpression of SERCA2 and diminished intracellular Ca^{2+} transients due to accelerated removal of cytoplasmic Ca^{2+} in response to T-cell receptor stimulation [174, 175]. Thus, in addition to metabolic regulation, HIF-1α limits excess intracellular accumulation of Ca^{2+}, which favors maintenance of the cell's viability.

15.3.2.5 **Cancer**
The fundamental properties common to many cancers, particularly the most aggressive ones, are the metabolic shift from OXPHOS to glycolysis (Warburg's effect), the tolerance of a hypoxic microenvironment, the avoidance of apoptosis [176–182], and the increased production of mitochondrial ROS with decreased levels of glutathione (GSH) [181–185]. All these phenomena are related to the altered function of mitochondria (Fig. 15.5).

Mechanisms of Metabolic Shift The metabolic transition of cancer cells is controlled by specific oncogenes such as RAS, Src, HER-2/Neu, myc, and p53 [186–190]. Activation of RAS, Src, and HER-2/Neu stabilizes HIF-1α, a protein that upregulates glycolytic enzymes but suppresses the PDH complex, whereas myc upregulates the LDH-A isoform [188]. On the contrary, mutations in p53 likely lead to suppression of COX 2 because p53 directly controls the expression of that enzyme [189]. As a result, the energetic metabolism of cancer cells changes from oxidative to glycolytic metabolism characterized by high rates of lactate production in aerobic conditions. This change, together with increased vascular growth induced by HIF-1α, confers the cancer cell with high tolerance to hypoxia (potentially developing in a core part of the tumor), whereas increased erythropoietin levels favor cancer cell migration [191]. TNF-α probably also participates in this type of metabolic transition, because by producing ROS it stabilizes HIF-1α in aerobic conditions [192]. Mitochondrial ROS most likely accumulates as a result of cancer-causing mitochondrial DNA mutations [193], impaired respiratory chain at the level of complexes I and III [194, 195], and abnormally high membrane potential, $\Delta\Psi$ [196]. Increased $\Delta\Psi$ in turn results from depressed activity of ATP synthase, which is due to increased expression of its inhibitor protein IF_1 [197–200], and downregulation of uncoupling protein (UCP) [201], all these changes diminishing utilization of $\Delta\Psi$. As an outcome, $\Delta\Psi$ stays at high levels that stimulate ROS production at the level of Q-cycle reactions [202].

Mechanisms of Antiapoptosis Notably, the status of OXPHOS in cancer cells may be taken as a marker of severity of tumor development: mitochondrial activities are strongly downregulated in very aggressive cancer forms, such as renal carcinomas, whereas relatively benign forms of cancer, e.g., oncocytomas, exhibit upregulation of mitochondrial membranes [203]. This observation suggests that cancer development requires mitochondria, at least in some of its stages. In this regard, an interesting finding is that breast and liver cancer cells overexpress hexokinase (HK) type II [204–207], due to stimulation of that process by HIF-1α [207]. HK II differs from other HK types by its ability to bind to the porin VDAC channel on the outer membrane; normally, this process takes place in brain cell mitochondria. However, in cancer cells HK II also binds to the mitochondria under stimulation by protein kinase B or Akt (PKB/Akt) [206, 208, 209]. As a result of HK II binding to mitochondria, ATP synthesized in mitochondria is transported via porin channels to active sites of HK II and is used to produce G-6-P. The latter then accelerates the glycolytic flux and can also be used for biosynthetic processes. Thus, coupling of HK II with mitochondrial ATP synthesis boosts both glycolytic ATP production and cancer growth (Fig. 15.5). On the other hand, inhibition of binding of HK II by 3-bromopyruvate significantly suppressed cellular growth and induced apoptosis via mitochondrial signaling cascades [207], which means that binding of HK II provides the cancer cells with an effective means for avoiding an apoptotic death. The underlying mechanisms may involve stabilization of the PT pore complex by HK II binding to VDAC, thereby suppressing the release of intermembrane proapoptotic proteins and/or blocking association of exogenous proapoptotic proteins (Bax) with the MOM [209]. It is also possible that association of HK II with VDAC suppresses the mitochondrial ROS production in cancer cells, as occurs in brain cells that should favor a closed state of the PT pore [73].

It should be considered that besides those antiapoptotic mechanisms that manifest at the level of HK II–VDAC interactions, cancer cells possess many other mechanisms aimed at achieving the same goals. Among them, upregulation of mitochondria-related antiapoptotic genes (Bcl-2, Bcl-x$_L$, survivin, XIAP) and downregulation of proapoptotic genes (Bax, Bak, Smac/Diablo) is characteristic of many types of human gastric cells [210–216]. This change can be attributed to the overexpression of PKB/Akt seen in many tumors that inhibits expression of death genes (Fas, Bim, and IGFBP-1) but promotes expression of antiapoptotic (Bcl2) and proliferation-supporting genes (cIAP1, cIAP2) through activation of NF-kB and CREB [178, 217]. In addition, other mechanisms exist that exert prosurvival/antiapoptotic effects in cancer cells (Fig. 15.5). First, the inflammatory mediators (e.g., cytokines, ROS, and NO) suppress apoptosis by causing mutations in Bcl2 and p53 proteins [217] or nitration of caspase 9 [218]. Second, OXPHOS can control Bax activation by inducing the PT pore and associated conformational changes in Bax and Bak [219], and this mechanism may be switched off if the cancer cell's metabolism shifts towards glycolysis. Third, given that mild acidosis protects mitochondria from PT pore opening and that acidosis accompanies intensive glycolysis in cancer cells, it may inhibit activation of Bax and Bak, thus favoring antiapoptosis in these cells [219]. Fourth, at least in breast cancer cells, estrogen, by binding to its mitochondrial receptors, upregulates mitochondrial MnSOD, which results in suppression of mitochondrial ROS production and blocking of apoptosis caused by mitochondrial ROS [185]. Fifth, in a variety of neoplastic cells, expression of the peripheral benzodiazepine receptor (PBR), a mitochondrial protein associated with the VDAC protein, is strongly upregulated [220]. Because the PBR exerts a strong protective effect against ROS damage [221], it supports cancer cell survival despite increased ROS loading. In light of these data, different pharmacological means for stimulating apoptosis in cancer cells are under investigation [222]. For example, it has been found that in many types of cancers appropriate chemo- and radiotherapy can recover the ability of mitochondria to release cytochrome c and activate apoptosis [223–226].

15.3.2.6 Inflammation

Inflammatory processes associate and complicate many pathological conditions – including cardiac ischemia and reperfusion, cardiac failure, neurodegenerative diseases (see also above), and diabetes mellitus – through promoting tissue hypoxia that is due to impaired endothelial function and through activation of inflammatory mediators and cytokines that are produced by circulating leukocytes, lymphocytes, and macrophages, as well as by endothelial cells, microglia cells, and other cells of host tissues.

Mechanisms of Activation of Mitochondrial ROS Production Increased production of ROS, which in turn induces the proinflammatory cytokines, is a well-known hallmark of inflammation [227]. Although cytokines such as IL-1 and TNF-α can increase ROS production via both membrane-bound NADPH oxidase and the mitochondrial respiratory chain [192, 227], increasing evidence suggests the primary importance of mitochondrial ROS. This was elegantly demonstrated in experiments using mice with targeted disruption of the UCP-2 gene ($Ucp2^{-/-}$), which resulted in greater macrophage phagocytotic activity, together with larger ROS production, greater expression of inducible NO synthase (iNOS), inflammatory cytokines (interferon-γ [IFN-γ] and tumor necrosis factor-α [TNF-α], and faster nuclear translocation of nuclear factor-κB (NF-κB) subunits in response to bacterial LPS challenge compared with that in wild-type ($Ucp2^{+/+}$) mice [235]. Because TNF-α inhibits mitochondrial oxidation of both NADH- and FADH$_2$-linked substrates in association with suppression of the respiratory chain complexes, it also increases ROS production [228–230]. ROS in turn increases the expression of proinflammatory cytokines, such as interleukin-2 (IL-2), TNF-α, and IL-10 [231, 232], and activates NF-κB, a common target for TNF-α and IL-1 [233]. Based on these facts, several groups have been seeking the mechanisms that might link the effects of TNF-α on mitochondrial function to activation of NF-κB (Fig. 15.6). Itoh et al. [230] found that Dok-4, one of the proteins downstream of tyrosine kinase (Dok), is involved in TNF-α–mediated ROS production, as Dok-4 recruits cytosolic c-Src protein kinase into mitochondria and causes its activation, these changes resulting in suppression of complex I and increased mitochondrial ROS production. Because activation of NF-κB in herpesvirus-infected macrophages is dependent on ROS and Ca^{2+} both originating from mitochondria, Mogensen et al. [234] have suggested that mitochondrial oxidative stress and mobilization of Ca^{2+} trigger a signaling system comprising a cascade of kinases (TAK1,

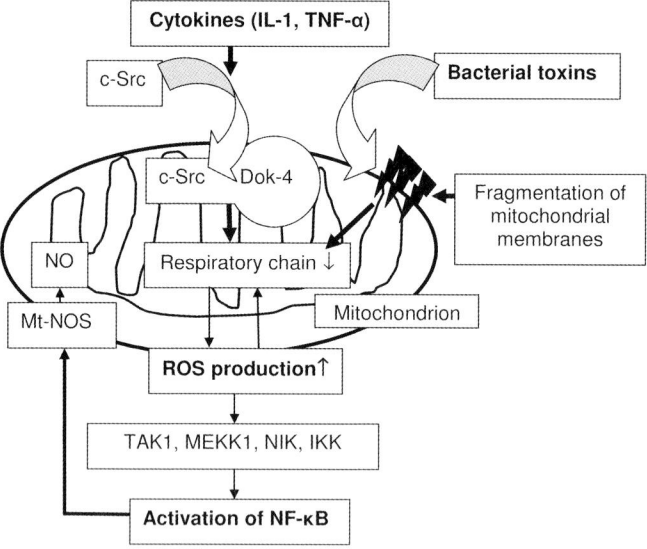

Fig. 15.6 Mechanisms of ROS production in mitochondria in conditions of inflammation.

MEKK1, NIK, and IκB kinase [IKK]) that eventually activates NF-κB. Together, these results strongly suggest that mitochondria can initiate and amplify a proinflammatory response to various stressors via generating ROS that activate NF-κB. Along with being the ROS producers, mitochondria are the source and target of reactive nitrogen species (RNS), because IL-1 and TNF-α stimulate expression of the iNOS [235] and mitochondrial NOS-l isoforms [236, 237] (Fig. 15.6). While accumulating in relatively low concentrations, NO reversibly inhibits respiration at the level of COX by competing with O_2; it also inhibits the activity of complex II and oxidizes ubiquinol [238, 239]. At higher concentrations it reacts with superoxide, producing the strong oxidant peroxynitrite ($ONOO^-$). Through S-nitrosylation and/or nitration, $ONOO^-$ irreversibly inhibits many mitochondrial proteins, such as the subunits of complexes I and II of the respiratory chain [236], thus suppressing the electron flow and ATP synthesis, and increasing ROS production [240–245]. In conditions of sepsis the endotoxin-induced impairment of mitochondrial function in heart and skeletal muscle of rabbits was demonstrated by decreased state 3 respiration rates caused by diminished activities of complex I+III [246] and by heavily decreased activities of complexes I+III and II+III in hearts of septic baboons [247]. Other evidence of impaired mitochondria in sepsis has been reviewed [248, 249].

It is important to consider that pathogenic bacteria can stimulate ROS production in mitochondria via direct effects of their components on mitochondrial membranes. For example, *Helicobacter pylori*, a major pathogen that causes inflammation of the gastric mucosa in humans, permeabilizes the MOM via translocation of the N-terminal, 34-kDa fragment of *H. pylori* vacA cytotoxin into the mitochondria [250]. This process is associated with depolarized mitochondrial membranes; depressed cellular respiration and ATP content; mitochondrial fragmentation [251, 252]; and increased production of ROS, NO, and ammonia, all of which secondarily exert cytotoxic and mitochondriotoxic effects [217, 253, 254].

Normally, the ROS generated in mitochondria are largely detoxified by mitochondrial Mn-dependent superoxide dismutase (MnSOD). However, in conditions of excess NO production, nitration of that protein inhibits its enzymatic activity as well [255]. Moreover, GSH generation becomes suppressed due to inactivation of $NADP^+$-dependent isocitrate dehydrogenase by $ONOO^-$ [256]. Through these mechanisms the positive feedback signaling mechanisms for maintaining inflammation in which the mitochondria play a central role can be formed: activation of inflammation is associated with mitochondrial production of ROS and RNS stimulated by proinflammatory cytokines, whereas mitochondria in turn promote expression of proinflammatory cytokines via ROS and RSN.

Mitochondrion – a Master Switch Between Inflammation and Repair   At present it is not clear why mitochondria are recruited by the cell to control the course of inflammation. Considering that the hypoxic microenvironment of the inflamed sites and activation of HIF-1α by inflammatory cytokines (e.g., IL-1 and TNF-α) should shift the cellular energy metabolism from OXPHOS towards glycolysis [227, 257], it is difficult to expect a significant role for mitochondria in regulation of the function of inflammatory cells. Nevertheless, several lines of evidence suggest that even if the mitochondrial number and/or their capacity to produce ATP in the cells decreased, they still maintain control over the cell's fate. In this regard, a remarkable observation is that many inflammatory cells that infiltrate the inflamed sites can function there for longer periods of time than observed for the same cells in the bloodstream. One reason for this is that the cytokines (e.g., IL-3, IL-5, granulocyte-macrophage colony-stimulating factors, and granulocyte colony-stimulating factors) inhibit mitochondria-dependent activation of apoptosis by suppressing many proapoptotic changes, such as translocation of Bax to the mitochondria, cytochrome *c* release, activation of caspases, and caspase-independent loss of ΔΨ, as seen in neutrophils and eosinophils [258, 259]. Notably, NO produced during inflammatory reactions also acts as an inhibitor of apoptosis, via suppression of caspases (S-nitration), PT pore opening, and stimulation of antiapoptotic Bcl2 [258]. A further example of restrained apoptosis comes from studies on peripheral blood lymphocytes (T cells) of patients with systemic lupus erythematosus (SLE) [260, 261], which showed that in contrast to the cells of healthy patients, exhibiting a transient increase in ΔΨ before entering apoptotic death, the T cells of SLE patients were characterized by persistent mitochondrial hyperpolarization associated with increased ROS production and by cellular ATP and GSH depletion leading to necrotic death in response to IL-10. It was considered that the necrotic mode of the cell's death is promoted by activation of dendritic cells through increased IFN-α production and inflammation in lupus patients [260, 261]. Thus, mitochondria serve as the targets for those intracellular signals that in parallel to retardation of apoptosis also sensitize the inflammatory cells to necrotic death, thereby aggravating the inflammatory tissue lesions and complicating the disease phenotype. Another suggestion for a central role of mitochondria is based on findings that disruption of mitochondrial function by either uncoupler FCCP or oligomycin in human neutrophils induced rapid changes in their shape and inhibited their ability for chemotaxis and late-phase respiratory burst. At the same time, impaired mitochondrial function had no effect on the rate of apoptosis. These findings mean that intact mitochondrial function is necessary for supporting the anti-inflammatory properties of the neutrophils, irrespective of its control over apoptotic processes [262]. Interestingly, the ways that mitochondria influence inflammatory processes varies depending on the type of inflammatory cells and their mode of activation. For example, during differentiation of eosinophils, mitochondria lost their capacity to respire and produce ATP via OXPHOS but retained their ability to generate $\Delta\Psi$ (at the expense of cytoplasmic/glycolytic ATP) and to induce apoptosis via mitochondrial cytochrome c release [263]. Because of these properties the eosinophils differ from other inflammatory cells, e.g., neutrophils and macrophages, in which the mitochondria contribute to inflammation not only by regulating apoptosis but also by production of ATP, ROS, and RNS. A very interesting novel observation is that activation of macrophages by either Th1 cytokines or Th2 cytokines may result in qualitatively different cellular functions. It was found that the Th1-derived cytokines (e.g., IFN-γ, TNF-α, LPS, and IL-1) activate the classical signaling pathways predominantly via HIF-1α, which leads to activation of glycolysis together with NO and ROS [227, 257, 264]. As an outcome, the macrophages produce proinflammatory cytokines (e.g., TNF-α, IL-1, IL-6, and IL-12) that besides amplifying inflammatory reactions such as the microbicidal effect (defense against invading pathogens) and the cellular immunity response also cause marked tissue damage. Alternatively, the Th2 cytokines (e.g., IL-4 and IL-13) stimulate oxidative mechanisms via activation of STAT6 (signal transducer and activator of transcription 6) and PGC-1β (PPARγ-co-activator-1β) that induce macrophage programs for fatty acid oxidation and mitochondrial biogenesis. Therefore, it appears that increased oxidative metabolism fuels the development of an anti-inflammatory phenotype, which via secretion of chitinases, chemokines, and collagen helps to limit inflammation and promote reparative processes, such as wound healing and granuloma formation [264, 265]. It is expected that shifting from an inflammatory phenotype into an anti-inflammatory/reparative phenotype of the inflammatory cells may represent a potential target for metabolic therapies in treating atherogenic or other types of inflammation [264].

15.4
Detection of Mitochondrial Dysfunction by Multiple Substrate Inhibitor Titration and Application of Metabolic Control Analysis

Because of the complexity of mitochondrial organization and function, it is usually difficult to identify important mitochondrial defects from a singular standard

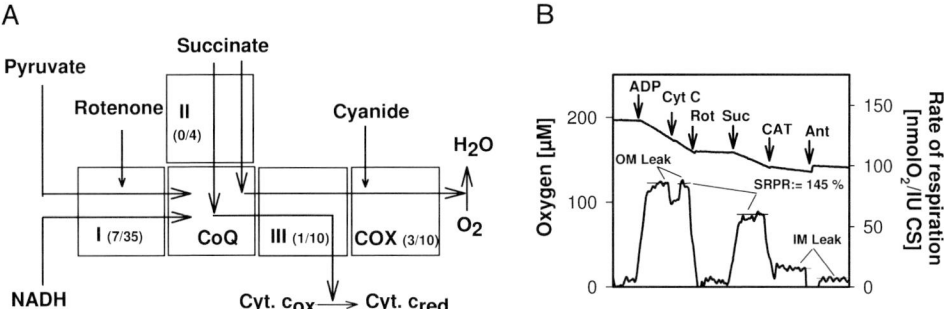

Fig. 15.7 Principle of multiple substrate inhibitor titration for detection of mitochondrial dysfunction.
(A) Metabolic scheme: I, II, III: complexes of the respiratory chain; COX: cytochrome *c* oxidase. The fractions given indicate the number of mitochondrial and the total number of subunits of the respective respiratory chain complex. SRPR can be calculated from complex I–dependent/ complex II–dependent respiration in percentages; succinate-related NADH oxidation (SRNO) can be calculated from the complex I+III:complex II+III enzyme activity ratio.
(B) Respirogram of mitochondrial respiration of hippocampal homogenates from 10-day-old normal rats incubated in a medium containing 5 mM MgCl$_2$, 75 mM mannitol,

25 mM sucrose, 100 mM KCl, 10 mM P$_i$, 0.5 mM EDTA, 20 mM TRIS, pH 7.4, and 10 mM pyruvate and 2 mM malate as substrates. Additions: ADP: 2 mM ADP; Cyt c: 10 µM cytochrome *c*; Rot: 10 µM rotenone; Suc: 10 mM succinate; CAT: 1 mM carboxyatractyloside; Ant: 10 µM antimycin A. In this experiment the SRPR equals to 145%. If after Cyt c addition the rate of respiration remains constant, there is no rate limitation by Cyt c, indicating an intact outer membrane (for comparison, see Fig. 15.8D, where a leaky MOM was detected by a Cyt c–caused increase in respiratory rate). The difference between the CAT- and Ant-inhibited rates of respiration allows the measurement of leak respiration, which increases at uncoupling and after PT.

experiment that considers only the maximal rates of respiration. Rather, it is necessary to apply a wider spectrum of methods and protocols, such as high-resolution respirometry combined with multiple substrate inhibitor titration [266–268] and the application of metabolic control analysis (MCA) [269–279]. In this context, we have specifically developed the following experimental protocols.

First, if suitable inhibitors are used in excess, different metabolic pathways can be investigated in the same measurement. The protocol shown in Fig. 15.7 is especially useful for detection of complex I–dependent impairments of mitochondrial function. In most tissues (except, e.g., liver and kidney) studied under these conditions, complex I respiration is higher than complex II respiration. We found in heart, muscle, and tumor tissues from different species (rat, rabbit, turkey, and human) that succinate-related pyruvate respiration (SRPR) significantly correlates with the activity ratio of complex I+III/complex II+III (SRNO). Figure 15.8 shows some examples of the identification of mitochondrial defects in human and animal tissues in different diseases. In a patient with chronic progressive external ophthalmoplegia (CPEO), we identified reduced state 3 respiration rates for

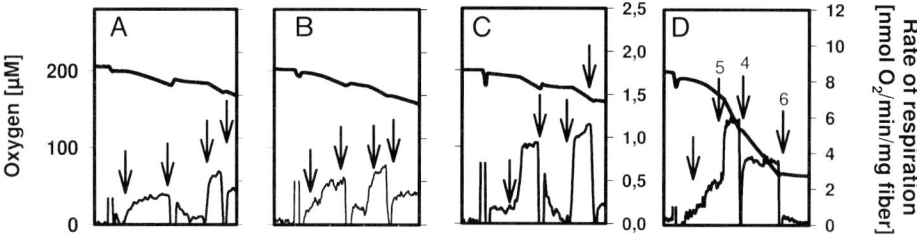

Fig. 15.8 Multiple substrate inhibitor titration protocol for detection of mitochondrial dysfunction in conditions of mitochondrial cytopathies and ischemia–reperfusion injury. Mitochondrial function was investigated in skinned fibers of patients with (A) CPEO due to 5.5 kb deletion at 34% heteroplasmy, (B) MELAS point mutation A3243G at 34% heteroplasmy, and (C) LHON mutation G11778A at homoplasmy and (D) in rabbit hearts perfused with a Langendorff system in conditions of 45 min ischemia followed by a normoxic reperfusion period of 90 min, incubated in a medium as described in Fig. 15.7B, with 10 mM pyruvate and 2 mM malate as complex I–dependent substrates. Sequential addition as marked by the non-numbered arrows: 2 mM ADP; 10 μM rotenone; 10 mM succinate; 1 mM carboxyatractyloside (also arrow 4). Arrow 5: 10 μM cytochrome c; arrow 6: 15 μM antimycin A.

both pyruvate/malate and succinate, indicating impaired mitochondrial function resulting from decreased activity of respiratory chain complexes (Fig. 15.8A). Moreover, the strongly reduced SRPR (58%), compared with 117% in control patients, indicates a massive defect in complex I [266]. In a patient suffering from combined mitochondrial myopathy, encephalopathy, lactic acidosis, and stroke-like episodes (MELAS) (Fig. 15.8B), the SRPR was not so markedly changed (73%), and the complex I–dependent state 3 respiration rates were somewhat higher than those in the CPEO patient. The smallest decrease in SRPR (82%) was found in a patient with Leber's hereditary optic neuropathy (LHON) (Fig. 15.8C). Obviously, all three mitochondrial diseases are characterized by a complex I–related dysfunction. Moreover, in CPEO and MELAS patients the activities of complexes III, IV, and V are diminished as well. The consequences of the mtDNA mutation depend on the level of heteroplasmy, which was the same in our patients with CPEO (34%) and MELAS (34%), but the LHON mutation occurred in homoplasmy (100%). The respirograms clearly show that the effect of mutations on mitochondrial dysfunction further depends on the type of mutation, on the order large deletions > MELAS > LHON. Deletions of the common deletion type [266] have the largest effect, because in this case three tRNA genes are missing, strongly reducing the activity of respiratory chain complexes I, III, IV, and V. On the other hand, the MELAS point mutation A3243G causes a dysfunction in only one tRNA, which modestly decreases the same respiratory chain complexes as in CPEO. The LHON point mutation G11778A affects only subunit ND4 of complex I, and therefore, although being homoplasmic, the consequences of that mutation on respiratory function are much smaller.

We further detected significantly reduced SRPR in endotoxemic rabbit hearts [246, 248], in human neck tumors (Kuhnt, T. and Gellerich, F. N., unpublished), in hearts and m. soleus of aged rats (Holtz, J., and Gellerich, F. N., unpublished), in skeletal muscle of transgenic HD R6/2 mice [114], and in antral and corpus mucosa specimens from patients with gastric diseases [280]. Numerous variations of this protocol are in use, e.g., the cytochrome *c* test allowing the detection of MOM leaks [268]. Figure 15.8D demonstrates stimulation of ADP-dependent respiration in heart mitochondria after ischemia–reperfusion by exogenous cytochrome *c*. It shows that the ischemia–reperfusion was associated with an impairment of the MOM that resulted in a loss of mitochondrial cytochrome *c*. The large difference between the CAT- and Ant-inhibited respiration rates as compared with normal mitochondria (Fig. 15.9A) demonstrates an increase in non-phosphorylating respiration due to an impairment of the mitochondrial inner membrane, probably caused by opening of the PT pores.

A second important approach is the stepwise addition of small amounts of inhibitors to allow the calculation of the flux control coefficient (Co) of the enzyme [269–279]. Co combines the properties of a single enzyme with the functional properties of a metabolic system [278, 279] as follows:

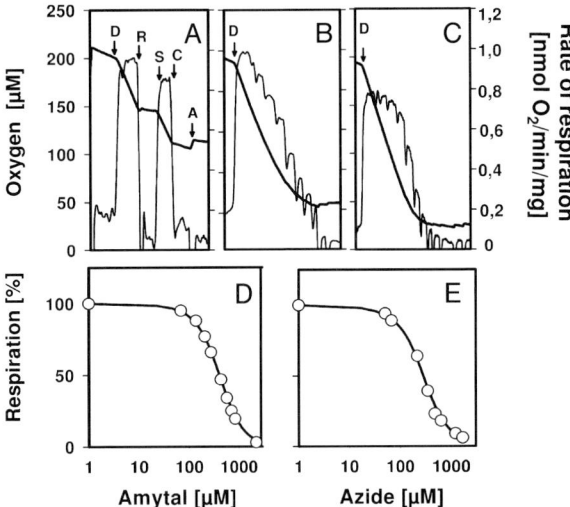

Fig. 15.9 Experimental determination of flux control coefficients of complex I and COX on mitochondrial state 3 respiration in skinned human skeletal muscle fibers. (A) Standard protocol for detection of SRPR, RCI, and the extent of leak respiration. Incubation of saponin skinned muscle fibers of a patient without mitochondrial diseases in a medium with additions as in Fig. 15.7B. SRPR equals to 110%. In healthy patients the mean was SRPR = 117 ± 15%, $n = 22$). (B, C) Experimental determination of flux control coefficients for complexes I and IV on mitochondrial respiration in muscle fibers of the same patient as in (A). Inhibitor titration of state 3$_{pyruvate/malate}$ with amytal (B) and azide (C). Indicated respiratory rates were normalized on the uninhibited state 3 and plotted versus the inhibitor concentrations (D, E). From these inhibitor titration curves, Co was calculated using a nonlinear regression program [269, 270]. For further details, see the text.

$$Co = dJ/dE \times E/J$$

where J is the mitochondrial rate of respiration under steady-state conditions and E is the activity of the enzyme under investigation. If the flux changes to the same extent as the enzyme activity, Co equals 1, but normally the Co of mitochondrial enzymes is much lower, between 0 and 0.3 [271, 273–275]. For the determination of Co, the enzyme activity must be diminished experimentally, e.g., by inhibitor titrations. In the case of irreversible inhibitors, the Co can be calculated from titration experiments by means of the following equation [269, 277]:

$$Co = dJ/dI \times I_{max}/J_o$$

Here, the rate of respiration is measured in dependence on inhibitor additions (dI) decreasing the enzyme activity. J_o is the flux in the absence of the inhibitor and J_{max} is the amount of inhibitor necessary for total inhibition. Figure 15.9 illustrates an example of inhibitor titrations of state 3 respiration with amytal and azide (inhibitors of complex I and COX) in human skeletal muscle fibers of a healthy patient. Under these conditions the Co values were 0.13 for complex I and 0.26 for COX. In CPEO patients, however, the Co values were significantly increased (Co = 0.33 for complex I, Co = 0.31 for COX) [271]. These increased Co values show that in CPEO, the activities of complex I and COX are decreased in relation to the total metabolic system of OXPHOS. The first detected example of increased Co in impaired mitochondria showed that the Co of succinate dehydrogenase of brain mitochondria was significantly enhanced after ischemic damage [272]. Later, further examples were observed, all showing that the functional consequence of reduced OXPHOS enzyme activity can be proven and quantified by its increased Co [271, 275, 276].

Acknowledgements

This work was supported by the European Huntington network, the DFG (Ge 664/11-2) and Estonian Science Foundations (Grants No. 5515 and 7117).

References

1 Warburg, O., Geissler, A. W., Lorenz, S. (1970) Genesis of tumor metabolism by vitamin B1 deficiency (thiamine deficiency). *Z. Naturforsch. B. 25*, 332–333.

2 Luft, R. (1992) Luft's disease revisited. Severe hypermetabolism of nonthyroid origin with a defect in the maintenance of mitochondrial respiratory control. *Mt. Sinai J. Med. 59*, 140–145.

3 Holt, I. J., Harding, A. E., Morgan-Hughes, J. A. (1988) Deletions of muscle mitochondrial DNA in patients with mitochondrial myopathies. *Nature 331*, 717–719.

4 Wallace, D. C., Singh, G., Lott, M. T., Hodge, J. A., Schurr, T. G., Lezza, A. M., Elsas, L. J., Nikoskelainen, E. K. (1988) Mitochondrial DNA mutation associated with Leber's

hereditary optic neuropathy. *Science 242*, 1427–1430.

5 Triepels, R. H., van Den Heuvel, L. P., Trijbels, J. M., Smeitink, J. A. (2001) Respiratory chain complex I deficiency. *Am. J. Med. Gen. 106*, 37–45.

6 Luft, R. (1995) The development of mitochondrial medicine. *Biochim. Biophys. Acta 1271*, 1–6.

7 Schlame, M., Ren, M. (2006) Barth syndrome, a human disorder of cardiolipin metabolism. *FEBS Lett. 80*, 5450–5455.

8 Quinzii, C. M., Dimauro, S., Hirano, M. (2006) Human coenzyme Q(10) deficiency. *Neurochem. Res.* (DOI 10.1007/s11064-006-9190-z).

9 Zeviani, M., Di Donato, S. (2004) Mitochondrial disorders. *Brain 127*, 2153–2172.

10 Smeitink, J. A., Zeviani, M., Turnbull, D. M., Jacobs, H. T. (2006) Mitochondrial medicine: a metabolic perspective on the pathology of oxidative phosphorylation disorders. *Cell. Metab. 3*, 9–13.

11 Morgan-Hughes, J. A., Hanna, M. G., (1999) Mitochondrial encephalomyopathies: the enigma of genotype versus phenotype. *Biochim. Biophys. Acta 1410*, 125–145.

12 Wiederkehr, A., Wollheim, C. B. (2006) Minireview: implication of mitochondria in insulin secretion and action. *Endocrinology 147*, 2643–2649.

13 Halestrap, A. P. (2006) Calcium, mitochondria and reperfusion injury: a pore way to die. *Biochem. Soc. Trans. 34*, 232–237.

14 Chan, P. H. (2004) Mitochondria and neuronal death/survival signaling pathways in cerebral ischemia. *Neurochem. Res. 29*, 1943–1949.

15 Galluzzi, L., Larochette, N., Zamzami, N., Kroemer, G. (2006) Mitochondria as therapeutic targets for cancer chemotherapy. *Oncogene 25*, 4812–4830.

16 Fantin, V. R., Leder, P. (2006) Mitochondriotoxic compounds for cancer therapy. *Oncogene 25*, 4787–4797.

17 Kalman, B., Leist, T. P. (2003) A mitochondrial component of neurodegeneration in multiple sclerosis. *Neuromolecular Med. 3*, 147–158.

18 Perl, A., Gergely, P. Jr, Banki, K. (2004) Mitochondrial dysfunction in T cells of patients with systemic lupus erythematosus. *Int. Rev. Immunol. 23*, 293–313.

19 Liu, H., Pope, R. M. (2003) The role of apoptosis in rheumatoid arthritis. *Curr. Opin. Pharmacol. 3*, 317–322.

20 Brand, M. D. (2005) The efficiency and plasticity of mitochondrial energy transduction. *Biochem. Soc. Trans. 33*, 897–904.

21 Gellerich, F. N., Trumbeckaite, S., Opalka, J. R., Chen, Y., Neuhoff, C., Schlag, H., Zierz, S. (2002) Mitochondrial dysfunction at sepsis: Evidences from bacteraemic baboons and endotoxaemic rabbits. *Bioscience Report 22*, 99–113.

22 Trumbeckaite, S., Opalka, J. R., Neuhof, C., Zierz, S., Gellerich, F. N. (2000) Different sensitivity of rabbit heart and skeletal muscle to endotoxin-induced impairment of mitochondrial function. *Eur. J. Biochem. 268*, 1422–1429.

23 Singer, M., Brealey, D. (1999) Mitochondrial dysfunction in sepsis. *Biochem. Soc. Symp. 66*, 149–166.

24 Skulachev, V. P. (2000) Mitochondria in the programmed death phenomena; a principle of biology: "it is better to die than to be wrong". *IUBMB Life 49*, 365–373.

25 Wallace, D. C. (2005) A mitochondrial paradigm of metabolic and degenerative diseases, aging, and cancer: a dawn for evolutionary medicine. *Annu. Rev. Genet. 39*, 359–407.

26 Gellerich, F. N., Laterveer, F. D., Zierz, S., Nicolay, K. (2002) The quantitation of ADP diffusion gradients across the outer membrane of heart mitochondria in the presence of macromolecules. *Biochim. Biophys. Acta 1554*, 48–56.

27 Parone, P. A., Martinou, J. C. (2006) Mitochondrial fussion and apoptosis: an ongoing trial. *Biochim. Biophys. Acta 1763*, 522–530.

28 Sickmann, A., Reinders, J., Wagner, Y., Joppich, C., Zahedi, R., Meyer, H. E., Schonfisch, B., Perschil, I., Chacinska, A., Guiard, B., Rehling, P., Pfanner, N., Meisinger, C. (2003) The proteome of Saccharomyces cerevisiae mitochondria.

Proc. Natl. Acad. Sci. USA. 100, 13207–13212.

29 Bredesen, D. E., Rammohan, V. R., Mehlen, P. (2006) Cell death in the nervous system. *Nature 443*, 796–802.

30 Vieira, K. L., Kroemer, G. (1999) Pathophysiology of mitochondrial cell death control. *Cell. Mol. Life Sci. 56*, 971–976.

31 Gellerich, F. N., Trumbeckaite, S., Müller, T., Chen, Y., Deschauer, M., Gizatullina, Z., Zierz, S. (2004) Energetic depression caused by mitochondrial dysfunction. *Mol. Cell. Biochem. 256/257*, 391–405.

32 Bernardi, P., Krauskopf, A., Basso, E., Petronelly, V., Blalchy-Dyson, E., Di Lisa, F., Forte, M. A. (2006) The mitochondrial permeability transition from *in vitro* artifact to disase target. *FEBS J. 273*, 2077–2099.

33 Garvan, C., Kane, Behfar, A., Dyer, R. B., Fearghas O'Cochlain, D., Liu, X.-K., Denice, M., Hodgson, D. M., Reyes, S., Miki, T., Seino, S., Terzic, A. (2006) *KCNJ11* gene knockout of the Kir6.2 K_{ATP} channel causes maladaptive remodeling and heart failure in hypertension. *Hum. Mol. Gen. 15*, 2285–2297.

34 Clarke, S. J., McStay, G. P., Halestrap, A. P. (2002) Sanglifehrin A acts as a potent inhibitor of the mitochondrial permeability transition and reperfusion injury of the heart by binding to cyclophilin-D at a different site from cyclosporin A. *J. Biol. Chem. 277*, 34793–34799.

35 Irwin, W. A., Bergamin, N., Sabatelli, P., Reggiani, C., Megighian, A., Merlini, L., Braghetta, P., Columbaro, M., Volpin, D., Bressan, G. M., Bernardi, P., Bonaldo, P. (2003) Mitochondrial dysfunction and apoptosis in myopathic mice with collagen VI deficiency. *Nat. Genet. 35*, 367–371.

36 Green, D. R., Kroemer, G. (2004) The pathophysiology of mitochondrial cell death. *Science 305*, 626–629.

37 Carafoli, E. (1986) Mitochondrial pathology: an overview. *Ann. N. Y. Acad. Sci. 488*, 1–18.

38 McCormack, J. G., Denton, R. M. (1986) The role of intramitochondrial Ca^{2+} in the regulation of oxidative phosphorylation in mammalian tissues. *TIBS 11*, 258–262

39 Jouaville, L. S., Ichas, F., Mazat, J. P. (1998) Modulation of cell calcium signals by mitochondria. *Mol. Cell. Biochem. 184*, 371–376.

40 Wussling, M. H. P., Krannich, K., Landgraf, G., Herrmann-Frank, A., Wiedenmann, D., Gellerich, F. N., Podhaisky, H. (1999) Sarcoplasmatic reticulum vesicles embedded in agarose gel exhibit propagating calcium waves. *FEBS Lett. 463*, 103–109.

41 Kranich, G., Gellerich, F. N., Wussling, M. H. P. (2004) Inhibitors of SERCA and mitochondrial Ca-uniporter decrease velocity of calcium waves in rat cardiomyocytes. *Mol. Cell. Biochem. 256/257*, 379–386.

42 Duszynski, J., Koziel, R., Brutkowski, W., Szczepanowska, J., Zablocki, K. (2006) The regulatory role of mitochondria in capacitive calcium entry. *Biochim. Biophys. Acta 1757*, 380–387.

43 Montero, M., Alonso, M. T., Carnicero, E., Cuchill-Ibanez, I., Albillos, A., Garcia, A. G., Garcua-Sancho, J., Alvarez, J. (2000) Chomaffin-cell stimulation triggers fast millimolar mitochondrial Ca^{2+} transients that modulate secretion. *Nat. Cell. Biol. 2*, 57–61.

44 Selivanov, V. A., Ichas, F., Holmuhamedov, E. L., Jouaville, L. S., Evtodienko, Y. V., Mazat, J. P. (1998) A model of mitochondrial Ca(2+)-induced Ca2+ release simulating the Ca2+ oscillations and spikes generated by mitochondria. *Biophys. Chem. 72*, 111–121.

45 Wasniewska, M., Karczmarewicz, E., Pronicki, M., Piekutowska Abramczak, K., Zablocki, K., Popowska, E., Pronicka, E., Duszynski, J. (2001) Abnormal calcium homeostasis in fibroblasts from patients with leigh disease. *Biophys. Biochem. Res. Comm. 283*, 687–693.

46 Seppet, E. K., Kaambre, T., Sikk, P., Tiivel, T., Vija, H., Tonkonogi, M., Sahlin, K., Kay, L., Appaix, F., Braun, U., Eimre, M., Saks, V. A. (2001) Functional complexes of mitochondria with Ca,MgATPases of myofibrils and sarcoplasmic reticulum in muscle cells. *Biochim. Biophys. Acta 1504*, 379–395.

47 Saks, V. A., Kaambre, T., Sikk, P., Eimre, M., Orlova, E., Paju, K., Piirsoo, A., Appaix, F., Kay, L., Regitz-Zagrosek, V., Fleck, E., Seppet, E. (2001) Intracellular energetic units in red muscle cells. *Biochem. J. 326*, 643–657.

48 Saks, V. A., Kuznetsov, A. V., Vendelin, M., Guerrero, K., Kay, L., Seppet, E. K. (2004) Functional coupling as a basic mechanism of feedback regulation of cardiac energy metabolism. *Mol. Cell. Biochem. 256/257*, 185–199.

49 Saks, V., Dzeja, P., Schlattner, U., Vendelin, M., Terzic, A., Wallimann, T. (2006) Cardiac system bioenergetics: metabolic basis of the Frank-Starling law. *J. Physiol. 571*, 253–273.

50 Anmann, T., Eimre, M., Kuznetsov, A. V., Andrienko, T., Kaambre, T., Sikk, P., Seppet, E., Tiivel, T., Vendelin, M., Seppet, E., Saks, V. A. (2005) Calcium-induced contraction of sarcomeres changes the regultaion of mitochondrial respiration in permeabilized cardiac cells. *FEBS J. 272*, 3145–3161.

51 Neubauer, S., Horn, M., Pabst, T., Godde, M., Lubke, D., Jilling, B., Hahn, D., Ertl, G. (1995) Contributions of ^{31}P-magnetic resonance spectroscopy to the understanding of dilated heart muscle disease. *Eur. Heart J. 16 (Suppl O)*, 115–118.

52 Ventura-Claper, R., Garnier, A., Veksler, V. (2003) Energy metabolism in heart failure. *J. Physiol. 555*, 1–13.

53 De Sousa, E., Veksler, V., Minajeva, A., Kaasik, A., Mateo, P., Mayoux, E., Hoerter, J., Bigard, X., Serrurier, B., Ventura-Clapier, R. (1999) Subcellular creatine kinase alterations. Implications in heart failure. *Circ. Res. 85*, 68–76.

54 Kaasik, A., Minajeva, A., De Sousa, E., Ventura-Clapier, R., Veksler, V. (1999) Nitric oxide inhibits cardiac energy production via inhibition of mitochondrial creatine kinase. *FEBS Lett. 414*, 75–77.

55 Nascimben, L., Ingwall, J. S., Pauletto, P., Friedrich, J., Gwathmey, J. K., Saks, V., Pessina, A. C., Allen, P. D. (1996) Creatine kinase system in failing and nonfailing human myocardium. *Circulation 94*, 1894–1901.

56 Ponticos, M., Lu, Q. L., Morgan, J. E., Hardie, D. G., Partridge, T. A., Carling, D. (1998) Dual regulation of the AMP-activated protein kinase provides a novel mechanism for the control of creatine kinase in skeletal muscle. *EMBO J. 17*, 1688–1699.

57 Dolder, M., Walzel, B., Speer, O., Schlattner, U., Wallimann, T. (2003) Inhibition of the mitochondrial permeability transition by creatine kinase substrates. Requirement for microcompartmentation. *J. Biol. Chem. 278*, 17760–17766.

58 Zoratti, M., Szabo, I. (1995) The mitochondrial permeability transition. *Biochim. Biophys. Acta 1241*, 139–176.

59 Gizatullina, Z. Z., Chen, Y., Zierz, S., Gellerich, F. N. (2005) Effects of extramitochondrial ADP on permeability transition of mouse liver mitochondria. *Biochim. Biophys. Acta 1706*, 98–104.

60 Dzeja, P. P., Zeleznikar, R. J., Goldberg, N. D. (1998) Adenylate kinase: Kinetic behavior in intact cells indicates it is integral to multiple cellular processes. *Mol. Cell. Biochem. 184*, 169–182.

61 Pucar, D., Janssen, E., Dzeja, P. P., Juranic, N., Macura, S., Wieringa, B., Terzic, A. (2000) Compromised energetics in the adenylate kinase AK1 gene knockout heart under metabolic stress. *J. Biol. Chem. 275*, 41424–41429.

62 Braun, U., Paju, K., Eimre, M., Seppet, E., Orlova, E., Kadaja, L., Trumbeckaite, S., Gellerich, F. N., Zierz, S., Jockusch, H., Seppet, E. K. (2001) Lack of dystrophin is associated with altered integration of the mitochondria and ATPases in slow-twitch muscle cells of MDX mice. *Biochim. Biophys. Acta 1505*, 258–270.

63 Kaasik, A., Veksler, V., Boehm, E., Novotova, M., Ventura-Clapier, R. (2003) From energy store to energy flux: a study in creatine kinase-deficient fast skeletal muscle. *FASEB J. 17*, 708–710.

64 Panov, A. V., Burke, J. R., Strittmatter, W. J., Greenamyre, J. T. (2003) *In vitro* effects of polyglutamine tracts on Ca2+-dependent depolarization of rat and human mitochondria: relevance to Huntington's disease. *Arch. Biochem. Biophys. 410*, 1–6.

65 Chang, T. W. D., Rintoul, L. G., Pandipati, S., Reynolds, I. J. (2006)

Mutant huntingtin aggregates impair mitochondrial movement and trafficking in cortical neurons. *Neurobiol. Dis. 22,* 388–400.

66 Casley, C. S., Land, J. M., Sharpe, M. A., Clark, J. B., Duchen, M. R., Canevari, L. (2002) Beta-amyloid fragment 25–35 causes mitochondrial dysfunction in primary cortical neurons. *Neurobiol. Dis. 10,* 258–267.

67 David, S., Shoemaker, M., Haley, B. E. (1998) Abnormal properties of creatine kinase in Alzheimer's disease brain: correlation of reduced enzyme activity and active site photolabeling with aberrant cytosol-membrane partitioning. *Brain Res. Mol. Brain Res. 54,* 276–287.

68 de Cerqueira Cesar, M., Wilson, J. E. (2002) Functional characteristics of hexokinase bound to the type a and type B sites of bovine brain mitochondria. *Arch. Biochem. Biophys. 397,* 106–112.

69 Wagner, G., Kovacs, J., Löw, P., Orosz, F., Ovadi, J. (2001) Tubulin and microtubule are potential targets for brain hexokinase binding. *FEBS Lett. 509,* 81–84.

70 Schlattner, U., Forstner, M., Eder, M., Stachowiak, O., Fritz-Wolf, K., Wallimann, T. (1998) Functional aspects of the X-ray structure of mitochondrial creatine kinase: a molecular physiology approach. *Mol. Cell. Biochem. 184,* 125–140.

71 Beutner, G., Rück, A., Riede, B., Brdiczka, D. (1998) Complexes between porin, hexokinase, mitochondrial creatine kinase and adenylate translocator display properties of the permeability transition pore. Implication for regulation of permeability transition by the kinases. *Biochim. Biophys. Acta 1368,* 7–18.

72 Speer, O., Bäck, N., Buerklen, T., Brdicka, D., Koretsky, A., Wallimann, T., Eriksson, O. (2005) Octameric mitochondrial creatine kinase induces and stabilizes contact sites between the inner and outer membrane. *Biochem. J. 385,* 445–450.

73 da-Silva, W. S., Gomez-Puyou, A., de Gomez-Puyou, M. T., Moreno-Sanchez, R., De Felice, F. G., de Meis, L., Oliveira, M. F., Galina, A. (2004) Mitochondrial bound hexokinase activity as a preventive antioxidant defense: steady-state ADP formation as a regulatory mechanism of membrane potential and reactive oxygen species generation in mitochondria. *J. Biol. Chem. 279,* 39846–39855.

74 Meyer, L. E., Machado, L. B., Santiago, A. P., da-Silva, W. S., De Felice, F. G., Holub, O., Oliveira, M. F., Galina, A. (2006) Mitochondrial creatine kinase activity prevents reactive oxygen species generation: Antioxidant role of mitochondrial kinase-dependent ADP re-cycling activity. *J. Biol. Chem. 281,* 37361–37371.

75 Wendt, S., Dedeoglu, A., Speer, O., Wallimann, T., Beal, M. F., Andreassen, O. (2002). Reduced creatine kinase activity in transgenic amyotrophic lateral sclerosis mice. *Free Rad. Biol. Med. 32,* 920–926.

76 Halliwell, B. (2006) Oxidative stress and neurodegeneration: where are we now? *J. Neurochem. 97,* 1634–1658.

77 Kwong, J. Q., Beal, M. F., Manfredi, G. (2006) The role of mitochondria in inherited neurodegenerative diseases. *J. Neurochem. 97,* 1659–1675.

78 Zhang, J., Asin-Cayuela, J., Fish, J., Michikawa, Y., Bonafe, M., Olivieri, F., Passarino, G., De Benedictis, G., Franceschi, C., Attardi, G. (2003) Strikingly higher frequency in centenarians and twins of mtDNA mutation causing remodeling of replication origin in leukocytes. *Proc. Natl. Acad. Sci. USA 100,* 1116–1121.

79 Simon, D. K., Lin, M. T., Ahn, C. H., Liu, G. J., Gibson, G. E., Beal, M. F., Johns, D. R. (2001) Low mutational burden of individual acquired mitochondrial DNA mutations in brain. *Genomics 73,* 113–116.

80 Chinnery, P. F., Taylor, G. A., Howell, N., Brown, D. T., Parsons, T. J., Turnbull, D. M. (2001) Point mutations of the mtDNA control region in normal and neurodegenerative human brains. *Am. J. Hum. Genet. 68,* 529–532.

81 Trifunovic, A., Wredenberg, A., Falkenberg, M., Spelbrink, J. N., Rovio, A. T., Bruder, C. E., Bohlooly, Y. M., Gidlof, S., Oldfors, A., Wibom, R.,

Tornell, J., Jacobs, H. T., Larsson, N. G. (2004) Premature ageing in mice expressing defective mitochondrial DNA polymerase. *Nature 429*, 417–423.

82 Trifunovic, A. (2006) Mitochondrial DNA and ageing. *Biochim. Biophys. Acta 1757*, 611–617.

83 Trifunovic, A., Hansson, A., Wredenberg, A., Rovio, A. T., Dufour, E., Khvorostov, I., Spelbrink, J. N., Wibom, R., Jacobs, H. T., Larsson, N. G. (2005) Somatic mtDNA mutations cause aging phenotypes without affecting reactive oxygen species production. *Proc. Natl. Acad. Sci. USA 102*, 17993–17998.

84 Lin, M. T., Beal, M. F. (2006) Mitochondrial dysfunction and oxidative stress in neurodegeneration. *Nature 443*, 787–795.

85 Sun, J., Folk, D., Bradley, T. J., Tower, J. (2002) Induced overexpression of mitochondrial Mn-superoxide dismutase extends the life span of adult Drosophila melanogaster. *Genetics 161*, 661–672.

86 Schriner, S. E., Linford, N. J., Martin, G. M., Treuting, P., Ogburn, C. E., Emond, M., Coskun, P. E., Ladiges, W., Wolf, N., Van Remmen, H., Wallace, D. C., Rabinovitch, P. S. (2005) Extension of murine life span by overexpression of catalase targeted to mitochondria. *Science 308*, 1909–1911.

87 Esposito, L. A., Melov, S., Panov, A., Cottrell, B. A., Wallace, D. C. (1999) Mitochondrial disease in mouse results in increased oxidative stress. *Proc. Natl. Acad. Sci. USA 96*, 4820–4825.

88 Hamilton, M. L., Van Remmen, H., Drake, J. A., Yang, H., Guo, Z. M., Kewitt, K., Walter, C. A., Richardson, A. (2001) Does oxidative damage to DNA increase with age? *Proc. Natl. Acad. Sci. USA 98*, 10469–10474.

89 Melov, S., Ravenscroft, J., Malik, S., Gill, M. S., Walker, D. W., Clayton, P. E., Wallace, D. C., Malfroy, B., Doctrow, S. R., Lithgow, G. J. (2000) Extension of life-span with superoxide dismutase/catalase mimetics. *Science 289*, 1567–1569.

90 Lee, C. K., Klopp, R. G., Weindruch, R., Prolla, T. A. (1999) Gene expression profile of aging and its retardation by caloric restriction. *Science 285*, 1390–1393.

91 Rubinsztein, D. C. (2006) The roles of intracellular protein-degradation pathways in neurodegeneration. *Nature 443*, 780–786.

92 Lansbury, P. T., Lashuel, H. A. (2006) A century-old debate on protein aggregation and neurodegeneration enters the clinic. *Nature 443*, 774–779.

93 Harjes, P., Wanker, E. E. (2003) The hunt for huntingtin function: interaction partners tell many different stories. *Trends. Biochem. Sci. 28*, 425–433.

94 Vonsattel, J. P., DiFiglia, M. J. (1998) Huntington disease. *J. Neuropathol. Exp. Neurol. 57*, 369–384.

95 The Huntington's Disease Collaborative Research Group (1993) A novel gene containing a trinucleotide repeat that is expanded and unstable on Huntington's disease chromosomes. *Cell 72*, 971–983.

96 Li, S. H., Schilling, G., Young 3rd, W. S., Li, X. mJ., Margolis, R. L., Stine, O. C., Wagster, M. V., Abbott, M. H., Franz, M. L., Ranen, N. G. et al. (1993) Huntington's disease gene (IT15) is widely expressed in human and rat tissues. *Neuron 11*, 985–993.

97 Langbehn, D. R., Brinkman, R. R., Falush, D., Paulsen, J. S., Hayden, M. R. (2004) International Huntington's Disease Collaborative Group. A new model for prediction of the age of onset and penetrance for Huntington's disease based on CAG length. *Clin. Gen. 65*, 276–277.

98 Browne, S. E., Beal, M. F. (2004) The energetics of Huntington's disease. *Neurochem. Res. 29*, 531–546.

99 Djousse, L., Knowlton, Cupples, L. A., Marder, K., Shoulson, I., Myers, R. H. (2002) Weight loss in early stage of Huntington's disease. *Neurologie 69*, 1325–1330.

100 Sanberg, P. R., Fibiger, H. C., Mark, R. C. (1981) Body weight and dietary factors in Huntington's disease patients compared with matched controls. *Med. J. Aust. 1*, 407–409

101 Morales, L. M., Estevez, J., Suarez, H., Villalobos, R., Chacin de Bonilla, L., Bonilla, E. (1989) Nutritional evaluation of Huntington disease patients. *Am. J. Clin. Nutr. 50*, 145–150.

102 Koroshetz, W. J., Jenkins, B. G., Rosen, B. R., Beal, M. F. (1997) Energy metabolism defects in Huntington's disease and effects of coenzyme Q10. *Ann. Neurol. 41*, 160–165.

103 Grunewald, T., Beal, M. F. (1999) Bioenergetics in Huntington's disease. *Ann. N.Y. Acad. Sci. 893*, 203–213.

104 Browne, S. E., Bowling, A. C., MacGarvey, U., Baik, M. J., Berger, S. C., Muqit, M. M., Bird, E. D., Beal, M. F. (1997) Oxidative damage and metabolic dysfunction in Huntington's disease: selective vulnerability of the basal ganglia. *Ann. Neurol. 41*, 646–653.

105 Tabrizi, S. J., Cleeter, M. W., Xuereb, J., Taanman, J. W., Cooper, J. M., Schapira, A. H. (1999) Biochemical abnormalities and excitotoxicity in Huntington's disease brain. *Ann. Neurol. 45*, 25–32.

106 Goebel, H. H., Heipertz, R., Scholz, W., Iqbal, K., Tellez-Nagel, I. (1978) Juvenile Huntington chorea: clinical, ultra-structural, and biochemical studies. *Neurology 28*, 23–31.

107 Tellez-Nagel, I., Johnson, A. B., Terry, R. D. (1974) Studies on brain biopsies with Huntington's chorea. *J. Neuropathol. Exp. Neurol. 33*, 308–332.

108 Milakovic, T., Johnson, G. V. (2005) Mitochondrial respiration and ATP production are significantly impaired in striatal cells expressing mutant huntingtin. *J. Biol. Chem. 280*, 30773–30782.

109 Ludolph, A. C., He, F., Spencer, P. S., Hammerstad, J., Sabri, M. (1991) 3-Nitropropionic acid-exogenous animal neurotoxin and possible human striatal toxin. *Can. J. Neurol. Sci. 18*, 492–498.

110 Brouillet, E., Jacquard, C., Bizat, N., Blum, D. (2005) 3-Nitropropionic acid: a mitochondrial toxin to uncover physiopathological mechanisms underlying striatal degeneration in Huntington's disease. *J. Neurochem. 95*, 1521–1540.

111 Strand, A. D., Aragaki, A. K., Shaw, D., Bird, T., Holton, J., Turner, C., Tapscott, S. J., Tabrizi, S. J., Schapira, A. H., Kooperberg, C., Olson, J. M. (2005) Gene expression in Huntington's disease skeletal muscle: a potential biomarker. *Hum. Mol. Genet. 14*, 1863–1876.

112 Lodi, R., Schapira, A. H., Manners, D., Styles, P., Wood, N. W., Taylor, D. J., Warner, T. T. (2000) Abnormal *in vivo* skeletal muscle energy metabolism in Huntington's disease and dentatorubro-pallidoluysian atrophy. *Ann. Neurol. 48*, 72–76.

113 Saft, C., Zange, J., Andrich, J., Muller, K., Lindenberg, K., Landwehrmeyer, B., Vorgerd, M., Kraus, P. H., Przuntek, H., Schols, L. (2005) Mitochondrial impairment in patients and asymptomatic mutation carriers of Huntington's disease. *Mov. Disord. 20*, 674–679.

114 Gizatullina, Z. Z., Chen, Y., Lindenberg, K. S., Harjes, P., Striggow, F., Zierz, S., Gellerich, F. N. (2006) Increased calcium sensitivity of respiration and of permeability transition in mitochondria of skeletal muscle in transgenic the R6/2mice with mouse model for Huntington disease. *Ann. Neurol. 59*, 407–411.

115 Schapira, A. H., Cooper, J. M., Dexter, D., Jenner, P., Clark, J. B., Marsden, C. D. (1989) Mitochondrial complex I deficiency in Parkinson's disease. *Lancet* 1(8649), 1269.

116 Schapira, A. H., Mann, V. M., Cooper, J. M., Dexter, D., Daniel, S. E., Jenner, P., Clark, J. B., Marsden, C. D. (1990) Anatomic and diseases specificity of NADH CoQ1 reductase (complex I deficiency) in Parkinson's disease. *J. Neurochem. 55*, 2142–2145.

117 Krige, D., Caroll, M. T., Cooper, J. M., Marsden, C. D., Schapira, A. H. (1992) Platelet mitochondrial function in Parkinson's disease. The Royal Kings and Queens Parkinson Disease Research Group. *Ann. Neurol. 32*, 782–788.

118 Parker, W. D. Jr, Boyson, S. J., Parks, J. K. (1989) Abnormalities of the electron transport chain in idiopathic Parkinson's disease. *Ann. Neurol. 26*, 719–723.

119 Winkler Stuck, K., Kirches, E., Mawrin, C., Dietzmann, K., Lins, H., Wallesch, C. W., Kunz, W. S., Wiedemann, F. R. (2005) Re-evaluation of the dysfunction of mitochondrial respiratory chain in skeletal muscle of patients with

Parkinson's disease. *J. Neural. Transm.* *112*, 499–518.

120 Tetrud, J. W., Langston, J. W. (1989) The effect of deprenyl (selegiline) on the natural history of Parkinson's disease. *Science 245*, 519–522.

121 Betarbet, R., Sherer, T. B., MacKenzie, G., Garcia-Osuna, M., Panov, A. V., Greenamyre, J. T. (2000) Chronic systemic pesticide exposure reproduces features of Parkinson's disease. *Nat. Neurosci. 3*, 1301–1306.

122 McCormack, A. L., Thiruchelvam, M., Manning-Bog, A. B., Thiffault, C., Langston, J. W., Cory-Slechta, D. A., Di Monte, D. A. (2002) Environmental risk factors and Parkinson's disease: selective degeneration of nigral dopaminergic neurons caused by the herbicide paraquat. *Neurobiol. Dis. 10*, 119–127.

123 Dawson, T. M., Dawson, V. L. (2002) Neuroprotective and neurorestorative strategies for Parkinson's disease. *Nat. Neurosci. 5*, 1058–1061.

124 Soto-Otero, R., Sanmartin-Suarez, C., Sanchez-Iglesias, S., Hermida-Ameijeiras, A., Sanchez-Sellero, I., Mendez-Alvarez, E. (2006) Study on the ability of 1,2,3,4-tetrahydropapaveroline to cause oxidative stress: Mechanisms and potential implications in relation to parkinson's disease. *J. Biochem. Mol. Toxicol. 20*, 209–220.

125 Simon, D. K., Pulst, S. M., Sutton, J. P., Browne, S. E., Beal, M. F., Johns, D. R. (1999) Familial multisystem degeneration with parkinsonism associated with the 11778 mitochondrial DNA mutation. *Neurology 53*, 1787–1793.

126 Luoma, P., Melberg, A., Rinne, J. O., Kaukonen, J. A., Nupponen, N. N., Chalmers, R. M., Oldfors, A., Rautakorpi, I., Peltonen, L., Majamaa, K., Somer, H., Suomalainen, A. (2004) Parkinsonism, premature menopause, and mitochondrial DNA polymerase gamma mutations: clinical and molecular genetic study. *Lancet 364*, 875–882.

127 Song, D. D., Shults, C. W., Sisk, A., Rockenstein, E., Masliah, E. (2004) Enhanced substantia nigra mitochondrial pathology in human alpha-synuclein transgenic mice after

treatment with MPTP. *Exp. Neurol. 186*, 158–172.

128 Martin, L. J., Pan, Y., Price, A. C., Sterling, W., Copeland, N. G., Jenkins, N. A., Price, D. L., Lee, M. K. (2006) Parkinson's disease alpha-synuclein transgenic mice develop neuronal mitochondrial degeneration and cell death. *J. Neurosci. 26*, 41–50.

129 Klivenyi, P., Siwek, D., Gardian, G., Yang, L., Starkov, A., Cleren, C., Ferrante, R. J., Kowall, N. W., Abeliovich, A., Beal, M. F. (2006) Mice lacking alpha-synuclein are resistant to mitochondrial toxins. *Neurobiol. Dis. 21*, 541–548.

130 Pesah, Y., Pham, T., Burgess, H., Middlebrooks, B., Verstreken, P., Zhou, Y., Harding, M., Bellen, H., Mardon, G. (2004) Drosophila parkin mutants have decreased mass and cell size and increased sensitivity to oxygen radical stress. *Development 131*, 2183–2194.

131 Palacino, J. J., Sagi, D., Goldberg, M. S., Krauss, S., Motz, C., Wacker, M., Klose, J., Shen, J. (2004) Mitochondrial dysfunction and oxidative damage in parkin-deficient mice. *J. Biol. Chem. 279*, 18614–18622.

132 Muftuoglu, M., Elibol, B., Dalmizrak, O., Ercan, A., Kulaksiz, G., Ogus, H., Dalkara, T., Ozer, N. (2004) Mitochondrial complex I and IV activities in leukocytes from patients with parkin mutations. *Mov. Disord. 19*, 544–548.

133 Darios, F., Corti, O., Lucking, C. B., Hampe, C., Muriel, M. P., Abbas, N., Gu, W. J., Hirsch, E. C., Rooney, T., Ruberg, M., Brice, A. (2003) Parkin prevents mitochondrial swelling and cytochrome *c* release in mitochondria-dependent cell death. *Hum. Mol. Genet. 12*, 517–526.

134 Canet-Aviles, R. M., Wilson, M. A., Miller, D. W., Ahmad, R., McLendon, C., Bandyopadhyay, S., Baptista, M. J., Ringe, D., Petsko, G. A., Cookson, M. R. (2004) The Parkinson's disease protein DJ-1 is neuroprotective due to cysteine-sulfinic acid-driven mitochondrial localization. *Proc. Natl. Acad. Sci. USA 101*, 9103–9108.

135 Kim, R. H., Smith, P. D., Aleyasin, H., Hayley, S., Mount, M. P., Pownall, S.,

Wakeham, A., You-Ten, A. J., Kalia, S. K., Horne, P., Westaway, D., Lozano, A. M., Anisman, H., Park, D. S., Mak, T. W. (2005) Hypersensitivity of DJ-1-deficient mice to 1-methyl-4-phenyl-1,2,3,6-tetrahydropyrindine (MPTP) and oxidative stress. *Proc. Natl. Acad. Sci. USA 102*, 5215–5220.

136 Petit, A., Kawarai, T., Paitel, E., Sanjo, N., Maj, M., Scheid, M., Chen, F., Gu, Y., Hasegawa, H., Salehi-Rad, S., Wang, L., Rogaeva, E., Fraser, P., Robinson, B., St George-Hyslop, P., Tandon, A. (2005) Wild-type PINK1 prevents basal and induced neuronal apoptosis, a protective effect abrogated by Parkinson disease-related mutations. *J. Biol. Chem. 280*, 34025–34032.

137 Deng, H., Jankovic, J., Guo, Y., Xie, W., Le, W. (2005) Small interfering RNA targeting the PINK1 induces apoptosis in dopaminergic cells SH-SY5Y. *Biochem. Biophys. Res. Commun. 337*, 1133–1138.

138 Hald, A., Lotharius, J. (2005) Oxidative stress and inflammation in Parkinson's disease: Is there a causal link? *Exp. Neuro. 193*, 179–290.

139 Evans, D. A., Funkenstein, H. H., Albert, M. S., Scherr, P. A., Cook, N. R., Chown, M. J., Hebert, L. E., Hennekens, C. H., Taylor, J. O. (1989) Prevalence of Alzheimer's disease in a community population of older persons. Higher than previously reported. *J. Am. Med. Assoc. 262*, 2551–2666.

140 Lee, V. M., Goedert, M., Trojanowski, J. Q. (2001) Neurodegenerative tauopathies.*Annu. Rev. Neurosci. 24*, 1121–1159.

141 McClure, R. J., Kanfer, J. N., Panchalingam, K., Klunk, W. E., Pettegrew, J. W. (1995) Magnetic resonance spectroscopy and its applica-tion to aging and Alzheimer's disease. *Neuroimag. Clin. N. Am. 5*, 69–86.

142 Alavi, A., Newberg, A. B., Souder, E., Berlin, J. A. (1993) Quantitative analysis of PET and MRI data in normal aging and Alzheimer's disease: atrophy weighted total brain metabolism and absolute whole brain metabolism as reliable discriminators. *J. Nucl. Med. 34*, 1681–1687.

143 Smith, C. D., Pettigrew, L. C., Avison, M. J., Kirsch, J. E., Tinkhtman, A. J., Schmitt, F. A., Wermeling, D. P., Wekstein, D. R., Markesberry, W. R. (1995) Frontal lobe phosphorus metabolism and neuropsychological function in aging and in Alzheimer's disease. *Ann. Neurol. 38*, 194–201.

144 Bosetti, F., Brizzi, F., Barogi, S., Mancuso, M., Siciliano, G., Tendi, E. A., Murri, L., Rapoport, S. I., Solaini, G. (2002) Cytochrome *c* oxidase and mitochondrial F_1F_0-ATPase (ATP synthase) activities in platelets and brain from patients with Alzheimer's disease. *Neurobiol. Aging. 23*, 371–366.

145 Swerdlow, R. H., Parks, J. K., Cassarino, D. S., Maguire, D. J., Maguire, R. S., Bennett, J. P. Jr, Davis, R. E., Parker, W. D. Jr. (1997) Cybrids in Alzheimer's disease: a cellular model of the disease? *Neurology 49*, 918–925.

146 Sheehan, J. P., Swerdlow, R. H., Parker, W. D., Miller, S. W., Davis, R. E., Tuttle, J. B. (1997) Altered calcium homeostasis in cells transformed by mitochondria from individuals with Parkinson's disease. *J. Neurochem. 68*, 1221–1233.

147 Ito, E., Oka, K., Etcheberrigaray, R., Nelson, T. J., McPhie, D. L., Tofel-Grehl, B., Gibson, G. E., Alkon, D. L. (1994) Internal Ca^{2+} mobilization is altered in fibroblasts from patients with Alzheimer disease. *Proc. Natl. Acad. Sci. U.S.A. 91*, 534–538.

148 Nunomura, A., Perry, G., Aliev, G., Hirai, K., Takeda, A., Balraj, E. K., Jones, P. K., Ghanbari, H., Wataya, T., Shimohama, S., Chiba, S., Atwood, C. S., Petersen, R. B., Smith, M. A. (2001) Oxidative damage is the earliest event in Alzheimer disease. *J. Neuropathol. Exp. Neurol. 60*, 759–767.

149 Elson, J. L., Herrnstadt, C., Preston, G., Thal, L., Morris, C. M., Edwardson, J. A., Beal, M. F., Turnbull, D. M., Howell, N. (2006) Does the mitochondrial genome play a role in the etiology of Alzheimer's disease? *Hum. Genet. 119*, 241–254.

150 Anandatheerthavarada, H. K., Biswas, G., Robin, M. A., Avadhani, N. G. (2003) Mitochondrial targeting and a novel transmembrane arrest of Alzheimer's amyloid precursor protein impairs mitochondrial function in neuronal cells. *J. Cell. Biol. 161*, 41–54.

151 Hansson, C. A., Frykman, S., Farmery, M. R., Tjernberg, L.O., Nilsberth, C., Pursglove, S. E., Ito, A., Winblad, B., Cowburn, R. F., Thyberg, J., Ankarcrona, M. (2004) Nicastrin, presenilin, APH-1, and PEN-2 form active gamma-secretase complexes in mitochondria. *J. Biol. Chem. 279*, 51654–5160.

152 Lustbader, J. W., Cirilli, M., Lin, C., Xu, H. W., Takuma, K., Wang, N., Caspersen, C., Chen, X., Pollak, S., Chaney, M., Trinchese, F., Liu, S., Gunn-Moore, F., Lue, L. F., Walker, D. G., Kuppusamy, P., Zewier, Z. L., Arancio, O., Stern, D., Yan, S. S., Wu, H. (2004) ABAD directly links aß to mitochondrial toxicity in Alzheimer's disease. *Science 304*, 448–452.

153 Crouch, P. J., Blake, R., Duce, J. A., Ciccotosto, G. D., Li, Q. X., Barnham, K. J., Curtain, C. C., Cherny, R. A., Cappai, R., Dyrks, T., Masters, C. L., Trounce, I. A. (2005) Copper-dependent inhibition of human cytochrome *c* oxidase by a dimeric conformer of amyloid-beta1-42. *J. Neurosci. 25*, 672–679.

154 Manczak, M., Anekonda, T. S., Henson, E., Park, B. S., Quinn, J., Reddy, P. H. (2006) Mitochondria are a direct site of A beta accumulation in Alzheimer's disease neurons: implications for free radical generation and oxidative damage in disease progression. *Hum. Mol. Genet. 15*, 1437–1449.

155 Casley, C. S., Canevari, L., Land, J. M., Clark, J. B., Sharpe, M. A. (2002) Beta-amyloid inhibits integrated mitochondrial respiration and key enzyme activities. *J. Neurochem. 80*, 91–100.

156 Gibson, G. E., Sheu, K. F., Blass, J. P., Baker, A., Carlson, K. C., Harding, B., Perrino, P. (1988) Reduced activities of thiamine-dependent enzymes in the brains and peripheral tissues of patients with Alzheimer's disease. *Arch. Neurol. 45*, 836–840.

157 Gotz, J., Ittner, L. M., Schonrock, N. (2006) Alzheimer's disease and frontotemporal dementia: prospects of a tailored therapy? *Med. J. Aust. 185*, 381–384.

158 Hutton, M., Lendon, C. L., Rizzu, P., Baker, M., Froelich, S., Houlden, H., Pickering-Brown, S., Chakraverty, S., Isaacs, A., Grover, A., Hackett, J., Adamson, J., Lincoln, S., Dickson, D., Davies, P., Petersen, R. C., Stevens, M., de Graaff, E., Wauters, E., van Baren, J., Hillebrand, M., Joosse, M., Kwon, J. M., Nowotny, P., Che, L. K., Norton, J., Morris, J. C., Reed, L. A., Trojanowski, J., Basun, H., Lannfelt, L., Neystat, M., Fahn, S., Dark, F., Tannenberg, T., Dodd, P. R., Hayward, N., Kwok, J. B., Schofield, P. R., Andreadis, A., Snowden, J., Craufurd, D., Neary, D., Owen, F., Oostra, B. A., Hardy, J., Goate, A., van Swieten, J., Mann, D., Lynch, T., Heutink, P. (1998) Association of missense and 5'-splice-site mutations in tau with the inherited dementia FTDP-17. *Nature 393*, 702–705.

159 Lewis, J., Dickson, D. W., Lin, W. L., Chisholm, L., Corral, A., Jones, G., Yen, S. H., Sahara, N., Skipper, L., Yager, D., Eckman, C., Hardy, J., Hutton, M., McGowan, E. (2001) Enhanced neurofibrillary degeneration in transgenic mice expressing mutant tau and APP. *Science 293*, 1487–1491.

160 David, D. C., Hauptmann, S., Scherping, I., Schuessel, K., Keil, U., Rizzu, P., Ravid, R., Drose, S., Brandt, U., Muller, W. E., Eckert, A., Gotz, J. (2005) Proteomic and functional analyses have revealed a mitochondrial dysfunction in P301L tau transgenic mice. *J. Biol. Chem. 280*, 23802–23814.

161 Hoppeler, H., Vogt, M., Weibel, E. R., Fluck, M. (2003) Response of skeletal muscle to hypoxia. *Exp. Physiol. 88*, 109–119.

162 Höckel, M., Vaupel, P. (2001) Tumor hypoxia: definitions and current clinical, biological, and molecular aspects. *J. Natl. Cancer. Inst. 93*, 266–276.

163 Guzy, R. D., Schumacker, P. T. (2006) Oxygen sensing by mitochondria at complex III: the paradox of increased reactive oxygen species during hypoxia. *Exp. Physiol. 91*, 807–819.

164 Chandel, N. S., McClintock, D. S., Feliciano, C. E., Wood, T. M., Melendez, J. A., Rodriguez, A. M., Schumacker,

P. T. (2000) Reactive oxygen species generated at mitochondrial complex III stabilize hypoxia-inducible factor-1α during hypoxia. *J. Biol. Chem. 275*, 25130–25138.

165 Brunelle, J. K., Bell, E. L., Quesada, N. M., Vercauteren, K., Tiranti, V., Zeviani, M., Scarpulla, R. C., Chandel, N. S. (2005) Oxygen sensing requires mitochondrial ROS but not oxidative phosphorylation. *Cell Metabolism 1*, 409–414.

166 Mansfield, K. D., Guzy, R. D., Pan, Y., Young, R. M., Cash, T. P., Schumacker, P. T., Simon, M. C. (2005) Mitochondrial dysfunction resulting from loss of cytochrome *c* impairs cellular oxygen sensing and hypoxic HIF-α activation. *Cell Metabolism 1*, 393–399.

167 Sanjuán-Pla, A., Cervera, A. M., Apostolova, N., Garcia-Bou, R., Victor, V. M., Murphy, M. P., McCreath, K. J. (2005) A targeted antioxidant reveals the importance of mitochondrial reactive oxygen species in the hypoxic signaling of HIF-1α. *FEBS Lett. 579*, 2669–2674.

168 Emerling, B. M., Platanias, L. C., Black, E., Nebreda, A. R., Davies, R. J., Chandel, N. S. (2005) Mitochondrial reactive oxygen species activation of p38 mitogen-activated protein kinase is required for hypoxia signaling. *Mol. Cell. Biol. 25*, 4853–4862.

169 Semenza, G. L. (2002) HIF-1 and tumor progression: pathophysiology and therapeutics. *Trends Mol. Med. 8*, S62–S67.

170 Iyer, N. V., Kotch, L. E., Agani, F., Leung, S. W., Laughner, E., Wenger, R. H., Gassmann, M., Gearhart, J. D., Lawler, A. M., Yu, A. Y., Semenza, G. L. (1998) Cellular and developmental control of O_2 homeostasis by hypoxia-inducible factor 1 alpha. *Genes Dev. 12*, 149–162.

171 Papandreou, I., Cairns, R. A., Fontana, L., Lim, A. L., Denko, N. C. (2006) HIF-1 mediates adaptation to hypoxia by actively downregulating mitochondrial oxygen consmption. *Cell Metabolism 3*, 187–197.

172 Kim, J.-W., Tchernyshyov, I., Semenza, G. L., Dang, C. V. (2006) HIF-1 mediated expression of pyruvate. *Metabolism 3*, 177–185.

173 Dahia, P. L. M., Ross, K. N., Wright, M. E., Hayashida, C. Y., Santagata, S., Barontini, M., Kung, A. L., Sanso, G., Powers, J. F., Tischler, A. S., Hodin, R., Heitritter, S., Moore, F. Jr, Dkuhy, R., Sosa, J. A., Ocal, I. T., Benn, D. E., Marsh, D. J., Robinson, B. G., Schneider, K., Garber, J., Arum, S. M., Korbonits, M., Grossman, A., Pigny, P., Toledo, S. P. A., Nosé, V., Li, C., Stiles, C. D. (2005) A HIFα regulatory loop links hypoxia and mitochondrial signals in pheochromas. *PloS Genetics 1*, 72–80.

174 Neumann, A. K., Yang, J., Biju, M., Joseph, S. K., Johnson, R. S., Haase, V. H., Freedman, B. D., Turka, L. A. (2005) Hypoxia inducible factor 1α regulates T cell receptor signal transduction. *Proc. Natl. Acad. Sci. USA 102*, 17071–17076.

175 Huang, Y., Hickey, R. P., Yeh, J. L., Liu, D., Dadak, A., Young, L. H., Johnson, R. S., Giordano, F. J. (2004) Cardiac myocyte-specific HIF-1α deletion alters vascularization, energy availabilty, calcium flux, and contractility in the normoxic heart. *FASEB J.* (DOI 10.1096/fj.04-1510fje. May 7, 2004).

176 Zörning, M., Hueber, A.-O., Baum, W., Evan, G. (2001) Apoptosis regulators and their role in tumorigenesis. *Biochim. Biophys. Acta 1551*, F1–F37.

177 Reed, J. C. (1999) Mechanisms of apoptosis avoidance in cancer. *Curr. Opinion. Oncol. 11*, 68–75.

178 Burgart, L. J., Zheng, J., Shu, Q., Shibata, D. (1995) Somatic mitochondrial mutation in gastric cancer. *Am. J. Pathol. 147*, 1105–1111.

179 Didelot, C., Barberi-Heyob, M., Bianchi, A., Becuwe, P., Mirjolet, J. F., Dauca, M., Merlin, J. L. (2001) Constitutive NF-kappaB activity influences basal apoptosis and radiosensitivity of head-and-neck carcinoma cell lines. *Int. J. Radiat. Oncol. Biol. Phys. 51*, 1354–1360

180 Takeda, Y., Togashi, H., Matsuo, T., Shinzawa, H., Takeda, Y., Takahashi, T. (2001) Growth inhibition and apoptosis of gastric cancer cell lines by Anemarrhena asphodeloides Bunge. *J. Gastroenterol. 36*, 79–90.

181 Eapen, C. E., Madesh, M., Balasubramanian, K., Pulimood, A.,

Mathan, M., Ramakrishna, B. S. (1998) Mucosal mitochondrial function and antioxidant defences in patients with gastric carcinoma. *Scand. J. Gastroenterol. 33*, 975–981.

182 Carretero, J., Obrador, E., Pellicer, J. A., Pascual, A., Estrela, J. M. (2000) Mitochondrial glutathione depletion by glutamine in growing tumor cells. *Free Rad. Biol. Med. 29*, 913–923.

183 Spitz, D. R., Sim, J. E., Ridnour, L. A., Galoforo, S. S., Lee, Y. J. (2000) Glucose deprivation-induced oxidative stress in human tumor cells. A fundamental defect in metabolism? *Ann. N.Y. Acad. Sci. 899*, 349–362.

184 Ahmad, I. M., Aykin-Burns, N., Sim, J. E., Walsh, S. A., Higashikubo, R., Buettner, G. R., Venkataraman, S., Mackey, M. A., Flanagan, S., Oberley, L. W., Spitz, D. R. (2005) Mitochondrial O2*- and H2O2 mediate glucose deprivation-induced stress in human cancer cells. *J. Biol. Chem. 280*, 4525–4563.

185 Pedram, A., Razandi, M., Wallace, D. C., Levin, A. R. (2006) Functional estrogen receptors in the mitochondria of breast cancer cells. *Mol. Biol. Cell 17*, 2125–2137.

186 Dang, C. V., Semenza, G. L. (1999) Oncogenic alterations of metabolism. *Trends. Biochem. 24*, 68–72.

187 Semenza, G. L., Artemov, D., Bedi, A., Bhujwalla, Z., Chiles, K., Feldser, D., Laughner, E., Ravi, R., Simons, J., Taghavi, P., Zhong, H. (2001) 'The metobolism of tumors': 70 years later. *Novartis Found. Symp. 240*, 251–260.

188 Fantin, V. R., St-Pierre, J., Leder, P. (2006) Attenuation of LDH-A expressino uncovers a link between glycolysis, mitochondrial physiology, and tumor maintenance. *Cancer Cell 9*, 426–434.

189 Matoba, S., Kang, J. G., Patino, W. D., Wragg, A., Boehm, M., Gavrilova, O., Hurley, P. J., Bunz, F., Hwang, P. M. (2006) p53 regulates mitochondrial respiration. *Science 312*, 1650–1653.

190 Shim, H., Dolde, C., Lewis, B. C., Wu, C. S., Dang, G., Jungmann, R. A., Dalla-Favera, R., Dang, C. V. (1997) C-Myc transactivation of LDH-A: implications for tumor metabolism and growth. *Proc. Natl Acad. Sci. USA 94*, 6658–6663.

191 Lester, R. D., Jo, M., Campana, W. M., Gonias, S. L. (2005) Erythropoietin promotes MCF-7 breast cancer cell migration by an ERK/mitogen activated protein kinase-dependent pathway and is primarily responsible for the increase in migration observed in hypoxia. *J. Biol. Chem. 280*, 39273–39277.

192 Haddad, J. J., Land, S. C. (2001) A non-hypoxic, ROS-sensitive pathway mediates TNF-α-dependent regulation of HIF-1α. *FEBS Lett. 505*, 269–274.

193 Petros, J. A., Baumann, A. K., Ruiz-Pesini, E., Amin, M. B., Sun, C. Q., Hall, J., Lim, S., Issa, M. M., Flanders, W. D., Hosseini, S. H., Marshall, F. F., Wallace, D. C. (2005) mtDNA mutations increase tumorigenicity in prostate cancer. *Proc. Natl. Acad. Sci. USA 102*, 719–724.

194 Boitier, F., Merad-Boudina, M., Guguen-Guillouzo, C., Defer, N., Ceballos-Picot, I., Leroux, J. P., Marsac, C. (1995) Impairment of the mitochondrial respiratory chain activity in diethyl-nitrosamine-induced rat hepatomas: possible involvement of oxygen free radicals. *Cancer Res. 55*, 3028–3035.

195 Ray, S., Ray, M. (1997) Does excessive adenosine 5′-triphosphate formation in cells lead to malignancy? A hypothesis on cancer. *Medical Hypotheses 48*, 473–476.

196 Chen, L. B. (1988) Mitochondrial membrane potential in living cells. *Annu. Rev. Cell Biol. 4*, 155–181.

197 Capuano, F., Varone, D., D'Eri, N., Russo, E., Tommasi, S., Montemurro, S., Prete, F., Papa, S. (1996) Oxidative phosphorylation and F(O)F(1) ATP synthase activity of human hepatocellular carcinoma. *Biochem. Mol. Biol. Int. 38*, 1013–1022.

198 Green, D. W., Grover, G. J. (2000) The IF(1) inhibitor protein of the mitochondrial F_1F_0-ATPase. *Biochim. Biophys. Acta 1458*, 343–355.

199 Bravo, C., Minauro-Sanmiguel, F., Morales-Rios, E., Rodriguez-Zavala, J. S., Garcia, J. J. (2004) Overexpression of the inhibitor protein IF(1) in AS-30D hepatoma produces a higher association

with mitochondrial F(1)F(0) ATP synthase compared to normal rat liver: functional and cross-linking studies. *J. Bioenerg. Biomembr.* 36, 257–264.

200 Bonora, E., Porcelli, A. M., Gasparre, G., Biondi, A., Ghelli, A., Carelli, V., Baracca, A., Tallini, G., Martinuzzi, A., Lenaz, G., Rugolo, M., Romeo, G. (2006) Defective oxidative phosphorylation in thyroid oncocytic carcinoma is associated with pathogenic mitochondrial DNA mutations affecting complexes I and III. *Cancer Res.* 66, 6087–6096.

201 Tan, M. G., Ooi, L. L., Aw, S. E., Hui, K. M. (2004) Cloning and identification of hepatocellular carcinoma down-regulated mitochondrial carrier protein, a novel liver-specific uncoupling protein. *J. Biol. Chem.* 279, 45235–45244.

202 Korshunov, S. S., Skulachev, V. P., Starkov, A. A. (1997) High protonic potential actuates a mechanism of production of reactive oxygen species in mitochondria. *FEBS Lett.* 416, 15–18.

203 Hervouet, E., Godinot, C. (2006) Mitochondrial disorders in renal tumors. *Mitochondrion* 6, 105–117.

204 Nakashima, R. A., Paggi, M. G., Scott, L. J., Pedersen, P. L. (1988) Purification and characterization of a bindable form of mitochondrial bound hexokinase from the high glycolytic AS-30D rat hepatoma cell line. *Cancer Res.* 48, 913–919.

205 Mathupala, S. P., Rempel, A., Pedersen, P. L. (1995) Glucose catabolism in cancer cells. Isolation, sequence, and activity of the promoter for type II hexokinase. *J. Biol. Chem.* 270, 16918–16925.

206 Mathupala, S. P., Ko, Y. H., Pedersen, P. L. (2006) Hexokinase II: cancer's double edged sword acting as both facilitator and gatekeeper of malignancy when bound to mitochondria. *Oncogene* 25, 4777–4786.

207 Gwak, G.-Y., Yoon, J.-H., Kim, K. M., Lee, H.-S., Chung, J. W., Gores, G. J. (2005) Hypoxia stimulates proliferation of human hepatoma cells through the induction of hexokinase II expression. *J Hepatology* 42, 358–364.

208 Nicholson, K. M., Anderson, N. G. (2002) The protein kinase B/Akt signalling pathway in human malignancy. *Cell Sign.* 14, 381–395.

209 Majewski, N., Nogueira, V., Robey, R. B., Hay, N. (2004) Akt inhibits apoptosis downstream of BID cleavage via a glucose-dependent mechanism involving mitochondrial hexokinases. *Mol. Cell. Biol.* 24, 730–740.

210 Chan, S.-L., Yu, V. C. (2004) Proteins of the Bcl-2 family in apoptosis signalling: from mechanistic insights to therapeutic opportunities. *Clin. Exp. Pharmacol. Physiol.* 31, 119–128.

211 Kirkin, V., Joos, S., Zörning, M. (2004) The role of Bcl-2 family members in tumorigenesis. *Biochim. Biophys. Acta* 1644, 229–249.

212 Sekimura, A., Konishi, A., Mizuno, K., Kobiyashi, Y., Sasaki, H., Yano, M., Fukai, I., Fujii, Y. (2004) Expression of Smac/DIABLO is a novel prognostic marker in lung cancer. *Oncol. Rep.* 11, 797–802.

213 Mizutani, Y., Nakanishi, H., Yamamoto, K., Li, Y. N., Matsubara, H., Mikami, K., Okihara, K., Kawauchi, A., Bonavida, B., Miki, T. (2004) Downregulation of Smac/DIABLO expression in renal cell carcinoma and its prognostic significance. *J. Clin. Oncol.* 23, 448–454.

214 Zörning, M., Hueber, A.-O., Baum, W., Evan, G. (2001) Apoptosis regulators and their role in tumorigenesis. *Biochim. Biophys. Acta* 1551, F1–F37.

215 Dohi, T., Beltrami, E., Wall, N. R., Plescia, J., Altieri, D. C. (2004) Mitochondrial survivin inhibits apoptosi and promotes tumorigenesis. *J. Clin. Invest.* 114, 1117–1127.

216 Hajra, K. M., Liu, J. R. (2004) Apoptosome dysfunction in human cancer. *Apoptosis* 9, 691–704.

217 Xia, H. H.-X., Talley, N. J. (2001) Apoptosis in gastric epithelium induced by Helicobacter pylori infection: Implications in gastric carcinogenesis. *Am. J. Gastroenterol.* 96, 16–26.

218 Török, N. J., Higuchi, H., Bronk, S., Goers, G. J. (2001) Nitric oxide inhibits apoptosis downstream of cytochrome *c* release by nitrosylating caspase 9. *Cancer Res* 62, 1648–1653.

219 Tomiyama, A., Serizawa, S., Tachibana, K., Sakurada, K., Samejima, H.,

Kuchino, Y., Kitanaka, C. (2006) Critical role for mitochondrial oxidative phosphorylation in the activation of tumor suppressors Bax and Bak. *J. National Cancer Inst.* 98, 1462–1473.

220 Galiègue, S., Casellas, P., Kramar, A., Tinel, N., Simony-Lafontaine, J. (2004) Immunohistochemical assessment of the peripheral benzodiazepine receptor in breat cancer and its relationship with survival. Clin. *Cancer Res. 10,* 2058–2064.

221 Carayon, P., Portier, M., Dussossoy, D., Bord, A., Petiprêtre, G., Canat, X., Le Fur, G., Casellas, P. (1996) Involvement of peripheral benzodiazepine receptors in the protection of hematopoietic cells against oxygen radical damage. *Blood 87,* 3170–3178.

222 Denicourt, C., Dowdy, S. (2004) Targeting apoptotic pathways in cancer cells. *Science 305,* 1411–1413.

223 Anderson, K. M., Harris, J. E. (2001) Is induction of type 2 programmed death in cancer cells from solid tumors directly related to mitochondrial mass? *Med. Hypotheses 57,* 87–90.

224 Grad, J. M., Cepero, E., Boise, L. H. (2001) Mitochondria as targets for established and novel anti-cancer agents. *Drug Resist. Updat. 4,* 85–91.

225 Zhang, L., Yu, J., Park, B. H., Kinzler, K. W., Vogelstein, B. (2000) Role of BAX in the apoptotic response to anticancer agents. *Science 290,* 989–992.

226 Zimmermann, K. C., Waterhouse, N. J., Goldstein, J. C., Schuler, M., Green, D. R. (2000) Aspirin induces apoptosis through release of cytochrome *c* from mitochondria. *Neoplasia (N.Y.) 2,* 505–513.

227 Haddad, J. J., Harb, H. L. (2005) Cytokines and the regulation of hypoxia-inducible factor (HIF)-1α. *Int. Immunopharm. 5,* 461–483.

228 Schulze-Osthoff, K., Bakker, A. C., Vanhasebroeck, B., Beyaert, R., Jacob, W. A., Fiers, W. (1992) Cytotoxic activity of tumor necrosis factor is mediated by early damage of mitochondrial functions. *J. Biol. Chem. 267,* 5317–5323.

229 Goossens, V., Stange, G., Moens, K., Pipeleers, D., Grooten, J. (1999) Regulation of tumor necrosis factor-induced, mitochondria- and reactive oxygen species-dependent cell death by the electron flux through the electron transport chain complex I. *Antioxid. Redox Signal. 1,* 285–295.

230 Itoh, S., Lemay, S., Osawa, M., Che, W., Duan, Y., Tompkins, A., Brookes, P. S., Sheu, S.-S., Abe, J.-i. (2005) Mitochondrial Dok-4 recruits Src kinase and regulates NF-kB activation in entothelial cells. *J. Biol. Chem. 280,* 26383–26396.

231 Le Moine, O., Louis, H., Stordeur, P., Collet, J. M., Goldman, M., Deviere, J. (1997) Role of reactive oxygen intermediates in interleukin 10 release after cold liver ischemia and reperfusion in mice. *Gastroenterology 113,* 1701–1706.

232 Llorente, L., Zou, W., Levy, Y., Richaud-Patin, Y., Wijdenes, J., Alcocer-Varela, J., Morel-Fourrier, B., Brouet, J. C., Alarcon-Segovia, D., Galanaud, P., Emilie, D. (1995) Role of interleukin 10 in the B lymphocyte hyperactivity and antibody production of human systemic lupus erythematosus. *J. Exp. Med. 181,* 839–844.

233 Marumo, T., Schini-Kerth, V. B., Fisslthaler, B., Busse, R. (1997) Platelet-derived growth factor-stimulated superoxide anion production modulates activation of transcription factor NF-kappaB and expression of monocyte chemoattractant protein 1 in human aortic smooth muscle cells. *Circulation 96,* 2361–2367.

234 Mogensen, T. H., Melchjorsen, J., Höllsberg, P., Paludan, S. R. (2003) Activation of NF-κB invirus-infected macrophages is dependent on mitochondrial oxidative stress and intracellular calcium: Downstream involvement of the kinases TGF-β-activated kinase 1, mitogen activated kinase/extracellular signal-regulated kinase kinase 1, and IκB kinase. *J. Immunol. 170,* 6224–6233.

235 Bai, Y., Onuma, H., Bai, X., Medvedev, A. V., Misukonis, M., Weinberg, J. B., Cao, W., Robidoux, J., Floering, L. M., Daniel, K. W., Collins, S. (2005) Persistent nuclear factor-κB activation in Ucp2−/− mice leads to enhanced nitric oxide and inflammatory cytokine production. *J. Biol. Chem. 280*, 19062–19069.

236 Boveris, A., Alvarez, S., Navarro, A. (2002) The role of mitochondrial nitric oxide synthase in inflammation and septic shock. *Free Rad. Biol. Med. 33*, 1186–1193.

237 Haynes, V., Elfering, S., Traaseth, N., Giulivi, C. (2004) Mitochondrial nitric-oxide synthase: enzyme expression, characterization, and regulation. *J Bioenerg Biomembr. 36*, 341–346.

238 Ramachandran, A., Levonen, A.-L., Brookes, P. S., Ceaser, E., Shiva, S., Barone, M. C., Darley-Usmar, V. (2002) Mitochondria, nitric oxide, and cardiovascular dysfunction. *Free Rad. Biol. Med. 33*, 1469–1474.

239 Borutaite, V., Moncada, S., Brown, G. C. (2005) Nitric oxide from inducible nitric oxide synthse sensitizes the inflamed aorta to hypoxic damage via respiratory inhibition. *Shock 23*, 319–323.

240 Frost, M. T., Wang, Q., Moncada, S., Singer, M. (2005) Hyoxia accelerates nitric oxide-dependent inhibition of mitochondrial complex I in activated macrophages. *Am. J. Physiol. 288*, 394–400.

241 Beltran, B., Orsi, A., Clementi, E., Moncada, S. (2000) Oxidative stress and S-nitrosylation of proteins in cells. *British J. Pharmacol. 129*, 953–960.

242 Borutaite, V., Budriunaite, A., Brown, G. C. (2000) Reversal nitric oxide-, peroxynitrite- and S-nitrosothiol-induced inhibition of mitochondrial respiration or complex I activity by light and thiols. *Biochim. Biophys. Acta 1459*, 405–412.

243 Clementi, E., Brown, G. C., Feelich, M., Moncada, S. (1998) Persistent inhibition of cellular respiration by nitic oxide: crucial role of S-nitrosylation of mitochondrial complex I and protective action of glutathione. *Proc. Natl. Acad. Sci. USA 95*, 7631–7636.

244 Moncada, S. (2000) Nitric oxide and cell respiration: physiology and Pathology. *Verh. K. Akad. Geneeskd. Belg. 62*, 171–179.

245 Takabayashi, A., Kawai, Y., Iwata, S., Kanai, M., Denno, R., Kawada, K., Obama, K., Taki, Y. (2000) Nitric oxide induces a decrease in the mitochondrial membrane potential of peripheral blood lymphocytes, especially in natural killer cells. *Antioxid. Redox. Signal 2*, 673–680.

246 Trumbeckaite, S., Opalka, J. R., Neuhof, C., Zierz, S., Gellerich, F. N. (2000) Different sensitivity of rabbit heart and skeletal muscle to endotoxin-induced impairment of mitochondrial function. *Eur. J. Biochem. 268*, 1422–1429.

247 Gellerich, F. N., Trumbeckaite, S., Hertel, K., Zierz, S., Müller-Werdan, U., Werdan, K., Redl, H., Schlag, G. (1999) Impaired energy metabolism in hearts of septic baboons: diminished activities of complex I and complex II of the mitochondrial respiratory chain. *Shock 11*, 336–341.

248 Gellerich, F. N., Trumbeckaite, S., Opalka, J. R., Chen, Y., Neuhoff, C., Schlag, H., Zierz, S. (2002) Mitochondrial dysfunction at sepsis: Evidences from bacteraemic baboons and endotoxaemic rabbits. *Bioscience Report 22*, 99–113.

249 Crouser, E. D. (2004) Mitochondrial dysfunction in septic shock and multiple organ dysfunction syndrome. *Mitochondrion. 4*, 729–741.

250 Galmiche, A., Rassow, J., Doye, A., Cagnol, S., Chambard, J. C., Contamin, S., de Thillot, V., Just, I., Ricci, V., Solcia, E., Van Obberghen, E., Boquet, P. (2000) The N-terminal 34 kDa fragment of Helicobacter pylori vacuolating cytotoxin targets mitochondria and induces cytochrome *c* release. *EMBO J. 19*, 6361–6370.

251 Ashktorab, H., Frank, S., Khaled, A. R., Durum, S. K., Kifle, B., Smoot, D. T. (2004) Bax translocation and mitochondrial fragmentation induced by Helicobacter pylori. *Gut 53*, 805–813.

252 Kimura, M., Goto, S., Wada, A., Yahiro, K., Niidome, T., Hatakeyama, T., Aoyagi, H., Hirayama, T., Kondo, T. (1999) Vacuolating cytotoxin purified from Helicobacter pylori causes mitochondrial damage in human gastric cells. *Microb. Pathog. 26*, 45–52.

253 Jung, H. K., Lee, K. E., Chu, S. H., Yi, S. I. (2001) Reactive oxygen species activity, mucosal lipoperoxidation and glutathione in Helicobacter pylori-infected gastric mucosa. *J. Gastroenterol. Hepatol. 16*, 1336–1340.

254 Kubota, Y., Kato, K., Dairaku, N., Koike, T., Iijima, K., Imatani, A., Sekine, H., Ohara, S., Matsui, H., Shimosegawa, T. (2004) Contribution of glutamine synthetase to ammonia-induced apoptosis in gastric mucosal cells. *Digestion 69*, 140–148.

255 Aulak, K. S., Koeck, T., Crabb, J. W., Stuehr, D. J. (2004) Dynamics of protein nitration in cells and mitochondria. *Am. J. Physiol. 286*, H30–H38.

256 Lee, J. H., Yang, E. S., Park, J.-W. (2003) Inactivation of $NADP^+$-dependent isocitrate dehydrogenase by peroxynitrite. *J. Biol. Chem. 278*, 52360–52371.

257 Cramer, T., Yamanishi, Y., Clausen, B. E., Förster, I., Pawlinski, R., Mackman, N., Haase, V. H., Jaenisch, R., Corr, M., Nizet, V., Firestein, G. S., Gerber, H.-P., Ferrara, N., Johnson, R. S. (2003) HIF-1α is essential for myeloid cell-mediated inflammation. *Cell 112*, 645–657.

258 Dewson, G., Cohen, G. M., Wardlaw, A. J. (2001) Interleukin-5 inhibits translocation of Bax to the mitochondria, cytochrome *c* release, and activation of caspoases in human eosinofils. *Blood 98*, 2239–2247.

259 Maianski, N. A., Mul, F. P. J., van Buul, J. D., Roos, D., Kuijpers, T. W. (2002) Granulocyte colony-stimulating factor inhibits the mitochondria-dependent activation of caspase-3 in neutrophils. *Blood 99*, 672–679.

260 Gergely, P. Jr, Niland, B., Gonchoroff, N., Pullmann, R. Jr, Phillips, P. E., Perl, A. (2002) Persistent mitochondrial hyperpolarization, increased reactive oxygen intermediate production, and cytoplasmic alkalization characterize altered IL-10 signalling in patients with systemic lupus erythematosus. *J. Immunol. 169*, 1092–1101.

261 Perl, A., Gergely, P. Jr, Nagy, G., Koncz, A., Banki, K. (2004) Mitochondrial hyperpolarization: a checkpoint of T-cell life, death and autoimmunity. *Trends Immunol. 25*, 360–367.

262 Fossati, G., Moulding, D. A., Spiller, D. G., Moots, R. J., White, M. R. H., Edwards, S. W. (2003) The mitochon-drial network of human neutrophils: Role in chemotaxis, phagocytosis, respiratory burst activation, and commitment of apoptosis. *J, Immunol, 170*, 1964–1972.

263 Peachman, K. K., Lyles, D. S., Bass, D. A. (2001) Mitochondria in eosinophils: Functional role in apoptosis but not respiration. *Proc. Natl. Acad. Sci. USA 98*, 1717–1722.

264 Vats, D., Mukundan, L., Odegaard, J. I., Zhang, L., Smith, K. L., Morel, C. R., Greaves, D. R., Murray, P. J., Chawla, A. (2006) Oxidative metabolism and PGC-1β attenuate macrophage-mediated inflammation. *Cell Metabolism 4*, 13–24.

265 Lacy-Hulbert, A., Moore, K. J. (2006) Designer macrophages: oxidative metabolism fuels inflammation repair. *Cell Metabolism, 4*, 7–8.

266 Gellerich, F. N., Deschauer, M., Chen, Y., Müller, T., Zierz, S. (2002) Functional impairment of mitochondria in skinned fibers of CPEO patients with single and multiple deletions of mt-DNA correlate with heteroplasmy. *Biochim. Biophys. Acta 1556*, 41–45.

267 Hutter, E., Unterluggauer, H., Garedew, A., Jansen-Durr, P., Gnaiger, E. (2006) High-resolution respirometry – a modern tool in aging research. *Exp. Gerontol. 41*, 103–109.

268 Gellerich, F. N., Trumbeckaite, S., Opalka, J. R., Seppet, E., Rasmussen, H. N., Neuhoff, C., Zierz, S. (2000) Function of the mitochondrial outer membrane as a diffusion barrier in health and disease. *Biochem. Soc. Trans. 28*, 164–169.

269 Gellerich, F. N., Bohnensack, R., Kunz, W. (1983) Control of mitochondrial respiration: The contribution of the adenine nucleotide translocator depends on the ATP and ADP consuming enzymes. *Biochem. Biophys. Acta 722*, 381–391.

270 Gellerich, F. N., Kunz, W. S., Bohnensack, R. (1990) Estimation of flux control coefficients from inhibitor titrations by non-linear regression. *FEBS Lett. 274*, 167–170.

271 Chen, Y. (2005) Analyse der Genotyp-Phänotyp-Beziehungen von Mito-chondrien in Muskelbiopsien und in

Cybrids mit "single deletions" mittels Metabolischer Kontrollanalyse. Thesis, Martin Luther Universität Halle Wittenberg.

272 Rigoulet, M., Averet, N., Mazat, J. P., Guerin, B., Cohadon, F. (1988) Redistribution of the flux-control coefficients in mitochondrial oxidative phosphorylations in the course of brain edema. *Biochim. Biophys. Acta 932*, 116–123.

273 Mazat, J. P., Reder, C., Letellier, T. (1996) Why are most flux control coefficients so small? *J. Theor. Biol. 182*, 253–258.

274 Letellier, T., Malgat, M., Rossignol, R., Mazat, J. P. (1988) Metabolic control analysis and mitochondrial pathologies. *Mol. Cell. Biochem. 184*, 409–417.

275 Kudin, A., Vielhaber, S., Elger, C. E., Kunz, W. S. (2002) Differences in flux control and reserve capacity of cytochrome *c* oxidase (COX) in human skeletal muscle and brain suggest different metabolic effects of mild COX deficiencies. *Mol. Biol. Rep. 29*, 89–92.

276 Kunz, W. S. (2001) Control of oxidative phosphorylation in skeletal muscle. *Biochim. Biophys. Acta 1504*, 12–19.

277 Groen, A. K., Wanders, R. J., Westerhoff, H. V., van der Meer, R., Tager, J. M. (1982) Quantification of the contribution of various steps to the control of mitochondrial respiration. *J. Biol. Chem. 257*, 2754–2757.

278 Kacser, H., Burns, J. A. (1979) Molecular democracy: who shares the controls? *Biochem. Soc. Trans. 7*, 1149–1160.

279 Heinrich, R., Rapoport, T. A. (1974) A linear steady-state treatment of enzymatic chains. General properties, control and effector strength. *Eur. J. Biochem. 42*, 89–95.

280 Gruno, M., Peet, N., Seppet, E., Kadaja, L., Paju, K., Eimre, M., Orlova, E., Peetsalu, M., Tein, A., Soplepmann, J., Schlattner, U., Peetsalu, A., Seppet, E. (2006) Oxidative phosphorylation and its coupling to mitochondrial creatine and adenylate kinases in human gastric mucosa. *Am. J. Physiol. 291*, R936–R946.

16
Tumor Cell Energetic Metabolome

Sybille Mazurek

Abstract

Tumor cells must survive in environments with varying oxygen and nutrient supplies depending upon their distance from blood vessels, which places special demands upon the energy metabolism of the cells. The tumor cell energetic metabolome is characterized by high glycolytic and glutaminolytic capacities. In glycolysis, the tetramer:dimer ratio of M2-PK – a special isoenzyme of pyruvate kinase that is found in all proliferating cells and is overexpressed in all tumor cells – determines whether glucose carbons are degraded to pyruvate and lactate under the production of ATP (highly active tetrameric form) or channeled into synthetic processes that debranch from glycolytic intermediates (inactive dimeric form). In contrast to mitochondrial respiration, glycolytic ATP production is independent of oxygen supply and enables tumor cells to survive under hypoxic conditions and to migrate into areas with poor vascularization. Glutaminolysis recruits reaction steps of the citric acid cycle, which is truncated in tumor cells due to an inhibition of aconitase by high levels of ROS. The glycolytic and glutaminolytic conversion rates are coordinated by the tetramer:dimer ratio of M2-PK, by the capacities of hydrogen shuttles, and by the composition of the glycolytic enzyme complex – an association of several glycolytic enzymes in the cytosol. Both glycolytic and glutaminolytic ATP production are highly sensitive to a reduction in NAD levels.

16.1
Introduction

Cell proliferation is a process that demands a large expenditure of energy. Furthermore, proliferating cells and especially tumor cells strongly depend on sufficient availability of cell building blocks, such as nucleic acids, phospholipids, amino acids, and C1-bodies. In contrast to normal proliferating cells, tumor cells must survive in environments with varying oxygen and nutrient supplies depending upon their distance from blood vessels. All together these circumstances

Molecular System Bioenergetics: Energy for Life. Edited by Valdur Saks
Copyright © 2007 WILEY-VCH Verlag GmbH & Co. KGaA, Weinheim
ISBN: 978-3-527-31787-5

place special demands upon the energy metabolism of tumor cells, resulting in the special metabolic phenotype of tumor cells termed the *tumor cell energetic metabolome* [1–3].

16.2
Otto Warburg's Discovery

Investigation of the tumor cell energetic metabolome was initiated in the 1920s when Otto Warburg first reported that tumor cells produce high levels of lactate even in the presence of oxygen [4]. The conversion of glucose to lactate yields two moles of ATP per mole of glucose, in comparison to 38 moles of ATP when glucose is completely degraded to CO_2 and H_2O. Thus, degradation of glucose to lactate yields only 5% of the energy available from glucose. This apparently senseless waste of energy prompted Warburg to postulate a defect in the respiration in tumor cells as a cause for increased *aerobic glycolysis*. Indeed, in tumor cells mutations in the mitochondrial genome, changes in the number and structure of mitochondria, a downregulation of respiratory chain activities, and expression of oxidative phosphorylation-dependent genes have been described [5–9]. In the same way, a decrease and defective regulation of pyruvate dehydrogenase, which catalyzes the oxidative decarboxylation of pyruvate to acetyl-CoA, are discussed as explanations for the increased aerobic glycolysis in tumor cells [10]. On the other hand, in human squamous cell carcinoma, the apoptotic effect of vanilloids has been correlated with an inhibition of mitochondrial respiration [11]. Until now, no consistent pattern had emerged, and it is still unclear whether the observed mitochondrial changes contribute to the increased aerobic glycolysis and pathogenesis of cancer or have taken place as secondary effects [5, 9]. Furthermore, in cell culture it has been shown that in tumor cells the mitochondrial biogenesis and oxidative metabolism can be stimulated solely by reducing glucose availability [5, 7, 12]. At the same time, normal proliferating and tumor cells may also use glutamine as an alternative energy substrate. Thus, glutamine oxidation – termed *glutaminolysis* – is a second source of energy transformation in tumor cells that requires an active oxidative phosphorylation for ATP production (see also Section 16.5) [5, 7, 12–19].

Furthermore, aerobic glycolysis is not limited to tumor cells, because it also can occur in normal proliferating cells [13]. On the other hand, cell proliferation is not generally linked to a high conversion rate of glucose to lactate. There are several tumor cell lines that are able to proliferate in the presence of low glucose supply [17, 20].

16.3
Glycolysis: A Bifunctional Pathway in Tumor Cells

Intensive metabolic characterizations have revealed that the increased glycolytic conversion rates observed in tumor cells correlate with an increase in total glyco-

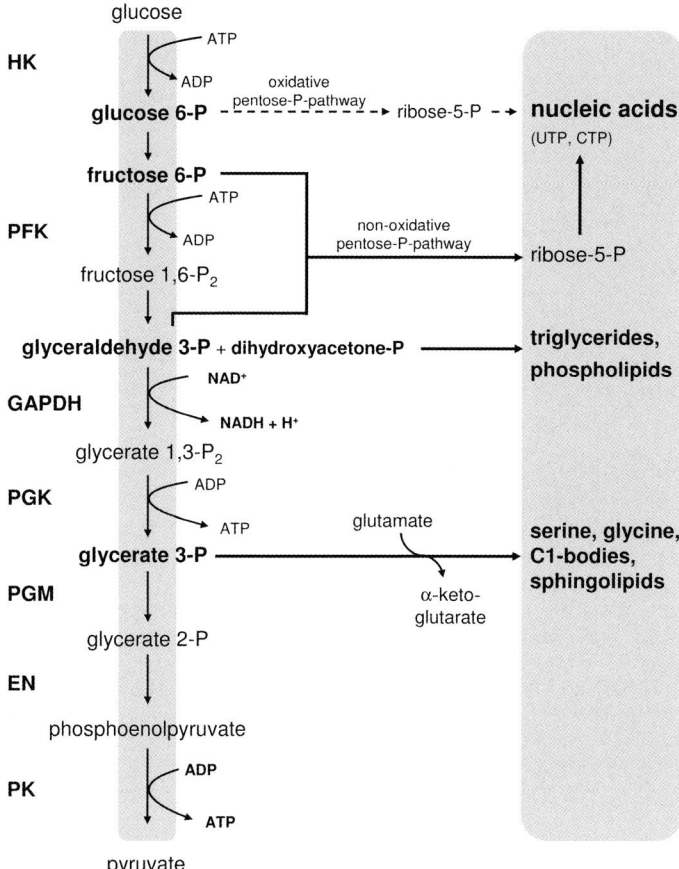

Fig. 16.1 Glycolysis with debranching synthetic processes. In glycolysis, first 2 moles of ATP are consumed for the synthesis of fructose 1,6-P2 and then are regained in the phosphoglycerate kinase (PGK)-catalyzed reaction. Net ATP synthesis occurs by the pyruvate kinase reaction. Furthermore, glycolytic intermediates are precursors for nucleic acid, phospholipid, and amino acid synthesis.

lytic enzyme activities (from 2- to more than 10-fold depending upon the enzyme and the tissue or cell model) as well as changes in the isoenzyme equipment and regulation mechanisms of the different enzymes (Fig. 16.1) [21, 22].

16.3.1
Compartmentalization of Glycolysis in the Cytosol by the Glycolytic Enzyme Complex

In the cytosol the different glycolytic enzymes (i.e., hexokinase, glyceraldehyde 3-P dehydrogenase, phosphoglycerate kinase, enolase, and pyruvate kinase) can be associated to the so-called *glycolytic enzyme complex*. In addition to the glycolytic

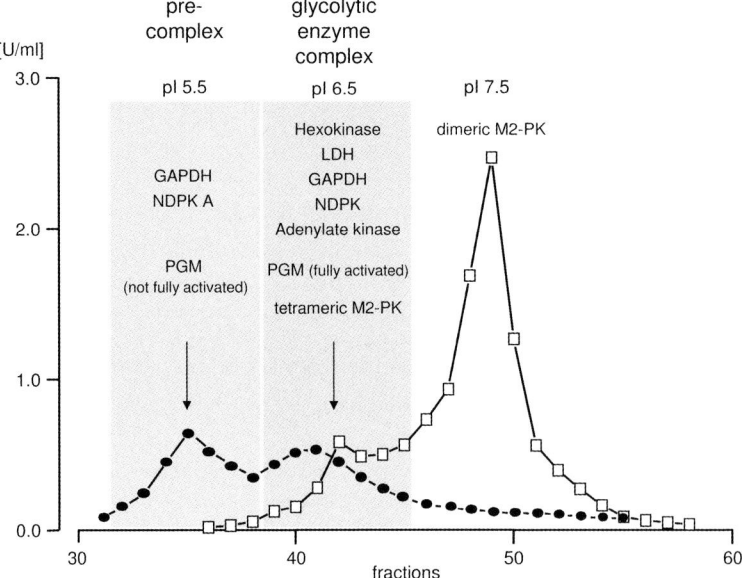

Fig. 16.2 Distribution of PGM and M2-PK in the glycolytic enzyme complex and pre-complex in rat liver oval cells; result of isoelectric focusing. The activated PGM and the tetrameric M2-PK focus together with other glycolytic enzymes within the glycolytic enzyme complex. The not-fully-activated PGM is found in the fractions of the pre-complex, whereas the dimeric form of M2-PK focuses outside both complexes at a more alkaline pH value (filled circles = PGM; squares = PK).

enzymes, other enzymes (e.g., nucleotide diphosphate kinase, adenylate kinase, glucose 6-P dehydrogenase, and 6-phosphogluconate dehydrogenase), components of the protein kinase cascade (i.e., RAF, MEK, and ERK), and AU-rich RNA also may be associated within this complex [1, 22–26]. The glycolytic enzyme complex can be demonstrated by free-flow isoelectric focusing (Fig. 16.2). Proteins associated within the complex focus at a common isoelectric point, which is different from the isoelectric point of the purified proteins. Migrations of proteins into or out of the glycolytic enzyme complex are reflected by shifts in their isoelectric points. The close spatial proximity of the glycolytic enzymes within the glycolytic enzyme complex allows a highly effective conversion of glucose to lactate.

16.3.2
Tumor Cell Characteristic Glycolytic Isoenzyme Profiles and Regulation Mechanisms

Hexokinase, the Gate to Glycolysis   Hexokinase (EC 2.7.1.1) controls the first reaction within the glycolytic sequence, the phosphorylation of glucose to glucose

6-P. Tumor cells are characterized by the expression of the hexokinase isoenzyme type II. This isoenzyme binds directly to the external mitochondria membrane, thereby profiting in an optimal manner from the ATP produced within the mitochondria. Furthermore, the product inhibition of isoenzyme type II by glucose 6-P is low, allowing a high glucose conversion rate [27].

6-Phosphofructo-1-kinase: A Key Regulator of Glycolysis in Differentiated Tissues   6-Phosphofructo-1-kinase (PFK; EC 2.7.1.11), which catalyzes the phosphorylation of fructose 6-P to fructose 1,6-P2, consists of four subunits that are encoded by different genes, termed types M, L, and P according to the organs where they are mainly expressed (muscle, liver, and platelets, respectively). The subtypes C and F are synonymous with subunit P. Tumor cells express mainly the subunit types L and P. Characterization of the kinetic properties of the different PFK isoenzymes has revealed no differences in the affinity to fructose 6-P. However, the control of PFK is passed from mitochondrial ATP to high fructose 1,6-P2 and fructose 2,6-P2 levels [28, 29].

Phosphoglycerate Mutase and Its Activation Mechanisms   Phosphoglycerate mutase (PGM; EC 5.4.2.1) catalyzes the conversion of glycerate 3-P to glycerate 2-P. In differentiated tissues PGM is activated by the cofactor glycerate 2,3-P2, which phosphorylates the enzyme in histidine. All differentiated cells, especially erythrocytes, are characterized by high glycerate 2,3-P2 levels, which may account for the fact that the PGM reaction is not a rate-limiting step of glycolysis in the majority of differentiated tissues [30].

Surprisingly, in tumor cells, which belong to the cells with the highest glycolytic conversion rates, glycerate 2,3-P2 levels are extremely low. Furthermore, tumor formation is correlated with a reduction of the glycerate 2,3-P2 synthesizing enzyme, glycerate 2,3-P2 mutase (EC 5.4.2.4), which points to another phosphate source for PGM activation [30, 31].

PGM can be associated with other glycolytic enzymes in the glycolytic enzyme complex. Part of the PGM protein also may focus outside the glycolytic enzyme complex at a more acidic IP in the so-called *pre-complex* (Fig. 16.2). Kinetic characterization has revealed that the PGM enzyme associated within the glycolytic enzyme complex is phosphorylated, fully activated, and independent of glycerate 2,3-P2, whereas the PGM enzyme associated within the pre-complex is not fully activated [22, 26, 32]. In the pre-complex, PGM is associated with nucleoside diphosphate kinase type A (NDPK; EC 2.7.4.6), which is also termed Nm23-H1. NDPK catalyzes the exchange of phosphate residues between different nucleotide tri- and diphosphates. In addition, NDPK also has protein kinase activity, and in MCF-7 cells it has been shown that NDPK A is able to phosphorylate and activate PGM with dCTP as phosphate donor when it is associated with glyceraldehyde 3-P dehydrogenase (GAPDH, EC 1.2.1.12) [32]. As glycolytic enzyme GAPDH is associated within the glycolytic enzyme complex but also colocalizes in small amounts together with NDPK and PGM in the pre-complex. The activation of PGM by NDPK A with nucleotides as phosphate donor opens the possibility of an optimal coordination of nucleic acid synthesis and glycolytic energy production.

Glycerate 3-P, the substrate of PGM, is the glycolytic precursor for serine synthesis, whereby the amino group derives from the glutamine derivative glutamate (Fig. 16.1). Accordingly, in tumor cell lines with PGM not associated within the glycolytic enzyme complex, a close correlation between glucose and glutamine consumption was found due to an increased channeling of glycerate 3-P into serine synthesis. However, the coordination of glycolysis and glutaminolysis by PGM is weakened in cell lines with an active glycerol 3-P shuttle (see Section 16.4.2) [17, 22, 26, 32].

Pyruvate Kinase: A Key Regulator of Glycolytic ATP Production and the Metabolic Budget System in Tumor Cells Pyruvate kinase (EC 2.7.1.40) catalyzes the dephosphorylation of phosphoenolpyruvate to pyruvate and is responsible for net ATP production within the glycolytic sequence. Depending on the metabolic functions of the tissues, different isoenzymes of pyruvate kinase are expressed. The pyruvate kinase isoenzyme type M1, which is characterized by the highest affinity to its substrate phosphoenolpyruvate (PEP), is found in tissues in which large amounts of ATP have to be rapidly provided, such as in muscle and brain. The pyruvate kinase isoenzyme with the lowest PEP affinity is the PK isoenzyme type L, which is characteristic of cells with high rates of gluconeogenesis, such as liver and kidney. The lung and all proliferating cells, such as normal proliferating cells, embryonic cells, adult stem cells, and especially tumor cells, express the pyruvate kinase isoenzyme type M2 (M2-PK). Accordingly, during tumorigenesis a shift in the isoenzyme composition of pyruvate kinase always takes place in such a manner that the tissue-specific isoenzymes, such as L-PK in liver or M1-PK in brain, disappear and M2-PK is expressed [15, 33–37].

The PEP affinity of the M2-PK isoenzyme depends on the quaternary structure of the enzyme. Whereas the other pyruvate kinase isoenzymes consist of four subunits (tetrameric quaternary structures), the M2-PK isoenzyme can occur in a tetrameric as well as a dimeric form. The tetrameric form has a high affinity to its substrate PEP and is highly active at physiological PEP concentrations, whereas the dimeric form has a low PEP affinity and is nearly inactive under physiological conditions (Fig. 16.3). Concerning the nucleotide diphosphate substrates, the tetrameric form has the highest affinity to ADP but also phosphorylates GDP, while affinity to UDP, CDP, and TDP is low [20, 33, 35].

Whereas in the lung the M2-PK isoenzyme is always in its highly active tetrameric form, in tumor cells the dimeric form of M2-PK is predominant and has therefore been termed *Tumor M2-PK* [33, 35, 37]. The tetrameric form is associated with other glycolytic enzymes within the glycolytic enzyme complex, while the dimeric form focuses outside the complex at a more alkaline pH value (Fig. 16.2) [1, 22, 26]. M2-PK is also located in the nucleus, where it has been found to participate in the phosphorylation of histone 1 [38].

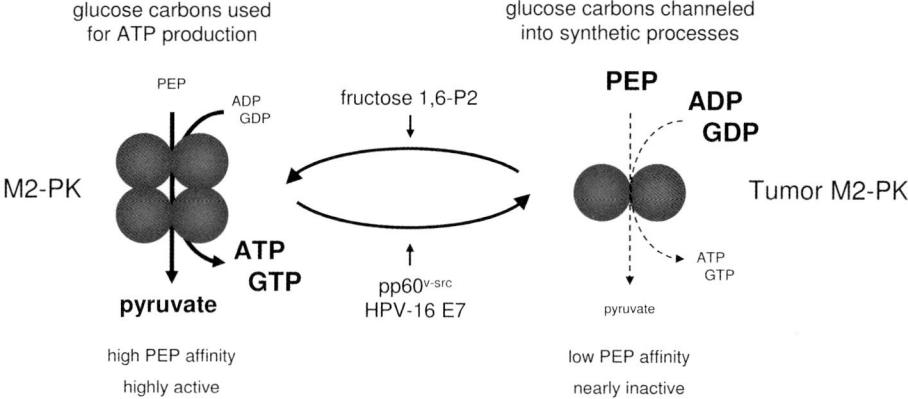

glucose carbons used
for ATP production

glucose carbons channeled
into synthetic processes

Fig. 16.3 Regulation and metabolic consequences of the quaternary structures of M2-PK. M2-PK may occur in a tetrameric or dimeric form and may use ADP or GDP for the dephosphorylation of PEP. When M2-PK is mainly in the highly active tetrameric form, glucose carbons are degraded to pyruvate. Accordingly, ADP and GDP levels are low and ATP and GTP levels are high. In tumor cells oncoproteins induce a dimerization of M2-PK. The nearly inactive dimeric form correlates with low ATP:ADP and GTP:GDP ratios as well as an increase in all phosphometabolites above the PK reaction, which are used as precursors of debranching synthetic processes (see also Fig. 16.1). High fructose 1,6-P2 levels induce a re-association of the dimeric form of M2-PK to the tetrameric form and are a main regulator of the oscillation cycle between the tetrameric and dimeric forms of M2-PK.

At first glance, the preponderance of the inactive dimeric form of M2-PK seems to be inconsistent with the increased conversion of glucose to lactate described for a wide variety of tumor cells. However, the inactive dimeric form of M2-PK has the metabolic advantage that all phosphometabolites above pyruvate kinase (e.g., PEP, glycerate 3-P, glyceraldehyde 3-P, fructose 1,6-P2, fructose 6-P, and glucose 6-P) accumulate. Thus, glycolytic phosphometabolites not only are intermediates within the ATP-producing degradation of glucose to lactate but also are important precursors of synthetic processes that debranch from glycolysis, such as nucleic acid, amino acid, and phospholipid synthesis (Fig. 16.1) [15, 33, 39]. In nucleic acid synthesis, ribose 5-P is synthesized mainly by transketolase (EC 2.2.1.1) and transaldolase (EC 2.2.1.2) as a result of inhibition of the oxidative pentose pathway by high fructose 1,6-P2 levels (Fig. 16.1) [40]. Accordingly, in tumor cells phosphometabolite levels are about 100 times higher than in differentiated tissues.

Cell proliferation is possible only when enough high-energetic phosphometabolites for synthetic processes and energy transformation are available. The underlying regulation mechanism is termed the *metabolic budget system* and depends on the tetramer:dimer ratio of M2-PK [35].

In tumor cells the tetramer:dimer ratio is not a stationary value, and it may oscillate between the highly active tetrameric and nearly inactive dimeric forms depending on the presence of distinct regulating signal metabolites. A key regulator within this oscillation cycle is the glycolytic intermediate fructose 1,6-P2 (Fig. 16.3). When the fructose 1,6-P2 levels reach a certain high value, the inactive dimeric form of M2-PK re-associates to the highly active tetrameric form [15, 20, 35, 41]. Consequently, glucose is converted to lactate with the production of ATP until the fructose 1,6-P2 levels drop below a minimum signal level. Then the tetrameric form dissociates to the dimeric form, and phosphometabolites increase and are used for synthesis of cell building blocks until the oscillation cycle begins again when fructose 1,6-P2 levels are high enough to induce a new tetramerization of M2-PK.

In addition to fructose 1,6-P2, the amino acid L-serine allosterically increases the affinity of M2-PK to its substrate PEP and reduces the amount of FBP necessary for M2-PK tetramerization. On the other hand, L-alanine, L-methionine, L-phenylalanine, L-valine, L-leucine, L-isoleucine, L-proline, and various fatty acids inhibit M2-PK (reviewed in Ref. [15]).

The inactive dimeric form of M2-PK correlates with low ATP and GTP levels and high ADP and GDP concentrations (Fig. 16.3) [1, 26]. Low ATP levels and low ATP:ADP ratios are correlated with a high degree of malignancy [42]. On the other hand, the dimeric form of M2-PK leads to an increase in nucleic acid synthesis whereby UTP and CTP increase (Fig. 16.1). Both nucleotides play important roles in synthetic processes. UTP is necessary for the synthesis of complex carbohydrates, while CTP is involved in phospholipid synthesis. Accordingly, the dimeric form of M2-PK as well as cell proliferation correlate with a low (ATP+GTP):(UTP+CTP) ratio [1, 26, 43].

Advantages and Disadvantages of High Glycolytic Capacities An increased glycolytic conversion rate has several advantages for tumor cells. First, in contrast to mitochondrial respiration, glycolytic ATP production is independent of oxygen supply. Therefore, the high glycolytic capacity allows survival of the cells at low oxygen levels and most importantly enables tumor cells to migrate into areas with poor vascularization. Secondly, glycolysis provides starting substances for synthetic processes that debranch from glycolytic intermediates, such as nucleic acid, phospholipid, and amino acid synthesis, whereby M2-PK – which can shift between an active tetrameric form and an inactive dimeric form – is an important metabolic sensor of tumor cells that allows adaptation of tumor metabolism to varying nutrient and oxygen supply conditions.

A high glycolytic flux rate may, however, hold potential danger for tumor cells. This danger results from the tumor-specific constellation of two highly active ATP-consuming enzymes (hexokinase and PFK) at the beginning of the glycolytic pathway and from a high level of the inactive dimeric M2-PK isoenzyme at the exit of glycolysis (Fig. 16.1). As a consequence, in glucose-starved tumor cells refeeding of glucose may lead to energetic imbalances when the ATP consumed by the increase of glycolytic phosphometabolites cannot be replaced in sufficient amounts by either tetramerization of M2-PK or glutaminolysis (see also Section 16.5) [15].

16.3.3
Influence of Transcription Factors and Oncoproteins

The expression of glycolytic enzymes is induced by different transcription factors and oncoproteins that react as transcription factors (i.e., C-ets for PFK; c-myc for LDH; and enolase, SP1, and SP3 for M2-PK expression) [44–47].

In normal cells as well as tumor cells, hypoxia leads to a stabilization of the transcription factor HIF-1 (hypoxia-inducing factor 1), which induces the transcription of several different genes that are important for survival under hypoxic conditions, including the glucose transporter GLUT1, LDH A, phosphoglycerate kinase 1, and PK type M (reviewed in [48]). The PK isoenzymes type M1 and type M2 are different splicing products of the M gene. The tumor suppressor protein pVHL (von Hippel Lindau protein) is involved in the degradation of the HIF-1α subunit in normoxia. Hereditary mutations of pVHL are associated with the development of highly vascularized tumors of the kidney and central nervous system [48].

Furthermore, different oncoproteins directly target glycolytic enzymes, thereby modulating their activities and kinetic properties. The pp60$^{v\text{-src}}$ kinase, the transforming principle of the Rous sarcoma virus, phosphorylates enolase, LDH, and M2-PK in tyrosine [33, 34, 49, 50]. The E7 oncoprotein of the human papillomavirus type 16, which is linked to malignant human cervix cancer, binds directly to M2-PK [1, 26]. Both RSV-induced tyrosine phosphorylation and E7 binding induce a dimerization of M2-PK, an increase in phosphometabolite levels, and an increased channeling of glucose carbons to synthetic processes. Ras, on the other hand, induces an increase and tetramerization of the M2-PK protein [1]. Activation of the PI3K/Akt/TOR pathway has been shown to induce an increase in aerobic glycolysis by an upregulation of glucose transporters and enhanced glucose phosphorylation by hexokinase [51].

16.4
Regeneration of Cytosolic NAD$^+$: An Achilles Heel in the Tumor Cell Energetic Metabolome

In glycolysis, NADH + H$^+$ is produced in addition to ATP. The corresponding reaction is catalyzed by GAPDH, whereby glyceraldehyde 3-P is oxidized to glycerate 1,3-P$_2$ (Fig. 16.1). Another source for cytosolic NADH production is serine synthesis. Consequently, glycolytic conversion rates as well as serine synthesis strongly depend on NAD$^+$ levels and the capability of the cells to recycle NAD$^+$ from NADH + H$^+$. One effective way to recycle NAD$^+$ is the transfer of the hydrogen to pyruvate and excretion as lactate. On the other hand, cytosolic hydrogen can be transported by hydrogen shuttles into the mitochondria and transferred to oxygen with the formation of H$_2$O and ATP. A lowering of NAD$^+$ levels or inhibition of NAD$^+$-regenerating enzymes may lead to a total inhibition of glycolysis (Fig. 16.4) (see also Sections 16.6 and 16.8).

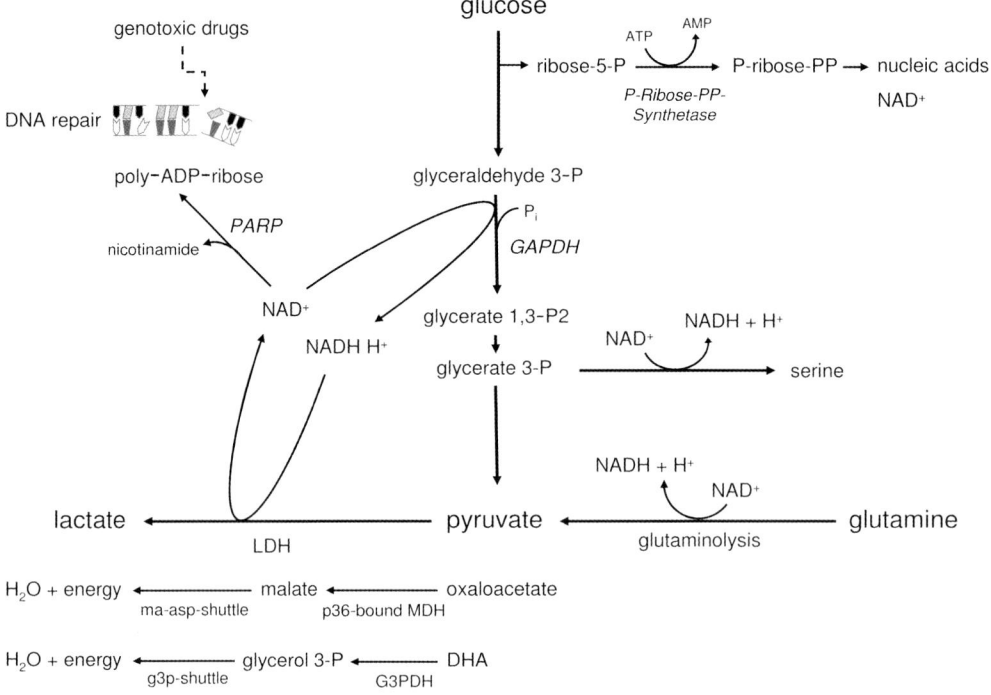

Fig. 16.4 NAD$^+$-regenerating and NAD$^+$-dependent pathways in the tumor metabolome. A reduction in NAD$^+$ levels, i.e., by inhibition of NAD$^+$ synthesis or by activation of poly-ADP-ribosyl polymerase (PARP), during DNA repair lowers glycolytic and glutaminolytic energy conversion and may lead to cell death. DHA = dihydroxyacetone-P; ma-asp-shuttle = malate–aspartate shuttle; g3P-shuttle = glycerol 3-P shuttle.

16.4.1
Lactate Dehydrogenase

Lactate dehydrogenase (LDH, EC 1.1.1.27) is a tetrameric protein composed of the subunits type H and type M. The LDH subunit type H is characteristic of tissues with good oxygen supply, whereas the subunit type M, which is characterized by a high affinity to its substrate pyruvate, dominates under anaerobic conditions [52]. Tumor cells are characterized by an overexpression of subunit type M (= product of LDH A gene) and are thereby equipped with an LDH isoenzyme that allows an optimal regeneration of NAD$^+$ for the cytosolic GAPDH reaction (Fig. 16.4). In Neu-initiated tumor cells, stable knockdown of LDH-A by expression of short hairpin RNAs (shRNA) complementary to LDH-A led to an increase in oxidative phosphorylation activity and a suppression of growth properties [53]. Lactate concentrations in tumors vary between low and extremely high, whereby high lactate levels correlate with a high incidence of distant metastasis [54]. Furthermore, the tumor-derived lactate lowers the pH value of the tumor environment and has been found to alter the phenotype and functional activity of dendritic cells in multicellular tumor spheroid models [55].

16.4.2
Glycerol 3-P Shuttle

In contrast to differentiated tissues, in most tumor cells the capacity of the glycerol 3-P-shuttle is reduced due to a reduction or total loss of cytosolic glycerol 3-P dehydrogenase (G3PDH; E.C. 1.1.1.8) activity (Fig. 16.4) [56].

16.4.3
Malate–Aspartate Shuttle

An alternative shuttle to transport cytosolic hydrogen into the mitochondria is the malate–aspartate shuttle [14]. The malate–aspartate shuttle consists of two enzymes: malate dehydrogenase (MDH, E.C. 1.1.1.37) and glutamate oxaloacetate transaminase (GOT, E.C. 2.6.1.1), which exist as both cytosolic and mitochondrial isoenzymes and are also part of the glutaminolytic pathway (Figs. 16.4 and 16.5).

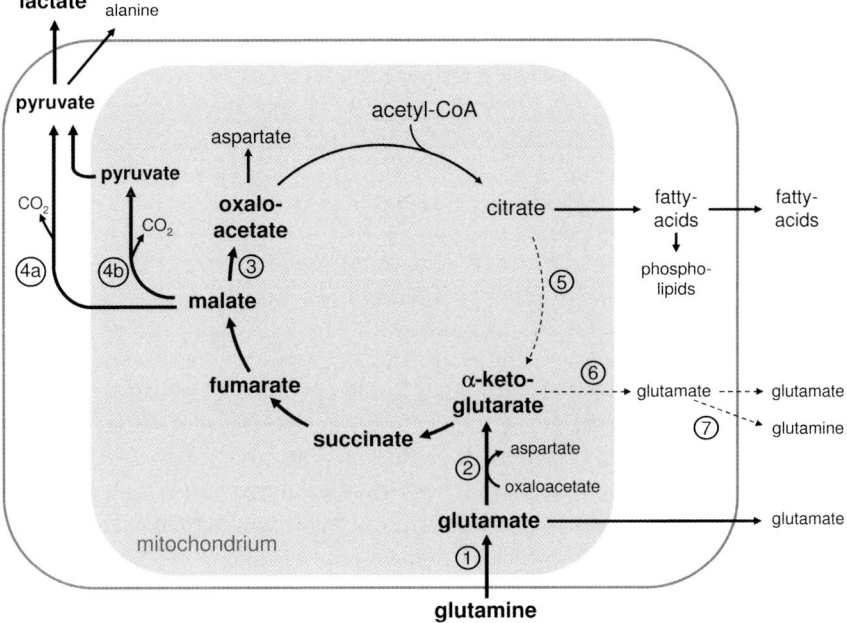

Fig. 16.5 Glutaminolysis with recruited steps of the truncated citric acid cycle. (1) glutaminase; (2) GOT; (3) MDH; (4a) cytosolic isoenzyme of malic enzyme; (4b) mitochondrial isoenzyme of malic enzyme; (5) aconitase; (6) glutamate dehydrogenase; (7) glutamine synthetase.

In differentiated tissues with a high capacity for gluconeogenesis, such as liver and kidney, the malate–aspartate shuttle is used to transport hydrogen from the mitochondria into the cytosol to provide enough hydrogen for gluconeogenesis. Accordingly, the mitochondrial isoenzyme of MDH has a high capacity for the conversion of $NADH + H^+$ to NAD^+, whereas the cytosolic isoenzyme is characterized by a higher capacity for the conversion of NAD^+ to $NADH + H^+$ [25]. As a prerequisite for the transport of hydrogen from the cytosol into the mitochondria, the precursor of the mitochondrial isoenzyme can be captured in the cytosol by binding to p36, which is found at high levels in tumor cells (Fig. 16.4) [17, 25]. In MCF-7 cells, high concentrations of the glycolytic intermediate fructose 1,6-P2 led to a dissociation of p36 from MDH and to an increase in the mitochondrial MDH isoenzyme, pointing to a direct influence of the glycolytic intermediate fructose 1,6-P2 on the capacity of the malate–aspartate shuttle [25].

16.5
Glutaminolysis: A Second Main Pillar for Energy Conversion in Tumor Cells

16.5.1
The Truncated Citric Acid Cycle

The citric acid cycle, an important source of energy conversion in differentiated cells, is truncated in tumor cells due to an inhibition of the conversion of citrate to isocitrate, which is catalyzed by aconitase (EC 4.2.1.3), an iron-sulfur–containing dehydratase whose activity is inhibited by superoxide radicals (Fig. 16.5). In normal cells, superoxide radicals are immediately inactivated by high activities of mitochondrial (Mn-dependent) and cytosolic (CuZn-dependent) superoxide dismutase and glutathione peroxidase, which are greatly reduced during tumor formation. Accordingly, overexpression of MnSOD enhances the activity of aconitase and inhibits growth of numerous tumor cell types [57]. The metabolic consequence of the truncated citric acid cycle is that the amount of pyruvate and acetyl-CoA infiltrated into the citric acid cycle is low and that acetyl-CoA is available for *de novo* synthesis of cholesterol and fatty acids in the cytosol [58]. The fatty acids synthesized in the cytosol can be used for phospholipid synthesis or can be released from tumor cells (Fig. 16.5) [2]. Fatty acids provide an effective hydrogen storage and are an effective way to eliminate surplus hydrogen. Furthermore, fatty acids are immunosuppressive and may protect tumor cells from immune system attacks [59].

16.5.2
The Glutaminolytic Pathway

Tumor cells overexpress phosphate-dependent glutaminase (EC 3.5.1.2) as well as $NAD(P)^+$-dependent malate decarboxylase (malic enzyme, EC 1.1.1.39 and 1.1.1.40), which in combination with the remaining reaction steps of the citric acid cycle from alpha-ketoglutarate to malate, oxaloacetate, and citrate impart the possibility of a new energy-transforming pathway: the degradation of the amino acid glutamine to glutamate, aspartate, pyruvate, CO_2, lactate, and citrate, which is termed *glutaminolysis* (Fig. 16.5) [1, 5, 7, 12–19, 26, 60, 61]. Glutaminase catalyzes the hydrolysis of the amino group of glutamine, yielding ammonium and glutamate. Transfection of Ehrlich ascites tumor cells with antisense glutaminase mRNA led to a reduction in total glutaminase protein content as well as glutaminase activity, which was accompanied by a longer cell-doubling time and a loss of the tumorigenic capacity of the cells [62]. Glutamine synthetase (EC 6.3.1.2), the enzyme that catalyzes the reverse reaction, the amination of glutamate to glutamine, is reduced in tumor cells (Fig. 16.5) [18].

Glutamate can be excreted or can be further metabolized to alpha-ketoglutarate. Because of low glutamate dehydrogenase (EC 1.4.1.2) and glutamate pyruvate transaminase (EC 2.6.1.2) activities, the hydrolysis of NH_4^+ from glutamate preferentially takes place via GOT-catalyzed transamination, whereby the amino group is transferred to oxaloacetate with the production of aspartate (Fig. 16.5) [18].

16.5.3
Energy Conversion in Glutaminolysis

In glutaminolysis, oxidation of 1 mole of glutamine can generate 1 GTP by direct phosphorylation of GDP, 2 ATP from oxidation of $FADH_2$, and 3 ATP at a time for the $NADH + H^+$ produced within the alpha-ketoglutarate dehydrogenase reaction, malate dehydrogenase, and malic enzyme reaction. In HeLa cells it was shown that about 65% of the ATP derived from glutamine degradation when the cells were cultured in the presence of glucose in the cultivation medium. When glucose was replaced by fructose or galactose, the contribution of glutamine to ATP production increased to 99.9% [5, 12].

16.5.4
Advantages and Disadvantages of Glutaminolysis

Glutamine is the most abundant amino acid in the plasma and can serve as an additional source of energy conversion, especially when M2-PK is in its inactive dimeric form. In addition, glutaminolysis provides glutamate and aspartate, which together with glutamine itself are precursors for nucleic acid and serine synthesis. Accordingly, high extracellular glutamine levels stimulate tumor cell proliferation and are essential for cell transformation, while a reduction in extracellular glutamine correlates with phenotypical and functional differentiation of the cells [19, 63].

In tumor cells a considerable amount of the glutamine consumed is released as glutamate immediately after hydrolysis of glutamine, thereby leaving the energy-transforming glutaminolytic pathway. However, this at first senseless-appearing waste of energy has some advantages for the tumor cells. First, the release of glutamate is an effective way to excrete hydrogen. Second, like fatty acids, glutamate is immunosuppressive and may protect tumor cells from immune attacks [64]. Furthermore, it is hypothesized that the pool of glutamate could be used to drive the endergonic uptake of other amino acids by system ASC [19]. The glutamate released from tumor cells can be converted to glucose in the liver and kidney, which may explain why tumor patients become cachectic without developing hypoglycemia or ketosis.

However, in contrast to glycolytic ATP production, glutaminolytic ATP production depends on oxygen supply and takes place only in small tumors with good vascularization and adequate oxygen levels [5, 12, 60, 65].

16.6
AMP: A Mediator Between Tumor Cell Energetic Metabolism and Cell Proliferation

In differentiated tissues high ATP and low ADP and AMP levels reflect a high capacity of the citric acid cycle and oxidative phosphorylation and lead to an inhibition of the glycolytic pathway due to an inhibition of the key glycolytic enzyme PFK (Fig. 16.1). This regulation mechanism is the basis of the downregulation of glycolysis in the presence of oxygen commonly known as the *Pasteur effect*. High AMP and low ATP levels, on the other hand, activate PFK and lead to a strong stimulation of the glycolytic flux rate.

In tumor cells high AMP levels also activate PFK but may have different effects on the glycolytic flux rate depending on the cell line and its metabolic characteristics. In the human breast carcinoma cell line MDA-MB-453, high AMP levels induce a slight activation of the glycolytic conversion rates, whereas in the human breast carcinoma cell line MCF-7, a total inhibition of glycolysis was found [17]. An increase in intracellular AMP levels can be induced in cell culture by the addition of AMP into the cultivation medium. Thereby, extracellular AMP is degraded by 5′-ectonucleotidase to adenosine, which is transported into the cells via an adenosine translocator. Intracellular adenosine is phosphorylated to AMP by adenosine kinase; high intracellular AMP levels induce an inhibition of cell proliferation [17, 66]. In tumor cells that lack adenosine kinase activity, such as Novikoff cells, AMP levels do not increase in the presence of extracellular AMP and cell proliferation is not inhibited [20].

In addition to the activation of PFK, high AMP levels also inhibit P-ribose-PP synthetase, the enzyme that catalyzes the synthesis of P-ribose-PP, which is the sugar component of all nucleic acids as well as of $NAD^+/NADH$ (Fig. 16.4). Accordingly, high AMP levels induce a decrease in $NAD^+/NADH$ levels, which has different effects on the glycolytic flux rate depending on the mechanisms of NAD^+ recycling. In MDA-MB-453 cells – which are characterized by a more differentiated metabolic phenotype and which contain cytosolic G3PDH – most of the hydrogen produced within the glycolytic GAPDH reaction is transported via the glycerol 3-P shuttle into mitochondria. In MCF-7 cells, which lack an active glycerol 3-P shuttle, the mass of cytosolic hydrogen is transferred onto pyruvate and is excreted as lactate. LDH is much more sensitive to reduction of intracellular NAD^+ and NADH levels than is G3PDH. Therefore, in AMP-treated MCF-7 cells NAD^+ cannot be recycled in sufficient amounts, and glycolysis is inhibited at the step of GAPDH [17].

AMP furthermore is the activator of AMP kinase, an important sensor of cellular energy status that regulates a wide variety of cell processes. Activation of AMP kinase downregulates fatty acid and cholesterol synthesis; diminishes the expression and half-lives of cyclin A, cyclin B, and p21 mRNA by a decrease in cytoplasmic HuR; and initiates cell cycle arrest by phosphorylation of p53 (reviewed in [67]).

16.7
The Tumor Cell Energetic Metabolome as a Tool for Tumor Diagnosis

From the various tumor-specific metabolite alterations, the high rate of glucose consumption is the basis of the clinical visualization of tumors by positron emission tomography (PET) using ^{18}F-2-deoxyglucose as glucose analogue. The expansion of phosphometabolite pools can be detected by ^{31}P-NMR spectroscopy.

From the tumor-specific enzymatic changes, neuron-specific enolase, LDH, and Tumor M2-PK (the dimeric form of M2-PK) are presently used for diagnostic purposes. The diagnostic tests are based on the fact that the metabolic markers are released from the tumors into the blood and/or stool of the tumor patients. Neuron-specific enolase is especially used for early detection and follow-up studies of small-cell lung carcinoma [68]. High serum LDH values have been shown to correlate with high malignancy and poor prognosis in cases of non-Hodgkin's lymphoma, myeloma, melanoma, lung adenocarcinoma, and colorectal carcinoma [69]. Tumor M2-PK increases in the plasma of patients with melanoma; renal cell carcinoma; and lung, breast, cervical, and gastrointestinal cancer and correlates with tumor size and stages. The determination of fecal tumor M2-PK is a sensitive tool for early detection of colorectal cancer [37, 70, 71].

16.8
The Tumor Cell Energetic Metabolome as a Target for Therapy

Because tumor cells use either glycolysis or glutaminolysis for production of ATP, glucose and glutamine themselves, and in principle all enzymes of both pathways as well as the hydrogen shuttles, are potential targets for tumor therapy. Accordingly, glucose analogues (i.e., 2-deoxyglucose), glutamine antagonists (i.e., 6-diazo-5-oxo-L-norleucine [DON], azaserine), as well as various inhibitors of glycolytic, glutaminolytic, and shuttle enzymes (i.e., LDH inhibitors: oxamate, disulfiram, colchicine, α-hydroxybutyrate; PGM inhibitors: peptide encompassing the PGM phosphorylation site; GOT inhibitors: cycloserine; α-ketoglutarate dehydrogenase inhibitors: valproic acid) have been shown to inhibit tumor cell proliferation [15, 32, 72, 73]. Recently, we showed that drugs that specifically induce a dimerization and migration of M2-PK out of the glycolytic enzyme complex, and that in the same way avoid the re-association of M2-PK to its tetrameric form, efficiently inhibit glycolytic ATP production and tumor cell proliferation [74]. Another approach to inhibit glycolytic and glutaminolytic energy conversion is the use of AMP analogues, which inhibit NAD^+ synthesis, or nicotinamide analogues, which are incorporated into NAD^+ and $NADP^+$ [75, 76]. Even genotoxic drugs in the end exert their cytotoxic effects by reducing NAD^+ and ATP levels, as NAD^+ is depleted for ADP-ribosylation during DNA repair (Fig. 16.4). NADPH produced by cytosolic isocitrate dehydrogenase (EC 1.1.1.42), glutaminolytic malic enzyme, or glucose 6-P dehydrogenase (EC 1.1.1.49) and 6-phosphogluconate dehydrogenase (EC 1.1.1.44) is a co-substrate in fatty acid and cholesterol synthesis as well as in GSH regeneration. Glutathione plays an important role in drug detoxification, i.e., cisplatin [15].

Besides the metabolic parameters that are consistently altered in the same way in all tumor cells, such as the isoenzyme shift of M2-PK or the high glutaminolytic capacity, there are still many metabolic differences in different tumors (e.g., distinct enzyme activities, hydrogen shuttle capacities, and metabolite levels, which influence the success of the respective therapy form). Therefore, an improvement in therapy will be attained by the metabolic individualization of the tumors and a true metabolite- and enzyme-guided tumor therapy. Cisplatin resistance, for example, has been shown to correlate with decreased levels of M2-PK protein and activity in 11 different gastric carcinoma cell lines [77]. Furthermore, for an optimal therapy the different energy-transforming pathways must be inactivated simultaneously, i.e., by combination of different drugs or by a drug design that allows interaction with targets in each pathway.

Acknowledgments

This chapter is dedicated to my scientific teacher Prof. Dr. Erich Eigenbrodt, who unexpectedly passed away in 2004.

References

1 Mazurek, S., Zwerschke, W., Jansen-Dürr, P., Eigenbrodt, E. (2001) Metabolic cooperation between different oncogenes during cell transformation: interaction between activated ras and HPV-16 E7. *Oncogene 20*, 6891–6898.

2 Mazurek, S., Grimm, H., Boschek, C. B., Vaupel, P., Eigenbrodt, E. (2002) Pyruvate kinase type M2: a crossroad in the tumor metabolome. *Brit. J. Nutr. 87*, S23–S29.

3 http://www.metabolic-database.com

4 Warburg, O., Poesener, K., Negelein, E. (1924) Über den Stoffwechsel der Tumoren. *Biochem. Z. 152*, 319–344.

5 Rossignol, R., Gilkerson, R., Aggeler, R., Yamagata, K., Remington, S. J., Capaldi, R. A. (2004) Energy substrate modulates mitochondrial structure and oxidative capacity in cancer cells. *Cancer Res. 64*, 985–993.

6 Cuezva, J. M., Krajewski, M., Lopez de Heredia, M., Krajewski, S., Santamaria, G., Kim, H., Zapata, J. M., Marusawa, H. et al. (2002) The bioenergetic signature of cancer: a marker of tumor progression. *Cancer Res. 62*, 6674–6681.

7 Weber, K., Ridderskamp, D., Alfert, M., Hoyer, S., Wiesner, R. J. (2002) Cultivation in glucose-deprived medium stimulates mitochondrial biogenesis and oxidative metabolism in HepG2 hepatoma cells. *Biol. Chem. 383*, 283–290.

8 Hervouet, E., Godinot, C. (2006) Mitochondrial disorders in renal tumors. *Mitochondrion 6*, 105–117.

9 Ohta, S. (2003) A multifunctional organelle mitochondrion is involved in cell death, proliferation and disease. *Curr. Med. Chem. 10*, 2485–2494.

10 Koukourakis, M. I., Giatromanolaki, A., Sividis, E., Gatter, K. C., Harris, A. L. (2005) Pyruvate dehydrogenase and pyruvate dehydrogenase kinase expression in non small cell lung cancer and tumor-associated stroma. *Neoplasia 7*, 1–6.

11 Hail, N., Lotan, R. (2002) Examining the role of mitochondrial respiration in vanilloid-induced apoptosis. *J. Natl. Cancer Inst. 94*, 1281–1292.

12 Reitzer, L. J., Wice, B. M., Kennell, D. (1979) Evidence that glutamine, not sugar, is the major energy source for cultured HeLa-cells. *J. Biol. Chem. 254*, 2669–2676.

13 McKeehan, W. L. (1982) Glycolysis, glutaminolysis and cell proliferation. *Cell Bio. Int. Rep. 6*, 635–650.

14 Matsuno, T. (1992) Oxidation of cytosolic NADH by the malate-aspartate shuttle in HuH13 human hepatoma cells. *Int. J. Biochem. 24*, 313–315.

15 Eigenbrodt, E., Gerbracht, U., Mazurek, S., Presek, P., Friis, R. (1994) Carbohydrate metabolism and neoplasia: New perspectives for diagnosis and therapy. In *Biochemical and Molecular Aspects of Selected Cancers*, Pretlow, T. G., Pretlow, T. P. (eds.), Academic Press INC., San Diego 2, 311–385.

16 Goossens, V., Grooten, J., Fiers, W. (1996) The oxidative metabolism of glutamine. *J. Biol. Chem. 271*, 192–196.

17 Mazurek, S., Michel, A., Eigenbrodt, E. (1997) Effect of extracellular AMP on cell proliferation and metabolism of breast cancer cell lines with high and low glycolytic rates. *J. Biol. Chem. 272*, 4941–4952.

18 Piva, T. J., McEvoy-Bowe, E. (1998) Oxidation of glutamine in HeLa cells: role and control of truncated TCA cycles in tumour mitochondria. *J. Cell. Biochem. 68*, 213–225.

19 Aledo, J. C. (2004) Glutamine breakdown in rapidly dividing cells: waste or investment? *BioEssays 26*, 778–785.

20 Mazurek, S., Grimm, H., Wilker, S., Leib, S., Eigenbrodt, E. (1998) Metabolic characteristics of different malignant cancer cell lines. *Anticancer Res. 18*, 3275–3292.

21 Board, M., Humm, S., Newsholme, E. A. (1990) Maximum activities of key enzymes of glycolysis, glutaminolysis, pentose phosphate pathway and tricarboxylic acid cycle in normal, neoplastic and suppressed cells. *Biochem. J. 265*, 503–509.

22 Mazurek, S., Eigenbrodt, E., Failing, K., Steinberg, P. (1999) Alterations in the glycolytic and glutaminolytic pathways after malignant transformation of rat liver oval cells. *J. Cell. Physiol. 181*, 136–146.

23 Hentze, M. W. (1994) Enzymes as RNA-binding proteins: a role for (di)nucleotide-binding domains? *Trends Biochem. Sci. 19*, 101–103.

24 Nagy, E., Rigby, W. F. (1995) Glyceraldehyde 3-phosphate dehydrogenase selectively binds AU-rich RNA in the NAD^+-binding region (Rossmann Fold). *J. Biol. Chem. 270*, 2755–2763.

25 Mazurek, S., Hugo, F., Failing, K., Eigenbrodt, E. (1996) Studies on associations of glycolytic and glutaminolytic enzymes in MCF-7 cells. Role of p36. *J. Cell. Physiol. 167*, 238–250.

26 Mazurek, S., Zwerschke, W., Jansen-Dürr, P., Eigenbrodt, E. (2001) Effects of the human papilloma virus HPV-16 E7 oncoprotein on glycolysis and glutaminolysis: role of pyuvate kinase type M2 and the glycolytic enzyme complex. *Biochem. J. 356*, 247–256.

27 Mathupala, S. P., Rempel, A., Pedersen, P. L. (2001) Glucose catabolism in cancer cells: identification and characterization of a marked activation response of the type II hexokinase gene to hypoxic conditions. *J. Biol. Chem. 276*, 43407–43412.

28 Staal, G. E. J., Kalff, A., Heesbeen, E. C., van Veelen, C. W., Rijksen, G. (1987) Subunit composition, regulatory properties, and phosphorylation of phosphofructokinase from human gliomas. *Cancer Res. 47*, 5047–5051.

29 Sanchez-Martinez, C., Estevez, A. M., Aragon, J. J. (2000) Phosphofructokinase C isozyme from ascites tumor cells: cloning, expression, and properties. *Biochem. Biophys. Res. Commun. 271*, 635–640.

30 Yeoh, G. C. (1980) Levels of 2,3-diphosphoglycerate in Friend leukaemia cells. *Nature 285*, 108–109.

31 Hegde, P., Qi, R., Gaspard, R., Abernathy, K., Dharap, S., Earle-Hughes, J., Gay, C., Nwokekeh, N. U. et al. (2001) Identification of tumor markers in models of human colorectal cancer using a 19.200-element complementary DNA microarray. *Cancer Res. 61*, 7792–7797.

32 Engel, M., Mazurek, S., Eigenbrodt, E., Welter, C. (2004) Phosphoglycerate mutase derived polypeptide inhibits glycolytic flux and induces cell growth arrest in tumor cell lines. *J. Biol. Chem. 279*, 35803–35812.

33 Eigenbrodt, E., Glossmann, H. (1980) Glycolysis – one of the keys to cancer? *Trends Pharmacol. Sci. 1*, 240–245.

34 Staal, G. E. J., Rijksen, G. (1991) Pyruvate kinase in selected human tumors. In *Biochemical and Molecular Aspects of Selected Cancers*, Pretlow, T. G., Pretlow, T. P. (eds.), Academic Press, San Diego 1, 313–337.

35 Eigenbrodt, E., Reinacher, M., Scheefers-Borchel, U., Scheefers, H., Friis, R. (1992) Double role for pyruvate kinase type M2 in the expansion of phospho-metabolite pools found in tumor cells. In *Critical Reviews in Oncogenesis*, Perucho, M. (ed.), CRC Press, Boca Raton, FL, 3, 91–115.

36 Hacker, H. J., Steinberg, P., Bannasch, P. (1998) Pyruvate kinase isoenzyme shift from L-type to M2-type is a late event in hepatocarcinogenesis induced in rats by a choline-deficient/DL-ethionine-supplemented diet. *Carcinogenesis 19*, 99–107.

37 Schneider, J., Neu, K., Grimm, H., Velcovsky, H.-G., Weisse, G., Eigenbrodt, E. (2002) Tumor M2-pyruvate kinase in lung cancer patients: Immunohisto-chemical detection and disease monitoring. *Anticancer Res. 22*, 311–318.

38 Ignacak, J., Stachurska, M. B. (2003) The dual activity of pyruvate kinase type M2 from chromatin extracts of neoplastic cells. *Comp. Biochem. Physiol. Part B 134*, 425–433.

39 Miccheli, A., Tomassini, A., Puccetti, C., Valerio, M., Peluso, G., Tuccillo, F., Calvani, M., Manetti, C. et al. (2006) Metabolic profiling by ^{13}C-NMR spectroscopy: [1,2-^{13}C$_2$] glucose reveals a heterogeneous metabolism in leukemia T cells. *Biochimie 88*, 437–448.

40 Boros, L. G., Cascante, M., Lee, W. N. (2002) Metabolic profiling of cell growth and death in cancer: applications in drug discovery. *Drug. Discov. Today 7*, 364–372.

41 Ashizawa, K., Willingham, M. C., Liang, C. M., Cheng, S. Y. (1991) *In vivo* regulation of monomer-tetramer conversion of pyruvate kinase subtype M2 by glucose is mediated via fructose 1,6-bisphosphate. *J. Biol. Chem. 266*, 16842–16846.

42 Jackson, R. C., Lui, M. S., Boritzki, T. J., Morris, H. P., Weber, G. (1980) Purine and pyrimidine nucleotide patterns of normal, differentiating, and regenerating liver and of hepatomas in rats. *Cancer Res. 40*, 1286–1291.

43 Ryll, T., Wagner, R. (1992) Intracellular ribonucleotide pools as a tool for monitoring the physiological state of *in vitro* cultivated mammalian cells during production processes. *Biotechnol. Bioengin. 40*, 934–946.

44 Dupriez, V. J., Darville, M. I., Antoine, I. V., Gegonne, A., Ghysdael, J., Rousseau, G. G. (1993) Characterization of a hepatoma mRNA transcribed from a tl promoter of a 6-phosphofructo2-kinase/fructose 2,3-bisphosphatase-encoding gene and controlled by ets oncogene-related products. *Proc. Natl. Acad. Sci USA 90*, 8244–8288.

45 Shim, H., Dolde, C., Lewis, B. C., Wu, C.-S., Dang, G., Jungmann, R. A., Dalla-

Favera, R., Dang, C. V. (1997) c-Myc transactivation of LDH-A: Implications for tumor metabolism and growth. *Proc. Natl. Acad. Sci. USA 94*, 6658–6663.

46 Dang, C. V., Semenza, G. L. (1999) Oncogenic alterations of metabolism. *Trends Biochem. Sci. 24*, 68–72.

47 Netzker, R., Weigert, C., Brand, K. (1997) Role of the stimulatory proteins Sp1 and Sp3 in the regulation of transcription of the rat pyruvate kinase M gene. *Eur. J. Biochem. 245*, 174–181.

48 Maxwell, P. H., Pugh, C. W., Ratcliffe, P. J. (2001) Activation of the HIF pathway in cancer. *Curr. Opin. Genet. Dev. 11*, 293–299.

49 Cooper, J. A., Esch, F. S., Taylor, S. S., Hunter, T. (1984) Phosphorylation sites in enolase and lactate dehydrogenase utilized by tyrosine protein kinases *in vivo* and *in vitro*. *J. Biol. Chem. 259*, 7835–7841.

50 Eigenbrodt, E., Mazurek, S., Friis, R. (1998) Double role of pyruvate kinase type M2 in the regulation of phosphometabolite pools. In *Cell Growth and Oncogenesis*, Bannasch, P., Kanduc, D., Papa, S., Tager, J. M. (eds.), Birkhäuser Verlag, Basel, 15–30.

51 Elstrom, R. L., Bauer, D. E., Buzzai, M., Karnauskas, R., Harris, M. H., Plas, D. R., Zhuang, H., Cinalli, R. M., et al. (2004) Akt stimulates aerobic glycolysis in cancer cells. *Cancer Res. 64*, 3892–3899.

52 Freitas, I., Bertone, V., Griffini, P., Accossato, P., Baronzio, G. F., Pontiggia, P., Stoward, P. J. (1991) *In situ* lactate dehydrogenase patterns as markers of tumour oxygenation. *Anticancer Res. 11*, 1293–1299.

53 Fantin, V. R., St-Pierre, J., Leder, P. (2006) Attenuation of LDH-A expression uncovers a link between glycolysis, mitochondrial physiology, and tumor maintenance. *Cancer Cell 9*, 425–434.

54 Walenta, S., Mueller-Klieser, W. F. (2004) Lactate: mirror and motor of tumor malignancy. *Semin. Radiat. Oncol. 14*, 267–274.

55 Gottfried, E., Kunz-Schughart, L. A., Ebner, S., Mueller-Klieser, W., Hoves, S., Andreesen, R., Mackensen, A., Kreutz, M. (2006) Tumor-derived lactic acid modulates dendritic cell activation and antigen expression. *Blood 107*, 2013–2021.

56 Sánchez-Jiménez, F., Martínez, P., Nunnez de Castro, I., Olavarría, J. S. (1985) The function of redox shuttles during aerobic glycolysis in two strains of Ehrlich ascites tumor cells. *Biochimie 67*, 259–264.

57 Kim, K. H., Rodriguez, A. M., Carrico, P. M., Melendez, J. A. (2001) Potential mechanisms for the inhibition of tumor cell growth by manganese superoxide dismutase. *Antioxid. Redox. Signal. 3*, 361–373.

58 Parlo, R. A., Coleman, P. S. (1984) Enhanced rate of citrate export from cholesterol-rich hepatoma mitochondria. *J. Biol. Chem. 259*, 9997–10003.

59 Grimm, H., Tibell, A., Norrlind, B., Blecher, C., Wilker, S., Schwemmle, K. (1994) Immunoregulation by parental lipids: impact of the n-3 to n-6 fatty acid ratio. *J. Parenter. Enteral. Nutr. 18*, 417–421.

60 Eigenbrodt, E., Kallinowski, F., Ott, M., Mazurek, S., Vaupel, P. (1989) Pyruvate kinase and the interaction of amino acid and carbohydrate metabolism in solid tumors. *Anticancer Res. 18*, 3267–3274.

61 Mazurek, S., Grimm, H., Oehmke, M., Weisse, G., Teigelkamp, S., Eigenbrodt, E. (2000) Tumor M2-PK and glutaminolytic enzymes in the metabolic shift of tumor cells. *Anticancer Res. 20*, 5151–5154.

62 Lobo, C., Ruiz-Bellido, M. A., Aledo, J. C., Marquez, J., Nunez De Castro, I., Alonso, F. J. (2000) Inhibition of glutaminase expression by antisense mRNA decreases growth and tumourigenicity of tumour cells. *Biochem. J. 348*, 257–261.

63 Spittler, A., Oehler, R., Goetzinger, P., Holzer, S., Reissner, C. M., Leutmezer, F., Rath, V., Wrba, F. et al. (1997) Low glutamine concentrations induce phenotypical and functional differentiation of U937 myelomonocytic cells. *J. Nutr. 127*, 2151–2157.

64 Eck, H. P., Drings, P., Dröge, W. (1989) Plasma glutamate levels, lymphocyte reactivity and death rate in patients with bronchial carcinoma. *J. Cancer Res. Clin. Oncol. 115*, 571–574.

65 Vaupel, P., Kallinowski, F., Okunieff, P. (1989) Blood flow, oxygen and nutrient supply, and metabolic microenvironment of human tumors. *Cancer Res.* 49, 6449–6465.

66 Rapaport, E. (1988) Experimental cancer therapy in mice by adenine nucleotides. *Eur. J. Cancer Clin. Oncol.* 24, 1491–1497.

67 Hardie, D. G. (2004) The AMP-activated protein kinase pathway – new players upstream and downstream. *J. Cell. Sci.* 117, 5479–5487.

68 Schneider, J., Philipp, M., Salewski, L., Velcovsky, H. G. (2003): Pro-gastrin-releasing peptide (ProGRP) and neuron specific enolase (NSE) in therapy control of patients with small-cell lung cancer. *Clin. Lab.* 49, 35–42.

69 Lin, J. T., Wang, W. S., Yen, C. C., Liu, J. H., Yang, M. H., Chao, T. C., Chen, P. M., Chiou, T. J. (2005) Outcome of colorectal carcinoma in patients under 40 years of age. *J. Gastroenterol. Hepatol.* 20, 900–905.

70 Mazurek, S., Lüftner, D., Wechsel, H. W., Schneider, J., Eigenbrodt, E. (2002) Tumor M2-PK: a marker of the tumor metabolome. In *Tumor Markers-Physiology, Pathobiology, Technology, and Clinical Applications*, Diamandis, E. P., Fritsche, H. A., Lilja, H., Chan, D. W., Schwartz, M. K. (eds.), AACC Press, Washington, 471–475.

71 Hardt, P. D., Mazurek, S., Toepler, M., Schlierbach, P., Bretzel, R. G., Eigenbrodt, E., Kloer, H. U. (2004) Faecal tumour M2 pyruvate kinase: a new, sensitive screening tool for colorectal cancer. *Brit. J. Cancer* 91, 980–984.

72 Luder, A. S., Parks, J. K., Frerman, F., Parker, W. D. (1990) Inactivation of beef brain α-ketoglutarate dehydrogenase complex by valproic acid and valproic acid metabolites. Possible mechanism of anticonvulsant and toxic actions. *J. Clin. Invest.* 86, 1574–1581.

73 Maschek, G., Savaraj, N., Priebe, W., Braunschweiger, P., Hamilton, K., Tidmarsh, G. F., De Young, L. R., Lampidis, T. J. (2004) 2-Deoxy-D-glucose increases the efficacy of adriamycin and paclitaxel in human osteosarcoma and non-small cell lung cancers *in vivo*. *Cancer Res.* 64, 31–34.

74 Muellner, S., Stark, H., Niskanen, P., Eigenbrodt, E., Mazurek, S., Fasold, H. (2006) From target to lead synthesis. In *Proteomics in Drug Research*, Hamacher, M., Marcus, K., Stühler, K., van Hall, A., Warscheid, B., Meyer, H. E. (eds.) Wiley-VCH, Weinheim, 28, 187–207.

75 Ghose, A. K., Viswanadhan, V. N., Sanghvi, Y. S., Nord, L. D., Willis, R. C., Revankar, G. R., Robins, R. K. (1989) Structural mimicry of adenosine by the antitumor agents 4-methoxy- and 4-amino-8-(beta-D-ribofuranosylamino)pyrimido[5,4-d]pyrimidine as viewed by a molecular modeling method. *Proc. Natl. Acad. Sci USA* 86, 8242–8246.

76 Nord, L. D., Stolfi, R. L., Colofiore, J. R., Martin, D. S. (1996) Correlation of retention of tumor methylmercaptopurine riboside-5′-phosphate with effectiveness in CD8F1 murine mammary tumor regression. *Biochem. Pharmacol.* 51, 621–627.

77 Yoo, B. C., Ku, J.-L., Hong, S. H., Shin, Y. K., Park, S. Y., Kim, H. K., Park, J. G. (2004) Decreased pyruvate kinase M2 activity linked to cisplatin resistance in human gastric carcinoma cell lines. *Int. J. Cancer* 108, 532–539.

17
AMPK and the Metabolic Syndrome

Benoit Viollet and Fabrizio Andreelli

Abstract

The metabolic syndrome may arise due to inadequate adaptation to environmental changes (e.g., imbalance between energy intake and energy expenditure) involving both genetic and environmental factors. In recent studies in humans and rodents, AMP-activated protein kinase (AMPK), a phylogenetically conserved serine/threonine protein kinase, has been described as an integrator of regulatory signals monitoring systemic and cellular energy status. Therefore, it has been proposed that AMPK could provide a link in metabolic defects underlying progression to the metabolic syndrome. In transgenic mouse models that are deficient in AMPK activity, several defects in the control of body weight, whole-body insulin sensitivity, glucose homeostasis, and lipid metabolism (all components of the metabolic syndrome) have been observed. Furthermore, it is now recognized that physical activity, metformin, and thiazolidinediones, all used during the management of the metabolic syndrome, activate AMPK. In consequence, AMPK probably participates both as a key factor in metabolic syndrome physiopathology and as a new treatment for this disease. In the present review, we update those topics and discuss new findings that suggest that AMPK might be a promising pharmacological target for the treatment of metabolic disorders associated with the metabolic syndrome.

17.1
Defining the Metabolic Syndrome

During the last 50 years there has been considerable effort to better identify risk factors contributing to cardiovascular disease, which is the major cause of morbidity and mortality in the developed world. It is now clearly recognized that cardiovascular disease depends on different risk factors often simultaneously present in the same patient. This risk factor clustering, and its association with insulin resistance, led investigators to propose the existence of a unique pathophysiolog-

Table 17.1 Principal criteria for the identification of the metabolic syndrome and threshold values, according to the World Health Organization (WHO), the European Group for the Study of Insulin Resistance (EGIR), the National Cholesterol Education Program in the United States (NCEP-ATP III), and the American Association of Clinical Endocrinologists (AACE).

	WHO 1998/1999	EGIR 1999	NCEP-ATP III 2001	AACE 2003
FG (mmol L^{-1})	≥6.1 and/or	≥6.1	≥6.1	6.1–6.9
G2h (mmol L^{-1})	≥6.1 and/or	–	–	≥7.8
Treatment	+	−(ND)	–	–
Triglycerides (mmol L^{-1})	≥1.7 and/or	≥2.0 and/or	≥1.7	≥1.7
HDL-C (mmol L^{-1})	<0.9 M/<1.0 F	<1.0 and/or	<1.04 M/<1.29 F	<1.04 M/<1.29 F
Treatment	–	+	–	–
SAP/DAP (mmHg)	≥140/90	≥140 and/or 90	≥130/85	≥130/85
Treatment	–	+	–	–
Waist/hips	>0.90 M/0.85 F and/or	–	–	–
BMI (kg m^{-2})	>30	–	–	≥25
Waist circumference (cm)	–	≥94 M/80 F	>102 M/88 F	–
Insulin	Clamp < Q1	FI > Q4	–	–
Other	UAE ≥ 20 µg mn^{-1} or Alb/creat ≥ 30 mg g^{-1}	–	–	–
Syndrome	I and/or G + 2 criteria (metabolic syndrome)	FI + 2 criteria (insulin resistance syndrome)	3 criteria (metabolic syndrome)	Left to the clinician's judgment

Abbreviations. FG: fasting glycemia; G2h: glycemia two hours after glucose loading; HDL-C: HDL-cholesterol; SAP: systolic arterial pressure; DAP: diastolic arterial pressure; BMI: body mass index; UAE: urinary albumin excretion; Alb/creat: urinary albumin-to-creatinine ratio; I: insulinemia; FI: fasting insulinemia; IR: insulin resistance. <Q1: 25th percentile > Q4: 75th percentile.

ical condition called the "metabolic syndrome" [1] or "insulin resistance" syndrome, first described by Reaven in 1988 [2]. Metabolic syndrome is the association in a given subject of various metabolic and hemodynamic factors resulting in a high risk of type 2 diabetes and/or cardiovascular disease. There are several definitions of this syndrome (WHO, EGIR, NCEP-ATP III, AACE) [3–5], but all include a collection of criteria (high glycemia or type 2 diabetes, arterial hypertension, low HDL cholesterol, high level of triglycerides, abdominal obesity), with risk increasing with the number of criteria met (Table 17.1).

The prevalence of the metabolic syndrome in adults varies from 10% to 30% in industrialized countries, depending on the definition used and the age of the subjects studied. In these countries, metabolic syndrome prevalence increases in parallel with the expansion of abdominal obesity. Whatever the definition used, the prevalence of the metabolic syndrome increases with age, and this syndrome is becoming increasingly frequent in women (linked in particular with the increase in the proportion of women with large waists). In addition, differences according to ethnic origin have been observed, both in terms of prevalence, with Hispanic Americans more frequently affected (about 30%), and in terms of the distribution of constitutive criteria. Another key finding is that 5% of subjects with metabolic syndrome have a strictly normal body mass index (BMI) but a central distribution of fat, suggesting that metabolic syndrome is more dependent on abdominal fat, and visceral fat in particular, than on BMI itself [6].

17.2
Pathophysiology of the Metabolic Syndrome

The underlying pathophysiology of the metabolic syndrome is the subject of debate. The different components of the disease (insulin resistance, lower HDL cholesterol, higher apolipoprotein B and triglyceride levels, smaller LDL particles, aortic stiffness, coronary artery calcification, and higher blood pressure) are often present simultaneously in the same patient. This finding opens the question of a common original process. The causes of the metabolic syndrome remain poorly understood but include both genetic and environmental factors. The genetic factors include those determining corpulence, fat distribution, hyperinsulinemia, and lipoprotein metabolism. Environmental factors also play an essential role, with sedentary lifestyle, cigarette smoking, and excess calorie intake being of particular importance. When the concept of the metabolic syndrome was first proposed, insulin resistance and/or hyperinsulinemia were initially thought to be the primary etiological mechanism [2, 7, 8]. Recently, it has been recognized that factors such as tissular lipotoxicity (due to an increased release of free fatty acids) [9], secretion of adipokines (such as adiponectin) [10, 11] or secretion of cytokines (such as tumor necrosis factor-α [TNF-α]) [12] play an additional role during the development of the disease and these different components (Table 17.2). Indeed, visceral adiposity and its inflammatory component are significant independent predictors of the insulin resistance [13–16], impaired glucose toler-

Table 17.2 Manifestations associated with insulin resistance and not included in the different definitions of the metabolic syndrome.

Inflammation	Lipoproteins	Pro-thrombosis	Other
Decreased adiponectin	Increased apo B	Increased fibrinogen	Non-alcoholic steatohepatitis
Increased interleukin 6	Decreased apo A-1	Increased plasminogen activator inhibitor 1	Polycystic ovary syndrome
Increased tumor necrosis factor α	Small dense LDL and HDL	Increased viscosity	Obstructive sleep apnea
Increased resistin	Increased apo C-III	Increased homocysteine	
Increased C-reactive protein			
Increased white blood cell count			

ance [17], elevated blood pressure [18–20], and dyslipidemia [15, 21–23] seen in the metabolic syndrome. Additionally, a strong association between fatty liver disease and hyperinsulinemic insulin resistance in patients has been observed. Almost a quarter of adults in many industrialized countries have excessive fat accumulation in the liver. Although the cause of fatty liver is not known, it is often associated with metabolic syndrome (Table 17.2). Fatty liver is now recognized as a major contributor to overall obesity-related morbidity and mortality. Although it appears that therapies focusing on dieting and exercise are helpful for management of the metabolic syndrome, there is little evidence that these approaches improve the outcome of fatty liver disease.

17.3
AMPK as a Strong Determinant of the Metabolic Syndrome

AMP-activated protein kinase (AMPK) is activated in response to environmental or nutritional stress factors that deplete intracellular ATP levels, including heat shock, hypoxia, starvation, or prolonged exercise (for a review on AMPK regulation and signaling pathways, see Chapter 9). Regardless, the result of AMPK activation is the inhibition of energy-consuming biosynthetic pathways (such as fatty acid synthesis in liver and adipocytes, cholesterol synthesis in liver, protein synthesis in liver and muscle, and insulin secretion from β-cells) and activation of ATP-producing catabolic pathways (such as fatty acid uptake and oxidation in multiple tissues, glycolysis in heart, and mitochondrial biogenesis in muscle).

AMPK can also modulate transcription of specific genes involved in energy metabolism, thereby exerting long-term metabolic control. AMPK causes genetic changes that assist the cell in making chronic adaptations, such as the increase in mitochondrial biogenesis and enzyme expression in muscle that accompany physical training. These effects are mediated, in part at least, by an AMPK-induced increase in the expression of the peroxisome proliferator–activated receptor γ co-activator 1α (PGC1α) [24]. Recent studies showed that AMPK is likely to be under both endocrine and autocrine control in rodents. Thus, in addition to exercise and starvation, AMPK is activated by the fat cell–derived hormones adiponectin and leptin [25–27] and interleukin-6 (IL-6) [28]. Conversely, AMPK activity is suppressed in muscle and liver by sustained hyperglycemia and in liver by re-feeding after starvation [29] and by increases in the plasma concentration of another adipocyte-derived hormone, resistin [30]. In addition to its role in the periphery, AMPK also regulates energy intake and body weight by mediating the opposing effects of anorexigenic and orexigenic signals in the hypothalamus [31–34]. It is thus apparent how a dysfunction in the AMPK signaling pathway may have sustained, deleterious effects at the systemic level and may contribute to the events that lead to the metabolic syndrome (Fig. 17.1). It is interesting to note that there is a strong correlation between a low activation state of AMPK and metabolic disorders associated with insulin resistance, obesity, and sedentary activities [35–37]. In addition, many therapies that are useful in treating the metabolic syndrome and associated disorders in humans, including thiazolidinediones

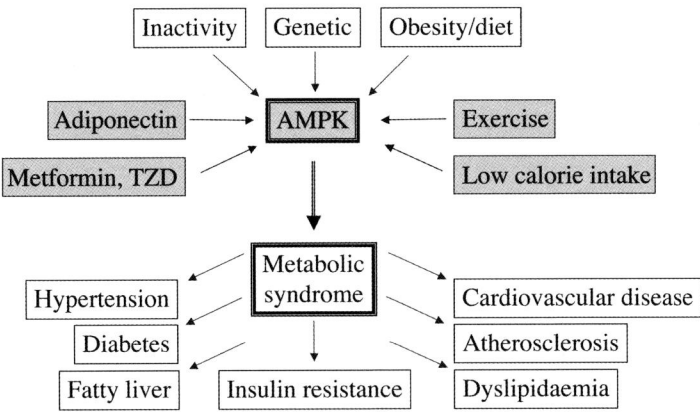

Fig. 17.1 Links between AMPK and the metabolic syndrome. Patients with the metabolic syndrome are characterized by insulin resistance; obesity; and a predisposition to hypertension, dyslipidemia, pancreatic β-cell dysfunction, type 2 diabetes, and premature atherosclerosis. A common feature linking these multiple abnormalities is dysregulation of AMPK and the signaling network. It is proposed that such dysregulation leads to alterations in cellular fatty acid metabolism, which in turn cause ectopic lipid accumulation, cellular dysfunction, and, ultimately, disease. Evidence is also presented that factors that activate AMP kinase might reverse these abnormalities or prevent them from occurring (shown by grey boxes).

Table 17.3 Genetic manipulation of AMPK subunits in mouse models.

Model	Phenotype	Reference
AMPKα1$^{-/-}$	Reduced isoproterenol-stimulated lipolysis in white adipose tissue and attenuated metformin-enhanced eNOS phosphorylation and activity in aortas	[64, 78]
AMPKα2$^{-/-}$	Insulin-resistance, increased catecholamine secretion in urine, no effect on muscle glucose uptake in response to contraction, more rapid onset of ischemic contracture in the heart, and increased fat mass following high-fat diet	[45, 65, 73]
AMPKγ3$^{-/-}$	Reduced glycogen synthesis following recovery from exercise	[43]
Skeletal and cardiac muscle-specific overexpression of α2 dominant negative (mAMPK-DN)	Decreased glucose uptake in skeletal muscle in response to contraction and hypoxia, decreased glucose uptake in heart during ischemia, decreased fatty acid oxidation following reperfusion, and impaired skeletal muscle mitochondrial biogenesis following chronic energy deprivation	[41, 97]

(TZDs), metformin, calorie deprivation, and exercise, have been shown to activate the AMPK system (Fig. 17.1).

Important progress has recently been made in the understanding of the physiological role of AMPK at both the cellular and organismal level. This is in part due to the development of transgenic and knockout (KO) mouse models, which have made it possible to study distinct physiological functions for AMPK isoforms (Table 17.3). AMPK is a heterotrimeric complex formed of a catalytic subunit (α) and two regulatory subunits (β and γ). Two isoforms have been identified for both the α-subunit (α1 and α2) and the β-subunit (β1 and β2), and three isoforms have been reported for the γ-subunit (γ1, γ2, and γ3; for the latter, the expression pattern is restricted to skeletal muscle), giving rise to a large variety of heterotrimeric combinations. A recent study investigated the isoform composition of AMPK complexes in human skeletal muscle and found that only 3 of the 12 theoretically possible AMPK complexes were present [38]. It is still unclear what cellular consequences the changes in AMPK isoform expression may have, but this study raised important questions about the function of each AMPK complex in relation to its particular subcellular localization and/or specific targets.

17.3.1
AMPK in the Control of Whole-body Insulin Sensitivity

Skeletal muscle is the major site of insulin-stimulated glucose disposal, and insulin resistance in this tissue is viewed as a contributing factor to the pathogenesis of the metabolic syndrome. One important finding is that pharmacological activation of AMPK by the use of 5-amino-imidazole-4-carboxamide ribonucleoside (AICAR) increases muscle glucose uptake concomitantly with glucose transporter 4 (GLUT4) translocation to the plasma membrane [39, 40]. This has also been illustrated in transgenic mice overexpressing a kinase-dead α2 isoform and in KO mouse models for the catalytic α2 or regulatory γ3 isoforms (Table 17.3) [41–43]. Interestingly, AMPK-induced glucose transport occurs through a mechanism distinct from that utilized by the classical insulin signaling pathway because it is not blocked by inhibitors of phosphatidylinositol 3-kinase and the effects of insulin and AMPK activators are additive [44]. Therefore, because the AMPK system is considered independent of the insulin pathway, one could expect that lack of AMPK activity cannot modify insulin sensitivity. Surprisingly, AMPKα2 KO mice exhibit decreased insulin sensitivity in muscle associated with a dramatic reduction of glycogen synthesis during the hyperinsulinemic euglycemic clamp procedure. In contrast, mice lacking the catalytic α1 isoform exhibited normal insulin sensitivity. These data are consistent with a more predominant role for the α2 catalytic subunit in controlling insulin action *in vivo* than the α1 isoform. The observation that transgenic mice overexpressing a dominant-inhibitory mutant of AMPK in skeletal muscle (mAMPK-DN) do not present a reduction in insulin sensitivity or a defect in muscle glycogen synthesis suggests that AMPKα2 could control insulin sensitivity through modification of autonomous nervous system activity [45]. Indeed, AMPKα2 KO mice exhibit a chronic sympathetic activation known to reduce insulin-stimulated glucose uptake in skeletal muscles and alter muscle glycogen synthesis. These data strongly suggest that partial or total inactivation of AMPK in skeletal muscle is probably unable *per se* to induce insulin resistance. However, even if the modulation of AMPK activity does not seem to be a direct cause of insulin resistance initiation, AMPK could be a potent target to treat insulin resistance.

17.3.2
AMPK Controls Glucose Homeostasis and Insulin Secretion

Insulin resistance and insulin secretion defects are major risk factors for type 2 diabetes [46]. The pathogenesis of type 2 diabetes is complex, comprising different degrees of β-cell failure relative to varying degrees of insulin resistance. A progressive decrease in β-cell function leads to glucose intolerance, which is followed by type 2 diabetes that inexorably aggravates with time [47]. Evidence from recent literature clearly demonstrated that AMPK regulates insulin secretion by direct or indirect pathways. AMPKα2 KO mice presented altered glucose regulation associated with a reduction in glucose-mediated insulin secretion [45]. This defect, observed only *in vivo* and not in isolated pancreatic islets of AMPKα2 KO mice, was dependent on an excess in sympathetic tone and was completely reversed when mice were treated with an α blockade drug. Because glucose-mediated insulin secretion was preserved *in vitro*, this model showed that the lack of pancreatic AMPKα2 subunit is not essential for the regulation of insulin secretion or might be compensated by the remaining pancreatic catalytic AMPKα1 subunit. In contrast, pharmacological activation of AMPK has been shown to inhibit insulin secretion *in vitro*, but only when chronically administered [48, 49]. These results have been confirmed using adenovirus-mediated overexpression of a constitutively active AMPK (AMPK-CA) in β-cells [50]. This inhibitory effect of AMPK on insulin secretion is associated with an increase in apoptotic index in β-cells.

One important point to be solved is the role of pancreatic AMPK during the adaptation of insulin secretion facing various degrees of insulin resistance *in vivo*. Indeed, the alterations of insulin secretion in human type 2 diabetes may result from changes in β-cell function, β-cell mass, or both. A decrease in β-cell mass is likely to play a role in the pathogenesis of human type 2 diabetes [51], as it does in rodent models of the disease [52, 53]. Pathways regulating β-cell turnover are also implicated in β-cell insulin secretory function. In consequence, a decrease in β-cell mass is not dissociable from an intrinsic secretory defect. Furthermore, because AMPK may act as a new regulator of β-cell apoptosis [54, 55], potential regulation of β-cell mass by AMPK requires further study.

17.3.3
AMPK Controls Lipid Metabolism

AMPK also has been involved in the regulation of lipid metabolism in order to acutely switch on alternative catabolic pathways and switch off anabolic pathways. Thus, once activated, AMPK phosphorylates and inactivates a number of metabolic enzymes involved in ATP-consuming cellular events, including cholesterol and fatty acid synthesis. 3-Hydroxy-3-methylglutaryl-coenzyme A reductase (HMG-CoA reductase) and acetyl-CoA carboxylase (ACC) – key enzymes in cholesterol and fatty acid synthesis, respectively – were the first enzymes shown to be phosphorylated and inactivated by AMPK. In the cholesterol synthesis pathway, AMPK blocks the conversion of HMG-CoA to mevalonate. One could expect detrimental effects on cholesterol homeostasis when AMPK activity is altered. In total and liver-specific AMPKα2 KO mice, plasma levels for total and HDL cholesterol are not statistically different compared with controls but have a tendency to be higher [45, 56]. This suggests that the remaining α1 subunit activity in AMPKα2 KO mice is sufficient to control hepatic cholesterol synthesis and that HMG-CoA reductase is a target for both catalytic isoforms of AMPK.

ACC is an important rate-controlling enzyme for the synthesis of malonyl-CoA, which is both a critical precursor for the biosynthesis of fatty acids and a potent inhibitor of mitochondrial fatty acid oxidation. Inhibition of ACC by AMPK leads to a drop in malonyl-CoA content and a subsequent decrease in fatty acid synthesis. Liver-specific AMPKα2 deletion leads to increased plasma triglyceride levels and enhanced hepatic lipogenesis [56]. Conversely, overexpression of AMPKα2 in the liver decreases plasma triglyceride levels [57]. Decreased triglyceride levels have also been observed during AICAR infusion in lean and obese rodents [58]. This emphasizes the critical role of the AMPKα2 subunit in the control of hepatic lipogenesis, which is a pathway not adequately controlled by the

AMPKα1 subunit. This suggests that the AMPKα2 subunit is critical for hepatic lipogenesis and that substitution by the AMPKα1 subunit is not possible.

17.3.4
AMPK and Ectopic Lipid Deposition in Tissue

Deposition of lipids in tissue is a hallmark defect of the metabolic syndrome in humans. Insulin resistance has been linked with excess lipids in tissue, resulting in the inhibition of insulin secretion and signaling (resistance to insulin-stimulated glucose transport in muscle and fat and insulin-stimulated suppression of glucose production in the liver). According to this lipotoxicity hypothesis, insulin resistance develops when excess lipids are deposited in insulin-sensitive cell types. In cells, a delicate balance controls whether fatty acids are transported into mitochondria and metabolized or are stored in the cytoplasm as triglycerides. This balance is regulated mainly by malonyl-CoA, generated by ACC, inhibiting transport of fatty acids into mitochondria via allosteric regulation of carnitine palmitoyltransferase-1 and thereby preventing them from being metabolized. Activated AMPK inhibits malonyl-CoA synthesis and shifts the balance towards mitochondrial fatty acid oxidation and away from fat storage. Interestingly, AMPK-induced ACC phosphorylation is impaired in hepatocytes deleted of both catalytic subunits [59], contributing to increased intracellular malonyl-CoA levels and triglyceride accumulation in the liver (author's unpublished results). This is not the case in AMPKα2 KO mice, which exhibit hepatic triglyceride content similar to that in control mice. This suggests that the remaining α1 subunit activity in AMPKα2 KO mice is sufficient to control hepatic lipogenesis and that key enzymes of lipogenesis are equal targets of both catalytic isoforms of the kinase. In skeletal muscle and pancreas, triglyceride content was similar in AMPKα2 KO and control mice. Similarly, lipid content was not increased in skeletal muscle from AMPKγ3 KO mice compared with control mice [43].

17.3.5
AMPK as a Regulator of White Adipose Tissue Physiology

It is well documented that changes in adipose tissue mass are frequently associated with alterations in insulin sensitivity [60]. Altered energy balance in the organism can cause a dramatic change in adipose tissue mass, leading to either obesity or lipoatrophy, disorders that are associated with pathologies such as type 2 diabetes and cardiovascular disease. Factors such as genetic background, diet, physical activity, and hormonal balance are involved in the control of fat mass. Adipose tissue is the major organ for storage of energy in the form of triglycerides, and an alteration in AMPK activity has been proposed as a factor contributing to the development of obesity. It is well known that AMPK regulates lipogenesis, mainly by phosphorylating ACC, as well as lipolysis in adipocytes [61]. Activation of AMPK in adipocytes is concomitant with a decreased expression of genes coding lipogenic enzymes [62] and leads to a decreased lipogenic flux and a decreased triglyceride synthesis [63, 64]. In adipocytes, AMPK activation using AICAR or overexpression of AMPK-CA has been shown to inhibit β-adrenergic–induced lipolysis [61, 63, 64]. A negative control of lipolysis by the AMPK system seems to contradict the paradigm that AMPK inhibits anabolism and stimulates catabolism. However, if fatty acids released by lipolysis are not removed from the cell, they recycle back into triglycerides with the consumption of ATP and generation of AMP, thus activating AMPK. The release of fatty acids from adipose tissue is controlled by inhibition of hormone sensitive lipase (HSL), a rate-limiting enzyme of lipolysis. Thus, inhibition of HSL by AMPK represents a mechanism for limiting this recycling and ensuring that the rate at which fatty acids are released by lipolysis does not exceed the rate at which they could be disposed of by export or by internal oxidation.

In agreement with these studies, AMPKα2 KO mice exhibit significantly higher free fatty acid levels than do controls in both the fasted and fed states, probably by chronically enhanced lipolysis. This suggests that the α2 isoform inhibits HSL activity in adipocytes. But we cannot exclude that lipolysis was enhanced in AMPKα2 KO mice because their insulin levels were chronically low or because sympathetic nervous activity was increased [45]. The consequences of lipid accumulation in adipose tissue of AMPKα2 KO mice fed a high-fat diet have been studied [65]. Microscopic analysis of gonadal adipose tissue from wild-type and AMPKα2 KO mice revealed that mice lacking the α2 isoform had larger adipocytes than did wild-type mice. In addition, a significantly higher triglyceride accumulation was observed in adipose depots of AMPKα2 KO mice compared with control animals. In AMPKα2 KO mice, an increase in adipose tissue mass may arise from an increase in triglyceride storage rather than from enhancement of pre-adipocyte differentiation into adipocytes. The higher body weight and the increased fat mass exhibited by AMPKα2 KO mice fed a high-fat diet are consistent with studies linking AMPK activity and alterations in fat mass. Indeed, sustained AMPK activation in obese Zucker rats by long-term administration of AICAR diminished the mass of epididymal and retroperitoneal fat pads up to 30–40%, although no difference in total body weight was observed in these studies [66]. Similarly, treatment of *ob/ob* mice with the antidiabetic drug metformin, which is reported to increase AMPK activity [67], also decreases epididymal adipose tissue mass [68].

In addition to its role in adipocyte metabolism, it has been reported that the treatment of 3T3-L1 or F442A pre-adipocytes with AICAR inhibits their differentiation into adipocytes and promotes apoptosis [69, 70]. However, in mice lacking either the α1 or α2 catalytic subunits, the differentiation potential does not seem to be grossly affected, because no changes in the expression of the mature adipocyte markers, PPARγ and C/EBPα, have been observed [64, 65]. It is possible that the observed effect in 3T3-L1 pre-adipocytes may not be caused by a specific activation of AMPK by AICAR but by its effect on other AMP-sensitive molecules or processes. It is thus not clear whether AMPK could have a physiological regulatory function in human adipocyte differentiation and this point remains to be addressed.

17.3.6
AMPK Alterations in Cardiovascular Pathology

The metabolic syndrome is associated with an increased risk of cardiovascular disease and coronary heart disease mortality. In humans, a variety of mutations in the $\gamma2$ subunit have been shown to produce a glycogen storage cardiomyopathy characterized by ventricular pre-excitation, conduction defects, and cardiac hypertrophy [71]. Although there is still controversy as to whether $\gamma2$ mutations are activating or inactivating mutations of AMPK, inappropriate regulation of AMPK can have profound morphological and functional consequences on the heart. The role of AMPK on cardiac energy metabolism is particularly relevant in the setting of cardiac ischemia and hypoxia. Total cessation of myocardial blood flow leads to rapid perturbations in myocardial metabolism. In a few seconds, oxidative phosphorylation and mitochondrial ATP production are seriously disturbed, with a consequent decrease in high-energy phosphate, creatine phosphate, and ATP levels. Liberation of free fatty acids is stimulated in myocardial ischemia by increased circulating catecholamines, but fatty acid oxidation and the tricarboxylic acid cycle are inhibited. This leads to cytosolic accumulation of free fatty acid CoA esters. Glycogenolysis and anaerobic glycolysis are stimulated, but accumulation of lactate and H^+ ions leads to inhibition of glycolysis and decreased anaerobic energy production [72]. Because AMPK regulates the balance between glucose and fatty acid metabolism at the cellular level, the metabolic response of the heart to global ischemia was studied in AMPK$\alpha2^{-/-}$ mice. These hearts displayed a more rapid onset of ischemic contracture, which was associated with a decrease in ATP content, in lactate production, in glycogen content, and in the phosphorylation state of ACC [73]. The importance of metabolic adaptation via AMPK activation during ischemia was also documented in another transgenic mouse model overexpressing a dominant-negative form of AMPK$\alpha2$ in the heart [74]. These studies indicate that the $\alpha2$ isoform of AMPK is required for the metabolic response of the heart to ischemia, suggesting that AMPK is cardioprotective. Several reports suggest a relationship between adiponectin and heart disease. For example, high blood levels of adiponectin are associated with a lower risk of heart attack and *vice versa* [75]. Additionally, adiponectin levels rapidly decline after the onset of acute myocardial infarction. Similarly, in mice, deletion of adiponectin induces increased heart damage after reperfusion that was associated with diminished AMPK signaling in the myocardium [76]. These findings clearly show that adiponectin has a cardioprotective role *in vivo* during ischemia through AMPK-dependent mechanisms.

Endothelial cell dysfunction, as manifested by impaired vascular relaxation or an increase in circulating vascular cell adhesion molecules, is present in patients with type 2 diabetes and the metabolic syndrome, and it is thought to be one component of the inflammatory process that initiates atherogenesis. That AMPK dysregulation could contribute to this process has been suggested by studies showing that AMPK activity was lower in aortic endothelium of obese rats [77] and that metformin-induced AMPK activation can increase endothelial nitric oxide synthase (eNOS) phosphorylation and activity in endothelium in an AMPK-dependent manner [78]. Interestingly, metformin was also shown to relax endothelium-denuded rat aortic rings pre-contracted with phenylephrine, showing that AMPK can induce vasorelaxation in an endothelium- and NOS-independent manner [79]. Thus, as a metabolic sensor, vascular AMPK also could be involved in metabolic regulation of vascular tone. AMPK activation in response to hypoxia or metabolic challenge can induce vasorelaxation of big vessels [80, 81], thereby favoring blood flow.

17.4
Management of the Metabolic Syndrome

During the past two decades, the metabolic syndrome has reached pandemic proportions in association with rising levels of obesity and inactivity. With the elevated risk of not only type 2 diabetes [60] but also cardiovascular disease from the metabolic syndrome [82, 83], there are major healthcare cost implications. Now, guidelines are urgently needed to prevent this emerging global epidemic. Although the metabolic syndrome appears to be more common in people who are genetically susceptible, acquired underlying risk factors such as overweight or obesity, sedentary lifestyle, and atherogenic diet commonly elicit clinical manifestations. Clinical management should first focus on controlling these underlying risk factors independent of an individual's risk status. There is currently no recognized drug treatment for the metabolic syndrome. But it is possible to treat each of the abnormalities making up this syndrome early and effectively by using drug-based treatment. It is obviously important to use drugs available for each condition that do not have deleterious effects or that even have beneficial effects associated with metabolic abnormalities.

The growing realization that AMPK regulates the coordination of anabolic (synthesis and storage of glucose and fatty acids) and catabolic (oxidation of glucose

Table 17.4 Effects of factors that activate AMPK on metabolic syndrome criteria.

Factor	Insulin resistance	β-cell dysfunction	Endothelial dysfunction	Fatty liver	Cardiovascular disease
Exercise	+	+	+	+	+
Calorie restriction	+	+	+	+	+
Adiponectin	+	+	+	+	+
Leptin	+	+	+	+	ND
AICAR	+	+	+	ND	ND
Metformin	+	+	+	+	+
TZDs	+	+	+	+	+

ND: not determined.

and fatty acids) metabolic processes presents an attractive therapeutic target for intervention in many conditions of disordered energy balance, including obesity, type 2 diabetes, and the metabolic syndrome (Table 17.4). Support for this idea came first from *in vivo* treatment with AICAR of various animal models of insulin resistance, causing improvement in most, if not all, of the metabolic abnormalities of these animals [58, 66, 84, 85]. Consistent with these results, two major classes of existing antidiabetic drugs – biguanides (metformin and phenformin) and TZDs (rosiglitazone, troglitazone, and pioglitazone) – have recently been reported to activate AMPK [67, 86, 87].

17.4.1
Management of the Metabolic Syndrome by Exercise

Current guidelines recommend practical, regular, and moderate regimens of physical activity [88]. Regular and sustained physical activity will improve all risk factors of the metabolic syndrome [89]. The combination of weight loss and exercise to reduce the incidence of type 2 diabetes in patients with glucose intolerance should not be dismissed [90]. It is well established that physical exercise is a prototypical AMPK activator, and it has been suggested that AMPK activation may recapitulate some of the exercise-induced adaptations and is likely to mediate beneficial effects of exercise on insulin sensitivity and glucose transport in skeletal muscle. Thus, it is expected that part of the effect of physical activity in preventing the development of diseases related to a sedentary lifestyle is due to activation of AMPK. Repetitive pharmacological activation of AMPK *in vivo* results in expression of specific muscle proteins mimicking some of the effects of exercise training. Thus, AICAR or chronic intake of the creatine analogue β-guanadinopropionic acid (β-GPA) in rodent increases muscle expression of GLUT4 and hexokinase II [91]. Increased expression of GLUT4 in skeletal muscle following AMPK activation also could be dependent on PGC-1α [92]. These gene expression effects are abolished in AMPKα2 KO and mAMPK-KD mice [93, 94]. In addition, chronic activation of AMPK with AICAR or β-GPA increases mitochondrial content and expression of mitochondrial proteins, leading to a mitochondrial biogenesis [95–97]. Favorable effects of AMPK activation on mitochondrial biogenesis involve PGC-1α, which is known to increase expression of the transcription factors nuclear respiratory factor (NRF) 1 and 2 and mitochondrial transcription factor A (mtTFA). In accordance with these data, AMPK seems to act as a key factor for the metabolic adaptation of skeletal muscle to exercise. For instance, AMPKα2 KO mice have a disturbed muscle energy balance during exercise, as indicated by a reduced ATP content [93]. In mAMPK-KD mice, mitochondrial biogenesis is completely abolished [97]. Interestingly, it has been proposed that the development of skeletal muscle insulin resistance may be partly linked to decreased mitochondrial density [98]. Surprisingly, even if mAMPK-DN mice showed a defect in skeletal muscle mitochondrial biogenesis, insulin sensibility was preserved in this model [45].

The Diabetes Prevention Program has recently shown that, similar to diet and exercise, metformin treatment reduces the risk of developing type 2 diabetes [99]. In addition, it has been demonstrated that long-term AICAR administration, like exercise training, can prevent the development of hyperglycemia in Zucker diabetic fatty (ZDF) rats. This is partly due to improved peripheral insulin sensitivity in skeletal muscles and the maintenance of β-cell function [100]. Activation of AMPK in response to physical exercise also has been observed in extra-muscular tissues such as liver and adipose tissue [101] and might account for additional metabolic benefits. Physical training increases circulating adiponectin and mRNA expression of its receptors in muscle, which may mediate the improvement of insulin resistance and the metabolic syndrome in response to exercise by activation of AMPK [102]. Interestingly, treatment of subjects with type 2 diabetes with metformin significantly increased AMPK activity in skeletal muscle, and this stimulation was accompanied by enhanced peripheral glucose disposal [103, 104]. Metformin was also able to restore glucose uptake stimulation in insulin-resistant cardiomyocytes, suggesting that AMPK activation could be a potential therapeutic approach to treat insulin resistance in diabetic hearts [105]. Thus, by mimicking the beneficial effects of physical activity, AMPK activators appear to be new therapeutic agents for the treatment of the metabolic syndrome.

17.4.2
Management of Obesity

Effective weight reduction improves all risk factors associated with the metabolic syndrome [106], and it further reduces the risk of type 2 diabetes [107]. Nevertheless, to date weight reduction drugs have not been particularly effective in the treatment of obesity. Weight reduction is best achieved by behavioral changes to reduce energy intake and by increasing physical activity to enhance energy expenditure. Therefore, the AMPK system may be an important pharmacological target for reducing fatty acid storage in adipocytes and for treating obesity. By inducing fatty acid oxidation within the adipocyte, activation of AMPK would reduce fat cell size and also prevent fatty acids from being exported to peripheral tissues and causing deleterious effects. Direct evidence linking AMPK activation to diminished adiposity was first obtained by chronic administration of AICAR to lean and obese rats [66, 95]. These studies suggest that the decrease in adiposity caused by AICAR is attributable, at least in part, to an increase in energy expenditure. A link between AMPK, fatty acid oxidation, and the anti-obesity hormone leptin has been described in skeletal muscle [25]. Induced hyperleptinemia in rats is able to activate AMPK activity in adipose tissue in association with increased uncoupling mitochondrial protein (UCP)-1 and UCP-2 expression, and it induces a depletion in body fat stores [62]. β3-Adrenoceptor (β3-AR) agonists were also found to have remarkable anti-obesity and antidiabetic effects in rodents. It appears that chronic β3-adrenergic stimulation in white adipose tissue increases the expression of mitochondrial uncoupling proteins (UCP-2 and UCP-3). The beneficial effects of β-adrenergic agonists on body weight control in obese patients might be explained by AMPK activation, because these compounds were found to stimulate AMPK in fat cells [108]. In addition, overexpression of UCP-1 in adipocytes leads to an increase in the AMP:ATP ratio, activation of AMPK, inactivation of ACC, and decreased lipogenesis [109]. Interestingly, these mice are resistant to nutrient-induced obesity. There is also a strong mitochondrial biogenesis in response to increased UCP-1 expression in adipocytes [62, 110], features that could enhance the fatty acid oxidation capacity of adipocytes in response to AMPK activation. Nevertheless, recent studies have reported that in contrast to skeletal muscle in which AMPK stimulates fatty acid oxidation to provide ATP as a fuel, in adipocytes AMPK activation inhibits both lipogenesis and fatty acid oxidation [111]. Therefore, the metabolic consequences and anti-obesity effects of activation of AMPK in white adipocytes must be carefully investigated.

17.4.3
Management of Lipotoxicity

Patients with the metabolic syndrome are insulin resistant and demonstrate impaired pancreatic β-cell function, dyslipidemia, and premature atherosclerosis. In addition, they are usually overtly obese or have more subtle manifestations of increased adiposity, such as an increase in visceral fat. These conditions are associated with excess triglyceride content in muscle, liver, and pancreatic β-cells, which has been shown to precede the onset of hyperglycemia and insulin resistance. What seems clear is that type 2 diabetes develops only when the metabolic syndrome is accompanied by a β-cell failure in insulin-resistant patients. In Zucker rats, triglyceride accumulation in the β-cell induces a decreased β-cell mass by apoptosis [112]. An emerging body of evidence suggests that metformin, TZD, and AICAR treatments favor fatty acid β-oxidation and prevent glucolipotoxicity-induced apoptosis in β-cells [113–115]. Lipotoxicity is also present in skeletal muscle from insulin-resistant patients. Several studies have shown that activation of AMPK with AICAR, α-lipoic acid, leptin, and adiponectin enhances muscle fatty acid β-oxidation [25, 26, 37, 40, 95]. Studies in transgenic animals support these observations, because expression of the activating $\gamma 3$ R225Q mutation in muscle increased fatty acid oxidation and protected against excessive triglyceride accumulation and insulin resistance in skeletal muscle [43]. Interestingly, recent data have shown that resistin lowers AMPK signaling in muscle cells and that this reduction is associated with a suppressed fatty acid oxidation [116].

17.4.4
Management of Glucose Homeostasis Alteration

Several reports indicate that metformin and TZDs can reduce the risk of type 2 diabetes in people with glucose intolerance [99, 117]. Interestingly, these two drugs have been reported to activate AMPK [67, 86]. There is now good evidence – from studies of mice that are deficient in the upstream AMPK kinase, LKB1, in the liver – that the blood glucose–lowering effect of metformin is mediated by AMPK activation [118]. Nevertheless, recent studies have shown that LKB1 phosphorylates and activates at least 12 AMPK-related kinases, and these studies raise the question of whether the glucose-lowering function of LKB1 is mediated by AMPK-related kinases rather than by AMPK itself. Metabolic and insulin-sensitizing effects of TZDs have been shown to be in part mediated through adiponectin-dependent activation of AMPK, because activation of AMPK by rosiglitazone treatment is diminished in adiponectin KO mice [119]. TZDs can markedly enhance the expression and secretion of adiponectin *in vitro* and *in vivo* through the activation of its promoter and also antagonize the suppressive effect of TNF-α on the production of adiponectin [120]. Interestingly, in human adipose tissue, AICAR has been shown to increase the expression of adiponectin [121, 122]. However, type 2 diabetic patients treated with metformin display no change in serum adiponectin concentration or adipocyte adiponectin content [123].

Circulating levels of adiponectin are decreased in individuals with obesity and insulin resistance, suggesting that its deficiency may have a causal role in the etiopathogenesis of these diseases. Therefore, adiponectin replacement in humans may represent a promising approach to preventing and/or treating obesity, insulin resistance, and the metabolic syndrome. The chronic effects of adiponectin on insulin resistance *in vivo* were investigated by generation of adiponectin transgenic mice. Globular adiponectin transgenic *ob/ob* mice showed partial amelioration of insulin resistance and diabetes [124], and full-length adiponectin showed suppression of insulin-mediated endogenous glucose production [125]. With respect to the molecular mechanisms underlying the insulin-sensitizing action of adiponectin, it has been demonstrated that full-length adiponectin activates AMPK in the liver, while globular adiponectin does so in both muscle and liver [26]. Blocking AMPK activation by the use of a dominant-negative mutant inhibited the action of full-length adiponectin on glucose hepatic production [26]. In addition, the lack of action of adiponectin on hepatic glucose production when the AMPKα2 catalytic subunit is missing strongly supports the concept that adiponectin's effects are strictly dependent on AMPK [56].

17.4.5
Management of Fatty Liver

Studies in humans and various animal models have suggested that efforts to enhance insulin sensitivity might improve fatty liver disease, a situation frequently observed in patients with the metabolic syndrome. The efficacy of metformin as a treatment for fatty liver disease has been confirmed in obese *ob/ob* mice, which develop hyperinsulinemia, insulin resistance, and fatty livers [68]. Similarly, adiponectin restores insulin sensitivity and decreases hepatic steatosis by lowering the triglyceride content in muscle and liver in obese mice [126, 127]. Metabolic improvement of adiponectin is linked to an activation of AMPK in the liver that decreases fatty acid biosynthesis and increases mitochondrial fatty acid oxidation [26]. This has been confirmed by a decrease in liver triglyceride content in lean and obese rodents during AICAR infusion [58] and treatment with small-molecule AMPK activators [128]. In addition, it was demonstrated recently that activation of AMPK by resveratrol, a polyphenolic compound found in red grapes or wine, protects against lipid accumulation in the liver of diabetic mice [129]. This effect has been correlated to increased mitochondrial number and decreased PCG-1α acetylation in the liver [130]. However, the role of the AMPK system in the treatment of fatty liver diseases remains to be clearly established in humans. Its importance is strongly indicated by recent successes in treating these disorders with therapies that activate AMPK, including diet, exercise, and TZDs [131, 132].

17.4.6
Management of Cardiovascular Disease

Patients with the metabolic syndrome have increased susceptibility to cardiovascular disorders. AMPK has emerged as a key regulator of energy metabolism in the heart. The high energy demands of the heart are primarily met by the metabolism of fatty acids and glucose, both processes being regulated by AMPK. Indeed, AMPK stimulates glycolysis and sustains energy supply during ischemic stress. Promotion of glucose oxidation or inhibition of fatty acid oxidation in ischemic/reperfused hearts could be a promising novel therapeutic approach to myocardial ischemic conditions. Such a mechanism has been demonstrated during the phenomenon called ischemic preconditioning. Brief episodes of myocardial ischemia render the heart more resistant to subsequent ischemic episodes [133]. This phenomenon, ischemic preconditioning, is known to induce endogenous protective mechanisms in the heart. Myocardial preconditioning activates AMPK in a PKC-dependent manner and promotes glucose utilization in myocardial cells, supporting resistance toward ischemic consequences [134]. AMPK mediates preconditioning in cardiac cells by regulating the activity and recruitment of sarcolemmal K(ATP) channels without being a part of the signaling pathway that regulates mitochondrial membrane potential [135]. Thus, AMPK activators could be of particular interest for the management of myocardial ischemia. Attractively, it has been reported that adiponectin protects the heart from ischemia by activating AMPK and increasing the energy supply to heart cells [76]. In addition, reduced adiponectin levels also have been involved in hypertrophic cardiomyopathy associated with diabetes and other obesity-related diseases [136]. Adiponectin deficiency leads to progressive cardiac remodeling in pressure-overloaded conditions mediated via decreasing AMPK signaling and impaired glucose metabolism [136, 137]. In this model, adenovirus-mediated supplementation of adiponectin attenuated cardiac hypertrophy in response to pressure overload through activation of AMPK signaling [136]. In another cellular model, AMPK was shown to mediate the suppressive effects of adiponectin on endothelin-1–induced hypertrophy in cultured cardiomyocytes [138]. Interestingly, AMPK-dependent adiponectin vascular effects also have been demonstrated for angiogenic repair in an ischemic hind limbs model [139]. These data indicate that adiponectin may be useful in the treatment of obesity-related vascular deficiency diseases.

Atherosclerosis and its clinical complications are dramatically accelerated in patients with the metabolic syndrome. This has been attributed to the effects on the vascular wall of dysglycemia, dyslipidemia, inflammation, and oxidant stress [140]. Attractively, polyphenols inhibit endothelial cell adhesion molecule expression and aortic atherogenesis events in LDL receptor-deficient mice in large part through their lipid-lowering effects, which are associated with their ability to stimulate hepatic AMPK activation [129]. Lipid accumulation in vascular endothelial cells also may play an important role in the pathogenesis of atherosclerosis in obese subjects. Statins are the most commonly used drugs to reduce mortality related to cardiovascular disease in patients with the metabolic syndrome. Statins have a number of vasoprotective effects to improve endothelial function outside of reductions in low-density lipoproteins and triglycerides. These cholesterol-independent actions have been found to downregulate vascular inflammation and promote cardioprotection against ischemic disorders and heart failure. Mechanisms of this protection include increases in endothelial nitric oxide synthase activity and a subsequent rise in nitric oxide bioavailability. Interestingly, recent data showed that AMPK activation can increase nitric oxide (NO) synthase (eNOS) phosphorylation and activity in the endothelium [78, 141, 142]. This suggests that the beneficial vaso- and cardioprotective effects of statins are expected by AMPK activation. In this respect, metformin has been proposed to improve endothelial functions in diabetes by favoring eNOS activation by AMPK activation [78], which may explain the reduction in cardiovascular events observed in the UKPDS study [143]. Similarly, by activating AMPK, α-lipoic acid improves vascular dysfunction by normalizing triglyceride and lipid peroxide levels and NO synthesis in endothelial cells from obese rats [77]. Attractively, adiponectin exhibits potent anti-atherosclerotic effects [124, 144] and suppresses endothelial cell proliferation induced by a low dose of oxidized low-density lipoprotein via AMPK activation [145]. The use of AICAR also could suppress vascular smooth muscle cell proliferation and inhibit cell cycle progression [146]. Endothelial dysfunction is also present in hypertensive patients. Observational studies have provided strong evidence that consuming a moderately reduced intake of sodium contributes to lowering blood pressure. Interestingly, long-term administration of AICAR reduced systolic blood pressure in an insulin-resistant animal model [66]. In this process, a potential role for AMPK could be the regulation of ion channels or sodium co-transporters, including ENaC and the Na-K-2Cl co-transporter [147, 148]. These data provide additional support to the hypothesis that AMPK activation might be a potential future pharmacological strategy for treating the cardiovascular risk factors linked to the metabolic syndrome.

17.5
Conclusions and Medical Perspectives

The AMPK system plays a major role in the regulation of glucose and lipid metabolism through its acute effects on energy metabolism pathways and its long-term effects involving changes in gene expression. By maintaining energy balance, both at the single-cell and the whole-body levels, AMPK appears to be an important player in the derangements of energy metabolism that occur in conditions such as the metabolic syndrome. Although there is no current evidence that mutations in or altered expression of AMPK is a common cause of the metabolic syndrome in humans, the latter is strongly correlated with obesity, a sedentary lifestyle, and a low activation state of AMPK in the periphery, due to overnutrition and lack of exercise, and may be a contributory factor in its onset. The relationship between AMPK activation and beneficial metabolic effects in insulin-resistant and diabetic rodent models has provided the rationale for the development of new therapeutic strategies based on pharmacological but also nutritional use of AMPK activators in order to prevent or reverse metabolic disorders linked to the metabolic syndrome.

The widespread and various cellular functions of AMPK make its selective targeting in therapeutics a difficult one, with simultaneous advantageous and deleterious consequences being possible. Indeed, an emerging concept is that the result of AMPK activation is context specific and can be either beneficial or deleterious, depending on the tissue, degree of stimulation, and conditions of activation. A therapeutic agent would ideally activate AMPK in peripheral tissues (to increase fatty acid oxidation and glucose uptake and to reduce gluconeogenesis) while inhibiting it in the hypothalamus (to reduce food intake and body weight). Furthermore, it is now becoming apparent that there are distinct expression patterns of AMPK isoforms across species and particularly between rodent and human tissues. These findings will have significant implications for the understanding of the physiological role of AMPK in different species as well as for the use of animal models in the development of AMPK activators as therapeutic agents. Another interesting area for future development of drugs targeted at AMPK is in the arena of cancer. It has recently been reported that type 2 diabetics treated with metformin have a lower incidence of cancer [149]. However, these findings leave open the question as to whether AMPK activation can be responsible for the therapeutic treatment of cancer.

References

1 Kahn, R., Buse, J., Ferrannini, E., Stern, M. (2005) The metabolic syndrome: time for a critical appraisal: joint statement from the American Diabetes Association and the European Association for the Study of Diabetes. *Diabetes Care 28*, 2289–2304.

2 Reaven, G. M. (1988) Banting lecture 1988. Role of insulin resistance in human disease. *Diabetes 37*, 1595–1607.

3 (1999) World Health Organization: Definition, Diagnosis, and Classification of Diabetes Mellitus and its Complications: Report of a WHO Consultation. Geneva, World Health Org.

4 (2001) Expert Panel on Detection, Evaluation, and Treatment of High Blood Cholesterol in Adults. Executive Summary of The Third Report of The National Cholesterol Education Program (NCEP) Expert Panel on Detection, Evaluation, And Treatment of High Blood Cholesterol In Adults (Adult

Treatment Panel III). *Jama 285*, 2486–2497.

5 Balkau, B., Charles, M. A., Drivsholm, T., Borch-Johnsen, K., Wareham, N., Yudkin, J. S., Morris, R., Zavaroni, I., et al. (2002) Frequency of the WHO metabolic syndrome in European cohorts, and an alternative definition of an insulin resistance syndrome. *Diabetes Metab 28*, 364–376.

6 Carr, D. B., Utzschneider, K. M., Hull, R. L., Kodama, K., Retzlaff, B. M., Brunzell, J. D., Shofer, J. B., Fish, B. E., et al. (2004) Intra-abdominal fat is a major determinant of the National Cholesterol Education Program Adult Treatment Panel III criteria for the metabolic syndrome. *Diabetes 53*, 2087–2094.

7 Ferrannini, E., Buzzigoli, G., Bonadonna, R., Giorico, M. A., Oleggini, M., Graziadei, L., Pedrinelli, R., Brandi, L., et al. (1987) Insulin resistance in essential hypertension. *N Engl J Med 317*, 350–357.

8 Ferrannini, E., Haffner, S. M., Mitchell, B. D., Stern, M. P. (1991) Hyperinsulinae-mia: the key feature of a cardiovascular and metabolic syndrome. *Diabetologia 34*, 416–422.

9 van Harmelen, V., Dicker, A., Ryden, M., Hauner, H., Lonnqvist, F., Naslund, E., Arner, P. (2002) Increased lipolysis and decreased leptin production by human omental as compared with subcutaneous preadipocytes. *Diabetes 51*, 2029–2036.

10 Yatagai, T., Nagasaka, S., Taniguchi, A., Fukushima, M., Nakamura, T., Kuroe, A., Nakai, Y., Ishibashi, S. (2003) Hypoadiponectinemia is associated with visceral fat accumulation and insulin resistance in Japanese men with type 2 diabetes mellitus. *Metabolism 52*, 1274–1278.

11 Cnop, M., Havel, P. J., Utzschneider, K. M., Carr, D. B., Sinha, M. K., Boyko, E. J., Retzlaff, B. M., Knopp, R. H., et al. (2003) Relationship of adiponectin to body fat distribution, insulin sensitivity and plasma lipoproteins: evidence for independent roles of age and sex. *Diabetologia 46*, 459–469.

12 Katsuki, A., Sumida, Y., Murashima, S., Murata, K., Takarada, Y., Ito, K., Fujii, M., Tsuchihashi, K., et al. (1998) Serum levels of tumor necrosis factor-alpha are increased in obese patients with noninsulin-dependent diabetes mellitus. *J Clin Endocrinol Metab 83*, 859–862.

13 Carey, D. G., Jenkins, A. B., Campbell, L. V., Freund, J., Chisholm, D. J. (1996) Abdominal fat and insulin resistance in normal and overweight women: Direct measurements reveal a strong relationship in subjects at both low and high risk of NIDDM. *Diabetes 45*, 633–638.

14 Cnop, M., Landchild, M. J., Vidal, J., Havel, P. J., Knowles, N. G., Carr, D. R., Wang, F., Hull, R. L., et al. (2002) The concurrent accumulation of intra-abdominal and subcutaneous fat explains the association between insulin resistance and plasma leptin concentrations: distinct metabolic effects of two fat compartments. *Diabetes 51*, 1005–1015.

15 Katsuki, A., Sumida, Y., Urakawa, H., Gabazza, E. C., Murashima, S., Maruyama, N., Morioka, K., Nakatani, K., et al. (2003) Increased visceral fat and serum levels of triglyceride are associated with insulin resistance in Japanese metabolically obese, normal weight subjects with normal glucose tolerance. *Diabetes Care 26*, 2341–2344.

16 Wagenknecht, L. E., Langefeld, C. D., Scherzinger, A. L., Norris, J. M., Haffner, S. M., Saad, M. F., Bergman, R. N. (2003) Insulin sensitivity, insulin secretion, and abdominal fat: the Insulin Resistance Atherosclerosis Study (IRAS) Family Study. *Diabetes 52*, 2490–2496.

17 Hayashi, T., Boyko, E. J., Leonetti, D. L., McNeely, M. J., Newell-Morris, L., Kahn, S. E., Fujimoto, W. Y. (2003) Visceral adiposity and the prevalence of hypertension in Japanese Americans. *Circulation 108*, 1718–1723.

18 Rattarasarn, C., Leelawattana, R., Soonthornpun, S., Setasuban, W., Thamprasit, A., Lim, A., Chayanunnu-kul, W., Thamkumpee, N., et al. (2003) Regional abdominal fat distribution in lean and obese Thai type 2 diabetic women: relationships with insulin sensitivity and cardiovascular risk factors. *Metabolism 52*, 1444–1447.

19 Kanai, H., Tokunaga, K., Fujioka, S., Yamashita, S., Kameda-Takemura, K. K., Matsuzawa, Y. (1996) Decrease in intra-abdominal visceral fat may reduce blood pressure in obese hypertensive women. *Hypertension 27*, 125–129.

20 Bacha, F., Saad, R., Gungor, N., Janosky, J., Arslanian, S. A. (2003) Obesity, regional fat distribution, and syndrome X in obese black versus white adolescents: race differential in diabetogenic and atherogenic risk factors. *J Clin Endocrinol Metab 88*, 2534–2540.

21 Pascot, A., Despres, J. P., Lemieux, I., Bergeron, J., Nadeau, A., Prud'homme, D., Tremblay, A., Lemieux, S. (2000) Contribution of visceral obesity to the deterioration of the metabolic risk profile in men with impaired glucose tolerance. *Diabetologia 43*, 1126–1135.

22 Pascot, A., Lemieux, I., Prud'homme, D., Tremblay, A., Nadeau, A., Couillard, C., Bergeron, J., Lamarche, B., et al. (2001) Reduced HDL particle size as an additional feature of the atherogenic dyslipidemia of abdominal obesity. *J Lipid Res 42*, 2007–2014.

23 Nieves, D. J., Cnop, M., Retzlaff, B., Walden, C. E., Brunzell, J. D., Knopp, R. H., Kahn, S. E. (2003) The athero-genic lipoprotein profile associated with obesity and insulin resistance is largely attributable to intra-abdominal fat. *Diabetes 52*, 172–179.

24 Suwa, M., Egashira, T., Nakano, H., Sasaki, H., Kumagai, S. (2006) Metformin increases the PGC-1{alpha} protein and oxidative enzyme activities possibly via AMPK phosphorylation in skeletal muscle *in vivo. J Appl Physiol.*

25 Minokoshi, Y., Kim, Y. B., Peroni, O. D., Fryer, L. G., Muller, C., Carling, D., Kahn, B. B. (2002) Leptin stimulates fatty-acid oxidation by activating AMP-activated protein kinase. *Nature 415*, 339–343.

26 Yamauchi, T., Kamon, J., Minokoshi, Y., Ito, Y., Waki, H., Uchida, S., Yamashita, S., Noda, M., et al. (2002) Adiponectin stimulates glucose utilization and fatty-acid oxidation by activating AMP-activated protein kinase. *Nat Med 8*, 1288–1295.

27 Tomas, E., Tsao, T. S., Saha, A. K., Murrey, H. E., Zhang Cc, C., Itani, S. I., Lodish, H. F., Ruderman, N. B. (2002) Enhanced muscle fat oxidation and glucose transport by ACRP30 globular domain: acetyl-CoA carboxylase inhibition and AMP-activated protein kinase activation. *Proc Natl Acad Sci USA 99*, 16309–16313.

28 Kelly, M., Keller, C., Avilucea, P. R., Keller, P., Luo, Z., Xiang, X., Giralt, M., Hidalgo, J., et al. (2004) AMPK activity is diminished in tissues of IL-6 knockout mice: the effect of exercise. *Biochem Biophys Res Commun 320*, 449–454.

29 Assifi, M. M., Suchankova, G., Constant, S., Prentki, M., Saha, A. K., Ruderman, N. B. (2005) AMP-activated protein kinase and coordination of hepatic fatty acid metabolism of starved/carbohydrate-refed rats. *Am J Physiol Endocrinol Metab 289*, E794–800.

30 Banerjee, R. R., Rangwala, S. M., Shapiro, J. S., Rich, A. S., Rhoades, B., Qi, Y., Wang, J., Rajala, M. W., et al. (2004) Regulation of fasted blood glucose by resistin. *Science 303*, 1195–1198.

31 Andersson, U., Filipsson, K., Abbott, C. R., Woods, A., Smith, K., Bloom, S. R., Carling, D., Small, C. J. (2004) AMP-activated protein kinase plays a role in the control of food intake. *J Biol Chem 279*, 12005–12008.

32 Kim, M. S., Park, J. Y., Namkoong, C., Jang, P. G., Ryu, J. W., Song, H. S., Yun, J. Y., Namgoong, I. S., et al. (2004) Anti-obesity effects of alpha-lipoic acid mediated by suppression of hypothalamic AMP-activated protein kinase. *Nat Med 10*, 727–733.

33 Kola, B., Hubina, E., Tucci, S. A., Kirkham, T. C., Garcia, E. A., Mitchell, S. E., Williams, L. M., Hawley, S. A., et al. (2005) Cannabinoids and ghrelin have both central and peripheral metabolic and cardiac effects via AMP-activated protein kinase. *J Biol Chem 280*, 25196–25201.

34 Minokoshi, Y., Alquier, T., Furukawa, N., Kim, Y. B., Lee, A., Xue, B., Mu, J., Foufelle, F., et al. (2004) AMP-kinase regulates food intake by responding to hormonal and nutrient signals in the hypothalamus. *Nature 428*, 569–574.

35 Luo, Z., Saha, A. K., Xiang, X., Ruderman, N. B. (2005) AMPK, the metabolic syndrome and cancer. *Trends Pharmacol Sci 26*, 69–76.

36 Martin, T. L., Alquier, T., Asakura, K., Furukawa, N., Preitner, F., Kahn, B. B. (2006) Diet-induced obesity alters AMP kinase activity in hypothalamus and skeletal muscle. *J Biol Chem 281*, 18933–18941.

37 Lee, W. J., Song, K. H., Koh, E. H., Won, J. C., Kim, H. S., Park, H. S., Kim, M. S., Kim, S. W., et al. (2005) Alpha-lipoic acid increases insulin sensitivity by activating AMPK in skeletal muscle. *Biochem Biophys Res Commun 332*, 885–891.

38 Wojtaszewski, J. F., Birk, J. B., Frosig, C., Holten, M., Pilegaard, H., Dela, F. (2005) 5′ AMP activated protein kinase expression in human skeletal muscle: effects of strength training and type 2 diabetes. *J Physiol 564*, 563–573.

39 Kurth-Kraczek, E. J., Hirshman, M. F., Goodyear, L. J., Winder, W. W. (1999) 5′ AMP-activated protein kinase activation causes GLUT4 translocation in skeletal muscle. *Diabetes 48*, 1667–1671.

40 Merrill, G. F., Kurth, E. J., Hardie, D. G., Winder, W. W. (1997) AICA riboside increases AMP-activated protein kinase, fatty acid oxidation, and glucose uptake in rat muscle. *Am J Physiol 273*, E1107–1112.

41 Mu, J., Barton, E. R., Birnbaum, M. J. (2003) Selective suppression of AMP-activated protein kinase in skeletal muscle: update on 'lazy mice'. *Biochem Soc Trans 31*, 236–241.

42 Jorgensen, S. B., Viollet, B., Andreelli, F., Frosig, C., Birk, J. B., Schjerling, P., Vaulont, S., Richter, E. A., et al. (2004) Knockout of the alpha2 but not alpha1 5′-AMP-activated protein kinase isoform abolishes 5-aminoimidazole-4-carboxamide-1-beta-4-ribofuranosidebut not contraction-induced glucose uptake in skeletal muscle. *J Biol Chem 279*, 1070–1079.

43 Barnes, B. R., Marklund, S., Steiler, T. L., Walter, M., Hjalm, G., Amarger, V., Mahlapuu, M., Leng, Y., et al. (2004) The 5′-AMP-activated protein kinase gamma3 isoform has a key role in carbohydrate and lipid metabolism in glycolytic skeletal muscle. *J Biol Chem 279*, 38441–38447.

44 Hayashi, T., Hirshman, M. F., Kurth, E. J., Winder, W. W., Goodyear, L. J. (1998) Evidence for 5′ AMP-activated protein kinase mediation of the effect of muscle contraction on glucose transport. *Diabetes 47*, 1369–1373.

45 Viollet, B., Andreelli, F., Jorgensen, S. B., Perrin, C., Geloen, A., Flamez, D., Mu, J., Lenzner, C., et al. (2003) The AMP-activated protein kinase alpha2 catalytic subunit controls whole-body insulin sensitivity. *J Clin Invest 111*, 91–98.

46 Kahn, S. E. (2003) The relative contributions of insulin resistance and beta-cell dysfunction to the pathophysiology of Type 2 diabetes. *Diabetologia 46*, 3–19.

47 (1995) U.K. prospective diabetes study 16. Overview of 6 years' therapy of type II diabetes: a progressive disease. U.K. Prospective Diabetes Study Group. *Diabetes 44*, 1249–1258.

48 Targonsky, E. D., Dai, F., Koshkin, V., Karaman, G. T., Gyulkhandanyan, A. V., Zhang, Y., Chan, C. B., Wheeler, M. B. (2006) alpha-Lipoic acid regulates AMP-activated protein kinase and inhibits insulin secretion from beta cells. *Diabetologia 49*, 1587–1598.

49 Leclerc, I., Woltersdorf, W. W., da Silva Xavier, G., Rowe, R. L., Cross, S. E., Korbutt, G. S., Rajotte, R. V., Smith, R., et al. (2004) Metformin, but not leptin, regulates AMP-activated protein kinase in pancreatic islets: impact on glucose-stimulated insulin secretion. *Am J Physiol Endocrinol Metab 286*, E1023–1031.

50 Richards, S. K., Parton, L. E., Leclerc, I., Rutter, G. A., Smith, R. M. (2005) Over-expression of AMP-activated protein kinase impairs pancreatic {beta}-cell function *in vivo*. *J Endocrinol 187*, 225–235.

51 Butler, A. E., Janson, J., Bonner-Weir, S., Ritzel, R., Rizza, R. A., Butler, P. C. (2003) Beta-cell deficit and increased beta-cell apoptosis in humans with type 2 diabetes. *Diabetes 52*, 102–110.

52 Rhodes, C. J. (2005) Type 2 diabetes-a matter of beta-cell life and death? *Science* *307*, 380–384.

53 Kaiser, N., Leibowitz, G., Nesher, R. (2003) Glucotoxicity and beta-cell failure in type 2 diabetes mellitus. *J Pediatr Endocrinol Metab 16*, 5–22.

54 Kefas, B. A., Heimberg, H., Vaulont, S., Meisse, D., Hue, L., Pipeleers, D., Van de Casteele, M. (2003) AICA-riboside induces apoptosis of pancreatic beta cells through stimulation of AMP-activated protein kinase. *Diabetologia 46*, 250–254.

55 Kefas, B. A., Cai, Y., Ling, Z., Heimberg, H., Hue, L., Pipeleers, D., Van de Casteele, M. (2003) AMP-activated protein kinase can induce apoptosis of insulin-producing MIN6 cells through stimulation of c-Jun-N-terminal kinase. *J Mol Endocrinol 30*, 151–161.

56 Andreelli, F., Foretz, M., Knauf, C., Cani, P. D., Perrin, C., Iglesias, M. A., Pillot, B., Bado, A., et al. (2006) Liver adenosine monophosphate-activated kinase-alpha2 catalytic subunit is a key target for the control of hepatic glucose production by adiponectin and leptin but not insulin. *Endocrinology 147*, 2432–2441.

57 Foretz, M., Ancellin, N., Andreelli, F., Saintillan, Y., Grondin, P., Kahn, A., Thorens, B., Vaulont, S., et al. (2005) Short-term overexpression of a constitutively active form of AMP-activated protein kinase in the liver leads to mild hypoglycemia and fatty liver. *Diabetes 54*, 1331–1339.

58 Bergeron, R., Previs, S. F., Cline, G. W., Perret, P., Russell, R. R., 3rd, Young, L. H., Shulman, G. I. (2001) Effect of 5-aminoimidazole-4-carboxamide-1-beta-D-ribofuranoside infusion on *in vivo* glucose and lipid metabolism in lean and obese Zucker rats. *Diabetes 50*, 1076–1082.

59 Guigas, B., Bertrand, L., Taleux, N., Foretz, M., Wiernsperger, N., Vertommen, D., Andreelli, F., Viollet, B., et al. (2006) 5-Aminoimidazole-4-Carboxamide-1-{beta}-D-Ribofuranoside and Metformin Inhibit Hepatic Glucose Phosphorylation by an AMP-Activated Protein Kinase-Independent Effect on Glucokinase Translocation. *Diabetes 55*, 865–874.

60 Eckel, R. H., Grundy, S. M., Zimmet, P. Z. (2005) The metabolic syndrome. *Lancet 365*, 1415–1428.

61 Corton, J. M., Gillespie, J. G., Hawley, S. A., Hardie, D. G. (1995) 5-aminoimidazole-4-carboxamide ribonucleoside. A specific method for activating AMP-activated protein kinase in intact cells? *Eur J Biochem 229*, 558–565.

62 Orci, L., Cook, W. S., Ravazzola, M., Wang, M. Y., Park, B. H., Montesano, R., Unger, R. H. (2004) Rapid transformation of white adipocytes into fat-oxidizing machines. *Proc Natl Acad Sci USA 101*, 2058–2063.

63 Sullivan, J. E., Brocklehurst, K. J., Marley, A. E., Carey, F., Carling, D., Beri, R. K. (1994) Inhibition of lipolysis and lipogenesis in isolated rat adipocytes with AICAR, a cell-permeable activator of AMP-activated protein kinase. *FEBS Lett 353*, 33–36.

64 Daval, M., Diot-Dupuy, F., Bazin, R., Hainault, I., Viollet, B., Vaulont, S., Hajduch, E., Ferre, P., et al. (2005) Anti-lipolytic action of AMP-activated protein kinase in rodent adipocytes. *J Biol Chem 280*, 25250–25257.

65 Villena, J. A., Viollet, B., Andreelli, F., Kahn, A., Vaulont, S., Sul, H. S. (2004) Induced adiposity and adipocyte hypertrophy in mice lacking the AMP-activated protein kinase-alpha2 subunit. *Diabetes 53*, 2242–2249.

66 Buhl, E. S., Jessen, N., Pold, R., Ledet, T., Flyvbjerg, A., Pedersen, S. B., Pedersen, O., Schmitz, O., et al. (2002) Long-term AICAR administration reduces metabolic disturbances and lowers blood pressure in rats displaying features of the insulin resistance syndrome. *Diabetes 51*, 2199–2206.

67 Zhou, G., Myers, R., Li, Y., Chen, Y., Shen, X., Fenyk-Melody, J., Wu, M., Ventre, J., et al. (2001) Role of AMP-activated protein kinase in mechanism of metformin action. *J Clin Invest 108*, 1167–1174.

68 Lin, H. Z., Yang, S. Q., Chuckaree, C., Kuhajda, F., Ronnet, G., Diehl, A. M. (2000) Metformin reverses fatty liver

disease in obese, leptin-deficient mice. *Nat Med 6*, 998–1003.

69 Dagon, Y., Avraham, Y., Berry, E. M. (2006) AMPK activation regulates apoptosis, adipogenesis, and lipolysis by eIF2alpha in adipocytes. *Biochem Biophys Res Commun 340*, 43–47.

70 Habinowski, S. A., Witters, L. A. (2001) The effects of AICAR on adipocyte differentiation of 3T3-L1 cells. *Biochem Biophys Res Commun 286*, 852–856.

71 Dyck, J. R., Lopaschuk, G. D. (2006) AMPK alterations in cardiac physiology and pathology: enemy or ally? *J Physiol 574*, 95–112.

72 Opie, L. H. (1976) Effects of regional ischemia on metabolism of glucose and fatty acids. Relative rates of aerobic and anaerobic energy production during myocardial infarction and comparison with effects of anoxia. *Circ Res 38*, 152–74.

73 Zarrinpashneh, E., Carjaval, K., Beauloye, C., Ginion, A., Mateo, P., Pouleur, A. C., Horman, S., Vaulont, S., et al. (2006) Role of the alpha2 isoform of AMP-activated protein kinase in the metabolic response of the heart to no-flow ischemia. *Am J Physiol Heart Circ Physiol*.

74 Russell, R. R., 3rd, Li, J., Coven, D. L., Pypaert, M., Zechner, C., Palmeri, M., Giordano, F. J., Mu, J., et al. (2004) AMP-activated protein kinase mediates ischemic glucose uptake and prevents postischemic cardiac dysfunction, apoptosis, and injury. *J Clin Invest 114*, 495–503.

75 Pischon, T., Girman, C. J., Hotamisligil, G. S., Rifai, N., Hu, F. B., Rimm, E. B. (2004) Plasma adiponectin levels and risk of myocardial infarction in men. *Jama 291*, 1730–1737.

76 Shibata, R., Sato, K., Pimentel, D. R., Takemura, Y., Kihara, S., Ohashi, K., Funahashi, T., Ouchi, N., et al. (2005) Adiponectin protects against myocardial ischemia-reperfusion injury through AMPK- and COX-2-dependent mechanisms. *Nat Med 11*, 1096–1103.

77 Lee, W. J., Lee, I. K., Kim, H. S., Kim, Y. M., Koh, E. H., Won, J. C., Han, S. M., Kim, M. S., et al. (2005) Alpha-lipoic acid prevents endothelial dysfunction in

obese rats via activation of AMP-activated protein kinase. *Arterioscler Thromb Vasc Biol 25*, 2488–2494.

78 Davis, B. J., Xie, Z., Viollet, B., Zou, M. H. (2006) Activation of the AMP-activated kinase by antidiabetes drug metformin stimulates nitric oxide synthesis *in vivo* by promoting the association of heat shock protein 90 and endothelial nitric oxide synthase. *Diabetes 55*, 496–505.

79 Majithiya, J. B., Balaraman, R. (2006) Metformin reduces blood pressure and restores endothelial function in aorta of streptozotocin-induced diabetic rats. *Life Sci 78*, 2615–2624.

80 Evans, A. M., Mustard, K. J., Wyatt, C. N., Peers, C., Dipp, M., Kumar, P., Kinnear, N. P., Hardie, D. G. (2005) Does AMP-activated protein kinase couple inhibition of mitochondrial oxidative phosphorylation by hypoxia to calcium signaling in O2-sensing cells? *J Biol Chem 280*, 41504–41511.

81 Rubin, L. J., Magliola, L., Feng, X., Jones, A. W., Hale, C. C. (2005) Metabolic activation of AMP kinase in vascular smooth muscle. *J Appl Physiol 98*, 296–306.

82 Hunt, K. J., Resendez, R. G., Williams, K., Haffner, S. M., Stern, M. P. (2004) National Cholesterol Education Program versus World Health Organization metabolic syndrome in relation to all-cause and cardiovascular mortality in the San Antonio Heart Study. *Circulation 110*, 1251–1257.

83 Lakka, H. M., Laaksonen, D. E., Lakka, T. A., Niskanen, L. K., Kumpusalo, E., Tuomilehto, J., Salonen, J. T. (2002) The metabolic syndrome and total and cardiovascular disease mortality in middle-aged men. *Jama 288*, 2709–2716.

84 Iglesias, M. A., Ye, J. M., Frangioudakis, G., Saha, A. K., Tomas, E., Ruderman, N. B., Cooney, G. J., Kraegen, E. W. (2002) AICAR administration causes an apparent enhancement of muscle and liver insulin action in insulin-resistant high-fat-fed rats. *Diabetes 51*, 2886–2894.

85 Song, X. M., Fiedler, M., Galuska, D., Ryder, J. W., Fernstrom, M., Chibalin, A. V., Wallberg-Henriksson, H., Zierath, J. R. (2002) 5-Aminoimidazole-4-

carboxamide ribonucleoside treatment improves glucose homeostasis in insulin-resistant diabetic (ob/ob) mice. *Diabetologia 45*, 56–65.

86 Fryer, L. G., Parbu-Patel, A., Carling, D. (2002) The Anti-diabetic drugs rosiglitazone and metformin stimulate AMP-activated protein kinase through distinct signaling pathways. *J Biol Chem 277*, 25226–25232.

87 Saha, A. K., Avilucea, P. R., Ye, J. M., Assifi, M. M., Kraegen, E. W., Ruderman, N. B. (2004) Pioglitazone treatment activates AMP-activated protein kinase in rat liver and adipose tissue *in vivo*. *Biochem Biophys Res Commun 314*, 580–585.

88 Thompson, P. D., Buchner, D., Pina, I. L., Balady, G. J., Williams, M. A., Marcus, B. H., Berra, K., Blair, S. N., et al. (2003) Exercise and physical activity in the prevention and treatment of atherosclerotic cardiovascular disease: a statement from the Council on Clinical Cardiology (Subcommittee on Exercise, Rehabilitation, and Prevention) and the Council on Nutrition, Physical Activity, and Metabolism (Subcommittee on Physical Activity). *Circulation 107*, 3109–3116.

89 Cook, S., Weitzman, M., Auinger, P., Nguyen, M., Dietz, W. H. (2003) Prevalence of a metabolic syndrome phenotype in adolescents: findings from the third National Health and Nutrition Examination Survey, 1988–1994. *Arch Pediatr Adolesc Med 157*, 821–827.

90 Tuomilehto, J., Lindstrom, J., Eriksson, J. G., Valle, T. T., Hamalainen, H., Ilanne-Parikka, P., Keinanen-Kiukaanniemi, S., Laakso, M., et al. (2001) Prevention of type 2 diabetes mellitus by changes in lifestyle among subjects with impaired glucose tolerance. *N Engl J Med 344*, 1343–1350.

91 Holmes, B. F., Kurth-Kraczek, E. J., Winder, W. W. (1999) Chronic activation of 5′-AMP-activated protein kinase increases GLUT4, hexokinase, and glycogen in muscle. *J Appl Physiol 87*, 1990–1995.

92 Michael, L. F., Wu, Z., Cheatham, R. B., Puigserver, P., Adelmant, G., Lehman, J. J., Kelly, D. P., Spiegelman, B. M.

(2001) Restoration of insulin-sensitive glucose transporter (GLUT4) gene expression in muscle cells by the transcriptional coactivator PGC-1. *Proc Natl Acad Sci USA 98*, 3820–3825.

93 Jorgensen, S. B., Wojtaszewski, J. F., Viollet, B., Andreelli, F., Birk, J. B., Hellsten, Y., Schjerling, P., Vaulont, S., et al. (2005) Effects of alpha-AMPK knockout on exercise-induced gene activation in mouse skeletal muscle. *Faseb J 19*, 1146–1148.

94 Holmes, B. F., Lang, D. B., Birnbaum, M. J., Mu, J., Dohm, G. L. (2004) AMP kinase is not required for the GLUT4 response to exercise and denervation in skeletal muscle. *Am J Physiol Endocrinol Metab 287*, E739–743.

95 Winder, W. W., Holmes, B. F., Rubink, D. S., Jensen, E. B., Chen, M., Holloszy, J. O. (2000) Activation of AMP-activated protein kinase increases mitochondrial enzymes in skeletal muscle. *J Appl Physiol 88*, 2219–2226.

96 Zong, H., Ren, J. M., Young, L. H., Pypaert, M., Mu, J., Birnbaum, M. J., Shulman, G. I. (2002) AMP kinase is required for mitochondrial biogenesis in skeletal muscle in response to chronic energy deprivation. *Proc Natl Acad Sci USA 99*, 15983–15987.

97 Bergeron, R., Ren, J. M., Cadman, K. S., Moore, I. K., Perret, P., Pypaert, M., Young, L. H., Semenkovich, C. F., et al. (2001) Chronic activation of AMP kinase results in NRF-1 activation and mitochondrial biogenesis. *Am J Physiol Endocrinol Metab 281*, E1340–1346.

98 Petersen, K. F., Befroy, D., Dufour, S., Dziura, J., Ariyan, C., Rothman, D. L., DiPietro, L., Cline, G. W., et al. (2003) Mitochondrial dysfunction in the elderly: possible role in insulin resistance. *Science 300*, 1140–1142.

99 Knowler, W. C., Barrett-Connor, E., Fowler, S. E., Hamman, R. F., Lachin, J. M., Walker, E. A., Nathan, D. M. (2002) Reduction in the incidence of type 2 diabetes with lifestyle intervention or metformin. *N Engl J Med 346*, 393–403.

100 Pold, R., Jensen, L. S., Jessen, N., Buhl, E. S., Schmitz, O., Flyvbjerg, A., Fujii, N., Goodyear, L. J., et al. (2005) Long-term AICAR administration and exercise prevents diabetes in ZDF rats. *Diabetes 54*, 928–934.

101 Park, H., Kaushik, V. K., Constant, S., Prentki, M., Przybytkowski, E., Ruderman, N. B., Saha, A. K. (2002) Coordinate regulation of malonyl-CoA decarboxylase, sn-glycerol-3-phosphate acyltransferase, and acetyl-CoA carboxylase by AMP-activated protein kinase in rat tissues in response to exercise. *J Biol Chem 277*, 32571–32577.

102 Bluher, M., Bullen, J. W., Jr., Lee, J. H., Kralisch, S., Fasshauer, M., Kloting, N., Niebauer, J., Schon, M. R., et al. (2006) Circulating adiponectin and expression of adiponectin receptors in human skeletal muscle: associations with metabolic parameters and insulin resistance and regulation by physical training. *J Clin Endocrinol Metab 91*, 2310–2316.

103 Musi, N., Hirshman, M. F., Nygren, J., Svanfeldt, M., Bavenholm, P., Rooyackers, O., Zhou, G., Williamson, J. M., et al. (2002) Metformin increases AMP-activated protein kinase activity in skeletal muscle of subjects with type 2 diabetes. *Diabetes 51*, 2074–2081.

104 Hojlund, K., Mustard, K. J., Staehr, P., Hardie, D. G., Beck-Nielsen, H., Richter, E. A., Wojtaszewski, J. F. (2004) AMPK activity and isoform protein expression are similar in muscle of obese subjects with and without type 2 diabetes. *Am J Physiol Endocrinol Metab 286*, E239–244.

105 Bertrand, L., Ginion, A., Beauloye, C., Hebert, A. D., Guigas, B., Hue, L., Vanoverschelde, J. L. (2006) AMPK activation restores the stimulation of glucose uptake in an *in vitro* model of insulin-resistant cardiomyocytes via the activation of protein kinase B. *Am J Physiol Heart Circ Physiol 291*, H239–250.

106 (1998) Clinical Guidelines on the Identification, Evaluation, and Treatment of Overweight and Obesity in Adults – The Evidence Report. National Institutes of Health. *Obes Res 6 Suppl 2*, 51S–209S.

107 Zimmet, P., Shaw, J., Alberti, K. G. (2003) Preventing Type 2 diabetes and the dysmetabolic syndrome in the real world: a realistic view. *Diabet Med 20*, 693–702.

108 Moule, S. K., Denton, R. M. (1998) The activation of p38 MAPK by the beta-adrenergic agonist isoproterenol in rat epididymal fat cells. *FEBS Lett 439*, 287–290.

109 Matejkova, O., Mustard, K. J., Sponarova, J., Flachs, P., Rossmeisl, M., Miksik, I., Thomason-Hughes, M., Grahame Hardie, D., et al. (2004) Possible involvement of AMP-activated protein kinase in obesity resistance induced by respiratory uncoupling in white fat. *FEBS Lett 569*, 245–248.

110 Rossmeisl, M., Barbatelli, G., Flachs, P., Brauner, P., Zingaretti, M. C., Marelli, M., Janovska, P., Horakova, M., et al. (2002) Expression of the uncoupling protein 1 from the aP2 gene promoter stimulates mitochondrial biogenesis in unilocular adipocytes *in vivo*. *Eur J Biochem 269*, 19–28.

111 Gaidhu, M. P., Fediuc, S., Ceddia, R. B. (2006) 5-Aminoimidazole-4-carboxamide-1-beta-D-ribofuranoside-induced AMP-activated protein kinase phosphorylation inhibits basal and insulin-stimulated glucose uptake, lipid synthesis, and fatty acid oxidation in isolated rat adipocytes. *J Biol Chem 281*, 25956–25964.

112 Pick, A., Clark, J., Kubstrup, C., Levisetti, M., Pugh, W., Bonner-Weir, S., Polonsky, K. S. (1998) Role of apoptosis in failure of beta-cell mass compensation for insulin resistance and beta-cell defects in the male Zucker diabetic fatty rat. *Diabetes 47*, 358–364.

113 El-Assaad, W., Buteau, J., Peyot, M. L., Nolan, C., Roduit, R., Hardy, S., Joly, E., Dbaibo, G., et al. (2003) Saturated fatty acids synergize with elevated glucose to cause pancreatic beta-cell death. *Endocrinology 144*, 4154–4163.

114 Lupi, R., Del Guerra, S., Fierabracci, V., Marselli, L., Novelli, M., Patane, G., Boggi, U., Mosca, F., et al. (2002) Lipotoxicity in human pancreatic islets and the protective effect of metformin. *Diabetes 51 Suppl 1*, S134–137.

115 Higa, M., Zhou, Y. T., Ravazzola, M., Baetens, D., Orci, L., Unger, R. H. (1999) Troglitazone prevents mitochondrial alterations, beta cell

destruction, and diabetes in obese prediabetic rats. *Proc Natl Acad Sci USA 96*, 11513–11518.

116 Palanivel, R., Sweeney, G. (2005) Regulation of fatty acid uptake and metabolism in L6 skeletal muscle cells by resistin. *FEBS Lett 579*, 5049–5054.

117 Buchanan, T. A., Xiang, A. H., Peters, R. K., Kjos, S. L., Marroquin, A., Goico, J., Ochoa, C., Tan, S., et al. (2002) Preservation of pancreatic beta-cell function and prevention of type 2 diabetes by pharmacological treatment of insulin resistance in high-risk hispanic women. *Diabetes 51*, 2796–2803.

118 Shaw, R. J., Lamia, K. A., Vasquez, D., Koo, S. H., Bardeesy, N., Depinho, R. A., Montminy, M., Cantley, L. C. (2005) The kinase LKB1 mediates glucose homeostasis in liver and therapeutic effects of metformin. *Science 310*, 1642–1646.

119 Nawrocki, A. R., Rajala, M. W., Tomas, E., Pajvani, U. B., Saha, A. K., Trumbauer, M. E., Pang, Z., Chen, A. S., et al. (2006) Mice lacking adiponectin show decreased hepatic insulin sensitivity and reduced responsiveness to peroxisome proliferator-activated receptor gamma agonists. *J Biol Chem 281*, 2654–2660.

120 Maeda, N., Takahashi, M., Funahashi, T., Kihara, S., Nishizawa, H., Kishida, K., Nagaretani, H., Matsuda, M., et al. (2001) PPARgamma ligands increase expression and plasma concentrations of adiponectin, an adipose-derived protein. *Diabetes 50*, 2094–2099.

121 Lihn, A. S., Jessen, N., Pedersen, S. B., Lund, S., Richelsen, B. (2004) AICAR stimulates adiponectin and inhibits cytokines in adipose tissue. *Biochem Biophys Res Commun 316*, 853–858.

122 Sell, H., Dietze-Schroeder, D., Eckardt, K., Eckel, J. (2006) Cytokine secretion by human adipocytes is differentially regulated by adiponectin, AICAR, and troglitazone. *Biochem Biophys Res Commun 343*, 700–706.

123 Phillips, S. A., Ciaraldi, T. P., Kong, A. P., Bandukwala, R., Aroda, V., Carter, L., Baxi, S., Mudaliar, S. R., et al. (2003) Modulation of circulating and adipose

tissue adiponectin levels by antidiabetic therapy. *Diabetes 52*, 667–674.

124 Yamauchi, T., Kamon, J., Waki, H., Imai, Y., Shimozawa, N., Hioki, K., Uchida, S., Ito, Y., et al. (2003) Globular adiponectin protected ob/ob mice from diabetes and ApoE-deficient mice from atherosclerosis. *J Biol Chem 278*, 2461–2468.

125 Combs, T. P., Pajvani, U. B., Berg, A. H., Lin, Y., Jelicks, L. A., Laplante, M., Nawrocki, A. R., Rajala, M. W., et al. (2004) A transgenic mouse with a deletion in the collagenous domain of adiponectin displays elevated circulating adiponectin and improved insulin sensitivity. *Endocrinology 145*, 367–383.

126 Yamauchi, T., Kamon, J., Waki, H., Terauchi, Y., Kubota, N., Hara, K., Mori, Y., Ide, T., et al. (2001) The fat-derived hormone adiponectin reverses insulin resistance associated with both lipoatrophy and obesity. *Nat Med 7*, 941–946.

127 Xu, A., Wang, Y., Keshaw, H., Xu, L. Y., Lam, K. S., Cooper, G. J. (2003) The fat-derived hormone adiponectin alleviates alcoholic and nonalcoholic fatty liver diseases in mice. *J Clin Invest 112*, 91–100.

128 Cool, B., Zinker, B., Chiou, W., Kifle, L., Cao, N., Perham, M., Dickinson, R., Adler, A., et al. (2006) Identification and characterization of a small molecule AMPK activator that treats key components of type 2 diabetes and the metabolic syndrome. *Cell Metab 3*, 403–416.

129 Zang, M., Xu, S., Maitland-Toolan, K. A., Zuccollo, A., Hou, X., Jiang, B., Wierzbicki, M., Verbeuren, T. J., et al. (2006) Polyphenols stimulate AMP-activated protein kinase, lower lipids, and inhibit accelerated atherosclerosis in diabetic LDL receptor-deficient mice. *Diabetes 55*, 2180–2191.

130 Baur, J. A., Pearson, K. J., Price, N. L., Jamieson, H. A., Lerin, C., Kalra, A., Prabhu, V. V., Allard, J. S., et al. (2006) Resveratrol improves health and survival of mice on a high-calorie diet.

131 Neuschwander-Tetri, B. A., Caldwell, S. H. (2003) Nonalcoholic steatohepatitis: summary of an AASLD Single Topic Conference. *Hepatology 37*, 1202–1219.

132 Carey, D. G., Cowin, G. J., Galloway, G. J., Jones, N. P., Richards, J. C., Biswas, N., Doddrell, D. M. (2002) Effect of rosiglitazone on insulin sensitivity and body composition in type 2 diabetic patients [corrected]. *Obes Res 10*, 1008–1015.

133 Murry, C. E., Jennings, R. B., Reimer, K. A. (1986) Preconditioning with ischemia: a delay of lethal cell injury in ischemic myocardium. *Circulation 74*, 1124–1136.

134 Nishino, Y., Miura, T., Miki, T., Sakamoto, J., Nakamura, Y., Ikeda, Y., Kobayashi, H., Shimamoto, K. (2004) Ischemic preconditioning activates AMPK in a PKC-dependent manner and induces GLUT4 up-regulation in the late phase of cardioprotection. *Cardiovascular research 61*, 610–619.

135 Sukhodub, A., Jovanovic, S., Du, Q., Budas, G., Clelland, A. K., Shen, M., Sakamoto, K., Tian, R., et al. (2007) AMP-activated protein kinase mediates preconditioning in cardiomyocytes by regulating activity and trafficking of sarcolemmal ATP-sensitive K(+) channels. *Journal of cellular physiology 210*, 224–236.

136 Shibata, R., Ouchi, N., Ito, M., Kihara, S., Shiojima, I., Pimentel, D. R., Kumada, M., Sato, K., et al. (2004) Adiponectin-mediated modulation of hypertrophic signals in the heart. *Nat Med 10*, 1384–1389.

137 Liao, Y., Takashima, S., Maeda, N., Ouchi, N., Komamura, K., Shimomura, I., Hori, M., Matsuzawa, Y., et al. (2005) Exacerbation of heart failure in adiponectin-deficient mice due to impaired regulation of AMPK and glucose metabolism. *Cardiovasc Res 67*, 705–713.

138 Fujioka, D., Kawabata, K., Saito, Y., Kobayashi, T., Nakamura, T., Kodama, Y., Takano, H., Obata, J. E., et al. (2006) Role of adiponectin receptors in endothelin-induced cellular hypertrophy in cultured cardiomyocytes and their expression in infarcted heart. *Am J Physiol Heart Circ Physiol 290*, H2409–2416.

139 Shibata, R., Ouchi, N., Kihara, S., Sato, K., Funahashi, T., Walsh, K. (2004) Adiponectin stimulates angiogenesis in response to tissue ischemia through stimulation of amp-activated protein kinase signaling. *J Biol Chem 279*, 28670–28674.

140 Miranda, P. J., DeFronzo, R. A., Califf, R. M., Guyton, J. R. (2005) Metabolic syndrome: evaluation of pathological and therapeutic outcomes. *American heart journal 149*, 20–32.

141 Chen, Z. P., Mitchelhill, K. I., Michell, B. J., Stapleton, D., Rodriguez-Crespo, I., Witters, L. A., Power, D. A., Ortiz de Montellano, P. R., et al. (1999) AMP-activated protein kinase phosphorylation of endothelial NO synthase. *FEBS Lett 443*, 285–289.

142 Morrow, V. A., Foufelle, F., Connell, J. M., Petrie, J. R., Gould, G. W., Salt, I. P. (2003) Direct activation of AMP-activated protein kinase stimulates nitric-oxide synthesis in human aortic endothelial cells. *J Biol Chem 278*, 31629–31639.

143 (1998) Effect of intensive blood-glucose control with metformin on complications in overweight patients with type 2 diabetes (UKPDS 34). UK Prospective Diabetes Study (UKPDS) Group. *Lancet 352*, 854–865.

144 Kubota, N., Terauchi, Y., Yamauchi, T., Kubota, T., Moroi, M., Matsui, J., Eto, K., Yamashita, T., et al. (2002) Disruption of adiponectin causes insulin resistance and neointimal formation. *J Biol Chem 277*, 25863–25866.

145 Motoshima, H., Wu, X., Mahadev, K., Goldstein, B. J. (2004) Adiponectin suppresses proliferation and superoxide generation and enhances eNOS activity in endothelial cells treated with oxidized LDL. *Biochem Biophys Res Commun 315*, 264–271.

146 Igata, M., Motoshima, H., Tsuruzoe, K., Kojima, K., Matsumura, T., Kondo, T., Taguchi, T., Nakamaru, K., et al. (2005) Adenosine monophosphate-activated protein kinase suppresses vascular smooth muscle cell proliferation through the inhibition of cell cycle progression. *Circ Res 97*, 837–844.

147 Carattino, M. D., Edinger, R. S., Grieser, H. J., Wise, R., Neumann, D.,

Schlattner, U., Johnson, J. P., Kleyman, T. R., et al. (2005) Epithelial sodium channel inhibition by AMP-activated protein kinase in oocytes and polarized renal epithelial cells. *J Biol Chem 280*, 17608–17616.

148 Fraser, S. A., Mount, P. F., Hill, R., Stapleton, D., Kemp, B. E., Power, D. A. (2003) Inhibition of the Na-K-2Cl cotransporter by novel interaction with the metabolic sensor AMP-activated protein kinase. *J Am Soc Nephrol 14*, 545A.

149 Evans, J. M., Donnelly, L. A., Emslie-Smith, A. M., Alessi, D. R., Morris, A. D. (2005) Metformin and reduced risk of cancer in diabetic patients. *Bmj 330*, 1304–1305.

18

A Systems Biology Perspective on Obesity and Type 2 Diabetes

Jolita Ciapaite, Stephan J.L. Bakker, Robert J. Heine, Klaas Krab, and Hans V. Westerhoff

Abstract

Type 2 diabetes, a disease characterized by impaired glucose homeostasis, is caused by defects in both insulin release from pancreatic β-cells and insulin action on peripheral tissues. Mitochondrial dysfunction is emerging as an important component of the cellular dysfunction that characterizes insulin-resistant states. In many cases overweight and obesity have been shown to accelerate the onset of the disease and to exacerbate it. The precise role of increased adiposity in the etiology of type 2 diabetes has not yet been established. The potential roles of various adipose tissue–derived molecules such as cytokines, hormones, and fatty acids are examples of the different individual molecular mechanisms that are being investigated intensely. Because of the high number of affected genes, enzymes, and pathways, the traditional, i.e., reductionist, molecular biology approach to systemic diseases such as obesity and type 2 diabetes may not lead to satisfactory understanding of underlying molecular mechanisms and the discovery of potential therapeutic targets. Here we take the tenet that type 2 diabetes is a biological systems disease affecting the entire glucose metabolism network of the human organism. We review the current knowledge of the molecular mechanisms underlying type 2 diabetes and discuss the usefulness of modular kinetic analysis and metabolic control analysis as tools for understanding complex diseases.

Abbreviations

DAG: diacylglycerol; GLUT: glucose transporter; LCAC: long-chain acyl-CoA esters; MCA: metabolic control analysis; NEFA: non-esterified fatty acids; PGC1α: peroxisome proliferator–activated receptor-γ co-activator 1α.

Molecular System Bioenergetics: Energy for Life. Edited by Valdur Saks
Copyright © 2007 WILEY-VCH Verlag GmbH & Co. KGaA, Weinheim
ISBN: 978-3-527-31787-5

18.1
Introduction

Diabetes mellitus is a metabolic disease described by defects in the secretion and action of insulin, a hormone that not only ensures normal glucose metabolism but also regulates fat and protein metabolism. This metabolic disease, characterized by hyperglycemia and long-term complications affecting the eyes, kidneys, nervous system, and large vessels, now affects up to 5–10% of the population in some countries. The number of adults with diabetes worldwide is expected to rise from 171 million in 2000 to 366 million in the year 2030 [1]. With the increasing incidence and prevalence of diabetes, especially in younger individuals, the economical burden to healthcare systems will be huge [2, 3].

Type 2 diabetes is considered to be a multifactorial, heterogeneous group of disorders with insulin resistance and β-cell dysfunction as the prime underlying disorders. Genetic and environmental factors determine one's susceptibility to diabetes. Major risk factors include age, obesity (especially visceral adiposity), and physical inactivity. Although type 2 diabetes is associated with a strong genetic predisposition (reviewed in [4]), it is the interaction with different environmental factors that is thought to determine the occurrence and phenotype of the disease.

Diabetes research is following various lines of investigation, including the genetics of diabetes and the role of adipose tissue, brain, and alterations in mitochondrial and β-cell function (see the special issue of *Science*, 21 January, Vol. 307, 2005). Most approaches to studying type 2 diabetes have been reductionist, i.e., researchers focused on one component of a biological system at a time (e.g., one gene, protein, or metabolic pathway). Although this approach has yielded a large amount of knowledge about the involvement of individual components, the studies are often performed in separation from the context, and it is difficult to reconstruct how different factors are connected at a system level (e.g., in the cell, tissue, organ, organ system, or whole organism). The situation is further complicated by the fact that many components of a system are affected and that the sequence of interactions between the observed phenomena is not always obvious. Taken together, this indicates that new, system-level approaches are needed for the elucidation of the etiology and effective treatment of the disease.

Here we review the current knowledge of the molecular mechanisms of type 2 diabetes and argue that modular kinetic analysis and metabolic control analysis may be helpful in understanding the complexity underlying this disease.

18.2
Glucose Homeostasis

Human blood glucose levels are maintained continuously at values of around 5 mM despite significant fluctuations in the amounts of incoming glucose. This homeostatic effect is the result of efficient communication between glucose-sensing cells (pancreatic α-cells [low glucose] and β-cells [high glucose]) and glucose-consuming (muscle, adipose tissue) and glucose-producing cells (liver) via two hormones, glucagon and insulin.

The pathway leading to glucose-stimulated insulin secretion in β-cells is now well characterized. Glucose is taken up into the cell by a low-affinity ($K_m = 15$– 20 mM) facilitated-diffusion glucose transporter (GLUT2) [5]. In this way the concentration of incoming glucose is proportional to the concentration of glucose in the blood. This allows β-cells to monitor blood glucose levels directly, and thereby regulate insulin secretion. Within a β-cell glucose undergoes glycolysis to form pyruvate, which is transported into the mitochondria where it enters the tricarboxylic acid cycle. Oxidation of pyruvate yields NADH and QH_2, and these provide electrons to the mitochondrial electron transfer chain upon their oxidation. The stimulation of electron transfer chain activity sequentially leads to an increase in mitochondrial membrane potential, increased production of ATP in the mitochondrial matrix, increased cytosolic ATP:ADP ratio, closure of ATP-sensitive K^+ channels in the plasma membrane, plasma membrane depolarization, opening of voltage-operated Ca^{2+} channels, Ca^{2+} influx, a rise in the cytoplasmic free Ca^{2+} concentration ($[Ca^{2+}]_i$), and activation of the exocytotic machinery (reviewed in [6]).

Liver is also an important organ in body glucose homeostasis. Hepatocytes express the same type of low-affinity glucose transporter (GLUT2) as β-cells, facilitating rapid glucose uptake when glucose is present in abundance. Depending on the physiological conditions, the liver deposits excess blood glucose in the form of glycogen (in the fed state) or releases glucose either through increased glycogenolysis or by gluconeogenesis (in the fasting state). The response of the liver to changes in blood glucose levels is regulated by insulin and glucagon, with the former stimulating glucose uptake and inhibiting hepatic glucose production and the latter exerting the opposite effects.

Skeletal muscle and adipose tissues, the main consumers of blood glucose, express an insulin-regulated glucose transporter, GLUT4 ($K_m = \sim5$ mM). In contrast to the other GLUT isoforms, which are primarily localized in the plasma membrane, GLUT4 transporter proteins remain in the cytosol under basal conditions, sequestered in specialized storage vesicles. The rise in blood glucose stimulates insulin secretion from β-cells. Upon binding to the insulin receptor, insulin activates several phosphorylation–dephosphorylation cascades. This ultimately leads to the mobilization of GLUT4 to the plasma membrane by promoting the fusion of vesicles containing the transporters with that membrane [7]. However, insulin is not required for glucose uptake in working skeletal muscle because GLUT4 is also mobilized by exercise (reviewed in [8]).

18.3
Obesity in Relation to Glucose Homeostasis

In genetically predisposed individuals, the occurrence of type 2 diabetes is associated with obesity [9], especially visceral adiposity [10]. In obesity, adipocytes exhibit an altered cytokine secretion profile with a tendency to secrete more of the inflammatory cytokines interleukin-6 and tumor necrosis factor-α [11, 12], which have been implicated in insulin resistance. Furthermore, adult-onset obesity in humans results in an increase in adipocyte size rather than cell number [13]. Larger adipocytes have been shown to exhibit a decreased sensitivity to insulin [14] and to have higher rates of both triacylglycerol synthesis and lipolysis, and therefore to have higher transmembrane fatty acid flux [15]. Although plasma non-esterified fatty acid (NEFA) levels can vary broadly even under physiological conditions (<0.5 mM on average, 10–20 μM under hyperinsulinemic conditions, and up to 3 mM under conditions of fasting or adrenergic stimulation) [16, 17], this variation is effectively regulated by the exquisite sensitivity of adipose tissue lipolysis to insulin [16] and catecholamines [18]. Ultimately, reduced sensitivity of adipose tissue to the antilipolytic effect of insulin results in increased plasma NEFA levels [19] and fluxes [20].

The half-life of circulating NEFA is only 3–4 min [21] due to fast tissue uptake. The process of cellular fatty acid uptake involves passive diffusion and protein-mediated transport (reviewed in [22]). Intracellular fatty acids are esterified to form acyl-CoA esters, after which these can be transported into mitochondria where they could undergo β-oxidation, providing energy for ATP synthesis.

In obesity, the increased availability of circulating NEFA leads to the buildup of large intracellular triacylglycerol stores in non-adipose tissues (muscle, liver, pancreas) [23–25] as well as to an increase in the concentrations of intermediates of lipid metabolism (acyl-CoA esters, diacylglycerol, ceramides), causing activation of signal transduction pathways that lead to impaired insulin secretion [26–29] and action [30–34].

18.4
Fatty Acid–induced Insulin Resistance

One of the main fundamental abnormalities determining disruption of glucose homeostasis is impaired insulin action in peripheral tissues (skeletal muscle, adipose tissue, and liver), a condition referred to as insulin resistance. Skeletal muscle is important in determining systemic insulin sensitivity, because under insulin-stimulated conditions it is a major consumer of blood glucose [35]. Under physiological conditions, skeletal muscle is able to switch from carbohydrate oxidation in the fed state (insulin stimulated) to increased fat oxidation in the fasting state. Under pathological conditions characteristic to insulin-resistant states, this metabolic flexibility becomes increasingly impaired [36].

Initially, based on the concept proposed by Randle and referred to as the "glucose fatty acid cycle," it was suggested that stimulation of β-oxidation during periods of lipid oversupply could underlie impaired utilization of carbohydrate as a fuel due to inhibition of pyruvate dehydrogenase and phosphofructokinase by increased concentrations of acetyl-CoA and citrate, respectively (reviewed in [37]). The inhibition of these enzymes is expected to result in reduced glycolytic flux

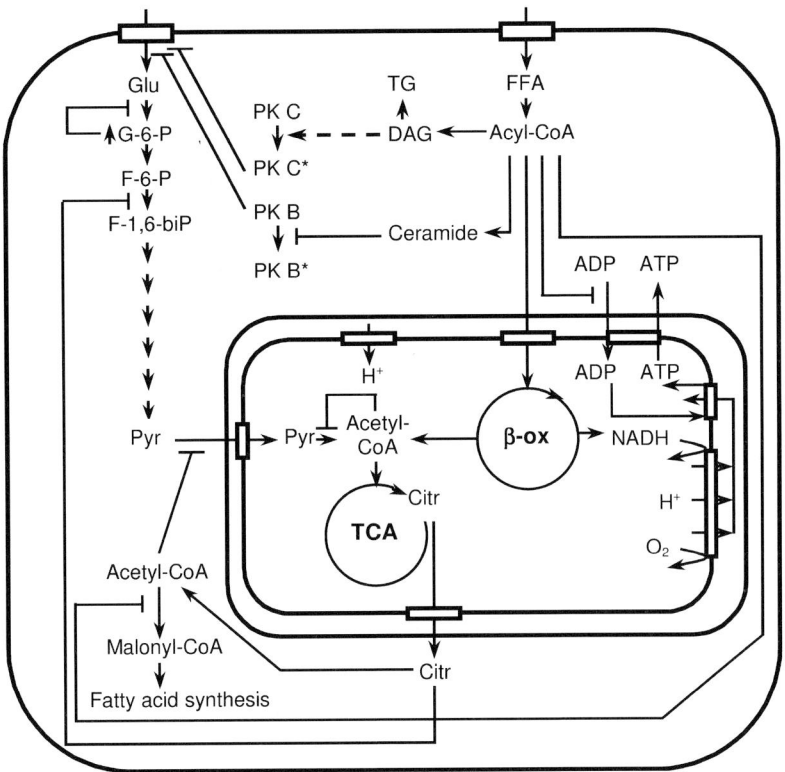

Fig. 18.1 Fatty acid–induced impairment of glucose metabolism in cells that have insulin-dependent glucose uptake. Oversupply of free fatty acids and the resulting stimulation of β-oxidation can cause accumulation of acetyl-CoA and citrate and lead to inhibition of pyruvate dehydrogenase and phospho-fructokinase, respectively, in turn causing accumulation of glucose-6-phosphate and inhibition of hexokinase and thus a decreased uptake of glucose (Randle cycle). Meanwhile, the accumulation of fatty acyl-CoA esters in the cytosol can lead to increased production of diacylglycerol and ceramide, causing stimulation of protein kinase C and inhibition of protein kinase B, respectively, and thus impairing insulin signaling and glucose uptake. Furthermore, fatty acyl-CoA esters can inhibit fatty acid synthesis and adenine nucleotide transport across the inner mitochondrial membrane. Dashed line: stimulation; Glu: glucose; G-6-P: glucose-6-phosphate; F-6-P: fructose-6-phosphate; F-1,6-biP: fructose-1,6-bisphosphate; Pyr: pyruvate; Citr: citrate; DAG: diacylglycerols; TG: triacylglycerols; FFA: free fatty acids; TCA: tricarboxylic acid cycle; β-ox: β-oxidation; PKC: protein kinase C; PKB: protein kinase B; PKC* and PKB*: active forms of protein kinases.

and accumulation of glucose 6-phosphate, which in turn can inhibit hexokinase and thereby decrease glucose uptake (Fig. 18.1).

Accumulating evidence indicates that this mechanism is likely to be just one of a number of mechanisms by which fatty acids influence insulin action and glucose metabolism. An important factor in insulin resistance appears to be the accumulation of long-chain acyl-CoA esters (LCAC) and diacylglycerol (DAG). DAG causes activation of protein kinase C, which in turn reduces the activity of insulin receptor substrate-1–associated phosphatidylinositol-3-kinase, an important step in the insulin signaling cascade [30, 31]. Consequently, the effect of insulin should be decreased. The accumulation of ceramides is emerging as another important cellular event that might impair insulin signaling by inhibiting the activation of protein kinase B (also known as Akt), a serine/threonine kinase implicated in the stimulation of glycogen synthesis and insulin-stimulated glucose transport (reviewed in [38]).

Impaired mitochondrial function is emerging as another important factor in insulin resistance, obesity, and type 2 diabetes [39–41]. However, it is not clear whether this mitochondrial dysfunction is the result of an inherent defect, a loss of mitochondria, a functional disturbance resulting from impaired glucose and lipid metabolism, or possibly a combination of these factors. It has been shown that obese individuals as well as individuals with type 2 diabetes have less skeletal muscle mitochondria, which also have reduced electron transfer chain activity [41]. Furthermore, a decreased expression of peroxisome proliferator–activated receptor-γ co-activator 1α (PGC1α), which also has been implicated in mitochondrial biogenesis [42], has been suggested to be responsible for the downregulation of PGC1α-dependent metabolic and mitochondrial genes observed in skeletal muscle of type 2 diabetes patients [43, 44]. Next to these abnormalities, it was demonstrated recently that β-oxidation is inhibited by the accumulation of malonyl-CoA in skeletal muscle of obese and type 2 diabetes patients, resulting in increased lipogenesis and the accumulation of LCAC and triacylglycerols in muscle [45]. It cannot be excluded that increased concentrations of intermediates of lipid metabolism also may exert acute effects on mitochondrial function (e.g., inhibition of mitochondrial adenine nucleotide translocator by LCAC [46]), leading to decreased substrate oxidation and phosphorylation rates.

18.5
Fatty Acid–induced β-cell Dysfunction

The second fundamental abnormality causing disruption of glucose homeostasis in type 2 diabetes is impairment of insulin secretion from pancreatic β-cells. Under physiological conditions the endocrine pancreas is able to adapt its β-cell mass depending on how much insulin is needed to assure optimal control of glucose homeostasis. A balance between β-cell growth (β-cell replication and neogenesis) and β-cell death (mainly by apoptosis) determines β-cell mass in adult mammals, while the disruption of this balance may lead to rapid and marked changes in β-cell mass (reviewed in [47]). Under the conditions of obesity and insulin resistance, this ability to adjust β-cell mass becomes increasingly impaired. Initially, β-cells have the capacity to respond to a glucose stimulus so that overproduction of insulin compensates for any insulin resistance–caused increase in blood glucose level. In this manner, higher functional demands are met by enhanced

β-cell proliferation, resulting in increased β-cell mass [48]. With the onset of type 2 diabetes in parallel to increasingly evident secretory defects, β-cells lose their proliferative potential and undergo apoptosis, leading to decreased β-cell mass and resulting in the release of inadequate amounts of insulin.

The process of fuel-stimulated (glucose in the fed state, fatty acids in the fasting state) insulin secretion from β-cells is well established, with mitochondrial metabolism playing a central role (see above, also [6]). This is further supported by the finding that β-cells depleted of mitochondrial DNA have impaired glucose-stimulated insulin secretion [49, 50]. The availability of a particular substrate (carbohydrate versus fat) determines which process will provide reducing equivalents for ATP synthesis: when glucose is abundant, glycolysis is stimulated and β-oxidation is inhibited by the accumulation of malonyl-CoA, a substance that inhibits carnitine palmitoyltransferase-1 and thus fatty acid transport into mitochondria. Conversely, increased availability of fatty acids may favor their oxidation, causing inhibition of glycolysis due to accumulation of acetyl-CoA and citrate (glucose–fatty acid cycle) [37].

Acute stimulation with fatty acids induces insulin secretion by increasing intracellular free Ca^{2+} concentration [51] and potentiates glucose-stimulated insulin secretion by increasing LCAC concentrations that may directly stimulate the exocytotic machinery or exert their effect indirectly via modulation of protein kinase C activity [52]. A chronic increase in fatty acid concentrations in the presence of high concentrations of glucose enhances basal insulin secretion possibly by the same mechanism but diminishes glucose-stimulated insulin release, presumably due to the inhibition of glycolysis [53]. Besides inhibiting glucose-stimulated insulin secretion, fatty acids inhibit insulin gene expression [54] and promote cell death by apoptosis [55–59]. Deleterious effects of fatty acids on β-cell function involve accumulation of intracellular triglycerides [56] and generation of lipid-derived cytosolic signals that adversely affect β-cell function (mainly ceramide) [57–59] and, potentially, oxidative stress, in the form of either reactive oxygen species [60, 61] or reactive nitrogen species [60]. A fatty acid–induced expression of uncoupling proteins, which leads to lower cytosolic ATP:ADP ratios [62], may be an additional mechanism contributing to impaired insulin release. Moreover, free and esterified fatty acids may interfere with mitochondrial oxidative phosphorylation due to a direct effect shown to occur *in vitro* [46, 63], resulting in lower cytosolic ATP:ADP ratios and thus the impairment of insulin release.

18.6
Type 2 Diabetes from the Systems Biology Point of View

Type 2 diabetes is a multifactorial disease in which insulin synthesis, release, and action are impaired at various levels and through a number of molecular mechanisms. Dysfunction at the cellular level leads to functional impairment at the level of virtually all tissues and organs of the human body. Complex interactions involving feedback regulation among all involved cell types (tissues) ensure that

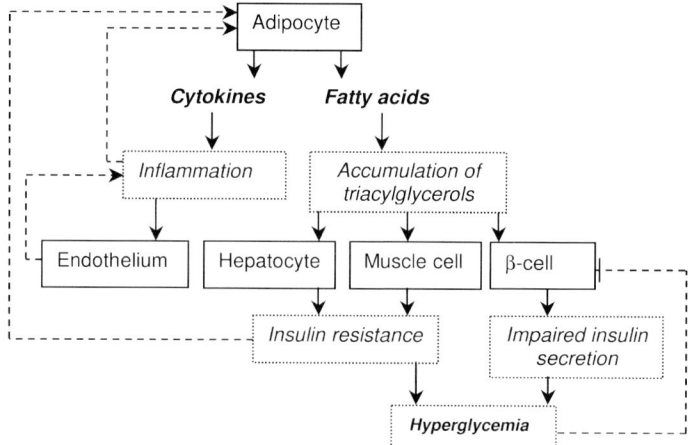

Fig. 18.2 Interactions among different cell types that underlie impaired glucose homeostasis in obesity. Dashed lines: routes of stimuli amplification.

impaired glucose homeostasis can persist for long periods of time (years) without becoming manifest (Fig. 18.2) and finally reach a certain point of "no return," which leads to manifestation of the disease. Because the disease reflects changes in many components in the network that regulates glucose homeostasis rather than just a single change in a single molecular mechanism, type 2 diabetes is best called a biological systems disease (similar to cancer) [64].

One can approach diseases caused by the failure of a single molecular function carried out by a single molecule by trying to eliminate the faulty molecule or adjust its properties. Often, however, identification of the faulty molecule is not simple because malfunctioning of that faulty molecule will permeate through the network and affect many physiological functions at the same time. This makes it difficult to identify the molecular origin of the disease on the basis of complex pathology. Moreover, other factors in the network will affect the pathology and may even appear to be additional causes thereof. This means that molecular diseases in biological systems are already biological systems diseases and therefore that their origin, even if simple, may be hard to identify. In addition, their treatment may be directed at the single molecular origin as well as at other steps in the network, which will alleviate the effects of the molecular malfunction: there will be many therapies for the disease, all of which will help a little, but perhaps none of which (except one, which may be inaccessible) will help completely.

How one should approach biological systems diseases is a major issue. Because the same malfunctioning of a network can be caused by many different molecular malfunctions, biological systems diseases can be caused by many different molecular defects (e.g., genetic or environmental). For this reason the traditional, i.e., fully reductionist, molecular biology approaches may not lead to satisfactory understanding of the underlying molecular mechanisms and of potential molecular therapies for the disease. The potential usefulness of systems approaches for understanding complex diseases such as obesity and type 2 diabetes has been pointed out before [65]. Systems biology strives to understand the functioning of complex biological systems [66, 67]. In contrast to traditional molecular biology – where a system is broken down to its components (e.g., genes, proteins) and then analyzed component by component in the hope of understanding the functioning of a system from understanding the functioning of its parts (a bottom-up approach) – systems biology focuses on the (regulatory) interactions between the components when they are connected into a system as a whole [68].

To achieve system-level understanding, two different approaches are usually taken: (1) mathematical modeling, where behavior of a system is described in terms of mathematics, and (2) experimentation using high-throughput methodologies (e.g., microarrays and proteomics), where changes in expression of genes and protein levels in a system in response to a perturbation are quantified. Both approaches have their strong and weak points: while a successful application of the former approach can be of great predictive value, it requires detailed preliminary information about the system of interest. The latter approach yields detailed quantitative information about the components of a system (e.g., DNA, mRNA, proteins) but fails to provide information on (regulatory) interactions among them.

Metabolic control analysis is one tool used in systems biology to quantitatively assess the contribution of individual system components (i.e., enzymes) to system behavior, and it does not require information about the kinetics of system components in advance. Metabolic control analysis allows the identification of system components that are crucial in the control of pathway flux or metabolite concentration and thus can be helpful for both understanding mechanisms underlying complex diseases and identification of potential drug targets [69, 70].

18.7
Metabolic Control Analysis

Metabolic control analysis (MCA), which was simultaneously developed by Kacser and Burns and by Heinrich and Rapoport in the early 1970s, allows a quantitative evaluation of the control of system variables (e.g., fluxes and metabolite concentrations) by enzymes in complex metabolic networks at steady state [71, 72] (the current terminology of MCA is given in [73]). With the help of MCA, system properties (control coefficients, see below) of an intact metabolic pathway can be explained in terms of the kinetic (i.e., local) properties of individual pathway constituents (elasticity coefficients). Since its introduction, some major developments have been achieved in the theory of MCA (e.g., [74–78], reviewed in [79]) that allow the control structure of metabolic pathways of any complexity to be determined. MCA was successfully applied experimentally to assess the control of different cellular processes in a number of organisms: dominance in *Neurospora crassa* [80], oxidative phosphorylation in isolated rat liver mitochondria and liver cells [81–88], oxidative phosphorylation in isolated potato tuber mitochondria [89], citrulline synthesis in isolated rat liver mitochondria [90], rat hepatocyte metabolism [91], threonine synthesis [92] and DNA supercoiling [93] in *Escherichia coli*, glycolysis in *Trypanosoma brucei* [94], insulin-stimulated glucose disposal in rat skeletal muscle [95], and many others.

The early theory is applicable to any metabolic pathway or network if it is at steady state, meaning that the production rate of each intermediate equals its consumption rate, the concentrations of all intermediates remaining constant in time [71]. Subsequent extensions included time-dependent phenomena, i.e., oscillations [96] and transients [97, 98].

18.7.1
Definitions of Metabolic Control Analysis

MCA focuses on the quantitative assessment of metabolic control within the system. Two types of control are distinguished: global and local. Control coefficients quantify global control; they are a relative measure of how an infinitesimal perturbation of a system parameter (e.g., enzyme activity) affects a system variable (e.g., flux J_k or intermediate concentration X_m) [71, 72]. When the system variable is a flux, the *flux control coefficient* is defined as the fractional change in steady-state system flux J_k in response to an infinitesimal change in the activity v of a step i directly caused by an external agent p_i:

$$ C_i^{J_k} = \left(\frac{dJ_k}{J_k} \middle/ \frac{\partial v_i}{v_i} \right)_{ss} = \left(\frac{d \ln J_k}{\partial \ln v_i} \right)_{ss} = \left(\frac{d \ln J_k}{dp_i} \right)_{ss} \middle/ \frac{\partial \ln v_i}{\partial p_i} \tag{1} $$

where the numerator in the final expression refers to the steady-state change in flux caused by a change in a parameter that affects only step i, and the denominator refers to the change in that process itself caused by that same parameter change, if all other factors around the process remained constant. The subscript *ss* refers to this steady-state condition and is often omitted, as are the parentheses. The enzyme activity is defined in terms of a proportional equal modulation of forward and reverse V_{max}.

When the system variable is a concentration of an intermediate X_m (or a ratio of concentrations), the *concentration control coefficient* is defined in the analogous manner as:

$$ C_i^{X_m} = \frac{\partial X_m}{X_m} \middle/ \frac{\partial v_i}{v_i} = \frac{\partial \ln X_m}{\partial \ln v_i} \tag{2} $$

The concentration control coefficient is also equal to the dependence of the chemical potential (μ) of the substance X_m on the modulation of the rate, normalized by the gas constant (R) times the absolute temperature (T) [99]:

$$C_i^{X_m} \equiv \frac{\partial \ln X_m}{\partial \ln v_i} = \frac{\partial \mu_{X_m}}{RT \partial \ln v_i} \tag{3}$$

Similar to the control coefficient, one can define the response of a system variable to change in an external parameter, e.g., concentration of an external effector. Thus, the *response coefficient* is defined as a fractional change in a system variable a (flux J_k or intermediate X_m) in response to an infinitesimal fractional change in the concentration of an external effector q [74]:

$$R_q^a = \frac{\partial a}{a} \Big/ \frac{\partial q}{q} = \frac{\partial \ln a}{\partial \ln q} \tag{4}$$

The *elasticity coefficient* quantifies local control and is a measure of how the rate of a reaction will respond to a change in concentration of an intermediate (e.g., the concentration of a substrate or product, also $\Delta\psi$, $\Delta\bar{\mu}_{H^+}$, concentration ratios) [71, 72, 99]. It is defined in terms of the fractional change in rate v through step i caused by a very small fractional change in the concentration of intermediate X_m:

$$\varepsilon_{X_m}^i = \left(\frac{\partial v_i}{v_i} \Big/ \frac{\partial X_m}{X_m}\right)_{\text{intermediates constant}} = \left(\frac{\partial \ln v_i}{\partial \ln X_m}\right)_{\text{intermediates constant}} \tag{5}$$

Here, the parentheses again refer to a partial differentiation condition, where the rate v is now considered to be a function of the concentrations of all metabolites that affect the enzyme, without the condition of a systemic steady state. All metabolites except the one that is varied explicitly are held constant in the differentiation.

The elasticity coefficient of step i to effector q is defined as the fractional change in rate v_i through that process caused by a very small fractional change in the concentration of an external effector q [74]:

$$\varepsilon_q^i = \frac{\partial v_i}{v_i} \Big/ \frac{\partial q}{q} = \frac{\partial \ln v_i}{\partial \ln q} \tag{6}$$

When an external effector acts on several processes in a network, the overall response coefficient of a system variable is determined by the control pattern of a network and the elasticity coefficient of each step to the external effector [74]:

$$R_q^a = \sum_{\text{all } i} C_i^a \cdot \varepsilon_q^i \tag{7}$$

Thus, to bring about significant changes in system fluxes, concentrations, or potentials, the external effector must act on a step that exerts appreciable control over these system variables and has the highest elasticity coefficient to the external effector.

The *co-response coefficients* quantify the ratio of responses of intermediate X_m and flux J_k after perturbation of any step i in the system [78]:

$$^iO_{J_k}^{X_m} = \frac{\partial \ln X_m}{\partial \ln v_i} \bigg/ \frac{\partial \ln J_k}{\partial \ln v_i} = \frac{\partial \ln X_m}{\partial \ln J_k} \tag{8}$$

18.7.2
Theorems of Metabolic Control Analysis

Summation theorems apply to both flux and concentration control coefficients. It has been shown that the sum of all flux control coefficients in the total network is 1 [71, 72]:

$$\sum_{i=1} C_i^{J_k} = 1 \tag{9}$$

This theorem shows that there need not be a rate-limiting enzyme; the rate limitation is a property that may be shared by the processes in a network. However, there can be a rate-limiting step in a special case where one of the enzymes has a control coefficient that is equal to 1 while the others have control coefficients that are equal to 0. Furthermore, the summation theorem implies that depending on the variation in the external conditions, the distribution of control within a network can change, i.e., if the flux control coefficient of one step decreases, the coefficient(s) of the other step(s) in a network must simultaneously increase. Thus, the summation theorem demonstrates that flux (also concentration, see below) control is a system property rather than a property of an isolated enzyme.

The sum of all concentration-control coefficients in a network is 0 [71, 72]:

$$\sum_{i=1} C_i^{X_m} = 0 \tag{10}$$

The sum of all (chemical or electrochemical) potential control coefficients in a network is also 0 [99].

The summation theorem for concentration control coefficients indicates that the control of the concentration of an intermediate is shared among the processes in a network in such a way that there cannot be a single step that determines a concentration. Instead, there must be processes that exert positive control and processes that exert negative control, and the sum of all positive controls and the sum of all negative controls in a network are of the same absolute magnitude. Thus, both positive and negative control are equally important in determining the concentration of an intermediate.

The system properties are determined by the local properties (i.e., elasticities), and this relationship is expressed by connectivity theorems. The connectivity for flux control coefficients reads [71]:

$$\sum_{i=1} C_i^{J_k} \varepsilon_{X_m}^i = 0 \qquad (11)$$

The connectivity for concentration control coefficients is [75, 76]:

$$\sum_{i=1} C_i^{X_m} \varepsilon_{X_n}^i = -\delta_{mn} \qquad (12)$$

where $\delta_{mn} = 1$ if $m = n$, and 0 otherwise.

The connectivity of (chemical or electrochemical) potential control coefficients is [99]:

$$\sum_{i=1} C_i^{\Delta G_{X_m}} \varepsilon_{\Delta G_{X_m}}^i = -\delta_{mn} \qquad (13)$$

It follows from the connectivity theorems that the lower the elasticity coefficient (i.e., sensitivity) of a step is, the stronger the flux, concentration, or potential controlled by that step is and *vice versa*.

If flux J branches into two fluxes J_1 and J_2, branching theorems apply at the branching point [76, 100]:

$$\frac{\sum_{branch\ 1} C_i^J}{J_1} + \frac{\sum_{branch\ 2} C_i^J}{J_2} = 0 \quad (J \text{ outside of branches}) \qquad (14)$$

$$\frac{\sum_{branch\ 1} C_i^{J_k}}{J_1} + \frac{\sum_{branch\ 2} C_i^{J_k}}{J_2} = \frac{1}{J_k} \quad (J_k \text{ inside a branch}) \qquad (15)$$

$$\frac{\sum_{branch\ 1} C_i^{X_m}}{J_1} + \frac{\sum_{branch\ 2} C_i^{X_m}}{J_2} = 0 \qquad (16)$$

The summation, connectivity, and branching theorems allow one to calculate all control coefficients in a network if the system fluxes at steady state and the elasticity coefficients of all constituents of a network are known [76, 77, 100].

18.7.3
Modular Metabolic Control Analysis

Conventional MCA considers individual enzymes. The control coefficient is determined from fractional change in a system flux or concentration of metabolite in

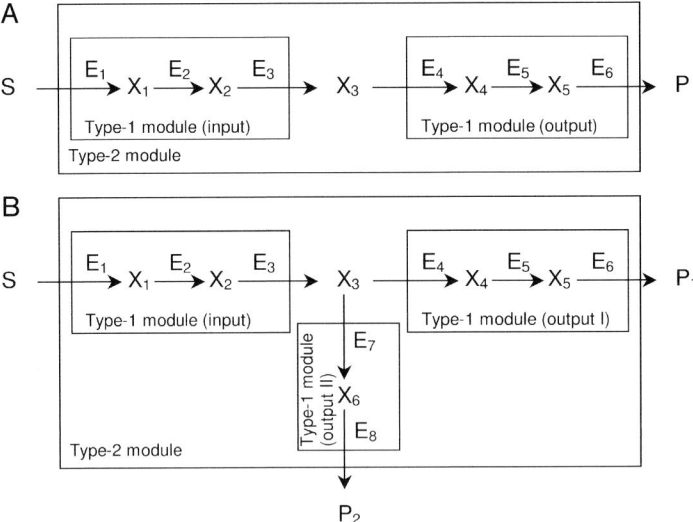

Fig. 18.3 Dissection of a pathway into modules. (A) Dissection of a linear pathway into two "black box" modules. Reaction intermediate X_3 allosterically inhibits the reaction catalyzed by enzyme E_1. (B) Dissection of a branched pathway into three "black box" modules. S: substrate; P: product; E_n: enzymes; X_m: reaction intermediates.

response to a fractional change in the activity of an enzyme of interest, which is achieved by genetic manipulations [71], by titrations with purified enzyme [101], or by use of a specific inhibitor [81–84]. However, if one is interested in resolving the control structure of a larger system with many enzymes, fluxes, and metabolites, such a "bottom-up" approach is not always feasible because it is not always possible to measure all enzymes, fluxes, and metabolites. A modular (or top-down) approach to MCA was developed to facilitate the assessment of the control structure in complex systems [99, 102–105] (reviewed in [106]). In this type of analysis, one considers a system of interest as a whole, where it is conceptually subdivided into a small number of functional modules interacting through a small number of connecting intermediates (Fig. 18.3). A module can contain cellular compartments (i.e., cytosol and mitochondria), enzymes, transporters, non-enzymatic processes, metabolites, feedback loops, and allosteric interactions and in the further analysis is treated as a single enzyme. It has been shown that modular MCA could be applied to a whole human body by treating each organ as a separate module that consumes and produces certain intermediates, although complications that can arise with such an approach (e.g., fulfillment of the steady-state condition) were indicated [107]. There are two types of modules: (1) type 1, or "black box" modules, where one is concerned only with the reactions and metabolites that link the modules, not with the internal reactions and metabolites; and (2) type 2 modules, where internal reactions and metabolites also are observed [103]. When the modules are assigned, one has to take care that (1) there are no conservation relations linking metabolites within different modules (i.e., ATP and ADP cannot belong to different modules) and (2) internal intermediates of one module do not act as effectors in another module (i.e., the corresponding elasticities are zero) [103, 106].

It has been demonstrated that the summation theorem (see above) also holds for the modular approach [103, 104]. The overall control coefficients of the modules (comprised of control coefficients of the module constituents) can be derived from the overall elasticity coefficients of the modules for the connecting intermediates. This way, the control exerted by a group of enzymes rather than one enzyme is quantified. If a more detailed analysis of the control structure is needed, the modules can be further decomposed and the control exerted by a smaller group of enzymes can be quantified.

If modules interact via one connecting intermediate, elasticities of the modules towards the connecting intermediate can be determined experimentally using titrations with module-specific modulators (i.e., inhibitors or activators). Consider a simple linear pathway that is subdivided into two modules (Fig. 18.3A). To obtain kinetics of the "input" module, one should titrate with output-specific modulator and measure the flux and concentration of an intermediate X_3. A plot of the flux against the concentration of X_3 then serves as a kinetic characterization of the input module (Fig. 18.4A). Kinetic characterization of the "output" module is obtained similarly by titrating with the input-specific modulator (Fig. 18.4B). The elasticity coefficients are then obtained from the slopes of kinetic curves at steady state. The assessment of module kinetics also in the presence of an external effector can yield information about the sites of action of that effector; this technique has been extensively used to study the effects of various agents on mitochondrial oxidative phosphorylation [108–111].

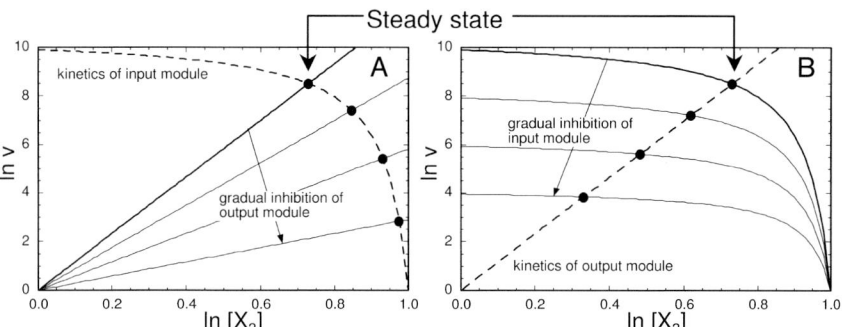

Fig. 18.4 Kinetic characterization of the modules. The linear pathway depicted in Fig. 18.3A is considered. (A) Kinetics of the input module determined by titration of the output module with specific inhibitor. (B) Kinetics of the output module determined by titration of the input module with specific inhibitor.

However, this procedure for determining elasticities cannot be used for complex pathways with more fluxes and connecting intermediates, because the obtained relationship of the flux and the level of intermediate is not unique (unless there is a way to clamp other fluxes and levels of intermediates at a constant value while measuring the dependence of a particular flux on the level of a particular intermediate [105]). In such a situation, the elasticities can be obtained by using a multiple modulation method [74, 112, 113] or a co-response analysis [78, 114].

Recently we demonstrated how modular MCA together with modular kinetic analysis can be used for testing a system-level hypothesis on a potential mechanism underlying cellular dysfunction in obesity and type 2 diabetes [115, 116]. The combination of these two methods allows one to determine whether an agent that causes a number of pathological changes acts on the network through a single mechanism or through multiple mechanisms. It also enables one to identify the mechanism. In this way we were able to show that the effect on a single molecular target (i.e., mitochondrial adenine nucleotide translocator) may result in a multitude of secondary effects leading to impaired mitochondrial function, which presumably results in impaired cellular function [117–119]. However, we applied this approach to a small system (i.e., mitochondrial oxidative phosphorylation), and therefore it would be useful to extend it to larger systems (cell, tissue, organ, and whole body). We hope that in the future the integration of two different systems biology approaches, i.e., mathematical modeling and high-throughput methodologies (e.g., microarrays, proteomics, metabolomics, and others), could result in large-scale, hypothesis-driven experimentation and lead to elucidation of the fundamental mechanisms underpinning the development and progression of type 2 diabetes.

Acknowledgments

This work was made possible by generous support from the Dutch Diabetes Research Foundation, from the BioSim (EU-FP6) network of excellence, from the NucSys (EU-FP7) training network, and from NWO and BBSRC.

References

1 Wild, S., Roglic, G., Green, A., Sicree, R., King H. (2004) Global prevalence of diabetes: estimates for the year 2000 and projections for 2030. *Diabetes Care 27*, 1047–1053.

2 Ray, J.A., Valentine, W.J., Secnik, K., Oglesby, A.K., Cordony, A., Gordois, A., Davey, P., Palmer, A.J. (2005) Review of the cost of diabetes complications in Australia, Canada, France, Germany, Italy and Spain. *Curr. Med. Res. Opin. 21*, 1617–1629.

3 Zhou, H., Isaman, D.J., Messinger, S., Brown, M.B., Klein, R., Brandle, M., Herman, W.H. (2005) A computer simulation model of diabetes progression, quality of life, and cost. *Diabetes Care 28*, 2856–2863.

4 Malecki, M.T. (2005) Genetics of type 2 diabetes mellitus. *Diabetes Res. Clin. Pract. 68*, S10–S21.

5 Shepherd, P.R., Kahn, B.B. (1999) Glucose transporters and insulin action – implications for insulin resistance and diabetes mellitus. *N. Engl. J. Med. 341*, 248–257.

6 Maechler, P. (2002) Mitochondria as the conductor of metabolic signals for insulin exocytosis in pancreatic beta-cells. *Cell. Mol. Life Sci. 59*, 1803–1818.

7 Saltiel, A.R., Kahn, C.R. (2001) Insulin signalling and the regulation of glucose and lipid metabolism. *Nature 414*, 799–806.

8 Dohm, G.L. (2002) Invited review: Regulation of skeletal muscle GLUT4 expression by exercise. *J. Appl. Physiol. 93*, 782–787.

9 Mokdad, A.H., Ford, E.S., Bowman, B.A., Dietz, W.H., Vinicor, F., Bales, V.S., Marks, J.S. (2003) Prevalence of obesity, diabetes, and obesity-related health risk factors, 2001. *JAMA. 289*, 76–79.

10 Nielsen, S., Guo, Z., Johnson, C.M., Hensrud, D.D., Jensen, M.D. (2004) Splanchnic lipolysis in human obesity. *J. Clin. Invest. 113*, 1582–1588.

11 Sopasakis, V.R., Sandqvist, M., Gustafson, B., Hammarstedt, A., Schmelz, M., Yang, X., Jansson, P.A., Smith, U. (2004) High local concentrations and effects on differentiation implicate interleukin-6 as a paracrine regulator. *Obes. Res. 12*, 454–460.

12 Hotamisligil, G.S., Shargill, N.S., Spiegelman, B.M. (1993) Adipose expression of tumor necrosis factor-alpha: direct role in obesity-linked insulin resistance. *Science 259*, 87–91.

13 Hirsch, J., Batchelor, B. (1976) Adipose tissue cellularity in human obesity. *Clin. Endocrinol. Metab. 5*, 299–311.

14 Salans, L.B., Knittle, J.L., Hirsch, J. (1968) The role of adipose cell size and adipose tissue insulin sensitivity in the carbohydrate intolerance of human obesity. *J. Clin. Invest. 47*, 153–165.

15 DiGirolamo, M., Howe, M.D., Esposito, J., Thurman, L., Owens, J.L. (1974) Metabolic patterns and insulin responsiveness of enlarging fat cells. *J. Lipid Res. 15*, 332–338.

16 Jensen, M.D., Caruso, M., Heiling, V., Miles, J.M. (1989) Insulin regulation of lipolysis in nondiabetic and IDDM subjects. *Diabetes 38*, 1595–1601.

17 Jensen, M.D., Haymond, M.W., Gerich, J.E., Cryer, P.E., Miles, J.M. (1987) Lipolysis during fasting. Decreased suppression by insulin and increased stimulation by epinephrine. *J. Clin. Invest. 79*, 207–213.

18 Galster, A.D., Clutter, W.E., Cryer, P.E., Collins, J.A., Bier, D.M. (1981) Epi-nephrine plasma thresholds for lipolytic effects in man: measurements of fatty acid transport with [l-13C]palmitic acid. *J. Clin. Invest. 67*, 1729–1738.

19 Camastra, S., Manco, M., Mari, A., Baldi, S., Gastaldelli, A., Greco, A.V., Mingrone, G., Ferrannini, E. (2005) beta-cell function in morbidly obese subjects during free living: long-term effects of weight loss. *Diabetes 54*, 2382–2389.

20 Horowitz, J.F., Coppack, S.W., Paramore, D., Cryer, P.E., Zhao, G., Klein, S. (1999) Effect of short-term fasting on lipid kinetics in lean and obese women. *Am. J. Physiol. 276*, E278–E284.

21 Heiling, V.J., Miles, J.M., Jensen, M.D. (1991) How valid are isotopic measurements of fatty acid oxidation? *Am. J. Physiol. 261*, E572–E577.

22 Koonen, D.P., Glatz, J.F., Bonen, A., Luiken, J.J. (2005) Long-chain fatty acid uptake and FAT/CD36 translocation in heart and skeletal muscle. *Biochim. Biophys. Acta 1736*, 163–180.

23 Boden, G., Lebed, B., Schatz, M., Homko, C., Lemieux, S. (2001) Effects of acute changes of plasma free fatty acids on intramyocellular fat content and insulin resistance in healthy subjects. *Diabetes 50*, 1612–1617.

24 Koyama, K., Chen, G., Lee, Y., Unger, R.H. (1997) Tissue triglycerides, insulin resistance, and insulin production: implications for hyperinsulinemia of obesity. *Am. J. Physiol. 273*, E708–E713.

25 Marchesini, G., Brizi, M., Bianchi, G., Tomassetti, S., Bugianesi, E., Lenzi, M., McCullough, A.J., Natale, S., Forlani, G.,

Melchionda, N. (2001) Nonalcoholic fatty liver disease: a feature of the metabolic syndrome. *Diabetes 50*, 1844–1850.

26 Hennes, M.M., Dua, A., Kissebah, A.H. (1997) Effects of free fatty acids and glucose on splanchnic insulin dynamics. *Diabetes 46*, 57–62.

27 Boucher, A., Lu, D., Burgess, S.C., Telemaque-Potts, S., Jensen, M.V., Mulder, H., Wang, M.Y., Unger, R.H., et al. (2004) Biochemical mechanism of lipid-induced impairment of glucose-stimulated insulin secretion and reversal with a malate analogue. *J. Biol. Chem. 279*, 27263–27271.

28 Briaud, I., Harmon, J.S., Kelpe, C.L., Segu, V.B., Poitout, V. (2001) Lipotoxicity of the pancreatic beta-cell is associated with glucose-dependent esterification of fatty acids into neutral lipids. *Diabetes 50*, 315–321.

29 Feng, D.D., Luo, Z., Roh, S.G., Hernandez, M., Tawadros, N., Keating, D.J., Chen, C. (2006) Reduction in voltage-gated K+ currents in primary cultured rat pancreatic beta-cells by linoleic acids. *Endocrinology 147*, 674–682.

30 Griffin, M.E., Marcucci, M.J., Cline, G.W., Bell, K., Barucci, N., Lee, D., Goodyear, L.J., Kraegen, E.W., et al. (1999) Free fatty acid-induced insulin resistance is associated with activation of protein kinase C theta and alterations in the insulin signaling cascade. *Diabetes 48*, 1270–1274.

31 Yu, C., Chen, Y., Cline, G.W., Zhang, D., Zong, H., Wang, Y., Bergeron, R., Kim, J.K., et al. (2002) Mechanism by which fatty acids inhibit insulin activation of insulin receptor substrate-1 (IRS-1)-associated phosphatidylinositol 3-kinase activity in muscle. *J. Biol. Chem. 277*, 50230–50236.

32 Powell, D.J., Turban, S., Gray, A., Hajduch, E., Hundal, H.S. (2004) Intracellular ceramide synthesis and protein kinase Czeta activation play an essential role in palmitate-induced insulin resistance in rat L6 skeletal muscle cells. *Biochem. J. 382*, 619–629.

33 Chung, S., Brown, J.M., Provo, J.N., Hopkins, R., McIntosh, M.K. (2005) Conjugated linoleic acid promotes human adipocyte insulin resistance through NFkappaB-dependent cytokine production. *J. Biol. Chem. 280*, 38445–38456.

34 Chen, X., Iqbal, N., Boden, G. (1999) The effects of free fatty acids on gluconeogenesis and glycogenolysis in normal subjects. *J. Clin. Invest. 103*, 365–372.

35 Basu, A., Basu, R., Shah, P., Vella, A., Johnson, C.M., Jensen, M., Nair, K.S., Schwenk, W.F., et al. (2001) Type 2 diabetes impairs splanchnic uptake of glucose but does not alter intestinal glucose absorption during enteral glucose feeding: additional evidence for a defect in hepatic glucokinase activity. *Diabetes 50*, 1351–1362.

36 Kelley, D.E. (2005) Skeletal muscle fat oxidation: timing and flexibility are everything. *J. Clin. Invest. 115*, 1699–1702.

37 Randle, P.J. (1998) Regulatory interactions between lipids and carbohydrates: the glucose fatty acid cycle after 35 years. *Diabetes Metab. Rev. 14*, 263–283.

38 Hajduch, E., Litherland, G.J., Hundal, H.S. (2001) Protein kinase B (PKB/Akt) – a key regulator of glucose transport? *FEBS Lett. 492*, 199–203.

39 Petersen, K.F., Befroy, D., Dufour, S., Dziura, J., Ariyan, C., Rothman, D.L., DiPietro, L., Cline, G.W., et al. (2003) Mitochondrial dysfunction in the elderly: possible role in insulin resistance. *Science 300*, 1140–1142.

40 Petersen, K.F., Dufour, S., Befroy, D., Garcia, R., Shulman, G.I. (2004) Impaired mitochondrial activity in the insulin-resistant offspring of patients with type 2 diabetes. *N. Engl. J. Med. 350*, 664–671.

41 Ritov, V.B., Menshikova, E.V., He, J., Ferrell, R.E., Goodpaster, B.H., Kelley, D.E. (2005) Deficiency of subsarcolemmal mitochondria in obesity and type 2 diabetes. *Diabetes 54*, 8–14.

42 Wu, Z., Puigserver, P., Andersson, U., Zhang, C., Adelmant, G., Mootha, V., Troy, A., Cinti, S., et al. (1999) Mechanisms controlling mitochondrial biogenesis and respiration through the

thermogenic coactivator PGC-1. *Cell 98*, 115–124.

43 Mootha, V.K., Lindgren, C.M., Eriksson, K.F., Subramanian, A., Sihag, S., Lehar, J., Puigserver, P., Carlsson, E., et al. (2003) PGC-1alpha-responsive genes involved in oxidative phosphorylation are coordinately downregulated in human diabetes. *Nat. Genet. 34*, 267–273.

44 Patti, M.E., Butte, A.J., Crunkhorn, S., Cusi, K., Berria, R., Kashyap, S., Miyazaki, Y., Kohane, I., et al. (2003) Coordinated reduction of genes of oxidative metabolism in humans with insulin resistance and diabetes: Potential role of PGC1 and NRF1. *Proc. Natl. Acad. Sci. U.S.A. 100*, 8466–8471.

45 Bandyopadhyay, G.K., Yu, J.G., Ofrecio, J., Olefsky, J.M. (2006) Increased malonyl-CoA levels in muscle from obese and type 2 diabetic subjects lead to decreased fatty acid oxidation and increased lipogenesis; thiazolidinedione treatment reverses these defects. *Diabetes 55*, 2277–2285.

46 Chua, B.H., Shrago, E. (1977) Reversible inhibition of adenine nucleotide translocation by long chain acyl-CoA esters in bovine heart mitochondria and inverted submitochondrial particles. *J. Biol. Chem. 252*, 6711–6714.

47 Rhodes, C.J. (2005) Type 2 diabetes-a matter of beta-cell life and death? *Science 307*, 380–384.

48 Bernard-Kargar, C., Ktorza, A. (2001) Endocrine pancreas plasticity under physiological and pathological conditions. *Diabetes 50*, S30–S35.

49 Soejima, A., Inoue, K., Takai, D., Kaneko, M., Ishihara, H., Oka, Y., Hayashi, J.I. (1996) Mitochondrial DNA is required for regulation of glucose-stimulated insulin secretion in a mouse pancreatic beta cell line, MIN6. *J. Biol. Chem. 271*, 26194–26199.

50 Tsuruzoe, K., Araki, E., Furukawa, N., Shirotani, T., Matsumoto, K., Kaneko, K., Motoshima, H., Yoshizato, K., et al. (1998) Creation and characterization of a mitochondrial DNA-depleted pancreatic beta-cell line: impaired insulin secretion induced by glucose, leucine, and sulfonylureas. *Diabetes 47*, 621–631.

51 Feng, D.D., Luo, Z., Roh, S.G., Hernandez, M., Tawadros, N., Keating, D.J., Chen, C. (2006) Reduction in voltage-gated K+ currents in primary cultured rat pancreatic beta-cells by linoleic acids. *Endocrinology 147*, 674–682.

52 Bratanova-Tochkova, T.K., Cheng, H., Daniel, S., Gunawardana, S., Liu, Y.J., Mulvaney-Musa, J., Schermerhorn, T., Straub, S.G., et al. (2002) Triggering and augmentation mechanisms, granule pools, and biphasic insulin secretion. *Diabetes 51*, S83–S90.

53 Zhou, Y.P., Grill, V.E. (1994) Long-term exposure of rat pancreatic islets to fatty acids inhibits glucose-induced insulin secretion and biosynthesis through a glucose fatty acid cycle. *J. Clin. Invest. 93*, 870–876.

54 Gremlich, S., Bonny, C., Waeber, G., Thorens, B. (1997) Fatty acids decrease IDX-1 expression in rat pancreatic islets and reduce GLUT2, glucokinase, insulin, and somatostatin levels. *J. Biol. Chem. 272*, 30261–30269.

55 Butler, A.E., Janson, J., Soeller, W.C., Butler, P.C. (2003) Increased beta-cell apoptosis prevents adaptive increase in beta-cell mass in mouse model of type 2 diabetes: evidence for role of islet amyloid formation rather than direct action of amyloid. *Diabetes 52*, 2304–2314.

56 Cnop, M., Hannaert, J.C., Hoorens, A., Eizirik, D.L., Pipeleers, D.G. (2001) Inverse relationship between cytotoxicity of free fatty acids in pancreatic islet cells and cellular triglyceride accumulation. *Diabetes 50*, 1771–1777.

57 Lupi, R., Dotta, F., Marselli, L., Del Guerra, S., Masini, M., Santangelo, C., Patane, G., Boggi, U., et al. (2002) Prolonged exposure to free fatty acids has cytostatic and pro-apoptotic effects on human pancreatic islets: evidence that beta-cell death is caspase mediated, partially dependent on ceramide pathway, and Bcl-2 regulated. *Diabetes 51*, 1437–1442.

58 Maedler, K., Spinas, G.A., Dyntar, D., Moritz, W., Kaiser, N., Donath, M.Y. (2001) Distinct effects of saturated and monounsaturated fatty acids on beta-cell

turnover and function. *Diabetes 50*, 69–76.

59 Shimabukuro, M., Zhou, Y.T., Levi, M., Unger, R.H. (1998) Fatty acid-induced beta cell apoptosis: a link between obesity and diabetes. *Proc. Natl. Acad. Sci. U.S.A. 95*, 2498–2502.

60 Maestre, I., Jordan, J., Calvo, S., Reig, J.A., Cena, V., Soria, B., Prentki, M., Roche, E. (2003) Mitochondrial dysfunction is involved in apoptosis induced by serum withdrawal and fatty acids in the β-cell line INS-1. *Endocrinology 144*, 335–345.

61 Wang, X., Li, H., De Leo, D., Guo, W., Koshkin, V., Fantus, I.G., Giacca, A., Chan, C.B., et al. (2004) Gene and protein kinase expression profiling of reactive oxygen species-associated lipotoxicity in the pancreatic beta-cell line MIN6. *Diabetes 53*, 129–140.

62 Joseph, J.W., Koshkin, V., Saleh, M.C., Sivitz, W.I., Zhang, C.Y., Lowell, B.B., Chan, C.B., Wheeler, M.B. (2004) Free fatty acid-induced beta-cell defects are dependent on uncoupling protein 2 expression. *J. Biol. Chem. 279*, 51049–51056.

63 Schonfeld, P., Wieckowski, M.R., Wojtczak, L. (2000) Long-chain fatty acid-promoted swelling of mitochondria: further evidence for the protonophoric effect of fatty acids in the inner mitochondrial membrane. *FEBS Lett. 471*, 108–112.

64 Hornberg, J.J., Bruggeman, F.J., Westerhoff, H.V., Lankelma, J. (2006) Cancer: a Systems Biology disease. *Biosystems 83*, 81–90.

65 Kitano, H., Oda, K., Kimura, T., Matsuoka, Y., Csete, M., Doyle, J., Muramatsu, M. (2004) Metabolic syndrome and robustness tradeoffs. *Diabetes 53*, S6–S15.

66 Kitano, H. (2002) Computational systems biology. *Nature 420*, 206–210.

67 Kitano, H. (2002) Systems biology: a brief overview. *Science 295*, 1662–1664.

68 Westerhoff, H.V., Palsson, B.O. (2004) The evolution of molecular biology into systems biology. *Nat. Biotechnol. 22*, 1249–1252.

69 Cascante, M., Boros, L.G., Comin-Anduix, B., de Atauri, P., Centelles, J.J.,

Lee, P.W. (2002) Metabolic control analysis in drug discovery and disease. *Nat. Biotechnol. 20*, 243–249.

70 Yarmush, M.L., Banta, S. (2003) Metabolic engineering: advances in modeling and intervention in health and disease. *Annu. Rev. Biomed. Eng. 5*, 349–381.

71 Kacser, H., Burns, J.A. (1973) The control of flux. *Symp. Soc. Exp. Biol. 27*, 65–104.

72 Heinrich, R., Rapoport, T.A. (1974) A linear steady-state treatment of enzymatic chains. General properties, control and effector strength. *Eur. J. Biochem. 42*, 89–95.

73 Burns, J.A., Cornish-Bowden, A., Groen, A.K., Heinrich, R., Kacser, H., Porteous, J.W., Rapoport, S.M., Rapoport, T.A., et al. (1985) Control of metabolic systems. *Trends Biochem. Sci. 10*, 16.

74 Kacser, H., Burns, J.A. (1979) Molecular democracy: who shares the controls? *Biochem. Soc. Trans. 7*, 1149–1160.

75 Westerhoff, H.V., Chen, Y.D. (1984) How do enzyme activities control metabolite concentrations? An additional theorem in the theory of metabolic control. *Eur. J. Biochem. 142*, 425–430.

76 Westerhoff, H.V., Kell, D.B. (1987) Matrix method for determining the steps most rate- limiting to metabolic fluxes in biotechnological processes. *Biotechnol. Bioeng. 30*, 101–107.

77 Reder, C. (1988) Metabolic control theory: a structural approach. *J. Theor. Biol. 135*, 175–201.

78 Hofmeyr, J.H., Cornish-Bowden, A., Rohwer, J.M. (1993) Taking enzyme kinetics out of control; putting control into regulation. *Eur. J. Biochem. 212*, 833–837.

79 Fell, D.A. (1992) Metabolic control analysis: a survey of its theoretical and experimental development. *Biochem. J. 286*, 313–330.

80 Kacser, H., Burns, J.A. (1981) The molecular basis of dominance. *Genetics 97*, 639–666.

81 Bohnensack, R., Kuster, U., Letko, G. (1982) Rate-controlling steps of oxidative phosphorylation in rat liver mitochondria. A synoptic approach of model and

experiment. *Biochim. Biophys. Acta. 680,*
271–280.

82 Groen, A.K., Wanders, R.J., Westerhoff,
H.V., van der Meer, R., Tager, J.M.
(1982) Quantification of the contribution of various steps to the control of
mitochondrial respiration. *J. Biol. Chem.*
257, 2754–2757.

83 Duszynski, J., Groen, A.K., Wanders,
R.J., Vervoorn, R.C., Tager, J.M. (1982)
Quantification of the role of the adenine
nucleotide translocator in the control of
mitochondrial respiration in isolated
rat-liver cells. *FEBS Lett. 146,* 262–266.

84 Tager, J.M., Groen, A.K., Wanders, R.J.,
Duszynski, J., Westerhoff, H.V., Vervoorn,
R.C. (1983) Control of mitochondrial re-
spiration. *Biochem. Soc. Trans. 11,* 40–43.

85 Gellerich, F.N., Bohnensack, R., Kunz,
W. (1983) Control of mitochondrial
respiration. The contribution of the
adenine nucleotide translocator depends
on the ATP- and ADP-consuming
enzymes. *Biochim. Biophys. Acta 722,*
381–391.

86 Borutaite, V., Mildaziene, V., Brown,
G.C., Brand, M.D. (1995) Control and
kinetic analysis of ischemia-damaged
heart mitochondria: which parts of the
oxidative phosphorylation system are
affected by ischemia? *Biochim. Biophys.*
Acta 1272, 154–158.

87 Westerhoff, H.V., Plomp, P.J., Groen,
A.K., Wanders, R.J., Bode, J.A., van
Dam, K. (1987) On the origin of the
limited control of mitochondrial
respiration by the adenine nucleotide
translocator. *Arch. Biochem. Biophys. 257,*
154–169.

88 Hafner, R.P., Brown, G.C., Brand, M.D.
(1990) Analysis of the control of
respiration rate, phosphorylation rate,
proton leak rate and protonmotive force
in isolated mitochondria using the "top-
down" approach of metabolic control
theory. *Eur. J. Biochem. 188,* 313–319.

89 Kesseler, A., Brand, M.D. (1994) Effects
of cadmium on the control and internal
regulation of oxidative phosphorylation
in potato tuber mitochondria. *Eur. J.*
Biochem. 225, 907–922.

90 Wanders, R.J., van Roermund, C.W.,
Meijer, A.J. (1984) Analysis of the
control of citrulline synthesis in isolated

rat-liver mitochondria. *Eur. J. Biochem.*
142, 247–254.

91 Ainscow, E.K., Brand, M.D. (1999)
Internal regulation of ATP turnover,
glycolysis and oxidative phosphorylation
in rat hepatocytes. *Eur. J. Biochem. 266,*
737–749.

92 Chassagnole, C., Fell, D.A., Rais, B.,
Kudla, B., Mazat, J.P. (2001) Control of
the threonine-synthesis pathway in
Escherichia coli: a theoretical and
experimental approach. *Biochem. J. 56,*
433–444.

93 Snoep, J.L., van der Weijden, C.C.,
Andersen, H.W., Westerhoff, H.V.,
Jensen, P.R. (2002) DNA supercoiling in
Escherichia coli is under tight and
subtle homeostatic control, involving
gene-expression and metabolic regulation of both topoisomerase I and DNA
gyrase. *Eur. J. Biochem. 269,* 1662–1669.

94 Bakker, B.M., Michels, P.A., Opperdoes,
F.R., Westerhoff, H.V. (1999) What
controls glycolysis in bloodstream form
Trypanosoma brucei? *J. Biol. Chem. 274,*
14551–14559.

95 Jucker, B.M., Barucci, N., Shulman, G.I.
(1999) Metabolic control analysis of
insulin-stimulated glucose disposal in
rat skeletal muscle. *Am. J. Physiol. 277,*
E505–E512.

96 Bier, M., Teusink, B., Kholodenko, B.N.,
Westerhoff, H.V. (1996) Control analysis
of glycolytic oscillations. *Biophys. Chem.*
62, 15–24.

97 Heinrich, R., Rapoport, T.A. (1975)
Mathematical analysis of multienzyme
systems. II. Steady state and transient
control. *Biosystems 7,* 130–136.

98 Hornberg, J.J., Bruggeman, F.J., Binder,
B., Geest, C.R., de Vaate, A.J., Lankelma,
J., Heinrich, R., Westerhoff, H.V. (2005)
Principles behind the multifarious
control of signal transduction. ERK
phosphorylation and kinase/phosphatase
control. *FEBS J. 272,* 244–258.

99 Westerhoff, H.V., van Dam, K. (1987)
Thermodynamics and control of free-energy
transduction, Elsevier, Amsterdam.

100 Fell, D.A., Sauro, H.M. (1985) Metabolic
control and its analysis. Additional
relationships between elasticities and
control coefficients. *Eur. J. Biochem. 148,*
555–561.

101 Torres, N.V., Mateo, F., Melendez-Hevia, E., Kacser, H. (1986) Kinetics of metabolic pathways. A system *in vitro* to study the control of flux. *Biochem. J. 234*, 169–174.

102 Brown, G.C., Hafner, R.P., Brand, M.D. (1990) A "top-down" approach to the determination of control coefficients in metabolic control theory. *Eur. J. Biochem. 188*, 321–325.

103 Schuster, S., Kahn, D., Westerhoff, H.V. (1993) Modular analysis of the control of complex metabolic pathways. *Biophys. Chem. 48*, 1–17.

104 Ainscow, E.K., Brand, M.D. (1995) Top-down control analysis of systems with more than one common intermediate. *Eur. J. Biochem. 231*, 579–586.

105 Ainscow, E.K., Brand, M.D. (1998) Control analysis of systems with reaction blocks that "cross-talk". *Biochim. Biophys. Acta 1366*, 284–290.

106 Schuster, S. (1999) Use and limitations of modular metabolic control analysis in medicine and biotechnology. *Metab. Eng. 1*, 232–242.

107 Brown, G.C. (1994) Control analysis applied to the whole body: control by body organs over plasma concentrations and organ fluxes of substances in the blood. *Biochem. J. 297*, 115–122.

108 Kesseler, A., Brand, M.D. (1994) Localisation of the sites of action of cadmium on oxidative phosphorylation in potato tuber mitochondria using top-down elasticity analysis. *Eur. J. Biochem. 225*, 897–906.

109 Dufour, S., Rousse, N., Cnioni, P., Diolez, P. (1996) Top-down control analysis of temperature effect on oxidative phosphorylation. *Biochem. J. 314*, 743–751.

110 Krab, K., Wagner, M.J., Wagner, A.M., Moler, I.M. (2000) Identification of the site where the electron transfer chain of plant mitochondria is stimulated by electrostatic charge screening. *Eur. J. Biochem. 267*, 869–876.

111 Mildaziene, V., Nauciene, Z., Baniene, R., Grigiene, J. (2002) Multiple effects of 2,2,5,5-tetrachlorbiphenyl on oxidative phosphorylation in rat liver mitochondria. *Toxicol. Sci. 65*, 220–227.

112 Giersch, C. (1994) Determining elasticities from multiple measurements of steady-state flux rates and metabolite concentrations: theory. *J. Theor. Biol. 169*, 89–99.

113 Ainscow, E.K., Brand, M.D. (1999) Top-down control analysis of ATP turnover, glycolysis and oxidative phosphorylation in rat hepatocytes. *Eur. J. Biochem. 263*, 671–685.

114 Hofmeyr, J.H., Cornish-Bowden, A. (1996) Co-response analysis: a new experimental strategy for metabolic control analysis. *J. Theor. Biol. 182*, 371–380.

115 Bakker, S.J., Ijzerman, R.G., Teerlink, T., Westerhoff, H.V., Gans, R.O., Heine, R.J. (2000) Cytosolic triglycerides and oxidative stress in central obesity: the missing link between excessive atherosclerosis, endothelial dysfunction, and β-cell failure? *Atherosclerosis 148*, 17–21.

116 Bakker, S.J., Gans, R.O., ter Maaten, J.C., Teerlink, T., Westerhoff, H.V., Heine, R.J. (2001) The potential role of adenosine in the pathophysiology of the insulin resistance syndrome. *Atherosclerosis 155*, 283–290.

117 Ciapaite, J., van Eikenhorst, G., Bakker, S.J., Diamant, M., Heine, R.J., Wagner, M.J., Westerhoff, H.V., Krab, K. (2005) Modular kinetic analysis of the adenine nucleotide translocator-mediated effects of palmitoyl-CoA on the oxidative phosphorylation in isolated rat liver mitochondria. *Diabetes 54*, 944–951.

118 Ciapaite, J., Bakker, S.J., Diamant, M., van Eikenhorst, G., Heine, R.J., Westerhoff, H.V., Krab, K. (2006) Metabolic control of mitochondrial properties by adenine nucleotide translocator determines palmitoyl-CoA effects. Implications for a mechanism linking obesity and type 2 diabetes. *FEBS J. 273*, 5288–5302.

119 Ciapaite, J., Bakker, S.J., van Eikenhorst, G., Wagner, M.J., Teerlink, T., Schalkwijk, C.G., Fodor, M., Ouwens, M.D., et al. (2007) Functioning of oxidative phosphorylation in liver mitochondria of high-fat diet fed rats. *Biochim. Biophys. Acta 1772*, 307–316.

Subject Index

Molecular System Bioenergetics: Energy for Life. Edited by Valdur Saks
Copyright © 2007 WILEY-VCH Verlag GmbH & Co. KGaA, Weinheim
ISBN: 978-3-527-31787-5